国外优秀数学著作
原　版　系　列

项目反应理论手册——第一卷，模型

Handbook of Item Response Theory—Volume one, Models

● [美] 维姆·J.范·德·林登（Wim J. van der Linden）主编

（英文）

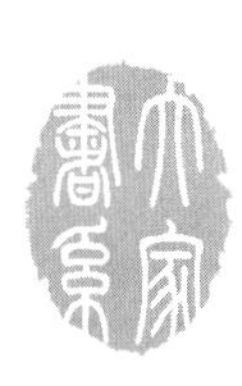

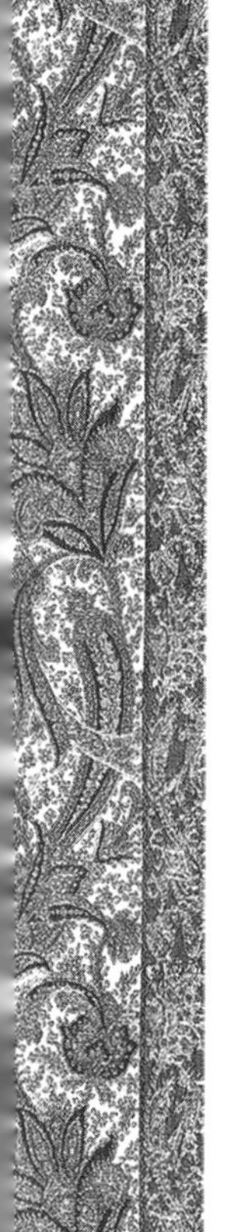

HITP
哈爾濱工業大學出版社
HARBIN INSTITUTE OF TECHNOLOGY PRESS

黑版贸审字 08-2019-189 号

图书在版编目(CIP)数据

项目反应理论手册. 第一卷,模型 = Handbook of Item Response Theory: Volume One, Models:英文/(美)维姆·J. 范·德·林登(Wim J. van der Linden)主编. —哈尔滨:哈尔滨工业大学出版社,2021. 11

ISBN 978 - 7 - 5603 - 9817 - 4

Ⅰ. ①项… Ⅱ. ①维… Ⅲ. ①统计学-手册-英文
Ⅳ. ①C8 - 62

中国版本图书馆 CIP 数据核字(2021)第 226175 号

Handbook of Item Response Theory: Volume One, Models/by Wim J. van der Linden/ISBN: 978 - 1 - 4665 - 1431 - 7

策划编辑　刘培杰　杜莹雪
责任编辑　刘春雷　李兰静
封面设计　孙茵艾
出版发行　哈尔滨工业大学出版社
社　　址　哈尔滨市南岗区复华四道街 10 号　邮编 150006
传　　真　0451-86414749
网　　址　http://hitpress.hit.edu.cn
印　　刷　哈尔滨市工大节能印刷厂
开　　本　787 mm×1 092 mm　1/16　印张 40　字数 670 千字
版　　次　2021 年 11 月第 1 版　2021 年 11 月第 1 次印刷
书　　号　ISBN 978 - 7 - 5603 - 9817 - 4
定　　价　138.00 元

(如因印装质量问题影响阅读,我社负责调换)

Contents

Contents for *Statistical Tools* v
Contents for *Applications* vii
Preface xi
Contributors xv

1. **Introduction** 1
   Wim J. van der Linden

Section I Dichotomous Models

2. **Unidimensional Logistic Response Models** 13
   Wim J. van der Linden

3. **Rasch Model** 31
   Matthias von Davier

Section II Nominal and Ordinal Models

4. **Nominal Categories Models** 51
   David Thissen and Li Cai

5. **Rasch Rating-Scale Model** 75
   David Andrich

6. **Graded Response Models** 95
   Fumiko Samejima

7. **Partial Credit Model** 109
   Geoff N. Masters

8. **Generalized Partial Credit Model** 127
   Eiji Muraki and Mari Muraki

9. **Sequential Models for Ordered Responses** 139
   Gerhard Tutz

10. **Models for Continuous Responses** 153
    Gideon J. Mellenbergh

Section III Multidimensional and Multicomponent Models

11. Normal-Ogive Multidimensional Models 167
Hariharan Swaminathan and H. Jane Rogers

12. Logistic Multidimensional Models 189
Mark D. Reckase

13. Linear Logistic Models 211
Rianne Janssen

14. Multicomponent Models 225
Susan E. Embretson

Section IV Models for Response Times

15. Poisson and Gamma Models for Reading Speed and Error 245
Margo G. H. Jansen

16. Lognormal Response-Time Model 261
Wim J. van der Linden

17. Diffusion-Based Response-Time Models 283
Francis Tuerlinckx, Dylan Molenaar, and Han L. J. van der Maas

Section V Nonparametric Models

18. Mokken Models 303
Klaas Sijtsma and Ivo W. Molenaar

19. Bayesian Nonparametric Response Models 323
George Karabatsos

20. Functional Approaches to Modeling Response Data 337
James O. Ramsay

Section VI Models for Nonmonotone Items

21. Hyperbolic Cosine Model for Unfolding Responses 353
David Andrich

22. Generalized Graded Unfolding Model 369
James S. Roberts

Section VII Hierarchical Response Models

23. Logistic Mixture-Distribution Response Models 393
Matthias von Davier and Jürgen Rost

24. Multilevel Response Models with Covariates and Multiple Groups 407
Jean-Paul Fox and Cees A. W. Glas

25. Two-Tier Item Factor Analysis Modeling 421
Li Cai

26. Item-Family Models 437
Cees A. W. Glas, Wim J. van der Linden, and Hanneke Geerlings

27. Hierarchical Rater Models 449
Jodi M. Casabianca, Brian W. Junker, and Richard J. Patz

28. Randomized Response Models for Sensitive Measurements 467
Jean-Paul Fox

29. Joint Hierarchical Modeling of Responses and Response Times 481
Wim J. van der Linden and Jean-Paul Fox

Section VIII Generalized Modeling Approaches

30. Generalized Linear Latent and Mixed Modeling 503
Sophia Rabe-Hesketh and Anders Skrondal

31. Multidimensional, Multilevel, and Multi-Timepoint Item Response Modeling 527
Bengt Muthén and Tihomir Asparouhov

32. Mixed-Coefficients Multinomial Logit Models 541
Raymond J. Adams, Mark R. Wilson, and Margaret L. Wu

33. Explanatory Response Models 565
Paul De Boeck and Mark R. Wilson

Index 581

编辑手记 597

Contents for *Statistical Tools*

Section I Basic Tools

1. **Logit, Probit, and Other Response Functions**
   James H. Albert

2. **Discrete Distributions**
   Jodi M. Casabianca and Brian W. Junker

3. **Multivariate Normal Distribution**
   Jodi M. Casabianca and Brian W. Junker

4. **Exponential Family Distributions Relevant to IRT**
   Shelby J. Haberman

5. **Loglinear Models for Observed-Score Distributions**
   Tim Moses

6. **Distributions of Sums of Nonidentical Random Variables**
   Wim J. van der Linden

7. **Information Theory and Its Application to Testing**
   Hua-Hua Chang, Chun Wang, and Zhiliang Ying

Section II Modeling Issues

8. **Identification of Item Response Theory Models**
   Ernesto San Martín

9. **Models With Nuisance and Incidental Parameters**
   Shelby J. Haberman

10. **Missing Responses in Item Response Modeling**
    Robert J. Mislevy

Section III Parameter Estimation

11. **Maximum-Likelihood Estimation**
    Cees A. W. Glas

12. **Expectation Maximization Algorithm and Extensions**
    Murray Aitkin

13. **Bayesian Estimation**
Matthew S. Johnson and Sandip Sinharay

14. **Variational Approximation Methods**
Frank Rijmen, Minjeong Jeon, and Sophia Rabe-Hesketh

15. **Markov Chain Monte Carlo for Item Response Models**
Brian W. Junker, Richard J. Patz, and Nathan M. VanHoudnos

16. **Statistical Optimal Design Theory**
Heinz Holling and Rainer Schwabe

Section IV Model Fit and Comparison

17. **Frequentist Model-Fit Tests**
Cees A. W. Glas

18. **Information Criteria**
Allan S. Cohen and Sun-Joo Cho

19. **Bayesian Model Fit and Model Comparison**
Sandip Sinharay

20. **Model Fit with Residual Analyses**
Craig S. Wells and Ronald K. Hambleton

Index

Contents for *Applications*

Section I Item Calibration and Analysis

1. **Item-Calibration Designs**
   Martijn P. F. Berger

2. **Parameter Linking**
   Wim J. van der Linden and Michelle D. Barrett

3. **Dimensionality Analysis**
   Robert D. Gibbons and Li Cai

4. **Differential Item Functioning**
   Dani Gamerman, Flávio B. Gonçalves, and Tufi M. Soares

5. **Calibrating Technology-Enhanced Items**
   Richard M. Luecht

Section II Person Fit and Scoring

6. **Person Fit**
   Cees A. W. Glas and Naveed Khalid

7. **Score Reporting and Interpretation**
   Ronald K. Hambleton and April L. Zenisky

8. **IRT Observed-Score Equating**
   Wim J. van der Linden

Section III Test Design

9. **Optimal Test Design**
   Wim J. van der Linden

10. **Adaptive Testing**
    Wim J. van der Linden

11. **Standard Setting**
    Daniel Lewis and Jennifer Lord-Bessen

12. **Test Speededness and Time Limits**
    Wim J. van der Linden

13. **Item and Test Security**
Wim J. van der Linden

Section IV Areas of Application

14. **Large-Scale Group-Score Assessments**
John Mazzeo

15. **Psychological Testing**
Paul De Boeck

16. **Cognitive Diagnostic Assessment**
Chun Wang and Hua-Hua Chang

17. **Health Measurement**
Richard C. Gershon, Ron D. Hays, and Michael Kallen

18. **Marketing Research**
Martijn G. de Jong and Ulf Böckenholt

19. **Measuring Change Using Rasch Models**
Gerhard H. Fischer

Section V Computer Programs

20. **IRT Packages in R**
Thomas Rusch, Patrick Mair, and Reinhold Hatzinger

21. **Bayesian Inference Using Gibbs Sampling (BUGS) for IRT models**
Matthew S. Johnson

22. **BILOG-MG**
Michele F. Zimowski

23. **PARSCALE**
Eiji Muraki

24. **IRTPRO**
Li Cai

25. **Xcalibre 4**
Nathan A. Thompson and Jieun Lee

26. **EQSIRT**
Peter M. Bentler, Eric Wu, and Patrick Mair

27. ACER ConQuest
Raymond J. Adam, Margaret L. Wu, and Mark R. Wilson

28. Mplus
Bengt Muthén and Linda Muthén

29. GLLAMM
Sophia Rabe-Hesketh and Anders Skrondal

30. Latent GOLD
Jeroen K. Vermunt

31. WinGen
Kyung (Chris) T. Han

32. Firestar
Seung W. Choi

32. jMetrik
J. Patrick Meyer

Index

Preface

Item response theory (IRT) has its origins in pioneering work by Louis Thurstone in the 1920s, a handful of authors such as Lawley, Mosier, and Richardson in the 1940s, and more decisive work by Alan Birnbaum, Frederic Lord, and George Rasch in the 1950s and 1960s. The major breakthrough it presents is the solution to one of the fundamental flaws inherent in classical test theory—its systematic confounding of what we measure with the test items used to measure it.

Test administrations are observational studies in which test takers receive a set of items and we observe their responses. The responses are the joint effects of both the properties of the items and abilities of the test takers. As in any other observational study, it would be a methodological error to attribute the effects to one of these underlying causal factors only. Nevertheless, it seems as if we are forced to do so. If new items are field tested, the interest is exclusively in their properties, and any confounding with the abilities of the largely arbitrary selection of test takers used in the study would bias our inferences about them. Likewise, if examinees are tested, the interest is in their abilities only and we do not want their scores to be biased by the incidental properties of the items. Classical test theory does create such biases. For instance, it treats the p-values of the items as their difficulty parameters, but these values depend equally on the abilities of the sample of test takers used in the field test. In spite of the terminology, the same holds for its item-discrimination parameters and definition of test reliability. On the other hand, the number-correct scores classical test theory typically is used for are scores equally indicative of the difficulty of the test as the abilities of test takers. In fact, the tradition of indexing such parameter and scores by the items or test takers only systematically hides this confounding.

IRT solves the problem by recognizing each response as the outcome of a distinct probability experiment that has to be modeled with separate parameters for the item and test taker effects. Consequently, its item parameters allow us to correct for item effects when we estimate the abilities. Likewise, the presence of the ability parameters allows us to correct for their effects when estimating the item parameter. One of the best introductions to this change of paradigm is Rasch (1960, Chapter 1), which is mandatory reading for anyone with an interest in the subject. The chapter places the new paradigm in the wider context of the research tradition found in the behavioral and social sciences with its persistent interest in vaguely defined "populations" of subjects, who, except for some random noise, are treated as exchangeable, as well as its use of statistical techniques as correlation coefficients, analysis of variance, and hypothesis testing that assume "random sampling" from them.

The developments since the original conceptualization of IRT have remained rapid. When Ron Hambleton and I edited an earlier handbook of item response theory (van der Linden and Hambleton, 1997), we had the impression that its 28 chapters pretty much summarized what could be said about the subject. But now, nearly two decades later, three volumes with roughly the same number of chapters each appear to be necessary. And I still feel I have to apologize to all the researchers and practitioners whose original contributions to the vast literature on IRT are not included in this new handbook. Not only have the original models for dichotomous responses been supplemented with numerous models for different response formats or response processes, it is now clear, for instance, that models for response times on test items require the same type of parameterization to account both for the item and test taker effects. Another major development has been the recognition of the need of deeper parameterization due to a multilevel or hierarchical structure of the response data. This development has led to the possibility to account for explanatory covariates, group structures with an impact on the item or ability parameters, mixtures of response processes, higher-level relationships between responses and response times, or special structures of the item domain, for instance, due to the use of rule-based item generation. Meanwhile, it has also become clear how to embed IRT in the wider development of generalized latent variable modeling. And as a result of all these extensions and new insights, we are now keener in our choice of treating model parameter as fixed or random. Volume 1 of this handbook covers most of these developments. Each of its chapters basically reviews one model. However, all chapters have the common format of an introductory section with some history of the model and a motivation of its relevance, and then continue with sections that present the model more formally, treat the estimation of its parameters, show how to evaluate its fit to empirical data, and illustrate the use of the model through an empirical example. The last section discusses further applications and remaining research issues.

As with any other type of probabilistic modeling, IRT depends heavily on the use of statistical tools for the treatment of its models and their applications. Nevertheless, systematic introductions and review with an emphasis on their relevance to IRT are hardly found in the statistical literature. Volume 2 is to fill this void. Its chapters are on such topics as commonly used probability distributions in IRT, the issue of models with both intentional and nuisance parameters, the use of information criteria, methods for dealing with missing data, model identification issues, and several topics in parameter estimation and model fit and comparison. It is especially in these last two areas that recent developments have been overwhelming. For instance, when the previous handbook of IRT was produced, Bayesian approaches had already gained some ground but were certainly not common. But thanks to the computational success of Markov chain Monte Carlo methods, these approaches have now become standard, especially for the more complex models in the second half of Volume 1.

The chapters of Volume 3 review several applications of IRT to the daily practice of testing. Although each of the chosen topics in the areas of item calibration and analysis, person fit and scoring, and test design have ample resources in the larger literature on test theory, the current chapters exclusively highlight the contributions that IRT has brought to them. This volume also offers chapters with reviews of how IRT has advanced such areas as large-scale educational assessments, psychological testing, cognitive diagnosis, health measurement, marketing research, or the more general area of measurement of change. The volume concludes with an extensive review of computer software programs available for running any of the models and applications in Volumes 1 and 3.

I expect this *Handbook of Item Response Theory* to serve as a daily resource of information to researchers and practitioners in the field of IRT as well as a textbook to novices. To better serve them, all chapters are self-contained. But their common core of notation and extensive cross-referencing allows readers of one of the chapters to consult others for background information without too much interruption.

I am grateful to all my authors for their belief in this project and the time they have spent on their chapters. It has been a true privilege to work with each of them. The same holds for Ron Hambleton who was willing to serve as my sparring partner during the conception of the plan for this handbook. John Kimmel, executive editor, *Statistics*,

Chapman & Hall/CRC has been a permanent source of helpful information during the production of this book. I thank him for his support as well.

Wim J. van der Linden
Monterey, CA

References

Rasch, G. 1960. *Probabilistic Models for Some Intelligence and Attainment Tests.* Copenhagen: Danish Institute for Educational Research.

van der Linden, W. J., and Hambleton, R. K. (Eds.) 1997. *Handbook of Modern Item Response Theory.* New York: Springer.

Contributors

Raymond J. Adams is an honorary professorial fellow of the University of Melbourne, and head of ACER's global educational monitoring center. Dr. Adams specializes in psychometrics, educational statistics, large-scale testing, and international comparative studies. He earned his PhD from the University of Chicago in 1989. He led the OECD PISA programme from its inception until 2013. Dr. Adams has published widely on the technical aspects of educational measurement and his item response modeling software packages are amongst the most widely used in educational and psychological measurement. He has served as chair of the technical advisory committee for the International Association for the Evaluation of Educational Achievement and as head of measurement at the Australian Council for Educational Research.

David Andrich is the Chapple professor of education at the University of Western Australia. He earned his PhD from the University of Chicago in 1973. In 1990, he was elected as a fellow of the Academy of Social Sciences of Australia for his contributions to measurement in the social sciences. He has contributed to the development of both single-peaked (unfolding) and monotonic (cumulative) response models, and is especially known for his contributions to Rasch measurement theory.

Tihomir Asparouhov earned his PhD in mathematics at the California Institute of Technology. He is a senior statistician and programmer at Muthén & Muthén, which develops and distributes the Mplus program. He has published on complex survey analysis, multilevel modeling, survival analysis, structural equation modeling, and Bayesian analysis.

Li Cai is a professor of education and psychology at UCLA, where he also serves as codirector of the National Center for Research on Evaluation, Standards, and Student Testing (CRESST). His methodological research agenda involves the development, integration, and evaluation of innovative latent variable models that have wide-ranging applications in educational, psychological, and health-related domains of study. A component on this agenda is statistical computing, particularly as related to item response theory (IRT) and multilevel modeling. He has also collaborated with substantive researchers at UCLA and elsewhere on projects examining measurement issues in educational games and simulations, mental health statistics, substance abuse treatment, and patient-reported outcomes.

Jodi M. Casabianca is an assistant professor of quantitative methods in the Department of Educational Psychology in the College of Education at the University of Texas, Austin. She earned a BA in statistics and psychology and an MS in applied and mathematical statistics from Rutgers University, and an MA and PhD in psychometrics from Fordham University. Her research interests are in psychometrics and educational measurement, specifically measurement models for rated data, measurement and computational methodology for large-scale testing and assessment, and evaluation of teaching quality and teacher professional development programs.

Paul De Boeck earned his PhD from the KU Leuven (Belgium) in 1977, with a dissertation on personality inventory responding. He has held positions at the KU Leuven as a professor of psychological assessment and at the University of Amsterdam (Netherlands) as a professor of psychological methods from 2009 to 2012. Since 2012 he is a professor of quantitative psychology at the Ohio State University. He is past section editor of ARCS *Psychometrika* and a past president of the Psychometric Society (2007–2008). His main research interests are explanatory item response models and applications in the domain of psychology and educational measurement.

Susan E. Embretson earned her PhD at the University of Minnesota in psychology in 1973. She was on the faculty of the University of Kansas for many years and has been a professor of psychology at the Georgia Institute of Technology, since 2004. She has served as president of the Psychometric Society (1999), the Society of Multivariate Psychology (1998), and American Psychological Association, Division 5 (1991). Embretson has received career achievement awards from NCME, AERA, and APA. Her current research interests include cognitive psychometric models and methods, educational test design, the measurement of change, and automatic item generation.

Jean-Paul Fox is a professor in the Department of Research Methodology, Measurement and Data Analysis, University of Twente, the Netherlands. His research is in the area of Bayesian item response modeling and he is the author of the monograph *Bayesian Item Response Modeling* (Springer, 2010). He is known for his work on multilevel IRT modeling, especially the integration of multilevel survey design and psychometric modeling. In 2001, he received the Psychometric Society Dissertation Award for his work on multilevel IRT modeling. He has also received two personal grants from the Netherlands Organisation for Scientific Research to develop psychometric models for large-scale survey research.

Hanneke Geerlings earned her PhD in psychometrics from the University of Twente, the Netherlands, where she is currently appointed as an assistant professor. Her PhD thesis was on multilevel response models for rule-based item generation.

Cees A. W. Glas is chair of the Department of Research Methodology, Measurement and Data Analysis, at the Faculty of Behavioral, Management and Social Sciences of the University of Twente in the Netherlands. He participated in numerous research projects including projects of the Law School Admission Council and the OECD international educational survey PISA. His published articles, book chapters, and supervised theses cover such topics as testing of fit to IRT models, Bayesian estimation of multidimensional and multilevel IRT models using MCMC, modeling with nonignorable missing data, concurrent modeling of item response and textual input, and the application of computerized adaptive testing in the context of health assessment and organizational psychology.

Margo G. H. Jansen earned her PhD in psychology from the University of Groningen in 1977 with a dissertation on applying Bayesian statistical methods in educational measurement. She has held positions at the Central Institute for Test Development (Cito) until 1979 and as an associate professor at the University of Groningen. Her current research interests are in educational measurement and linguistics.

Rianne Janssen earned her PhD on componential IRT models from the KU Leuven (Belgium) in 1994. She has been an associate professor at the same university since 1996.

Her research interests are in nearly every aspect of educational measurement. She is currently responsible for the national assessments of educational progress in Flanders (Belgium).

Brian W. Junker is a professor of statistics and associate dean for academic affairs in the Dietrich College of Humanities and Social Sciences at Carnegie Mellon University. Dr. Junker has broad interests in psychometrics, education research, and applied statistics, ranging from nonparametric and Bayesian item response theory, to Markov Chain Monte Carlo and other computing and estimation methods, and rating protocols for teacher quality, educational data mining, social network analysis, and mixed membership modeling. He earned a BA in mathematics from the University of Minnesota, and an MS in mathematics and PhD in statistics from the University of Illinois.

George Karabatsos earned his PhD from the University of Chicago in 1998, with specialties in psychometric methods and applied statistics. He has been a professor at the University of Illinois-Chicago since 2002. He received a New Investigator Award from the Society for Mathematical Psychology in 2002, and is an associate editor of *Psychometrika*. His current research focuses on the development and use of Bayesian nonparametrics, especially the advancement of regression and psychometrics.

Geoff N. Masters is chief executive officer and a member of the Board of the Australian Council for Educational Research (ACER)—roles he has held since 1998. He is an adjunct professor in the Queensland Brain Institute. He has a PhD in educational measurement from the University of Chicago and has published widely in the fields of educational assessment and research. Professor Masters has served on a range of bodies, including terms as president of the Australian College of Educators; founding president of the Asia–Pacific Educational Research Association; member of the Business Council of Australia's Education, Skills, and Innovation Taskforce; member of the Australian National Commission for UNESCO; and member of the International Baccalaureate Research Committee.

Gideon J. Mellenbergh earned his PhD in psychology from the University of Amsterdam in The Netherlands. He is professor emeritus of psychological methods, the University of Amsterdam, former director of the Interuniversity Graduate School of Psychometrics and Sociometrics (IOPS), and emeritus member of the Royal Netherlands Academy of Arts and Sciences (KNAW). His research interests are in the areas of test construction, psychometric decision making, measurement invariance, and the analysis of psychometrical concepts.

Dylan Molenaar earned his PhD in psychology from the University of Amsterdam in the Netherlands in 2012 (cum laude). His dissertation research was funded by a personal grant from the Netherlands Organization for Scientific Research and he was awarded the Psychometric Dissertation Award in 2013. As a postdoc he studied item response theory models for responses and response times. In addition, he has been a visiting scholar at Ohio State University. Currently, he is an assistant professor at the University of Amsterdam. His research interests include item response theory, factor analysis, response time modeling, intelligence, and behavior genetics.

Ivo W. Molenaar (PhD, University of Amsterdam) is professor emeritus of statistics and measurement, University of Groningen, The Netherlands. He is a past president of the Psychometric Society, a former editor of *Psychometrika*, and past president of

The Netherlands Society for Statistics and Operations Research (VvS). His research is in measurement models for abilities and attitudes (Rasch models and Mokken models), Bayesian methods (prior elicitation, robustness of model choice), and behavior studies of the users of statistical software. Together with Gerard H. Fischer, he coedited a monograph on Rasch models in 1995.

Eiji Muraki is a professor emeritus, Tohoku University, School of Educational Informatics, and a research advisor for the Japan Institute for Educational Measurement. He earned a PhD in measurement, evaluation, and statistical analysis from the University of Chicago and has developed several psychometric software programs, including PARSCALE, RESGEN, BILOG-MG (formerly BIMAIN), and TESTFACT. His research interests in psychometrics have been in polytomous and multidimensional response models, item parameter drift, and the method of marginal maximum-likelihood estimation. Recently, he has added computer-based testing, web-based education, and instructional design to his research agenda.

Mari Muraki is an education data consultant and was a Code for America Fellow in 2015. She currently builds technology to help schools use their student data efficiently. Previously, she led the Stanford University Center for Education Policy Analysis data warehouse and district data partnerships across the United States. She earned a BA in mathematics and statistics from the University of Chicago and an MS in statistics, measurement, assessment and research technology from the University of Pennsylvania.

Bengt Muthén obtained his PhD in statistics at the University of Uppsala, Sweden and is professor emeritus at UCLA. He was the president of the Psychometric Society from 1988 to 1989 and the recipient of the Psychometric Society's Lifetime Achievement Award in 2011. He has published extensively on latent variable modeling and many of his procedures are implemented in Mplus.

Richard J. Patz is a chief measurement officer at ACT, with responsibilities for research and development. His research interests include statistical methods, assessment design, and management of judgmental processes in education and assessment. He served as president of the National Council on Measurement in Education from 2015 to 2016. He earned a BA in mathematics from Grinnell College, and an MS and a PhD in statistics from Carnegie Mellon University.

Sophia Rabe-Hesketh is a professor of education and biostatistics at the University of California, Berkeley. Her previous positions include professor of social statistics at the Institute of Education, University of London. Her research interests include multilevel, longitudinal, and latent variable modeling and missing data. Rabe-Hesketh has over 100 peer-reviewed articles in *Psychometrika, Biometrika,* and *Journal of Econometrics,* among others. Her six coauthored books include *Generalized Latent Variable Modeling* and *Multilevel and Longitudinal Modeling Using Stata*. She has been elected to the National Academy of Education, is a fellow of the American Statistical Association, and was president of the Psychometric Society.

James O. Ramsay is professor emeritus of psychology and an associate member in the Department of Mathematics and Statistics at McGill University. He earned a PhD from Princeton University in quantitative psychology in 1966. He served as chair of the department from 1986 to 1989. Dr. Ramsay has contributed research on various topics in psychometrics, including multidimensional scaling and test theory. His current research focus is on functional data analysis, and involves developing methods for analyzing samples of curves and images. He has been the president of the Psychometric Society and the Statistical Society of Canada. He received the Gold Medal of the Statistical Society of Canada in 1998 and the Award for Technical or Scientific Contributions to the Field of Educational Measurement of the National Council on Measurement in Education in 2003, and was made an honorary member of the Statistical Society of Canada in 2012.

Mark D. Reckase is a university distinguished professor emeritus at Michigan State University in East Lansing, Michigan. He earned his PhD in psychology from Syracuse University in Syracuse, New York. His professional interests are in the areas of advanced item response theory models, the design of educational tests, setting standards of performance on educational tests, computerized adaptive testing, and statistical methods for evaluating the quality of teaching. He is the author of the book *Multidimensional Item Response Theory*. He has been the president of the National Council on Measurement in Education and the vice president of division D of the American Educational Research Association (AERA). He has received the E. F. Lindquist Award from AERA for contributions to educational measurement.

James S. Roberts earned his PhD in experimental psychology from the University of South Carolina in 1995 with a specialty in quantitative psychology. He subsequently completed a postdoctoral fellowship in the division of statistics and psychometric research at Educational Testing Service. He is currently an associate professor of psychology at the Georgia Institute of Technology and has previously held faculty positions at the Medical University of South Carolina and the University of Maryland. His research interests focus on the development and application of new model-based measurement methodology in education and the social sciences.

H. Jane Rogers earned her bachelor's and master's at the University of New England in Australia and her PhD in psychology at the University of Massachusetts Amherst. Her research interests are applications of item response theory, assessment of differential item functioning, and educational statistics. She is the coauthor of a book on item response theory and has published papers on a wide range of psychometric issues. She has consulted on psychometric issues for numerous organizations and agencies as well as on projects funded by Educational Testing Service, Law School Admissions Council, Florida Bar, and National Center for Educational Statistics.

Jürgen Rost earned his PhD from the University of Kiel in Germany in 1980. He became a professor at its Institute of Science Education and led its methodology department until 2005. He has authored more than 50 papers published in peer-reviewed journals on Rasch models and latent class models. He developed his mixture-distribution Rasch model in 1990. He edited two volumes on latent class and latent trait models. In addition, he authored a textbook on test theory and is the founding editor of *Methods of Psychological Research*, the first online open-access journal on research methodology.

Fumiko Samejima earned her PhD in psychology from Keio University, Japan, in 1965. She has held academic positions at the University of New Brunswick, Canada, and the University of Tennessee. Although her research is wide-ranging, it is best known for her pioneering work in polytomous item response modeling. Dr. Samejima is a past president of the Psychometric Society.

Klaas Sijtsma earned his PhD from the University of Groningen in the Netherlands in 1988, with a dissertation on the topic of nonparametric item response theory. Since 1981, he has held positions at the University of Groningen, Vrije Universiteit in Amsterdam, and Utrecht University, and has been a professor of methods of psychological research at Tilburg University since 1997. He is a past president of the Psychometric Society. His research interests encompass all topics with respect to the measurement of individual differences in psychology. Together with Ivo W. Molenaar, he published a monograph on nonparametric item response theory, in particular Mokken models, in 2002.

Anders Skrondal earned his PhD in statistics from the University of Oslo for which he was awarded the Psychometric Society Dissertation Prize. He is currently a senior scientist, Norwegian Institute of Public Health, adjunct professor, Centre for Educational Measurement, University of Oslo, and adjunct professor, Graduate School of Education, University of California, Berkeley. Previous positions include professor of statistics and director of the Methodology Institute, London School of Economics, and adjunct professor of biostatistics, University of Oslo. His coauthored books include *Generalized Latent Variable Modeling* and *The Cambridge Dictionary of Statistics.* His research interests span topics in psychometrics, biostatistics, social statistics, and econometrics. Skrondal is currently president-elect of the Psychometric Society.

Hariharan Swaminathan earned his BS (hon.; mathematics) from Dalhousie University, Halifax, Canada, and MS (mathematics) and MEd and PhD from the University of Toronto specializing in psychometrics, statistics, and educational measurement/evaluation. He is currently a professor of education at the University of Connecticut. His research interests are in the areas of Bayesian statistics, psychometrics, item response theory, and multivariate analysis. He has more than 300 papers, chapters, technical reports, and conference presentations to his credit. Professor Swaminathan is the coauthor of two books (with Hambleton and Rogers) on item response theory and a fellow of the American Educational Research Association. He has received outstanding teacher and mentoring awards from both the University of Massachusetts and the American Psychological Association as well as the Governor's award for outstanding contribution to the state of Connecticut by a naturalized citizen for his work with its department of education.

David Thissen is a professor of psychology at the University of North Carolina at Chapel Hill and the L.L. Thurstone Psychometric Laboratory. He earned his PhD from the University of Chicago in 1976. He was previously at the University of Kansas. His research interests include statistical models and estimation in item response theory, test scoring, test linking, and graphical data analysis. He published *Test Scoring* (with Howard Wainer) in 2001. He is a past president of the Psychometric Society, and has received a career achievement award from the NCME.

Francis Tuerlinckx earned his PhD in psychology from the University of Leuven in Belgium in 2000. He held a research position at the Department of Statistics of Columbia University, New York. Since 2004, he is a professor of quantitative psychology at the KU Leuven, Belgium. His research interests are item response theory, response time modeling in experimental psychology and measurement, Bayesian statistics, and time series analysis.

Gerhard Tutz is a professor of statistics at the Ludwig Maximilian University (LMU), Munich and a former director of the Department of Statistics at LMU. His research interests include categorical data, latent trait models, survival analysis, multivariate statistics, and regularization methods. He has authored several books, including *Regression for Categorical Data* and *Multivariate Statistical Modelling Based on Generalized Linear Models* (with Ludwig Fahrmeir).

Wim J. van der Linden is a distinguished scientist and director of research innovation, Pacific Metrics Corporation, Monterey, California, and professor emeritus of measurement and data analysis, University of Twente, the Netherlands. He earned his PhD in psychometrics from the University of Amsterdam in 1981. His research interests include test theory, computerized adaptive testing, optimal test assembly, parameter linking, test equating, and response-time modeling, as well as decision theory and its application to problems of educational decision making. He is a past president of the Psychometric Society and the National Council on Measurement in Education and has received career achievement awards from NCME, ATP, and the Psychometric Society, as well as the E. F. Lindquist award from AERA.

Han L. J. van der Maas earned his PhD in developmental psychology in 1993 (cum laude), with dissertation research on methods for the analysis of phase transitions in cognitive development. After a five-year KNAW fellowship, he joined the faculty of the Developmental Group of the University of Amsterdam, first as an associate professor and in 2003 as a professor. In 2005, he became professor and chair of the Psychological Methods Group at the University of Amsterdam. Since 2008, he is also director of the Graduate School of Psychology at the University of Amsterdam. His current research includes network models of general intelligence, new psychometric methods, and adaptive learning systems for education.

Matthias von Davier currently holds the position of senior research director, global assessment, at Educational Testing Service, Princeton, New Jersey, USA. He earned his PhD from the University of Kiel in Germany in 1996, specializing in psychometric methods. He serves as the editor of the *British Journal of Mathematical and Statistical Psychology* and is a coeditor of the *Journal of Large Scale Assessments in Education*. He received the ETS Research Scientist Award in 2006 and the Bradley Hanson Award for contributions to Educational Measurement in 2012. His research interests are item response theory, including extended Rasch models and mixture distribution models for item response data, latent structure models, diagnostic classification models, computational statistics, and developing advanced psychometric methods for international large-scale surveys of educational outcomes.

Mark R. Wilson is a professor of education at the University of California, Berkeley. He earned his PhD in psychometrics from the University of Chicago in 1984. His research interests include item response modeling, especially extensions of Rasch models, test, and instrument design based on cognitive models, philosophy of measurement, and the development of psychometric models for use in formative assessment contexts. In recent years he was elected president of the Psychometric Society, he became a member of the U.S. National Academy of Education, and he has published three books: *Constructing Measures: An Item Response Modeling Approach*, *Explanatory Item Response Models: A Generalized Linear and Nonlinear Approach* (with Paul De Boeck), and *Towards Coherence between Classroom Assessment and Accountability*.

Margaret L. Wu is a professor at the Work-based Education Research Centre, Victoria University, Melbourne. Her background is in statistics and educational measurement. She completed her PhD on the topic of the application of item response theory to problem solving. Since 1992, she has worked as a psychometrician at the Australian Council for Educational Research (ACER), the Assessment Research Centre at the University of Melbourne, and Victoria University, where she teaches educational measurement and supervises higher-degree research students. Dr. Wu has written numerous book chapters and published papers in refereed journals. She has conducted workshops and carried out consultancies within Australia and overseas. She has also coauthored an item response modeling software program, ConQuest which has been used in major projects such as OECD PISA and Australian national and statewide testing programs.

1

Introduction

Wim J. van der Linden

CONTENTS

1.1 Alfred Binet 1
1.2 Louis Thurstone 3
1.3 Frederic Lord and George Rasch 5
1.4 Later Contributions 6
1.5 Unifying Features of IRT 8
References 9

The foundations of item response theory (IRT) and classical test theory (CTT) were laid in two journal articles published only one year apart. In 1904, Charles Spearman published his article "The proof and measurement of two things," in which he introduced the idea of the decomposition of an observed test score into a true score and a random error generally adopted as the basic assumption of what soon became known as the CTT model for a fixed test taker. The role of Alfred Binet as the founding father of IRT is less obvious. One reason may be the relative inaccessibility of his 1905 article "Méthodes nouvelles pour le diagnosis du niveau intellectuel des anormaux" published with his coworker Théodore Simon in *l'Année Psychologie*. But a more decisive factor might be the overshadowing of his pioneering statistical work by something attributed to him as more important in the later psychological literature—the launch of the first standardized psychological test. Spearman's work soon stimulated others to elaborate on his ideas, producing such contributions as the definition of a standard error of measurement, methods to estimate test reliability, and the impact of the test lengthening on reliability. But, with the important exception of Thurstone's original work in the 1920s, Binet's contributions hardly had any immediate follow-up.

1.1 Alfred Binet

What exactly did Binet set in motion? In order to understand his pioneering contributions, we have to view them against the backdrop of his predecessors and contemporaries, such as Fechner, Wundt, Ebbinghaus, Quetelet, and Galton. These scientists worked in areas that had already experienced considerable growth, including anthropometrics, with its measurement of the human body, and psychophysics, which had successfully studied the relationships between the strength of physical stimuli as light and sound and human sensations of them.

Binet was fully aware of these developments and did appreciate them. But the problem he faced was fundamentally different. The city of Paris had asked him to develop a test that would enable its schools to differentiate between students that were mentally retarded, with the purpose of assigning them to special education, and those that were just unmotivated. Binet had already thought deeply about the measurement of intelligence and he was painfully aware of its problems. In 1898, he wrote (*Revue Psychologique*; cf. Wolf, 1973, p. 149):

> There is no difficulty in measurement as long as it is a question of experiments on... tactile, visual, or auditory sensations. But if it is a question of measuring the keenness of intelligence, where is the method to be found to measure the richness of intelligence, the sureness of judgment, the subtlety of mind?

Unlike the earlier anthropometricians and psychophysicists with their measurement and manipulation of simple physical quantities, Binet realized he had to measure a rather complex variable that could be assumed to be "out there" but to which we have no direct access. In short, something we now refer to as a *latent variable*.

Binet's solution was innovative in several respects. First, he designed a large variety of tasks supposed to be indicative of the major mental functions, such as memory, reasoning, judgment, and abstraction, believed to be included in intelligence. The variety was assumed to cover the "richness of intelligence" in his above quote. Second, he used these tasks in what he became primarily known for—a fully standardized test. Everything in it, the testing materials, administration, and scoring rules, was carefully protocolled. As a result, each proctor independently administering the test had to produce exactly the same results for the same students. But although Binet was the first to do so, the idea of standardization was not original at all. It was entirely in agreement with the new methodological tradition of the psychological experiment with its standardization and randomization, which had psychology fully in its grip since Wundt opened his laboratory in Leipzig in 1897. Binet had been in communication with Wundt and had visited his laboratory.

Third, Binet wanted to scale his test items but realized there was no natural scale for the measurement of intelligence. His solution was equally simple as ingenious; he chose to use the chronological age of his students to determine scale values for his items. During a pretest, he tried out all items with samples of students from each of the age groups 3–11 and assigned as scale value to each item the chronological age of the group for which it appeared to be answered correctly by 75% of its students. These scale values were then used to estimate the mental age at which each student actually performed. (Six years later, William Stern proposed to use the ratio of mental and chronological age as intelligence quotient [IQ]. A few more years later, Lewis Terman introduced the convention of multiplying this IQ by 100. Ever since, the mean IQ for a population has invariably been set at 100.)

This author believes that we should honor Binet primarily for the introduction of his idea of scaling. He dared to measure something that did not exist as a directly observable or *manifest variable*, nevertheless felt the necessity to map both his items and students on a single scale for it, estimated the scale values of his items using empirical data collected in a pretest, and then scored his students using the scale values of the items—exactly the same practice as followed by any user of IRT today. In fact, Binet's trust in the scaling of his test items was so strong that he avoided the naïve idea of the need for a standardized test to administer exactly the same items to each of his students. Instead, the items were selected adaptively. The protocol of his intelligence test included a set of rules that precisely prescribed how the selection of the items had to move up and down along the scale as a function of the student's achievements. We had to wait for nearly a century before Binet's idea of adaptive testing became generally accepted as the benchmark of efficient testing!

In Binet's work, several notions were buried which a modern statistician would have picked up immediately. For instance, his method of scaling involved the necessity to estimate population distributions. And with his intuitive ideas about the efficiency of testing, he entered an area now generally known as optimal design in statistics. But, remarkably, Spearman did not recognize the importance of Binet's work. He remained faithful to his linear model with a true score and measurement error and became the father of factor analysis. (And probably needed quite a bit of his time for a fight with Karl Pearson over his correction of the product–moment correlation coefficient for unreliability of measurement.)

1.2 Louis Thurstone

The one who definitely did recognize the importance of Binet's work was Louis Thurstone. In 1925, he introduced a method of scaling with the intention to remove the necessity to use age as a manifest substitute for the latent intelligence variable. He did so by exactly reversing what Binet had treated as given and to be estimated: Binet used the chronological age of his students as a given quantity and subsequently estimated the unknown shape of the empirical curves of the proportion of correct answers as a function of age for each of his items to identify their scale values. Thurstone, on the other hand, assumed an unknown, latent scale for the items but did impose a known shape on these curves—that of the cumulative normal distribution function, using their estimated location parameters as scale values for the items. As a result, he effectively disengaged intelligence from age, giving it its own scale (and fixing an unforeseen consequence of Terman's definition of IQ, namely, an automatic decrease of it with chronological age). Figure 1.1 shows the location of a number of Binet's test items on his new scale, with the zero and unit fixed at the mean and standard deviation of the normalized distribution of the scale values for the 3.5-year-old children in his dataset (Thurstone, 1925). Thurstone also showed how his assumption of normality could be checked empirically, and thus basically practiced the

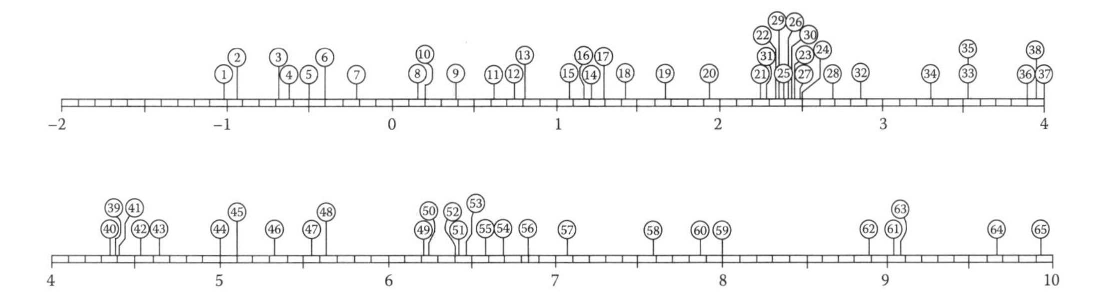

An absolute scale of Binet test questions
Linear unit: Standard deviation of Binet test intelligence of 3.5-year-old children
Origin: The mean of Binet test intelligence of 3.5-year-old children

FIGURE 1.1
Thurstone's scaling of the Binet test items. (Reproduced from Thurstone, L. L. 1925. *Journal of Educational Psychology*, 16, 433–451.)

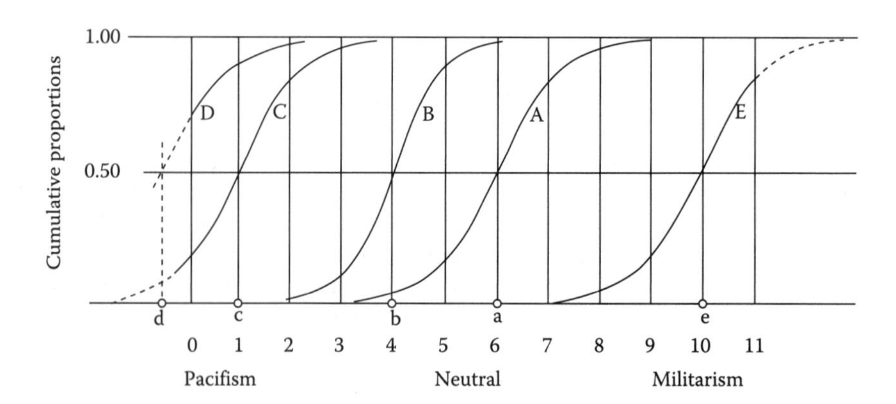

FIGURE 1.2
A few items from Thurstone's scale for pacifism–militarism. (Reproduced from Thurstone, L. L. 1928. *American Journal of Sociology*, 23, 529–554.)

idea of a statistical model as a hypothesis to be tested for its fit to reality. A few years later, he expanded the domain of measurement by showing how something hitherto vague as an attitude could be measured. The only two things necessary to accomplish this were his new method of scaling and a set of agree–disagree responses to a collection of statements evaluating the object of the attitude. Figure 1.2 shows a few of his response functions for the items in a scale for the measurement of the attitude dimension of pacifism–militarism.

In spite of his inventiveness and keen statistical insights, Thurstone's work was plagued by one consistent aberration—the confusion between the use of the normal ogive as the mathematical shape of a response function and as the distribution function for the score in some population. His confusion was no doubt due to the general idea spread by Quetelet that the "true" distribution of data properly collected for any unrestricted population had to be normal (Stigler, 1986, Chapter 5). This platonic belief was so strong in his days that when an observed distribution did not look normal, it was just normalized to uncover the "true" scale for the test scores, which automatically was supposed to have "equal measurement units."

In hindsight, it looks as if Thurstone wanted to reconcile this hypothesis of normality with the well-established use of the normal ogive as response function in psychophysics. Anyhow, the impact of psychophysical scaling on the early history of test theory is unmistakable. As already noted, psychophysicists study the relationships between the strength of physical stimuli and the psychological sensations they invoke in human subjects. Their main method was a fully standardized psychological experiment in which a stimulus of varying strength was administered, for instance, a flash of light of varying clarity or duration, and the subjects were asked to report whether or not they had perceived it. As the probability of perception obviously is a monotonically increasing function of the strength of the stimulus, the choice of a response function with a shape identical to that of the normal distribution function appeared obvious. Thurstone made the same choice but for a latent rather than a manifest variable, with the statistical challenges of having to fit it just to binary response data and the need to evaluate its goodness of fit.

Mosier (1940, 1941) was quite explicit in his description of such parallels between psychophysics and psychometrics. In his 1940 publication, he provided a table with the different terms used to describe analogous quantities in the two disciplines. Others using the same normal-ogive model as response functions for test items at about the same time include

Ferguson (1942), Lawley (1943), Richardson (1936), and Tucker (1946). These authors differed from Thurstone, however, in that they fell back completely upon the psychophysical method, using the normal-ogive function just as a nonlinear regression model for the response to the individual items on the observed score on the entire test. Ferguson even used it to regress an external, dichotomous success criterion on the observed scores on several test forms with different difficulties.

1.3 Frederic Lord and George Rasch

The first not to suffer in any way from the confusion between distributions functions and their use as response functions were Lord (1952) and Rasch (1960). In his two-parameter normal-ogive model, Lord directly formulated the normal-ogive function as a mathematical model for the probability of a correct response given the unknown ability θ measured by it. He also used the model to describe the bivariate distributions of observed item scores and the latent ability, explore the limiting frequency distributions of number-correct scores on large tests, and derived the bivariate distributions of number-correct scores on two tests measuring the same ability.

Rasch was more rigorous in his approach and introduced his model as an attempt to change the paradigm for social and behavioral research more fundamentally. Unlike the typical researchers in this field, he rejected the notion of a population as entirely undesirable. His best-known model involved two different types of parameters, ability parameters representing the ability of the persons that are measured and difficulty parameters for the test items. Although we would need more than one person and item to estimate these parameters, their datasets are basically ad hoc; the estimates they produce are no descriptions of some population but of the intrinsic properties of individual persons and items only.

Rasch formulated several arguments in support of his proposed change of paradigm. Interestingly, one of his first was that his concepts were "more akin to the psychophysical measurements where the concern is with individual subjects, each observed on several occasions" (Rasch, 1960, p. 4). Indeed, psychophysics never has been interested in the normal hypothesis as a postulate for some population of subjects and always tried to verify its laws at the level of individual subjects. Rasch's main argument, however, was that the concepts of ability and difficulty have exactly the same status, for instance, as the basic concepts of force and mass in physics. And indeed, the parallels are convincing. In Maxwell's analysis of the role of force and mass in the theory of motion, reviewed in Rasch (1960, Section 7.2), both represent quantities that cannot be observed directly—our only access to them is through observing the acceleration produced by the force when applied to an object of a certain mass, as described by Newton's second law. In other words, force and mass are latent quantities as well. And just as Newton's law describes what happens when an individual force is exerted on an individual physical object, Rasch's model is for the encounter between an individual subject and an individual item.

The usual presentation of Rasch's model with these two basic parameters is in the form of the ratio of the ability parameter over the sum of the ability and difficulty parameters (Rasch, 1960, Section 5.10; Volume One, Chapter 3, Equation 3.3). Although Rasch was aware of the fact that it could be represented in a form close to the normal-ogive function used by Lord (1952) by using the logistic transform (Rasch, 1960, Section 6.8), he hardly ever used it.

In the subsequent years, the developments for the normal-ogive model basically came to a halt. Lord lost much of his interest in it, mainly because it did not account for what he considered as an inherent aspect of test items, the likelihood of a correct answer just as a result of guessing. The developments for the Rasch model were impressive though. In his 1960 book, Rasch introduced two other models with a similar parameter structure, a Poisson model for reading errors and a gamma model for reading time. Immediately after it, he proposed a general unidimensional model for items with multiple response categories that included his dichotomous model as a special case (Rasch, 1961). He also laid the statistical foundation for parameter estimation and statistical testing of the goodness of fit of his models.

Through his many contacts with the international statistical community, Rasch must have been aware of the Pitman–Koopman–Darmois theorem about the exponential family of distributions admitting sufficient statistics for its parameters published in the mid-1930s as well as the possibility of conditional inference for this family (e.g., Lehmann, 1959, Section 2.7). Each of Rasch's models was intentionally formulated to belong to the exponential family, and his statistical treatment of them was based exactly on the type of conditional inference it admits. His initial statistical work immediately stimulated others to develop an impressive toolkit for parameter estimation and goodness-of-fit analysis for his models, including major advances made, for instance, by Andersen (1970, 1971), Fischer (1974), and Wright (1968) and his students.

As for Lord's two-parameter model, the role of Alan Birnbaum was decisive. Birnbaum was a statistician who worked in relative isolation on IRT in the late 1950s, mainly through a series of internal research reports. But his work has become well known through his contribution to Lord and Novick (1968, Chapters 17–20). Birnbaum was aware of the tradition of modeling dose–response curves in bioassay and the change it had made from the normal-ogive to the mathematically more tractable logistic response function mainly through the work by Berkson (1944, 1951). He suggested the same change for Lord's model. In addition, he suggested the addition of a guessing parameter to it, and basically launched what has become known as the three-parameter logistic (3PL) model. Birnbaum's formidable chapters in Lord and Novick (1968) laid the statistical foundation for maximum-likelihood estimation of the item and ability parameters in this model. They also introduced Fisher's information measure as a quantity key to IRT, and demonstrated its applicability to an optimal design approach to test construction. Another significant contribution by Birnbaum was his derivation of optimal weights for use in test scoring and two-point classification problems.

1.4 Later Contributions

Later landmark contributions to IRT include those by Bock (1972) and Samejima (1969, 1972), who, continuing Rasch's (1961) early explorations of IRT for test items with polytomous scoring, proposed alternative models for this response format. Especially, Samejima's 1972 monograph has turned out to be prophetic; she already suggested several types of models for nominal, graded, and rating responses that were picked up by a next generation of psychometricians.

During the following two decades of IRT, the field basically matured. Its scope became much wider and the number of active researchers devoting their time to it increased considerably. Alternative models were developed, for instance, for ordinal data and multidimensional abilities, some of them being nonparametric. At the same time, IRT ventured on more explanatory models for response processes or mixtures thereof, and addressed the possibility of nonmonotone response functions. Also, though still somewhat hesitatingly, the first efforts were made to model response times on test items outside the domain of reading speed already charted by Rasch (1960). The number of empirical applications of IRT remained still rather small though, mainly because of limited computational resources. A rather comprehensive summary of all developments in this period can be found in a previous handbook edited by van der Linden and Hambleton (1997) with some 28 chapters each highlighting one of the models developed in these two decades.

The growth in the number of IRT models developed in the last two decades has been even more dramatic. It is hardly possible to open a new issue of any of the main measurement journals and not find an article with a modification of an existing model or a proposal for an entirely new one. Several of these modifications or new models represent rather fundamental changes. While the very first authors emphasized the applicability of their models at the level of an single subject responding to a single item, with Rasch's (1960) explicit attempt at changing the existing paradigm as a prime example, many of the newer attempts embed these models as lowest-level models in a framework with a multilevel or hierarchical structure. As a result, IRT models now enable us to introduce explanatory covariates for response processes, deal with group structures with systematic differences between ability parameters, and address mixtures of response processes. They also allow us to account for higher-level relationships between responses and response times, or for special structures in item domains, such as family structures due to the use of rule-based item generation. This proliferation of models for responses with multiple layers of parameters is definitely not something that has emerged solely in the field of IRT, but has been part of a much wider development in probability modeling and statistics known as generalized linear and nonlinear latent variable modeling.

One of most effective changes, however, has been in the computational and statistical treatment of all these new models. As any other type of probabilistic modeling, applications of IRT have always depended heavily on the efficiency of their parameter estimation methods and the power of statistical tests of model fit. But, with a few obvious exceptions, its initial use of statistics did not always profit from the rigor that could have been realized. The current landscape is entirely different, though; the use of statistics and computational techniques in IRT has become quite advanced. A substantial part of the change is no doubt due to the ample availability of the extreme computational power produced by the revolution in information technology. A similarly overwhelming development has taken place in Bayesian parameter estimation and model comparison. Bayesian approaches had already gained some ground in the early days of IRT, but were certainly not common at the time. But thanks to the computational success of Markov chain Monte Carlo methods, these approaches have now become dominant, especially for the more complex hierarchical models.

IRT has now also penetrated nearly every daily practice of testing, not only in the original areas of educational and psychological measurement but also in health measurement, marketing research, large-scale international surveys, and cognitive diagnosis. In each of the areas, IRT is used to calibrate items, assess their model fit, assemble tests according to optimal design principles, score subjects, set time limits, link scores and parameters for different test forms, run adaptive testing programs, set standards for mastery or admission to programs or treatments, and detect item or test security breaches.

1.5 Unifying Features of IRT

Given all these different developments, one may wonder what still unifies IRT. In the opinion of this author, three clear principles demarcate the field. The first principle is a focus on the responses by human subjects to test items rather than an *a priori* chosen score on the entire test, as in more traditional test theory. The second is the recognition of the random nature of these responses and the acknowledgment of the need of a probabilistic model to explain their distribution. The third principle is the presence of separate parameters for the effects of the subjects' abilities, skills, or attitudes and the properties of the items. This parameter separation is the ultimate defining characteristic of IRT. It distinguishes IRT from CTT, which models an expected *a priori* chosen score by a subject on a set of items as a linear composition of a true score and an error, without any further decomposition of this true score into separate effects.

The principle of parameter separation is illustrated by the two causal diagrams in Figure 1.3. The first represents the separation for a single-level IRT model. Its two arrows represent the separate effects due to the latent ability of the subject (first oval) and the properties of the item (second oval) on the observed score (rectangle). This diagram underlies the original models for objectively scored dichotomous items by Birnbaum Lord, and Rasch, as well as several of their later generalizations to polytomous items. The second diagram is for the case of a response to an item that requires human rating, for instance, a short answer to a constructive-response item or an essay written in response to a given prompt. As the diagram reveals, we now have to apply the principle of parameter separation twice—once for the response produced by the subject in reply to the item and again for the score produced by the rater in reply to the subject's response. The diagram, which underlies the hierarchical rater models reviewed by Casabianca, Junker, and Patz (Volume One, Chapter 27), shows how quickly we may have to move from the realm of single-level models to models with multiple levels of parameters.

Binet's tentative application of the principle of parameter separation to the responses by Parisian students to the items in his intelligence test a century ago started a tradition of

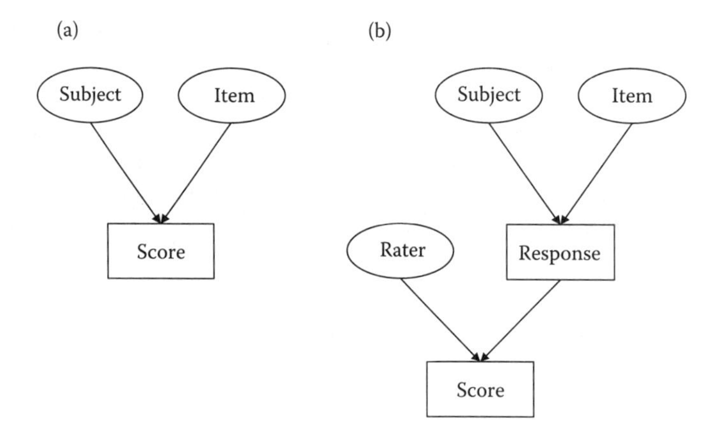

FIGURE 1.3
Examples of the two causal diagrams underlying IRT models for the scores on (a) objectively scored test items and (b) constructive-response items that involve human rating of the response.

response modeling he impossibly could have foreseen. Given the multitude of recent contributions collected in this handbook, it seems fair to admit that we are unable to assess where its future developments will lead us, or if it ever will reach a more stable state.

References

Andersen, E. B. 1970. Asymptotic properties of conditional maximum likelihood estimators. *Journal of the Royal Statistical Society B*, 32, 283–301.

Andersen, E. B. 1971. Asymptotic properties of conditional likelihood ratio tests. *Journal of the American Statistical Association*, 66, 630–633.

Berkson, J. A. 1944. Applications of the logistic function to bio-assay. *Journal of the American Statistical Association*, 39, 357–365.

Berkson, J. A. 1951. Why I prefer logits to probits. *Biometrics*, 7, 327–329.

Binet, A. and Simon, Th. A. 1905. Méthodes nouvelles pout le diagnostic du niveau intellectuel des anormaux. *l'Année Psychologie*, 11, 191–336.

Birnbaum, A. 1968. Some latent trait models and their use in inferring an examinee's ability. In F. M. Lord and M. R. Novick (Eds.), *Statistical Theories of Mental Test Scores* (pp. 19–20). Reading, MA: Addison-Wesley.

Bock, R. D. 1972. Estimating item parameters and latent abilities when responses are scored in two or more latent categories. *Psychometrika*, 43, 561–573.

Ferguson, G. A. 1942. Item selection by the constant process. *Psychometrika*, 7, 19–29.

Fischer, G. H. 1974. *Einführung in die Theorie Psychologischer Tests*. Bern, Switzerland: Huber.

Lawley, D. N. 1943. On problems connected with item selection and test construction. *Proceedings of the Royal Society of Edinburgh*, 61, 73–287.

Lehmann, E. L. 1959. *Testing Statistical Hypotheses*. New York: Wiley.

Lord, F. M. 1952. A theory of test scores. *Psychometric Monograph No. 7*. Richmond, VA: Psychometric Corporation. Retrieved from: http://www.psychometrika.org/journal/online/MN07.pdf

Lord, F. M. and Novick, M. R. 1968. *Statistical Theories of Mental Test Scores*. Reading, MA: Addison-Wesley.

Mosier, C. I. 1940. Psychophysics and mental test theory: Fundamental postulates and elementary theorems. *Psychological Review*, 47, 355–366.

Mosier, C. I. 1941. Psychophysics and mental test theory. II. The constant process. *Psychological Review*, 48, 235–249.

Rasch, G. 1960. *Probabilistic Models for Some Intelligence and Attainment Tests*. Copenhagen, Denmark: Danish Institute for Educational Research.

Rasch, G. 1961. On general laws and meaning of measurement in psychology. In *Proceedings of the Fourth Berkeley Symposium on Mathematical Statistics and Probability* (Volume 4) (pp. 321–333).

Richardson, M. W. 1936. The relationship between the difficulty and the differential validity of a test. *Psychometrika*, 1, 33–49.

Samejima, F. 1969. Estimation of ability using a response pattern of graded scores. *Psychometric Monograph No. 17*. Richmond, VA: Psychometric Corporation. Retrieved from: http://www.psychometrika.org/ journal/online/MN07.pdf

Samejima, F. 1972. A general model for free-response data. *Psychometric Monograph No. 18*. Richmond, VA: Psychometric Corporation. Retrieved from: http://www.psychometrika.org/journal/online/MN07.pdf

Spearman, C. 1904. The proof and measurement of association between two things. *American Journal of Psychology*, 15, 72–101.

Stigler, S. M. 1986. *The History of Statistics: The Measurement of Uncertainty Before 1900*. Cambridge, MA: Harvard University Press.

Thurstone, L. L. 1925. A method of scaling psychological and educational tests. *Journal of Educational Psychology*, 16, 433–451.

Thurstone, L. L. 1928. Attitudes can be measured. *American Journal of Sociology*, 23, 529–554.

Tucker, L. R. 1946. Maximum validity of a test with equivalent items. *Psychometrika*, 11, 1–13.

van der Linden, W. J. and Hambleton, R. K. (Eds.) 1997. *Handbook of Modern Item Response Theory*. New York: Springer.

Wolf, T. H. 1973. *Alfred Binet*. Chicago: The University of Chicago Press.

Wright, B. D. 1968. Sample-free test calibration and person measurement. In *Proceedings of the 1967 Invitational Conference on Testing Problems*. Princeton, NJ: Educational Testing Service.

Section I

Dichotomous Models

2

Unidimensional Logistic Response Models

Wim J. van der Linden

CONTENTS

2.1 Introduction 13
2.2 Presentation of the Models 14
 2.2.1 Fixed-Effects Models 14
 2.2.2 Random-Effects Models 17
 2.2.3 Model Identifiability 19
 2.2.4 Parameter Linking 21
 2.2.5 Model Interpretation 21
2.3 Parameter Estimation 25
2.4 Model Fit 26
2.5 Empirical Example 26
2.6 Discussion 28
Acknowledgments 29
References 29

2.1 Introduction

The family of models discussed in this chapter is for responses to dichotomous items. This family of models has a long tradition of successful applications in educational and psychological testing as well as several other areas of behavioral and cognitive measurement. In fact, it is by far the most frequently used family of models for these applications.

The most general member of the family, known as the three-parameter logistic (3PL) model, was introduced by Birnbaum (1968). It seems fair to ascribe the origins of a special version of it, known as the two-parameter logistic (2PL) model, to Lord (1952). Although he actually used the normal-ogive instead of the logistic response function, his model had a parameter structure identical to that of the 2PL model. The one-parameter logistic (1PL) model is a special case of the 2PL model. It can be shown to be equivalent to the Rasch model. Rasch (1960, Section 6.8) was aware of this option but invariably used the simpler representation of it with exponential versions of the difficulty and ability parameters in this chapter. However, as the Rasch model is an exponential family model from which it borrows special statistical properties, it deserves a separate review (Volume One, Chapter 3). In this chapter, we therefore review the 1PL model only to the extent that it shares statistical properties with the other two models.

2.2 Presentation of the Models

All three models exist as versions with fixed-effects and random-effects parameters. Historically, the introduction of the former preceded the latter. The main reason for the introduction of the latter was to overcome the computational issues associated with the fixed-effects models discussed later in this chapter.

2.2.1 Fixed-Effects Models

The distributions addressed by the fixed-effects versions of the three models are for the dichotomous responses $U_{pi} = 0,1$ by test takers $p = 1, \ldots, P$ on items $i = 1, \ldots, I$. The prime examples of this type of responses are items scored as correct or incorrect. The distributions are Bernoulli with probability functions

$$f(u_{pi};\pi_{pi}) = \pi_{pi}^{u_{pi}}(1-\pi_{pi})^{1-u_{pi}}, \quad p=1,\ldots,P; \quad i=1,\ldots,I, \tag{2.1}$$

where $\pi_{pi} \in [0,1]$ are the success parameters for the distributions; that is, the probabilities of a response $U_{pi} = 1$ by the test takers on each of the items (Casabianca and Junker, Volume Two, Chapter 2).

Making the usual assumption of independence between the responses by the same test taker ("local independence"), and assuming they all worked independently, the probability function of the joint distribution of a complete response matrix, $\mathbf{U} \equiv (U_{pi})$, is the product of each of these Bernoulli distributions:

$$f(\mathbf{u};\pi) = \prod_{p=1}^{P}\prod_{i=1}^{I}\pi_{pi}^{u_{pi}}(1-\pi_{pi})^{1-u_{pi}} \tag{2.2}$$

with parameter vector $\boldsymbol{\pi} \equiv (\pi_{11}, \ldots, \pi_{1I}, \ldots, \pi_{P1}, \ldots, \pi_{PI})$.

The 3PL model explains each π_{pi} as a function of parameters for the effects of the test taker's ability and the properties of the item. More specifically, let θ_p denote the parameters for the effects of the individual abilities of the test takers and a_i, b_i, and c_i the effects of the items generally interpreted as representing their difficulties, discriminating power, and success probabilities when guessing randomly on them, respectively. For a given response matrix, the 3PL model equations are

$$\pi_{pi} = c_i + (1-c_i)\frac{\exp[a_i(\theta_p - b_i)]}{1+\exp[a_i(\theta_p - b_i)]}, \quad p=1,\ldots,P; \quad i=1,\ldots,I \tag{2.3}$$

with $\theta_p \in (-\infty, \infty)$, $a_i \in (0, \infty)$, $b_i \in (-\infty, \infty)$, and $c_i \in [0,1]$ as ranges for the values of their parameters. The model thus consists of $P \times I$ nonlinear equations, one for each of the success parameters. In other words, rather than a single-level probabilistic model, it is a system of second-level mathematical equations. (The statistical literature is somewhat ambiguous in its assignments of the number of levels to a model; some sources count the levels of parameters as we do here, while others count the levels of randomness that are modeled.)

It is common to introduce the 3PL model graphically instead of as the system of equations in Equation 2.3, emphasizing the shape of the success probabilities π_{pi} as a function of a mathematical variable θ. Figure 2.1 shows these response functions for 40 arithmetic items estimated under the 3PL model. For each of these functions, the c_i parameter represents the height of the lower asymptote to it. More formally, these parameters are defined as

$$c_i = \lim_{\theta_p \to -\infty} \pi_{pi}. \tag{2.4}$$

Naively, for a multiple-choice item with A alternatives, one might expect to find $c_i = 1/A$. But in practice, guessing turns out to be somewhat more complicated than a test taker just picking one of the alternatives completely at random; empirical estimates of the c_i parameters typically appear to be slightly lower than $1/A$. The b_i parameters represent the location of the curves on the scale for θ. Formally, they are the values of θ with success probability

$$\pi_{pi} = (1 + c_i)/2, \tag{2.5}$$

that is, the probability halfway between their maximum of 1 and minimum of c_i. Finally, the α_i parameters can be shown to be proportional to the slope of the response functions at $\theta = b_i$, at which point the slopes take the value

$$\frac{\partial \pi_{pi}}{\partial \theta_p} = 0.25a_i(1 - c_i). \tag{2.6}$$

Although graphs of response functions as in Figure 2.1 definitely add to our understanding of the 3PL model, they are also potentially misleading. This happens especially

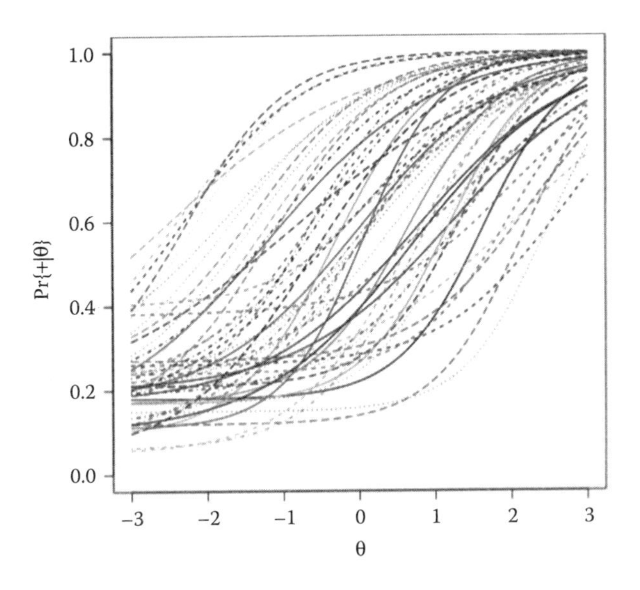

FIGURE 2.1
Response functions for a set of 40 arithmetic items estimated under the 3PL model.

when, mainly for didactic reasons, the model is introduced by explaining the graph of the response function for one fictitious item only. For the case of one item, the system of equations is not identified. Consequently, as its parameters cannot be estimated, the model does not have any empirical meaning (see below). For the case of multiple items, as in Figure 2.1, the graph nicely reveals the relative values of the parameters of each of the I items in Equation 2.3 but still hides those of the P ability parameters.

Use of Equation 2.3 as the representation of the model also reveals that, in spite of its wide acceptance, the adoption of the qualifier "3PL" in the name of the model is potentially confusing as well. It suggests a restriction of the parameter count to the parameters of one item only, ignoring both those of all other items and the ability parameters. However, replacing it by the more appropriate qualifier "$(P+3I)$PL" would still leave us with a misnomer. The shapes of the response functions shown in Figure 2.1 are *not* those of a logistic function. Only the 2PL and 1PL models below have logistic response functions.

Finally, it is still not uncommon to find a version of the model with a scale constant $D = 1.7$ added to it, giving it $1.7a_i(\theta_p - b_i)$ as its core structure. The original purpose of this practice was to bring the shape of the response functions close to that of the normal-ogive model, generally believed to be the "true model" that would give us a scale with "equal units of measurement" in the early days of test theory (Volume One, Chapter 1). However, as discussed below, the scale for the θ_p parameters is arbitrary up to a nonmonotonic transformation, and the practice of bringing it close to the scale of a normal ogive is a relict from a distant past, bound to create communication problems between users of item response theory (IRT) rather than add to their understanding of it.

The 2PL model follows from Equation 2.3 upon the choice of $c_i = 0$ for all items. It has

$$\pi_{pi} = \frac{\exp[a_i(\theta_p - b_i)]}{1+\exp[a_i(\theta_p - b_i)]}, \quad p=1,\ldots,P; \quad i=1,\ldots,I \tag{2.7}$$

as its system of model equations. Observe that this model does assume the logistic function

$$\Psi(x) = \frac{\exp(x)}{1+\exp(x)}$$

as response function. For this choice, location parameters b_i simplify to the point on the θ scale with success probability $\pi_{pi} = 0.5$ for each item, which now also is the point at which the slope of the response functions in Equation 2.6 specializes to $0.25a_i$. A more substantive assumption underlying the 2PL model is thus absence of any guessing—an assumption that may hold when there is nothing at stake for the test takers. However, the model is sometimes applied in cases where the assumption is clearly untenable, typically with the claim that the c_i parameters are generally difficult to estimate. But even if this claim were valid (it is not), it would empirically be more defensible to set these parameters equal to a value somewhat larger than zero to better account for the effects of guessing.

The 1PL model follows upon the additional assumption of $a_i = 1$ for all items, yielding

$$\pi_{pi} = \frac{\exp(\theta_p - b_i)}{1+\exp(\theta_p - b_i)}, \quad p=1,\ldots,P; \quad i=1,\ldots,I \tag{2.8}$$

as its system of model equations. As already alluded to, reparameterizing the model through substitution of

$$\begin{aligned} \theta_p &= \ln(\tau_p), \\ b_i &= \ln(\xi_i), \ \tau_p, \xi_i > 0 \end{aligned} \tag{2.9}$$

into it gives us

$$\pi_{pi} = \frac{\tau_p}{\tau_p + \xi_i}, \quad p = 1,\ldots,P; \quad i = 1,\ldots,I, \tag{2.10}$$

which is the version typically used as representation of the Rasch model (Volume One, Chapter 3).

Attempts have been made to add an upper asymptote to the model in Equation 2.3 to make it robust against careless response behavior in addition to guessing (Barton and Lord, 1981), but this "4PL" model just has not made it. An empirically more successful model may be the Rasch model extended with the structure for the guessing parameters in Equation 2.3 added to it; for details on this 1PL-G model, see San Martín et al. (2006). Finally, the success of the family of logistic models in this chapter has led to its generalization to items with polytomous scoring and multidimensional ability parameters; examples of such generalizations are reviewed by Masters (Volume One, Chapter 7), Muraki and Muraki (Volume One, Chapter 8), Tutz (Volume One, Chapter 9), and Reckase (Volume One, Chapter 12).

2.2.2 Random-Effects Models

In spite of the convincing nature of the logistic models, their initial number of applications was extremely low. For the Rasch model, the favorite method of parameter estimation was maximum conditional likelihood (CML) estimation (Volume One, Chapter 3), but due to the limited computational power available in the 1960–1970s, it appeared impossible to deal with its elementary symmetric functions for longer tests, even for Fischer's efficient summation algorithm (van der Linden, Volume Two, Chapter 6). The typical estimation methods in use for the 3PL and 2PL models were maximum joint likelihood (JML) methods. It is still not fully clear why these methods occasionally failed to show convergence, even for larger datasets and shorter test lengths. Although lack of convergence may be caused by lack of model identifiability, it is now clear that the major computer programs imposed effective additional restrictions to prevent this from happening (see below). In hindsight, these disturbing results might have been just due to limited numerical precision and/or the presence of the ability parameters as incidental parameters during item calibration (Volume Two, Chapter 9).

One of the first attempts to neutralize the possible role of incidental parameters was Bock and Lieberman's (1970) reformulation of the fixed-effects three-parameter normal-ogive model as a model with random ability parameters. Their reformulation applies equally well to the logistic models.

The reformulation involves the adoption of a different probability experiment. Instead of a set of P fixed test takers, it assumes these test takers to be randomly and independently sampled from the same probability distribution. Let Θ be the random ability parameter assumed to have a common density $f(\theta)$ for the test takers and $\pi_i(\theta)$ be the probability of a correct response to item i by a test taker with realization $\Theta = \theta$.

The focus is on the random response vector to be observed for each random test taker (observe that we now have two levels of randomness). We use $\mathbf{u}_v \equiv (u_{v1}, \ldots, u_{vI})$, $v = 1, \ldots, 2^I$ to denote each possible realization of this vector. Because of local independence, the conditional probability of observing $\mathbf{U}_v = \mathbf{u}_v$ given $\Theta = \theta$ is equal to

$$\pi_v(\theta) = \prod_{i=1}^{I} \pi_i(\theta)^{u_{vi}} [1 - \pi_i(\theta)]^{1-u_{vi}}. \tag{2.11}$$

Continuing the argument, the marginal probabilities of observing $\mathbf{U}_v = \mathbf{u}_v$ are equal to

$$\pi_v = \int \prod_{i=1}^{I} \pi_i(\theta)^{u_{vi}} [1 - \pi_i(\theta)]^{1-u_{vi}} f(\theta) d\theta, \quad v = 1, \ldots, 2^I. \tag{2.12}$$

A random sample of P test takers amounts to an equal number of independent draws from the space of all possible response vector. Let X_v denote the number of times vector v is observed, with $\Sigma_v X_v = P$. The probability model for this experiment is a multinomial distribution with probability function

$$f(\mathbf{x}; \boldsymbol{\pi}) = \frac{P!}{x_1! \ldots x_{2^I}!} \prod_{v=1}^{2^I} \pi_v^{x_v}, \tag{2.13}$$

where $\mathbf{x} \equiv (x_v)$ and now $\boldsymbol{\pi} \equiv (\pi_v)$.

Treating $p = 1, \ldots, P$ as index for the order in which the test takers are sampled, and using $\mathbf{U} \equiv (U_{pi})$ again to denote the response matrix, the function can also be written as

$$f(\mathbf{u}; \boldsymbol{\pi}) = \prod_{p=1}^{P} \int \prod_{i=1}^{I} \pi_{pi}^{u_{pi}} (1 - \pi_{pi})^{1-u_{pi}} f(\theta_p) d\theta_p \tag{2.14}$$

with π_{pi} given by Equation 2.3, which is the form usually met in the IRT literature. This form hides the crucial difference between the fixed-effects and random-effects models as a consequence of their different sampling distributions, though. The former is the second-level system of $P \times I$ equations for the Bernoulli parameters π_{pi} in Equation 2.3. The latter is a system at the same level but now with the 2^I equations for the multinomial parameters π_v in Equation 2.12.

Somewhat later, Bock and Aitkin (1981) introduced an application of the expectation–maximization (EM) algorithm for the estimation of the item parameters in the 3PL model based on the marginal likelihood associated with Equation 2.14, which has become one of the standard procedures for item calibration in IRT. A description of the logic underlying the EM algorithm is provided by Aitkin (Volume Two, Chapter 12). For an application of the algorithm, it is not necessary for the test takers to actually be randomly sampled; in principle, a statistical model can produce good results even if not all of its assumptions are met. And extensive parameter recovery studies of item calibration with maximum marginal likelihood (MML) estimation with the EM algorithm have certainly proven the random-effects models to do so, in spite of their imposition of a mostly arbitrary common density $f(\theta)$ on the ability parameters. But for the issue of model identifiability addressed in the next section, it is crucial to be aware of the precise sampling model underlying it.

It is not necessary to treat the ability parameters as the only random parameters in the model, leaving the item parameters fixed. Several advantages may be related to the adoption of random-effects parameters for the items as well (De Boeck, 2008). A natural application of random-item IRT is modeling of the distributions of the item parameters in the different families of items produced by different settings of a rule-based item generator; for details, see Glas, van der Linden and Geerlings (Volume One, Chapter 26).

2.2.3 Model Identifiability

The problem of lack of model or parameter identifiability arises when different combinations of parameter values are observationally equivalent, that is, imply the same distribution for the observed data. As it is generally impossible to use identically distributed data to distinguish between different parameter values, any attempt at statistical inference with respect to the parameters then breaks down.

Fortunately, the product of Bernoulli functions in Equation 2.2 is an example of a fully identified sampling model. For any shift in any combination of the components of its parameter vector $\boldsymbol{\pi} \equiv (\pi_{11}, \ldots, \pi_{1I}, \ldots, \pi_{P1}, \ldots, \pi_{PI})$, the model yields a distinct joint distribution of the responses. Conversely, for any distinct response distribution, it is thus possible to infer a unique combination of values for this vector. This last statement has only theoretical meaning though; in the practice of educational and psychological testing, due to memory or learning effects, it is mostly impossible to get more than one response per test taker to the same item without changing the ability parameters, leaving us with the extreme values of zero and one as direct estimates of each success parameter.

It is at this point, however, that second-level systems of model equations as in Equations 2.3, 2.7, and 2.8 show favor. Basically, they enable us to pool the data from multiple test takers and responses into estimates of each of the success parameters. Because of their feature of parameter separation ("different parameters for every p and i in the joint index of π_{pi}") as well as their cross-classified nature ("multiple equations with the same parameters"), these systems enable us to adjust the data on the test-taker parameters for the differences between the items, and conversely. As a result, we effectively have multiple data points per ability and item parameter, and are able to combine their estimates into estimates of each of the individual success parameters. (The problem of incidental parameters is intentionally ignored here.) As discussed below, these success parameters are key parameters in item response modeling which provide us with the empirical interpretation of the test takers' scores.

These systems of equations are only able to do their job, however, if they are identified themselves; that is, when, for the 3PL model, each distinct combination of values for the model parameters $(\theta_p, a_i, b_i, c_i)$ corresponds with a distinct value of π_{pi}. Generally, mathematical systems of equations are only identifiable if the number of equations is at least as great as the number of unknowns. Thus, at a minimum, it should hold for this model that $P \times I \geq P + 3I$—a condition which implies, for example, that for a five-item test we are unable to produce any fixed parameter estimates unless the number of test takers is at least equal to four.

However, the condition is not sufficient. For one thing, as is well known, no matter the numbers of test takers and items, it is always possible for some subset of the model parameters to compensate for certain changes in the others. More specifically, the addition of the same constant to all b_i and θ_p parameters in Equation 2.3 does not lead to a change in any of the π_{pi}. Likewise, multiplying all a_i parameters by a constant has no effect on these success parameters when all θ_p and b_i parameters are divided by it as well. The identifiability problem is more fundamental, though. Deeper analysis has revealed cases for which the c_i parameters are not identifiable either (Maris, 2002; van der Linden and Barrett, 2016). A more comprehensive review of identifiability issues in IRT is provided by San Martín (Volume Two, Chapter 8).

For a response matrix with fixed dimensions, the only way to make the system of model equations identifiable is by adding more equations to it. A recent result by van der Linden and Barrett (2016, Volume Three, Chapter 2, Theorem 3) can be used as a check on the sufficiency of such identifiability restrictions for the logistic models in this chapter. The result is a formal characterization of the class of observationally equivalent parameters values for the 3PL model in the form of a mapping φ which, for an arbitrary test taker and item, for any given solution to the model equations gives us all other solutions. The mapping can be shown to be the vector function

$$\varphi(\theta,a,b,c)=(u\theta+v,u^{-1}a,ub+v,c) \tag{2.15}$$

with $u \equiv [\varphi_\theta(\theta)-\varphi_\beta(\beta)]/(\theta-\beta)$, $\theta \neq b$, and $v \equiv \varphi(b)-ub=\varphi(\theta)-u\theta$.

The critical quantities in this result are the (unknown) parameters u and v. Given any arbitrary combination of values for $(\theta_p, a_i, b_i, c_i)$, they index all other combinations with identical values for $(\pi_{11}, \ldots, \pi_{1I}, \, , \pi_{P1}, \ldots, \pi_{PI})$ in Equation 2.3.

Observe that for the choice of $(u,v)=(1,0)$ the function just returns its input. It follows that adding equations to the system that effectively restrict u and v to this pair of values does restrict it to have only one solution as well—in other words, makes the model identifiable. As an example, suppose we set $\theta_p=\kappa_1$ and $\theta_{p'}=\kappa_2$ for two arbitrary test takers p and p' and constants κ_1 and κ_2. The first component of Equation 2.15 can then only take the form

$$\theta_p = u\theta_p + v$$

$$\theta_{p'} = u\theta_{p'} + v$$

for these two test takers, leaving $(u,v)=(1,0)$ as the only admissible values.

These two restrictions are definitely not the only option. More generally, it can be shown that in order to make the fixed-effect 3PL model identifiable, it is sufficient to add (i) one linear restriction on one of the a_i parameters in combination with another on one of the b_i or θ_p parameters or (ii) two independent linear restrictions on the θ_p and/or b_i parameters (van der Linden, 2016, Theorem 1).

The sets of identifiability restrictions for the 3PL and 2PL model are identical. The restrictions for the 1PL model follow if we ignore the a_i parameters. As for the c_i parameters, the fact that they do not need to be restricted does not imply that they are automatically identifiable though. The mapping in Equation 2.15 is a vector function with components that hold only simultaneously; we should not isolate one of the components and declare it to hold independently of the others. The correct conclusion from Equation 2.15 is that the c_i parameters are identifiable once all other model parameters are. In fact, our earlier references to specific cases for which the c_i parameters have shown to lack identifiability already implied the necessity of additional conditions.

A similar analysis for the random-effects versions of the logistic models has not resulted in any final conclusions yet. Remember that their sampling distribution is multinomial with the vector of marginal probabilities $\boldsymbol{\pi}=(\pi_1,\ldots,\pi_{2^I})$ in Equation 2.12 as parameters.

The good news is that the multinomial model is identifiable as well; its family of distributions is known to have distinct members for distinct vectors of success probabilities. The problematic part, however, is the more complicated system of equations for these success probabilities as a function of the model parameters in Equation 2.12, which now includes the unknown parameters of the density $f(\theta)$ as well.

A common practice for the 3PL and 2PL models is to set the mean and variance of the ability parameters equal to zero and one, respectively:

$$\mu_\theta = 0 \quad \text{and} \quad \sigma_\theta^2 = 1. \tag{2.16}$$

For the fixed-effects versions of these models, these restrictions were typically implemented by renorming all ability parameters estimates upon each iteration step in the JML estimation procedure. It is easy to show that these restrictions pass our earlier check on the validity of the identifiability restrictions for these models (van der Linden, 2016). For the random-effects versions, the restrictions are typically implemented through the choice of the standard normal density for $f(\theta)$ in Equation 2.14. Although, a formal proof is still missing, numerous parameter recovery studies have invariably supported the validity of this practice.

2.2.4 Parameter Linking

One of the consequences of the fundamental lack of identifiability of the logistic response models is the necessity to link parameter values obtained in different calibration studies through common items and/or test takers. This is necessary even when formally identical identifiability restrictions are imposed.

A simple case illustrates the necessity. Suppose the same test taker responds to different items in two calibration studies, both with Equation 2.16 as identifiability restrictions. Owing to the presence of other test takers, the restrictions have a differential effect on the ability parameter for this common test taker. For example, if the test taker would be among the most able in the first study but the least able in the second, the use of $\mu_\theta = 0$ would force his parameter value to be positive in the former but negative in the latter. A similar analysis is possible for common items in different calibration studies.

Observe that the issue has nothing to do with the impact of estimation error; all changes are in what commonly is referred to as true parameter values. "True" should not be taken to mean unique, though. Lack of identifiability means that each parameter has an entire class of observationally equivalent values. Identifiability restrictions are necessary to reduce each of these classes to a unique value. These restrictions do not operate on each parameter in isolation but, as an integral part of the system of model equations, have a joint effect on all of them.

The close relationship between the lack of model identifiability and the necessity of parameter linking suggests an equally important role of the characterization of the equivalent parameter values in Equation 2.15 with respect to the latter. Indeed, the linking functions required to map the unique parameters values in one calibration onto those in another have to be derived from Equation 2.15 as well; details are provided in van der Linden and Barrett (Volume Three, Chapter 2).

2.2.5 Model Interpretation

Several times so far, we have emphasized the formal nature of the response models as a system of equations for the test takers' success probabilities on the items. These success probabilities are the "empirical reality" explained by the models. Conversely, in order to empirically interpret the features of the model, we have to restrict ourselves to the given set of probabilities. Any appeal to a reality outside of it is speculative at best.

It follows that formal features of the response models that can be changed without changing the success probabilities for the test takers are meaningless. Only features that remain invariant under such changes can have a valid empirical interpretation. A prime example of features of the logistic models that are not invariant are the values of their parameters. As just noted, each of them can be made to vary as a result of the infinitely many choices for the identifiability restrictions that have to be imposed on the models. Hence, statements as "John's ability score on this test is equal to 50" or "the difficulty of this item is equal to –2" should not be taken to have any absolute meaning.

Actually, the same lack of invariance holds for the entire parameter structure of the model. As already demonstrated, it is possible to use Equation 2.9 to reparameterize the 1PL model into the Rasch model. Both representations have a completely different structure. But empirically, it is impossible to tell one from the other; for the same test takers and items, they imply the same probabilities of success and thus identical response distributions. This observation forces us to extend our earlier notion of observational equivalence to include equivalence across reparameterizations of the model as well.

Note that it is possible to go back and forth between the 1PL and Rasch models because the logarithmic transformation in Equation 2.9 has the exponential as its inverse. More generally, it is always possible to reparameterize probabilistic models provided the vectors of the old and new parameters are of the same dimension and have a reversible (bijective) relationship. The logistic models are monotone, continuous functions in their parameters (the only exception to the monotonicity is when $\theta \neq b$, which case we exclude). If we want to keep these features, the required relationship reduces to a similar monotone function between the old and new parameters (van der Linden and Barrett, 2016, theorem 3). Hence, the following conclusion: For the logistic models, it is always possible to replace their parameters by new parameters provided the old and new parameters are monotone continuous functions of each other.

For each of these infinitely many possible reparameterizations, the response probabilities for the test takers remain the same but the response functions look entirely different. Figure 2.2 illustrates the impact of a few reparameterizations of the Rasch model. Its first plot shows the response functions for 20 items for the original version of the model in Equation 2.10. Following S. S. Stevens' (1946) classification of different levels of measurement, the choice is sometimes claimed to yield an absolute zero for the ability scale, and thus measurement on a ratio scale. The second plot shows the shapes of the response functions for the same items for the standard parameterization of the 1PL model. The fact that the curves now run "parallel" seems to suggest equal measurement units along the ability scale and this feature has led to claims of measurement on an interval rather than a ratio scale. These two claims are already inconsistent. But actually, almost every claim of special features for the scale of ability parameters can be "reparameterized away." The third plot shows the shift in response functions obtained for an alternative parameterization of the Rasch model, using

$$\begin{aligned} \tau_p &= \lambda_p - 1, \lambda_p > 1, \\ \xi_i &= \beta_i, \beta_i > 0 \end{aligned} \tag{2.17}$$

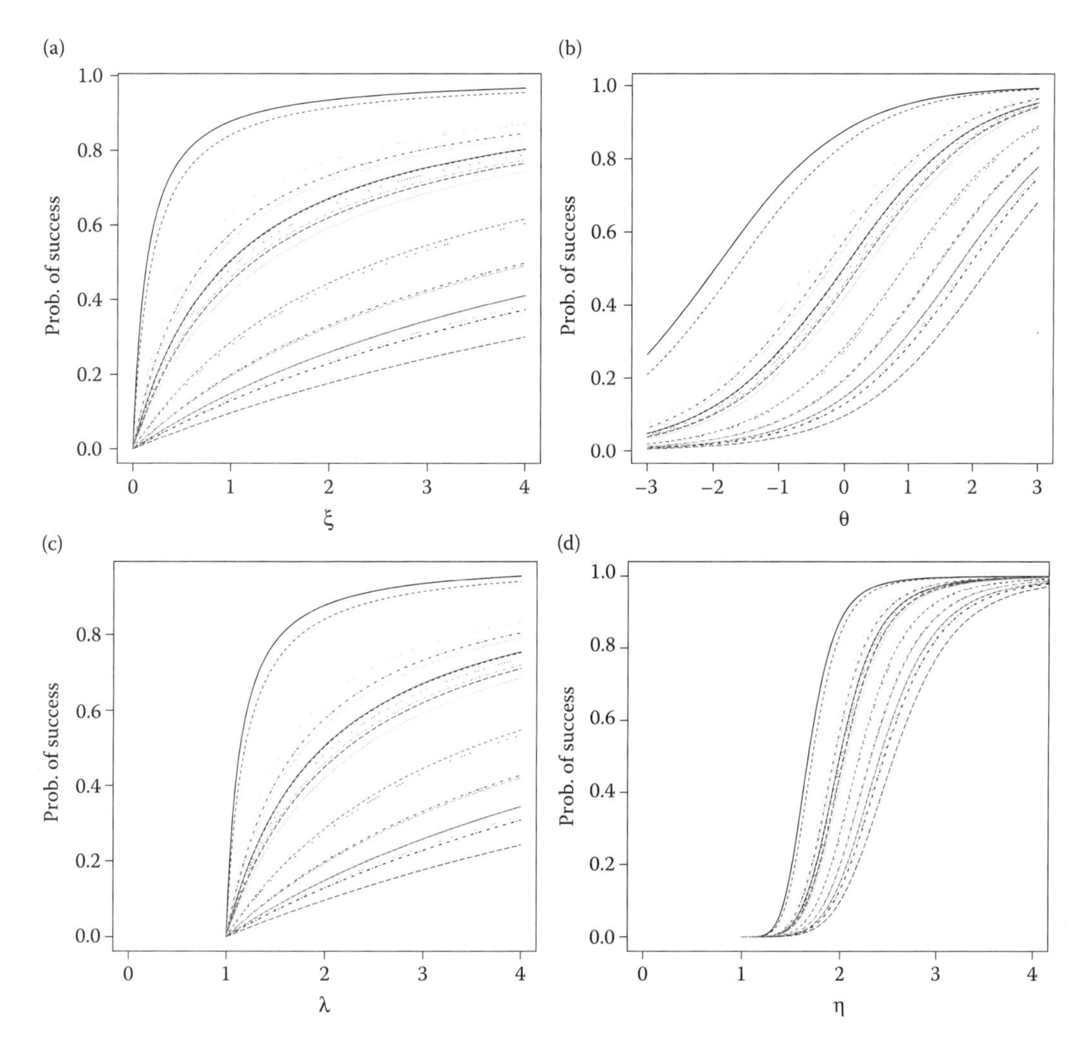

FIGURE 2.2
Four different parameterizations of the Rasch model, each with the same goodness of fit to the same response data. (a) Equation 2.10; (b) Equation 2.8; (c) Equation 2.18; (d) Equation 2.20.

for all p and i. The result is

$$\pi_{pi} = \frac{\lambda_p - 1}{\lambda_n + \beta_i - 1}, \tag{2.18}$$

which now has a scale for the ability parameters with the earlier "zero" at the point $\lambda = 1$. Our last example is

$$\begin{aligned} \tau_p &= \eta_p^5 - 1, \ \eta_p > 1, \\ \xi_i &= \varepsilon_i, \ \varepsilon_i > 0 \end{aligned} \tag{2.19}$$

for all p and i, which gives us

$$\pi_{pi} = \frac{\eta_p^5 - 1}{\eta_p^5 + \varepsilon_i - 1} \tag{2.20}$$

as another representation of the Rasch model. The result has the same shift of "zero" but now gives us ogive-shaped response functions squeezed together at the lower end of range of possible values for the ability parameters. The squeeze clearly illustrates how the same interval of values for the ability parameters at different positions along the scale is associated with different changes in success probability. (Actually, the same observation could already have been made directly for the standard 1P model. The "parallelness" of its response functions in the second plot is just an optical illusion. How could monotone curves each mapping the real line onto [0,1] ever be made to run completely parallel?)

Figure 2.2 illustrates how dangerous it is to ascribe absolute meaning to apparent features of the logistic response models. The possibilities of reparameterization are virtually endless. Other choices could have been made with more surprising changes in the shapes of the response functions, especially if we would include other transformations of the item parameters than with the identify functions above. The same conclusions hold for the 2PL and 3PL models, with even a larger variety of changes due to the possibility to play with their a_i and c_i parameters as well.

What, if any, features of the logistic models are invariant across all possible reparameterizations and thus allow for absolute interpretation? There are three. The first is the order of the ability parameters. Although their scale does have an arbitrary zero and unit, the models do order all ability parameters identically across all possible parameterizations—a fundamental prerequisite for measurement of a unidimensional variable.

The second is the quantile rank of the test takers in *any* norm group for which the test is in use. The feature follows from a simple statistical fact. Let $q_p \in [0,1]$ be the rank of test taker p with score θ_p in a norm group with any distribution $F_\Theta(\cdot)$ defined as the proportion of test takers below this person; that is, $F_\Theta(\theta_p) = q_p$. For any monotone reparameterization $\vartheta(\cdot)$, it holds that

$$F_{\vartheta(\Theta)}(\vartheta(\theta_p)) = F_\Theta(\theta_p). \tag{2.21}$$

Thus, $F_{\vartheta(\Theta)}(\vartheta(\theta_p)) = q_p$ as well, and p has the same quantile rank in the new as in the old distribution.

The final feature, not surprisingly, is the invariant link between each of the test takers and their probabilities of success on the items. No matter how we reparameterize the model, for the same items, it always assigns the same probability of success to each of the test takers. The feature can be used to map the content of the items along the current ability scale, providing it with a point-by-point empirical interpretation. Graphically, each reparameterization of the model amounts to local stretching or shrinking of the scale, but the order of all points remains the same and so does their empirical interpretation. Item maps have been shown to be key to IRT-based test score interpretation (Hambleton and Zenisky, Volume Three, Chapter 7), standard setting (Lewis and Lord-Bessen, Volume Three, Chapter 11), and more generally item writing and test construction (Wilson, 2005). In our example below, we use a logistic response model to construct an item map that illustrates the behavioral meaning of the variable of body height.

In educational testing, it is common to distinguish between norm-referenced and criterion-referenced interpretations of test scores. The former refers to the quantile ranks of test scores in the distributions of the norm groups for which the test is in use; the latter to empirical information on what test takers with a given score can and cannot perform. From the preceding discussion, we are able to conclude that the logistic response models in this chapter provide us with measurement of a unidimensional variable on scale that, except for nonmonotone transformation, is entirely arbitrary but nevertheless has absolute norm- and criterion-referenced interpretations.

The fact that the parameter structure of the logistic response models is arbitrary should not lead to relativism or carelessness. As highlighted by Hand (2004), measurement in any science has both a representational and pragmatic aspect. The pragmatic aspect is the result of the necessity to make arbitrary choices with respect to such steps as the design of the measurement instrument, the actual measurement operations, definition of the data, and auxiliary assumptions required to infer the measurements from them. These choices are entirely separate from the empirical features the measurements present, but are necessary to map them on numbers.

As for the item parameters, the standard parameterization of the 3PL model in Equation 2.3 involves parameters (a_i, b_i, c_i) that, strictly speaking, do not have any meaning beyond their formal definitions in Equations 2.4 through 2.6. Our references to them as parameters for the discriminating power, difficulty, and probabilities of successful guessing on an item are just empirical metaphors. Although fundamentally arbitrary, these metaphors are well established and have helped us to carefully communicate differences between test items. Any other choice of parameterization would lead to loss of these metaphors and seriously disrupt our communications.

2.3 Parameter Estimation

The most frequently used methods for estimating the parameters in the logistic models are Bayesian and MML methods, where the former have taken the lead due to the current popularity of its Markov chain Monte Carlo (MCMC) methods. This handbook has several other chapters exclusively devoted to these methods, which use one or more of our logistic models to illustrate their ideas, equations, and algorithms, and highlight important details of their implementation. It is pointless to duplicate their materials in this chapter.

A general introduction to Bayesian statistical methods, explaining their logic of inference with its focus on the joint posterior distribution of all model parameters, is provided by Johnson and Sinharay (Volume Two, Chapter 13). As a continued example, these authors use the one-parameter normal-ogive model. The treatment of their example easily transfers to the 1PL model though. Junker, Patz and VanHoudnos (Volume Two, Chapter 15) should be consulted for a comprehensive introduction to the theory and practice of MCMC methods for sampling posterior distributions. Specifically, they introduce the history of general ideas underlying these methods, address all choices that have to be made when implementing them, and use the 2PL model extensively to illustrate how to use a Metropolis–Hastings algorithm to sample the parameters in the 2PL model. Readers interested in *R* code to run the algorithm find it in their example.

Glas (Volume Two, Chapter 11) offers an extensive introduction to MML estimation of item parameters in IRT models, including the logistic models in this chapter, based on Fisher's identity as organizing principle. The identity equates the derivatives of the log of the marginal likelihood in Equation 2.14 with respect to the parameters to their posterior predictive expectation across the missing ability parameters. The resulting systems of estimation equations can be solved numerically or using an EM algorithm. The use of the EM algorithm in this context is carefully explained in Aitkin (Volume Two, Chapter 12). Unlike the Bayesian methods, MML estimation only provides us with estimates of the item parameters. Estimates of the ability parameters are typically obtained either through a second maximum-likelihood step treating the item parameters as known constants or in a Bayesian fashion that accounts for their remaining uncertainty.

Almost every computer program in an extensive section with reviews of programs in *Handbook of Item Response Theory, Volume 3: Applications* can be used to estimate the current logistic models, including programs that embed them in larger modeling frameworks, such as *Mplus*, generalized linear latent modeling (GLLAM), and *Latent GOLD.*

2.4 Model Fit

Bayesian model fit methodology for item response models is reviewed in Sinharay (Volume Two, Chapter 19). The methodology enables us to check on such features of the models as monotonicity of their response functions, local independence between the responses, differential item function across subgroups of test takers, or an assumed shape for the ability distribution (given the current parameterization). The methods include Bayesian residual analysis (i.e., evaluation of the size of the observed u_{pi} under the posterior distribution of π_{pi}) and predictive checks on a variety of test quantities given the prior or posterior distributions of the parameters. The chapter also shows how easy it is to implement these methods for MCMC sampling of the posterior distribution of the parameters, and extensively demonstrates an implementation for the 1PL model.

Glas (Volume Two, Chapter 17) shows how to test the validity of each logistic models against a variety of alternative hypotheses using a coherent framework of Lagrange multiplier (LM) tests. His chapter also reviews alternatives based on likelihood-ratio and Wald tests.

For a more descriptive analysis of model fit based on the classical residuals $U_{pi} - \pi_{pi}$ aggregated across the test takers and/or items, the reader should consult Wells and Hambleton (Volume Two, Chapter 20). This review is built entirely around a demonstration of the application of their methods to an empirical dataset for the 3PL model.

The logistic models in this chapter are nested in the sense that a version with fewer parameters is obtained constraining some of the parameters in a version with more. For comparison with alternative models outside this nested family, the information criteria in Cohen and Cho (Volume Two, Chapter 18) or Bayes factors in Sinharay (Volume Two, Chapter 19) can be used. The chapter by Cohen and Cho has an entire section reviewing the applications of information criteria for model comparison throughout the most recent IRT literature.

2.5 Empirical Example

The purpose of the example is to illustrate the use of a logistic response model for the measurement of body height. The application is somewhat unusual in that body height seems a

I bump my head quite often
For most people, my shoes would be too large
When a school picture was taken, I was always asked to stand in the first row
In bed, I often suffer from cold feet
When walking down the stairs, I usually take two steps at a time
I think I would do reasonably well in a basketball team
As a police officer, I would not make much of an impression
In most cars, I sit uncomfortably
I literally look up to most of my friends
I often have to stand on my toes to look in the mirror
Seats in theaters are usually too small for me

FIGURE 2.3
Sample items from 32-item test of body height.

physical variable only to be measured by a yardstick. However, differences in body height do have behavioral consequences. Conversely, it is thus perfectly possible to treat these consequences as indicators of body height and infer measurements from them.

The measurement instrument used to collect the current dataset was a test of 32 dichotomous items, each formulating a different behavioral consequence. Examples of the items are given in Figure 2.3. The subjects were 214 students in a class on test theory at Free University, Amsterdam, who were asked to respond to each of the items indicating whether or not they endorsed the experience formulated in them. As nothing was at stake for the test takers, guessing was not expected to play any role, and the responses were analyzed using the 2PL model. The method of estimation was MML estimation with the EM algorithm as implemented in *BILOG-MG*, version 3 (Volume Three, Chapter 23; Zimowski et al., 2003), using its default option of a standard normal distribution for the θ_p parameters.

Three of the items yielded values for the residual-based chi-square type statistic in *BILOG-MG* with probabilities of exceedance lower than 0.10. The items were "I often sit uncomfortably in the back seat of a car" ($p = 0.08$), "When people were chosen for a school basketball team, I was usually chosen last" ($p = 0.04$), and "When I sit at a table, I usually have trouble with legroom" ($p = 0.05$). It is tempting to speculate about the reasons of misfit—Dutch students generally bike? Not too many Dutch schools with a basketball team? Other people at the same table creating the problem?—but we just accepted the lack of fit of these items as a statistical fact. In addition, we removed two items with rather skewed response distributions from the analysis. One had only seven zeroes among its 214 responses, which resulted in an estimate of its b_i parameters of −4.62 with a standard error of estimation equal to 1.34. The other had only 15 ones, with a b_i estimate of 6.18 and a standard error equal to 2.19. These extreme values were taken to point at poor identifiability of the parameters from the given data.

The estimates of the b_i parameters of the remaining 27 items varied across [−2.23, 3.55] with standard errors in the range of [0.09, 0.90]. Likewise, the estimates of the a_i parameters ran between [0.36, 2.94] with standard errors of [0.10, 0.76].

Figure 2.4 shows the item map for a selection from the remaining items with brief content descriptions at their points of 0.50 probability of endorsement. Our selection is largely arbitrary; in an electronic version of the map, we would have put all items in it, with the options to zoom in on certain portions of the map, click on them to see full items or an interval of uncertainty about their location, etc. But the current version already illustrates how much richer a plain variable as body height becomes when it is presented lined with

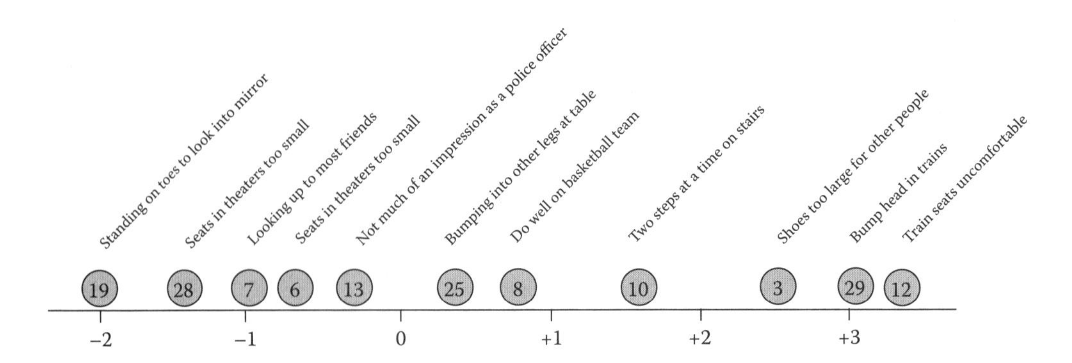

FIGURE 2.4
Example of an item map for a test of body height (locations of the items at 0.50 probability of their endorsement).

behavioral consequences rather than a single number for each subject. Unfortunately, our dataset did not include physical measurements of the body heights of its subjects. Otherwise we could have used the arbitrariness of the scale of the 2PL model to introduce a transformation of it that would have mapped the item content along the physical scale used to obtain the measurements.

2.6 Discussion

The logistic models in this chapter belong to the best-researched and most frequently used models in the field of IRT. At times, however, their use has led to debates between supporters of the 1PL and 3PL model pleading an *a priori* preference for their model. The former have typically supported their plea by pointing at such unique properties of the 1PL model as the presence of sum scores as sufficient statistics for its model parameter, measurement on interval or ratio scales, and similar ordering of the response functions across the ability scale. The latter have pointed at a higher likelihood of fit of their model to empirical response data and its better accounting for the presence of guessing on test items, especially in high-stakes testing.

Either position has both something against and in favor of it. The 1PL model belongs to the exponential family and consequently does have sufficient statistics for its parameters, a feature that gives us the option of conditional inference about them. Except for the elegance of its estimation equations and the expressions of its goodness-of-fit statistics, conditional reference does not have any serious advantages though. It certainly does not have any computational advantages, a criterion that, given the overwhelming computational power currently at our disposal, has lost its significance anyhow. More importantly, the 2LP and 3PL models have sufficient statistics too—the complete response vectors—that contain all statistical information available about their ability and item parameters. In fact, if we condition on sum scores for the 1PL model, some of the information in the response vectors (not much) is lost because their conditional distributions still depends on the parameters we try to eliminate (Eggen, 2000).

As for the unique scale properties claimed to hold for the 1PL model, Figure 2.2 illustrates that none of the logistic models has any of them. The common thing they provide are measurements of a unidimensional variable on a scale with an invariant order of the test takers.

The 1PL model does have one invariant property in addition to those shared with the other logistic models though—similar order of its response functions across all possible reparameterizations. This feature of noncrossing functions, which is *not* a prerequisite for the identical ordering of the test takers, definitely allows for an easier choice of the response probability necessary to create an item map, in the sense that each choice leads to the same ordering of the item content in the map. For the other two logistic models, the order may change locally as a result of crossing response functions, a fact sometimes difficult to understand by consumers with lack of statistical training (for more on this issue, see Lewis and Bessen-Lord, Volume Three, Chapter 11).

The 3PL does offer free parameters to account for the possibility of guessing on test items. However, though empirical estimates of these parameters do a better job of catching the effects of guessing than versions of the model with these parameters set to zero, the assumption of knowledge-or-random guessing on which they are based is definitely too simple to represent all the subtle processes that may go on in test takers who are uncertain about their answers (von Davier, 2009).

As the 3PL is more flexible than the two other logistic models, it does have a higher likelihood of fitting a given response matrix. But it does not automatically follow that its ability parameter estimates are therefore always better. For smaller datasets (short test lengths and/or few test takers), its greater flexibility actually make it adjust to random peculiarities in the data that better be avoided. The statistical issue at play here is the bias-accuracy trade-off fundamental to all statistical modeling: For a given dataset, if we increase the flexibility of a model by adding more parameters, the likelihood of bias in their estimates as a result of misfit decreases, but always at the price of a greater inaccuracy due to a smaller number of observations per parameter. For small datasets, it may therefore pay off to choose a more constrained model as the 1PL model. Lord caught this point exactly when he gave his 1983 article the title "Small *N* Justifies Rasch Model."

Acknowledgments

The idea of a test of body height has been circulating for quite some time in the Netherlands. As far as I have been able to trace back, the original idea is due to Frits Zegers, who introduced an early version of the test as a teaching tool in his test theory class in the 1970. Klaas Sijtsma adjusted and expanded the set of items in the 1980s. Henk Kelderman was so generous to give me a copy of the dataset, which he had collected more recently using students in one of his test theory classes as subjects. Although classical item analyses for the test have been distributed, I am not aware of any IRT analysis of it. I thank all my colleagues for the opportunity they have given me to use the items and dataset for the example in this chapter.

References

Barton, M. A. and Lord, F. M. 1981. An upper asymptote for the three-parameter logistic item-response model. Research Bulletin, 81–20. Princeton, NJ: Educational Testing Service.

Birnbaum, A. 1968. Some latent trait models and their use in inferring an examinee's ability. In F. M. Lord and M. R. Novick, *Statistical Theories of Mental Test Scores* (pp. 392–479). Reading, MA: Addison-Wesley.

Bock, R. D. and Aitkin, M. 1981. Marginal maximum likelihood estimation of item parameters: Application of an EM algorithm. *Psychometrika*, 46, 443–459.

Bock, R. D. and Lieberman, M. 1970. Fitting a response model for *n* dichotomously scored items. *Psychometrika*, 35, 179–197.

De Boeck, P. 2008. Random item IRT models. *Psychometrika*, 73, 533–599.

Eggen, T. J. H. M. 2000. On the loss of information in conditional maximum likelihood estimation of item parameters. *Psychometrika*, 65, 337–362.

Hand, D. J. 2004. *Measurement Theory and Practice: The World through Quantification*. London: Oxford University Press.

Lord, F. M. 1952. A theory of test scores. *Psychometric Monograph No. 7*. Richmond, VA: Psychometric Corporation. Retrieved from: http://www.psychometrika.org/journal/online/MN07.pdf

Lord, F. M. 1983. Small *N* justifies Rasch model. In D. J. Weiss (Ed.), *New Horizons in Testing: Latent Trait Test Theory and Computerized Adaptive Testing* (pp. 51–61). New York: Academic Press.

Maris, G. 2002. Concerning the identification of the 3PL model (Measurement and Research Department Reports 2002-3). Arnhem, The Netherlands: Cito.

Rasch, G. 1960. *Probabilistic Models for Some Intelligence and Attainment Tests*. Chicago: University of Chicago Press.

Rasch, G. 1960. Probabilistic Models for Some Intelligence and Attainment Tests. Chicago: University of Chicago Press.

San Martín, E., Del Pino, G., and De Boeck, P. 2006. IRT models for ability-based guessing. *Applied Psychological Measurement*, 30, 183–203.

Stevens, S. S. 1946. On the theory of scales of measurement. *Science*, 103, 677–680.

van der Linden, W. J. 2016. Identifiability restrictions for the fixed-effects 3PL model. *Psychometrika*, 81, in press.

van der Linden, W. J. and Barrett, M. D. 2016. Linking item response model parameters. *Psychometrika*, 81, in press (doi: 20.1007/211336-015-9496-6).

von Davier, M. 2009. Is there a need for the 3PL model? Guess what? *Measurement: Interdisciplinary Research and Perspectives*, 7, 110–114.

Wilson, M. 2005. *Constructing Measures: An Item Response Modeling Approach*. Mahwah, NJ: Erlbaum.

Zimowski, M. F., Muraki, E., Mislevy, R., and Bock, R. D. 2003. *BILOG-MG: Multiple Group IRT Analysis and Test Maintenance for Binary Items* (3rd ed.) [Computer software]. Lincolnwood, IL: Scientific Software International, Inc.

3

Rasch Model

Matthias von Davier

CONTENTS

3.1 Introduction 31
3.2 Presentation of the Model 32
3.3 Parameter Estimation 35
3.4 Model Fit 41
3.5 Empirical Example 42
3.6 Discussion 44
References 45

3.1 Introduction

The Rasch model (Rasch, 1960, 1966a,b) is widely used and highly praised by some researchers but criticized by others who prefer more highly parameterized models for item response data. The reason behind both of these conflicting viewpoints is that the Rasch model can be derived from a small number of simple assumptions. Proponents say these assumptions form a mathematically elegant foundation of the model, while opponents say they are rather restrictive.

Even though this divide exists, many researchers in educational measurement, psychometrics, patient-reported outcomes (including quality-of-life research), and other domains are at least familiar with the name of the Rasch model and are aware that it is an important approach in the field that has inspired many applications and methodological developments.

The amount of research on the Rasch model is almost impossible to summarize in a single chapter. For more information, readers are referred to edited volumes on the Rasch model, its extensions, and diagnostics (e.g., Fischer and Molenaar, 1995; von Davier and Carstensen, 2007) or the detailed chapter by Fischer (2006). This chapter focuses on an introduction of the model, highlighting its derivation and links it to other models. In doing so, it touches only lightly on some of the most important results on specific objectivity and parameter separability (e.g., Rasch, 1960, 1966a,b), as well as on consistent estimation (e.g., Kiefer and Wolfowitz, 1956; Andersen, 1972; Haberman, 1977; de Leeuw and Verhelst, 1986; Follmann, 1988; Lindsay et al., 1991), fit diagnostics, model testing (e.g., Andersen, 1973; Molenaar, 1983; Glas, 1988a, 2007; Rost and von Davier, 1994, 1995), and attempts to put the effects of model misspecification in context (Molenaar, 1997).

3.2 Presentation of the Model

Tasks in which examinees are asked to read sentences aloud are an example of the type of items for which the Rasch model was originally developed, as is a series of problems requiring the use of essentially the same mathematical operation but with different numbers. In the motor domain, the Rasch model could be used to analyze repeated attempts of trials, such as the high jump with different heights of the bar or free throws with different distances from the basket. With the advent of the computer, we could also add repeated matches of chess players against computer programs of different strengths to this list. The commonality is that these examples confront a test taker (or athlete or player) with a series of very similar trials or items to be mastered, with success or failure expected to depend only on a specific skill of the examinee and stable common characteristics of the different items.

The following notational conventions are used: Let $\boldsymbol{U} = (U_1, \ldots, U_I)$ denote a vector of discrete observed variables, with $U_i \in \{0,1\}$ and 1 indicating success and 0 indicating failure. For each test taker $p = 1, \ldots, P$, let $U_{pi} \in \{0,1\}$ denote the test taker's response in terms of success or failure on item i. Rasch (1966a) introduces the model using the following assumptions:

1. The probability that an examinee p succeeds in trial i is

$$Pr\{U_i = 1 \mid p\} = \frac{\lambda_{pi}}{1+\lambda_{pi}}, \tag{3.1}$$

which is equivalent to

$$Pr\{U_i = 0 \mid p\} = \frac{1}{1+\lambda_{pi}}. \tag{3.2}$$

2. Parameter $\lambda_{pi} = (Pr\{U_i = 1 \mid p\} / Pr\{U_i = 0 \mid p\})$ can be written as

$$\lambda_{pi} = \frac{\tau_p}{\xi_i}, \tag{3.3}$$

which implies that $Pr\{U_i = 1|p\} = (\tau_p/\xi_i + \tau_p)$ and $Pr\{U_i = 0|p\} = (\xi_i/\xi_i + \tau_p)$.

3. Stochastic independence of these probabilities holds for examinee p attempting multiple trials $i = 1, \ldots, I$.

Note that with the transformation $\theta_p = \ln(\tau_p)$ and $\beta_i = \ln(\xi_i)$, we obtain an equivalent form of the Rasch model:

$$Pr\{U_i = u \mid p\} = \frac{\exp(u(\theta_p - \beta_i))}{1+\exp(\theta_p - \beta_i)}, \tag{3.4}$$

which relates more readily to other logistic models (Volume One, Chapter 2).

Parameters τ_p and ξ_i are unknown quantities that describe the level of skill or proficiency of examinee p and the difficulty of item i, respectively. Presenting the Rasch model as in Equations 3.1, 3.2, or 3.3 may be less familiar but helps to relate the model to a wider range of approaches for the analysis of (multivariate) binary data. Also, the fractions in Equation 3.3 relate the Rasch model in a straightforward way to the simple estimate of success probabilities based on sample frequency of successes and failures for repeatable trials. Further, note that the ratio of parameters τ_p and ξ_i in the second assumption are the (betting) odds for a binary variable, that is, the ratio of the probabilities of success and failure. A few very important properties of the Rasch model (as well as some of more general item response theory, or IRT, models) can conveniently be illustrated based on this representation of the model. If the Rasch model holds for all examinees p in a given population and for items $i = 1, \ldots, I$, it can be shown that

- Comparisons of examinees can be carried out independent of the items involved. For any item i, the ratio of parameters for two examinees p and p′ equals

$$\frac{\lambda_{ip}}{\lambda_{ip'}} = \frac{\tau_p}{\tau_{p'}}$$

  independently of which item i was chosen.
- Similarly, comparisons of items i and i' can be carried out independently of the examinees involved. For any examinee p, we have

$$\frac{\lambda_{ip}}{\lambda_{i'p}} = \frac{\xi_{i'}}{\xi_i}.$$

- Likewise, a direct comparison of item difficulties or examinee proficiencies is possible even when two sets of observations overlap only partially, or if two sets of observations are connected only through a third set of observations (items or examinees). For example, if examinee p took only items i and i' while examinee q took items i' and i'', how do we compare between item i and i''? The answer is easily obtained if the different λ_{ip} parameters are known. This follows because

$$\frac{(\lambda_{ip}/\lambda_{i'p})}{(\lambda_{i''q}/\lambda_{i'q})} = \frac{(\xi_{i'}/\xi_i)}{(\xi_{i'}/\xi_{i''})} = \frac{\xi_{i''}}{\xi_i},$$

  even though the two items have never been taken together, provided that there is a item i' taken by two (groups of) examinees that links items i and i''. This feature can be utilized for comparisons among items or test takers in the same way. As an example, Glas (1988b) shows how the Rasch model can be applied to multistage testing with a large number of different test forms assigned to different examinees.

There is only one issue: The parameters λ_{ij} are unknown, and naïve estimators are not available because each test-taker-by-item interaction is, in a typical testing situation, observed only once. The solution to this problem is an approach that capitalizes on the sufficiency (parameter separability) properties of the Rasch model. For any two items i and i', consider the conditional probability of solving i given that exactly one of the two items was solved:

$$\Pr\{U_{ip} = 1 \mid U_{ip} + U_{i'p} = 1\}.$$

If the Rasch model holds, local stochastic independence can be used. Then, applying Equation 3.3, the probability can be written as

$$\frac{\Pr\{U_{ip} = 1\}\Pr\{U_{i'p} = 0\}}{\Pr\{U_{ip} = 1\}\Pr\{U_{i'p} = 0\} + \Pr\{U_{ip} = 0\}\Pr\{U_{i'p} = 1\}} = \frac{(\tau_p/\xi_i + \tau_p)(\xi_{i'}/\xi_{i'} + \tau_p)}{(\tau_p/\xi_i + \tau_p)(\xi_{i'}/\xi_{i'} + \tau_p) + (\xi_{i*}/\xi_i + \tau_p)(\tau_p/\xi_{i'} + \tau_p)}$$

which, upon some simplification, yields

$$\Pr\{U_{ip} = 1 \mid U_{ip} + U_{i'p} = 1\} = \frac{\xi_{i'}}{\xi_{i'} + \xi_i} \tag{3.5}$$

for any examinee p. This is a central result, which we will discuss for the case of more than two items below. For now, we only note that Equation 3.5 shows that by conditioning on the total score on any two items, the examinee parameter τ_p is effectively eliminated from the conditional probability.

Now assume that $p = 1, \ldots, P$ test takers with parameters τ_p were sampled randomly from the population for which the Rasch model holds. Consider the sample estimate of the marginal conditional probability of observing success on item i given that exactly one of the two items i and i' was solved:

$$\widehat{\Pr}\{U_i = 1 \mid U_i + U_{i'} = 1\} = \frac{1}{P}\sum_{p=1}^{P} 1_{\{U_{ip}=1 \mid U_{ip}+U_{i'p}=1\}}. \tag{3.6}$$

Taking the expectation yields

$$E\left(\frac{1}{P}\sum_{p=1}^{P} 1_{\{U_{ip}=1 \mid U_{ip}+U_{i'p}=1\}}\right) = \sum_{p=1}^{P}\frac{1}{P}E(1_{\{U_{ip}=1 \mid U_{ip}+U_{i'p}=1\}}).$$

Inserting Equation 3.4 for each of the expected values inside the sum and simplifying yields

$$E(\widehat{\Pr}\{U_i = 1 \mid U_i + U_{i'} = 1\}) = \frac{\xi_{i'}}{\xi_{i'} + \xi_i}. \tag{3.7}$$

Note that this is the expected value of the sample estimate of the marginal conditional probability $\widehat{\Pr}\{U_i = 1 \mid U_i + U_{i'} = 1\}$ of observing success on item i given that exactly one of the two items i and i' was solved. By symmetry, the ratio of this estimator and its complement yields

$$\frac{\widehat{\Pr}\{U_i = 1 \mid U_i + U_{i'} = 1\}}{\widehat{\Pr}\{U_{i'} = 1 \mid U_i + U_{i'} = 1\}} \approx \frac{\xi_{i'}}{\xi_i}. \tag{3.8}$$

This means that the ratio of any two item parameters can be compared by marginal two-way counts independent of the sample of test takers that was used.

As Rasch (1966a) stated:

> Our aim is to develop probabilistic models in the application of which the population can be ignored. It was a discovery of some mathematical significance that such models could be constructed, and it seemed remarkable that data collected in routine psychological testing could be fairly well represented by such models (p. 89).

The results presented in Equations 3.7 and 3.8 allow comparisons of item difficulties without looking at properties of the sample of examinees that were used to obtain the sample statistics in Equation 3.6. In a testing situation, we typically have many more test takers than items in the test. These results are particularly useful in this situation: If the Rasch model holds, the only requirement to be fulfilled is a large enough sample size from the population of interest in order to obtain estimates of desired accuracy for Equations 3.6 and 3.8. By symmetry, in a situation in which there are few test takers but a large number of randomly chosen items, examinees can be compared independent of the sample of items they took.

This particular feature of the Rasch model is sometimes referred to as "parameter separability" or "specific objectivity." The same feature is not found in most other higher parametric item response models, with only a few exceptions (e.g., Verhelst and Glas, 1995). The feature, which is more generally based on the fact that the Rasch model is an exponential family model with simple sufficient statistics, is important in derivations of conditional maximum likelihood (CML) estimation methods that can be established simultaneously for all item parameters without introducing bias by nuisance parameters or assumptions about the distribution of unobserved test-taker parameters (e.g., Andersen, 1972; Verhelst and Glas, 1995). The feature can also be used to estimate pairwise relationships between item parameters (Zwinderman, 1995). Moreover, this feature puts the Rasch model—or better, its ability to compare two items (or two test takers) independent of the others—in the context of exponential family models for paired comparisons in incomplete designs (Zermelo, 1929; Bradley and Terry, 1952). The connection to paired comparison models also allows stating conditions for the existence of maximum likelihood estimators (Fischer, 1981). It is of particular importance for linking or equating multiple test forms using the Rasch model as it allows comparisons across a large set of items or test takers without the requirement that all test takers have responded to all the items, for example with multistage tests (Glas, 1988b).

3.3 Parameter Estimation

Several estimation methods are available for the Rasch model. This chapter focuses on maximum likelihood methods and presents a short description of each method while providing more detail on one that can be considered as most directly utilizing the mathematical properties of the Rasch model. Commonly discussed estimation methods for the Rasch model are

1. Joint maximum likelihood (JML): This method attempts to estimate all parameters τ_p and ξ_i for items $i = 1, \ldots, I$ and respondents $p = 1, \ldots, P$ simultaneously. There are well-known results that show that a joint estimation of parameters leads to biased estimates if $I \ll P$ (Kiefer and Wolfowitz, 1956). For the Rasch model, the main result given by Andersen (1972) and Haberman (1977) is that the bias is of order $(I/(I-1))$. Haberman (1977, 2004) notes that besides this bias of JML estimates, JML estimation is often problematic because (finite) estimates do not exist (see also Fischer, 1981) when there are respondents with perfect scores or with a vector of all incorrect responses. It should be noted that alternatives exist that do not share these issues, so there is no need to use JML methods in typical testing situations.
2. Maximum marginal likelihood (MML): This method is, among maximum likelihood methods, the most commonly used approach for higher parameterized item response models. In order to apply MML estimation, an assumption concerning the distribution $\phi(\theta)$ of examinee parameters is necessary to enable the calculation of the marginal probability:

$$Pr\{U_i = 1\} = \int_{-\infty}^{+\infty} Pr\{U_i = 1 \mid \theta\}\phi(\theta)d\theta. \tag{3.9}$$

   While MML is a commonly used approach in IRT also available for the Rasch model (Thissen, 1982), there is an issue of choice of the distribution $\phi(\theta)$. For example, many software packages allow only normal distributions here and also approximate the integral in the equation using numerical quadrature. Note, however, that the distribution does not need to be normal (Andersen and Madsen, 1977; Thissen, 1982) and that the MML approach can be used with more general distributions (e.g., Xu and von Davier, 2008; Xu and Jia, 2011).
3. Pairwise (conditional) estimation: As shown in Equations 3.7 and 3.8, simple ratios of probabilities can be used to eliminate the parameters associated with respondents to obtain estimates of parameters τ_p and ξ_i (e.g., Zwinderman, 1995; Andrich, 2001).
4. Nonparametric or semi-nonparametric estimation using located latent classes (LLCA): de Leeuw and Verhelst (1986), Lindsay, Clogg and Grego (1991), Follmann (1988), and Formann (1994) describe an estimation method that is based on approximating the integral in Equation 3.9 by a sum over a small number of unobserved groups instead. This approach estimates the marginal probability as a finite sum:

$$Pr\{U_i = 1\} = \sum_{g=1}^{G} Pr\{U_i = 1 \mid \theta_g\}\pi_g \tag{3.10}$$

   over $g = 1 \ldots G$ latent groups with locations θ_g and relative sizes π_g with $\sum_g \pi_g = 1$. Heinen (1996) describes these approaches for the Rasch model in more detail, and

Haberman et al. (2008) describe an extension of this approach to multidimensional item response models, while Kelderman (1984) describes log-linear-type Rasch models.

5. CML: This is the estimation approach that can be considered most fundamentally based on the existence of simple sufficient statistics (i.e., what Rasch called specific objectivity) in the Rasch model. That is, the CML approach is based on a general version of result presented in Equation 3.5 capitalizing on the fact that the sum of correct responses on the number of items $i = 1, \ldots, I$ to which an examinee p responds is a sufficient statistic for his or her parameter θ_p (e.g., Andersen, 1977; Bickel and Doksum, 1977). Consequently, utilizing the factorization theorem of sufficiency allows the construction of estimation equations for $\beta_1, \ldots, \beta_I$ that do not contain any examinee parameters. Fischer (1981) provides necessary and sufficient conditions for the existence and uniqueness of CML estimates in the Rasch model, while Eggen (2000) describes minimal loss of information due to the conditioning on this sufficient statistic in CML relative to JML and MML. It is important to note that, unlike the approach in Equations 3.9 and 3.10, CML allows item parameter estimation without making any assumption about the distribution of θ.

The remainder of this section presents the CML estimation method in more detail. For the Rasch model, using local stochastic independence, the probability of a vector of responses $\boldsymbol{u} = (u_1, \ldots, u_I)$ is

$$\Pr\{\boldsymbol{u} \mid p\} = \prod_{i=1}^{I} \frac{\exp(u_i(\theta_p - \beta_i))}{1 + \exp(\theta_p - \beta_i)}. \tag{3.11}$$

Let $S = \sum_{i=1}^{I} u_i$ denote the sum of correct responses for the vector. Then, we have

$$\Pr\{\boldsymbol{u} \mid p\} = \exp\{S\theta_p\} \prod_{i=1}^{I} \frac{\exp(-\beta_i u_i)}{1 + \exp(\theta_p - \beta_i)}. \tag{3.12}$$

Further, let $\Pr\{S \mid p\}$ denote the probability of score S when testing examinee p, which can be written as

$$\Pr\{S \mid p\} = \sum_{\left\{(v_1, \ldots, v_I): \sum_{i=1}^{I} v_i = S\right\}} \exp(S\theta_p) \prod_{i=1}^{I} \frac{\exp(-\beta_i v_i)}{1 + \exp(\theta_p - \beta_i)}. \tag{3.13}$$

The conditional probability of the response vector given the sum score S is defined as

$$\Pr\{\boldsymbol{u} \mid S, p\} = \frac{\Pr\{u_1, \ldots, u_I \mid p\}}{\Pr\{S \mid p\}}. \tag{3.14}$$

Inserting Equations 3.12 and 3.13 into Equation 3.14 and rearranging the result, we obtain

$$\Pr\{\boldsymbol{u} \mid S, p\} = \frac{\exp\left(\sum_{i=1}^{I} -\beta_i u_i\right)}{\sum_{\left\{v:\sum_{i=1}^{I} v_i = S\right\}} \exp\left(\sum_{i=1}^{I} -\beta_i v_i\right)}, \tag{3.15}$$

which does not contain the examinee parameter θ_p. Introducing the symmetric functions (e.g., Andersen, 1972; Gustafsson, 1980)

$$\gamma_S(\boldsymbol{\beta}) = \sum_{\left\{v:\sum_{i=1}^{I} v_i = S\right\}} \exp\left(-\sum_{i=1}^{I} \beta_i v_i\right) \tag{3.16}$$

and marginalizing over the sample distribution of the examinees, we obtain

$$\Pr\{\boldsymbol{u} \mid S\} = \frac{\exp\left(-\sum_{i=1}^{I} \beta_i u_i\right)}{\gamma_S(\boldsymbol{\beta})} \tag{3.17}$$

for the conditional probability of a response vector $\boldsymbol{u} = (u_1, \ldots, u_I)$ given total score S. Note that $\Pr\{0, \ldots, 0 \mid 0\} = \Pr\{1, \ldots, 1 \mid I\} = 1$ holds for the extreme score groups 0 and I because there is a unique single pattern of responses that produces each of these scores.

It is important to note that all items are not required to be taken by all respondents: The same approach to conditional estimation can be applied using a generalized version of the symmetric functions and Equation 3.17 in which an indicator vector $J_p = (j_{p1}, \ldots, j_{pI})$, with $j_{pi} \in \{0, 1\}$ associated with each test taker p showing the subset of items taken by this person. We can then define

$$\Pr\{\boldsymbol{u} \mid S_{J_p}, p\} = \frac{\exp\left(\sum_{i=1}^{I} -\beta_i j_{pi} u_i\right)}{\sum_{\left\{v:\sum_{i=1}^{I} v_i = S_{Jp}\right\}} \exp\left(\sum_{i=1}^{I} -\beta_i j_{pi} v_i\right)} \tag{3.18}$$

with $S_{J_p} = \sum_{i=1}^{I} j_{pi} u_i$ defining the total score based on the subset of items actually taken by p. All subsequent calculations can be adjusted similarly to realize this more general approach, with the advantage of, for instance, estimation of all parameters across multiple test forms in incomplete block designs on the same scale. Minimal requirements for this type of model-based linking were published as early as 1929 by Zermelo.

Note that the symmetric functions in Equation 3.16 can be written in a recursive manner as

$$\gamma_S(\boldsymbol{\beta}) = \exp(-\beta_i)\gamma^{(i)}{}_{S-1}(\boldsymbol{\beta}) + \gamma^{(i)}{}_S(\boldsymbol{\beta}), \tag{3.19}$$

where $\gamma_s^{(i)}(\boldsymbol{\beta})$ is defined as the symmetric function calculated under omission of item i (e.g., Andersen, 1973; Gustafsson, 1980; van der Linden, Volume Two, Chapter 6). Moreover, application of some basic calculus shows that the derivative of $\gamma_S(\boldsymbol{\beta})$ with respect to β_i is

$$\frac{\partial \gamma_S(\boldsymbol{\beta})}{\partial \beta_i} = -\exp(-\beta_i)\gamma^{(i)}{}_{S-1}(\boldsymbol{\beta}). \tag{3.20}$$

The conditional likelihood function for $p = 1, \ldots, P$ independently observed respondents, each responding to $i = 1, \ldots, I$ items, can be written as

$$\ln L_M(\boldsymbol{\beta} \mid \boldsymbol{u}_1, \ldots, \boldsymbol{u}_P) = \ln\left(\prod_{p=1}^{P}\left[\sum_{T} \Pr\{\boldsymbol{u}_p \mid T\}\pi_T\right]\right) = \sum_{p=1}^{P} \ln(\Pr\{\boldsymbol{u}_p \mid S_p\}\pi_{S_p}) \tag{3.21}$$

with score probabilities $\Pr\{S\} = \pi_S$ that can be based on relative sample frequencies H_s/P of the examinees' total scores $S_p = 0, \ldots, I$. The right-hand side of Equation 3.21 holds because $\Pr\{\boldsymbol{u} \mid T\} = 0$ if $T \neq S_p$. Dropping the terms $\ln\pi_S$ that do not contain any item parameters and inserting Equation 3.17 into Equation 3.21 yields

$$\ln L(\boldsymbol{\beta} \mid \boldsymbol{u}_1, \ldots, \boldsymbol{u}_P) = \sum_{p=1}^{P} \ln\left[\frac{\exp\left(-\sum_{i=1}^{I} \beta_i u_{pi}\right)}{\gamma_{S_p}(\boldsymbol{\beta})}\right]. \tag{3.22}$$

Collecting the terms in Equation 3.22 into score groups $S = 1, \ldots, I-1$ (the terms for the extreme score groups vanish), we obtain

$$\ln L(\boldsymbol{\beta} \mid \boldsymbol{u}_1, \ldots, \boldsymbol{u}_P) = \sum_{S=1}^{I-1} H_S[-\ln \gamma_S(\boldsymbol{\beta})] - \left[\sum_{i=1}^{I} \beta_i F_i\right] \tag{3.23}$$

with F_i denoting the count of the correct responses for item i in score groups $S = 1, \ldots, I-1$, and H_S denoting the frequency of score S in the sample.

The partial derivatives of Equation 3.23 with respect to β_i are

$$\frac{\partial \ln L(\boldsymbol{\beta} \mid \boldsymbol{u}_1, \ldots, \boldsymbol{u}_P)}{\partial \beta_i} = \left[-\sum_{S=1}^{I-1} H_S \frac{\partial \ln \gamma_S(\boldsymbol{\beta})}{\partial \beta_i}\right] - F_i, \tag{3.24}$$

which upon transformation, simplification, and insertion of Equation 3.19 into Equation 3.24 yield

$$\frac{\partial \ln L(\boldsymbol{\beta} \mid \boldsymbol{u}_1, \ldots, \boldsymbol{u}_P)}{\partial \beta_i} = \left[\sum_{S=1}^{I-1} H_S \frac{\exp(-\beta_i)\gamma^{(i)}{}_{S-1}(\boldsymbol{\beta})}{\gamma_S(\boldsymbol{\beta})}\right] - F_i. \tag{3.25}$$

Note that this is a result to be expected for exponential families (Bickel and Doksum, 1977; Andersen, 1980; Haberman, Volume Two, Chapter 4) because

$$\Pr\{U_i = 1 \mid S\} = \frac{\exp(-\beta_i)\gamma^{(i)}{}_{S-1}(\boldsymbol{\beta})}{\gamma_S(\boldsymbol{\beta})} \tag{3.26}$$

holds for the conditional form of the Rasch model. The derivatives in Equation 3.25 can be used either for the Newton–Raphson algorithm or with gradient (von Davier and Rost, 1995) or quasi-Newton methods (e.g., Dennis and Schnabel, 1996) to obtain maximum likelihood estimates of $\beta_1, \ldots, \beta_I$. When Newton methods are used, the second-order derivatives of Equation 3.25 are easily derived. Solving Equation 3.19 for $\exp(-\beta_i)\gamma^{(i)}{}_{S-1}(\boldsymbol{\beta})$ and inserting the result into Equation 3.26 yields

$$\frac{\gamma_S(\boldsymbol{\beta}) - \gamma^{(i)}{}_S(\boldsymbol{\beta})}{\gamma_S(\boldsymbol{\beta})} = \frac{\exp(-\beta_i)\gamma^{(i)}{}_{S-1}(\boldsymbol{\beta})}{\gamma_S(\boldsymbol{\beta})} = \Pr\{U_i = 1 \mid S\},$$

which implies

$$\frac{\gamma^{(i)}{}_S(\boldsymbol{\beta})}{\gamma_S(\boldsymbol{\beta})} = 1 - \Pr\{U_i = 1 \mid S\}.$$

Using the quotient rule for differentiation, along with some collecting of the terms and further simplifying, provides

$$\frac{\partial L(\boldsymbol{\beta} \mid \boldsymbol{u}_1, \ldots, \boldsymbol{u}_P)}{\partial^2 \beta_i} = \sum_{S=1}^{I-1} H_S \left[\frac{\left(\gamma_S(\boldsymbol{\beta}) - \gamma^{(i)}{}_S(\boldsymbol{\beta})\right)\left(\gamma^{(i)}{}_S(\boldsymbol{\beta})\right)}{[\gamma_S(\boldsymbol{\beta})]^2} \right]. \tag{3.27}$$

Alternatively, applying some of the properties of exponential families to the Rasch model (Bickel and Doksum, 1977; Haberman, 1977; Andersen, 1980) in order to find the second-order derivative of Equation 3.24 with respect to β_i will produce the equivalent result

$$\frac{\partial L(\boldsymbol{\beta} \mid \boldsymbol{u}_1, \ldots, \boldsymbol{u}_P)}{\partial^2 \beta_i} = \left[\sum_{S=1}^{I-1} H_S \Pr\{U_i = 1 \mid S\}(1 - \Pr\{U_i = 1 \mid S\}) \right]. \tag{3.28}$$

Once item parameter estimates $\hat{\beta}_1, \ldots, \hat{\beta}_I$ are available, they can be plugged into the joint likelihood equations. Solving the equations while holding $\hat{\beta}_1, \ldots, \hat{\beta}_I$ constant provides estimates of the examinee parameters. Alternatively, bias-corrected estimators can be found by maximizing a penalized likelihood function while holding the item parameters fixed (Warm, 1989; Firth, 1993). Finally, Bayesian estimates of the examinee parameters can be generated using their posterior modes or expectations, for instance, by adopting some "weakly informative" (Gelman et al., 2003; Gelman, 2007) prior distributions for them.

3.4 Model Fit

Numerous approaches to test the fit of the Rasch model are available (e.g., Andersen, 1973; Martin-Löf, 1973; van den Wollenberg, 1982; Molenaar, 1983; Glas, 1988a). Only a selection of tests specifically developed for this particular model will be discussed. More recent fit measures based on generalizations of Pearson χ^2 statistics have been developed for the Rasch model by Glas (1988a, 2007) and Glas and Verhelst (1995). For a more comprehensive overview of these methods, refer to Glas (Volume Two, Chapter 10).

Many of the early attempts to test the fit of the Rasch model were based on the fact that if the model holds, its item parameters can be estimated consistently regardless of the population involved (Rasch, 1966a; Andersen, 1973). These tests typically split the sample into two or more groups and estimate the parameters of the conditional model in each of them separately. The rationale behind these tests is that if parameter estimates differ substantially in different subsamples, then the Rasch model cannot adequately describe the whole sample and hence has to be rejected for the population from which the sample was drawn. Rasch (1960) utilized a graphical representation in the form of a simple scatterplot to determine whether separately estimated parameters line up without any obvious deviations. This approach is sometimes referred to as Rasch's graphical model test.

Andersen (1973) proposed a more rigorous statistical test of the Rasch model. This approach requires splitting the sample into groups of respondents according to their sum score S and estimating the Rasch model in each of the (nonextreme score) groups separately. Note that CML estimation is possible even if all respondents have the same (nonextreme) sum score, because the likelihood function in Equation 3.23 maximized with respect to the item parameters does not depend on the actual total scores observed. If we split the sample by score groups, we may define

$$\ln L_S(\boldsymbol{\beta} \mid \mathbf{u}'_1, \ldots, \mathbf{u}'_{H_s}) = H_S[-\ln \gamma_S(\boldsymbol{\beta})] - \left[\sum_{i=1}^{I} \beta_i F_{iS}\right], \tag{3.29}$$

where $F_{iS} = \sum_{p|S} u_{pi}$ denotes the count of the correct responses for item i only for respondents with sum score S and H_S is defined as above.

Maximizing $\ln L_S(\boldsymbol{\beta} \mid \mathbf{u}'_1, \ldots, \mathbf{u}'_{H_s})$ with respect to the item parameters for each score group $S = 1, \ldots, I-1$ separately yields score-group-based estimates $\hat{\boldsymbol{\beta}}_s = (\hat{\beta}_{1s}, \ldots, \hat{\beta}_{Is})$. As a test for the Rasch model, Andersen (1973) proposed evaluation of the likelihood ratio

$$LR_{Andersen} = \frac{L_S(\hat{\boldsymbol{\beta}} \mid \mathbf{u}_1, \ldots, \mathbf{u}_P)}{\prod_{S=1}^{I-1} L_S(\hat{\boldsymbol{\beta}}_s \mid \mathbf{u}'_1, \ldots, \mathbf{u}'_{H_s})}. \tag{3.30}$$

This expression evaluates the gain in likelihood based on separate estimation of item parameters in score groups, $\hat{\boldsymbol{\beta}}_s$, against single estimates based on the total sample, $\hat{\boldsymbol{\beta}}$.

Alternatively, the model can be tested splitting the sample into groups using background variables, such as level of education (e.g., years of schooling in broad categories) or gender. Rost and von Davier (Rost and von Davier, 1995; Volume One, Chapter 23) proposed using a mixture distribution Rasch model as the alternative hypothesis in a test of parameter invariance, introducing a split of the sample of examinees into latent groups.

Another test of the Rasch model is the Martin-Löf test (e.g., Glas and Verhelst, 1995; Verguts and De Boeck, 2000). This test is one of the few that checks on the unidimensionality assumption for the Rasch model (Verhelst, 2001). Unlike the previous tests, which split the examinee sample into subsamples to check on its homogeneity, the Martin-Löf test splits the item set into two subsets and then addresses their dimensionality.

Some other fit measures discussed in the literature are based on response residuals:

$$r_{pi} = u_{pi} - \Pr\{U_i = 1 \mid p\}.$$

These residuals are aggregated either by first squaring and then adding them or by first adding them, and then squaring the result. Aggregation across items yields a measure of person fit; across persons, it yields a measure of item fit. Depending on the type of aggregation, different types of standardization have been proposed (Wright and Stone, 1979); however, the exact null distribution of these standardized statistics is unknown (Wells and Hambleton, Volume Two, Chapter 20).

Fit measures based on CML inference have also been presented (e.g., Rost and von Davier, 1994). More recent approaches to model fit include the work by von Davier and Molenaar (2003) on person fit indices for Rasch models and its extensions to multiple populations and mixture distributions. Klauer (1991) developed a uniformly most powerful test for the person parameter in the Rasch model, while Ponocny (2000) presented an exact person fit index. Verhelst (2001) compared the Martin-Löf test with van den Wollenberg's (1982) as well as Molenaar's (1983) splitter item technique. He also proposed a new test statistic that addresses some of the statistical issues involved with the latter.

An important issue is the effect of misfit of the Rasch model on its use. Molenaar (1997) outlined a rationale to evaluate whether the increase in model complexity involved in replacing the Rasch model by a more highly parametrized model is justified. Also, person parameters estimates will still correlate highly, often substantially above 0.9, between different unidimensional IRT models applied to the same data. Finally, if the Rasch model is found not to fit the data well, more general models are available that preserve many of its characteristics, including the one-parameter logistic model (OPLM) (Verhelst and Glas, 1995) and logistic mixture models (Rost and von Davier, 1995; Volume One, Chapter 23).

3.5 Empirical Example

This empirical example is based on 10 items taken from a cognitive skills test for children ages 9–13. The dataset is available through the exercises in the textbook by Rost (2004). In the current example, the focus is on parameter recovery using different estimation methods. The item parameter estimates from the original dataset were treated as the generating parameters for a simulated dataset with known characteristics. These generating parameters (in Table 3.1 denoted by "Truth") served as the targets that were to be recovered by the different estimation methods. The dataset was simulated with 10 item responses for each of 15,000 test takers drawn from a normal ability distribution with mean 0.0 and standard deviation 1.5.

The item parameters were estimated utilizing the CML and JML estimation methods in WINMIRA 2001 (von Davier, 1994, 2001) and BIGSTEPS (Wright and Linacre, 1991),

TABLE 3.1

Generating Parameters (Truth) and CML, LLCA, and JML Item Parameter Estimates for the Example Data

Item	Truth	CML	S.E.	LLCA	S.E.	JML	S.E.
1	−0.941	−0.921	0.021	−0.921	0.020	−1.030	0.020
2	−0.457	−0.461	0.020	−0.462	0.020	−0.520	0.020
3	0.265	0.241	0.020	0.242	0.019	0.270	0.020
4	0.919	0.917	0.021	0.917	0.020	1.030	0.020
5	1.335	1.350	0.022	1.351	0.021	1.520	0.020
6	−1.223	−1.235	0.022	−1.235	0.021	−1.390	0.020
7	0.022	0.033	0.020	0.033	0.019	0.040	0.020
8	−0.457	−0.429	0.020	−0.430	0.020	−0.480	0.020
9	0.832	0.816	0.021	0.817	0.020	0.910	0.020
10	−0.295	−0.310	0.020	−0.311	0.020	−0.350	0.020

respectively, and a semiparametric approach (Heinen, 1996). The latter was implemented specifying four different latent classes (LLCA) and estimating the model with the mdltm software (von Davier, 2008).

The comparison focused primarily on the differences among the CML, JML, and LLCA methods for the estimation of item parameters and between JML and weighted likelihood estimation (WLE; Warm, 1989) for the estimation of the person parameters. Table 3.1 shows the generating item parameters and their CML, LLCA, and JML estimates along with the standard errors as provided by the programs.

The estimated standard errors for the three methods were comparable, while the estimates for CML and LLCA were virtually identical and appeared to be closer to the true values of the parameters than the JML parameter estimates. Since CML and LLCA are virtually identical, we focus our discussion and compare the CML and JML estimates only.

Figure 3.1 shows the differences between the CML and JML estimates and the true item parameters.

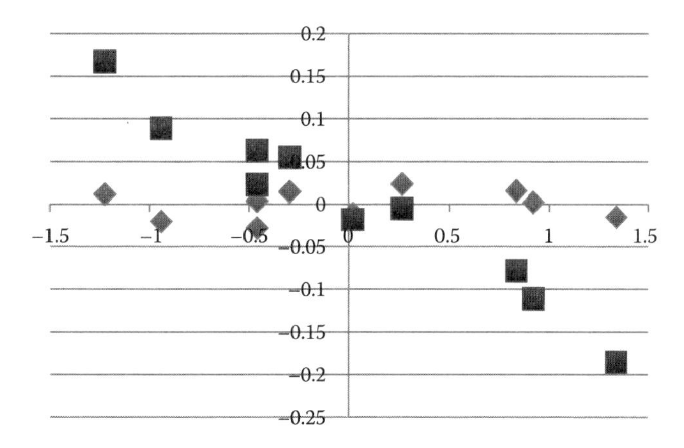

FIGURE 3.1

Differences between CML and true parameters (diamonds) and JML and true parameters (squares) for the dataset.

TABLE 3.2

Person Parameter Estimates Obtained Using Three Different Methods

Score	Frequency	MLE	S.E.	WLE	S.E.	JMLE	S.E.
0	780			−3.348	1.564	(−3.27)	(1.47)
1	1162	−2.429	1.079	−2.094	0.961	−2.49	1.08
2	1493	−1.565	0.824	−1.415	0.795	−1.61	0.83
3	1556	−0.970	0.730	−0.894	0.721	−1.00	0.74
4	1637	−0.472	0.689	−0.440	0.687	−0.49	0.70
5	1702	−0.007	0.678	−0.009	0.678	−0.01	0.69
6	1621	0.459	0.692	0.424	0.690	0.47	0.70
7	1591	0.964	0.735	0.886	0.726	0.99	0.75
8	1510	1.567	0.830	1.417	0.801	1.61	0.84
9	1198	2.442	1.084	2.109	0.969	2.50	1.09
10	750			3.377	1.574	(3.29)	(1.48)

Note: JML estimates and standard errors for the extreme-score groups were not estimated but extrapolated.

As Figure 3.1 shows, the bias of JML estimates correlated negatively with the true parameter, while much smaller differences were seen for CML with no visible correlation with the true parameter values.

Table 3.2 shows the person parameter estimates obtained using subsequent maximum likelihood estimation (MLE) with the CML item difficulties as fixed constants (MLE), WLE, and the estimates when jointly maximizing the item and person parameters (joint maximum likelihood estimation, or JMLE).

While it is possible to calculate means and variances of the JML and ML estimates across score groups 1–9, it is questionable whether anything is gained given that 1530 cases in the data were found to be in extreme score groups for which (finite) estimates do not exist (Haberman, 1977, 2004; Fischer, 1981). Also, note that the entries in the table for JML and score groups 0 and 10 were extrapolated—not estimated—values, so it is meaningless to evaluate their average and variance as well. The mean estimate for the person distribution for the LLCA approach was 0.001, with a standard deviation of 1.423. The average of the WLEs was 0.001 and their standard deviation was 1.567, respectively. These values need to be compared to the generating values of 0.00 and 1.5, respectively.

3.6 Discussion

This chapter presented an introduction to the Rasch model and the estimation of its parameters using a customary method that capitalized on the parameter separability offered by the existence of simple sufficient statistics for this exponential family model. It was pointed out that, while there are alternative unbiased estimation methods, only CML estimation makes use of this property to the fullest extent. However, the conditional approach is only applicable to the Rasch model and a few of its extensions, while the LLCA approach presented above can also be used for more highly parameterized IRT models (e.g., Heinen, 1996; Haberman et al., 2008; von Davier, 2008).

The Rasch model is used operationally in international as well as national test and assessment programs. A variety of software programs for consistent estimation (e.g.,

Pfanzagl, 1994) of its parameters are available, including Conquest (Wu et al., 1997; Adams et al., Volume Three, Chapter 27), LPCM-WIN (Fischer and Ponocny-Seliger, 1998), Logimo (Kelderman and Steen, 1988), Multira (Carstensen and Rost, 2003), RSP (Glas and Ellis, 1994), RUMM2010 (Andrich, 2001), and WINMIRA (von Davier, 1994, 2001). Most of these software packages also offer model extensions such as the ability to estimate Rasch models for polytomous data (e.g., Volume One, Chapters 1 and 7), models with covariates of person parameters (Volume One, Chapters 32 and 33), and (dichotomous as well as polytomous) models for multiple populations (Volume One, Chapter 23). In addition, there exist add-ons to general-purpose statistical software such as R (R Core Team 2012), STATA, and SAS; for example, the eRm (Mair and Hatzinger, 2007) and ltm packages (Rizopoulos, 2006). These add-ons, due to their nature of being implemented as interpreted scripts in a general-purpose software package, are typically somewhat less suitable for very large datasets as they offer lower computational performance than dedicated programs.

The Rasch model will remain the model of choice for some researchers, while it will remain the "too restrictive" IRT model "that never fits the data" (de Leeuw, 1997) for others. While it is often found that a more general IRT model will fit the data better, Molenaar (1997) reminds us that even with some misfit, use of the Rasch model may still be warranted when the conclusions drawn from this model differ little from a higher parameterized unidimensional IRT model. This does not mean that the issue of model fit should be ignored. As with any other model, it is imperative to examine whether the Rasch model fits the data satisfactorily.

References

Andersen, E. B. 1972. The numerical solution to a set of conditional estimation equations. *Journal of the Royal Statistical Society, Series B*, 34, 283–301.

Andersen, E. B. 1973. A goodness of fit test for the Rasch model. *Psychometrika*, 38, 123–140.

Andersen, E. B. 1977. Sufficient statistics and latent trait models. *Psychometrika*, 42, 69–81.

Andersen, E. B. 1980. *Discrete Statistical Models with Social Science Applications*. Amsterdam, The Netherlands: North-Holland Publishing Company.

Andersen, E. B., and Madsen, M. 1977. Estimating the parameters of the latent population. *Psychometrika*, 42, 357–374.

Andrich, D. 2001. *RUMM2010: Rasch Unidimensional Measurement Models*. Perth, Australia: RUMM Laboratory.

Bickel, P. J., and Doksum, K. A. 1977. *Mathematical Statistics: Basic Ideas and Selected Topics*. San Francisco, CA: Holden-Day, Incorporated.

Bradley, R. A., and Terry, M. E. 1952. Rank analysis of incomplete block designs. I. The method of paired comparisons. *Biometrika*, 39, 324–345.

Carstensen, C. H., and Rost, J. 2003. *Multira [Computer Software]*. Bamberg, Germany: University of Bamberg. Retrieved from http://www.multira.de/.

de Leeuw, J. 1997. Review of Rasch models—Foundations, recent developments, and applications. *Statistics in Medicine*, 16, 1431–1434.

de Leeuw, J., and Verhelst, N. 1986. Maximum-likelihood-estimation in generalized Rasch models. *Journal of Educational Statistics*, 11, 183–196.

Dennis, J. E., and Schnabel, R. B. 1996. Numerical methods for unconstrained optimization and nonlinear equations. *SIAM Classic in Applied Mathematics*, 16, 1996.

Eggen, T. J. H. M. 2000. On the loss of information in conditional maximum likelihood estimation of item parameters. *Psychometrika*, 65(3), 337–362.

Firth, D. 1993. Bias reduction of maximum likelihood estimates. *Biometrika*, 80(1), 27–38.

Fischer, G. H. 1981. On the existence and uniqueness of maximum-likelihood estimates in the Rasch model. *Psychometrika*, 46, 59–77.

Fischer, G. H. 2006. Rasch models. In C. R. Rao and S. Sinharay (Eds.), *Handbook of Statistics, Vol. 26: Psychometrics* (pp. 515–586). Amsterdam, The Netherlands: Elsevier Science Publishers.

Fischer, G. H., and Molenaar I. W. 1995. *Rasch Models: Foundations, Recent Developments and Applications*. New York, NY: Springer.

Fischer, G. H., and Ponocny-Seliger, E. 1998. *Structural Rasch Modeling. Handbook of the Usage of LPCM-WIN 1.0*. Groningen, The Netherlands: Institute ProGamma.

Follmann, D. 1988. Consistent estimation in the Rasch model based on nonparametric margins. *Psychometrika*, 53, 553–562.

Formann, A. K. 1994. Measuring change in latent subgroups using dichotomous data: Unconditional, conditional, and semiparametric maximum-likelihood-estimation. *Journal of the American Statistical Association*, 89, 1027–1034.

Gelman, A. 2007. http://andrewgelman.com/2007/05/11/weakly_informat/

Gelman, A., Carlin, J. B., Stern, H. S., and Rubin, D. B. 2003. *Bayesian Data Analysis* (2nd ed.). New York, NY: Chapman and Hall/CRC Texts in Statistical Science.

Glas, C. A. W. 1988a. The derivation of some tests for the Rasch model from the multinomial distribution. *Psychometrika*, 53, 525–546.

Glas, C. A. W. 1988b. The Rasch model and multistage testing. *Journal of Educational Statistics*, 13, 45–52.

Glas, C. A. W. 2007. Testing generalized Rasch models. In M. von Davier and C. H. Carstensen (Eds.), *Multivariate and Mixture Distribution Rasch Models: Extensions and Applications* (pp. 37–55). New York, NY: Springer.

Glas, C. A. W., and Ellis, J. L. 1994. *RSP: The Rasch Scaling Package [Computer Program]*. Groningen, The Netherlands: Institute ProGamma.

Glas, C. A. W., and Verhelst, N. D. 1995. Testing the Rasch model. In G. H. Fischer and I. W. Molenaar (Eds.), *Rasch Models: Their Foundations, Recent Developments and Applications* (pp. 69–96). New York, NY: Springer.

Gustafsson, J.-E. 1980. A solution of the conditional estimation problem for long tests in the Rasch model for dichotomous items. *Educational and Psychological Measurement*, 40, 327–385.

Haberman, S. J. 1977. Maximum likelihood estimates in exponential response models. *Annals of Statistics*, 5, 815–841.

Haberman, S. J. 2004. *Joint and Conditional Maximum Likelihood Estimation for the Rasch Model for Binary Responses* (Research Report No. RR-04-20). Princeton, NJ: Educational Testing Service.

Haberman, S. J., von Davier, M., and Lee, Y. 2008. *Comparison of Multidimensional Item Response Models: Multivariate Normal Ability Distributions versus Multivariate Polytomous Ability Distributions*. (Research Report No. RR-08-45). Princeton, NJ: Educational Testing Service.

Heinen, T. 1996. *Latent Class and Discrete Latent Trait Models: Similarities and Differences*. Thousand Oaks, CA: Sage.

Kelderman, H. 1984. Loglinear Rasch model tests. *Psychometrika*, 49, 223–245.

Kelderman, H., and Steen, R. 1988. *LOGIMO: A Program for Loglinear IRT Modeling*. Enschede, The Netherlands: University of Twente, Department of Education.

Kiefer, J., and Wolfowitz, J. 1956. Consistency of the maximum likelihood estimator in the presence of infinitely many incidental parameters. *Annals of Mathematical Statistics*, 27, 887–906.

Klauer, K. C. 1991. An exact and optimal standardized person fit test for assessing consistency with the Rasch model. *Psychometrika*, 56(2), 213–228.

Lindsay, B., Clogg, C. C., and Grego, J. 1991. Semiparametric estimation in the Rasch model and related exponential response models, including a simple latent class model for item analysis. *Journal of the American Statistical Association*, 86, 96–107.

Mair, P., and Hatzinger, R. 2007. Extended Rasch modeling: The eRm package for the application of IRT models in R. *Journal of Statistical Software*, 20, 1–20.

Martin-Löf, P. 1973. Statistiska Modeller. *Anteckningar fran seminarier läsaret 1969–70 utarbetade av Rolf Sundberg. Obetydligt ändrat nytryk*. Stockholm, Sweden: Institutet för försäkringsmatematik och matematisk statistik vid Stockholms universitet.

Molenaar, I. W. 1983. Some improved diagnostics for failure in the Rasch model. *Psychometrika*, 48, 49–72.

Molenaar, I. W. 1997. Lenient or strict application of IRT with an eye on practical consequences. In J. Rost and R. Langeheine (Eds.), *Applications of Latent Trait and Latent Class Models in the Social Sciences* (pp. 38–49). Münster, Germany: Waxmann.

Pfanzagl, J. 1994. On item parameter estimation in certain latent trait models. In G. H. Fischer and D. Laming (Eds.), *Contributions to Mathematical Psychology, Psychometrics, and Methodology* (pp. 249–263). New York, NY: Springer.

Ponocny, I. 2000. Exact person fit indexes for the Rasch model for arbitrary alternatives. *Psychometrika*, 65, 29–42.

R Core Team 2012. *R: A Language and Environment for Statistical Computing*. Vienna, Austria: R Foundation for Statistical Computing. Retrieved from http://www.R-project.org/.

Rasch, G. 1960. *Probabilistic Models for Some Intelligence and Attainment Tests*. Chicago, IL: University of Chicago Press.

Rasch, G. 1966a. An individualistic approach to item analysis. In P. F. Lazarsfeld and N. W. Henry (Eds.), *Readings in Mathematical Social Science* (pp. 89–107). Chicago, IL: Science Research Associates.

Rasch, G. 1966b. An item analysis which takes individual differences into account. *British Journal of Mathematical and Statistical Psychology*, 19, 49–57.

Rizopoulos, D. 2006. ltm: An R package for latent variable modelling and item response theory analyses. *Journal of Statistical Software*, 17, 1–25.

Rost, J. 2004. *Lehrbuch Testtheorie—Testkonstruktion* [Textbook test theory—Test construction]. Bern, Germany: Huber.

Rost, J., and von Davier, M. 1994. A conditional item fit index for Rasch models. *Applied Psychological Measurement*, 18, 171–182.

Rost, J., and von Davier, M. 1995. Mixture distribution Rasch models. In G. H. Fischer and I. W. Molenaar (Eds.), *Rasch Models: Foundations, Recent Developments and Applications* (pp. 257–268). New York, NY: Springer.

Thissen, D. 1982. Marginal maximum likelihood estimation for the one-parameter logistic model. *Psychometrika*, 47, 175–186.

van den Wollenberg, A. L. 1982. Two new test statistics for the Rasch model. *Psychometrika*, 47, 123–139.

Verguts, T., and De Boeck, P. 2000. A note on the Martin-Löf test for unidimensionality. *Methods of Psychological Research Online*, 5, 77–82. Retrieved from http://ppw.kuleuven.be/okp/_pdf/Verguts2000ANOTM.pdf

Verhelst, N. 2001. Testing the unidimensionality assumption of the Rasch model. *Methods of Psychological Research Online 2001*, 6(3), 231–271. Retrieved from http://136.199.86.12/fachgruppen/methoden/mpr-online/issue15/art2/verhelst.pdf

Verhelst, N. D., and Glas, C. A. W. 1995. The generalized one-parameter model: OPLM. In G. H. Fischer and I. W. Molenaar (Eds.), *Rasch Models: Foundations, Recent Developments and Applications* (pp. 215–238). New York, NY: Springer.

von Davier, M. 1994. *WINMIRA—A Windows Program for Analyses with the Rasch Model, with the Latent Class Analysis and with the Mixed Rasch Model*. Kiel, Germany: Institute for Science Education.

von Davier, M. 2001. *WINMIRA 2001: Software for Estimating Rasch Models, Mixed and Hybrid Rasch Models, and the Latent Class Analysis*. Manual retrieved from http://www.von-davier.com

von Davier, M. 2008. A general diagnostic model applied to language testing data. *British Journal of Mathematical and Statistical Psychology*, 61, 287–307.

von Davier, M., and Carstensen. C. H. (Eds.) 2007. *Multivariate and Mixture Distribution Rasch Models: Extensions and Applications*. New York, NY: Springer.

von Davier, M., and Molenaar, I. W. 2003. A person-fit index for polytomous Rasch models, latent class models, and their mixture generalizations. *Psychometrika*, 68, 213–228.

von Davier, M., and Rost, J. 1995 Polytomous mixed Rasch models. In G. H. Fischer and I. W. Molenaar (Eds.), *Rasch Models: Foundations, Recent Developments and Applications* (pp. 371–379). New York, NY: Springer.

Warm, T. A. 1989. Weighted likelihood estimation of ability in the item response theory. *Psychometrika*, 54, 427–450.

Wright, B. D., and Linacre, J. M. 1991. *BIGSTEPS Computer Program for Rasch Measurement*. Chicago, IL: MESA Press.

Wright, B. D., and Stone, M. H. 1979. *Best Test Design: Rasch Measurement*. Chicago, IL: MESA Press.

Wu, M. L., Adams, R. J., and Wilson, M. R. 1997. *ConQuest: Multi-Aspect Test Software*. Camberwell, Australia: Australian Council for Educational Research.

Xu, X., and Jia, Y. 2011. *The Sensitivity of Parameter Estimates to the Latent Ability Distribution* (Research Report No. RR-11-40). Princeton, NJ: Educational Testing Service.

Xu, X., and von Davier, M. 2008. *Comparing Multiple-Group Multinomial Loglinear Models for Multidimensional Skill Distributions in the General Diagnostic Model* (Research Report No. RR-08-35). Princeton, NJ: Educational Testing Service.

Zermelo, E. 1929. Die Berechnung der Turnier-Ergebnisse als ein Maximumproblem der Wahrscheinlichkeitsrechnung [The calculation of tournament results as a maximum problem of probability calculus]. *Mathematische Zeitschrift*, 29, 436–460.

Zwinderman, A. H. 1995. Pairwise estimation in the Rasch models. *Applied Psychological Measurement*, 19, 369–375.

Section II

Nominal and Ordinal Models

4

Nominal Categories Models

David Thissen and Li Cai

CONTENTS

4.1 Introduction ... 51
4.2 Presentation of the Model ... 52
4.2.1 Bock's (1972) Original Nominal Categories Model ... 53
4.2.2 Thissen et al.'s (2010) General-Purpose Multidimensional Nominal Model ... 54
4.2.2.1 Fourier Version for Linear Effects and Smoothing ... 55
4.2.2.2 Identity-Based **T** Matrix for Equality Constraints ... 55
4.2.2.3 Relationship of Nominal Model with Other IRT Models ... 56
4.3 Parameter Estimation ... 56
4.4 Model Fit ... 57
4.5 Empirical Examples ... 58
4.5.1 Unidimensional Example: Testlet Scores on a Reading Comprehension Test ... 58
4.5.2 Multidimensional Example: Response Alternatives on a Quality-of-Life Scale with Item Clusters ... 62
4.6 Discussion and Conclusions ... 65
4.6.1 Other Uses of the Nominal Model ... 65
4.6.1.1 Response Pattern Testlets ... 65
4.6.1.2 Determination of Response Category Order: Additional Examples ... 66
4.6.1.3 Checking the Behavior of Item Responses ... 68
4.6.2 Current Research ... 70
Acknowledgments ... 70
References ... 71

4.1 Introduction

The nominal categories model (Bock, 1972, 1997) was originally proposed soon after Samejima (1969, 1997) described the first general item response theory (IRT) model for polytomous responses. Samejima's graded models (in normal ogive and logistic form) had been designed for item responses that have some *a priori* order as they relate to the latent variable being measured (θ) (Volume One, Chapter 6). The nominal model was designed for responses with no predetermined order; it uses the multivariate logistic distribution as the model for the probability of the selection of each of $A > 2$ response alternatives, conditional on the value of a latent individual differences variable. The model was originally developed as a logistic approximation to an individual differences version of the extension of Thurstone's (1927) method of paired comparisons to first choices among three or more objects. It can also be viewed as an individual differences extension of Luce's (1959) choice model, or less theoretically as the use of the multivariate logistic function as a link function in a standard nonlinear latent variable regression model; for a more extensive discussion of the theoretical underpinnings of the model, see Thissen et al. (2010).

Thissen et al. (2010) proposed a new multidimensional parameterization of the nominal model that separates the *scoring function values* associated with each nominal category from the overall item-discrimination parameter(s). In so doing, the unidimensional version of the new parameterization is similar to unidimensional models that were described by Andersen (1977), Andrich (1978), Masters (1982), and Muraki (1992), and the multidimensional version to extensions of these models that were proposed by Kelderman and Rijkes (1994), Adams et al. (1997), and Yao and Schwarz (2006), except that in all of those earlier models, the scoring function values are fixed, whereas in the nominal model, they may be estimated from the data.

Takane and de Leeuw (1987) described the theoretical development of a less restricted parameterization of the multidimensional nominal model, as an IRT approach to the factor analysis of nominal variables. The unrestricted version of the multidimensional nominal model has almost never been used; however, variants with other restrictions on the parameter set have been applied for specific purposes: Bolt and Johnson's (2009) study of extreme responding used a version of the multidimensional nominal model that did not separate the scoring function values from the slope parameters; other constraints were used. Johnson and Bolt (2010) used a minimally constrained multidimensional nominal model and other variants. They observed that while the unconstrained multidimensional nominal model may be technically identified it "may not be well identified empirically for a given data set. This is often evident from convergence problems or exceptionally large standard errors" (Johnson and Bolt, 2010, p. 11).

4.2 Presentation of the Model

Unlike many other IRT models, the nominal categories model is best regarded as a general template with more potentially free parameters than are used in any given application. A very general form of the model is

$$f(U_{pi}=a;\boldsymbol{\theta}_p,\boldsymbol{\beta}_{1i},\mathbf{S}_i,\boldsymbol{\beta}_{0i})=f(a)=\frac{\exp(z_a)}{\sum_h \exp(z_h)}, \tag{4.1}$$

in which U_{pi} is the response of person p to item i and

$$z_a=\left[\boldsymbol{\beta}_{1i}\circ \mathbf{s}_{i(a+1)}\right]'\boldsymbol{\theta}_p+\beta_{0i(a+1)} \tag{4.2}$$

and $\circ$ represents the Hadamard (or Schur) product of the vector $\boldsymbol{\beta}'_{1i}$ with the ath column of the matrix $\mathbf{S}_i$, while $f(U_{pi}=a;\boldsymbol{\theta}_p,\boldsymbol{\beta}_{1i},\mathbf{S}_i,\boldsymbol{\beta}_{0i})$ is the probability of a response in category a as a function of the D-dimensional latent variable $\boldsymbol{\theta}_p$. In what follows, we will often shorten this notation to $f(a)$. The response alternatives are numbered $a=0,1,\ldots,A-1$ for an item with A response categories. $\boldsymbol{\beta}_{1i}$ is a D-dimensional vector of slope or discrimination parameters; as in many IRT models, values of the elements of $\boldsymbol{\beta}_{1i}$ reflect the degree to which the item's responses are related to each of the latent variables. $\mathbf{S}_i$ is a $D \times A$ matrix of scoring function values, which indicate the value or order in which the response alternatives are related to each θ. $\boldsymbol{\beta}_{0i}$ is an a-dimensional vector of intercept parameters.

The choice of the notation $\boldsymbol{\beta}_{1i}$ and $\boldsymbol{\beta}_{0i}$ has been made to harmonize the presentation of the nominal model here with the common notation of this volume. However, we would note that in other literature on the nominal model, the slope parameters in $\boldsymbol{\beta}_{1i}$ are most often referred to as $\mathbf{a}_i$, and the intercept parameters in $\boldsymbol{\beta}_{0i}$ are most often referred to as $\mathbf{c}_i$; $\mathbf{a}_i$ and $\mathbf{c}_i$ are also the notation used by software fitting the model. We would also note that $\boldsymbol{\beta}_{1i}$ and $\boldsymbol{\beta}_{0i}$ are not the same as "slope" and "intercept" parameters in other models that may use that notation. The only common element is that they represent the slope and intercept of a linear form; when embedded in other models, the notation $\boldsymbol{\beta}$ has other numerical meanings.

As of the time of this writing, there has been no implementation in general-purpose IRT software of this very general form of the model; however, many models in the literature are subsets of the model described by Equations 4.1 and 4.2.

4.2.1 Bock's (1972) Original Nominal Categories Model

Bock's (1972) original formulation of the nominal model was

$$f(U_{pi} = a; \theta_p, \boldsymbol{\beta}_{1i}^{[o]}, \boldsymbol{\beta}_{0i}) = f(a) = \frac{\exp(z_a)}{\sum_h \exp(z_h)}, \tag{4.3}$$

the multivariate logistic function, with arguments

$$z_a = \beta_{1i(a+1)}^{[o]}\theta_p + \beta_{0i(a+1)}, \tag{4.4}$$

in which z_a is a response process (value) for category a, expressed as a (linear) function of θ with slope parameter $\beta_{1(a+1)}^{[o]}$ and intercept $\beta_{0(a+1)}$; here, the superscript [o] denotes "original." $f(U_i = a; \theta, \boldsymbol{\beta}_{1i}^{[o]}, \boldsymbol{\beta}_{0i})$ is the curve tracing the probability that the item response U is in category a; it is a function of a unidimensional latent variable θ with vector parameters $\boldsymbol{\beta}_{1i}^{[o]}$ and $\boldsymbol{\beta}_{0i}$. Relative to the more general formulation of Equations 4.1 and 4.2, in Equations 4.3 and 4.4, the latent variable θ is unidimensional, and the scoring functions values $\mathbf{s}_i$, here a vector because $D = 1$, are concatenated with the (scalar) slope parameter β_{1i} to yield $\boldsymbol{\beta}_{1i}^{[o]}$. The intercept parameters $\boldsymbol{\beta}_{0i}$ are the same in both formulations.

As stated in Equations 4.3 and 4.4, the model is not identified: The addition of any constant to either all of the $\beta_{1i(a+1)}^{[o]}$ or all of the $\beta_{0i(a+1)}$ yields different parameters but the same values of $f(a)$. As identification constraints, Bock (1972) suggested

$$\sum_{a=0}^{A-1} \beta_{1i(a+1)}^{[o]} = \sum_{a=0}^{A-1} \beta_{0i(a+1)} = 0,$$

implemented by reparameterizing, and estimating the parameter vectors $\boldsymbol{\alpha}_i^{[o]}$ and $\boldsymbol{\gamma}_i$ using

$$\boldsymbol{\beta}_{1i}^{[o]} = \mathbf{T}\boldsymbol{\alpha}_i^{o} \quad \text{and} \quad \boldsymbol{\beta}_{1i}^{[o]} = \mathbf{T}\boldsymbol{\gamma}_i, \tag{4.5}$$

in which "deviation" contrasts from the analysis of variance were used:

$$\underset{A\times(A-1)}{\mathbf{T}_{\mathbf{Dev}}} = \begin{bmatrix} \frac{1}{A} & \frac{1}{A} & \cdots & \frac{1}{A} \\ \frac{1}{A}-1 & \frac{1}{A} & \cdots & \frac{1}{A} \\ \frac{1}{A} & \frac{1}{A}-1 & \cdots & \frac{1}{A} \\ \vdots & \vdots & & \vdots \\ \frac{1}{A} & \frac{1}{A} & \cdots & \frac{1}{A}-1 \end{bmatrix}. \tag{4.6}$$

With the **T** matrices defined as in Equation 4.6, the vectors $\boldsymbol{\alpha}_i^o$ and $\boldsymbol{\gamma}_i$ (of length $A - 1$) may take any value and yield vectors $\boldsymbol{\beta}_{1i}^{[o]}$ and $\boldsymbol{\beta}_{0i}$ with elements that sum to zero.

Other contrast (**T**) matrices may be used as well (see Thissen and Steinberg [1986] for examples). Some other formulations identify the model with the constraints

$$\beta_{1i1}^{[o]} = \beta_{0i1} = 0$$

instead of the zero-sum constraint.

4.2.2 Thissen et al.'s (2010) General-Purpose Multidimensional Nominal Model

Thissen et al. (2010) proposed a parameterization of the nominal model that is a more general subset of the model formulation of Equations 4.1 and 4.2, designed to facilitate item factor analysis (or multidimensional IRT [MIRT] analysis) for items with nominal responses. This nominal MIRT model is

$$f(U_{pi} = a; \boldsymbol{\theta}_p, \boldsymbol{\beta}_{1i}, \mathbf{s}_i, \boldsymbol{\beta}_{0i}) = f(a) = \frac{\exp(z_a)}{\sum_h \exp(z_h)}, \tag{4.7}$$

in which the logit values for each response category a are

$$z_a = \left[\boldsymbol{\beta}_{1i} s_{i(a+1)}\right]' \boldsymbol{\theta}_p + \beta_{0i(a+1)}. \tag{4.8}$$

The difference between this model and that of Equations 4.1 and 4.2 is that $\mathbf{s}_i$ is a vector instead of a matrix because one set of scoring function values is used for all dimensions; this is equivalent to a constraint that the D rows of the matrix $\mathbf{S}_i$ in Equation 4.1 are equal. The model remains nominal in the sense that the scoring functions may be estimated from the data, so in any direction in the $\boldsymbol{\theta}$ space, any cross section of the trace surfaces $f(a)$ may take the variety of shapes provided by the unidimensional nominal model.

Making use of the separation of the overall item-discrimination parameters from the scoring function values in $\mathbf{s}_i$, this multidimensional nominal model has a vector of discrimination parameters $\boldsymbol{\beta}_{1i}$, one value indicating the slope in each direction of the $\boldsymbol{\theta}$ space. This vector of discrimination parameters taken together indicates the direction of highest discrimination of the item, which may be along any of the θ axes or between them.

The following restrictions are imposed for identification:

$$s_{11} = 0, \quad s_{1(A-1)} = A-1, \quad \text{and} \quad \beta_{0i1} = 0.$$

This is accomplished by reparameterizing, and estimating the parameters $\boldsymbol{\alpha}$ and $\boldsymbol{\gamma}$ in

$$\mathbf{s}_i = \mathbf{T}\boldsymbol{\alpha}_i \quad \text{and} \quad \boldsymbol{\beta}_{0i} = \mathbf{T}\boldsymbol{\gamma}_i.$$

4.2.2.1 Fourier Version for Linear Effects and Smoothing

To provide a smooth parametric transition from arbitrarily ordered to graded responses, a Fourier basis, augmented with a linear column, may be used as the $\mathbf{T}$ matrix:

$$\underset{A\times(A-1)}{\mathbf{T}_{\mathbf{F}}} = \begin{bmatrix} 0 & 0 & \cdots & 0 \\ 1 & t_{22} & \cdots & t_{2(A-1)} \\ 2 & t_{32} & \cdots & t_{3(A-1)} \\ \vdots & \vdots & & \vdots \\ A-1 & 0 & \cdots & 0 \end{bmatrix}$$

in which $t_{ak} = \sin[\pi(k-1)(a-1)/(A-1)]$ and $\alpha_1 \equiv 1$.

The Fourier $\mathbf{T}$ matrix provides several useful variants of the nominal model: When β_1, $\{\alpha_2, \ldots, \alpha_{A-1}\}$, and γ are estimated parameters, the result is the "full rank" nominal model. If $\{\alpha_2, \ldots, \alpha_{A-1}\}$ are restricted to be equal to zero, the result is a reparameterized version of Muraki's (1992, 1997) generalized partial credit (GPC) model. The Fourier basis provides a way to create models "between" the GPC and nominal model, as suggested by Thissen and Steinberg (1986) and used by Thissen et al. (1989), Wainer et al. (1991), and others.

4.2.2.2 Identity-Based T Matrix for Equality Constraints

An alternative parameterization that facilitates some models involving equality constraints uses $\mathbf{T}$ matrices for $\mathbf{s}_i$ of the form

$$\underset{A\times(A-1)}{\mathbf{T}_{\mathbf{Is}}} = \begin{bmatrix} 0 & \underline{\mathbf{0}}'_{A-2} \\ \underline{\mathbf{0}}_{A-2} & \mathbf{I}_{A-2} \\ A-1 & \underline{\mathbf{0}}'_{A-2} \end{bmatrix}$$

with the constraint that $\alpha_1 \equiv 1$.

This arrangement provides for the following variants of the nominal model, among others: When β_1, $\{\alpha_2, \ldots, \alpha_{A-1}\}$, and γ are estimated parameters, the result is again the full rank nominal model. If $\alpha_i = i-1$ for $\{\alpha_2, \ldots, \alpha_{A-1}\}$, the result is another reparameterized version of the GPC model. The restriction $s_1 = s_2$ may be imposed by setting $\alpha_2 = 0$. The restriction

$s_{A-1} = s_A$ is imposed by setting $\alpha_{(A-1)} = A - 1$. For the other values of **s**, the restriction $s_{a'} = s_a$ may be imposed by setting $\alpha_a' = \alpha_a$.

If it is desirable to impose equality constraints on the β_0s, we use the following **T** matrix:

$$\underset{A\times(A-1)}{\mathbf{T}_{\mathbf{I}\beta_0}} = \begin{bmatrix} \mathbf{0}'_{A-1} \\ \mathbf{I}_{A-1} \end{bmatrix}.$$

4.2.2.3 Relationship of Nominal Model with Other IRT Models

Imposition of various restrictions on the parameters of the nominal model yields many widely used IRT models. If the scoring function values (s) are equal to the integer category values 0, …, $A - 1$, the nominal model is a model for graded data, equivalent to Muraki's (1992, 1997) GPC model (Volume One, Chapter 8). If in addition, the slope (β_1) parameters are constrained equal across items, the model is a variant of Masters's (1982) partial credit model (Volume One, Chapter 7). Andrich's (1978) rating scale model (Volume One, Chapter 5), or "rating scale" a parameterizations of the GPC model, may be obtained with additional suitable constraints on the intercept (β_0) parameters; see Thissen et al. (2010) or Thissen and Steinberg (1986) for details about how that is done using various **T** matrices.

While this is all interesting from a theoretical point of view, the most important single fact about the nominal model is this: If the scoring function values (s) are equal to the integer category values 0, …, $A - 1$, the "nominal" model is a model for graded item response data. To the extent that the scoring function values (s) deviate from successive integers, that describes some other, arbitrary, order of the response options. This fact is used in the examples in the remainder of this chapter.

4.3 Parameter Estimation

Most applications of the nominal model have made use of maximum likelihood (ML) estimation for the parameters, sometimes referred to as "marginal maximum likelihood" (Bock and Aitkin, 1981) or more correctly as "maximum marginal likelihood" (Ramsay and Winsberg, 1991) because the process involves integrating over or "marginalizing" the latent variable(s) (Glas, Volume Two, Chapter 11). A general-purpose implementation of ML estimation of the parameters of Bock's (1972) original unidimensional nominal model, with some additional **T** matrices suggested by Thissen and Steinberg (1984), was in the software *Multilog* (du Toit, 2003) using the Bock and Aitkin (1981) EM algorithm (Aitkin, Volume Two, Chapter 12). That implementation has been superseded by several algorithms for ML estimation of the parameters of the multidimensional new parameterization that are implemented in the software *IRTPRO* (Cai et al., 2011; Cai, Volume Three, Chapter 25), including adaptive quadrature (Schilling and Bock, 2005) and the MH-RM algorithm (Cai, 2010a, 2010b) as well as the EM algorithm.

In their study of extreme responding, Bolt and Johnson (2009) used an implementation of ML estimation for a multidimensional nominal model in the *Latent GOLD* software that combines EM and direct Newton–Raphson approaches (Vermunt and Magidson, 2005, 2006, 2008) (Vermunt, Volume Three, Chapter 31). Johnson and Bolt (2010) estimated parameters of multidimensional nominal model variants using *SAS PROC MIXED*, which provides direct estimation using adaptive quadrature. These more general-purpose systems for parameter estimation are more flexible than the current generation of dedicated IRT software, but not as efficient, so the number of items that can be analyzed using these systems is somewhat limited.

4.4 Model Fit

Because the nominal model is really a template, or family of models, in many applications, alternatives are considered with more, fewer, or different constraints on the parameters, and various statistics may be used for model comparison or selection. When the parameters are estimated using the method of ML, the value of –2 log likelihood is available for each model; for hierarchically nested models, the likelihood-ratio test, the difference between these values for the more and less restricted models, is distributed as a χ^2 statistic with *df* equal to the difference in the number of parameters. A significant likelihood-ratio χ^2 value indicates that the additional parameters in the less restricted model are different than would be expected by chance if their true values were zero, so the less restricted model is preferred.

For any sets of models, the Akaike information criterion (AIC) (Akaike, 1974) or the Bayesian information criterion (BIC) (Schwarz, 1978) may be used for model selection (Cohen and Cho, Volume Two, Chapter 18). AIC and BIC correct –2 log likelihood with different penalties for using additional parameters; both are used by selecting the model with the smallest value. For nested models, when all three procedures can be used, significance as judged by the likelihood-ratio test usually suggests using a relatively more highly parameterized model, BIC tends to suggest a more restricted model, with the model with the best value of AIC between the previous two.

In most cases, an overall or absolute goodness-of-fit statistic for any IRT model is a chimera. The IRT model is fitted to the table of response-pattern frequencies; in most applied situations, that table has more cells than there are persons in the sample. Under those circumstances, there is no statistical theory about the way the persons are distributed into the cells of that very large table—each cell (response pattern) has some very small probability, but each person must be placed in some cell. For small numbers of dichotomous items, or even smaller numbers of polytomous items, the likelihood-ratio G^2 or Pearson's χ^2 goodness-of-fit statistic may be used. However, the nominal model is designed for polytomous items; with even a few such items, the response pattern cross classification becomes so large that the modeled expected values in each cell are too small for the likelihood-ratio G^2 or Pearson's χ^2 goodness-of-fit statistics to follow their nominal χ^2 distribution.

Limited information goodness-of-fit statistics, such as M_2 (Maydeu-Olivares and Joe, 2005, 2006; Cai et al., 2006), provide an index of model fit that may be useful under some circumstances. The M_2 statistic is computed from the observed and expected frequencies in the one- and two-way marginal tables of the response pattern classification; if the model fits exactly, it follows a χ^2 distribution with *df* computed using methods described by Maydeu-Olivares and Joe (2005, 2006). A nonsignificant value would indicate perfect model fit. However, this statistic, like similar χ^2 statistics obtained in the context of structural equation modeling, is generally unrealistic because there will be some error in any strong parametric model; the associated value of the root mean squared error of approximation (RMSEA) has been suggested to aid the interpretation of statistics such as M_2 (Browne and Cudeck, 1993). If one follows Browne's and Cudeck's (1993) suggested rules of thumb, values of RMSEA of .05 or less indicate close fit, values of .08 or less indicate reasonable fit, and values greater than .1 indicate poor fit of the model.

The M_2 statistic cannot be used in all applications of the nominal model. If there are many items with many response categories, there are too many pairwise marginal tables, and they are too large to compute M_2 as originally defined. If the responses are ordered or graded, that fact can be used to compress those large tables further into summary statistics and a value of M_2 can be computed using them (Cai and Hansen, 2013). However, the nominal model may be fitted to data with large numbers of response categories that are not graded or ordered; no solution has yet been offered for that situation.

If the item responses are graded or ordered, or close enough to ordered so that a summed score is meaningful, then a generalization to the polytomous case of the $S\text{-}\chi^2$ statistic (Orlando and Thissen, 2000, 2003) may be a useful diagnostic statistic for item fit. This statistic is computed for each item by constructing a special kind of table that is marginal to the cross classification of response patterns; the table has rows corresponding to each summed score and columns corresponding to the response alternatives. An approximately χ^2 statistic is computed comparing the observed and expected frequencies. A nonsignificant result is an indicator of adequate model fit. Significant values suggest that the underlying table of frequencies should be examined to determine the reason the statistic is large, which may or may not be lack of model fit. Because even these two-way tables may become very large for large numbers of items (and therefore summed scores) or for large numbers of response categories, even with a sophisticated cell-collapsing algorithm, the $S\text{-}\chi^2$ statistic is sometimes significant merely because there are too many cells with small expected values.

4.5 Empirical Examples

4.5.1 Unidimensional Example: Testlet Scores on a Reading Comprehension Test

Thissen et al. (1989) described the use of the nominal model to fit data from a test made up of testlets, to avoid local dependence among item responses within testlets that might otherwise present problems for the use of unidimensional IRT models. Subsequently, more sophisticated MIRT models have been proposed for testlet-based tests; see Thissen and Steinberg (2010) for a brief summary, or Wainer et al. (2007) on testlet response theory, or Gibbons and Hedeker (1992), Gibbons et al. (2007), Cai (2010c), and Cai et al. (2011) on bifactor or two-tier models, which are contemporary alternatives. Nevertheless, the nominal model remains useful in some contexts to fit testlet data, in which the testlet scores are summed scores or response patterns across several of the actual item responses; the testlet scores then become the "item" data fitted with the nominal model.

To illustrate several properties of the nominal model, we use the data considered by Thissen et al. (1989): the responses of 3866 examinees to a four-passage, 22-item test of reading comprehension. The reading passages were of varying lengths, and were followed by varying numbers of multiple-choice items: seven, four, three, and eight questions for the four passages. Each of the four passages was about different topics. Thissen et al. proposed suppression of multidimensionality that could be due to individual difference variables related to the passage topics by rerepresenting the test as a four-"item" (passage) test, with polytomous scores 0–7, 0–4, 0–3, and 0–8—the number-correct scores for the questions following each passage. This change of representation defines θ in the unidimensional item response model fitted to the passage scores as "reading comprehension proficiency"—the latent variable that explains the observed covariation among number-correct scores across passages. A graded polytomous IRT model might be useful for the testlet scores; however, the fact that the original questions are multiple choice might mean that guessing could lead to the reflection of similar levels of proficiency by low scores (i.e., zero and one item correct might both represent low θ, with the only difference being more or less successful guessing), whereas higher scores could reflect increasingly larger amounts of proficiency. The nominal model is capable of showing this pattern.

Thissen et al. (1989) used the implementation of the original nominal model in the software *Multilog* (du Toit, 2003), with polynomial **T**-matrices. In this section, we describe parallel results obtained with the implementation in *IRTPRO* (Cai et al., 2011) of the new parameterization of the nominal model with Fourier **T**-matrices. We followed the general procedure described by Thissen et al. (1989), by first fitting the nominal model with no constraints (which involved 44 parameters), and then a more constrained model (with 21 parameters). The constrained model was developed using our prior belief that the response categories should be almost ordered, and the proportions choosing each response (passage summed score) should be a fairly smooth distribution. Table 4.1 shows goodness-of-fit statistics for the two models; the M_2 statistics indicate that both models provide satisfactory overall fit. The likelihood-ratio test of the goodness of fit of the constrained versus the unconstrained model, $G^2(23) = 49.23$, was significant, $p = 0.001$. However, the trace lines were not very different between the two models; see Figure 4.1, which shows the trace lines for the constrained (solid lines) and unconstrained model (dotted lines). The AIC statistic suggested the unconstrained model was better, but the BIC suggested the opposite (Table 4.1). Table 4.2 shows the values of the S-χ^2 item-level goodness-of-fit diagnostic statistics for both models; both models provide a satisfactory fit for all four items, which is not surprising given the similarity of the trace lines shown in Figure 4.1. The trace lines differ only at the extremes of θ, where there are very few examinees.

TABLE 4.1

Goodness of Fit Statistics for the Four-Passage Reading Comprehension Data

	Unconstrained Model	Constrained Model
Statistics Based on Log Likelihood		
−2 log likelihood	46119.55	46168.78
Akaike information criterion	46207.55	46210.78
Bayesian information criterion	46482.99	46342.24
Statistics Based on One- and Two-Way Marginal Tables		
M_2	137.01	186.47
df	151	174
p	0.79	0.25
RMSEA	0.0	0.00

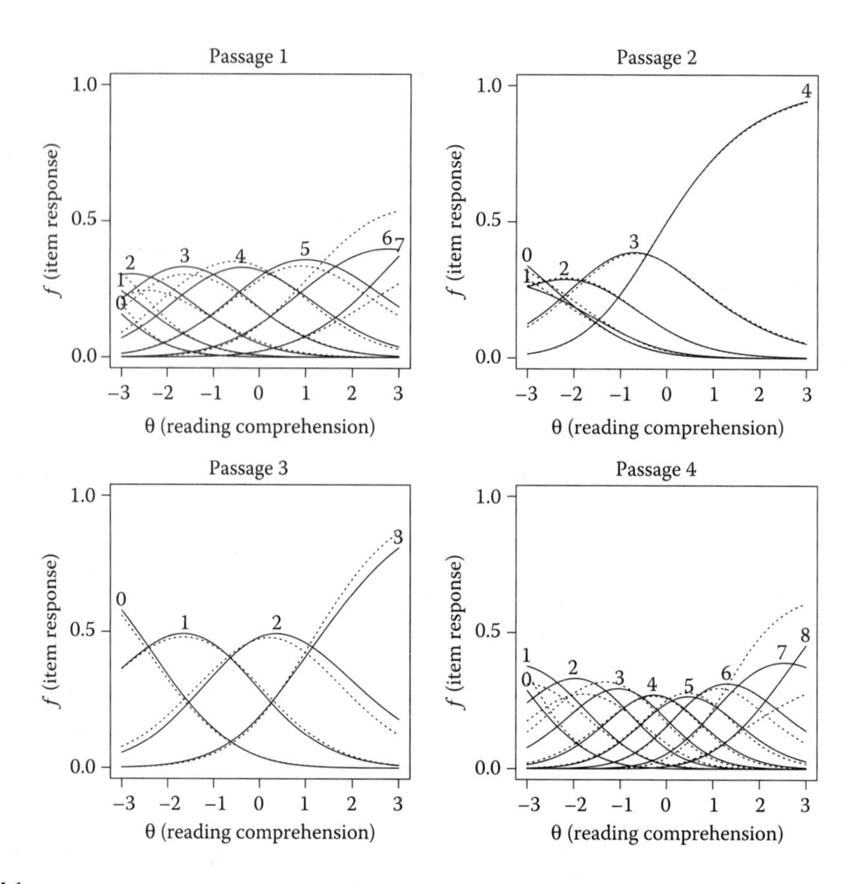

FIGURE 4.1
Trace lines for the four passages' scores; solid lines are the constrained model, dotted lines are the unconstrained model.

The largest difference between the two models is that the unconstrained model imposes no constraints on the scoring function parameters s_k, while the constrained model restricts those parameters to be equal to the score for Passages 1, 3, and 4, and uses a rank-2 Fourier fit for Passage 3. Figure 4.2 shows a plot of the unconstrained $s_{i(a+1)}$ values as a function of the summed score (response category) for each passage, with vertical bars indicating twice each parameter's standard error; the constrained model's straight-line fit for Passages 1, 3, and 4 and the curve for Passage 2 are also shown. Thissen et al. (1989) found that only Passage 2 exhibited the expected guessing effect in the parameters, with

TABLE 4.2
S-χ^2 Item-Level Diagnostic Statistics for the Four-Passage Reading Comprehension Data

	Unconstrained Model			Constrained Model		
Passage	S-χ^2	*df*	*p*	S-χ^2	*df*	*p*
Passage 1	60.18	75	0.89	82.24	84	0.52
Passage 2	45.81	52	0.72	45.90	54	0.78
Passage 3	47.25	44	0.34	57.62	46	0.12
Passage 4	81.81	75	0.28	100.27	83	0.10

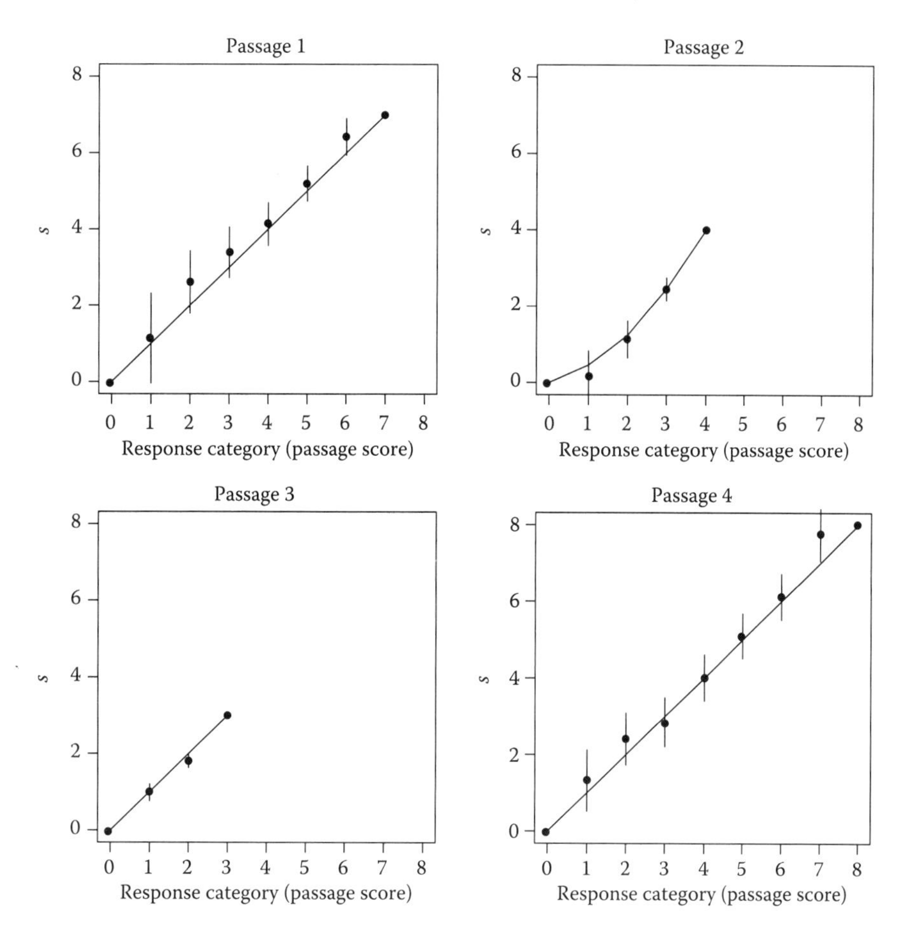

FIGURE 4.2
Unconstrained scoring function values (s) for each of the four passages, and the constrained values as smooth curves. The vertical bars indicate plus or minus twice the standard error of each parameter estimate.

scores of 0 and 1 reflecting very similar values of θ; see Figure 4.1 for the trace lines and Figure 4.2 for the scoring function values. However, Figure 4.1 shows that for Passages 1 and 4, the trace lines for responses of 0 and 1 were nearly proportional at the left side of the plots; the region involved in any guessing effect on low scores for those two passages was so far to the left in the distribution of θ that it would be unlikely to show in the parameters. Table 4.3 shows the frequencies in each score group for the four passages; Passages 1 and 4 have only 20 and 48 zeroes, respectively, which was not a lot of data on which to base a distinction between the relationships of Category 0 and 1 with θ. Passage 3 had too few items to make it very likely to obtain a score of one instead of zero purely by chance.

Parsimony suggests the use of the constrained model. Table 4.4 shows the parameters of the Thissen et al. (2010) parameterization of the constrained model; Table 4.5 shows the corresponding α and γ contrast coefficients. Parameters constrained to be 0.0 or 1.0 are shown with standard errors of "—" in Table 4.5. This example illustrates the idea that the nominal model may be treated as a template in which only a subset of parameters are used for any particular application—the nominal model differs from virtually all other item response models in this respect.

TABLE 4.3

Frequencies for Each Score for the Four Passages on the Reading Comprehension Test

Score	Passage 1	Passage 2	Passage 3	Passage 4
0	20	145	281	48
1	78	186	1132	167
2	292	462	1565	433
3	684	1181	888	649
4	1096	1892		739
5	970			730
6	579			592
7	147			382
8				126

TABLE 4.4

Constrained Item Parameters for the Four-Passage-Score Reading Comprehension Data

Parameter		Passage 1		Passage 2		Passage 3		Passage 4	
β_1	(*s.e.*)	0.55	(0.03)	0.54	(0.04)	0.80	(0.04)	0.68	(0.04)
s_1	β_{01}	0.0	0.00	0.00	0.00	0.0	0.00	0.0	0.00
s_2	β_{02}	1.0	2.08	0.47	0.49	1.0	1.94	1.0	2.31
s_3	β_{03}	2.0	3.98	1.24	1.75	2.0	2.45	2.0	3.91
s_4	β_{04}	3.0	5.28	2.47	2.96	3.0	1.52	3.0	4.80
s_5	β_{05}	4.0	5.84	4.00	3.32			4.0	5.16
s_6	β_{06}	5.0	5.76					5.0	5.07
s_7	β_{07}	6.0	4.86					7.0	4.64
s_8	β_{08}	7.0	3.12					7.0	3.59
s_9	β_{09}							8.0	1.74

TABLE 4.5

Constrained Contrast Parameters for the Four-Passage-Score Reading Comprehension Data

	Passage 1		Passage 2		Passage 3		Passage 4	
Contrast	α (*s.e.*)	γ (*s.e.*)	α (*s.e.*)	γ (*s.e.*)	α (*s.e.*)	γ (*s.e.*)	α (*s.e.*)	γ (*s.e.*)
1	1.0 —	0.45 (0.04)	1.0 —	0.83 (0.04)	1.0 —	0.51 (0.03)	1.0 —	0.22 (0.03)
2	0.0 —	4.18 (0.17)	–0.76 (0.18)	0.09 (0.10)	0.0 —	1.65 (0.06)	0.0 —	4.55 (0.21)
3	0.0 —	–0.26 (0.07)	0.0 —	–0.40 (0.04)	0.0 —	0.0 —	0.0 —	0.07 (0.05)
4	0.0 —	0.10 (0.04)	0.0 —	0.0 —			0.0 —	0.32 (0.04)
5	0.0 —	–0.07 (0.03)					0.0 —	–0.03 (0.03)
6	0.0 —	0.0 —					0.0 —	0.05 (0.03)
7	0.0 —	0.0 —					0.0 —	0.0 —
8							0.0 —	0.0 —

4.5.2 Multidimensional Example: Response Alternatives on a Quality-of-Life Scale with Item Clusters

Gibbons et al. (2007) described the use of a bifactor (MIRT) rating-scale version of the graded model to fit data from a 35-item quality-of-life scale (Lehman, 1988). Their focus was on the fact that there were seven clusters of 4–6 items each on this scale, and a bifactor

MIRT model (Gibbons and Cai, Volume Three, Chapter 3) with a general factor and orthogonal second-tier factors for each cluster fitted the data better than a unidimensional model. In this example, we used a similarly constructed nominal MIRT model to examine the performance of the seven response alternatives for this *QOL* scale. Each of the 35 items had a stem indicating a topic (like satisfaction with "free time," or "emotional well-being"), and a seven-point response scale with the following labels: "terrible" (0), "unhappy" (1), "mostly dissatisfied" (2), "mixed, about equally satisfied and dissatisfied" (3), "mostly satisfied" (4), "pleased" (5), and "delighted" (6). While previous analyses of these data have treated the seven response categories as if they are ordered, it is not clear that, for example, "unhappy" (1) is always a lower response than "mostly dissatisfied" (2).

The data considered by Gibbons et al. (2007), and in this example, are the responses of 586 chronically mentally ill patients. We fitted several more or less constrained nominal models, but all used a similar bifactor structure to that used by Gibbons et al. (2007), except we used only seven second-tier factors, leaving the unclustered global item to be associated only with the general factor, whereas Gibbons et al. (2007) had put that on a second-tier factor of its own.

As a first overall test of the gradedness of the response categories, we fitted the unconstrained nominal model and the GPC model. The likelihood-ratio test of the difference between the fit of those two models is $G^2(175) = 606.27$, $p < 0.0001$, so we conclude that the response alternatives are not as ordered and equally spaced as required by the GPC (or, by implication, the graded) model. Further investigation was suggested by the pattern of the scoring function value estimates, which are displayed graphically in Figure 4.3, using graphics similar to those used by Liu and Verkuilen (2013). (One use of the "Identity" **T** matrices in *IRTPRO* is to identify the estimated contrast parameters with the scoring function values plotted on such graphics, so standard errors are provided.)

Perusal of those values indicated that the most obvious violation of the expected ordering of the response alternatives involved "unhappy" (1) and "mostly dissatisfied" (2): For eight of the 35 items, the scoring function value estimates were in the wrong order. The small-multiple plots in Figure 4.3 are ordered (left to right, and then top to bottom) by the value of the difference between the scoring function value estimates for "unhappy" (1) and "mostly dissatisfied" (2). The order has the effect of showing that there were a number of items on the scale (in the top row, and part of the second) for which the scoring function values were essentially linear (and so the GPC would fit), but that there were also some items (in the bottom row, and part of the next row up) for which the scoring function values were not in the expected order, and very unequally spaced.

The fact that there were eight items for which the scoring function value estimates for "unhappy" (1) and "mostly dissatisfied" (2) were in the wrong order suggests the question: Can we infer that, for those questions, "mostly dissatisfied" was a lower response than "unhappy"? That question can be answered, in a hypothesis-testing framework, by using as null hypotheses equality for the two scoring function values. In estimation, the hypothesis can be implemented by using "identity" **T** matrices for the $\mathbf{s} - \boldsymbol{\alpha}$ parameters, and imposing constraints that set $s_2 = s_3$ for those items. Doing so, we found that the likelihood-ratio test of the difference between the model thus constrained and the unconstrained model was $G^2(8) = 3.46$, $p = 0.9$; so we concluded that there were insufficient data to establish the direction of those two alternatives. However, the same result could lead a data analyst to wonder if the two alternatives "unhappy" (1) and "mostly dissatisfied" (2) might be indistinguishable for all of the items. So we also fitted a model that constrained $s_2 = s_3$ for all items; the likelihood-ratio test of the difference between that model and the unconstrained model was $G^2(35) = 105.23$, $p < 0.0001$. So one concludes that for *some* items "unhappy" (1) < "mostly dissatisfied" (2).

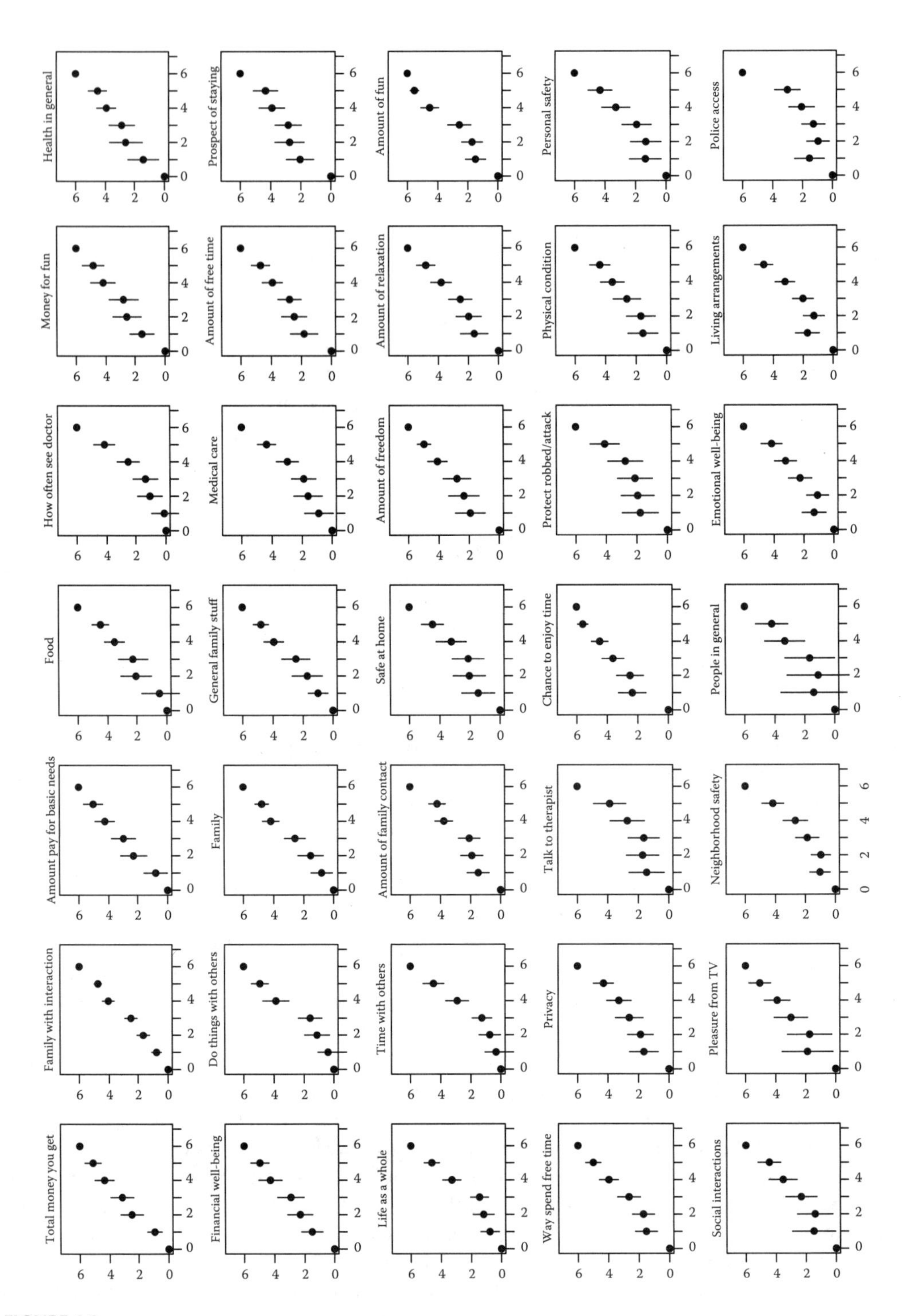

FIGURE 4.3
Scoring function value estimates for the 35 item QOL scale fitted with an unconstrained 8-factor (bifactor) MIRT model; the vertical bars indicate ±2 standard errors.

This example illustrates how the nominal model can be used to examine the expected, or empirical, ordering of response alternatives, even in the context of a MIRT model.

4.6 Discussion and Conclusions

4.6.1 Other Uses of the Nominal Model

4.6.1.1 Response Pattern Testlets

If sets of only two or three items with small numbers of response categories exhibit local dependence in the context of some larger scale, it may be feasible and useful to combine those items into testlets with as many categories as there are patterns of response to the original items. The nominal model may then be used for the analysis of the resulting data, because it may be very difficult to know the order of the response patterns *a priori*. For example, for two dichotomous items with responses coded 0 and 1, one would expect response pattern 00 to be lowest and 11 to be highest; however, it is difficult to guess the positions of 01 and 10 relative to each other or to 00 or 11.

Thissen and Steinberg (1988) illustrated this concept with data obtained from a set of four items measuring numerical knowledge at the preschool level. The four items required the children to identify numerals (3 or 4), and to match those numerals with the corresponding number of blocks. An item-level 2PL model fitted the data for those four items poorly, because they were two locally dependent pairs (*identifying* and *matching*). Fitting the nominal model to the same data relabeled as arising from two four category testlets, each with response categories 00, 01, 10, and 11, yielded better fit and interesting trace lines describing the relations of the response-pattern categories with underlying numerical knowledge; see Thissen and Steinberg (1988), Thissen (1993), or Thissen and Steinberg (2010) for more detailed discussions. Hoskens and De Boeck (1997) used these data as one of the examples in their development of explanatory models for local dependence. They found that two of their explanatory models were equivalent to the constrained nominal models selected by Thissen and Steinberg (1988).

Huber (1993) used these concepts to evaluate potential scoring rules for locally dependent pairs and a triplet of items on the *short portable mental status questionnaire* (SPMSQ), a brief diagnostic instrument used to screen for dementia. One of Huber's locally dependent pairs involved the items that asked for the respondent's age and date of birth. Thissen et al. (2010) presented a graphic illustrating Huber's fitted nominal model that suggests that either response pattern with one correct and one incorrect answer is intermediate between both wrong and both correct. Steinberg and Thissen (1996) used the nominal model to fit response patterns for locally dependent pairs and a triplet of items on scales from Klassen and O'Connor's (1989) *Violence Risk Assessment Study*, and showed how the results of those analyses could be used to clarify the meaning of the latent variables; Thissen and Steinberg (2010) also discussed one of those examples. For the final data analysis for this study, the nominal IRT model fitting was used to guide the selection of item scores used in summed scores on the measurement instruments, illustrating the idea that the IRT model may be used as part of a data analytic strategy without necessarily requiring all aspects of IRT to be used.

Another example of the use of the nominal model for response-pattern testlets in such a supporting role was described by Panter and Reeve (2002). Their primary goal was to fit a unidimensional IRT model to a set of five items that indicated "positive utility beliefs"

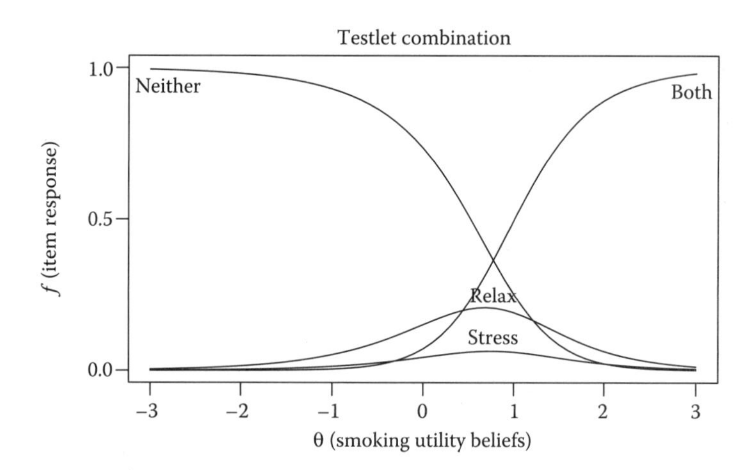

FIGURE 4.4
Trace lines for the four response patterns for two dichotomous items about the utility of smoking: "Helps people relax" and "Helps reduce stress."

about smoking. However, the data suggested that two of the items, expressing the ideas that smoking "helps people relax" and "reduces stress" were locally dependent. Panter and Reeve fitted the four response patterns to those two items with the nominal model, with the other three items on the scale using the 2PL model. The nominal model trace lines for the four response patterns, shown in Figure 4.4, suggested that the two-item response pattern testlet could usefully be scored dichotomously, 0 for 00 and 1 for any of {01, 10, or 11}. In the end, Panter and Reeve (2002) used a 2PL IRT model for the (now) four-item scale, in which one of the items was that dichotomously scored combination of two questions. Although the nominal model does not appear in the final scoring; its use was an essential part of the process, just as was the case in the analyses Steinberg and Thissen (1996) did for the Klassen and O'Connor (1989) study.

Liu and Verkuilen (2013) used the nominal model for testlet-like combinations of presence–severity items, like the pair "In the past week, did you have dry mouth," followed by "If yes, how much did it bother you" with graded responses "not at all," "a little," "somewhat," and "a lot." Liu and Verkuilen combined each such pair of presence–severity items on the *Memorial symptom assessment scale—short form* (MSAS-SF) (Chang et al., 2000) into a single five-category nominal testlet ("item"). They fitted those combinations with the nominal model, and used graphics very much like that shown in Figure 4.3 to evaluate the degree to which the five-category responses for the presence–severity combinations were graded or ordered.

4.6.1.2 Determination of Response Category Order: Additional Examples

While it is usually the case that response alternatives for individual items have some scoring order that is known *a priori*, or they are collapsed into meaningful dichotomous scores (such as *correct* and *incorrect* for multiple choice items on educational tests), sometimes there is some questions about the order of scoring even within a single item. As described in the *QOL* example in the previous section, analysis with the nominal model can be used to provide answers.

Thissen (1993) described the use of the nominal model to fit trace lines for the five response alternatives for one of the items on the *BULIT*, a 36-item scale created to identify individuals with, or at risk for developing bulimia (Smith and Thelen, 1984). Most of the items on the scale had five clearly ordered response alternatives, but one item provided five apparently completely nominal alternatives about where the respondent preferred to eat. Analysis of this item with the nominal model suggested that the response alternatives were not in the order originally proposed to score the instrument.

Huber (1993) used the nominal model to fit an apparently dichotomous item on the *SPMSQ* as three response categories: correct, incorrect, and no-response (or refusal to answer). While it may not often be wise to consider no-response a response that depends solely on the same latent variable as the other responses, in this case, it made sense: The *SPMSQ* item asked the respondents to "count down from 20 by 3s"; standard (non-IRT) scoring for the *SPMSQ* counted nonresponse to this item as incorrect, assuming that respondents declined to answer because they could not do it. Huber's (1993) analysis with the nominal model suggested that nonresponse was actually very similar to wrong in this case (Thissen et al., 2010).

More recent uses of the nominal model to suggest scoring rules have involved data obtained with multiple-choice situational judgment tests (SJTs), which may be useful for the measurement of workforce readiness or for preemployment testing. In the latter context, Thissen-Roe and Mangos (2012) described an *SJT* that presented a situation in which heavy supplies needed to be moved, but required equipment was not available. Respondents were asked to choose among (A) attempting to move the supplies on their own, (B) waiting for the equipment and moving the supplies later, (C) asking a coworker to help, and (D) leaving a note saying the move would be done the next day. Fitting the nominal model to this item, along with several others like it, yielded trace lines like those shown in the upper panel of Figure 4.5. The scoring function values for this item, $\mathbf{s}' = [0.0, 1.30, 2.08, 3.00]$ with the alternatives in the order A, C, D, B, suggest graded scoring in that order for this item.

Zu and Kyllonen (2012) described fitting the nominal model to items on two assessments made up of SJTs, some of which they described as having "ambiguous keys." The nominal

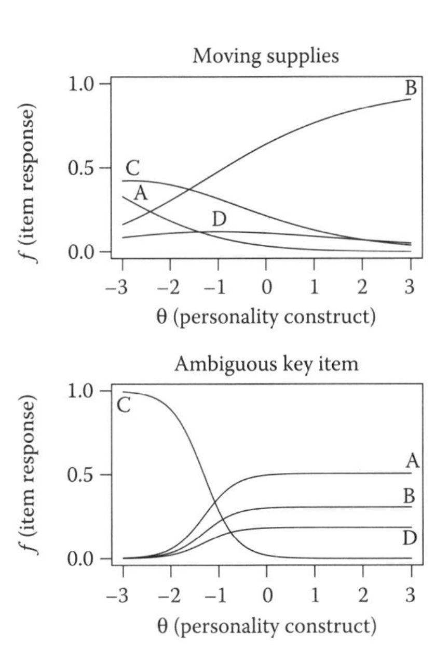

FIGURE 4.5
Trace lines for two multiple-choice situational judgment items.

model trace lines for one such item is shown in the lower panel of Figure 4.5. The situation for this item told the respondent that they were nervous about a speech they were to give at school and worried that some would not understand it. They were asked what they would do in this situation, choosing among (A) working on the speech to make it easier to understand, (B) practicing the speech in front of family or friends, (C) just giving the speech, or (D) being positive and confident. Zu and Kyllonen (2012) pointed out that the nominal model analysis suggests that alternatives A, B, and D are all positive responses, and C is not—which suggests a dichotomous scoring key for this item with three right answers!

It is probably not a good idea, in general, to apply the nominal model to data obtained from multiple-choice items, even though such items were used for the illustrations of its performance in Bock's (1972) original presentation, because the nominal model does not include components for the guessing behavior that is ubiquitous on selected response items on educational assessments. Thissen and Steinberg (1984) proposed a modification of the nominal model that includes guessing parameters and is more suitable for educational multiple-choice items. However, SJTs in multiple-choice format may be different, in that with well-constructed alternatives, each alternative may be attractive to examinees in some range of the latent variable, including the lowest range. In any event, the nominal model appears to provide the IRT analysis of choice for SJTs, to provide empirical guidance or checks on item scoring if not the basis for the scores.

4.6.1.3 Checking the Behavior of Item Responses

Even when items have response alternatives one might expect to be graded, analysis with the nominal model may show that some of them are not always in the expected order, or their spacing or relative discrimination varies. Sharp et al. (2012) used the nominal model to analyze data obtained with the *Problem Gambling Severity Index* (PGSI), a scale intended to assess characteristics of gambling behavior, severity, and consequences. The scale includes items like "Has gambling caused you any health problems, including stress, or anxiety?" and "Have you felt that you might have a problem with gambling?," with the (apparently) graded response alternatives "never" (0), "sometimes" (1), "most of the time" (2), and "almost always" (3). Sharp et al. (2012) found that the use of the nominal model with these items fitted the data satisfactorily and provided more information than dichotomized scoring. However, some of the items' trace lines were less ordered than one might expect from the response alternative labels. Figure 4.6 shows the trace lines for two of the items. For the item about health problems (upper panel), the difference between a response of 0 and any of the higher responses was large, while the difference among the other responses was much smaller; $\mathbf{s}' = [0.0, 2.05, 2.45, 3.00]$. For the "problem with gambling" item (lower panel), the responses were as though it was a three-category graded item, with "sometimes" (1) and "most of the time" (2) reflecting similar levels of severity; $\mathbf{s}' = [0.0, 1.08, 1.36, 3.00]$.

Preston et al. (2011) proposed the use of a derivative set of parameters of the nominal model, category boundary discrimination (CBD) parameters, to evaluate response categories. In terms of the new parameterization, the CBD parameters are the differences between the scoring function values for adjacent response categories multiplied by the item slope (β_{1i}) for unidimensional models. If the item responses are graded, or would be fitted well with the GPC, all of the CBD parameters within an item would be equal. Preston et al. (2011) analyzed data obtained in the development of the *PROMIS* emotional distress scales, which used five response alternatives: "never," "rarely," "sometimes," "often," and "always." The highest two categories were collapsed to yield "often/always" because those two categories were used rarely. Preston et al. found that it was often the case that the CBD

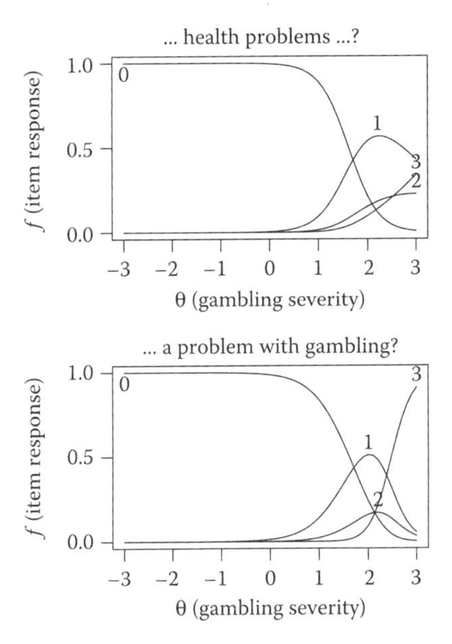

FIGURE 4.6
Trace lines for two items from the problem gambling severity index.

parameter for the first ("never" vs. "rarely") and last ("sometimes" vs. "often/always") response distinctions were higher than those for the remaining distinction ("rarely" vs. "sometimes"). They also found that even though these differences in discrimination were significant, they appeared to yield only small differences in scores. Nevertheless, Preston et al. (2011, p. 548) concluded that the nominal model "may be useful for (a) testing the assumption of ordered response categories inherent in most psychological scales, (b) testing the assumptions of equal CBD parameters inherent in more restricted models such as the GPC, (c) identifying poorly functioning response categories that should be collapsed, (d) conducting a fine-grained study of cross-cultural differences in category usage, and (e) studying the effects of reverse wording."

In addition to these considerations, sometimes analysis with the nominal model to check the behavior of item response alternatives may suggest that the sample would not be best for a purpose like item calibration. Revicki et al. (2009) used the nominal model to analyze data from the item tryout for the *PROMIS Pain Behavior* scale, which had items like "When I was in pain I screamed" and used a 6-point response scale, with the response "had no pain" prepended to a Likert-type response scale ("never," "rarely," "sometimes," "often," "always"). Revicki et al. found that the difference between the scoring function values for the first and second responses (i.e., between "had no pain" and the other Likert-type responses) was very large relative to all the other differences. This difference among the response category discriminations made the use of a graded model with those data problematic. This result becomes easier to understand when it is juxtaposed with Liu and Verkuilen's (2013) use of the nominal model with presence–severity items: The pain behavior items could be viewed as presence–severity items in which the "presence" item *was always the same question!* Repeatedly asking the same question (implicitly: "Did you have pain?") induced extreme local dependence, with the result that the latent class distinction pain/no-pain induced "θ theft" (Thissen and Steinberg, 2010, p. 131) and the slope for that trace line in this context became nearly vertical.

Revicki et al. ultimately concluded that the problem was that, while the items made little sense for persons who had no pain in the designated 7-day response window, they had been administered to an unselected population, many of whom selected "had no pain" for most, if not all, items. The final *PROMIS* pain scale calibration was done with a subsample selected to include mostly persons with some pain, because the scale will probably only be used in practice with persons with medical conditions that involve some degree of pain. Samejima's (1969) graded model was used with those data. But the use of the nominal model was an important contribution to the detection of a property of the data that might otherwise have introduced other problems in their analysis.

In a very different context, Thissen-Roe and Thissen (2013) found that fitting a unidimensional nominal model to personality items with Likert-type response scales like "strongly disagree," "disagree," "agree," and "strongly agree" may yield a phenomenon in which the response scale appears to "fold" into the order "disagree"–"agree"–"strongly disagree"–"strongly agree" when "extreme response set" is a strong second dimension of individual differences indicated by the items, along with the personality trait the questions are intended to measure. When a unidimensional nominal model is fitted to such two-dimensional data, the single latent variable may become an oblique combination of the intended trait (disagree on the left and agree on the right) with extreme response set (not "strongly" at the bottom but "strongly" at the top). Then the nominal model reorders the responses from those that are associated with low trait and low extreme response set (lower left) to high trait and high extreme response set (upper right). The solution in this case was not to change the data, but rather to change the model. However, in keeping with the theme of this section, this is another example in which analysis with the nominal model reveals unexpected behavior of item responses that requires attention.

4.6.2 Current Research

Since its introduction 40 years ago, the nominal categories model has found many uses in data analysis based on IRT. Most of these uses involved the unidimensional nominal model, and some of them have been described in the preceding sections of this chapter.

Parameter estimation for multidimensional variants of the nominal model has only recently become feasible. Many open research questions remain about the MIRT nominal model: What constraints are required for identification? To what extent might the model parameters show poor "empirical identifiability?" Will the restricted MIRT nominal model proposed by Thissen et al. (2010) be widely useful when applied to data? Or will it be the case that to fit nominal responses with an MIRT model, the scoring function values will often need to vary from dimension to dimension? Using variants of the nominal model that do *not* have the scoring function values equal for all dimensions, Bolt and Johnson (2009) and Johnson and Bolt (2010) studied extreme responding and other response sets. How widely will those models be useful? These questions, and others, will no doubt be topics of active research in the next decade or two.

Acknowledgments

We thank R. Darrell Bock for ideas leading to the development of the new parameterization of the nominal model, and Ying Liu, Anne Thissen-Roe, and Jiyun Zu for helpful contributions to the preparation of this chapter; any errors it contains are, of course, our own. Li Cai's research has been partially supported by the Institute of Education Sciences (R305B080016 and R305D100039).

References

Adams, R., Wilson, M., and Wang, W. 1997. The multidimensional random coefficients multinomial logit model. *Applied Psychological Measurement*, 21, 1–23.

Akaike, H. 1974. A new look at the statistical model identification. *IEEE Transactions on Automatic Control*, 19, 716–723.

Andersen, E. B. 1977. Sufficient statistics and latent trait models. *Psychometrika*, 42, 69–81.

Andrich, D. 1978. A rating formulation for ordered response categories. *Psychometrika*, 43, 561–573.

Bock, R. D. 1972. Estimating item parameters and latent ability when responses are scored in two or more latent categories. *Psychometrika*, 37, 29–51.

Bock, R. D. 1997. The nominal categories model. In: W. van der Linden and R. K. Hambleton (Eds.), *Handbook of Modern Item Response Theory* (p. 33–50). New York: Springer.

Bock, R. D., and Aitkin, M. 1981. Marginal maximum likelihood estimation of item parameters: An application of the EM algorithm. *Psychometrika*, 46, 443–459.

Bolt, D., and Johnson, T. 2009. Applications of a MIRT model to self-report measures: Addressing score bias and DIF due to individual differences in response style. *Applied Psychological Measurement*, 33, 335–352.

Browne, M., and Cudeck, R. 1993. Alternative ways of assessing model fit. In: K. Bollen and J. Long (Eds.), *Testing Structural Equation Models* (p. 136–162). Newbury Park, CA: SAGE.

Cai, L. 2010a. High-dimensional exploratory item factor analysis by a Metropolis–Hastings Robbins–Monro algorithm. *Psychometrika*, 75, 33–57.

Cai, L. 2010b. Metropolis–Hastings Robbins–Monro algorithm for confirmatory item factor analysis. *Journal of Educational and Behavioral Statistics*, 35, 307–335.

Cai, L. 2010c. A two-tier full-information item factor analysis model with applications. *Psychometrika*, 75, 581–612.

Cai, L., and Hansen, M. 2013. Limited-information goodness-of-fit testing of hierarchical item factor models. *British Journal of Mathematical and Statistical Psychology*, 66, 245–276.

Cai, L., Maydeu-Olivares, A., Coffman, D., and Thissen, D. 2006. Limited information goodness-of-fit testing of item response theory models for sparse 2^p tables. *British Journal of Mathematical and Statistical Psychology*, 59, 173–194.

Cai, L., Thissen, D., and du Toit, S. 2011. IRTPRO version 2: Flexible, multidimensional, multiple categorical IRT modeling [Computer software manual]. Chicago, IL.

Cai, L., Yang, J., and Hansen, M. 2011. Generalized full-information item bifactor analysis. *Psychological Methods*, 16, 221–248.

Chang, V., Hwang, S., Feuerman, M., Kasimis, B., and Thaler, H. 2000. The Memorial symptom assessment scale short form (MSAS-SF). *Cancer*, 89, 1162–1171.

du Toit, M. (Ed.). 2003. *IRT from SSI: BILOG-MG MULTILOG PARSCALE TESTFACT.* Lincolnwood, IL: Scientific Software International.

Gibbons, R., Bock, R., Hedeker, D., Weiss, D., Segawa, E., Bhaumik, D. et al. 2007. Full-information item bifactor analysis of graded response data. *Applied Psychological Measurement*, 31, 4–19.

Gibbons, R., and Hedeker, D. 1992. Full-information item bi-factor analysis. *Psychometrika*, 57, 423–436.

Hoskens, M., and De Boeck, P. 1997. A parametric model for local dependence among test items. *Psychological Methods*, 2, 261–277.

Huber, M. 1993. An item response theoretical approach to scoring the short portable mental status questionnaire for assessing cognitive status of the elderly. Unpublished master's thesis, Department of Psychology, University of North Carolina, Chapel Hill, NC.

Johnson, T., and Bolt, D. 2010. On the use of factor-analytic multinomial logit item response models to account for individual differences in response style. *Journal of Educational and Behavioral Statistics*, 35, 92–114.

Kelderman, H., and Rijkes, C. 1994. Loglinear multidimensional IRT models for polytomously scored items. *Psychometrika*, 59, 149–176.

Klassen, D., and O'Connor, W. 1989. Assessing the risk of violence in released mental patients: A cross-validation study. *Psychological Assessment*, 1, 75–81.

Lehman, A. 1988. A quality of life interview for the chronically mentally ill. *Evaluation and Program Planning*, 11, 51–62.

Liu, Y., and Verkuilen, J. 2013. Item response modeling of presence-severity items: Application to measurement of patient-reported outcomes. *Applied Psychological Measurement*, 37, 58–75.

Luce, R. D. 1959. *Individual Choice Behavior.* New York: Wiley.

Masters, G. N. 1982. A Rasch model for partial credit scoring. *Psychometrika*, 47, 149–174.

Maydeu-Olivares, A., and Joe, H. 2005. Limited and full information estimation and testing in 2^n contingency tables: A unified framework. *Journal of the American Statistical Association*, 100, 1009–1020.

Maydeu-Olivares, A., and Joe, H. 2006. Limited information goodness-of-fit testing in multidimensional contingency tables. *Psychometrika*, 71, 137–732.

Muraki, E. 1992. A generalized partial credit model: Application of an EM algorithm. *Applied Psychological Measurement*, 16, 159–176.

Muraki, E. 1997. A generalized partial credit model. In: W. van der Linden and R. K. Hambleton (Eds.), *Handbook of Modern Item Response Theory* (p. 153–164). New York: Springer.

Orlando, M., and Thissen, D. 2000. Likelihood-based item fit indices for dichotomous item response theory models. *Applied Psychological Measurement*, 24, 50–64.

Orlando, M., and Thissen, D. 2003. Further investigation of the performance of s-x^2: An item fit index for use with dichotomous item response theory models. *Applied Psychological Measurement*, 27, 289–298.

Panter, A., and Reeve, B. 2002. Assessing tobacco beliefs among youth using item response theory models. *Drug and Alcohol Dependence*, 68, S21–S39.

Preston, K., Reise, S., Cai, L., and Hays, R. 2011. Using the nominal response model to evaluate response category discrimination in the PROMIS emotional distress item pools. *Educational and Psychological Measurement*, 71, 523–550.

Ramsay, J. O., and Winsberg, S. 1991. Maximum marginal likelihood estimation in semiparametric data analysis. *Psychometrika*, 56, 365–380.

Revicki, D., Chen, W., Harnam, N., Cook, K., Amtmann, D., Callahan, L. et al. 2009. Development and psychometric analysis of the PROMIS pain behavior item bank. *Pain*, 146, 158–169.

Samejima, F. 1969. Estimation of latent ability using a response pattern of graded scores. *Psychometric Monographs*, No. 18.

Samejima, F. 1997. Graded response model. In: W. van der Linden and R. K. Hambleton (Eds.), *Handbook of Modern Item Response Theory* (p. 85–100). New York: Springer.

Schilling, S., and Bock, R. D. 2005. High-dimensional maximum marginal likelihood item factor analysis by adaptive quadrature. *Psychometrika*, 70, 533–555.

Schwarz, G. 1978. Estimating the dimension of a model. *Annals of Statistics*, 6, 461–464.

Sharp, C., Steinberg, L., Yaroslavsky, I., Hofmeyr, A., Dellis, A., Ross, D. et al. 2012. An item response theory analysis of the problem gambling severity index. *Assessment*, 19, 167–175.

Smith, M., and Thelen, M. 1984. Development and validation of a test for bulimia. *Journal of Consulting and Clinical Psychology*, 52, 863–872.

Steinberg, L., and Thissen, D. 1996. Uses of item response theory and the testlet concept in the measurement of psychopathology. *Psychological Methods*, 1, 81–97.

Takane, Y., and de Leeuw, J. 1987. On the relationship between item response theory and factor analysis of discretized variables. *Psychometrika*, 52, 393–408.

Thissen, D. 1993. Repealing rules that no longer apply to psychological measurement. In: N. Frederiksen, R. Mislevy, and I. Bejar (Eds.), *Test Theory for a New Generation of Tests* (p. 79–97). Hillsdale, NJ: Lawrence Erlbaum Associates.

Thissen, D., Cai, L., and Bock, R. D. 2010. The nominal categories item response model. In: M. Nering and R. Ostini (Eds.), *Handbook of Polytomous Item Response Theory Models* (p. 43–75). New York: Routledge.

Thissen, D., and Steinberg, L. 1984. A response model for multiple-choice items. *Psychometrika*, 49, 501–519.

Thissen, D., and Steinberg, L. 1986. A taxonomy of item response models. *Psychometrika*, 51, 567–577.

Thissen, D., and Steinberg, L. 1988. Data analysis using item response theory. *Psychological Bulletin*, 104, 385–395.

Thissen, D., and Steinberg, L. 2010. Using item response theory to disentangle constructs at different levels of generality. In: S. Embretson (Ed.), *Measuring Psychological Constructs: Advances in Model-Based Approaches* (p. 123–144). Washington, DC: American Psychological Association.

Thissen, D., Steinberg, L., and Mooney, J. A. 1989. Trace lines for testlets: A use of multiple-categorical-response models. *Journal of Educational Measurement*, 26, 247–260.

Thissen-Roe, A., and Mangos, P. 2012. *Plays Well with SJTs: Building a Mixed-Format Item Pool for Personality CAT.* Presentation at the annual meeting of the Society for Industrial and Organizational Psychology, San Diego, CA, April 26–28.

Thissen-Roe, A., and Thissen, D. 2013. A two-decision model for Likert item responses. *Journal of Educational and Behavioral Statistics*, 38, 522–547.

Thurstone, L. L. 1927. A law of comparative judgment. *Psychological Review*, 34, 278–286.

Vermunt, J., and Magidson, J. 2005. *Technical Guide for Latent GOLD 4.0: Basic and Advanced.* Belmont, MA: Statistical Innovations, Inc.

Vermunt, J., and Magidson, J. 2006. *Latent GOLD 4.0 and IRT Modeling.* Available from http://www.statisticalinnovations.com/products/LGIRT.pdf

Vermunt, J., and Magidson, J. 2008. *LG-Syntax User's Guide: Manual for Latent GOLD 4.5 Syntax Module.* Belmont, MA: Statistical Innovations, Inc.

Wainer, H., Bradlow, E., and Wang, X. 2007. *Testlet Response Theory and Its Applications.* New York: Cambridge University Press.

Wainer, H., Thissen, D., and Sireci, S. G. 1991. DIFferential testlet functioning: Definitions and detection. *Journal of Educational Measurement*, 28, 197–219.

Yao, L., and Schwarz, R. D. 2006. A multidimensional partial credit model with associated item and test statistics: An application to mixed-format tests. *Applied Psychological Measurement*, 30, 469–492.

Zu, J., and Kyllonen, P. 2012. Item response models for multiple-choice situational judgment tests. In: J. P. Bertling (Chair), *Advanced Psychometric Models for Situational Judgment Tests.* Symposium presented at the meeting of the National Council on Measurement in Education, Vancouver, British Columbia, Canada, April 13–17.

5

*Rasch Rating-Scale Model**

David Andrich

CONTENTS

5.1 Introduction 75
5.2 Presentation of the Models 76
5.2.1 Class of Rasch Models 76
5.2.2 Noncollapsed Adjacent Categories 77
5.2.3 Empirical Ordering of Categories 78
5.2.4 Parameter Estimation 82
5.2.4.1 Item Parameters 82
5.2.4.2 Person Parameters 85
5.3 Model Fit 85
5.3.1 Item Fit 86
5.3.2 Person Fit 86
5.4 Empirical Example 87
5.4.1 Threshold Order 90
5.4.2 Interpretation of Response Structure and Threshold Disorder 91
5.4.3 Interpreting Threshold Distances 91
5.5 Discussion 92
References 93

5.1 Introduction

The Rasch rating-scale model (RSM) is a member of the class of Rasch models (Rasch, 1961; Andersen, 1977; Andrich, 2005). Its application is particularly relevant when responses to items are in ordered categories. Table 5.1 shows three such formats, each in four categories. Also known as a Likert (1932), format it is an extension of the dichotomous response format and is used when more information may be obtained than from a dichotomous format. In elementary applications, the categories in the intended order are scored with successive integers, beginning with zero (0), and the scores of a person on multiple items of a test or questionnaire summed to characterize a person. In more-advanced applications, a probabilistic model, such as the RSM is applied.

In the RSM, the successive categories are also scored with integers beginning with zero, and if the responses conform to the model, then, given the total score, there is no further information for the person parameter in the responses to the items. Two further properties of the RSM are distinctive: first, that adjacent categories cannot be collapsed in the usual

* The work on the chapter was funded in part by the Australian Research Council and Pearson.

TABLE 5.1

Examples of Ordered Category Assessments in a Rating-Scale Format

Strongly disagree	Disagree	Agree	Strongly agree
Never	Sometimes	Often	Always
Fail	Pass	Credit	Distinction

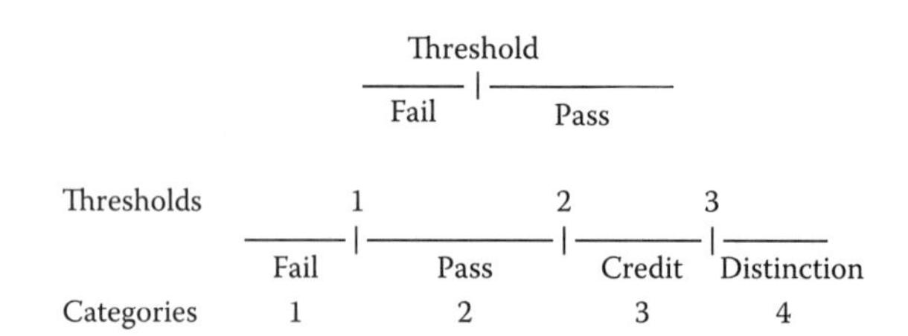

FIGURE 5.1
Graphical representation of contiguous intervals for ordered categories.

way arbitrarily; second, that the model permits making the important empirical test of whether or not the ordering of the categories is working as required.

Figure 5.1 shows the rating extension of the dichotomous case for the last of the three illustrative formats in Table 5.1. The dichotomous format has the continuum partitioned at one point, called a *threshold*; the format with four categories has the continuum partitioned at three thresholds. This relationship between the categories and thresholds is that if one is located in a particular category, then, all the thresholds before that category are exceeded, and all those following the category have not been exceeded. This relationship is analogous to that in a typical measuring instrument. Although seemingly obvious, because the values of the thresholds in the model are taken as a hypothesis and may be disordered in a data set, it is relevant to make the relationship between the thresholds and the categories they define explicit.

5.2 Presentation of the Models

5.2.1 Class of Rasch Models

Instruments in social measurement generally have more than one item where the responses are discrete to increase both the precision and validity of measurement. Having more than two categories, as in a rating scale, is intended to increase precision relative to a dichotomous response. However, it is necessary to check empirically both that the items are consistently working with each other to reflect a single continuum, and that the ordering of the categories within the items is functioning as required.

The Rasch model for a response in one of the *two* ordered categories takes the form

$$\Pr\{U_{pi} = a; \theta_p, b_i\} = [\exp a(\theta_p - b_i)]/\gamma_{pi}, \tag{5.1}$$

where θ_p, b_i are respectively the locations of person p and item i on the hypothesized latent continuum, $a = 0$, 1 are respectively the positive and negative responses, and

$\gamma_{pi} = 1 + \exp(\theta_p - b_i)$ is a normalizing factor that is the sum of the terms in the numerator across the two responses, which ensures that their probabilities sum up to 1 (Volume One, Chapter 3). In proficiency assessment, b_i is referred to as the *difficulty* and Equation 5.1 is referred to in this chapter as the simple logistic model (SLM).

Rasch's (1961) most general formulation for a polytomous response in $A + 1$ categories effectively is

$$\Pr\{U_{pi} = a; \boldsymbol{\theta}_p, \mathbf{b}_i\} = \exp(\theta_{na} - d_{ia})/\gamma_{ni}, a > 0, \tag{5.2}$$

where $\Pr\{U_{pi} = 0;\ \boldsymbol{\theta}_p,\ \mathbf{b}_i\} = 1/\gamma_{ni}$ and where $\boldsymbol{\theta}_p = (\theta_{p1},\ \theta_{p2}, \ldots \theta_{pa}, \ldots,\ \theta_{pA})$, $\mathbf{b}_i = (d_{i1},\ d_{i2}, \ldots d_{ia}, \ldots d_{iA})$ are respectively vectors of person and item locations in A dimensions and $\gamma_{pi} = 1 + \sum_{k=1}^{A} \exp(\theta_{pk} - b_{ik})$ is a normalizing factor. Rasch (1961) specialized the model of Equation 5.2 immediately into a unidimensional structure, and in the modern notation that is a generalization of the dichotomous case, it was equivalent to

$$\Pr\{U_{pi} = a; \theta_p, b_i, \kappa_a, \phi_a\} = [\exp(\varphi_a(\theta_p - b_i) + \kappa_a)]\gamma_{pi}, \tag{5.3}$$

where $a = 0, 1, \ldots, A$ index the categories, φ_a and κ_a are called scoring functions and category coefficients, respectively, and the normalizing factor is $\gamma_{ni} = 1 + \sum_{k=1}^{A} \exp(\varphi_k(\theta_p - b_i) + \kappa_k)$. Rasch (1966) observed that if the probabilities of adjacent categories were summed to provide a new category, then, the invariance property was destroyed. The observation implied that if the response to an item were in, say, five categories, and if the responses fitted the model, then, if the frequencies in a pair of adjacent categories were collapsed into one category, the responses would not fit the model. Although Rasch noted that this might come as a surprise, he did not elaborate further.

Andersen (1977) had the important insight, and then formally proved that if Equation 5.3 was to have sufficient statistics for a unidimensional person parameter, the scoring functions were restricted to

$$\varphi_{a+1} - \varphi_a = \varphi_a - \varphi_{a-1}. \tag{5.4}$$

He also proved that only if two adjacent categories a and $a - 1$ had the same scoring function, $\varphi_a = \varphi_{a-1}$, could they be collapsed into one category.

5.2.2 Noncollapsed Adjacent Categories

Andersen did not interpret the scoring functions and category coefficients with familiar parameters. Andrich (1978) did so in terms of thresholds, and discriminations at these thresholds. This formulation is different from the already-familiar graded-response model (Thurstone and Chave, 1929) and is elaborated in Samejima (1969; Volume One, Chapter 6). While this earlier model is based on partitioning a single-probability distribution with thresholds, the RSM is based on partitioning the continuum with different kinds of thresholds. These two models are contrasted in Andrich (1978, 1995a).

Specifically, the coefficients and scoring functions for the RSM were resolved as

$$\begin{aligned} &\kappa_a = -c_1\tau_1 - c_2\tau_2 - \cdots - c_a\tau_a; \quad \kappa_0 = \kappa_A = 0, \\ &\varphi_a = c_1 + c_2 + \cdots + c_a, \end{aligned} \tag{5.5}$$

where τ_a is the threshold at which the probability of a response in two adjacent categories $a-1$ and a are equal, and c_a is the discrimination at threshold a. First, it is evident that if $c_a > 0$, then, $\varphi_{a+1} > \varphi_a$, which provides for an increasing scoring function for adjacent categories. This is a necessary, but not sufficient condition, for the categories to be ordered empirically.

Second, if $c_a = 0$, then $\varphi_a = \varphi_{a-1}$. Thus, the condition identified by Andersen for collapsing two successive categories is explained as a zero discrimination between them. Explained in this way, the condition that adjacent categories cannot be combined arbitrarily does not seem surprising. Nevertheless, it did generate some debate (Jansen and Roskam, 1986; Andrich, 1995b; Roskam, 1995).

Third, from Equation 5.5, if $c_1 = c_2 = \ldots c_a = c$, then, $\varphi_a = ac$ and Equation 5.4 is satisfied. Thus, the condition of Equation 5.4 identified by Andersen can be explained in terms of *equal* discriminations at the thresholds. This identity of discriminations, which provides sufficient statistics, is characteristic of the Rasch class of models. The common discriminations c may be defined by $c = 1$ simplifying Equation 5.5 to

$$\begin{aligned} \kappa_a &= -\tau_1 - \tau_2 - \cdots - \tau_a, \\ \varphi_a &= 1 + 1 + \cdots + 1 = a \end{aligned} \tag{5.6}$$

giving an integer-scoring function $\varphi_a = a$. The model in Equation 5.3 then specializes to one of the familiar forms of the RSM

$$\Pr\{U_{pi} = a; \theta_p, b_i, \tau\} = [\exp(a(\theta_p - b_i) - \sum_{k=0}^{a} \tau_k)]/\gamma_{pi}, \tag{5.7}$$

where $\tau_0 \equiv 0$ and is introduced for notational efficiency. In this form, an identifying constraint is $\sum_{k=0}^{a} \tau_k \equiv 0$. If the response is dichotomous, then, Equation 5.7 specializes to Equation 5.1.

Sufficiency of the total score is demonstrated as part of the exposition of estimation.

5.2.3 Empirical Ordering of Categories

The RSM can be used simply to model ratings descriptively. However, as indicated above, it can also be used to test the hypothesis that the empirical ordering of the categories is as required. The hypothesis may be rejected if, for example, there are too many categories for raters to distinguish, if the descriptions of categories are not clearly distinguished, if the response structure across categories involves more than one dimension, and so on. With the ubiquitous application of rating scales in the social sciences, the facility to test the empirical ordering of categories is critical. Otherwise, decisions, ranging from those of policy to individual diagnoses, may be misleading. A rationale, Andrich (1978, 2013), which enables testing the empirical ordering of categories using the RSM is summarized below.

Consider illustratively the four ordered categories of Fail, Pass, Credit, and Distinction of Figure 5.1. This is referred to as the *standard design*. Now, consider a design in which a single judgment in one of the *four* ordered categories in Figure 5.1 is *resolved into three independent dichotomous* judgments. This *resolved design* may be considered as a thought experiment for each item (currently we do not subscript the items).

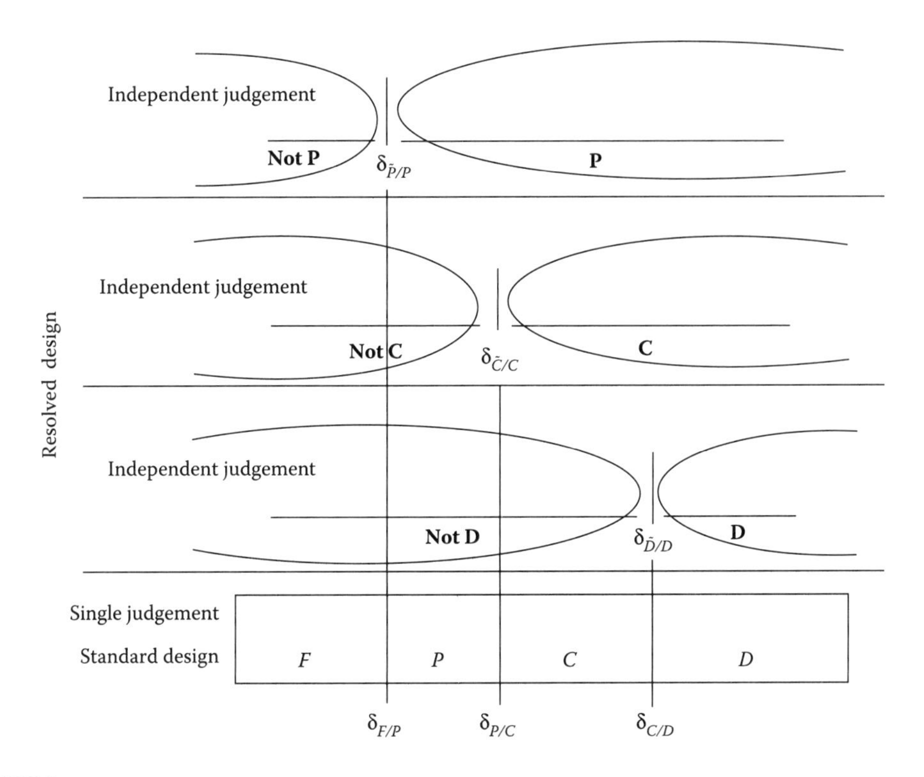

FIGURE 5.2
Resolved-independent design constructed from the standard-ordered designs.

Each judgment in the resolved design is whether or not the performance of the assessment of person p is successful in achieving the standard at only one of the thresholds, Pass, Credit, or Distinction, independently of each other. The design is summarized in Figure 5.2. Its last row repeats the standard design in relation to the resolved design.

Let the independent judgments be modeled by the SLM and let the difficulty of the thresholds be notated as $b_{\tilde{P}/P}, b_{\tilde{C}/C}$, and $b_{\tilde{D}/D}$ respectively. The threshold values $b_{F/P}$, $b_{P/C}$, and $b_{C/D}$ between the adjacent categories in the standard design in the last row of Figure 5.2 are interpreted shortly.

With three independent judgments in the resolved design, there are 2^3 possible response vectors for each performance. Let the space of these response vectors be Ω. Further, let the Bernoulli random variable

$$Y_{pk} = y \equiv \{0,1\}, \quad P_{pk} = \Pr\{Y_{pk} = 1 \mid \Omega\}; \quad Q_{pk} = 1 - P_{pk}, \tag{5.8}$$

characterize the response of person p at threshold k

$$b_k,\ k = \{1,2,3\} \leftrightarrow \{\tilde{P}/P,\ \tilde{C}/C,\ \tilde{D}/D\}, \tag{5.9}$$

where $y = 0$, 1 are the unsuccessful and successful responses, respectively. It is stressed that it is the *continuum*, and not a single-probability distribution as in the graded-response model, that is partitioned at different thresholds. This partitioning is emphasized in Figure 5.2 by the incomplete ovals that envelop the descriptors on either side of a threshold.

Then, the response at each threshold is modeled by

$$\Pr\{Y_{pk} = 1; \theta_p, b_k \mid \Omega\} = [\exp(\theta_p - b_k)]/\gamma_{pk}, \tag{5.10}$$

where $\gamma_{pk} = 1 + \exp(\theta_p - b_k)$ is a normalizing factor.

The resolved design makes it clear that the required threshold order is $b_{k-1} < b_k$; that is, less proficiency is required at b_{k-1} than at b_k. However, there is no artificial constraint that forces success rates at thresholds to take this order.

The standard design is reconnected to the resolved design by invoking the required threshold order as a hypothesis. With this hypothesis, the subspace, $\Omega^G \subset \Omega$, is a sequence of successful responses followed by a sequence of unsuccessful ones at the thresholds—this is referred to as a Guttman (1950) space within which each vector is referred to as a Guttman vector. Both Ω^G and Ω are illustrated in Table 5.2 for a format with four categories. Within Ω^G, the total score $a_p = \Sigma_{k=1}^{3} y_{pk} | \Omega^G$ can be identified with a response in the standard design.

For example, the vector (1, 1, 0) that is a success at Pass and at Credit, but not at Distinction, can be identified with Credit. This identification is also shown in Table 5.2. In contrast, the vector (1, 0, 1) that implies a success at Pass and Distinction but not at Credit, is inconsistent with the hypothesized order. Currently, these vectors are ignored and the latent structure of the RSM focuses on the subspace Ω^G.

Because the probabilities of Ω^G in Ω do not sum up to 1, Ω^G is not a probability space. It is made up by normalizing the probabilities in Ω^G by their sum. Appling (i) the SLM at the thresholds in Ω, (ii) taking only those responses in the space of Ω^G, (iii) recognizing that each vector in Ω^G uniquely identifies the score $a_p = \Sigma_{k=1}^{A} y_{pk}$ and in turn the response in the ordered category, and (iv) normalizing within Ω^G, gives

$$\Pr\{a_p; \theta_p, \mathrm{b} \mid \Omega^G\} = \prod_{k=1}^{a} P_{pk} \prod_{k=a+1}^{A} Q_{pk} / H_p, \tag{5.11}$$

where $H_p = \Sigma_{a=1}^{A} [\Pi_{k=1}^{a} P_{pk} \Pi_{k=a+1}^{A} Q_{pk}]$ is the normalizing factor.

TABLE 5.2

Experimentally Independent Space Ω, the Guttman Subspace

$a_p = \sum_{k=1}^{3} y_{pk} \mid \Omega^G$	y_{p1}	y_{p2}	y_{p3}		
0 (Fail)	0	0	0		
1 (Pass)	1	0	0		
2 (Credit)	1	1	0	Ω^G	
3 (Distinction)	1	1	1		
	0	1	0		Ω
	0	0	1		
	1	0	1		
	0	1	1		

$\sum_{\Omega} \Pr\{y_{p1}, y_{p2}, y_{p3}\} = 1, \sum_{\Omega^G} \Pr\{y_{p1}, y_{p2}, y_{p3} \mid \Omega^G\} = 1$

After further simplification

$$\Pr\{a_p;\theta_p,\mathbf{b}\,|\,\Omega^G\} = [\exp(a\theta_p - \sum_{k=1}^{a} b_k)]/\gamma_p, \tag{5.12}$$

where $\gamma_p = 1+\sum_{a=1}^{A}\exp(a\theta_p - \sum_{k=1}^{a} b_k)$ is the normalizing factor, and is one common form of the RSM. We continue with this form but for completeness, we show below that it is identical to Equation 5.7. Thus, consider, $i = 1, 2, \ldots, I$ items, where each item is imagined to have been resolved and then reconnected to the standard design. Then, Equation 5.12 is expanded to

$$\Pr\{a_{pi};\theta_p,b_i\,|\,\Omega^G\} = [\exp(a\theta_p - \sum_{k=1}^{a} b_{ki})]/\gamma_{pi}. \tag{5.13}$$

Let $b_i = (\sum_{k=1}^{A} b_{ki})/A$ be the average of the thresholds for item i. Further, let $\tau_k = b_i - b_{ki}$. Then, $b_{ki} = b_i - \tau_k$ and $\sum_{k=1}^{A}\tau_k = 0$. Substituting $b_i - \tau_k$ for b_{ki} in Equation 5.13, gives Equation 5.7.

Identical parameters estimated by the SLM and the RSM in the respective spaces. The first result of Equation 5.12 is that none of the parameters have changed their meaning. This implies that an analysis of responses in Ω using the SLM, and an analysis of the subset of responses in the space Ω^G using the RSM estimates identical parameters.

Identical threshold parameters in the respective spaces. A second result from Equation 5.12 is that the conditional probability of the response in the higher of two adjacent categories in the standard design is identical to the probability of success at the corresponding threshold in the resolved design. Thus, from Equation 5.11, for a single item and upon simplification

$$\begin{aligned}&\Pr\{a_p;\theta_p,\mathbf{b}\ |(a-1)_p \cup a_p;\Omega^G\} = P_{pa}/(P_{p(a-1)} + P_{pa})\\ &= [\exp(\theta_p - b_k)/\gamma_{pk},\end{aligned} \tag{5.14}$$

where $\gamma_{pk} = 1 + \exp(\theta_p - b_k)$. Equation 5.14 is identical to Equation 5.10; that is, the conditional probability of a latent success in the higher of two adjacent categories in the subspace Ω^G is just the SLM probability of success at threshold b_k in the resolved space Ω.

In Figure 5.2, the thresholds in the space Ω are $b_k, k = \{1,2,3\} \leftrightarrow \{b_{\tilde{P}/P}, b_{\tilde{C}/C}, b_{\tilde{D}/D}\}$. In the space Ω^G, which characterizes the standard design, the thresholds in Equation 5.14 may be identified as b_k, $k = \{1,2,3\} \leftrightarrow \{b_{F/P}, b_{P/C}, b_{C/D}\}$. Thus, the two sets of thresholds are identical: $b_k, k = \{1,2,3\} \leftrightarrow \{b_{\tilde{P}/P}, b_{\tilde{C}/C}, b_{\tilde{D}/D}\} \leftrightarrow \{b_{F/P}, b_{P/C}, b_{C/D}\}$.

It is possible to construct the RSM and inferences by beginning with Equation 5.14, and taking the implied latent, conditional response in two adjacent categories, in which the higher of the two categories is taken as the relatively successful response (Luo, 2005; Andrich, 2013). This construction demonstrates that the structure and the identity of the thresholds is necessary and sufficient for the RSM.

Mathematical derivations are tautologies. However, they can provide genuinely new insights. The identity of Equations 5.10 and 5.14 is one of them. Because the thresholds in the resolved design are required to have a particular order, they are required to take the same order in the standard design. Without the rationale for the RSM, this required order is not obvious. The implications are illustrated in the example below. It is stressed that the relationship between the resolved and standard designs is mathematical and structural, and does not depend on any real resolved design. Thus, suppose that responses of objects to A stimuli in the space Ω follow the SLM. Then, for any hypothesized difficulty order of the stimuli, an analysis of the subset of responses in the space Ω^G with the RSM gives estimates of exactly the same parameters as the original parameters that follow the SLM (Andrich, 2013). Reciprocally, when responses in ordered categories are analyzed by the RSM, an inference can be made that the threshold estimates from the space Ω^G are the same as if the responses came from an inferred, hypothesized full space Ω, of which the space Ω^G is a compatible subspace, and which were analyzed using the SLM.

5.2.4 Parameter Estimation

5.2.4.1 Item Parameters

Andersen (1973a) demonstrated conditional estimation of the parameters using Equation 5.2, the general form of the Rasch model and in Andersen (1997), he showed the conditional estimation in the specific case equivalent to the now-familiar form of Equation 5.7. Estimation that exploits the sufficiency of the total score is demonstrated efficiently using Equation 5.3 in which the definition of the scoring function $\varphi_a = a$ of Equation 5.6 is applied; that is,

$$\Pr\{U_{pi} = a; \theta_p, d_i, \kappa_a, \phi_a\} = \exp(a(\theta_p - d_i) + \kappa_a)/\gamma_{pi}. \tag{5.15}$$

From the estimates of κ_a, the estimates of the thresholds are readily obtained from Equation 5.6:

$$\tau_a = \kappa_a - \kappa_{a+1}. \tag{5.16}$$

To demonstrate sufficiency of the total score of a person across items, and to introduce the conditional equation for estimating the item parameters independently of the person parameters, consider the responses of a single person p to $i = 1, 2, \ldots, I$ items. On the assumption of statistical independence of the responses, their joint probability is

$$\begin{aligned}\Pr\{\mathbf{U}_p = \mathbf{a}; \theta_p, \mathbf{d}, \boldsymbol{\kappa}\} &= \prod_{i=1}^{I} [\exp(a_{pi}(\theta_p - b_i) + \kappa_a)]/\gamma_{pi} \\ &= \left[\exp\left[X_p\theta_p - \left(\sum_{i=1}^{I} a_{pi}b_i\right) + \sum_{a=0}^{m} f_{pa}\kappa_a\right]\right] \Big/ \prod_{i=1}^{I} \gamma_{pi},\end{aligned} \tag{5.17}$$

where $X_p = \sum_{i=1}^{I} a_{pi}$ is the total score of person p, and where f_{pa} is the frequency of κ_a. Different total scores give possible different frequencies of any response a.

The sum of the probabilities of all the possible ways of obtaining the score X is

$$\sum_{(\mathbf{a}|X)} \Pr\{\mathbf{U}_p = \mathbf{a}; \theta_p, \mathbf{b}, \boldsymbol{\kappa}\} = \sum_{(\mathbf{a}|X)} \left[\exp\left[X_p\theta_p - \left(\sum_{i=1}^{I} a_{pi}b_i\right) + \sum_{a=0}^{A} f_{pa}\kappa_a\right]\right] \Big/ \prod_{i=1}^{I} \gamma_{pi}, \tag{5.18}$$

where $\sum_{(\mathbf{a}|X)}$ indicates the sum of all the possible ways of obtaining a total score of X.

The conditional probability of the vector of responses, $(\mathbf{a}|X)$, of person p given X, is given by the ratio of Equations 5.17 and 5.18. On simplification

$$P\{\mathbf{U}_p = \mathbf{a}; \mathbf{d}, \boldsymbol{\kappa} \mid X\} = \frac{\exp\left[X_p\theta_p - \left(\sum_{i=1}^{I} a_{pi}b_i\right) + \sum_{a=0}^{A} f_{pa}\kappa_a\right] \Big/ \prod_{i=1}^{I} \gamma_{pi}}{\sum_{\mathbf{a}|r} \exp\left[X_p\theta_p - \left(\sum_{i=1}^{I} a_{pi}b\right) + \sum_{a=0}^{A} f_{pa}\kappa_a\right] \Big/ \prod_{i=1}^{I} \gamma_{pi}}$$

$$= \frac{\exp\left[-\left(\sum_{i=1}^{I} a_{pi}b_i\right) + \sum_{a=0}^{A} f_{pa}\kappa_a\right]}{\Gamma_X}, \tag{5.19}$$

where

$$\Gamma_X = \sum_{\mathbf{a}|X} \exp\left[-\left(\sum_{i=1}^{I} a_{Xi}b_i\right) + \sum_{a=0}^{A} f_{Xa}\kappa_a\right], \tag{5.20}$$

is the elementary symmetric function of all possible ways of obtaining a score of X (van der Linden, Volume Two, Chapter 6). Note that because it does not represent a particular response of a person, but one of the possible scores $a = 0, \ldots, A$ that is compatible with a total score of X across all items, the response a for item i in Equation 5.3 is subscripted in the form of a_{Xi}, and its frequency is subscripted in the form of f_{Xa}.

Clearly, Equation 5.19 is independent of θ_p. This results from the sufficiency of the total score X_p for θ_p, which also means that all the information for θ_p is contained in X_p. It is also clear that if the responses to all items are either 0 or m, in which case there is only one vector of responses that gives the total scores of 0 and I_m, respectively, then, Equation 5.19 has the value of 1 and has no information for the estimation of the item parameters. Accordingly, the relevant possible scores are $X = 1, \ldots, X'$, where $X' = \text{Im} - 1$.

For the persons with a score $X = 1, \ldots, X'$, let N_X be the number of persons. Then, the log likelihood of responses of these persons is

$$\ln L_X = -\left(\sum_{p=1}^{N_X}\sum_{i=1}^{I} a_{pi}b_i\right) + \sum_{p=1}^{N_X}\sum_{a=0}^{A} f_{pa}\kappa_a] - N_X \ln \Gamma_X$$
$$= \sum_{i=1}^{I}\left(\sum_{p=1}^{N_X} a_{pi}b_i\right) + \sum_{a=0}^{A}\sum_{p=1}^{N_X} f_{pa}\kappa_a - N_X \ln \Gamma_X. \tag{5.21}$$

However, for the score X, $\sum_{p=1}^{N_X} a_{pi}b_i$ is the sum of the scores of all persons on item i and $\sum_{p=1}^{N_X} f_{pa}$ is the frequency of the score a. Let $S_{Xi} = \sum_{p=1}^{N_X} a_{pi}b_i$ and $F_{Xa} = \sum_{p=1}^{N_X} f_{pa}$. Then, Equation 5.21 reduces to

$$\ln L_X = -\sum_{i=1}^{I} S_{Xi} b_i + \sum_{a=0}^{A} F_{Xa} \kappa_a - N_X \ln \Gamma_X. \tag{5.22}$$

Differentiating Equation 5.22 with respect to each b_i and each κ_a, equating the result to 0 to obtain the maximum, gives, on simplifying, the solution equations

$$S_{Xi} = N_X \sum_{\mathbf{a}|X} a_{Xi} P\{\mathbf{U}_p = \mathbf{a}; \mathbf{d}, \boldsymbol{\kappa} \mid X\}, \quad i = 1, 2, \ldots, I \tag{5.23}$$

and

$$F_{Xa} = N_X \sum_{\mathbf{a}|X} f_{Xa} P\{\mathbf{U}_p = \mathbf{a}; \mathbf{d}, \boldsymbol{\kappa} \mid X\}, \quad a = 1, 2, \ldots, A-1. \tag{5.24}$$

Recall that $\kappa_0 = \kappa_m \equiv 0$. The above equations are solved iteratively. Only one identification constraint needs to be imposed, generally taken to be $\sum_{i=1}^{I} \hat{b}_i = 0$.

Andersen has shown that the estimates based on conditional distributions, as shown above, are consistent. This estimation method directly follows from Rasch's measurement theory regarding the separation of item and person parameters. In particular, in estimating the item parameters, no assumptions need to be made about the person distribution.

Other more general methods of estimation have been implemented that appears in the solution of Equations 5.23 and 5.24. One method, which retains conditional estimation, has consistent estimates, and readily allows for missing data, is the conditional pairwise method (Andrich and Luo, 2003) that is implemented in *RUMM2030* (Andrich et al., 2014). Other methods include joint maximum likelihood (JML) also known as unconditional estimation (Wright and Masters, 1982; Jansen et al., 1988; Wright, 1988) that is implemented, for example, in the *WINSTEPS* software (Linacre, 2002), marginal maximum likelihood (MML) (Bock and Aitken, 1981), which is implemented in *ConQuest* (Adams, Wu, and Wilson, Volume Three, Chapter 27; Wu et al., 2007), and the expected a posteriori (EAP) approach (Bock and Mislevy, 1982), which is implemented in even more general software that takes the Rasch class of models as special cases. JML estimation does not require assumptions regarding the person distribution, but the estimates are inconsistent and for relatively small numbers of items need to be corrected. The MML approach may or may not make assumptions regarding the person distribution.

The RSM is generally applied using software, in which it is a special case of the more general form known as the partial credit parameterization

$$P\{U_{pi} = a; \theta_p, d_i, \tau\} = \left[\exp(a(\theta_p - d_i)) - \sum_{k=0}^{a} \tau_{ki} \right] \Big/ \gamma_{pi}, \tag{5.25}$$

in which the thresholds are subscripted to denote each item giving τ_{ki}. Although each item can have a different number of categories, the interpretation of the response structure of the models of Equations 5.25 and 5.7 at the level of a single item are identical.

5.2.4.2 Person Parameters

Although it is possible to estimate the person parameters by conditioning out vectors of all item parameters, with large numbers of items and persons, it is generally considered not feasible. Nevertheless, Andrich (2010) has shown that conditioning out the vector of item parameters results in an elegant equation from which conditional pairwise estimates are feasible, even when the model is applied to test scores that may have maximum scores ranging from 0 to 100 and where many scores have zero frequencies.

In general, when item parameters are estimated by conditioning out the person parameters, the person parameters are estimated taking the item parameters as known. Effectively, each person parameter is estimated individually. From Equation 5.15, and given the item parameters, the likelihood of the responses of a single person is

$$\begin{aligned} L &= \prod_{i=1}^{I} P\{U_{pi} = a_{pi}; \theta_p, \mathbf{b}, \boldsymbol{\kappa}\} = \prod_{i=1}^{I} \exp(a_{pi}(\theta_p - b_i) + \kappa_a)/\gamma_{pi} \\ &= [\exp \sum_{i=1}^{I} (a_{pi}(\theta_p - b_i) + \kappa_a)] / \prod_{i=1}^{I} \gamma_{pi}. \end{aligned} \tag{5.26}$$

The maximum likelihood estimate is obtained by taking the logarithm of Equation 5.26, differentiating with respect to θ_p, and equating the result to zero. On simplification

$$\begin{aligned} \ln L &= \sum_{i=1}^{I} (a_{pi}(\theta_p - b_i) + \kappa_a) - \sum_{i=1}^{I} \ln \gamma_{pi} \\ \frac{\partial \ln L}{\partial \theta_p} &= \sum_{i=1}^{I} a_{pi} - \sum_{i=1}^{I} \partial \left[\left[\sum_{i=1}^{I} \sum_{k=0}^{A} \exp(k(\theta_p - b_i) + \kappa_k) \right] \gamma_{pi}^{-1} \right] / \partial \theta_p \\ &= X_p - \sum_{i=1}^{I} E[U_p] = 0. \end{aligned} \tag{5.27}$$

Equation 5.27 simply equates the observed total score to its expected value for those items to which person p has responded. Thus, it readily accounts for either structural or random missing data. However, it is known that, especially when the persons are close to the minimum or maximum total scores, these estimates are biased by respectively being smaller or larger than they should be (Bock and Jones, 1968). These estimates can be made less biased by including weighting in the estimating equations (Warm, 1989). Methods of estimation that do condition out the person parameters, estimate the person parameters along with the item parameters, including persons with extreme total scores.

5.3 Model Fit

Various tests of fit between the data and the model have been devised. All of them involve some kind of check on the invariance of the estimates of the item parameters with respect to different subsets of persons, or person parameters with respect to different subsets of items. Because of the emphasis on scale construction, and because there are generally many more persons than items, the majority of tests of fit focus on the former.

5.3.1 Item Fit

Andersen considered two kinds of tests of fit, although both seem not to be routinely implemented in well-known computer software such as those referenced earlier. One reason is that they imply having complete data. Both involve estimating the item parameters for every total score group, and for the sample as a whole. In the first, the test of fit involves conducting a likelihood ratio test to check if the former estimates are significantly different from the latter Andersen (1973b). In the second, standard errors of the differences in estimates of all item parameters are estimated and checked for the significance of difference (Andersen, 1995). This method permits a residual analysis that focuses on deviations between observations and expectation at the item level. Andersen's tests of fit were both based on conditional distributions that excluded person parameters.

Often, when there are many score groups, some with small frequencies, class intervals based on person estimates are formed and the observed total score and the sum of the expected scores for the persons in each class interval are compared. Formally, for class intervals $g = 1, \ldots, G$, and for each item $i = 1, \ldots, I$, the residual z_{gi} is formed:

$$z_{gi} = \sum_{n \in g} a_{pi} - \sum_{n \in g} E[U_{pi}] \Bigg/ \sqrt{\sum_{n \in g} V[U_{pi}]}. \tag{5.28}$$

Then

$$\chi_i^2 = \sum_{g=1}^{G} z_{gi}^2 \tag{5.29}$$

is often taken to be an approximate Pearson chi-square statistic on $G - 1$ degrees of freedom. However, because the parameters are not estimated from the same pooled data, no formal proof justifying the claim exists. Thus, these statistics are best interpreted as statistics ordering the items from best-to-worst fitting where the order can be used to detect patterns of misfit. In addition, it is possible to use standard procedures to uncover patterns in the residuals for each person to each item. The application in Equation 5.29, together with its graphical counterpart, is shown in the example.

5.3.2 Person Fit

In addition to the fit of items with each other, it is possible to check each person's response profile relative to that expected under the model. This can be done conditionally, given a person's total score. Thus, for each total score, each pattern of responses to the items has a probability defined by Equation 5.19. In the SLM for dichotomous items, and for a specified order of item difficulty, the vector with the greatest probability is the Guttman vector. By analogy, in the RSM, the vector with the greatest probability is also defined as the Guttman vector (Andrich, 1985). Given a total score across items, the probabilities of all the possible patterns can be ranked and the probabilities from the one with the smallest probability to the one being considered are summed. If this summed probability is less than some specified value, say 0.05, it indicates that the profile does not fit the model at the specified level of significance with the usual interpretation of significance. This method of testing person fit is cumbersome, and most computer programs implement a simpler method using a function of the standardized residuals that are based on the person's proficiency estimate and each item's threshold parameters.

5.4 Empirical Example

The example concerned ratings of muscle tone according to the clinician-rated Ashworth (1964) spasticity scale that was divided into five ostensibly ordered categories in the structure of Table 5.1: *(0) limbs rigid, (1) increased tone—restricting movement, (2) increased tone—easily flexed, (3) "catch," and (4) normal tone.* The ratings, reordered for this analysis so that the greater the score the better the tone, were generated by a multicenter, randomized controlled, clinical trial of cannabinoids treatment of multiple sclerosis (Zajicek et al., 2003). In the example, data from eight ratings of the parts of the lower limbs (hip adduction, knee extension, knee flexion, and foot plantar flexion) for each side were made, thus resulting in eight items. The total score was taken as a summary of the muscle tone of the lower limbs for each person. In the example, analyses of ratings of 660 people with responses to all items only from their first visit are reported. Although the responses came from different clinicians, these were not identified and the threshold estimates of the items could therefore not be separated from any of their effects. An earlier analysis of the same data using the partial credit parameterization of Equation 5.25 was given in Andrich (2011). Here, the analysis is with the RSM, while *RUMM2030* was used for the analysis.

In parallel to Figure 5.1, Figure 5.3 explicitly shows the implied continuum on this scale. Four points are noted immediately. First, this figure highlights the interpretation of the standard to be met at each threshold. For example, Thresholds 2 and 3 show the standard of *restricting movement* and *easily flexed*, respectively. Then, an assessment that *restricting movement* has been exceeded, and *easily flexed* has not, gives the category response of *increased tone (restricting movement).* Second, although successive threshold estimates do not have to be equidistant on the continuum, they are required, *a priori,* to be ordered. Third, because any significance test depends on the sample size, a statistically significant distance between a pair of successive thresholds is not enough. How this might be decided independently of a traditional significance test is suggested in Section 5.5. Fourth, because

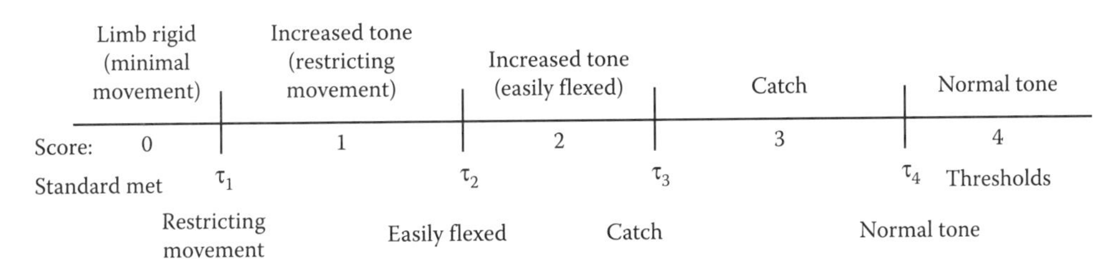

FIGURE 5.3
Implied rating continuum in the example.

TABLE 5.3

Frequencies of Responses in All Categories to All Items, and Fit Probability

	Category Frequencies						Fit	
Description	**0**	**1**	**2**	**3**	**4**	$\hat{b}_i$	**SE**	**Prob**
1. L hip adduction	38	227	221	91	76	0.106	0.047	0.678
2. R hip adduction	38	231	216	94	74	0.115	0.047	0.404
3. L knee extension	47	214	234	79	79	0.132	0.047	0.440
4. R knee extension	45	200	247	94	67	0.107	0.047	0.699
5. L knee flexion	37	138	232	111	135	−0.373	0.046	0.204
6. R knee flexion	37	129	233	127	127	−0.385	0.046	0.482
7. L foot plantar	63	207	211	86	86	0.134	0.047	0.201
8. R foot plantar	62	213	209	87	82	0.163	0.047	0.181
						0.000	Joint	0.325

the test of threshold order is one tailed, $\hat{\tau}_a > \hat{\tau}_{a-1}$, if thresholds are reversed, no matter how small the magnitude of the reversal, the hypothesis of the correct order is rejected.

Table 5.3 shows the frequency of responses in the categories of all items. It also shows the estimate of the locations $\hat{b}_i$, their standard errors, and the fit probability of the χ^2 in Equation 5.29. It is evident that all categories of all items have reasonably high frequencies, with greater frequencies in the higher than the lower ones, suggesting that persons tend to be more at the *normal tone* end of the continuum. Figure 5.4, which shows the distribution of both the person and threshold estimates of all items, confirms this suggestion. Figure 5.4 also shows that the majority of persons and the majority of the thresholds overlapped, which is also where the information for person estimates was maximal. Because all persons in the sample responded to all items, it was meaningful to consider the transformation of the total score to a location estimate. Table 5.4 shows their frequency and cumulative frequency, together with location estimates and the standard errors, for

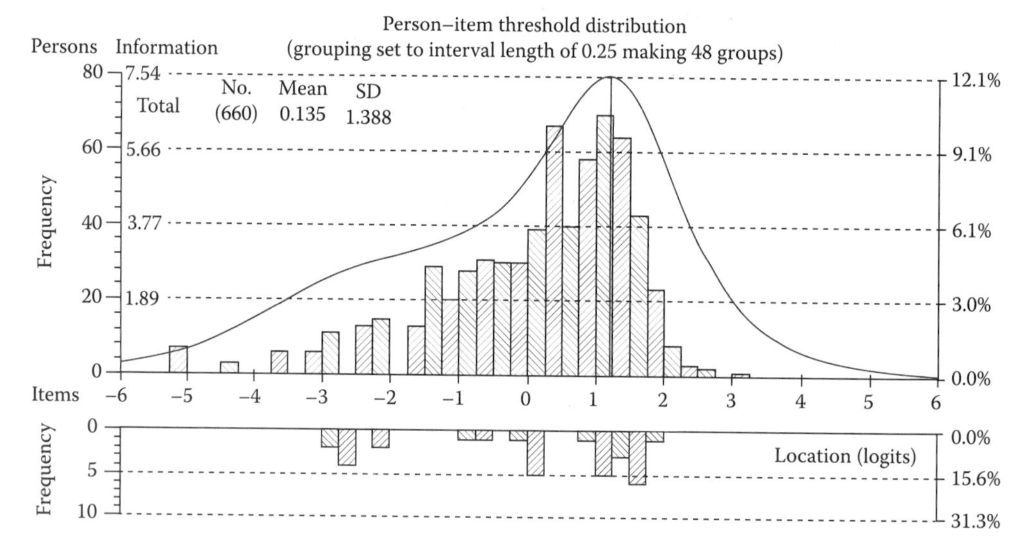

FIGURE 5.4
Person and threshold distributions.

TABLE 5.4

Distribution of Person and Location Estimates and Standard Errors for Selected Total Scores

Score	F	Cumulative F	Cumulative%	Location	Std Error
0	7	7	1.1	–5.110	1.286
1	3	10	1.5	–4.257	0.929
…	…	…	…	…	…
8	29	103	15.6	–1.398	0.555
…	…	…	…	…	…
16	38	348	52.7	0.448	0.399
…	…	…	…	…	…
24	27	607	92.0	1.519	0.378
…	…	…	…	…	…
31	1	660	100.0	3.060	0.743
32	0	660	100.0	3.677	1.061

selected total scores. The estimates for the extreme scores 0 and 32 were extrapolated ensuring that $\hat{\theta}_{X'+1} - \hat{\theta}_{X'} > \hat{\theta}_{X'} - \hat{\theta}_{X'-1}$ where $X' + 1$ is the maximum score, and similarly that $|\hat{\theta}_0 - \hat{\theta}_1| > |\hat{\theta}_1 - \hat{\theta}_2|$. Table 5.4 shows that seven persons had a score of 0 and none of them had a maximum score of 32.

Table 5.3 shows that all items had a very satisfactory fit and that Items 7 and 8, concerned with knee flexion, had a similar location that was relatively easier than the other items.

For completeness, Figure 5.5 shows the expected value curve (EVC) and the observed means of 10 class intervals for Item 8, which according to the fit statistic is the worst-fitting item. It is evident that the observed means were close to their expected values. The slope at every point on the continuum is a function of the relative distances between the thresholds, and is simply given by $V[U_a]$. This is also the information for the person parameter at that location. The closer the thresholds, the steeper the EVC. In Figure 5.5, the value of the slope is shown at the location estimate $\hat{b}_i$. Because all thresholds have the same relative

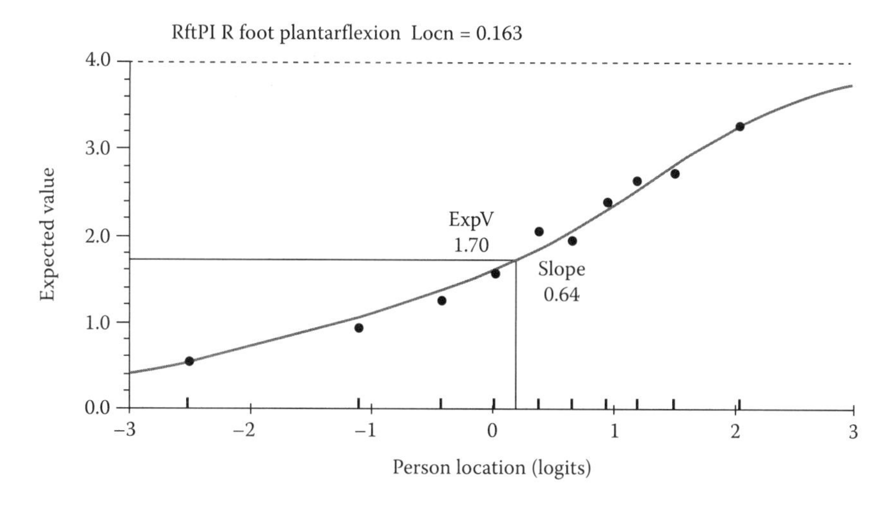

FIGURE 5.5
EVC and the observed means of 10 class intervals for Item 8.

values, with only the location of the items being different, the EVCs for all items had the same shape.

5.4.1 Threshold Order

Because the reversed threshold estimates reflect a property of the responses, and then the reversed threshold estimates are used to recover the responses, a test of fit of the kind used above is irrelevant to the assessment of threshold order. Thus, even when the thresholds are reversed, the responses may fit the model. Table 5.5 shows the threshold estimates using the rating-scale parameterization of Equation 5.7. Andrich (2011) shows the estimates with the partial credit parameterization of Equation 5.25 and the estimates are indeed similar for all items, justifying the RSM parameterization. Although the first three thresholds were ordered and there was a substantial distance between these thresholds, the estimates of Thresholds 3 and 4 were reversed. Similarly, the magnitude of the reversal was small, and for all practical purposes, Thresholds 3 and 4 could be taken as equivalent. However, because the hypothesis that they are correctly ordered is one tailed meant that no significance test was necessary to reject the hypothesis of order.

To appreciate the implications of this small reversal, Figure 5.6 shows the category characteristic curves (CCCs) for Item 3, whose location was closest to zero, together with its latent threshold characteristic curves (TCCs). These TCCs were obtained from Equation 5.14 and show the estimated conditional probability of a successful response

TABLE 5.5

Threshold Estimates Assuming All Thresholds the Same for All Items

	Threshold Estimates				
Mean	**1**	**2**	**3**	**4**	$\hat{\tau}_4 - \hat{\tau}_3$
0.000	−2.618	−0.115	1.522	1.212	−0.311

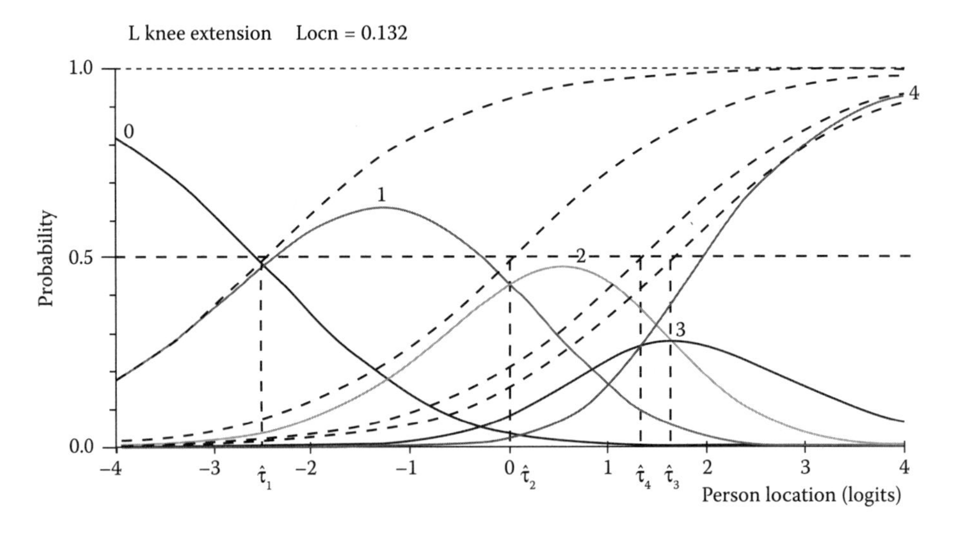

FIGURE 5.6
Category and TCCs for Item 3.

between two adjacent categories. Because the response was latent, the TCCs are shown in dotted lines.

5.4.2 Interpretation of Response Structure and Threshold Disorder

In the development of the model from the rating rationale, it was established that the Guttman subspace Ω^G with respect to the *a priori*-required order of the thresholds was at the core of the model. However, this Guttman space can be considered as a *subspace* of a hypothetical, compatible, full-space Ω, and $\Omega^G \subset \Omega$, of dichotomous responses in the resolved design with statistically independent responses. Thus, an analysis of responses in the subspace Ω^G according to the RSM is identical to one of the dichotomous responses in the compatible, inferred full-space Ω according to the SLM.

Therefore, the TCCs in Figure 5.6 would be the same as if they arose from an SLM analysis of responses in the resolved design, in which responses at the thresholds of each item were experimentally independent and compatible with the responses available for the RSM analysis. And, in this resolved design, it was clear that having the estimate of the threshold at the standard or *normal tone* that was equivalent, or perhaps a little easier, to achieve than the threshold at the standard of *catch*, was an anomaly—*normal tone* was intended to reflect better muscle tone than *catch*. Some aspect of the assessment in this region of the continuum had not been working as required.

Thus, although the model accounted for the data according to the test of fit, it identified that the clinicians effectively judged the standards of *catch* and *normal tone* to be indistinguishable. The only solution to deal with the anomaly is to reconsider, in conjunction with the clinicians, the structure, definition, operationalization, and administration of the rating scale. Without correcting the disclosed anomaly substantively, incorrect decisions, ranging from those at the individual clinical level to those in large-scale evaluation studies, are likely to be misleading. The logic and opportunity to disclose such lack of empirical ordering of categories is a major distinctive feature of the RSM and the more general partial credit parameterization in Equation 5.25.

5.4.3 Interpreting Threshold Distances

It was indicated above that because the hypothesis of threshold order was one tailed, and that the estimates were reversed, there was no need for a significance test. However, suppose Thresholds 3 and 4 in the above example were correctly ordered but the magnitude of the distances between them was the same as in the above example. With a large-enough sample size, they could be deemed as significantly different. However, that would in no way obviate the concern that the clinicians were not able to distinguish satisfactorily between the standards at *catch* and *normal tone*.

A useful concrete interpretation of threshold distances can be made by considering a person of sufficient muscle tone to be located at threshold 2 (*easily flexed*, $\hat{\tau}_2 = -0.115$) and whose probability of exceeding this threshold was 0.50. Simultaneously, this person had the probability of 0.92 for exceeding the next lower standard *restricting movement*. This pair of probabilities seems consistent with the requirements of ordered categories. This consistency between the probabilities is a reflection of the distance between the two thresholds. The same consistency is present between Thresholds 2 and 3.

However, with Thresholds 3 and 4, a person with sufficient muscle tone to have met the standard at Threshold 4 (*normal tone* $\hat{\tau}_2 = -0.115$ with a probability of 0.5, simultaneously had only a 0.42 probability of meeting the lower standard at threshold 3 (*catch*, $\hat{\tau}_4 = 1.212$ This reversal of probabilities of success at the successive thresholds reflected the reversed threshold estimates and was clearly an anomaly.

However, suppose that Thresholds 3 and 4 were correctly ordered, with their location estimates simply reversed; that is, $\hat{\tau}_3 = 1.212$ and $\hat{\tau}_4 = 1.522$. Then, someone located at Threshold 4 who had 0.50 probability of meeting the standard *normal tone*, has 0.58 probability of meeting the standard *catch*. Although the ordering of thresholds and probabilities is now correct, it might be considered that a probability of 0.58 of meeting the standard *catch* at threshold 3 was not sufficiently greater than the probability of 0.50 of meeting the standard *normal tone* at Threshold 4.

The choice of a sufficient distance between successive thresholds inevitably rests on the substantive variable and practical implications of anomalous assessments. If one required that a proficiency that resulted in a probability of success at one threshold had at least a probability of success of 0.8 at a previous threshold, then, the distance between the two thresholds should have been $2\ln 2 = 1.386 \approx 1.4$. The number is only an observation, and not necessarily a recommendation of the required distance between the thresholds.

5.5 Discussion

To interpret parameters meaningfully in probabilistic models, it is necessary that the data are consistent with the model. But even when data and a model are incompatible, the study of misfit can lead to insights. However, in the case of the RSM, there are two distinctive features that go beyond fit and an analysis of residuals.

First, the case for the model being applied is not that it fits a data set but that if the data fit the model, then, the item parameters can be estimated independently of the person parameters, and vice versa, and the estimates from different subsets of data will be statistically equivalent (Duncan, 1984). This separation of the parameters in estimation is rendered possible by the existence of sufficient statistics, a construction of R. A. Fisher (1934). On the basis of the conditions of invariance and sufficiency, the development of the RSM rested on three successive sets of theoretical derivations—Rasch (1961), Andersen (1977), and Andrich (1978). Second, and independent of the test of fit between the model and the data, is that it permits the ready checking of the empirical ordering of the categories of a rating scale against the expectations. Because rating scales are ubiquitous and used in important assessments, this facility is very powerful in improving rating scales that may not be working as required. Although the construction of sufficiency by Fisher and its role in Rasch's conceptualization of invariance in probabilistic models is now well known, it is less well known that Fisher considered it important that the operational ordering of categories was a property of the responses, and not of the model used to analyze them. In commenting on the analysis of a data set by a method based on finding weights for the successive categories when maximizing the sum of squares for a relative to the total sum of squares for a dependent variable with ordered categories in a contingency table, Fisher noted the following:

> It will be observed that the numerical values … lie … in the proper order of increasing reaction. This is not a consequence of the procedures by which they have been obtained, but a property of the data examined (Fisher, 1958, p. 294).

An analysis of Fisher's data set with the RSM showed that the threshold estimates were indeed ordered (Andrich, 1996). It seems that despite this comment and method considered by Fisher, the importance of checking the empirical ordering of the categories did not become standard. The advent of the RSM, where evidence of the empirical ordering of the categories is immediately available, is a major contribution to the analysis of rating scales. The same property of testing the empirical ordering of categories follows for the partial credit parameterization of the model, shown in Equation 5.25. In contrast, the thresholds defined in the graded-response model are ordered as a property of the model, irrespective of the property of the data.

References

Andersen, E. B. 1973a. Conditional inference for multiple choice questionnaires. *British Journal of Mathematical and Statistical Psychology*, 26, 31–44.

Andersen, E. B. 1973b. A goodness of fit test for the Rasch model. *Psychometrika*, 38, 123–240.

Andersen, E. B. 1977. Sufficient statistics and latent trait models. *Psychometrika*, 42, 69–81.

Andersen, E. B. 1995. Residual analysis in the polytomous Rasch model. *Psychometrika*, 60, 375–393.

Andersen, E. 1997. The rating scale model. In: van der Linden, W. J. and Hambleton, R. K. (Eds.), *Handbook of Modern Item Response Theory* (pp. 67–84). New York: Springer.

Andrich, D. 1978. A rating formulation for ordered response categories. *Psychometrika*, 43, 561–574.

Andrich, D. 1985. An elaboration of Guttman scaling with Rasch models for measurement. In: Brandon-Tuma, N. (Ed.), *Sociological Methodology* (pp. 33–80). San Francisco: Jossey-Bass.

Andrich, D. 1995a. Distinctive and incompatible properties of two common classes of IRT models for graded responses. *Applied Psychological Measurement*, 19, 101–119.

Andrich, D. 1995b. Models for measurement, precision and the non-dichotomization of graded responses. *Psychometrika*, 60, 7–26.

Andrich, D. 1996. Measurement criteria for choosing among models for graded responses. In: von Eye, A. and Clogg, C. C. (Eds.), *Analysis of Categorical Variables in Developmental Research* (pp. 3–35). Orlando, FL: Academic Press.

Andrich, D. 2005. Georg Rasch: Mathematician and statistician. In: Kempf-Leonard, K. (Ed.), *Encyclopedia of Social Measurement* (Volume 3) (pp. 299–306). Amsterdam, The Netherlands: Academic Press.

Andrich, D. 2010. Sufficiency and conditional estimation of person parameters in the polytomous Rasch model. *Psychometrika*, 75, 292–308.

Andrich, D. 2011. Rating scales and Rasch measurement. *Expert Review of Pharmacoeconomics and Outcomes Research*, 11, 571–585.

Andrich, D. 2013. An expanded derivation of the threshold structure of the polytomous Rasch rating model which dispels any "threshold disorder controversy". *Educational and Psychological Measurement*, 73, 78–124.

Andrich, D. and Luo, G. 2003. Conditional estimation in the Rasch model for ordered response categories using principal components. *Journal of Applied Measurement*, 4, 205–221.

Andrich, D., Sheridan, B. S. and Luo, G. 2014. *RUMM2030: Rasch Unidimensional Models for Measurement*. Perth, Western Australia: RUMM Laboratory.

Ashworth, B. 1964. Preliminary trial of carisoprodol in multiple sclerosis. *Practitioner*, 192, 540–542.

Bock, R. D. and Aitken, M. 1981. Marginal maximum likelihood estimation of item parameters: Application of an EM algorithm. *Psychometrika*, 46, 443–459.

Bock, R. D. and Jones, L. V. 1968. *The Measurement and Prediction of Judgement and Choice*. San Francisco: Holden Day.

Bock, R. D. and Mislevy, R. J. 1982. Adaptive EAP estimation of ability in a microcomputer environment. *Applied Psychological Measurement*, 6, 431–444.
Duncan, O. D. 1984. Rasch measurement in survey research: Further examples and discussion. In: Turner, C. F. and Martin, E. (Eds.), *Surveying Subjective Phenomena* (Volume 2) (pp. 367–403). New York: Russell Sage Foundation.
Fischer, G. H. 1981. On the existence and uniqueness of maximum-likelihood estimates in the Rasch model. *Psychometrika*, 46, 59–77.
Fisher, R. A. 1934. Two new properties of mathematical likelihood. *Proceedings of the Royal Society A*, 144, 285–307.
Guttman, L. 1950. The basis for scalogram analysis. In: Stouffer, S. A., Guttman, L., Suchman, E. A., Lazarsfeld, P. F., Star, S. A., and Clausen, J. A. (Eds.), *Measurement and Prediction* (pp. 60–90). New York: Wiley.
Jansen, P. G. W. and Roskam, E. E. 1986. Latent trait models and dichotomization of graded responses. *Psychometrika*, 51, 69–91.
Jansen, P. W. G., van den Wollenberg, A. L., and Wierda, F. W. 1988. Correcting unconditional parameter estimates in the Rasch model for inconsistency. *Applied Psychological Measurement*, 12, 297–306.
Likert, R. 1932. A technique for the measurement of attitudes. *Archives of Psychology*, 140, 5–55.
Linacre, J. M. 2002. *Winsteps Rasch Measurement Computer Program*. Chicago: Winsteps.com.
Luo, G. 2005. The relationship between the rating scale and partial credit models and the implication of disordered thresholds of the Rasch models for polytomous responses. *Journal of Applied Measurement*, 6, 443–455.
Rasch, G. 1961. On general laws and the meaning of measurement in psychology. In: Neyman, J. (Ed.), *Proceedings of the Fourth Berkeley Symposium on Mathematical Statistics and Probability IV* (pp. 321–334). Berkeley, CA: University of California Press.
Rasch, G. 1966. An individualistic approach to item analysis. In: Lazarsfeld, P. F. and Henry, N. W. (Eds.), *Readings in Mathematical Social Science* (pp. 89–108). Chicago: Science Research Associates.
Roskam, E. E. 1995. Graded responses and joining categories: A rejoinder to Andrich' models for measurement, precision, and the non-dichotomization of graded responses. *Psychometrika*, 60, 27–35.
Thurstone, L. L. and Chave, E. J. 1929. *The Measurement of Attitude*. Chicago: University of Chicago Press.
Warm, T. A. 1989. Weighted likelihood estimation of ability in item response theory. *Psychometrika*, 54, 427–450.
Wright, B. D. 1988. The efficacy of unconditional maximum likelihood bias correction: Comment on Jansen, van den Wollenberg, and Wierda. *Applied Psychological Measurement*, 12, 315–324.
Wright, B. D. and Masters, G. N. 1982. *Rating Scale Analysis: Rasch Measurement*. Chicago: MESA Press.
Wu, M. L., Adams, R. J., Wilson, M. R., and Haldane, S. A. 2007. *ACERConQuest Version 2: Generalised Item Response Modelling Software* [Computer program]. Camberwell: Australian Council for Educational Research.
Zajicek, J., Fox, P., Sanders, H., Wright, D., Vickery, J., Nunn, A., and Thompson, A. 2003. Cannabinoids for treatment of spasticity and other symptoms related to multiple sclerosis (CAMS study): Multicentre randomised placebo-controlled trial. *The Lancet*, 362, 8, 1517–1526.

6

Graded Response Models

Fumiko Samejima

CONTENTS

6.1 Introduction 95
6.2 Presentation of the Models 95
    6.2.1 General Graded-Response Model 96
    6.2.2 Homogeneous Case 98
    6.2.3 Heterogeneous Case 100
6.3 Parameter Estimation 102
6.4 Goodness of Fit 104
6.5 Heterogeneous Case Based on the Logistic Positive Exponent Family Models 105
6.6 Discussion 106
References 107

6.1 Introduction

The graded-response model represents a family of mathematical models that deals with ordered polytomous categories. These ordered categories include a rating such as letter grading, A, B, C, D, and F, used in the evaluation of students' performance; strongly disagree, disagree, agree, and strongly agree, used in attitude surveys; or partial credit given in accordance with an examinee's degree of attainment in solving a problem.

6.2 Presentation of the Models

Let θ be the latent trait, or ability, which represents a hypothetical construct underlying certain human behavior, such as problem-solving ability, and this ability is assumed to take on any real number. Let i denote an item, which is the smallest unit for measuring θ. Let U_i be a random variable to denote the graded-item response to item i, and $u_i = 0, 1, \ldots, m_i$ denote the actual responses. The score category response function, $P_{u_i}(\theta)$, is the probability with which an examinee with ability θ receives a score u_i, that is,

$$P_{u_i}(\theta) \equiv \Pr\{U_i = u_i \mid \theta\}.$$

$P_{u_i}(\theta)$ will be assumed to be 5 times differentiable with respect to θ. For convenience, u_i will be used both for a specific discrete response and for the event $U_i = u_i$, and a similar usage is applied for other symbols.

For a set of n items, a response pattern, denoted by V, indicates a sequence of U_i for $i = 1, 2, \ldots, n$, and its realization υ, can be written as

$$\upsilon = \{u_1, u_2, \ldots, u_i, \ldots, u_n\}'.$$

It is assumed that local independence (Lord and Novick, 1968) holds, so that within any group of examinees with the same value of ability θ, the distributions of the item responses are independent of each other. Thus, the conditional probability $P_\upsilon\theta$ given θ, for the response pattern υ can be written as

$$P_\upsilon(\theta) \equiv \Pr\{V = \upsilon \mid \theta\} = \prod_{u_i \in \upsilon} P_{u_i}(\theta), \tag{6.1}$$

which is also the likelihood function, $L(\upsilon|\theta)$, for $V = \upsilon$.

Samejima (1969) proposed the first graded-response models: the normal-ogive model and the logistic model for graded-response data (i.e., ordered polytomous categories). Later, she proposed a broader framework for graded-response models, distinguishing the *homogeneous* case, to which the normal-ogive and logistic models belong, and the *heterogeneous* case (Samejima, 1972).

6.2.1 General Graded-Response Model

Suppose, for example, that a cognitive process, such as problem solving, contains a finite number of steps. The graded-item score u_i should be assigned to the examinees who successfully complete up to step u_i but fail to complete the step $(u_i + 1)$. Let $M_{u_i}(\theta)$ be the *processing function* (Samejima, 1995) of the graded-item score u_i which is the probability with which the examinee completes the step u_i successfully, under the joint conditions that (i) the examinee's ability level is θ and (ii) the steps up to $(u_i - 1)$ have already been completed successfully. Let $(m_i + 1)$ be the next graded-item score above m_i. Since everyone can at least obtain the item score 0, and no one is able to obtain the item score $(m_i + 1)$, it is reasonable to set

$$M_{u_i}(\theta) = \begin{cases} 1 & \text{for } u_i = 0 \\ 0 & \text{for } u_i = m_i + 1, \end{cases} \tag{6.2}$$

for all θ. For each of the other u_i's, it is assumed that $P_{u_i}(\theta)$ is nondecreasing in θ. This assumption is reasonable considering that each item has some direct and positive relation to the ability measured. Thus, the category response function, $P_{u_i}(\theta)$, of the graded-item score u_i is given by

$$P_{u_i}(\theta) = \left[\prod_{s \le u_i} M_s(\theta)\right][1 - M_{(u_i+1)}(\theta)]. \tag{6.3}$$

This provides the fundamental framework for the general graded-response model.

Let $P^*_{u_i}(\theta)$ denote the conditional probability with which an examinee of ability θ completes the cognitive process successfully up to step u_i, or further. Then

$$P^*_{u_i}(\theta) = \prod_{s \le u_i} M_s(\theta). \tag{6.4}$$

This function is called *cumulative score category response function* (Samejima, 1995), although cumulation is actually the usual convention. From Equations 6.3 and 6.4, the score category response function can also be expressed as

$$P_{u_i}(\theta) = P^*_{u_i}(\theta) - P^*_{(u_i+1)}(\theta). \tag{6.5}$$

Note that $P^*_{u_i}(\theta)$ becomes the response function, $P_i(\theta)$, of the positive response to item i, when the graded-item score u_i is changed to the binary score, assigning 0 to all scores less than u_i and 1 to those score categories greater than or equal to u_i. It is obvious from Equations 6.2 and 6.4 that $P^*_{u_i}(\theta)$ is also nondecreasing in θ, and assumes unity for $u_i = 0$ and zero for $u_i = m_i + 1$ for the entire range of θ (see, e.g., Samejima, 2004).

The general graded-response model includes many specific mathematical models. In an effort to select a model interpretable within a particular psychological framework, it will be wise to examine whether the model has certain desirable features. Among others, a model should be examined as to whether (i) the principle behind the model and the set of assumptions agree with the psychological reality in question; (ii) it satisfies the unique maximum condition (Samejima, 1969, 1972); (iii) it provides the ordered modal points of the score category response functions in accordance with the item scores; (iv) additivity of the score category response functions (Samejima, 1995) holds; and (v) the model can be naturally generalized to a continuous-response model.

The unique maximum condition is satisfied if the *basic function*, $A_{u_i}(\theta)$, defined by

$$A_{u_i}(\theta) \equiv \frac{\partial}{\partial\theta}\log P_{u_i}(\theta) = \sum_{s \le u_i} \frac{\partial}{\partial\theta}\log M_s(\theta) + \frac{\partial}{\partial\theta}\log[1 - M_{(u_i+1)}(\theta)] \tag{6.6}$$

is strictly decreasing in θ, its upper asymptote is nonnegative and its lower asymptote is nonpositive (Samejima, 1969, 1972). Satisfaction of this condition ensures that the likelihood function of *any* response pattern consisting of such response categories has a unique local or terminal maximum. Using this basic function, a sufficient, though not necessary, condition for the strict orderliness of the modal points of the category response functions is that $A_{(u_i-1)}(\theta) < A_{u_i}(\theta)$ for all θ for $u_i = 1, 2, \ldots, m_i$.

Additivity of a model will hold if the score category response functions still belong to the same mathematical model under finer recategorizations and combinations of two or more categories together, implying that the unique maximum condition will be satisfied by the resulting score category response functions if it is satisfied by those of the original u_i's. Graded-item scores, or partial credits, are more or less incidental. For example, sometimes, letter grades, A, B, C, D, and F, are combined to pass–fail grades. Also, with the advancement of computer technologies, more abundant information can be obtained from an examinee's performance in computerized experiments, and thus, finer recategorizations of the whole cognitive process are possible. Additivity of response categories and generalizability to a continuous model are, therefore, important criteria in evaluating models.

6.2.2 Homogeneous Case

The homogeneous case of the graded-response model represents a family of models with functions $P^*_{u_i}(\theta)$, $u_i = 1, \ldots, m_i$, that are identical in shape and positioned alongside the ability continuum in accordance with the item score u_i. Thus, it is obvious that additivity of the category response functions always holds for mathematical models that belong to the homogeneous case.

The *asymptotic basic function* $\tilde{A}_{u_i}(\theta)$ has been defined in the homogeneous case for $u_i = 1, 2, \ldots, m_i$ by

$$\tilde{A}_{u_i}(\theta) \equiv \lim_{\lambda_{(u_i+1)} \to \lambda_{u_i}} A_{u_i}(\theta) = \frac{\partial}{\partial\theta} \log P_{u_i}(\theta) = \frac{\partial}{\partial\theta} \log\left[\frac{\partial}{\partial\theta} P^*_{u_i}(\theta)\right] - \frac{\frac{\partial^2}{\partial\theta^2} M_1(\theta - \lambda_{u_i})}{\frac{\partial}{\partial\theta} M_1(\theta - \lambda_{u_i})} \tag{6.7}$$

with λ_{u_i} being zero for $u_i = 1$ and increases with u_i, which is identical in shape for all $u_i = 1, 2, \ldots, m_i$ except for the positions alongside the dimension θ (Samejima, 1969, 1972). Using this asymptotic basic function, Samejima (1972, 1995) demonstrated that, in the homogeneous case, the unique maximum condition is simplified: (i) the lower and upper asymptotes of $M_{u_i}(\theta)$ for $u_i = 1$ are zero and unity, respectively, (ii) $\partial/\partial\theta\ \tilde{A}_{u_i}(\theta) < 0$ for the entire range of θ for an arbitrarily selected u_i, and (iii) for this specific u_i, the upper and lower asymptotes of $\tilde{A}_{u_i}(\theta)$ have some positive and negative values, respectively.

When the unique maximum condition is satisfied, the mathematical model can be represented by

$$P^*_{u_i}(\theta) = \int_{-\infty}^{a_i(\theta - b_{u_i})} \psi(t)dt, \tag{6.8}$$

where the item discrimination parameter a_i is finite and positive, and the difficulty or location parameters b'_{u_i}s, satisfy

$$-\infty = b_0 < b_1 < b_2 < \cdots < b_{m_i} < b_{m_i+1} = \infty$$

and $\psi(\cdot)$ denotes a density function that is four times differentiable with respect to θ, and is unimodal with zero as its two asymptotes as θ tend to negative and positive infinities, and with a first derivative that does not assume zero except at the modal point.

In the homogeneous case, satisfaction of the unique maximum condition also implies

1. A strict orderliness among the modal points of $P_{u_i}(\theta)$'s, for it can be shown that

$$A_{(u_i-1)}(\theta) < \tilde{A}_{u_i}(\theta) < A_{u_i}(\theta)$$

   for $u_i = 1, 2, \ldots, m_i$, throughout the whole range of θ (Samejima, 1972, 1995).
2. Additivity of the score category response functions, for, even if two or more adjacent graded-item scores are combined, or if a response category is more finely

recategorized, the $\tilde{A}_{u_i}(\theta)$'s for the remaining u_i's will be unchanged, and those of the newly created response categories will have the same mathematical form as that of the $\tilde{A}_{u_i}(\theta)$'s for the original response categories.

3. A natural expansion of the model to a continuous-response model is obtained by replacing u_i in Equation 6.8 by z_i, which denotes a continuous item response to item i and assumes any real number between 0 and 1, and by defining the *operating density characteristic*

$$H_{z_i}(\theta) = \lim_{\Delta z_i \to 0} \frac{P^*_{z_i}(\theta) - P^*_{(z_i+\Delta z_i)}(\theta)}{\Delta z_i} = a_i \psi(a_i(\theta - b_{z_i}))\left[\frac{db_{z_i}}{d_{z_i}}\right],$$

where b_{z_i} is the difficulty parameter for the continuous response z_i and is a strictly increasing function of z_i (Samejima, 1973). The basic function $A_{z_i}(\theta)$, in the continuous-response model is identical with the asymptotic base function $\tilde{A}_{u_i}(\theta)$ defined by Equation 6.7 on the graded-response level, with the replacement of u_i by z_i, and thus, for these models, the unique maximum condition is also satisfied on the continuous-response level.

It is wise, therefore, to select a specific model from those that satisfy the unique maximum condition, if the fundamental assumptions in the model agree with the psychological reality reflected in the data.

Samejima (1969, 1972) demonstrated that both the normal-ogive model and the logistic model belong to this class of models. In these models, the score category response function is given by

$$P_{u_i}(\theta) = \frac{1}{[2\pi]^{1/2}} \int_{a_i(\theta - b_{u_i+1})}^{a_i(\theta - b_{u_i})} \exp\left[\frac{-t^2}{2}\right] dt \tag{6.9}$$

and

$$P_{u_i}(\theta) = \frac{\exp\left[-D_{a_i}\left(\theta - b_{u_i+1}\right)\right] - \exp\left[-D_{a_i}\left(\theta - b_{u_i}\right)\right]}{\left[1 + \exp\left[-D_{a_i}\left(\theta - b_{u_i}\right)\right]\right]\left[1 + \exp\left[-D_{a_i}\left(\theta - b_{u_i+1}\right)\right]\right]}, \tag{6.10}$$

respectively. In both models, $M_{u_i}(\theta)$ is a strictly increasing function of θ with unity as its upper asymptote; the lower asymptote is zero in the former model, and $\exp[-D_{a_i}(b_{u_i} - b_{u_i} - 1)]$ in the latter (see Figure 5.2.1 in Samejima, 1972). The upper asymptote to the basic function $A_{u_i}(\theta)$ for $u_i = 1, 2, \ldots, m_i$ and the lower asymptote for $u_i = 1, 2, \ldots, m_i - 1$, are positive and negative infinities in the normal-ogive model, and D_{a_i} and $-D_{a_i}$ in the logistic model (Samejima, 1969, 1972).

Figure 6.1 illustrates the score category response functions in the normal-ogive model for $m_i = 4$, $a_i = 1.0$, and b_{u_i}'s equal to –2.0, –1.0, 0.7, and 2.0 for $u_i = 1, 2, 3$, and 4, respectively. The corresponding score category response functions in the logistic model with $D = 1.7$ are very similar to those in the normal-ogive model with the identical set of modal points; however, those curves are a little more peaked.

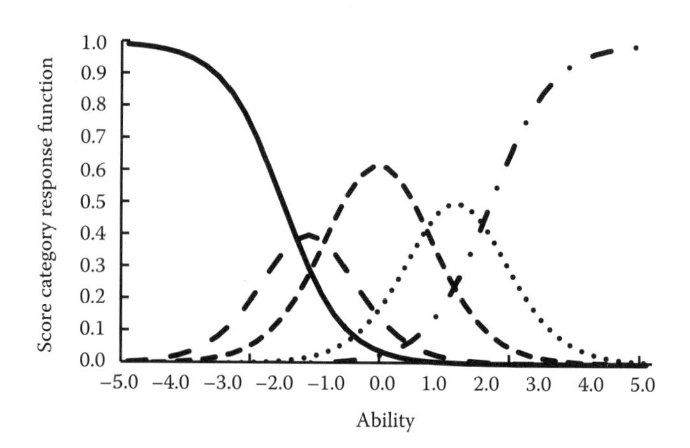

FIGURE 6.1
Example of a set of score category response functions in the normal-ogive model for the item scores 0, 1, 2, 3, and 4.

In the normal-ogive model for continuous-response data, there exists a simple sufficient statistic, $t(\upsilon)$, for a specific response pattern υ provided by

$$t(\upsilon) = \sum_{z_i \in \upsilon} a_i^2 b_{z_i}, \tag{6.11}$$

so that the maximum likelihood estimate (MLE) $\hat{\theta}$ is directly obtained from this simple sufficient statistic (Samejima, 1973). This fact suggests that, when m_i is substantially large, the normal-ogive model for continuous-response data can be used as an approximation to the model for graded-response data. In doing so, the value of m_i should be perhaps nine or greater.

6.2.3 Heterogeneous Case

The heterogeneous case of the graded-response model represents all mathematical models that provide a set of cumulative score category response functions $P_{u_i}^*(\theta)$'s, *not all* of which are identical in shape, that is, those that do not belong to the homogeneous case. One example is Bock's nominal response model (Bock, 1972; Volume One, Chapter 4), represented by

$$P_{h_i}(\theta) = \frac{\exp[\alpha_{h_i}\theta + \beta_{h_i}]}{\sum_{s \in H_i} \exp[\alpha_s \theta + \beta_s]}, \tag{6.12}$$

where h_i denotes a nominal response to item i and H_i is the set of all h_i's, and $\alpha_{h_i} (> 0)$ and β_{h_i} are item score category parameters. Samejima (1972) demonstrated that Bock's nominal response model can be considered as a graded-response model in the heterogeneous case, if the nominal response h_i in Equation 6.12 is replaced by the graded-item response u_i and the parameter α_{u_i} satisfies

$$\alpha_0 \le \alpha_1 \le \alpha_2 \ldots \le \alpha_{m_i}, \tag{6.13}$$

where a strict inequality should hold for at least one pair of α values. Two examples are Masters' (1982; Volume One, Chapter 7) partial credit model (PCM), and Muraki's (1992; Volume One, Chapter 8) generalized partial credit model (GPCM). This family of nominal response models satisfies the unique maximum condition, and the perfect orderliness of the modal points of the category response functions is realized with strict inequality held between every pair of α'_{u_i}s in Equation 6.13. However, it does not have additivity or generalizability to a continuous-response model, and thus applicabilities are limited because of the assumption used in Bock's model (Samejima, 1995).

Samejima (1995) proposed the *acceleration model* that belongs to the heterogeneous case of the graded-response model. Consider a situation, such as problem solving, that requires a number of subprocesses to be completed correctly before attaining the solution. It is assumed that there is more than one step in the process and the steps are observable. Graded-item scores, or partial credits, 1 through m_i, are assigned for the successful completion of these separate observable steps. The processing function for each $u_i = 1, 2, \ldots, m_i$ is given by

$$M_{u_i}(\theta) = [\Psi_{u_i}(\theta)]^{\xi_{u_i}}, \tag{6.14}$$

where $\xi_{u_i} > 0$ is called the *step acceleration parameter*. The acceleration model is a family of models in which $\Psi_{u_i}(\theta)$ is specified by a strictly increasing, five times differentiable function of θ with zero and unity as its two asymptotes. Here, a specific model in this family will be introduced, in which $\Psi_{u_i}(\theta)$ is given by

$$\Psi_{u_i}(\theta) = \frac{1}{1+\exp[-\alpha_{u_i}(\theta-\beta_{u_i})]}, \tag{6.15}$$

where $\alpha_{u_i} > 0$ and β_{u_i} are the discrimination and location parameters, respectively. It is assumed that the process leading to a problem solution consists of a finite number of *clusters*, each containing one or more steps, and within each cluster, the parameters α and β in the logistic distribution function are common. Thus, if two or more adjacent u_i's belong to the same cluster, then, the parameters α_{u_i}'s and β_{u_i}'s are the same for these u_i's, and, otherwise, at least one parameter is different.

It can be seen from Equations 6.14 and 6.15 that the roles of the step acceleration parameter ξ_{u_i} are to control (i) the general shape of the curve representing $M_{u_i}(\theta)$, (ii) the steepness of the curve, and (iii) the position of the curve alongside the ability continuum. When $\Psi_{u_i}(\theta)$ is given by Equation 6.15, $M_{u_i}(\theta)$ is point symmetric only when $\xi_{u_i} = 1$, and point asymmetric otherwise. Let θ_{stp} denote the point of θ at which $M_{u_i}(\theta)$ is steepest. This is provided by

$$\theta_{\text{stp}} = \Psi_{u_i}^{-1}\left[\frac{\xi_{u1}}{1+\xi_{u_i}}\right],$$

which increases with ξ_{u_i}. This maximum slope at $\theta = \theta_{\text{stp}}$ equals $a_i\left[\xi_{u_i}/1+\xi_{u_i}\right]^{\xi_{u_i}+1}$, which is also an increasing function of ξ_{u_i}. The value of $M_{u_i}(\theta)$ at $\theta = \theta_{\text{stp}}$ is $\left[\xi_{u_i}/1+\xi_{u_i}\right]^{\xi_{u_i}}$, which decreases with ξ_{u_i}. The general position of the curve representing $M_{u_i}(\theta)$ is shifted to the positive side alongside the continuum as ξ_{u_i} increases, and in the limiting situation when ξ_{u_i} tends to positive infinity, $M_{u_i}(\theta)$ approaches 0 for all θ, whereas in the other limiting situation where ξ_{u_i} tends to zero, $M_{u_i}(\theta)$ approaches 1 for all θ.

Let w denote a subprocess, which is the smallest unit in the cognitive process. Thus, each step contains one or more w's. Let $\xi_w > 0$ represent the *subprocess acceleration* parameter, and then, the step acceleration parameter ξ_{u_i} for each of $u_i = 1, 2, \ldots, m_i$, is given as the sum of ξ_w's over all $w \in u_i$. The name, *acceleration parameter,* derives from the fact that, within each step, separate subprocesses contribute to accelerate the value of θ at which the discrimination power is maximal (Samejima, 1995).

Note that any grading system is arbitrary. If our experimental setting is improved and allows observation of the examinee's performance in more finely graded steps, then m_i will become larger. It is obvious from Equation 6.14 and the definition of ξ_{u_i} that the resulting category response functions still belong to the acceleration model that partially satisfies the additivity criterion. It is also obvious that the model can be generalized to a continuous-response model as the limiting situation in which the number of steps approaches infinity.

From Equations 6.4 and 6.14, the cumulative category response function, $P^*_{u_i}(\theta)$, is given by

$$P^*_{u_i}(\theta) = \prod_{s=0}^{u_i} [\Psi_s(\theta)]^{\xi_s}$$

and from Equations 6.3 and 6.14, the cumulative score category response function is obtained such that

$$P^*_{u_i}(\theta) = \prod_{s=0}^{u_i} [\Psi_s(\theta)]^{\xi_s} \left[1 - [\Psi_{(u_i+1)}(\theta)]^{\xi_{u_i+1}}\right]. \tag{6.16}$$

The basic function, $A_{u_i}(\theta)$, in this model is obtained from Equations 6.6, 6.14, and 6.15, so that

$$A_{u_i}(\theta) = D\left[\sum_{s \le u_i} \xi_s \alpha_s [1 - \Psi_s(\theta)] - \xi_{u_i+1}\alpha_{u_i+1} \frac{[\Psi_{(u_i+1)}(\theta)]^{\xi_{u_i+1}} [1 - \Psi_{(u_i+1)}(\theta)]}{1 - [\Psi_{(u_i+1)}(\theta)]^{\xi_{u_i+1}}}\right] \tag{6.17}$$

for $u_i = 1, 2, \ldots, m_i - 1$, and for $u_i = 0$ and $u_i = m_i$, the first and second terms on the right-hand side of Equation 6.17 disappear, respectively. The upper and lower asymptotes of this basic function are $D\sum_{s \le u_i} \xi_s \alpha_s (> 0)$ for $u_i = 1, 2, \ldots, m_i$, and $-D\alpha_{u_i+1} (< 0)$ for $u_i = 1, 2, \ldots, m_i - 1$, respectively, and both the upper asymptote for $u_i = 0$ and the lower asymptote for $u_i = m_i$ are zero. It has been demonstrated (Samejima, 1995) that $A_{u_i}(\theta)$ in Equation 6.17 is strictly decreasing in θ. Thus, the unique maximum condition is satisfied for $u_i = 1, 2, \ldots, m_i$. It has also been shown (Samejima, 1995) that the orderliness of the modal points of the category response functions usually holds, except for cases in which the unidimensionality assumption is questionable, and additivity of the category response functions practically holds.

6.3 Parameter Estimation

In estimating the item parameters of the normal-ogive model in the homogeneous case, if our data are collected for a random sample from an unscreened population of examinees, assuming a normal ability distribution, we can (i) compute the polychoric correlation coefficient for each pair of items, (ii) factor analyze the resulting correlation matrix and confirm the unidimensionality, and (iii) estimate a_i and b_{u_i} for $u_i = 1, 2, \ldots, m_i$, using the same method adopted by Lord (1952) and Samejima (1995). In the logistic model, item parameter estimation can be done by using the expectation–maximization (EM) algorithm. Thissen wrote *Multilog* for this purpose, as well as for Bock's nominal response model, among others, with facilities for fixing parameter values at priors and imposing equality constraints (Thissen, 1991).

Multilog is based on the EM solution of the marginal likelihood equations. Bock and Aitkin (1981) proposed a solution for dichotomous responses, and *Multilog* uses a direct expansion to handle graded responses in the homogeneous case. Unlike the joint likelihood solution, the marginal likelihood solution treats the parameter θ_j of the examinee j as incidental, and integrates it out from the likelihood function. Thus, the marginal likelihood function, $L(a_i, b_{u_i})$ can be written as

$$L(a_i, b_{u_i}) = \prod_{p=1}^{P} \int_{-\infty}^{\infty} g(\theta_p) P_{\upsilon_p}(\theta_p) d\theta_p = \prod_{p=1}^{P} P_{\upsilon_j} = \prod_{V} P_{\upsilon}^{r_\upsilon}, \tag{6.18}$$

where P is the number of examinees in the sample, $g(\theta_p)$ is the ability density function, υ_p is the response pattern obtained by an examinee p, $P_{\upsilon_p}(\theta)$ given by Equation 6.1 for an examinee p equals the joint likelihood function, $L(\theta_p, a_i, b_{u_i}), P_{\upsilon_p}$ is the marginal probability for the response pattern υ_p of an examinee p, r_υ is the frequency of a specific response pattern υ, and $\prod_\upsilon$ indicates the product over all possible response patterns. Actually, the continuous ability θ is replaced by q discrete latent classes, each characterized by θ_k, $k = 1, 2, \ldots, q$, and homogeneity is assumed within each class.

Let $\tilde{P}_\upsilon$ be an approximation for P_υ, which is given by

$$\tilde{P}_\upsilon = \sum_{k=1}^{q} P_\upsilon(\theta_k) G(\theta_k), \tag{6.19}$$

where $G(\theta_k)$ is the Gauss–Hermite quadrature weight. The expected frequency, $\tilde{r}_{u_i k}$, of the graded response u_i to item i in the latent class k is provided by

$$\tilde{r}_{u_i k} = \frac{\sum_\upsilon r_\upsilon x_{\upsilon u_i} P_\upsilon(\theta_k) G(\theta_k)}{\tilde{P}_\upsilon}, \tag{6.20}$$

where $x_{\upsilon u_i}$ (= 0, 1) is the indicator variable that assumes 1 if $u_i \in \upsilon$, and 0, otherwise. The expected sample size, $\tilde{P}_k$, of the latent class k is given by

$$\tilde{P}_k = \frac{\sum_\upsilon r_\upsilon x_{\upsilon u_i} P_\upsilon(\theta_k) G(\theta_k)}{\tilde{P}_\upsilon}. \tag{6.21}$$

Thus, in the E-step of the EM algorithm, $P_\upsilon(\theta_k)$ is computed for provisional a_i and b_{u_i}'s and then, $\tilde{P}_\upsilon$, $\tilde{r}_{u_i k}$, and $\tilde{P}_k$ are obtained, using Equations 6.19 through 6.21, respectively. In the M-step, improved estimates of a_i and b_{u_i}'s are obtained by maximizing the approximated likelihood function, which is given by the last expression of Equation 6.18, with P_υ replaced by $\tilde{P}_\upsilon$. The cycles continued until the estimates become stable to the required number of places. Muraki and Bock (1993) wrote *Parscale* (Muraki, Volume Three, Chapter 23), which includes the logistic model for graded-response data, based on essentially the same EM algorithm.

Currently, software for the acceleration model is not available. Parameter estimation can be done, however, using the following method when Equation 6.15 is adopted for $\Psi_{ui}(\theta)$. Suppose we have used a nonparametric estimation method such as Levine's (1984) or Samejima's (1983, 1993), and estimated category response functions, $P_{u_i}(\theta)$'s, and they have tentatively been parameterized using a very general semiparametric method (see, e.g., Ramsay and Wang, 1993). From these results, $\hat{M}_{u_i}(\theta)$ and its partial derivative with respect to θ can be obtained by means of Equations 6.4 and 6.5. Then, select three arbitrary probabilities, P_1, P_2, and P_3, which are in an ascending order, and determine θ_1, θ_2, and θ_3 at which $\hat{M}_{u_i}(\theta)$ equals P_1, P_2, and P_3, respectively. From Equations 6.14 and 6.15, the estimated acceleration parameter $\hat{\xi}_{u_i}$ is obtained as the solution of

$$\frac{\theta_3-\theta_2}{\theta_2-\theta_1}=\frac{\log\left[(P_2)^{-1/\xi_{ui}}-1\right]-\log\left[(P_3)^{-1/\xi_{ui}}-1\right]}{\log\left[(P_1)^{-1/\xi_{ui}}-1\right]-\log\left[(P_2)^{-1/\xi_{ui}}-1\right]}. \tag{6.22}$$

The estimate, $\hat{\beta}_{u_i}$, is given as the solution of

$$\hat{M}_{u_i}(\beta_{u_i})=\left[\frac{1}{2}\right]^{\xi_{ui}} \tag{6.23}$$

and from these results, the estimate of α_{u_i} is obtained by

$$\hat{\alpha}_{u_i}=\frac{2^{\hat{\xi}_{ui}+1}}{D\hat{\xi}_{u_i}}\frac{\partial}{\partial\theta}\hat{M}_{u_i}(\theta)\quad\text{at}\quad\theta=\hat{\beta}_{u_i}. \tag{6.24}$$

Note that this method can be applied for any curve as long as $\frac{\partial}{\partial\theta}\hat{M}_{u_i}(\theta)(>0)$ is available at $\theta=\hat{\beta}_{u_i}$. Actually, the parameters in the acceleration model were obtained by this method from the $\hat{M}_{u_i}(\theta)$'s in Masters' or Muraki's model as the solutions of Equations 6.22 through 6.24, setting $p_1=0.25$, $p_2=0.50$, and $p_3=0.75$ in Equation 6.22.

6.4 Goodness of Fit

For small numbers of items, the observed frequencies for each response pattern may be tabulated; *Multilog* computes the expected frequencies for each response pattern and then the likelihood-ratio statistic as a measure of the goodness of fit of the model. *Parscale* adopts the likelihood-ratio chi-square statistic as a measure of fit for each item, and the sum of these chi-square statistics provides the likelihood-ratio chi-square statistic for the whole test. When the polychoric correlation matrix is used for parameter estimation in the normal-ogive model, goodness of fit can be examined by the chi-square statistic using bivariate normal frequency for each of the $(m_i+1)(m_h+1)$ cells for a pair of items, i and

h, as the theoretical frequency; when most of the $n(n-1)/2$ chi-square values are large, then, we must conclude that one or more assumptions are violated; when large chi-square values are clustered around a relatively small number of items, exclusion of these items will usually improve the fit, although the psychological meaningfulness of the remaining items must be checked to be such that the construct has not been distorted or changed (Samejima, 1994). Note that goodness of fit of the curves should not be used as the *sole* criterion in accepting or rejecting a specified model; there are many models that are based on quite different principles and yet produce similar sets of curves. More on the topic of model fit will be addressed in Volume Two, Chapters 17 through 20, of this handbook.

6.5 Heterogeneous Case Based on the Logistic Positive Exponent Family Models

Another example of the heterogeneous case of the graded-response model (Samejima, 1969, 1972, 1997) can be seen in the graded-response model expanded from the logistic positive exponent family (LPEF) of models for dichotomous responses (Samejima, 2004). This family of models was defined from the following observations.

In ordering the examinees' MLEs of θ, there are two conceivable principles, that is,

- Penalizing failure in solving an easier item
- Greater credit is given for solving a more difficult item

With point-symmetric item response functions, exemplified by the normal-ogive model, neither of these two opposing principles consistently works (Samejima, 2000).

The response function of an LPEF model is given by

$$P_i(\theta) = [\Psi_i(\theta)]^{\xi_i}, \tag{6.25}$$

where $\Psi_i(\theta)$ is the logistic function

$$\Psi_i(\theta) = \frac{1}{1+\exp[-a_i(\theta - b_i)}. \tag{6.26}$$

The third item parameter $\xi_i > 0$, called the acceleration parameter, has a critical role in this family of models. When $0 < \xi_i < 1$, the first principle works consistently, whereas the second principle works whenever $\xi_i > 1$. When $\xi_i = 1$, Equation 6.25 becomes Equation 6.26, namely, the logistic function, and it does not follow either of the two opposing principles. Thus, it can be interpreted as *transition* from one principle to the other within this family of models. In this specific case, neither of the above two principles works.

It is noted from Equations 6.25 and 6.26 that the response functions of an LPEF model are strictly increasing in θ, and also for two items h and i having the same discrimination parameter a and difficulty parameter b, the relationship $P_h(\theta) < P_i(\theta)$ holds for any given θ, if, and only if, $\xi_h > \xi_i$. This implies that the response function defined by Equation 6.1 can be the cumulative operating characteristic, $P^*_{x_k}(\theta)$, of a polytomous item k, where $x_k = 0, 1, \ldots, m_k$ represents the graded-item score.

The fundamental equations of the general graded-response model (LPEFG) are specified as follows. Defining $\xi_{-1} = 0$, the processing function is given by

$$M_{x_i}(\theta) = [\Psi_i(\theta)]^{\xi_{x_i} - \xi_{x_i - 1}} \tag{6.27}$$

for $x_i = 0, 1, \ldots, m_i, m_{i+1}$, where $\Psi_i(\theta)$ is defined by Equation 6.26, and

$$0 = \xi_0 < \xi_1 < \xi_2 < \ldots < \xi_{m_i} < \xi_{m_{i+1}} = \infty. \tag{6.28}$$

Substituting Equation 6.27 into Equation 6.4, the cumulative score category response function in LPEFG can be written as

$$P^*_{x_i}(\theta) = [\Psi_i(\theta)]^{\xi_{x_i}} \tag{6.29}$$

This outcome is of the same form as the response functions of a model in LPEF. Substituting this into Equation 6.5, we obtain

$$P_{x_i}(\theta) = [\Psi_i(\theta)]^{\xi_{x_i}} - [\Psi_i(\theta)]^{\xi_{x_i+1}} > 0 \tag{6.30}$$

(Samejima, 2008).

The reader is directed to Samejima (2008), for the basic functions and the information functions for the LPEFG models, and for the proof that these models satisfy the unique maximum condition and other characteristics of the models.

Note that LPEFG can be considered as a special case of the acceleration model. The way the word *acceleration parameter* is used is different from that in the acceleration model, however. Notably, the LPEFG models satisfy all the other four criteria for desirable characteristics of ordered polytomous-graded response models (Samejima, 1996) that include two types of additivity, generalizability to a continuous-response model, and order of modal points of the item score category response functions.

If one wants to use the normal-ogive function instead of the logistic function in Equation 6.29, that can be done without difficulty. The author might point out that the idea of LPEF using the normal-ogive model has been worked out as early as in Samejima (1972).

6.6 Discussion

It has been observed that, in the homogeneous case, the models have features that satisfy the criteria for good models, while, in the heterogeneous case, fulfillment of these criteria becomes more difficult. The heterogeneous case provides greater varieties in the configuration of the cumulative score category response functions, and therefore model fit will be better. In certain situations, such as Likert-scale attitude surveys, it will be reasonable to apply a model in the homogeneous case, which assumes invariance of the discrimination power of an item for every possible redichotomization of the graded categories. This is supported by the results of Koch (1983). In general, however, it will be wise to start with a nonparametric estimation of the cumulative score category response functions, and let the data determine which of these cases is more appropriate.

References

Bock, R. D. 1972. Estimating item parameters and latent ability when responses are scored in two or more nominal categories. *Psychometrika, 37,* 29–51.

Bock, R. D. and Aitkin, M. 1981. Marginal maximum likelihood estimation of item parameters: Application of an EM algorithm. *Psychometrika, 46,* 443–459.

Koch, W. R. 1983. Likert scaling using the graded response latent trait model. *Applied Psychological Measurement, 7,* 15–32.

Levine, M. 1984. *An Introduction to Multilinear Formula Scoring Theory* (Office of Naval Research Report, 84–4). Champaign, IL: Model-Based Measurement Laboratory, Education Building, University of Illinois.

Lord, F. M. 1952. A theory of mental test scores. *Psychometric Monograph No. 7.* Richmond, VA: Psychometric Corporation. Retrived from: http://www.psychometrika.org/journal/online/MN07.pdf

Lord, F. M. and Novick, M. R. 1968. *Statistical Theories of Mental Test Scores.* Reading, MA: Addison-Wesley.

Masters, G. N. 1982. A Rasch model for partial credit scoring. *Psychometrika, 47,* 149–174.

Muraki, E. 1992. A generalized partial credit model: Application of an EM algorithm. *Applied Psychological Measurement, 16,* 159–176.

Muraki, E. and Bock, R. D. 1993. *Parscale.* Chicago, IL: Scientific Software.

Ramsay, J. O. and Wang, X. 1993. Hybrid IRT models. Paper presented at the *Meeting of the Psychometric Society,* Berkeley, CA.

Samejima, F. 1969. Estimation of ability using a response pattern of graded scores. *Psychometrika Monograph No. 17.* Richmond, VA: Psychometric Corporation. Retrived from: http://www.psychometrika.org/journal/online/MN17.pdf

Samejima, F. 1972. A general model for free-response data. *Psychometrika Monograph No. 18.* Richmond, VA: Psychometric Corporation. Retrived from: http://www.psychometrika.org/journal/online/MN18.pdf

Samejima, F. 1973. Homogeneous case of the continuous response model. *Psychometrika, 38,* 203–219.

Samejima, F. 1983. Some methods and approaches of estimating the operating characteristics of discrete item responses. In H. Wainer and S. Messick (Eds.), *Principals of Modern Psychological Measurement: A Festschrift for Frederic M. Lord* (pp. 159–182). Hillsdale, NJ: Lawrence Erlbaum.

Samejima, F. 1993. Roles of Fisher type information in latent trait models. In H. Bozdogan (Ed.), *Proceedings of the First U.S./Japan Conference on the Frontiers of Statistical Modeling: An Informational Approach* (pp. 347–378). Netherlands: Kluwer Academic Publishers.

Samejima, F. 1994. Nonparametric estimation of the plausibility functions of the distractors of vocabulary test items. *Applied Psychological Measurement, 18,* 35–51.

Samejima, F. 1995. Acceleration model in the heterogeneous case of the general graded response model. *Psychometrika, 60,* 549–572.

Samejima, F. 1996. Evaluation of mathematical models for ordered polytomous responses. *Behaviormetrika, 23,* 17–35.

Samejima, F. 1997. Graded response model. In W. J. van der Linden and R. K. Hambleton (Eds.), *Handbook of Modern Item Response Theory* (pp. 85–100). New York: Springer.

Samejima, F. 2000. Logistic positive exponent family of models: Virtue of asymmetric item characteristic curves. *Psychometrika, 65,* 319–335.

Samejima, F. 2004. Graded response model. In K. Kempf-Leonard (Ed.), *Encyclopedia of Social Measurement* (Volume 2, pp. 145–153). Amsterdam: Elsevier.

Samejima, F. 2008. Graded response model based on the logistic positive exponent family of models for dichotomous responses. *Psychometrika, 73,* 561–575.

Thissen, D. 1991. *Multilog User's Guide—Version 6.* Chicago, IL: Scientific Software.

7

Partial Credit Model

Geoff N. Masters

CONTENTS

7.1 Introduction 109
7.2 Presentation of the Model 111
  7.2.1 Response Functions 112
  7.2.2 Parameter Interpretation 113
  7.2.3 Relations to Other Models 114
7.3 Parameter Estimation 115
  7.3.1 CML Estimation 115
  7.3.2 Marginal Maximum-Likelihood Estimation 116
7.4 Goodness of Fit 118
  7.4.1 Unweighted Mean-Square ("Outfit") 118
  7.4.2 Weighted Mean-Square ("Infit") 119
7.5 Empirical Example 119
  7.5.1 Measuring Essay Quality 120
  7.5.2 Calibrating Markers 120
7.6 Discussion 123
References 123

7.1 Introduction

The partial credit model (PCM) is a model for constructing measures using items with two or more ordered response categories; that is, items scored polytomously. The model is a simple application of Rasch's model for dichotomies (Volume One, Chapter 3). Specifically, if responses to item i are recorded in categories a = 0, 1, ... A_i, then the probability of a person's response to item i being in category x rather than in category $x - 1$ is modeled by the PCM, as being governed by Rasch's model for dichotomies.

The PCM was developed for a similar purpose to several other models in this Handbook, including Samejima's graded response model (Volume One, Chapter 6) and the generalized PCM (Volume One, Chapter 8). However, the PCM differs from these models in that, like the Rasch rating scale model (Volume One, Chapter 5), which is a particular case of the PCM, it belongs to the Rasch family of models and so shares the distinguishing characteristics of that family: separable person and item parameters and sufficient statistics (Andersen, 1977). These features enable "specifically objective" comparisons of persons and items (Rasch, 1977) and allow each set of model parameters to be conditioned out of the estimation procedure for the other. Among polytomous item response models, the PCM and its special cases is the only model for which the sum of the item scores is a sufficient statistic for the person parameter. It is also the only model that permits the stochastic ordering of persons by item-sum score (Hemker et al., 1996, 1997; Ligtvoet, 2012).

Because it is based on the application of Rasch's model for dichotomies to each pair of adjacent categories in an ordered sequence, the PCM (Masters, 1982, 1987, 1988a,b; Masters and Wright, 1996) is the simplest of all polytomous item response models. It contains only two sets of parameters—one for persons and one for items—and all parameters are locations on the underlying measurement variable. This feature distinguishes the PCM from other models that include item "discrimination" or "dispersion" parameters, which qualify locations and complicate the substantive interpretation of variables.

The simplicity of the PCM's formulation makes it easy to implement in practice, and the model has been incorporated into a range of software packages (e.g., Mair and Hatzinger, 2007; Adams et al., 2012; Linacre, 2012). The model can be applied in any situation in which performances on a test item or assessment task are recorded in two or more ordered categories and there is an intention to combine results across items/tasks to obtain measures on some underlying variable. Early applications of the model included applications to ratings of infant performance (Wright and Masters, 1982); computerized patient simulation problems (Julian and Wright, 1988); ratings of writing samples (Pollitt and Hutchinson, 1987; Harris et al., 1988); measures of critical thinking (Masters and Evans, 1986); ratings of second language proficiency (Adams et al., 1987); computer adaptive testing (Dodd and Koch, 1987; Koch and Dodd, 1989; Dodd et al., 1993, 1995; Chen et al., 1998); answer-until-correct scoring (Wright and Masters, 1982); measures of conceptual understanding in science (Adams et al., 1991); embedded figures tests (Pennings, 1987); applications of the SOLO taxonomy (Wilson and Iventosch, 1988); the diagnosis of mathematics errors (Adams, 1988); measures of conceptual understanding in social education (Doig et al., 1994); the construction of item banks (Masters, 1984); and statewide testing programs (Titmanis et al., 1993).

Since the turn of the century, the PCM has been used routinely for the scaling of polytomously scored items in a number of statewide, national, and international assessment programs. These programs include the first five cycles of the *Programme for International Student Assessment* (PISA) (Organisation for Economic Cooperation and Development, 2012), the *Civic Education Study of the International Association for the Evaluation of Educational Achievement* (IEA) (Torney-Purta et al., 2001), the IEA's *International Civic and Citizenship Education Study* (Schulz et al., 2011) and national assessment programs in Civics and Citizenship (Gebhardt et al., 2011), and ICT Literacy (Gebhardt et al., 2012).

The PCM also has been applied to a variety of practical measurement problems in health and medicine, including the construction of mental health scales (Bode, 2001); the assessment of complex decision-making by nurses (Fox, 1999); quality of life measures (Sebille et al., 2007; Jafari et al., 2012); and health self-report surveys (Tandon et al., 2003).

Research studies have investigated the usefulness of the PCM in a wide range of contexts and for varying forms of assessment, including computer adaptive testing (Gorin et al., 2005), performance assessments (Fitzpatrick et al., 1996), "testlets" (Verhalst and Verstralen, 2008), multistage testing (Kim et al., 2013), and hybrid tests consisting of multiple-choice and fixed response items (Sykes and Yen, 2000).

7.2 Presentation of the Model

When responses to an item are recorded in only two-ordered categories (i.e., $U_{pi} = 0, 1$), the probability of scoring 1 rather than 0 is expected to increase with the ability being measured. In Rasch's model for dichotomies, this expectation is modeled as

$$\frac{\Pr\{U_{pi} = 1\}}{\Pr\{U_{pi} = 0\} + \Pr\{U_{pi} = 1\}} = \frac{\exp(\theta_p - b_i)}{1 + \exp(\theta_p - b_i)} \tag{7.1}$$

where θ_p is the ability of person p, and b_i is the difficulty of item i defined as the location on the measurement variable at which a score of 1 on item i is as likely as a score of 0. The model is written here as a conditional probability to emphasize that it is a model for the probability of person p scoring 1 rather than 0 (i.e., given one of only two possible outcomes and conditioning out all other possibilities of the person-item encounter such as "missing" and "answered but not scorable").

When an item provides more than two response categories (e.g., three ordinal categories scored 0, 1, and 2), a score of 1 is not expected to be increasingly likely with increasing ability because score of 2 becomes a more probable result beyond some point. Nevertheless, it follows from the intended order $0 < 1, \ldots, < A_i$ of a set of categories that the conditional probability of scoring a rather than $a - 1$ on an item should increase monotonically throughout the ability range. In the PCM, this expectation is modeled using Rasch's model for dichotomies:

$$\frac{\Pr\{U_{pi} = a\}}{\Pr\{U_{pi} = a-1\} + \Pr\{U_{pi} = a\}} = \frac{\exp(\theta_p - b_{ia})}{1 + \exp(\theta_p - b_{ia})} \quad (a = 1, 2, \ldots, A_i) \tag{7.2}$$

where θ_p is the ability of person p, and b_{ia} is an item parameter governing the probability of scoring a rather than $a - 1$ on item i.

By conditioning a pair of adjacent categories and so eliminating all other response possibilities from consideration, Equation 7.2 focuses on the local comparison of categories $a - 1$ and a. This local comparison is the heart of the PCM and, as Masters and Wright (1984) and Masters (2010) demonstrate, is an essential and defining feature of all members of the Rasch family of models.

To apply the PCM, it is conventional to rewrite the model as the unconditional probability of each possible outcome $0, 1, \ldots, A_i$ of person p's attempt at item i. This reexpression of the model requires only that person p's performance on item i is assigned one of the $A_i + 1$ available scores for that item, that is,

$$\sum_{a=0}^{A_i} \Pr\{U_{pi} = a\} = 1 \tag{7.3}$$

The probability of person p scoring a on item i can then be written as

$$\Pr\{U_{pi} = a\} = \frac{\exp \sum_{k=0}^{a} (\theta_p - b_{ik})}{\sum_{h=0}^{Ai} \exp \sum_{k=0}^{h} (\theta_p - b_{ik})} \quad (a = 0, 1, \ldots, Ai) \tag{7.4}$$

where, for notational convenience,

$$\sum_{k=0}^{0} (\theta_p - b_{ik}) \equiv 0 \quad \text{and} \quad \sum_{k=0}^{h} (\theta_p - b_{ik}) \equiv \sum_{k=1}^{h} (\theta_p - b_{ik})$$

For an item with only two categories, Equation 7.4 is simply Rasch's model for dichotomies.

7.2.1 Response Functions

When the PCM is applied, it provides estimates of the item parameters $(b_{i1}, b_{i2}, \ldots, b_{iA_i})$ for each item *i*. When these estimates are substituted into Equation 7.4, the estimated probabilities of scoring 0, 1, …, A_i on item *i* are obtained for any specified ability θ.

Figure 7.1a shows the model probabilities of scoring 0, 1, 2, and 3 on a particular 4-category item calibrated with the PCM. Notice that the modes of the response curves are in

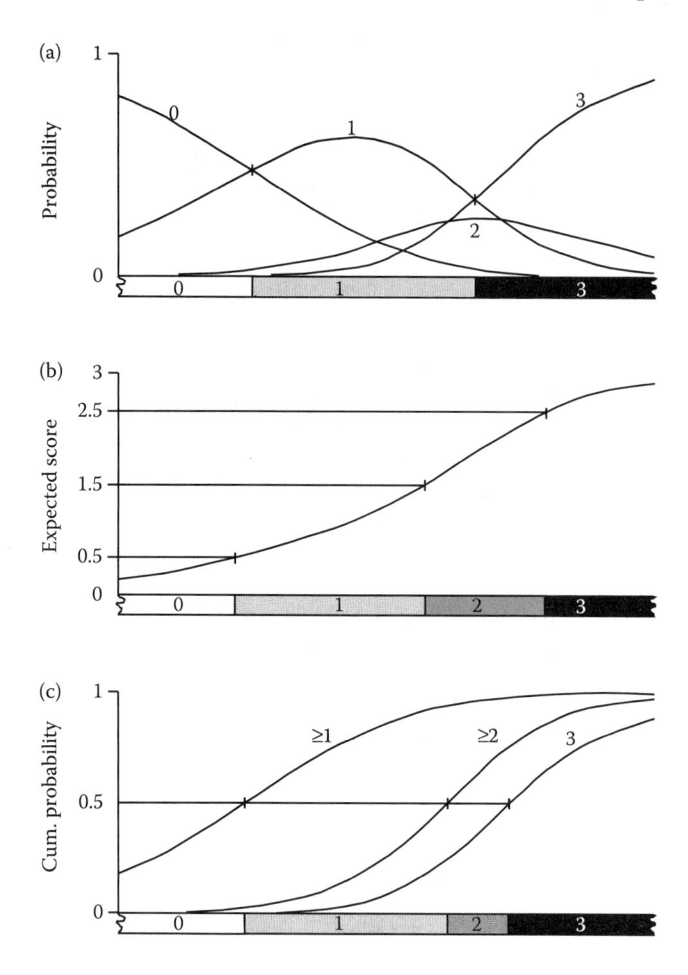

FIGURE 7.1
Three approaches to defining regions of a measurement continuum using the PCM. Measurement continuum regions defined in terms of (a) single most probable response, (b) expected score, and (c) cumulative probability.

the order $0 < 1 < 2 < 3$ from left to right. This is a basic feature of the model; the response curve modes are always ordered $0 < 1 < 2 < \ldots < A_i$ on the measurement variable.

The item parameters $(b_{i1}, b_{i2}, \ldots, b_{iA_i})$ have a simple interpretation in this picture. Each parameter b_{ia} corresponds to the position on the measurement variable at which a person has the same probability of responding in category a as in category $a - 1$. The parameter estimates for the item in Figure 7.1a are thus at the intersections of the response curves for categories 0 and 1, 1 and 2, and 2 and 3.

7.2.2 Parameter Interpretation

Person ability parameter θ_p in the PCM is the modeled location of person p on the variable being measured. For each item i scored 0, 1, ..., A_i, the PCM defines a set of A_i item parameters. These parameters, all of which are locations on the measurement variable, can be used to map the qualitative meaning of the variable and to interpret person parameters.

The kind of open-ended and performance tasks the PCM is designed to analyze are usually intended to be accessible to a wide-range of abilities and to differentiate among test takers on the basis of their levels of response. Response categories for each item capture this response diversity and thus provide the basis for the qualitative mapping of measurement variables and the substantive interpretation of ability estimates. For items with more than two response categories, however, the mapping of response categories on to measurement variables is a little less straightforward than for right/wrong scoring.

Because each item parameter in the PCM is defined locally with respect to just two adjacent categories (rather than globally taking into account all categories simultaneously), the item parameters in the model can take *any* order. For the item in Figure 7.1a they are ordered $b_{i1} < b_{i3} < b_{i2}$. Because they have this property, these parameters are not always, in themselves, helpful in marking out regions of an underlying variable. Three useful alternatives are illustrated in Figure 7.1a–c.

The first method, illustrated in Figure 7.1a, identifies regions of the "single most probable response" to an item. The three shaded regions at the bottom of Figure 7.1a indicate the single most probable response (0, 1, or 3) for the range of abilities shown.

A disadvantage of this method is that, in this particular case, it defines no region for a score of 2. In fact, it gives the impression that category 2 has "disappeared" entirely. This is not the case, a score of 2 on this item is a quite likely result ($p = 0.28$) for an ability at the junction of curves 1 and 3. When PCM item parameters are in the order $b_{i1} < b_{i2} < b_{i3}$, the parameters mark out regions of "single most probable response" for *all* categories. However, as can be seen from Figure 7.1a, other orders of PCM item parameters are possible.

Some caution is required in the interpretation of "most probably" statements. Toward the far right of the shaded region labelled "1" in Figure 7.1a, it is *not* the case that persons will most probably score 1. Because the sum of the probabilities for categories 2 and 3 exceeds the probability of a response in category 1, persons with abilities in this part of the continuum are more likely to score at least 2 than to score only 1. This observation makes it clear that the most useful method of mapping response categories on to underlying measurement variables is not obvious when items are scored in more than two categories.

The second method, illustrated in Figure 7.1b, is based on the calculation of the expected score of person p on item i as

$$E_{pi} = \sum_{a=0}^{A_i} h \Pr\{U_{pi} = a; \theta_p, b_{i1}, b_{i2}, \ldots, b_{iA_i}\} \tag{7.5}$$

In Figure 7.1b, the abilities at which $E_{pi} = 0.5, 1.5$, and 2.5 are identified. These are used to mark out an alternative set of response regions along the horizontal variable.

An attraction of the method in Figure 7.1b is that it is consistent with procedures for testing model-data fit that contrast a person's observed score on an item with his/her expected score. For expected scores between, say 1.5 and 2.5, the observed score that minimizes the observed–expected residual is a score of 2. In this sense it might be argued that the "2 region" of the measurement variable is best defined as that range of abilities for which the expected score is between 1.5 and 2.5. A disadvantage of this method is that it provides category regions that may seem implausibly wide.

The third method, illustrated in Figure 7.1c, sums the curves in Figure 7.1a. The curve labelled " ≥1" is the result of summing curves 1, 2, and 3 in Figure 7.1a. The curve labelled " ≥2" is the result of summing curves 2 and 3. These cumulative ogives give the probability of scoring "1 or better", "2 or better," and "3" on this item. The abilities at which the cumulative probabilities equal 0.5 have been used to mark out a third set of regions. Persons with abilities in the "2 region" in Figure 7.1c will most probably (i.e., $p > 0.5$) score *at least* 2 on this item, but will most probably ($p > 0.5$) *not* score 3.

An attractive feature of the method in Figure 7.1c is that it parallels the interpretation of variables constructed from dichotomously scored items. On a variable defined by dichotomous items, a person's estimated location places them above items they will most probably ($p > 0.5$) pass and below items they will most probably ($p > 0.5$) fail. Method (c) similarly interprets an individual's location by reference to thresholds they will most probably ($p > 0.5$) pass and thresholds they will most probably ($p > 0.5$) fail.

The method also has the advantage of being consistent with Thurstone's approach to mapping response categories on to underlying measurement variables. Thurstone used cumulative probabilities to define category boundaries or thresholds on a latent continuum. The distance between cumulative ogives Thurstone described as the width of the corresponding response category (e.g., Edwards and Thurstone, 1952).

7.2.3 Relations to Other Models

The difference between the PCM and Samejima's graded response model (GRM) (Volume One, Chapter 6) can be understood in terms of the cumulative ogives in Figure 7.1c. In the PCM, these cumulative probabilities are not modeled directly, but are the result of summing the category response functions in Figure 7.1a, which in turn are the result of applying Rasch's model for dichotomies separately to each pair of adjacent categories. In the graded response model, cumulative probabilities are modeled directly. Thus, while the PCM gives the probability of person p scoring a on item i, the GRM models the probability of person p scoring *a or better* on item i. As a result, the PCM is not a case of Samejima's graded response model (e.g., with equal item discriminations).

As a member of the Rasch family of item response models, the PCM is closely related to other members of that family. Masters and Wright (1984) describe several members of this family and show how each has as its essential element Rasch's model for dichotomies. Andrich's (1978; Volume One, Chapter 5) model for rating scales, for example, can be thought of as a version of the PCM with the added assumption that the response categories are defined and function in the same way for each item in an instrument. With this added assumption, rather than modeling a set of A_i parameters for each item, a single parameter b_i is modeled for item i, and a set of A parameters ($\tau_1, \tau_2, \ldots, \tau_A$) are proposed for the common response categories. To obtain the rating scale version of the PCM, each item parameter in the model is redefined as $b_{ia} = b_i + \tau_a$. Wilson (1992) and

Muraki (Volume One, Chapter 8) have proposed generalized versions of the PCM for other kinds of multi-category data.

7.3 Parameter Estimation

Conditional and joint maximum-likelihood procedures for estimating PCM parameters are described by Masters (1982) and Wright and Masters (1982), who also describe a procedure based on the pairwise comparison of responses to items (PAIR), and a simplified procedure (PROX) based on the assumption that the effects of the person sample on item calibration and of the test on person measurement can be summarized by means and standard deviations on the variable. Joint maximum-likelihood estimates for the model have been further investigated by Bertoli-Barsotti (2005).

7.3.1 CML Estimation

The conditional maximum-likelihood (CML) procedure begins with the conditional probability of the response vector $\boldsymbol{a}_i$ given test score r

$$\Pr\{\mathbf{a}_i;\mathbf{b}_{ik}\mid r\}=\frac{\exp\left(-\sum_{i}^{n}\sum_{k=0}^{a_j}b_{ik}\right)}{\sum_{\mathbf{a}_q}^{r}\exp\left(-\sum_{i}^{n}\sum_{k=0}^{a_q}b_{ik}\right)} \tag{7.6}$$

where $b_{i0}\equiv 0$ and $\sum_{\mathbf{a_q}}^{r}$ is the sum over all response vectors $\mathbf{a}_q$, which produce the score r.

The conditional probability of responding in category h of item i given score r is

$$P_{irh}^{*}=\frac{\exp\left(-\sum_{k=0}^{h}b_{ik}\right)\sum_{(a_{q\neq i})}^{r-h}\exp\left(-\sum_{q\neq i}^{n}\sum_{k=0}^{a_q}b_{qk}\right)}{\sum_{g=0}^{A_i}\left\{\exp\left(-\sum_{k=0}^{g}b_{ik}\right)\sum_{(a_{q\neq i})}^{r-g}\exp\left(-\sum_{q\neq i}^{n}\sum_{k=0}^{a_q}b_{qk}\right)\right\}}=\frac{\exp\left(-\sum_{k=0}^{h}b_{ik}\right)\gamma_{r-h,i}}{\gamma_r} \tag{7.7}$$

where $\sum_{a_{q\neq i}}^{r-h}$ is the sum over all response vectors $a_{q\neq i}$, which exclude item i and produce the score $r-h$.

The conditional likelihood over N persons with various scores is

$$\Lambda=\prod_{j}^{N}\left[\frac{\exp\left(-\sum_{i}^{n}\sum_{k=0}^{a_{ij}}b_{ik}\right)}{\gamma_r}\right]=\exp\frac{\left[-\sum_{j}^{N}\sum_{i}^{n}\sum_{k=0}^{a_{ij}}b_{ik}\right]}{\prod_{r}^{M-1}(\gamma_r)^{N_r}} \tag{7.8}$$

where $M=\sum_{i}^{n}A_i$ is the maximum possible score on the instrument, $\prod_{j}^{N}\gamma_r=\prod_{r}^{M-1}(\gamma_r)^{N_r}$ and N_r is the number of persons with a particular score r.

The log-likelihood can then be written as

$$\lambda \equiv \log \Lambda = -\sum_{i}^{n}\sum_{k=1}^{A_i} S_{ik}b_{ik}\sum_{r}^{M-1} N_r \log \gamma_r \tag{7.9}$$

where S_{ik} is the number of persons scoring *k or better* on item *i*.

The estimation equations for the conditional maximum-likelihood procedure require the first and second derivatives of the log-likelihood with respect to b_{ia}. These equations are

$$\frac{\partial \lambda}{\partial b_{ia}} = -S_{ia} - \sum_{r}^{M-1} \frac{N_r}{\gamma_r}\left(\frac{\partial \gamma_r}{\partial b_{ia}}\right) = -S_{ih} + \sum_{r}^{M-1} N_r \sum_{k=0}^{A_i} P_{irk} = -S_{ih} + \sum_{r}^{M-1} N_r \sum_{k=a}^{A_i} P_{irk} \tag{7.10}$$

and

$$\frac{\partial^2 \lambda}{\partial b_{ih}^2} = -\sum_{r}^{M-1} N_r \left(\sum_{k=a}^{A_i} P_{irk}\right)\left(1 - \sum_{k=h}^{A_i} P_{irk}\right) \tag{7.11}$$

where $\sum_{k=a}^{A} P_{irk}$ is the probability of a person with score *r* scoring *a* or better on item *i*.

In the implementation of partial credit analysis, it is possible to estimate all parameters for an item only if observations occur in each of the available response categories. Wilson and Masters (1993) have developed a procedure for automatically reparameterizing the model to provide JML estimates of a smaller number of item parameters when one or more response categories for an item are unused (i.e., "null").

7.3.2 Marginal Maximum-Likelihood Estimation

Marginal maximum-likelihood (MML) estimation was developed as an alternative to the joint estimation of person and item parameters. In models which do not have simple sufficient statistics for θ, and so do not permit conditional maximum-likelihood estimates, the number of person parameters to be estimated increases with the number of persons, meaning that joint maximum likelihood may not provide consistent estimates of item parameters. MML treats person parameters as nuisance parameters and removes them from the likelihood function by assuming that persons are sampled randomly from a population in which ability is distributed according to some density function g(θ).

In the PCM, the person score $r_p = \sum_i^n a_{ip}$ is a sufficient statistic for θ and so the number of person parameters to be estimated is $\sum_i^n A_i - 1$, regardless of sample size. CML removes θ from the likelihood by conditioning on the person score, and so provides consistent estimates of the item parameters. Although the process leads to some information on the item parameter estimates, as CML does not require an assumption about the population distribution, it is more robust and is in general the procedure of choice.

An MML procedure for the PCM is developed in detail by Glas and Verhelst (1989) and Wilson and Adams (1993). The procedure follows Bock and Aitkin (1981).

If the probability of person *p* scoring *a* on item *i* is P_{ipa}, then the probability of person *p*'s observed response vector $\boldsymbol{a}_p$ given θ_p and the set of item parameters $\boldsymbol{b}$ is

$$\Pr\{\mathbf{a}_p \mid \mathrm{p}, \mathbf{b}\} = \prod_{i=1}^{n} P_{ipa_i} \tag{7.12}$$

For a person sampled randomly from a population with a continuous density function $g(\theta)$, the unconditional probability of response vector $\boldsymbol{a}_p$ is

$$\Pr\{\mathbf{a}_p \mid \mathbf{b}, \boldsymbol{\alpha}\} = \int_{\theta} \Pr\{p \mid \theta_p, \mathbf{b}\} g(\theta \mid \boldsymbol{\alpha})\, d\theta \tag{7.13}$$

where $\boldsymbol{\alpha}$ is a set of population parameters, for example, $\boldsymbol{\alpha} = (\mu, \sigma^2)$. When N persons are drawn at random from this population, the likelihood of their observed set of response patterns is

$$\Lambda = \prod_{p=1}^{N} \Pr\{\mathbf{a}_p \mid \mathbf{b}, \boldsymbol{\alpha}\} \tag{7.14}$$

in which the nuisance parameters θ have been eliminated.

MML finds the values of the structural item and population parameters that maximize the log-likelihood, $\lambda = \log \Lambda$, by solving the likelihood equations

$$\frac{\partial \lambda}{\partial \mathbf{b}} = 0 \tag{7.15}$$

and

$$\frac{\partial \lambda}{\partial \boldsymbol{\alpha}} = 0 \tag{7.16}$$

Rather than solving Equations 7.19 and 7.20 directly, it is possible, following Bock and Aitkin (1981), to apply an EM algorithm (Aitkin, Volume Two, Chapter 12). When applied to the PCM, this approach is particularly straightforward because of the existence of sufficient statistics for the model parameters.

If it is assumed that, in addition to the response data, the person parameters θ are known, then an expression can be written for the complete likelihood $\Lambda(\mathbf{b}, \boldsymbol{\beta} \mid \boldsymbol{\alpha}, \boldsymbol{\theta})$ and log-likelihood $\lambda = \log \Lambda(\mathbf{b}, \boldsymbol{\beta} \mid \boldsymbol{\alpha}, \boldsymbol{\theta})$. The complete log-likelihood cannot be maximized directly because it depends on a knowledge of θ. An alternative is to maximize its posterior predictive expectation $E[\lambda]$ given current estimates of θ and the data. First, the marginal posterior density of θ_p given a_p and the data is estimated using provisional estimates of the item and population parameters (the E-step). Then the expected value of the complete log-likelihood is maximized, that is, the equations

$$\frac{\partial E[\lambda]}{\partial \mathbf{b}} = 0$$

$$\frac{\partial E[\lambda]}{\partial \boldsymbol{\alpha}} = 0$$

are solved, to obtain improved estimates of the item and population parameters (the M-step): These two steps are repeated until convergence.

7.4 Goodness of Fit

Tests of fit to the PCM are conducted at several different levels. Tests of item fit identify problematic items. These may be items that do not work to define quite the same variable as the other items in an instrument, or items that are ambiguous or flawed. Tests of person fit identify persons with responses that do not follow the general pattern of responses. Tests of global fit indicate the overall fit of a set of data to the PCM.

Most PCM computer programs address the goodness-of-fit question at all three levels. In addition to statistical tests of fit, most also provide graphical displays of model-data fit.

Some widely used PCM software programs provide tests of item and person fit as weighted and unweighted mean-squares. Although the exact distributional properties of these statistics are unknown, weighted and unweighted mean-squares may still be useful in practice for diagnosing sources of model-data misfit. These two statistics are explained briefly here; for more details, see Wright and Masters (1982, Chapter 5).

When person p with ability θ_p responds to item i, that person's response a_{ip} takes a value between 0 and the maximum possible score on item i, A_i. Under the PCM, the probability of person p scoring a on item i is denoted P_{ipa}. The expected value and variance of a_{ip} are

$$E_{ip} \equiv E[a_{ip}] = \sum_{k=0}^{A_i} kP_{ipk}$$

and

$$W_{ip} \equiv V[a_{ip}] = \sum_{k=0}^{A_i} (k - E_{ip})^2 P_{ipk}$$

The standardized difference between person p's observed and expected response to item i is

$$z_{ip} = \frac{(a_{ip} - E_{ip})}{\sqrt{W_{ip}}}$$

7.4.1 Unweighted Mean-Square ("Outfit")

For each item i, an unweighted mean-square is calculated as

$$u_i = \frac{\sum_p^N z_{ip}^2}{N} \tag{7.17}$$

Analogously, an unweighted mean-square can be calculated for each person p:

$$u_p = \frac{\sum_{i}^{n} z_{ip}^2}{n} \tag{7.18}$$

In practice, these unweighted item and person mean squares can be overly sensitive to outliers (i.e., unexpected high scores on difficult items or unexpected low scores on easy items). To reduce the influence of outliers, a weighted mean-square can be calculated by weighting z_{ip}^2 by the information available about person p from item i. This is the basis of the "infit" mean-square.

7.4.2 Weighted Mean-Square ("Infit")

For each item i, a weighted mean-square is calculated as

$$v_i = \frac{\sum_{p}^{N} W_{ip} z_{ip}^2}{\sum_{p}^{N} W_{ip}} \tag{7.19}$$

For each person p, a weighted mean-square is calculated as

$$v_p = \frac{\sum_{i}^{n} W_{ip} z_{ip}^2}{\sum_{i}^{n} W_{ip}} \tag{7.20}$$

A considerable volume of research has investigated methods for testing the fit of data to members of the Rasch family of models. Much of this research has sought diagnostic goodness-of-fit tests with known null distributions. The fact that Rasch models are members of the family of exponential models has facilitated the development of methods for testing model-data fit (Glas, Volume Two, Chapter 17).

7.5 Empirical Example

To illustrate the use of partial credit analysis, an application to the assessment of student essay writing is described. These data were collected to investigate the consistency with which 70 markers were able to make holistic judgements of the quality of student essays using a five-point scale. The 70 markers had applied for positions as essay markers in a major testing program. After initial training in the use of the five-point grade scale (E, D, C, B, A), all markers were given the same set of 90 essays to mark. Analysis was undertaken of the grades allocated by the aspiring markers and the results were used in appointing markers to the program.

In this analysis the quality θ_j of each student essay j was measured by treating the 70 markers as "items," each providing information about θ_j. In this way, all 70 markers were calibrated. This approach allowed each marker to have their own interpretation of the five-point grade scale and provided a basis for studying differences in marker behavior.

7.5.1 Measuring Essay Quality

The first result of the partial credit analysis was a set of estimates $(\hat{\theta}_1, \hat{\theta}_2, \ldots, \hat{\theta}_{90})$ for the 90 student essays. The estimates were plotted along the line of the variable in Figure 7.2. The essay with the highest $\hat{\theta}$ (about +2.0 logits) was Essay 61. The essay with the lowest $\hat{\theta}$ (about −2.6 logits) was Essay 53.

The grades awarded to five of the 90 essays are shown on the right of Figure 7.2. Column heights indicate the number of markers giving each grade. For Essay 61, the most commonly assigned grade was an A. Many markers assigned this essay a B, some gave it a C. A few assigned it a grade of D. At the bottom of Figure 7.2, Essay 21 was most often assigned a grade of E. None of the 70 markers rated Essay 21 higher than C.

Figure 7.2 provides insight into the basis of the $\hat{\theta}$ estimates. When the grade awarded by each marker was replaced by a score so that 0 = E, 1 = D, 2 = C, 3 = B, and 4 = A, and this score was averaged over the 70 markers, then the ordering of essays by their average score was identical to their order by estimated θ shown in Figure 7.2. (This is the case in this example because all 70 markers rated all 90 essays.)

The dotted lines on the right of Figure 7.2 show the expected number of markers giving each grade to these essays under the PCM. This is obtained by calculating the model probability of each marker giving a particular grade to an essay and then summing these probabilities over the 70 markers. The essays on the right of Figure 7.2 all show good fit to the expectations of the model.

7.5.2 Calibrating Markers

A second outcome of the partial credit analysis is a set of marker calibrations. Figure 7.3 illustrates the results of this process for four markers, #57, #25, #43, and #24.

On the left of Figure 7.3, measures of essay quality are reproduced from Figure 7.2. Each essay's identification number has been replaced here by the grade most commonly assigned to that essay. From this distribution it is possible to see the regions of the essay quality continuum that these 70 markers, as a group, associate with the five grades.

On the right of Figure 7.3, the calibrations of the four markers are shown. For each marker, the partial credit analysis had provided a set of threshold estimates, which had been used to mark out regions (shaded) reflecting that marker's use of the five-point grade scale. If an essay of quality $\hat{\theta}$ is located within a marker's "C" region, for example, then under the PCM, that marker will probably ($p > 0.5$) rate that essay at least C, but probably not ($p > 0.5$) rate it better than C.

The four markers in Figure 7.3 have been selected because of their different uses of the grade scale. Marker 57 was a typical marker. His interpretation of the five grade points matches quite closely the distribution on the left of Figure 7.3. Marker 25, on the other hand, was a harsh marker. This marker typically gave Bs to essays that most commonly received As, and Cs to essays that most commonly received Bs. Marker 25 had higher than usual thresholds for all grades.

At the other extreme, Marker 43 was atypically lenient. But as Figure 7.3 shows, this leniency applied only at lower grades: Marker 43 was not lenient in allocating As but was reluctant to award Es and allocated only Ds to very poor essays that would typically be assigned a grade of E.

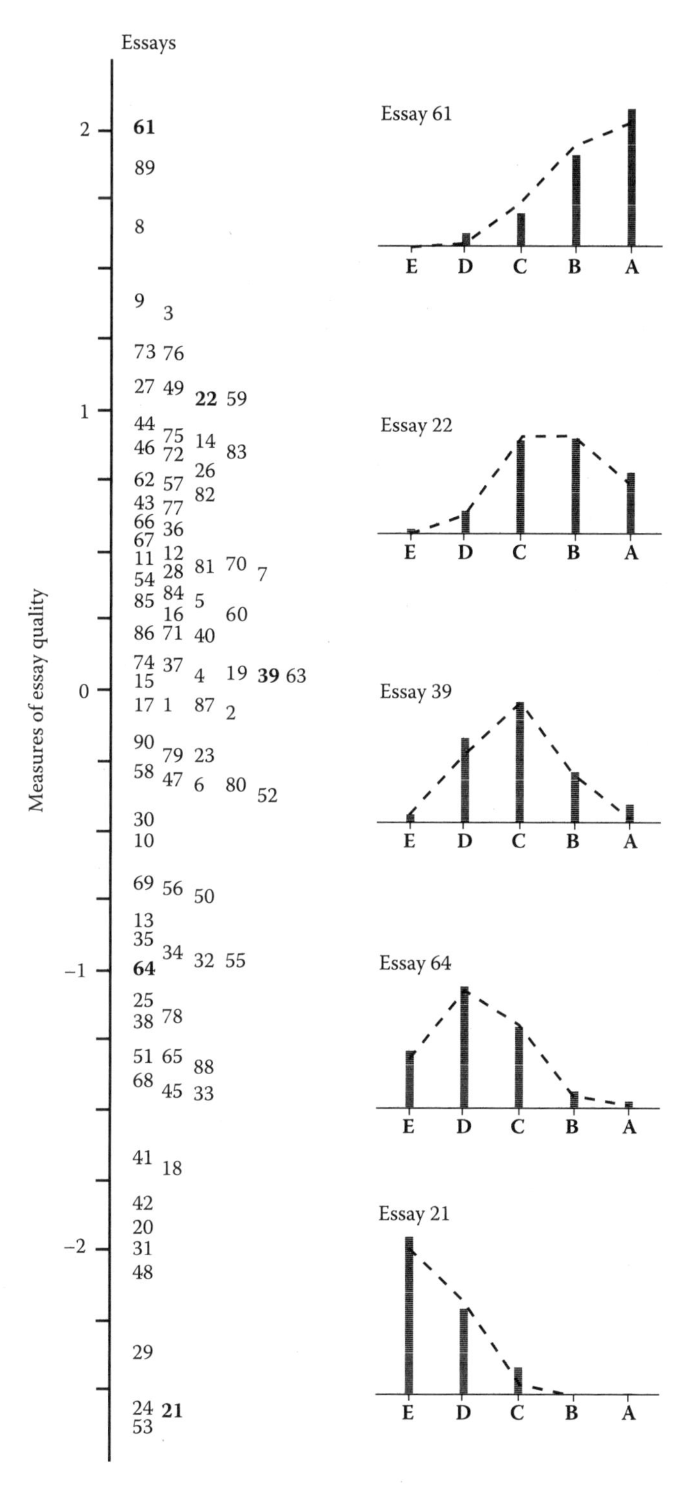

FIGURE 7.2
Partial credit measures of essay quality.

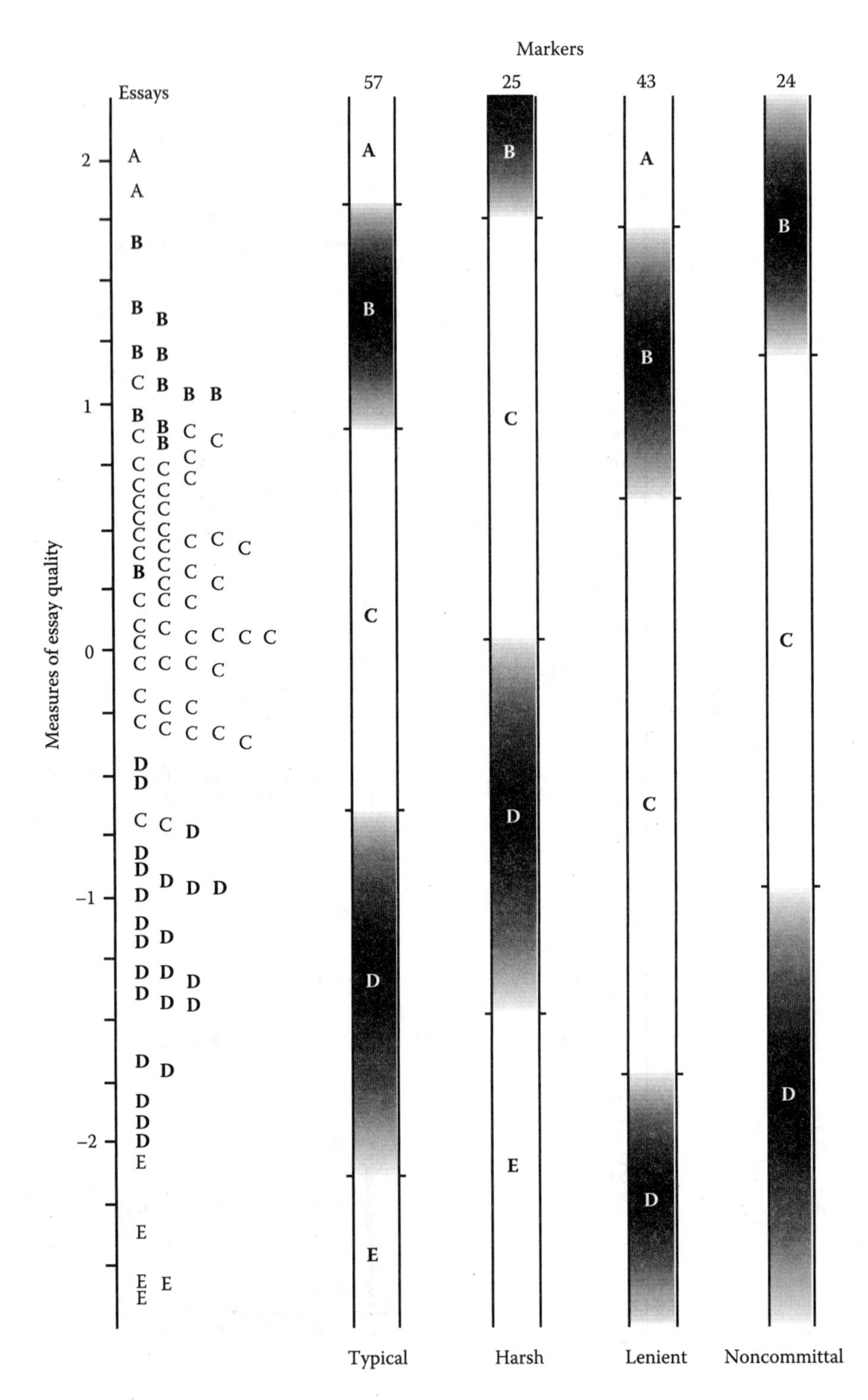

FIGURE 7.3
Partial credit calibrations of markers.

Finally, Marker 24 displayed an unusual tendency not to assign the highest grade, A, or the lowest grade, E. This marker showed a tendency to be noncommittal by making unusual use of the middle grades.

Each of these four markers displayed good fit to the PCM. Their ratings were in good agreement with the general ordering of these essays on the variable. Their different uses of the A to E grades, reflected in their different threshold estimates in Figure 7.3, are not an indication of misfit, but suggest that they have different interpretations of these grades.

Partial credit analysis can be used in this way to identify differences among markers in their use of the provided grade points. Once markers are calibrated, differences in marker behavior could be taken into account automatically in future marking exercises.

7.6 Discussion

The strength of the PCM resides in its simplicity. It is simply Rasch's (1960) model for dichotomies applied to each pair of adjacent categories in an ordered sequence of response alternatives. The use of Rasch's model for dichotomies to model the probability that person p's response to item i will be in category a rather than in category $a - 1$ means that the PCM has only two sets of parameters—one for persons and one for items; all parameters in the model are locations on the measurement variable; and, because it can be "repeated without modification in different parts of the measurement continuum" (Thurstone, 1931, p. 257), Rasch's simple logistic expression defines a unit for the PCM that is maintained over the range of the variable.

As a result of its algebraic formulation, the PCM shares the distinguishing statistical properties of the Rasch family of item response models: separable person and item parameters, and hence the possibility of "specifically objective" comparisons of persons and items (Rasch, 1977), and sufficient statistics for all model parameters. These features make the PCM easy to implement. Partial credit analysis is now a standard feature of many IRT software packages.

A number of recent applications of the PCM have been to the analysis of alternative assessments such as performances, projects, and portfolios. In these contexts, assessors typically rate student work either holistically or against a set of prespecified criteria. Measures of achievement are constructed by aggregating ratings on different tasks or across assessment criteria. The PCM is ideally suited to the analysis of assessments of this kind.

References

Adams, R. J. 1988. Applying the partial credit model to educational diagnosis. *Applied Measurement in Education*, 4, 347–362.

Adams, R. J., Doig, B. A., and Rosier, M. R. 1991. *Science Learning in Victorian Schools: 1990 (Monograph 41).* Melbourne, Victoria: Australian Council for Educational Research.

Adams, R. J., Griffin, P. E., and Martin, L. 1987. A latent trait method for measuring a dimension in second language proficiency. *Language Testing*, 4, 9–27.

Adams, R. J., Wu, M., and Wilson, M. 2012. *ACER ConQuest 3.0.1 (Computer Program)*. Melbourne: Australian Council for Educational Research.

Andersen, E. B. 1977. Sufficient statistics and latent trait models. *Psychometrika*, 42, 69–81.

Andrich, D. 1978. A rating formulation for ordered response categories. *Psychometrika*, 43, 561–573.

Bertoli-Barsotti, L. 2005. On the existence and uniqueness of JML estimates for the partial credit model. *Psychometrika*, 70, 517–531.

Bock, R. D., and Aitkin, M. 1981. Marginal maximum likelihood estimation of item parameters: Application of an EM algorithm. *Psychometrika*, 46, 443–459.

Bode, R. K. 2001. Partial credit model and pivot anchoring. *Journal of Applied Measurement*, 2, 78–95.

Chen, S., Hou, L., and Dodd, B. G. 1998. A comparison of maximum likelihood estimation and expected a posteriori estimation in CAT using the partial credit model. *Educational and Psychological Measurement*, 58, 569–595.

Dodd, B. G., De Ayala, R. J., and Koch, W. R. 1995. Computerized adaptive testing with polytomous items. *Applied Psychological Measurement*, 19, 5–22.

Dodd, B. G., and Koch, W. R. 1987. Effects of variations in step values on item and test information for the partial credit model. *Applied Psychological Measurement*, 11, 371–384.

Dodd, B. G., Koch, W. R., and De Ayala, R. J. 1993. Computerized adaptive testing using the partial credit model: Effects of item pool characteristics and different stopping rules. *Educational and Psychological Measurement*, 53, 61–77.

Doig, B. A., Mellor, S., Piper, K., and Masters, G. N. 1994. *Conceptual Understanding in Social Education*. Melbourne, Victoria: Australian Council for Educational Research.

Edwards, A. L., and Thurstone, L. L. 1952. An internal consistency check for scale values determined by the method of successive intervals. *Psychometrika*, 17, 169–180.

Fitzpatrick, A. R., Link, V. B., Yen, W. M., Burket, G. R., Ito, K., and Sykes, R. C. 1996. Scaling performance assessments: A comparison of one-parameter and two-parameter partial credit models. *Journal of Educational Measurement*, 33, 291–314.

Fox, C. 1999. An introduction to the partial credit model for developing nursing assessments. *Journal of Nursing Education*, 38, 340–346.

Gebhardt, E., Fraillon, J., Schulz, W., O'Malley, K., Freeman, C., Murphy, M., and Ainley, J. 2012. National Assessment Program: ICT Literacy Technical Report, Sydney: Australian Curriculum Assessment and Reporting Authority.

Gebhardt, E., Fraillon, J., Wernert, N., and Schulz, W. 2011. National Assessment Program: Civics and Citizenship Technical Report, Sydney: Australian Curriculum Assessment and Reporting Authority.

Glas, C. A. W., and Verhelst, N. D. 1989. Extensions of the partial credit model. *Psychometrika*, 53, 525–546.

Gorin, J. S., Dodd, B. G., Fitzpatrick, S. J., and Shieh, Y. Y. 2005. Computerized adaptive testing with the partial credit model: Estimation procedures, population distributions, and item pool characteristics. *Applied Psychological Measurement*, 29, 433–456.

Harris, J., Laan, S., and Mossenson, L. T. 1988. Applying partial credit analysis to the construction of narrative writing tests. *Applied Measurement in Education*, 4, 335–346.

Hemker, B. T., Sijtsma, K., Molenaar, I. W., and Junker, B. W. 1996. Polytomous IRT models and monotone likelihood ratio of the total score. *Psychometrika*, 61, 679–693.

Hemker, B. T., Sijtsma, K., Molenaar, I. W., and Junker, B. W. 1997. Stochastic ordering using the latent trait and the sum score in polytomous IRT models. *Psychometrika*, 62, 331–347.

Jafari, P., Bagheri, Z., and Safe, M. 2012. Item and response-category functioning of the Persian version of the Kidscreen-27: Rasch partial credit model. *Health and Quality of Life Outcomes*, 10, 1–6.

Julian, E. R., and Wright, B. D. 1988. Using computerized patient simulations to measure the clinical competence of physicians. *Applied Measurement in Education*, 4, 299–318.

Kim, J., Chung, H., Park, R., and Dodd, B. G. 2013. A comparison of panel designs with routing methods in the multistage test with the partial credit model. *Behavior Research Methods*, 45, 1087–1098.

Koch, W. R., and Dodd, B. G. 1989. An investigation of procedures for computerized adaptive testing using partial credit scoring. *Applied Measurement in Education*, 2, 335–357.

Ligtvoet, R. 2012. An isotonic partial credit model for ordering subjects on the basis of their sum scores. *Psychometrika*, 77, 479–494.

Linacre, J.M. 2012. A User's Guide to *WINSTEPS:* Multiple-choice, rating scale and partial credit Rasch analysis, Version 3.75.0 http://www.winsteps.com/a/winsteps-manual.pdf accessed 12/3/14.

Mair, P. and Hatzinger, R. 2007. Extended Rasch modelling: The eRm package for the application of IRT models in R. *Journal of Statistical Software*, 20, 1–20.

Masters, G. N. 1982. A Rasch model for partial credit scoring. *Psychometrika*, 47, 149–174.

Masters, G. N. 1984. Constructing an item bank using partial credit scoring. *Journal of Educational Measurement*, 21, 19–32.

Masters, G. N. 1987. Measurement models for ordered response categories. In Langeheine, R., and Rost, J. (Eds.), *Latent Trait and Latent Class Models*, pp. 11–29. New York: Plenum Publishing Corporation.

Masters, G. N. 1988a. Partial credit model. In Keeves, J.P. (Ed.) *Educational Research Methodology, Measurement, and Evaluation*, pp. 292–296. Oxford: Pergamon Press.

Masters, G. N. 1988b. The analysis of partial credit scoring. *Applied Measurement in Education*, 1, 279–298.

Masters, G. N. 2010. Partial credit model. In Nering, M., and Ostini, R. (Eds.), *Handbook of Polytomous Item Response Theory Models: Development and Applications*, pp. 109–122. New York: Routledge Academic.

Masters, G. N., and Evans, J. 1986. Banking non-dichotomously scored items. *Applied Psychological Measurement*, 10, 355–367.

Masters, G. N., and Wright, B. D. 1984. The essential process in a family of measurement models. *Psychometrika*, 49, 529–544.

Masters, G. N., and Wright, B. D. 1996. The partial credit model. In Hambleton, R. K., and van der Linden, W. J. (Eds.), *Handbook of Modern Item Response Theory*, pp. 101–119. New York: Springer.

Muraki, E. 1992. A generalized partial credit model: Application of an EM algorithm. *Applied Psychological Measurement*, 16, 159–176.

Organisation for Economic Cooperation and Development. 2012. PISA 2009 Technical Report. Paris: Programme for International Student Assessment, OECD Publishing.

Pennings, A. 1987. The diagnostic embedded figures test. Paper presented at the Second Conference on Learning and Instruction of the European Association of Learning and Instruction, University of Tübingen, Germany.

Pollitt, A., and Hutchinson, C. 1987. Calibrating graded assessments: Rasch partial credit analysis of performance in writing. *Language Testing*, 4, 72–92.

Rasch, G. 1960. *Probabilistic Models for Some Intelligence and Attainment Tests.* Copenhagen: Denmarks Paedagogiske Institut.

Rasch, G. 1977. On specific objectivity: An attempt at formalizing the request for generality and validity of scientific statements. *Danish Yearbook of Philosophy*, 14, 58–94.

Samejima, F. 1969. Estimation of latent ability using a response pattern of graded scores. *Psychometrika Monograph Supplement*, 34(4, Pt. 2, Monograph No. 17), 1–100.

Sebille, V., Challa, T., and Mesbah, M. 2007. Sequential analysis of quality of life measurements with the mixed partial credit model. In Auget, J.-L., Balakrishan, N., Mesbah, M., and Molenberghs, G. (Eds.). *Advances in Statistical Methods for the Health Sciences*, pp. 109–125. Boston: Birkhauser.

Schulz, W., Ainley, J., Friedman, T., and Lietz, P. 2011. *ICCS 2009 Latin American Report: Civic Knowledge and Attitudes among Lower-Secondary Students in Six Latin American Countries.* Amsterdam, The Netherlands: IEA.

Sykes, R.C., and Yen, W. M. 2000. The scaling of mixed-item-format tests with the one-parameter and two-parameter partial credit models. *Journal of Educational Measurement*, 37, 221–244.

Tandon, A., Murray, C. J. L., Salomon, J. A., and King, G. 2003. Statistical models for enhancing cross-population comparability. In Murray, J. L., and Evans, D. B. (Eds.) *Health Systems Performance Assessment: Debates, Methods and Empiricism*, pp. 727–746. Geneva: World Health Organisation.

Thurstone, L. L. 1931. Measurement of social attitudes. *Journal of Abnormal and Social Psychology*, 26, 249–269.

Titmanis, P., Murphy, F., Cook, J., Brady, K., and Brown, M. 1993. *Profiles of Student Achievement: English and Mathematics in Western Australian Government Schools, 1992*. Perth: Ministry of Education.

Torney-Purta, J., Lehmann, R., Oswald, H., and Schulz, W. 2001. *Citizenship and Education in Twenty-Eight Countries: Civic Knowledge and Engagement at Age Fourteen*. Amsterdam, The Netherlands: IEA.

Verhalst, N. D., and Verstralen, H. H. F. M. 2008. Some considerations on the partial credit model. *Psicologica*, 29, 229–254.

Wilson, M. R. 1992. The ordered partition model: An extension of the partial credit model. *Applied Psychological Measurement*, 16, 309–325.

Wilson, M. R., and Adams, R. J. 1993. Marginal maximum likelihood estimation for the ordered partition model. *Journal of Educational Statistics*, 18, 69–90.

Wilson, M. R., and Iventosch, L. 1988. Using the partial credit model to investigate responses to structured subtests. *Applied Measurement in Education*, 1, 319–334.

Wilson, M. R., and Masters, G. N. 1993. The partial credit model and null categories. *Psychometrika*, 58, 87–99.

Wright, B. D., and Masters, G. N. 1982. *Rating Scale Analysis*. Chicago: MESA Press.

8

Generalized Partial Credit Model

Eiji Muraki and Mari Muraki

CONTENTS

8.1 Introduction 127
8.2 Presentation of the Model 127
8.3 Parameter Estimation 130
 8.3.1 E-Step 131
 8.3.2 M-Step 132
8.4 Goodness of Fit 133
8.5 Empirical Example 134
8.6 Discussion 136
References 136

8.1 Introduction

A generalized partial credit model (GPCM) was formulated by Muraki (1992). Based on Masters' (Volume One, Chapter 7) partial credit model (PCM), the GPCM relaxes the assumption of the uniform discriminating power of the test items. However, the difference between these two types of item response models is not only the parameterization of item characteristics but also the basic assumption about the latent variable. An item response model is viewed here also as a member of a family of latent variable models, which includes such models as the linear or nonlinear factor analysis model, the latent class model, and the latent profile model as well (Bartholomew, 1987).

As a member of the Rasch family of polytomous item response models, Masters' PCM assumes the discrimination power to be constant across all items. The GPCM is a generalization of the PCM with the inclusion of varying discrimination power for each item. The GPCM not only can attain some of the objectives that the Rasch model achieves but also provides more information about the characteristics of test items.

8.2 Presentation of the Model

The PCM (Volume One, Chapter 7; Muraki, 1992) is formulated for items $i = 1, \ldots, n$ with response categories $a = 1, \ldots, A_i$. It is based on the assumption that the probability of choosing the response category a of item i over $a - 1$ is defined by the logistic dichotomous response function

$$C_{ia} = P_{ia|a-1,a}(\theta) = \frac{P_{ia}(\theta)}{P_{i,a-1}(\theta) + P_{ia}(\theta)} = \frac{\exp[Z_{ia}(\theta)]}{1 + \exp[Z_{ia}(\theta)]} \tag{8.1}$$

Equation 8.1 can be written as

$$P_{ia}(\theta) = \frac{C_{ia}}{1 - C_{ia}} P_{i,a-1}(\theta) = \exp[Z_{ia}(\theta)] P_{i,a-1}(\theta) \tag{8.2}$$

Note that $C_{ia}/(1 - C_{ia})$ in Equation 8.2 is the ratio of two conditional probabilities and can be described as the odds of choosing the ath instead of the $(a - 1)$th category, given two available choices, $a - 1$ and a. Its log transformation, $Z_{ia}(\theta)$, is called a logit.

After normalizing each $P_{ia}(\theta)$ within an item such that $\sum P_{ia}(\theta) = 1$, the GPCM is written as

$$P_{ia}(\theta) = \frac{\exp\left[\sum_{v=1}^{a} Z_{iv}(\theta)\right]}{\sum_{c=1}^{A_i} \exp\left[\sum_{v=1}^{c} Z_{iv}(\theta)\right]} \tag{8.3}$$

and

$$Z_{ia}(\theta) = Da_i(\theta - b_{ia}) = a_i(\theta - b_i + d_a) \tag{8.4}$$

where a_i is a slope parameter, b_{ia} is an item-category parameter, b_i is an item-location parameter, and d_a is a category parameter. The slope parameter indicates the degree to which the categorical responses vary among items as the θ level changes. This concept of item discriminating power is closely related to that of the item-reliability index in classical test theory (Muraki and Wang, 1992). This parameter captures information about differential discriminating power of the different items. P_{ia} in Equation 8.3 is also called the item-category response function (ICRF) of the GPCM.

As the number of response categories is A, only $A_i - 1$ item-category parameters can be identified. Consequently, any of the A_i category threshold parameters can be arbitrarily set to a fixed value. Observe that the terms that include the parameter in the numerator and denominator of the model cancel (Muraki, 1992). We arbitrarily define $b_{i1} \equiv 0$.

As shown in Equations 8.2 and 8.4, the item-category parameters b_{ia} are the points on the θ scale at which the plots of $P_{i,a-1}(\theta)$ and $P_{ia}(\theta)$ intersect. These two ICRFs intersect only once, and the intersection can occur anywhere along the θ scale. Thus, under the assumption $a_i > 0$,

$$\begin{aligned} &\text{if} \quad \theta = b_{ia} \quad \text{then} \quad P_{ia}(\theta) = P_{i,a-1}(\theta) \\ &\text{if} \quad \theta > b_{ia} \quad \text{then} \quad P_{ia}(\theta) > P_{i,a-1}(\theta) \end{aligned} \tag{8.5}$$

and

$$\text{if} \quad \theta < b_{ia} \quad \text{then} \quad P_{ia}(\theta) < P_{i,a-1}(\theta)$$

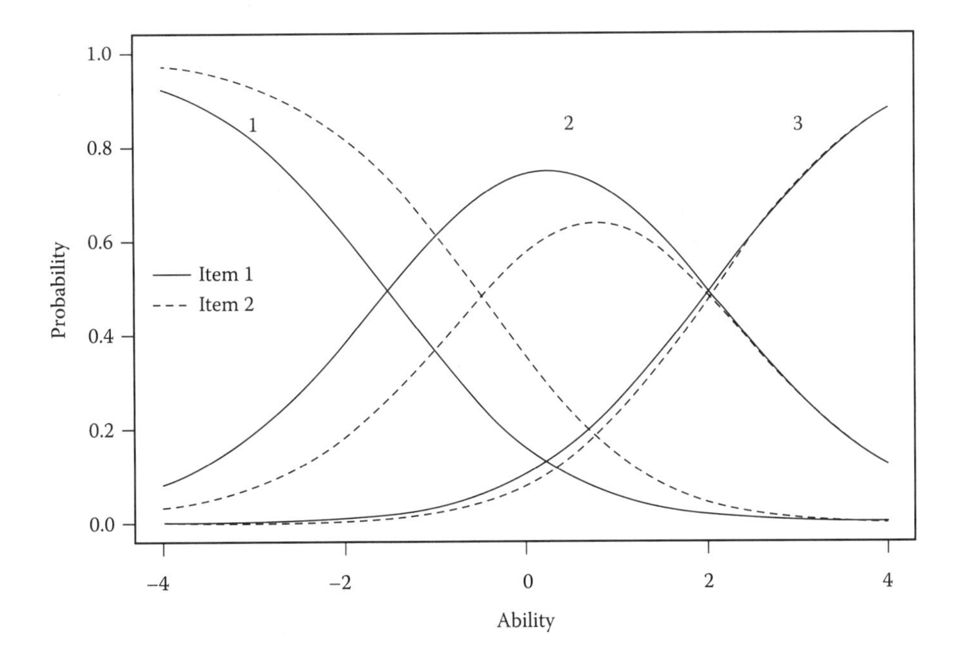

FIGURE 8.1
ICRF of items 1 and 2.

It should be noted that values of the b_{ia} parameters are not necessarily ordered with the answer categories within each item; they only represent the relative magnitude of the adjacent probabilities $P_{i,a-1}(\theta)$ and $P_{ia}(\theta)$. Figure 8.1 shows the ICRFs for two GPCM items, each having three categorical responses. For item 1 with $a_1 = 1.0$, $b_{11} = 0.0$, $b_{12} = -1.5$, and $b_{13} = 2.0$, the ICRFs of $P_{11}(\theta)$ and $P_{12}(\theta)$ intersect at $\theta = -1.5$, and the ICRFs of $P_{12}(\theta)$ and $P_{13}(\theta)$ intersect at $\theta = 2.0$. If the second item-category parameter changes from −1.5 to −0.5 (item 2), the intersection of the first and the second ICRFs moves from −1.5 to −0.5 (Figure 8.1). Since the probability of a response in the first category becomes dominant over a wider range of θ values, the expected frequency of its responses increases. Consequently, the ICRF of the second category is depressed. If the slope parameter is decreased, the intersections of ICRFs remain the same but the curves of the ICRFs become flatter. As a result, the second category probability, $P_{22}(\theta)$, is further depressed, and the responses of the examinees become more likely to fall into the first or third category. The cumulative response frequencies for these two items were simulated for 10,000 examinees with ability parameters sampled from the standard normal distribution. For these generated data, the response frequencies of the first, second, and third categories were 1449, 7709, and 842 for item 1 and 3685, 4973, and 1342 for item 2, respectively.

In the GPCM, following Andrich's (1978, Volume One, Chapter 5) rating scale model, the item-category parameters b_{ia} may be additively decomposed as $b_{ia} = b_i - d_a$. Because the values of b_{ia} are not necessarily ordered within each item, the category parameters (d_a) are also not necessarily sequentially ordered within an item. Again, parameter d_a should be interpreted as the relative difficulty of category a relative to the other categories for the item, or as the deviation of each categorical threshold from the item location, b_i.

For the polytomous item response models, there is an indeterminacy between the set of category parameters and location parameters in each block of items that share the same set of category parameters. To obtain a unique set of parameters, a location constraint must be imposed on the category parameters, that is, the mean of category parameters within a categorical scale must be set equal to a constant over blocks. The following location constraint can be imposed across blocks to eliminate an indeterminacy:

$$\sum_{A=2}^{A_i} d_a = 0 \tag{8.6}$$

The sum of the $Z_{iv}(\theta)$ for item i in the GPCM can be written as

$$Z_{ia}^{+}(\theta) = \sum_{v=1}^{a} Z_{iv}(\theta) = a_i[T_a(\theta - b_i) + K_a] \tag{8.7}$$

where

$$K_a = \sum_{v=1}^{a} d_v \tag{8.8}$$

Andrich (Volume One, Chapter 5) called T_a and K_{ah} the scoring function and the category coefficient, respectively. For the PCM, the scoring function T_a is a linear integer scoring function ($T_a = a$). The PCM becomes a model for ordered categorical responses only when the scoring function is increasing, that is, $T_a > T_{a-1}$ for any a within item i and $a_i > 0$.

The expected value of the scoring function is computed by

$$\bar{T}_i(\theta) = \sum_{a=1}^{A_i} T_a P_{ia}(\theta) \tag{8.9}$$

$\bar{T}_i(\theta)$, in Equation 8.9, is called the item response function (IRF) for a polytomously scored item with a given scoring function (T_a, $a = 1, 2, \ldots, A$). It can be viewed as a regression of the item score onto the ability scale (Lord, 1980). For dichotomous IRT models, the IRF coincides with the ICRF. In the GPCM, the IRF is the conditional mean of item scores at a given θ (Muraki, 1993). Chang and Mazzeo (1944) proved that, if two polytomously scored items have the same IRFs with the linear integer scoring function, then the model parameters must be identical.

8.3 Parameter Estimation

Estimating both person and item parameters simultaneously by maximum joint likelihood (JML) estimation (Lord, 1980), and treating the former as fixed effects, fails to produce numerically and statistically stable estimates (Neyman and Scott, 1948), especially when either the number of examinees or the number of items is very large. Similarly, as the model has no sufficient statistics other than the response vectors themselves, the maximum conditional likelihood (CML) estimation method (Andersen, 1973; Fisher and Parzer, 1991; Glas, Volume Two, Chapter 11; Wright and Masters, 1982) cannot be used either.

For this reason, Bock and Lieberman (1970) proposed the maximum marginal likelihood (MML) method for the case of dichotomously scored items. In their method, the latent trait is treated as a random component (Bartholomew, 1987) and integrated out of the likelihood assuming an ability distribution. The Bock and Lieberman method, however, is computationally feasible only when the number of items is very small because of the necessity of inversion of the $2n \times 2n$ information matrix (n is the number of items). Bock and Aitkin (1981) reformulated the MML method as an application of the EM algorithm (Dempster et al., 1977; Aitkin, Volume Two, Chapter 12). The MML–EM estimation method consists of two steps. The first is the expectation step (E-step) where provisional expected frequencies $\bar{r}_{iaf}$, and provisional expected sample sizes $\bar{N}_f$ ($\bar{N}_{if}$ when the sample size differs for each item) at quadrature points indexed by $f = 1, \ldots, F$ are computed. Then, in the maximization step (M-step), the MML estimates are obtained by Fisher's scoring method (Kendall and Stuart, 1973). Both the E-step and the M-step are repeated (EM cycles) until all estimates are stable. More specifically, the EM algorithm for the GPCM is as follows.

8.3.1 E-Step

We denote the number of examinees with response pattern $l = 1, 2, \ldots, S$ by r_l, where $S \leq \min(N, J)$, $N = \Sigma r_l$, and $J = \Pi A_i$. For the lth response pattern, let $U_{ial} = 1$ denote a response to item i is in the ath category; otherwise $U_{ial} = 0$. The provisional expected frequency of the ath categorical response of item i, $\bar{r}_{iaf}$, and the provisional expected sample size $\bar{N}_f$, at each quadrature point X_f are computed from the following respective equations:

$$\bar{r}_{iaf} = \sum_{l=1}^{S} \frac{r_l L_l(X_f) A(X_f) U_{ial}}{\bar{P}_l} \tag{8.10}$$

$$\bar{N}_f = \sum_{l=1}^{S} \frac{r_l L_l(X_f) A(X_f)}{\bar{P}_l} \tag{8.11}$$

where

$$\bar{P}_l = \sum_{f=1}^{F} L_l(X_f) A(X_f) \tag{8.12}$$

and

$$L_l(X_f) = \prod_{i=1}^{n} \prod_{a=1}^{A_i} [P_{ia}(X_f)]^{U_{ial}} \tag{8.13}$$

Note that Equation 8.12 is a numerical approximation of the posterior predictive probability of the observed response pattern l, where the prior distribution is assumed to be standard normal. $A(X_f)$ is the weight of the Gauss–Hermite quadrature, and X_f is the quadrature point (Stroud and Secrest, 1966). The quadrature weight $A(X_f)$ is approximately the standard normal probability density at the point X_f, such that across all quadrature points:

$$\sum_{f=1}^{F} A(X_f) = 1 \tag{8.14}$$

Empirical weights can also be used in place of $A(X_f)$ as long as the condition in Equation 8.14 is met.

8.3.2 M-Step

The qth cycle of the iterative process can be expressed as

$$\boldsymbol{v}_q = \boldsymbol{v}_{q-1} + \boldsymbol{V}^{-1}\boldsymbol{t} \tag{8.15}$$

where $\boldsymbol{v}_q$ and $\boldsymbol{v}_{q-1}$ are the vectors with the parameter estimates of the qth and $q-1$th cycles, respectively, $\boldsymbol{V}^{-1}$ is the inverse of the information matrix, and $\boldsymbol{t}$ is the gradient vector.

The elements of the gradient vector and the (2×2) information matrix for the slope and item-location parameters for the item i are

$$t_{a_i} = a_i^{-1} \sum_{f=1}^{F} \sum_{a=1}^{A_i} \bar{r}_{iaf} [Z_{ia}^{+}(X_f) - \bar{Z}_i^{+}(X_f)] \tag{8.16}$$

$$t_{b_i} = a_i \sum_{f=1}^{F} \sum_{a=1}^{A_i} \bar{r}_{iaf} [-T_a + \bar{T}_i(X_f)] \tag{8.17}$$

$$V_{a_i a_i} = a_i^{-2} \sum_{f=1}^{F} \bar{N}_f \sum_{a=1}^{A_i} P_{ia}(X_f) [Z_{ia}^{+}(X_f) - \bar{Z}_i^{+}(X_f)]^2 \tag{8.18}$$

$$V_{b_i b_i} = a_i^{2} \sum_{f=1}^{F} \bar{N}_f \sum_{a=1}^{A_i} P_{ia}(X_f) [-T_a + \bar{T}_i(X_f)]^2 \tag{8.19}$$

$$V_{a_i b} = \sum_{f=1}^{F} \bar{N}_f \sum_{a=1}^{A_i} P_{ia}(X_f) [Z_{ia}^{+}(X_f) - \bar{Z}_i^{+}(X_f)][-T_a + \bar{T}_i(X_f)] \tag{8.20}$$

where $\bar{T}_i(X_f)$ was already defined in Equation 8.9 and

$$\bar{Z}_i^{+}(X_f) = \sum_{a=1}^{A_i} Z_{ia}^{+}(X_f) P_{ia}(X_f)] \tag{8.21}$$

The entry of the gradient vector for estimation of the ath category parameter is

$$t_{d_a} = \sum_{f=1}^{F}\sum_{i=1}^{n'} a_i \sum_{k=a}^{A_i}\left[\bar{r}_{ikf} - P_{ik}(X_f)\sum_{c=1}^{A_i}\bar{r}_{icf}\right] \tag{8.22}$$

The entry of the information matrix for $a' \leq a$ is

$$\begin{aligned} V_{d_a d_{a'}} &= \sum_{f=1}^{F}\bar{N}_f \sum_{i=1}^{n'}\sum_{k=1}^{A_i} \frac{1}{P_{ik}(X_f)} \frac{\partial P_{ik}(X_f)}{\partial d_a} \frac{\partial P_{ik}(X_f)}{\partial d_{a'}} \\ &= \sum_{f=1}^{F}\bar{N}_f \sum_{i=1}^{n'} a_i^2 \left[\sum_{k=a}^{A_i} P_{ik}(X_f)\right]\left[1 - \sum_{k=a'}^{A_i} P_{ik}(X_f)\right] \end{aligned} \tag{8.23}$$

where n' is the number of items in a given block. When a set of category parameters is unique for each item, $n' = 1$ and the summations drop from Equations 8.22 and 8.23. Since d_i has been set equal to 0, the orders of the gradient vector $\boldsymbol{t}$ and the information matrix $\boldsymbol{V}$ are $A - 1$ and $(A - 1) \times (A - 1)$, respectively.

In the M-step, the provisional expected values computed in the E-step are used to calculate Equations 8.16 through 8.20, and Equations 8.22 and 8.23, and the parameters for each item are estimated individually. The iterative process is repeated over the n items until the estimates of all item parameters are stable at a specified level of precision.

The optimal number of quadrature points for the calibration must be chosen considering the numbers of items and categories relative to the available computing resources. For polytomous response models, more quadrature points are generally necessary than for the calibration of dichotomous items. However, the number of EM cycles required to reach convergence and the computer memory necessary for the computation increase exponentially with the number of quadrature points. Although 30–40 quadrature points seem to be adequate, further empirical research is necessary.

The MML–EM estimation method described above is implemented in the *PARSCALE* program (Muraki, Volume Three, Chapter 24; Muraki and Bock, 1991). The *PARSCALE* program is also capable of estimating parameters of Samejima's normal and logistic models for graded responses (Volume One, Chapter 6). The program is particularly suitable for large-scale studies of rating data or performance assessments, and operates both at the individual level and the level of scoring other groups of examinees (e.g., schools).

8.4 Goodness of Fit

The goodness of fit of the GPCM can be tested item by item. If the test is sufficiently long, the method of the *PC-BILOG3* program (Mislevy and Bock, 1990) can be used with minor modifications. In this method, the examinees in a sample of size N are assigned to intervals of the θ-continuum. The expected a posteriori (EAP) is used as the estimator of each examinee's ability score. The EAP estimate is the mean of the posterior distribution of θ, given the observed response pattern l (Bock and Mislevy, 1982). The EAP score of the response pattern l is approximated by

$$\hat{\theta}_l \cong \frac{\sum_{f=1}^{F} X_f L_l(X_f) A(X_f)}{\sum_{f=1}^{F} L_l(X_f) A(X_f)} \tag{8.24}$$

After all examinees' EAP scores are assigned to any one of the predetermined W intervals on the θ-continuum, the observed frequency of the ath categorical responses to item i in interval w, r_{wia}, and the number of examinees assigned to item i in the wth interval, N_{wi}, are computed. Thus, the W by A_i contingency table for each item is obtained. Then, the estimated θs are rescaled so that the variance of the sample distribution equals that of the latent distribution on which the MML estimation of the item parameters was based. For each interval w, its mean $\bar{\theta}_w$, based on the rescaled θs, and the value of the fitted ICRF function, $P_{ia}(\bar{\theta}_w)$, are computed. Finally, a likelihood-ratio chi-square statistic for each item is computed by

$$G_i^2 = 2\sum_{w=1}^{W_i}\sum_{a=1}^{A_i} r_{wia} \ln \frac{r_{wia}}{N_{wi} P_{ia}(\bar{\theta}_w)} \tag{8.25}$$

where W_i is the number of intervals remaining after neighboring intervals are merged, if necessary to avoid expected values $N_{wi}P_{ih}(\bar{\theta}_w) < 5$. The number of degrees of freedom is equal to the number of intervals W_i multiplied by $A_i - 1$. The likelihood-ratio chi-square statistic for the test as a whole is simply the sum of the statistic in Equation 8.25 over items. Similarly, the number of degrees of freedom is the sum of the degrees of freedom for each item.

The method discussed above is quite sensitive to the number of classification intervals. If the number is too large, the expected values of the contingency table become fragmented and the fit statistics are inflated. Further research of the goodness-of-fit index for IRT models in general is needed.

8.5 Empirical Example

The *National Assessment of Educational Progress* (NAEP) (Beaten and Zwick, 1992; Mazzeo, Volume Three, Chapter 14) is an ongoing survey designed to measure students' academic knowledge in various subject areas. The GPCM was applied to calibrate the items of the 1992 cross-sectional writing assessment of Grades 8 and 12 (Grima and Johnson, 1992).

Six of the nine writing items for the Grade 8 assessment were common with Grade 12 (linking items). Students' writings were scored by trained readers on a five-point scale. Omitted responses were treated as a response in the lowest category (scored as 1). For some items, the two highest categories were combined so that the observed frequency for any category was at least 10.

Two analyses were conducted using the *PARSCALE* computer program. For both analyses, 33 quadrature points were used for normal prior distributions, and the precision level for the convergence of the parameter estimates was set at 0.001. First, a set of item parameters was estimated based on the pooled response data of the Grade 8 and 12 assessments (pooled calibration). For the second analysis, the calibrations were done separately for each grade group (separate calibrations).

The total number of response vectors used in the calibration for the Grade 8 and 12 samples were 10,167, and 9344, respectively. A total of 76 EM cycles was required for the pooled calibration, while 102 cycles were needed for the separate calibration of the Grade 8 assessment. For the Grade 8 assessment, the goodness-of-fit statistics were computed based on 10 intervals. They were equal to 4293.741 ($df = 123$) for the pooled calibration and 3863.961 ($df = 124$) for the separate calibration. The substantial reduction of 429.780 in the chi-square statistic suggests that the separate calibration estimates should be used.

Figures 8.2 and 8.3 present the ICRFs and IRFs of items 5 and 6, respectively. The IRFs was computed from Equation 8.9 based on the scoring function, T = (0, 1, 2, …) and divided by the number of categories minus 1, so that both ICRFs and IRFs could be plotted in the same figures. For the computation of the ICRF, there is no difference as long as the successive increments of the scoring function are equal to one; that is, using either T = (0, 1, 2, …), T = (1, 2, 3, …), or T = (101, 102, 103, …) produces the same probabilities for the GPCM.

Both items have five response categories, and their mean response for the Grade 8 sample was about 2.70. Based upon this classical item statistic, these two items seemed to be equally difficult. However, their ICRFs did show differences. For instance, the difference between the observed response frequency of the third category for items 5 and 6 was considerable, 757 versus 1055. Figure 8.2 also shows that the third ICRF of item 2 was quite dominant relative to that of item 6. Another distinctive difference was that the ICRFs of

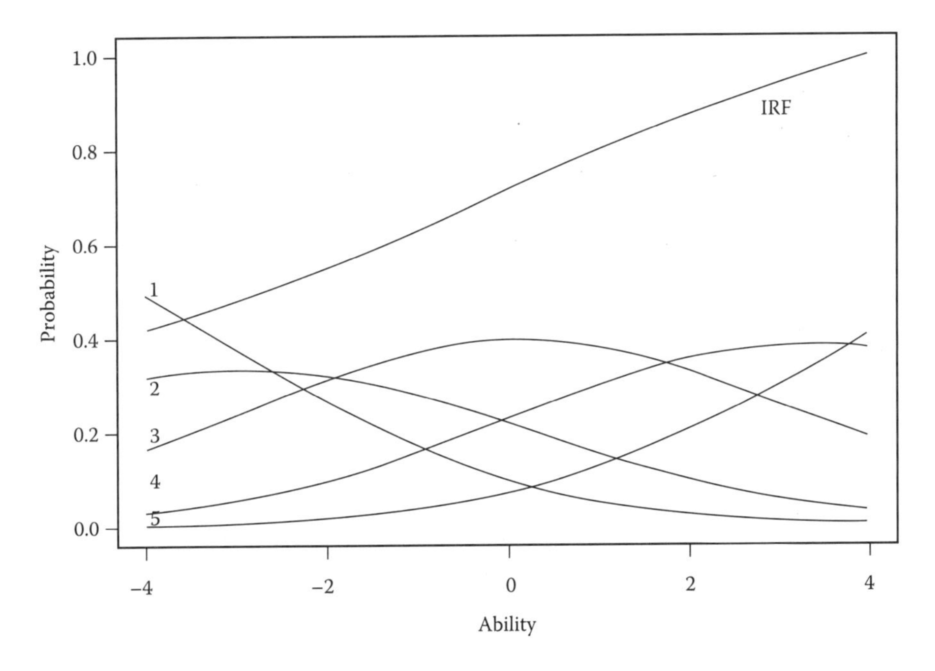

FIGURE 8.2
IRF and ICRF of item 5.

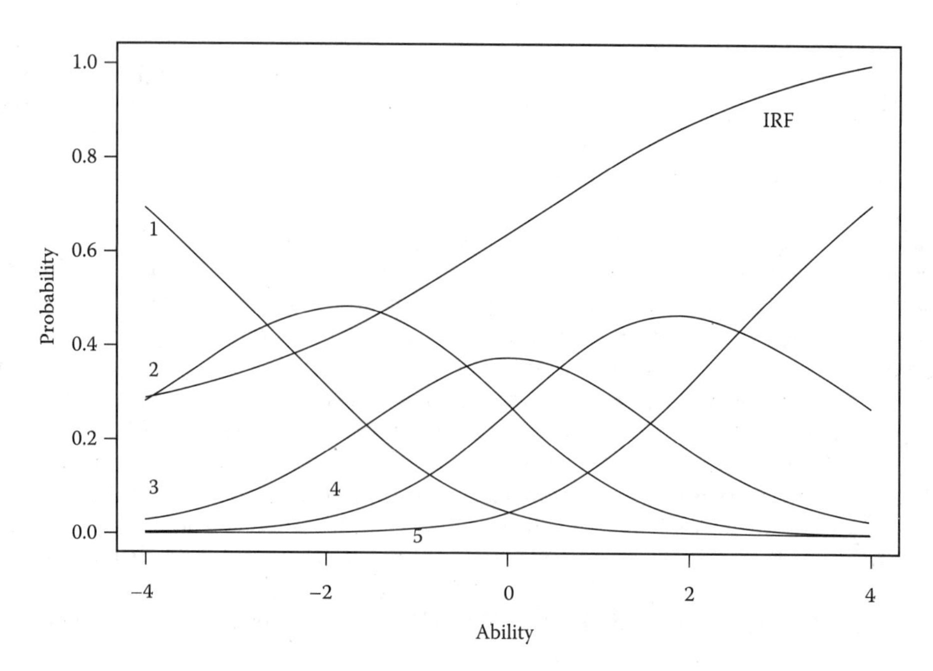

FIGURE 8.3
IRF and ICRF of item 6.

item 5 were relatively flat compared to those of item 6. The polyserial correlations between the scores on items 5 and 6 and the total scores were equal to 0.768 and 0.853, respectively. The difference between these classical item discrimination indices reflects on the slope of IRFs as well as the shapes of ICRFs. The IRF of item 6 was considerably steeper than that of item 5. Both items covered a wide range of the θ scale.

8.6 Discussion

A major advantage of the use of polytomously scored items is that their general coverage of a wider range of the ability scale with a sufficient information relative to an assessment with the same number of dichotomous items. For polytomous response models, the parameter values should be interpreted with the aid of graphical presentations (Masters, 1982; Muraki, 1992, 1993). Inspecting the plots of ICRFs, IRF, and item information functions for each item is an essential item analysis procedure (Muraki, 1993).

References

Andersen, E. B. 1973. Conditional inference for multiple-choice questionnaires. *British Journal of Mathematical and Statistical Psychology*, 26, 31–44.

Andrich, D. 1978. A rating formulation for ordered response categories. *Psychometrika*, 43, 561–573.

Bartholomew, D. J. 1987. *Latent Variable Models and Factor Analysis*. London: Charles Griffin and Company.

Beaten, A. E. and Zwick, R. 1992. Overview of the national assessment of educational progress. *Journal of Educational Statistics*, 17, 95–109.

Bock, R. D. and Aitkin, M. 1981. Marginal maximum likelihood estimation of item parameters: Application of an EM algorithm. *Psychometrika*, 41, 443–459.

Bock, R. D. and Lieberman, M. 1970. Fitting a response model for n dichotomously scored items. *Psychometrika*, 35, 179–197.

Bock, R. D. and Mislevy, R. J. 1982. Adaptive EAP estimation of ability in a microcomputer environment. *Applied Psychological Measurement*, 6, 431–444.

Chang, H.-H. and Mazzeo, J. 1944. The unique correspondence of item response function and item category response functions in polytomously scored item response models. *Psychometrika*, 59, 391–404.

Dempster, A. P., Laird, N. M., and Rubin, D. B. 1977. Maximum likelihood from incomplete data via the EM algorithm. *Journal of the Royal Statistical Society, Series B*, 39, 1–38.

Fisher, G. H. and Parzer, P. 1991. An extension of the rating scale model with an application to the measurement of change. *Psychometrika*, 56, 637–651.

Grima, A. M. and Johnson, E. G. 1992. Data analysis for the writing assessment. In: Johnson, E. G. and Allen, N. L. (Eds.), *The NAEP 1990 Technical Report*. Princeton, New Jersey: Educational Testing Service.

Kendall, M. G. and Stuart, A. 1973. *The Advanced Theory of Statistics* (Volume 2). New York: Hafner Publishing Company.

Lord, F. M. 1980. *Applications of Item Response Theory to Practical Testing Problems*. Hillsdale, New Jersey: Erlbaum.

Masters, G. N. 1982. A Rasch model for partial credit scoring. *Psychometrika*, 47, 149–174.

Mislevy, R. J. and Bock, R. D. 1990. *BILOG 3: Item Analysis and Test Scoring with Binary Logistic Models* [Computer program]. Chicago, Illinois: Scientific Software, Inc.

Muraki, E. 1990. Fitting a polytomous item response model to Likert-type data. *Applied Psychological Measurement*, 14, 59–71.

Muraki, E. 1992. A generalized partial credit model: Application of an EM algorithm. *Applied Psychological Measurement*, 16, 159–176.

Muraki, E. 1993. Information functions of the generalized partial credit model. *Applied Psychological Measurement*, 17, 351–363.

Muraki, E. and Bock, R. D. 1991. *PARSCALE: Parameter Scaling of Rating Data* [Computer program]. Chicago, Illinois: Scientific Software, Inc.

Muraki, E. and Wang, M. 1992. Issues relating to the marginal maximum likelihood estimation of the partial credit model. Paper presented at the *Annual Meeting of the American Educational Research Association*, San Francisco, California.

Neyman, J. and Scott, E. L. 1948. Consistent estimates based on partially consistent observations. *Econometrika, 16*, 1–32.

Samejima, F. 1969. Estimating of latent ability using a response pattern of graded scores. *Psychometrika Monograph Supplement*, No. 17.

Stroud, A. H. and Secrest, D. 1966. *Gaussian Quadrature Formulas*. Englewood Cliffs, New Jersey: Prentice-Hall.

Wright, B. D. and Masters, G. 1982. *Rating Scale Analysis*. Chicago, Illinois: MESA Press.

9

Sequential Models for Ordered Responses

Gerhard Tutz

CONTENTS

9.1 Introduction 139
9.2 Presentation of the Model 140
  9.2.1 Item Steps and Response: The Sequential Mechanism 140
  9.2.2 Modeling of Steps 141
  9.2.3 An Alternative Concept of Steps: Partial Credit Model 143
  9.2.4 The Sequential Model and the Graded Response Model 144
9.3 Parameter Estimation 144
  9.3.1 Joint Maximum Likelihood 145
  9.3.2 Marginal Maximum Likelihood 146
9.4 Model Fit 147
9.5 Empirical Example 148
9.6 Discussion 149
References 150

9.1 Introduction

Early item response models were concerned with the simple dichotomous case where items were either solved or not. But items often carry more information when the response is in ordered categories with higher categories meaning better performance. Various models that exploit the information in ordered categories have been proposed in recent years. The model considered here is appropriate for items that are solved in a stepwise manner such that higher levels can be reached only if previous levels were reached earlier. Items of this structure are quite common, in particular, in ability tests. A simple example that illustrates items of this type is as follows.

Wright and Masters (1982) consider the item $\sqrt{9.0/0.3-5}=?$ Three levels of performance may be distinguished: no subproblem solved (level 0), $9.0/0.3=30$ solved (level 1), $30-5=25$ solved (level 2), and $\sqrt{25}=5$ (level 3). The important feature is that each level in a solution to the problem can be reached only if the previous level has been reached.

The main feature of the *sequential* or *step model* is that the stepwise manner of solving an item is exploited. The transition from one level to the next is modeled and item parameters are interpreted directly in terms of the difficulty of this transition. It should be emphasized that for sequential models, steps are *consecutive steps* in problem solving; the next step in the solution to a problem can be performed successfully only if the previous steps were completed successfully. Item difficulties explicitly refer to the difficulty of solving single steps.

An early reference to the explicit modeling of transitions in item response theory is Molenaar (1983). In Tutz (1990), a parametric model was introduced under the name "sequential model"; Verhelst et al. (1997) called the corresponding model a "step model." The model is strongly related to the general class of continuation ratio models, which are in common use in regression modeling with categorical ordered response; see, for example, Agresti (2002) and Tutz (2012). A quite different approach to the modeling of steps is the partial credit model (Volume One, Chapter 7), which is related to the adjacent categories approach. Differences between the two types of models will be discussed later.

9.2 Presentation of the Model

Let the graded response for item $i \in \{1, \ldots I\}$ and person $p \in \{1, \ldots P\}$ be given by the response variable $U_{pi} \in \{0, \ldots A_i\}$. Thus, item i has levels $0, \ldots, A_i$, where 0 stands for the lowest and A_i for the highest level of performance.

9.2.1 Item Steps and Response: The Sequential Mechanism

The basic assumption is that each item is solved step by step. Let U_{pia}, $a = 1, \ldots, A_i$, denote the step from level $a - 1$ to level a, where $U_{pia} = 1$ stands for successful transition and $U_{pia} = 0$ denotes an unsuccessful transition. Let us consider the response process step by step.

Step 1. The process always starts at level 0. If the transition to level 1 fails, that is, $U_{pi1} = 0$, the process stops and the person's score is $U_{pi} = 0$. If the transition is successful, the person's score is at least level 1. Thus, one has

$$U_{pi} = 0 \quad \text{if } U_{pi1} = 0.$$

Step 2. If the first step was successful, that is, $U_{pi} = 1$, the person tries to make the second step, from level 1 to level 2. Successful transition yields a response $U_{pi} \geq 2$. If the person fails at the second step, the process stops. Thus, one obtains

$$\{U_{pi} = 1 \text{ given } U_{pi} \geq 1\} \quad \text{if } U_{pi2} = 0.$$

The conditioning on $U_{pi} \geq 1$ is essential because the second step would become relevant only if the first step was successful ($U_{pi1} = 1$).

Step $a + 1$. Step $a + 1$ is under consideration only if all previous steps were successful, that is, $U_{pi1} = \ldots = U_{pia} = 1$, or equivalently, $U_{pi} \geq a$. In the same way as for previous steps, the final response is a when the transition to $a + 1$ fails:

$$\{U_{pi} = a \text{ given } U_{pi} \geq a\} \quad \text{if } U_{pi,a+1} = 0. \tag{9.1}$$

Equation 9.1 gives the sequential response mechanism, which is based solely on the assumption of a step-by-step process. Figure 9.1 illustrates the process for an item with categories 0–4.

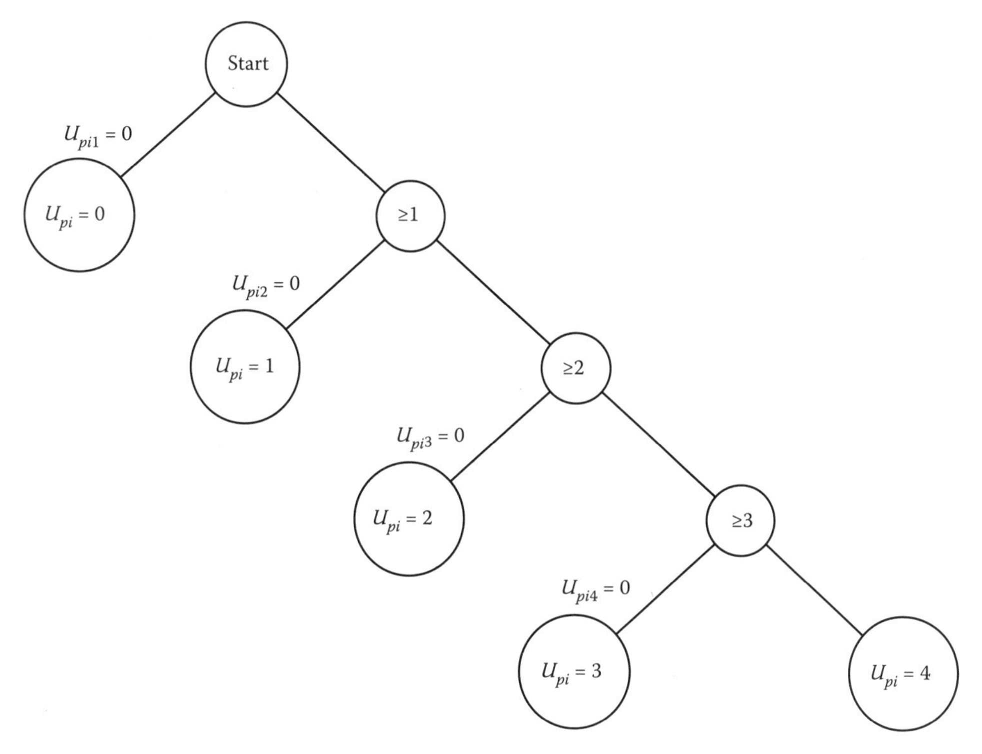

FIGURE 9.1
Tree for item with Categories 0, 1, 2, 3, 4.

9.2.2 Modeling of Steps

To obtain a model that links observed scores and abilities, the steps themselves are modeled parametrically. Since the transition is a binary process, one option is to use for the conditional transition to category $a + 1$, given that level a has been reached the binary model

$$\Pr\{U_{pi,a+1} = 1\} = F(\theta_p - b_{i,a+1}), \quad a = 0, \ldots, A_i - 1,$$

where θ_p is the ability of the person p, $b_{i,a+1}$ is the item difficulty for the step $a + 1$, and $F(.)$ is a response function, typically chosen to be a distribution function (Albert, Volume Two, Chapter 1). It is noteworthy that the model contains one parameter for the ability of the person, but $A_i - 1$ item difficulties, one for each step. The item difficulty b_{i1} represents the difficulty for the first step, that is, the transition from category 0 to category 1, the item difficulty b_{i2} represents the difficulty for the second step, etc.

If F is the logistic function $F(x) = \exp(x)/(1 + \exp(x))$, each step is modeled by the dichotomous Rasch model. However, using alternative response functions one obtains a much larger class of models.

Using the equivalence of $U_{pi,a+1} = 1$ and $U_{pi} > a$ given $U_{pi} \geq a$, the sequential model (with response function F) has the closed form

$$\Pr\{U_{pi} > a | U_{ij} \geq a\} = F(\theta_p - b_{i,a+1}), \quad a = 0, \ldots, A_i - 1. \tag{9.2}$$

The probability of response in category a is given by

$$\Pr\{U_{pi}=a\}=\begin{cases}\prod_{s=0}^{a-1}F(\theta_p-b_{i,s+1})(1-F(\theta_p-b_{i,a+1})), & \text{for } a=0,\ldots,A_i-1\\ \prod_{s=0}^{A_i-1}F(\theta_j-b_{i,s+1}) & \text{for } a=A_i.\end{cases} \tag{9.3}$$

Equation 9.3 has a simple interpretation: the person's response is in category $a < A_i$ if the a steps $U_{pi1}, \ldots, U_{pia}$, were successful but the $(a + 1)$th step was not. Category A_i is obtained if all of the steps, $U_{pi1}, \ldots, U_{pi,Ai}$ were successful. If one chooses the logistic distribution function F (i.e., the Rasch model) for the transitions, the model is called a *sequential* or *stepwise Rasch model*. Figure 9.2 shows the response functions in the case of four categories in an item. The first picture shows the probabilities for scores of 0, 1, 2, and 3 in the case of equal item (step) difficulties $b_{i1} = b_{i2} = b_{i3} = 1.0$, which means a person with ability $\theta_p = 1.0$ has a

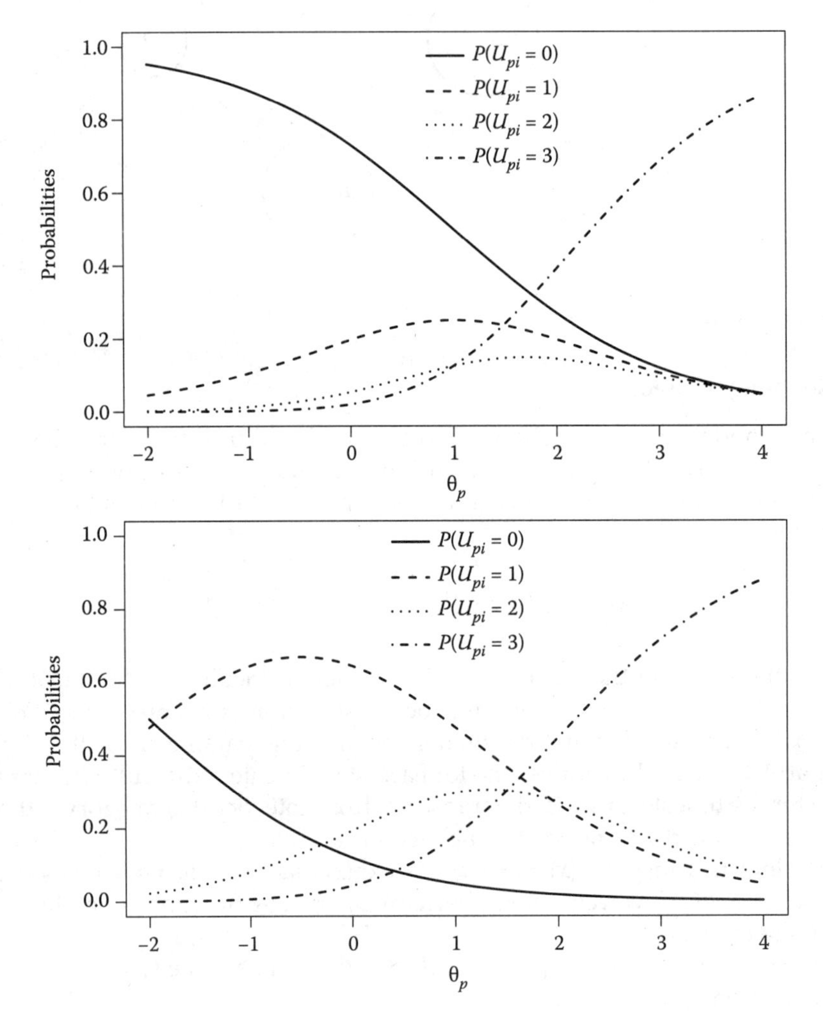

FIGURE 9.2
Category probabilities for two items, the first with $b_{i1} = b_{i2} = b_{i3} = 0$, the second with $b_{i1} = 0.0$, $b_{i2} = -1.0$, $b_{i3} = 2.0$ (numbers denote the category).

probability of 0.5 in each step. As is seen, for this person $\Pr\{U_{pi} = 0\} = 0.5$ and probabilities for the higher score levels of 2 and 3 are very low. In the second picture, the first step is easy with $b_{i1} = -2.0$ and the third step is more difficult to perform with $b_{i3} = 1.5$. The second threshold is unchanged ($b_{i2} = 1.0$). Consequently, probabilities for categories 1 and 2 are much higher now. For category 3, the probability is rather similar to that in the first picture, the reason being that, although the threshold b_{i3} is higher, now the increased probability of categories 1 and 2 have to be taken into account.

Interpretation of the model parameters is simple and straightforward. Since the model is derived from a stepwise mechanism, interpretation may be done in terms of dichotomous response models, which are used to model the transitions. For example, in the stepwise Rasch model, θ_p is the usual ability of a person that determines the solving of the problems behind each step. If the item difficulty of the specific step b_{ia} is small its corresponding subproblem is easily solved, whereas for large b_{ia} the specific step is hard to perform. It should be noted that the model is unidimensional. There is a latent scale for measuring abilities and item difficulties; a unidimensional ability is measured. The ordered response is used to measure that ability more precisely than is possible with dichotomously scored items.

9.2.3 An Alternative Concept of Steps: Partial Credit Model

Masters (Volume One, Chapter 7) considered the so-called partial credit model. It can be given in several forms. A form that makes the underlying concept of "steps" most obvious is

$$\Pr\{U_{pi} = a | U_{pi} \in \{a-1, a\}\} = \exp\frac{(\theta_p - b_{ia})}{(1+\exp(\theta_p - b_{ia}))},$$

where U_{pi} denotes the response, θ_p the ability of person p, and b_{ia} the difficulty of "step a." It is seen that the response in category a, given either category $a-1$ or a is taken, is specified as a dichotomous Rasch model. However, the difficulty b_{ia} is *not* the difficulty of a Rasch model for the step from $a-1$ to a given at least level $a-1$ is reached, which is the difficulty in sequential models. The step modeled by the partial credit model is a local step: it is conditional on the response being in category $a-1$ or a. Therefore, it is not suitable to model sequential solution strategies.

Suppose a person solves $9.0/0.3 = 30$ (level 1) in the example in the introduction. The next step is solving the second subproblem, $(30 - 5 = 25)$. Assuming a model like the Rasch model for this subproblem corresponds to the sequential approach, whereas the partial credit model assumes a Rasch model given that the first and the second subproblem but not the third are solved. Therefore, not the next step (solving the second subproblem) is determined by a Rasch Model. Although Masters (1982) refers to the binary decision as "how likely it is that a person will make a 2 rather than a 1," he omits that the decision is conditional on the response being in category 1 or 2. Actually, a conditional step is modeled in the partial credit model framework and this should be clearly distinguished from modeling the process of the sequential solving of subproblems.

In the regression context, logit type models that condition on adjacent categories as the partial credit model does are known as *adjacent categories logits* (Agresti, 2002). For the partial credit model the adjacent categories logits have the form

$$\log\left(\frac{\Pr\{U_{pi} = a\}}{\Pr\{U_{pi} = a-1\}}\right) = \theta_p - b_{ia},$$

which shows that the log-rate of adjacent categories is modeled instead of the next step. The partial credit model is well suited for the analysis of rating scales but less appropriate for modeling the process of the subsequent solving of subproblems. However, for items like $\sqrt{9.0/0.3-5}=?$ the sequential model does more adequately capture the underlying process; see also Verhelst and Verstralen (2008).

9.2.4 The Sequential Model and the Graded Response Model

Samejima (Volume One, Chapter 6) proposed the graded response model. A special form that aims at equal discrimination (Andrich, 1978) is

$$\Pr\{U_{pi} \geq a\} = F(\theta_p - \tilde{b}_{ia}), \quad a = 1,\ldots,A_i,$$

where $F(.)$ is a distribution function and the item parameters are ordered as $-\infty = \tilde{b}_{i0} < \cdots < \tilde{b}_{iA_i}$. Again, $F(.)$ is typically assumed to be of a logistic form, while item parameters may be interpreted as thresholds on the ability scale. Response $U_{pi} = a$ is observed when the ability, randomized by the addition of a noise variable, lies between thresholds b_{ia} and $b_{i,a+1}$. Regression models of this form are known as *cumulative probability models* and have been propagated in particular by McCullagh (1980).

There is an interesting relation between the sequential model and the cumulative type model. Let the response function in the sequential model in Equation 9.2 be given as the extreme-maximal-value distribution function $F(x) = \exp(-\exp(-x))$. Simple derivation shows that the sequential model

$$\Pr\{U_{pi} > a | U_{pi} \geq a\} = \mathrm{F}(\theta_p - b_{i,a+1})$$

is equivalent to the cumulative model

$$\Pr\{U_{pi} \geq a\} = F(\theta_p - \tilde{b}_{ia}).$$

The transformation of item parameters is given by $\tilde{b}_{ia} = \log(e^{b_{i1}} + \cdots + e^{b_{ia}})$, and the transformed item parameters $\tilde{b}_{ia}$ may again be interpreted as thresholds on a latent continuum for which $-\infty = \tilde{b}_{i0} < \cdots < \tilde{b}_{iA_i}$ holds. Thus, for the special case of the extreme-maximal value distribution function (and only for this case), the stepwise approach is equivalent to the cumulative approach (Tutz, 1991). A formal proof that the equivalence of the sequential and the graded response model holds only for the extreme-maximal-value distribution was given by Bechger and Akkermans (2001).

9.3 Parameter Estimation

In the following, two estimation methods are considered namely joint maximum-likelihood estimation and marginal estimation based on a prior distribution. An alternative method often used for dichotomous Rasch models is conditional maximum-likelihood estimation. Although this method may be used with some modification (Tutz, 1990), it is not considered here.

9.3.1 Joint Maximum Likelihood

The joint (or unconditional) likelihood is equivalent to the likelihood of conditionally independent dichotomous responses following the dichotomous Rasch model. If person p attempts item i, a response in one of the categories 0, …, A_i is obtained. Usually, multicategorical responses are given in dummy coding with

$$x_{pia} = \begin{cases} 1 & \text{response of person } p \text{ to item } i \text{ in category } a \\ 0 & \text{otherwise,} \end{cases}$$

and $a = 0, \ldots A_i$. Thus, the vector $x_{pi}^T = (x_{pio}, \ldots, x_{piA_i})$ with just one '1' as component is obtained. The kernel of the log-likelihood contribution from item i and person p has the common form

$$l_{pi} = \sum_{a=0}^{A_i} x_{pia} \log \pi_{pia}, \tag{9.4}$$

where $\pi_{pia} = \Pr\{U_{pi} = a\}$ is the probability of a response in category a. Here, l_{pi} is an abbreviation for $\log(L_{pi})$, where L_{pi} is the likelihood contributed by person p attempting item i. For the sequential model, it is useful to recode the responses in terms of the transition process given by $U_{pi1}, \ldots$ The essential term in Equation 9.2 is

$$\gamma_{\text{pi,a+1}} = \Pr\{U_{pi} > a | \, U_{pi} \geq a\} = \Pr\{U_{pi,a+1} = 1 | \, U_{pi1} = \cdots = U_{pia} = 1\},$$

which represents the probability of the transition to category $a + 1$ given all previous transitions were successful. The probability in Equation 9.4 can be written as

$$\pi_{pia} = \Pr\{U_{pi} = a | \, U_{pi} \geq a\} \prod_{s=0}^{a-1} P(U_{pi} > s | \, U_{pi} \geq s) = (1 - \gamma_{pi,a+1}) \prod_{s=1}^{a} \gamma_{pis}.$$

Alternatively, instead of $(u_{pi0} \ldots, u_{piA_i})$ the coding of transitions can be given by

$$u_{pia} = 1 - (x_{pi0} + \cdots + x_{pi,a-1}), \quad a = 1, \ldots, A_i.$$

It thus holds that $u_{pia} = 1$ when a transition to category a takes place and $u_{pia} = 0$ otherwise. For example, a response in category 3 is recorded as $x_{pi} = (0, 0, 0, 1, 0, \ldots, 0)$ in observation coding, but becomes $u_{pi} = (1, 1, 1, 0, \ldots, 0)$ in transition coding. If $A_i = 3$ and the response is in category 3, one has $x_{pi} = (0, 0, 0, 1)$ in observation coding and $u_{pi} = (1, 1, 1)$ when coding transition data. Using this recoding, upon some algebra, it can be shown that the log-likelihood in Equation 9.4 has the form

$$\begin{aligned} l_{pi} &= \sum_{s=0}^{A_i-1} x_{pis} \log(1 - \gamma_{pi,s+1}) + (1 - x_{pi0} - \cdots - x_{pis}) \log \gamma_{pi,s+1} \\ &= \sum_{s=1}^{\min(r_{pi}+1, A_i)} u_{pis} \log \gamma_{pis} + (1 - u_{pis}) \log(1 - \gamma_{pis}), \end{aligned} \tag{9.5}$$

where r_{pi} is the observed response category for individual p to item i. Since

$$\gamma_{\text{pis}} = \Pr\{U_{pi} > s-1 \mid U_{pi} \geq s-1\} = F(\theta_{\text{p}} - \text{b}_{\text{is}}) \tag{9.6}$$

represents the probability that the s-th step is successful (transition from $s-1$ to s), l_{pi} is equivalent to the contribution to the log-likelihood of the dichotomous response model in Equation 9.6 when person p meets the pseudo-items transition from $s-1$ to s in item i for $s = 1, \ldots, s_{pi} = \min(r_{pi}+1, A_i)$. The number of pseudo-items is $s_{pi} = \min(r_{pi}+1, A_i)$; if $r_{pi} < A_i$ the r_{pi} successful transitions and the last failing transition have to be included; if $r_{pi} = A_i$ all the A_i transitions were successful and each of them has to be included. Thus, the total log-likelihood for persons $p = 1, \ldots, P$ and items $i = 1, \ldots, I$ given by

$$\log L = \sum_{pi} l_{pi}$$

is equivalent to the dichotomous transition model in Equation 9.6 where person p has observations $u_{pi1}, \ldots, u_{pis_{pi}}$, $i = 1, \ldots, I$. It should be noted that the items that have to be included in the log-likelihood depend on the person and the design is thus incomplete. The existence and uniqueness of estimates of incomplete designs was investigated in Fischer (1981). This embedding into the framework of dichotomous response models makes things easier. By considering transitions and the corresponding pseudo-items, standard software and inference procedures developed for dichotomous models can be used.

To use programs for regression models, the regressor term $\theta_p - b_{is}$ should have the usual linear form $w_{pis}^T\beta$, where β is an unknown parameter. The design vector corresponding to observation u_{pis} is given by $w_{pis}^T = (1(p)^T, z_{is})$ where the unit vector $1(p) = (0, \ldots, 1, \ldots, 0)^T$ with 1 in column p contains the coding of person p and the item is coded by $z_{is}^T = \left(z_{11}^{is}, \ldots, z_{1A_1}^{is}, z_{21}^{is}, \ldots, z_{IA_I}^{is}\right)$, where $z_{lm}^{is} = 1$ if $(i, s) = (l, m)$ and $z_{lm}^{is} = 0$, otherwise.

9.3.2 Marginal Maximum Likelihood

The basic idea of marginal maximum-likelihood estimation is to consider the person ability θ_{p} as a random effect drawn from a distribution with density $g(\theta)$. Then integration over $g(\theta)$ yields the marginal likelihood.

Using Equation 9.5, the marginal likelihood for person p is given by

$$L_p^m = \int \prod_{i=1}^{I} \prod_{s=1}^{s_{pi}} F(\theta - b_{is})^{u_{pis}} (1 - F(\theta - b_{is}))^{1-u_{pis}} g(\theta) d\theta. \tag{9.7}$$

The marginal likelihood is again given in the form of dichotomous response models, where instead of items $1, \ldots, I$ the pseudo-items corresponding to transitions $u_{pi1}, \ldots, u_{pis_{pi}}$, $i = 1, \ldots, I$, are considered.

In the same way as for the joint estimates in the sequential Rasch model the methods for the dichotomous model may be applied (Bock and Aitkin, 1981; Thissen, 1982). A helpful device is the embedding of the estimation procedure in the framework of generalized random-effect regression models, which allows for the use of the new methods and programs developed in this area.

The parameters to be estimated may be summarized as $b^{\mathrm{T}} = \left(b_1^{\mathrm{T}},\ldots,b_I^{\mathrm{T}}\right)$, where $b_i^{\mathrm{T}} = (b_{il},\ldots,b_{iA_i})$ are the difficulty parameters associated with the transition from one level to another for item i. The corresponding design vector when transition from level $s-1$ to level s for item i is considered may be chosen as $z_{is}^T = \left(z_{11}^{is},\ldots,z_{1A_1}^{is},z_{21}^{is},\ldots,z_{IA_I}^{is}\right)$, where $z_{lm}^{is}=1$ if $(i, s) = (l, m)$ and $z_{lm}^{is}=0$, otherwise.

Consequently, the term in Equation 9.7 that corresponds to the linear influence term in regression models may be written as $\theta - b_{is} = \theta + z_{is}^T b$. This assumption means for person p with random effect θ_p we have for $s = 1, \ldots, s_{ip}$, $i = 1, \ldots, I$, the dichotomous response u_{pis} and linear regression term $\theta_p + z_{is}^T b$, and therefore the generalized random effects model $P(U_{ips}=1) = F\left(\theta_p + z_{is}^T b\right)$, $\theta_p \sim N(0, \sigma^2)$ holds.

For the maximization of the corresponding marginal likelihood

$$l = \sum_p \log L_p^m$$

with respect to the unknown parameters b (item difficulties) and σ (heterogeneity of individuals) various procedures are available. The integral in L_p^m can be solved by Gauss–Hermite integration or Monte Carlo methods. Procedures for the estimation of nonrandom generalized linear models may be effectively used for the construction of the maximization algorithm. For details of the Gauss–Hermite procedure, see Anderson and Aitkin (1985) and Anderson and Hinde (1988). A procedure that may reduce the number of quadrature points is adaptive Gauss–Hermite quadrature (Liu and Pierce, 1994; Pinheiro and Bates, 1995). Alternatively, estimation of b and σ may be based on their posterior modes implying the use of an EM type algorithm (Stiratelli et al., 1984) or Laplace approximation (Breslow and Clayton, 1993); both yield the same estimates of item and person parameters. Details of these methods may be found, for example, in Fahrmeir and Tutz (Chapter 7). A general framework for mixed models in item response theory was given by Rijmen et al. (2003). The decomposition into binary data also allows one to use programs that are designed for generalized linear mixed models; see De Boek et al. (2011).

9.4 Model Fit

Goodness of fit checks may be based on the dichotomous nature of transitions to higher levels. The basic idea is to consider only persons who have reached at least level $a_1 \ldots, a_I$ for items $1 \ldots, I$. Further, assume that the transitions from a_1 to $a_1+1, \ldots, a_n$ to a_n+1 are determined by a dichotomous Rasch model. For this dichotomous model, its local goodness of fit may be checked by all methods given in Hambleton and Swaminathan (1985), Andersen (1973), and Kelderman (1984). As an example, consider the log-likelihood ratio statistic, which is the basic statistic investigated in Kelderman (1984). It has the form

$$G^2(m,m^*) = 2\sum m^* \log\left(\frac{m^*}{m}\right),$$

where m and m^* stand for the expected counts of model M and M^*, the former being a submodel of the latter. Currently, M stands for the Rasch model and M^* may be chosen as the saturated model, which does not restrict the data.

The log-likelihood ratio statistic gives a measure for the goodness of fit of the transitions from a_i to $a_i + 1$, $i = 1, \ldots, I$. Comparison with the asymptotic chi-square distribution shows if the model is appropriate for these steps. Of course, testing all the possible combinations of binary transitions yields $A_1 \ldots A_I$ tests. The resulting multiple test procedure makes it necessary to adjust the significance level, for example, by controlling its false discovery rate (Benjamini and Hochberg, 1995). By considering all the possible combinations of transitions each step is tested several times. Therefore, one may consider only specific groups of transitions. For example, one may build a group for the transitions from categories 0 to 1 for all items, from 1 to 2 for all items, etc. Then, the number of transition groups is the maximal number of item steps. If $A_i = A$, one has only A goodness-of-fit statistics G^2 corresponding to the steps in the items. This procedure is also used in the example given in Section 9.5.

9.5 Empirical Example

The example refers to the solution of problems in statistics. Six items with three levels were presented to 99 students. For most of the items the first step was to find the proper statistical tool (e.g., Wilcoxon test) and the second step was to carry out the computations and interpret the results. For some of the items finding a statistical procedure and using it was considered as the first step, the second step being some additional question, that is, concerning the invariance properties of the used procedure. The fitted model was the sequential Rasch model with F being the logistic distribution function. The parameters involved were the abilities θ_p, $p = 1, \ldots, 99$ and the item difficulties b_{i1}, b_{i2}, $i = 1, \ldots, 6$. Instead of joint maximum-likelihood estimation, the current focus is on marginal estimation where only the 12-item parameters need to be estimated. The estimation procedures considered were posterior mode estimation and Gauss–Hermite integration with six and eight quadrature points (GH(6) and GH(8), respectively). The underlying assumption was that the abilities were drawn from $N(0, \sigma^2)$. Table 9.1 shows the estimated item difficulties with standard errors given in brackets.

The results of the marginal estimation procedures were quite similar, indicating that six–eight quadrature points seemed to be enough. As is seen from the estimated standard deviation $\hat{\sigma}$, heterogeneity across individuals was quite large: consider the first item with $b_{11} = -0.527$ and $b_{12} = -0.279$. Since $b_{11} < b_{12}$, the difficulty of the first step was lower than that of the second step. For a person with $\theta = 0$ the probability of performing the first step successfully was $F(0.527) = 0.845$, while the probability for the second step was $F(0.279) = 0.570$. The resulting probabilities of categories 0, 1, 2 for this person were 0.155, 0.363, 0.489, respectively. The principle that the first step is easier than the second step held for items 1, 3, 4, 5, 6, but not for item 2. The difficulty of the second step for item 2 was very low and the probability of successful performance is $F(1.448) = 0.810$. From the estimated standard error of parameters, it is seen that $\sigma(b_{i1}) < \sigma(b_{i2})$ for each item. This is a natural consequence of the stepwise solution of items. All of the individuals faced the first step whereas the second step was considered only by individuals who successfully completed

TABLE 9.1

Marginal Estimates Based on Gauss–Hermite Quadrature (GH(6) and GH(8)) and Posterior Modes

	GH(6)	GH(8)	Posterior Mode
b_{11}	−0.527 (0.276)	−0.530 (0.294)	−0.504
b_{12}	−0.279 (0.367)	−0.285 (0.394)	−0.297
b_{21}	−0.365 (0.295)	−0.369 (0.315)	−0.357
b_{22}	−1.448 (0.471)	−0.452 (0.503)	−0.385
b_{31}	−0.365 (0.288)	−0.369 (0.307)	−0.357
b_{32}	−0.421 (0.414)	−0.424 (0.439)	−0.426
b_{41}	−0.692 (0.292)	−0.695 (0.313)	−0.655
b_{42}	0.080 (0.344)	0.077 (0.367)	0.036
b_{51}	−0.748 (0.290)	−0.751 (0.314)	−0.706
b_{52}	0.958 (0.351)	0.955 (0.369)	0.838
b_{61}	−0.979 (0.322)	−0.983 (0.347)	−0.917
b_{62}	0.233 (0.330)	0.233 (0.352)	0.181
$\hat{\sigma}$	1.212 (0.140)	−0.209 (0.151)	1.135

step 1. Consequently, the precision of item parameter estimates for the second step was lower.

Goodness of fit was checked for the two steps separately. First, all individuals were included and only the first steps for the items were considered. For this dichotomous Rasch model, the likelihood-ratio test versus the saturated model yielded a value of 36.199 on 56-degrees of freedom (Kelderman, 1984). Second, for the 22 individuals who succeeded in the first step of each item, the second step was checked yielding the likelihood-ratio statistic 17.720. However, the use of the asymptotic distribution of the likelihood-ratio test is not trustworthy for only 99 individuals. Therefore, the mixed Rasch model approach (Rost, 1991) was used as additional check. For the assumption of two latent classes, the likelihood-ratio statistics were 4.420 (first steps) and 4.553 (second steps). The difference between the one group model and the two-class model was 31.779 (first steps) and 12.767 (second steps) on eight-degrees of freedom. Hence, the deviances with respect to the two-class model suggested that it was doubtful if the Rasch model held for the first steps of all items. All computations of the goodness of fit checks as well as conditional estimates were done with MIRA (Rost and Davier, 1992).

9.6 Discussion

The sequential model is suited for the special type of items that are successively solved. This stepwise character of items should be taken seriously and directly modeled by considering the conditional probabilities of transitions to higher levels. The reduction to dichotomous transitions makes much of the theory and computational tools already developed for this case accessible. The accessibility compensates for the lack of sufficient statistics, which makes conditional maximum-likelihood estimates harder to obtain.

References

Agresti, A. 2002. *Categorical Data Analysis*. New York: Wiley.

Andersen, E. B. 1973. A goodness of fit test for the Rasch model. *Psychometrika*, 38, 123–139.

Anderson, D. A. and Aitkin, M. 1985. Variance component models with binary responses: Interviewer variability. *Journal of the Royal Statistical Society B*, 47, 203–210.

Anderson, D. A. and Hinde, J. 1988. Random effects in generalized linear models and the EM algorithm. *Communication in Statistics-Theory and Methods*, 17, 3847–3856.

Andrich, D. A. 1978. A rating formulation for ordered response categories. *Psychometrika*, 43, 561–573.

Benjamini, Y. and Hochberg, Y. 1995. Controlling the false discovery rate: A practical and powerful approach to multiple testing. *Journal of the Royal Statistical Society*, 57, 289–300.

Bechger, T. M. and Akkermans, W. 2001. A note on the equivalence of the graded response model and the sequential model. *Psychometrika*, 66, 461–463.

Bock, R. D. and Aitkin, M. 1981. Marginal maximum likelihood estimation of item parameters: Application of an EM algorithm. *Psychometrika*, 46, 443–459.

Breslow, N. E. and Clayton, D. G. 1993. Approximate inference in generalized linear mixed model. *Journal of the American Statistical Association*, 88, 9–25.

De Boeck, P., Bakker, M., Zwitser, R., Nivard, M., Hofman, A., Tuerlinckx, F., and Partchev, I. 2011. The estimation of item response models with the lmer function from the lme4 package in R. *Journal of Statistical Software*, 39, 1–28.

Fahrmeir, L. and Tutz, G. 2001. *Multivariate Statistical Modelling Based on Generalized Linear Models*. New York: Springer.

Fischer, G. H. 1981. On the existence and uniqueness of maximum-likelihood estimates in the Rasch model. *Psychometrika*, 46, 59–77.

Hambleton, R. K. and Swaminathan, H. 1985. *Item Response Theory: Principles and Applications*. Boston: Kluwer Academic Publishing.

Kelderman, H. 1984. Loglinear Rasch model tests. *Psychometrika*, 49, 223–245.

Liu, Q. and Pierce, D. A. 1994. A note on Gauss-Hermite quadrature. *Biometrika*, 81, 624–629.

Masters, G. N. 1982. A Rasch model for partial credit scoring. *Psychometrika*, 47, 149–174.

McCullagh, P. 1980. Regression models for ordinal data. *Journal of the Royal Statistical Society A*, 135, 370–384.

Molenaar, I. W. 1983. Item steps (Report No. HB-83-630-EX). Heymans Bulletins Psychologische Instituten R. U. Groningen.

Pinheiro, J. C. and Bates, D. M. 1995. Approximations to the log-likelihood function in the nonlinear mixed-effects model. *Journal of Computational and Graphical Statistics*, 4, 12–35.

Rost, J. 1991. A logistic mixture distribution model for polychotomous item responses. *British Journal of Mathematical and Statistical Psychology*, 44, 75–92.

Rost, J. and Davier, M.v. 1992. *MIRA–A PC Program for the Mixed Rasch Model* (user Manual). Kiel, Germany: IPN-Institute for Science Education.

Rijmen, F., Tuerlinckx, F., De Boeck, P., and Kuppens, P. 2003. A nonlinear mixed model framework for item response theory. *Psychological Methods*, 8, 185–204.

Samejima, F. 1969. Estimation of latent ability using a response pattern of graded scores. *Psychometrika Monograph Supplement No. 17.*

Stiratelli, R., Laird, N., and Ware J. H. 1984. Random-effects models for serial observation with binary response. *Biometrics*, 40, 961–971.

Thissen, D. M. 1982. Marginal maximum likelihood estimation for the one-parameter logistic model. *Psychometrika*, 47, 175–186.

Tutz, G. 1990. Sequential item response models with an ordered response. *British Journal of Mathematical and Statistical Psychology*, 43, 39–55.

Tutz, G. 1991. Sequential models in categorical regression. *Computational Statistics and Data Analysis*, 11, 275–295.

Tutz, G. 2012. *Regression for Categorical Data*. Cambridge: Cambridge University Press.
Verhelst, N. D., Glas, C., and de Vries, H. 1997. A steps model to analyse partial credit. In W. J. van der Linden and R. Hambleton (Eds.), *Handbook of Modern Item Response Theory*. New York: Springer.
Verhelst, N. D. and Verstralen, H. 2008. Some considerations on the partial credit model. *Psicologica*, 29, 229–254.
Wright, B. D. and Masters, G. N. 1982. *Rating Scale Analysis*. Chicago: MESA Press.

10

Models for Continuous Responses

Gideon J. Mellenbergh

CONTENTS

10.1 Introduction ... 153
10.2 Presentation of the Models ... 154
10.3 Parameter Estimation ... 157
 10.3.1 Item Parameters ... 157
 10.3.2 Person Parameters ... 157
 10.3.3 Precision of Estimates ... 157
 10.3.4 Observed Test-Score Precision ... 158
10.4 Model Fit ... 159
10.5 Empirical Example ... 159
10.6 Discussion ... 160
References ... 162

10.1 Introduction

Test takers give their responses to items on a response scale. This scale can be discrete or continuous. Examples of discrete scales are the dichotomous scale, which has two categories (e.g., correct or incorrect and yes or no), and the ordered polytomous scale that has more than two ordered categories (e.g., a five-point Likert scale). Examples of continuous response scales are a continuous line segment and time to complete a task.

Usually, continuous response scales are bounded. For example, a 10-cm line is bounded by 0 and 10 cm, and a time scale is bounded by 0 s and an upper bound of, for example, 120 s. The test taker's response is either the distance from a bound to his (her) mark on the line segment or the time he or she needed.

Two types of latent variable models were developed for discrete item responses. Latent trait models assume that one or more continuous latent variables causally affect test takers' responses to discrete items, while latent class models assume that discrete latent classes causally affect test takers' item responses.

Latent trait models for dichotomous item responses were generalized to models for discrete polytomous item responses. For example, Andrich (1978; Volume One, Chapter 5) generalized Rasch's (1960) model for dichotomous item responses to his rating scale model for ordinal-polytomous responses, and Samejima (1969; Volume One, Chapter 6) generalized Birnbaum's (1968) two-parameter logistic model to her graded response model for ordered polytomous responses.

In this tradition of latent variable models for discrete item responses, models for continuous item responses have been developed. Rasch (1960; see also Volume One, Chapter 15), presented a model for continuous reading time. Samejima (1973) extended her graded response model to responses to an open line segment, that is, a line segment, where the test taker is not allowed to mark the bounds of the line. Müller (1987) extended Andrich's rating scale model to responses to a closed line segment, where the test taker is allowed to mark the bounds of the line. Both Samejima and Müller used the same thought experiment to derive their models. The line segment is split into halves, each of these halves is halved again, and so on. The models for the continuous line segments are the limiting case of the graded response and rating scale model, respectively, where the halving process goes to infinity.

But, latent variable models for the analysis of continuous variables have also been developed in other traditions. In the beginning of the twentieth century, factor analytic models were developed for the analysis of correlations and covariances between continuous variables (Bartholomew, 1995). The application of factor analytic models to continuous item responses yields latent variable models for continuous items (McDonald, 1982; see also Volume One, Chapter 11). In the middle of the twentieth century, latent class models, known as latent profile models, were developed for the analysis of continuous variables (Gibson, 1959; Lazarsfeld and Henry, 1968, Chapter 8).

Thus, (continuous) latent variable and (discrete) latent class models exist for discrete as well as continuous item responses. Both have a number of common elements (Mellenbergh, 1994a, 2011, Chapter 7): First, they assume that one or more latent variables causally affect test takers' item scores; for a discussion of this assumption see Borsboom (2005, Chapter 3). Latent trait models can assume one or more than one latent trait. Models that assume one trait (or factor) are unidimensional response (or one-factor) models, and models that assume more than one latent trait are multidimensional (or multiple-factor) models.

Second, it is assumed that test taker p's response to an item has an item response distribution. For example, the Rasch (1960) and Birnbaum (1968) models for dichotomous items assume that the responses by test taker $p = 1, \ldots, P$ to items $i = 1, \ldots, n$ are Bernoulli distributed with expected value (probability of giving the correct or yes answer) π_{pi} and variance $\sigma_{pi}^2 = \pi_{pi}(1-\pi_{pi})$. This distribution is not observed, but can be constructed in different ways (Borsboom, 2005, Chapter 2; Holland, 1990). For example, Lord and Novick (1968, Section 2.2) used Lazarsfeld's thought experiment, where an item is assumed to be repeatedly administered to the same person and the person is brainwashed between test administrations, to construct the test taker's item response (propensity) distribution.

Third, it is assumed that test taker p's responses to the n items of a test are conditionally independent given the latent parameter (local independence). Some item response models make weaker assumptions, for example, they assume that the conditional item covariances of p's responses are equal to zero, whereas some other models specify specific relations between p's item responses.

Finally, an assumption is made on the regression of the test takers' responses on the latent variable(s). For example, the Rasch (1960; Volume One, Chapter 3) and Birnbaum (1968; Volume One, Chapter 2) models for dichotomous items assume that the regression of the observed item responses on the latent trait follows a logistic function (for alternative link functions, see Albert, Volume Two, Chapter 1).

10.2 Presentation of the Models

The models in this chapter specify the common elements of latent variable models in the previous section for the case of continuous item scores. First, it is assumed that D latent variables ($\Theta_1, \Theta_2, \ldots, \Theta_D$) causally affect test taker p's continuous item score. Second, it is assumed that test taker p's responses to the ith item are normally distributed with expected value τ_{pi} and variance σ_{pi}^2. Third, it is assumed that the test taker's responses are conditionally independently distributed, given $(\Theta_1 = \theta_{p1}, \ldots, \Theta_D = \theta_{pD})$. Finally, it is assumed that the regression of the item score on the latent variables can be modeled as a linear function of the latter:

$$E(U_{pi} \mid \Theta_1 = \theta_{p1}, \Theta_2 = \theta_{p2}, \ldots, \Theta_D = \theta_{pD}) = \tau_{pi} = b_i + a_{i1}\theta_{p1} + a_{i2}\theta_{p2} + \cdots + a_{iD}\theta_{pD} \quad (10.1)$$

where U_{pi} is the random score of the ith item by test taker p, b_i is the intercept, and $a_{i1}, a_{i2}, \ldots, a_{iD}$ are the slopes of the regression function of the ith item, $\theta_{p1}, \theta_{p2}, \ldots, \theta_{pD}$ are test taker p's values on the D latent variables, and τ_{pi} is test taker p's true item score.

Factor analytic and latent profile models are obtained from Equation 10.1 by specifying the latent variables and additional making assumptions on test takers' item score variances. The D-factor model is obtained by specifying the D latent variables as continuous latent traits and assuming that the item score variance is homogeneous with respect to test takers (i.e., $\sigma_{pi}^2 = \sigma_i^2$). The $(D + 1)$-class latent profile model is obtained from Equation 10.1 by specifying the D latent variables as latent dummy variables that indicate the test takers' latent class membership, assuming that the item score variance is homogeneous within each of the latent classes.

Spearman's one-factor model is the special case where only one latent trait is assumed. Setting $D = 1$ in Equation 10.1 and suppressing the latent variable subscript, yields the regression function of the one-factor model,

$$E(U_{pi} \mid \Theta = \theta_p) = \tau_{pi} = b_i + a_i\theta_p \quad (10.2)$$

with homogenous variance with respect to test takers,

$$Var(U_{pi} \mid \Theta = \theta_p) = \sigma_i^2 \quad (10.3)$$

The model is both Spearman's one-factor model for continuous variables and Jöreskog's (1971) model for congeneric measurements. In the remainder of this chapter, the model is called the congeneric model. θ_p is test taker p's factor score or latent trait parameter. The b parameters are comparable to the difficulty parameters of IRT models for intelligence and achievement test items, or the attractiveness parameters of IRT models for attitude and personality questionnaire items. If the latent trait is standardized with mean zero, the b parameter is equal to the mean item score. The a parameters are item factor loadings in factor analysis. They are comparable to the discrimination parameters of IRT models because they are the slopes of the regression functions of the item scores on the latent trait.

Special cases of the congeneric model are obtained by constraining the item parameters. Setting the a, b, and σ^2 parameters of the test items equal to each other ($a_i = a$, $b_i = b$, $\sigma_i^2 = \sigma^2$, $i = 1, 2, \ldots, n$) yields the version of the model for parallel items; setting only the a and b parameters equal to each other ($a_i = a$, $b_i = b$, $i = 1, 2, \ldots, n$) yields the version for tau-equivalent items, while equality of just the a parameters ($a_i = a$, $i = 1, 2, \ldots, n$) yields the model for essentially tau-equivalent items (Lord and Novick, 1968, Section 2.13). Also, an equality constraint on the a and σ^2 parameters ($a_i = a$, $\sigma_i^2 = \sigma^2$, $i = 1, 2, \ldots, n$) yields a model for continuous item scores that is the counterpart of the Rasch model for dichotomous item responses (Mulaik, 1972).

The two-class latent profile model is obtained from Equation 10.1 by specifying one latent dummy variable (Θ) and assuming that test takers' item score variances are homogeneous within either class: setting $D = 1$ and specifying $\Theta = 0, 1$ for the two classes yields

$$E(U_{pi} \mid \Theta = 1) = b_i + a_i = b_i' \tag{10.4}$$

and

$$E(U_{pi} \mid \Theta = 0) = b_i \tag{10.5}$$

Moreover, it is assumed that test takers' item score variances are homogeneous within each of the two latent classes; that is,

$$Var(U_{pi} \mid \Theta = 1) = \sigma_{1i}^2 \tag{10.6}$$

and

$$Var(U_{pi} \mid \Theta = 0) = \sigma_{0i}^2 \tag{10.7}$$

Equations 10.4 and 10.5 show that the model allows the mean scores on the ith item to differ between the two latent classes. Consequently, the profile of the means of the n test item scores can also differ between the two latent classes, which motivates the name "latent profile model."

A striking relation exists between factor analytic and latent profile models. For instance, McDonald (1967) noted the similarity of the one-factor model and the two-class latent profile model, while Bartholomew (1987, Section 2.4), Molenaar and Von Eye (1994), and Loken and Molenaar (2008) discussed more specific relationships between factor analytic and latent profile models. These authors showed that a D-factor model can be rewritten as a $(D + 1)$-class latent profile model with the same means, variances, and covariances as the factor model, and vice versa.

Factor analytic models are more common in the behavioral sciences than latent profile models. Therefore, this chapter focuses on factor analytic models. McDonald (1982) discussed a general class of factor analytic models for continuous item scores. The discussion of this chapter is restricted to the congeneric model because it is comparable to Birnbaum's (1968) two-parameter logistic model for dichotomous responses. Under the Birnbaum model, the logit transformation of test taker p's probability of giving the correct (yes) answer to the ith item yields

$$\ln\frac{\pi_{pi}}{1-\pi_{pi}} = a_i(b_i - \theta_p) = b_i^* + a_i^*\theta_p \tag{10.8}$$

where ln denotes the natural logarithm, $b_i^* = a_i b_i$, and $a_i^* = -a_i$. The two-parameter logistic model in Equation 10.8 has a similar structure as the congeneric model Equation 10.2. Both the equations are linear functions of one latent variable with the same parameterization.

The θ parameter of the congeneric model Equation 10.2 is not identified. A convenient way to fix the scale is to set the mean and variance of the latent trait in a population of persons equal to

$$E(\Theta) = 0 \tag{10.9}$$

and

$$Var(\Theta) = 1 \tag{10.10}$$

respectively. These scaling constraints are applied in this chapter, but other constraints can be used as well.

10.3 Parameter Estimation

The parameters of the congeneric model are assumed to be estimated from a sample of N_s test takers. The item parameters can be estimated from the item-score means, variances, and covariances, while the test takers latent values are to be estimated from their responses to the items.

10.3.1 Item Parameters

It follows from Equations 10.2 and 10.9 that the b parameters can be estimated from the sample means of the item scores. The a and σ^2 parameters are then estimated from the sample variances and inter-item covariances using, for example, the maximum-likelihood or generalized least-squares methods (Mulaik, 2009, Chapter 7).

10.3.2 Person Parameters

Test taker p's latent parameter is estimated from his or her n item scores. The maximum-likelihood estimator of the parameter is

$$\hat{\theta}_p = \frac{\sum_{i=1}^{n} a_i(u_{pi} - b_i)/\sigma_i^2}{\sum_{i=1}^{n} (a_i^2/\sigma_i^2)} \tag{10.11}$$

(Mellenbergh, 1994b), which is equal to Bartlett's least-squares estimator of p's factor score in a one-factor model (McDonald, 1982). It follows from Equations 10.2 and 10.11 that $\hat{\theta}_p$ is an unbiased estimator of θ_p.

10.3.3 Precision of Estimates

The precision of the person parameter estimate has two aspects (Mellenbergh, 1996). The first is the within-person variance of the test taker's parameter estimate, given his (her) latent-trait value. Second, when distinguishing between different persons it makes sense to evaluate their between-person precision.

The former is asymptotically equal to

$$Var(\hat{\theta}_p \mid \Theta = \theta_p) = \frac{1}{\sum_{i=1}^{n} (a_i^2/\sigma_i^2)} = \frac{1}{I} \tag{10.12}$$

(Mellenbergh, 1994b), where $I = \sum_{i=1}^{n}(a_i^2/\sigma_i^2) = \sum_{i=1}^{n} I_i$ is the Fisher's information in the test and $I_i = a_i^2\sigma_i^2$ the information in the item (Chang and Zhing, Volume Two, Chapter 7). These functions are constants independent of the latent parameter that is estimated. Observe that the test information is just the sum of the item information values.

The between-persons precision is typically assessed using a reliability coefficient for the population of test takers. In classical test theory, the reliability coefficient is defined as the squared product–moment correlation between observed and true test scores in a population of persons (Lord and Novick, 1968, Section 3.4). Analogously, the reliability of the latent-trait estimates can be defined as the squared product–moment correlation between estimated and true person parameter values in a population of persons:

$$Rel(\hat{\Theta}) = [Cor(\hat{\Theta}, \Theta)]^2 = \frac{I\,Var(\Theta)}{1 + I\,Var(\Theta)} \tag{10.13}$$

where *Rel* and *Cor* denote reliability and product–moment correlation, respectively (Mellenbergh, 1994b).

The reliability of the person parameter estimates contains the term $Var(\Theta)$, which can be set equal to one (Equation 10.10) as part of the identification of the model. However, the reliability coefficient, being a squared product–moment correlation, is actually invariant under a linear transformation of the scale of the parameter. Therefore, setting $Var(\Theta)$ equal to one does not change its value. However, as Equation 10.13 shows, the reliability may differ between populations or subpopulations. If the item parameters of the congeneric model are invariant across different populations, but the variance of the person parameter values differs between them, their reliability will differ as well. Generally, the reliability is smaller for populations with smaller variance.

The within-person variance and reliability are estimated by inserting estimates of the item parameters into Equations 10.12 and 10.13, respectively. The estimated within-person variance can be used to compute confidence intervals for the test takers' parameter estimates.

10.3.4 Observed Test-Score Precision

In the practice of testing, test taker p's observed test score (X_p) is sometimes used to estimate his (her) classical true test score (τ_p), where X_p is taken to be the unweighted sum of the item scores (i.e., $X_p = \sum_{i=1}^{n} u_{pi}$). The within-person variance of the observed score is its squared standard error of measurement, and its reliability is the squared product–moment correlation of observed and true scores.

The within-person variance and the reliability of the observed test score can be expressed in terms of the congeneric model parameters. It follows from Equation 10.3 and the local independence assumption that the within-person variance of p's observed score is

$$Var(X_p \mid \Theta = \theta_p) = \sum_{i=1}^{n} \sigma_i^2 \tag{10.14}$$

Similarly, the reliability of the observed score is

$$Rel(X) = \frac{\left(\sum_{i=1}^{n} a_i\right)^2 Var(\Theta)}{\sum_{i=1}^{n} \sigma_i^2 + \left(\sum_{i=1}^{n} a_i\right)^2 Var(\Theta)} \tag{10.15}$$

which is equal to McDonald's (1999, p. 89) coefficient omega for $Var(\Theta) = 1$.

The within-person variance and the reliability of the observed score can be estimated by inserting estimates of the item parameters into Equations 10.14 and 10.15, respectively. The estimated within-person variance can be used to compute confidence intervals of p's true test score.

10.4 Model Fit

If the item scores are from a multivariate normal distribution and the parameters are estimated by the maximum-likelihood or generalized least-squares methods, the fit of the congeneric model can be tested (Kaplan, 2000, Section 3.4). The null hypothesis of the congeneric model holding for the item variance–covariance matrix is tested against the alternative hypothesis of this matrix, being an arbitrary symmetric positive definite matrix, using an asymptotically chi-square distributed test statistic with $n(n-3)/2$ degrees of freedom. Moreover, a number of descriptive indices can be used to assess the fit of the model (Hu and Bentler, 1995; Kaplan, 2000, Section 6.1).

10.5 Empirical Example

Use of the congeneric model is illustrated for a subtest of the Dutch translation of Gough and Heilbrun's (1980) *Adjective Check List* (ACL). The items of the ACL are adjectives that describe different personality attributes. The original dichotomous response scale was replaced by a 57-millimeter line segment that runs from a lower endpoint (labeled "not at all relevant") to an upper end point (labeled "completely relevant"). This adapted version was administered to 236 psychology freshmen at the University of Amsterdam. The test takers' responses were recorded in millimeters.

The data of the 10-item *Aggression* subtest were used. Seven of the 10 items are indicative of aggression and three items are contraindicative. In the analysis, the scale of the latter was just reversed. Van den Berg (2002) fitted the congeneric model to the data using the maximum-likelihood estimation method in the *LISREL* program (Jöreskog and Sörbom, 1999). The mean and variance of the person parameters were set at 0 and 1, respectively.

The parameter estimates are given in Table 10.1, and are used to illustrate the computation of the precision indices.

The congeneric model was rejected by the chi-square statistic, indicating its lack of fit of to the data. Inspection of the residual covariances (i.e., differences between observed and fitted covariances) shows that item 1 (Patient) and 5 (Impatient) and item 2 (Calm) and 7

TABLE 10.1

Maximum-Likelihood Parameter Estimates, Congeneric Model, Aggression Subtest of the ACL ($N_s = 236$)

Item	Estimate			
	$\hat{b}$	$\hat{a}$	$\hat{\sigma}^2$	$\hat{a}^2/\hat{\sigma}^2$
1. Patient[a]	26.7	7.69	135.49	0.44
2. Calm[a]	23.9	6.28	182.30	0.22
3. Squabbling	15.5	7.37	101.33	0.54
4. Unpleasant	9.4	3.15	81.51	0.12
5. Impatient	26.7	8.96	140.14	0.57
6. Irritable	22.1	9.90	118.09	0.83
7. Quiet[a]	26.4	5.49	191.92	0.16
8. Argumentative	11.1	6.67	79.32	0.56
9. Easily annoyed	21.4	10.04	112.92	0.89
10. Vengeful	10.0	4.71	123.54	0.18
Sum		70.26	1266.56	4.51

[a] The response scale of these items was reversed.

(Quiet) had the largest residual covariances. Items 1 and 5 were antonyms and items 2 and 7 were synonyms, which was taken to account for a large part of the misfit.

Using Equations 10.12 and 10.13, the within-person variance and reliability of the latent-trait estimate were estimated to be $\hat{Var}(\hat{\theta}_p | \Theta = \theta_p) = 1/4.51 = 0.22$ and $\hat{Rel}(\hat{\Theta}) = 4.51/(1 + 4.51) = 0.82$, respectively. Similarly, using Equations 10.14 and 10.15, the within-person variance and reliability of the observed test score were estimated to be $\hat{Var}(X_p | \Theta = \theta_p) = 1266.56$ and $REL(X) = 70.26^2/(1266.56 + 70.26^2) = 0.80$, respectively.

10.6 Discussion

As remarked before, the parameter structure of the congeneric model for continuous item scores is similar to that of the slope–intercept version of Birnbaum's (1968) two-parameter logistic model for dichotomous item responses (cf. Equations 10.2 and 10.8). This similarity also holds for the generalizations of these two models. The generalization of the congeneric model is the multiple-factor model Equation 10.1; the generalization of the two-parameter logistic model is the multidimensional item response model (Reckase, 2009, Section 4.1; Volume One, Chapter 12). Applying the logit transformation to test taker p's probability of giving a correct answer to the ith item under the multidimensional item response model for dichotomous responses yields Equation 10.1 of the multiple-factor model for continuous item scores.

A difference between the logistic models for dichotomous responses and the factor analytic models for continuous item scores is that the former have a and b parameters only, whereas the latter have a, b, and σ^2 parameters. This difference follows from the logistic model assumption that test taker p's responses to an item are Bernoulli distributed. The variance of the Bernoulli distribution is a function of test taker p's probability of giving the correct (yes) answer (i.e., $\sigma^2_{pi} = \pi_{pi}[1 - \pi_{pi}]$), whereas factor analytic models need a separate σ^2 parameter to characterize p's normal item response distribution.

As observed earlier, a D-factor model can be rewritten as a $(D + 1)$-class latent profile model with the same means, variances, and covariances. Therefore, means, variances, and covariances cannot give conclusive evidence on the distinction between dimensional and typological theories in the behavioral sciences (Loken and Molenaar, 2008; Molenaar and Von Eye, 1994). For example, the congeneric model for the *Aggression* subtest of the ACL assumes that a person's aggression can be represented by a continuous latent dimension. However, the congeneric model can be rewritten as a two-class latent profile model with the same means, variances, and covariances as the congeneric model, and this latent profile model distinguishes two types of test takers, aggressive and nonaggressive persons.

The congeneric model is built on strong assumptions. First, the model assumes that the item score given the person's latent parameter value is normally distributed. Second, the model assumes that the item response function is linear (Equation 10.2). Third, the model assumes that item parameters a, b, and σ^2 are homogeneous with respect to the person parameter, that is, do not vary with it. Finally, in order to use the maximum-likelihood method to estimate the parameters, the model requires that the item scores be multivariate normally distributed, which implies that the latent person parameters are normally distributed as well. These assumptions are easily violated for empirical test data. For example, the response distributions of test takers who respond near a bound on the continuous response scale may have a smaller variance than those of test takers who respond in the middle of the scale, which means violation of the homogeneous variance assumption (Equation 10.3). Hessen and Dolan (2009) presented a one-factor model with heteroskedastic variance. The parameters of this model can be estimated using the marginal maximum-likelihood method, while its homogeneous variance assumption can be tested with a likelihood-ratio test. Molenaar, Dolan, and Verhelst (2010) extended this model to a model with a and σ^2 parameters that can vary between test takers, and a nonnormal person parameter distribution. The model can be estimated using the marginal maximum-likelihood method, and both the assumptions of invariant a and σ^2 parameters and a normal person parameter distribution can be tested using likelihood-ratio tests.

Several methods for the construction and analysis of discrete item responses have been developed for discrete response models (Hambleton and Swaminathan, 1985; Lord, 1980; Mellenbergh, 2011). Most of these methods can be adapted to continuous responses under the congeneric model. An exception is adaptive testing (van der Linden, Volume Three, Chapter 10), which cannot be conducted under a congeneric model. Adaptive testing is typically based on the use of maximization of Fisher's information as item selection criterion; that is, it selects the items to have maximum information at the interim ability estimates. However, the congeneric model is linear in the person parameters and, consequently, Fisher's information is a constant independent of it (as is clear from the constant variance in Equation 10.3). Therefore, under the congeneric model it is impossible to adapt the item selection to a test taker's parameter estimates.

Tests are typically constructed selecting items from a larger pool. Discrete item response models have been used for optimal selection of n items from the pool (van der Linden, 2005; Volume Three, Chapter 10). The test is optimal in the sense that it optimizes an objective under a given number of constraints. Optimal test design methods can also be applied to continuous response items. The congeneric model provides objectives, such as maximization of the reliability of the person parameter estimates that can be used for optimal test construction.

Methods are available to detect differentially functioning discrete items under item response models (Millsap, 2011, Chapters 7 and 8). Under the congeneric model differential item functioning (DIF) (Gamerman, Gonçalves, and Soares, Volume Three, Chapter 4) between subpopulations is absent if the item parameters in the congeneric model are invariant across the subpopulations. DIF can be studied under the congeneric model using multiple-group factor analytic methods (Millsap, 2011, Chapter 4).

The response behavior of some test takers may deviate from the behavior of the majority of them. A person-fit statistic assesses whether a test taker's responses to the n test items deviate from his (her) responses that are expected under an item response model. For a review of person-fit statistics for discrete item responses; see Glas (Volume Two, Chapter 17) or Meijer and Sijtsma (2001). A person-fit statistic for continuous items based on the congeneric model was introduced by Ferrando (2007).

Earlier, we discussed the evaluation of within-person and between-persons precision both for person parameter estimates ($\hat{\theta}_p$) and the observed test-score (X_p). In real-world testing, both types of precision are easily confused. But Equations 10.12 and 10.13 should be used only for latent-trait parameter estimates, whereas Equations 10.14 and 10.15 should be applied when the interest is in the precision of observed test-scores.

References

Andrich, D. 1978. Application of a psychometric rating model to ordered categories which are scored with successive integers. *Applied Psychological Measurement*, 2, 581–594.

Bartholomew, D. J. 1987. *Latent Variable Models and Factor Analysis.* London: Griffin.

Bartholomew, D. J. 1995. Spearman and the origin and development of factor analysis. *British Journal of Mathematical and Statistical Psychology*, 48, 211–220.

Birnbaum, A. 1968. Some latent trait models and their use in inferring an examinee's ability. In: Lord, F. M. and Novick, M. R. (Eds.), *Statistical Theories of Mental Test Scores*, pp. 397–479. Reading, MA: Addison-Wesley.

Borsboom, D. 2005. *Measuring the Mind.* Cambridge, United Kingdom: Cambridge University Press.

Ferrando, P. J. 2007. Factor-analytic procedures to model misspecification and model types. *Multivariate Behavioral Research*, 42, 481–507.

Gibson, W. A. 1959. Three multivariate models: Factor analysis, latent structure analysis, and latent profile analysis. *Psychometrika*, 24, 229–252.

Gough, H. G. and Heilbrun, A. B. 1980. *The Adjective Check List Manual.* Palo Alto, CA: Consulting Psychologists Press.

Hambleton, R. K. and Swaminathan, H. 1985. *Item Response Theory.* Boston, MA: Kluwer-Nijhoff.

Hessen, D. J. and Dolan, C. V. 2009. Heteroscedastic one-factor models and marginal maximum likelihood estimation. *British Journal of Mathematical and Statistical Psychology*, 62, 57–77.

Holland, P. W. 1990. On the sampling theory foundations of item response theory models. *Psychometrika*, 55, 577–601.

Hu, L. -T. and Bentler, P. M. 1995. Evaluating model fit. In: Oyle, R. H. (Ed.), *Structural Equation Modeling: Concepts, Issues, and Applications*, pp. 76–99. Thousand Oaks, CA: Sage.

Jöreskog, K. G. 1971. Statistical analysis of sets of congeneric tests. *Psychometrika*, 36, 109–133.

Jöreskog, K. G. and Sörbom, D. 1999. *LISREL (8.30)* [Computer program]. Chicago, IL: Scientific Software International.

Kaplan, D. 2000. *Structural Equation Modeling: Foundations and Extensions.* Thousand Oaks, CA: Sage.

Lazarsfeld, P. F. and Henry, N. W. 1968. *Latent Structure Analysis.* New York: Houghton Mifflin.

Loken, E. and Molenaar, P. 2008. Categories or continua? The correspondence between mixture models and factor models. In: Hancock, G. R. and Samuelsen, K. M. (Eds.), *Advances in Latent Variable Mixture Models*, pp. 277–297. Charlotte, NC: Information Age Publishing.

Lord, F. M. 1980. *Applications of Item Response Theory to Practical Testing Problems.* Hillsdale, NJ: Erlbaum.

Lord, F. M. and Novick, M. R. 1968. *Statistical Theories of Mental Test Scores*. Reading, MA: Addison-Wesley.
McDonald, R. P. 1967. Nonlinear factor analysis. *Psychometric Monograph*, No. 15.
McDonald, R. P. 1982. Linear versus nonlinear models in item response theory. *Applied Psychological Measurement*, 4, 379–396.
McDonald, R. P. 1999. *Test Theory: A Unified Treatment*. Mahwah, NJ: Erlbaum.
Meijer, R. R. and Sijtsma, K. 2001. Methodology review: Evaluating person fit. *Applied Psychological Measurement*, 25, 107–135.
Mellenbergh, G. J. 1994a. Generalized linear item response theory. *Psychological Bulletin*, 115, 300–307.
Mellenbergh, G. J. 1994b. A unidimensional latent trait model for continuous item responses. *Multivariate Behavioral Research*, 29, 223–236.
Mellenbergh, G. J. 1996. Measurement precision in test score and item response models. *Psychological Methods*, 1, 293–299.
Mellenbergh, G. J. 2011. *A Conceptual Introduction to Psychometrics: Development, Analysis, and Application of Psychological and Educational Tests*. The Hague, the Netherlands: Eleven International Publishing.
Millsap, R. E. 2011. *Statistical Approaches to Measurement Invariance*. New York: Routledge.
Molenaar, D., Dolan, C. V., and Verhelst, N. D. 2010. Testing and modelling non-normality within the one-factor model. *British Journal of Mathematical and Statistical Psychology*, 63, 293–317.
Molenaar, P. C. M. and Von Eye, A. 1994. On the arbitrary nature of latent variables. In: von Eye, A. and Glogg, C. C. (Eds.), *Latent Variables Analysis*, pp. 226–242. Thousand Oaks, CA: Sage.
Mulaik, S. A. 1972. March. A mathematical investigation of some multidimensional Rasch models for psychological tests. Paper presented at the *Meeting of the Psychometric Society*, Princeton, NJ.
Mulaik, S. A. 2009. *Linear Causal Modeling with Structural Equations*. Boca Raton, FL: Chapman and Hall.
Müller, H. 1987. A Rasch model for continuous ratings. *Psychometrika*, 52, 165–181.
Rasch, G. 1960. *Probabilistic Models for Some Intelligence and Attainment Tests*. Copenhagen, Denmark: The Danish Institute for Educational Research.
Reckase, M. D. 2009. *Multidimensional Item Response Theory*. New York: Springer.
Samejima, F. 1969. Estimation of latent ability using a response pattern of graded scores. *Psychometrika Monograph*, No. 17.
Samejima, F. 1973. Homogeneous case of the continuous response model. *Psychometrika*, 38, 203–219.
van den Berg, R. G. 2002. Psychometrics report: Analysis of the aggression scale of the ACL. Unpublished manuscript, Psychological Methods, Department of Psychology, University of Amsterdam, the Netherlands.
van der Linden, W. J. 2005. *Linear Models for Optimal Test Design*. New York: Springer.

Section III

Multidimensional and Multicomponent Models

11

Normal-Ogive Multidimensional Models[*]

Hariharan Swaminathan and H. Jane Rogers

CONTENTS

11.1 Introduction 167
11.2 Presentation of the Model 170
11.3 Parameter Estimation 177
  11.3.1 Asymptotic Distribution of Estimators 178
11.4 Goodness of Fit 182
11.5 Empirical Example 183
11.6 Conclusion 184
References 185

11.1 Introduction

The normal probability density function is central to statistical modeling in the natural, social, and physical sciences. The cumulative normal distribution function, or the normal ogive, is thus a natural choice for modeling probabilities (Albert, Volume Two, Chapter 1). Gustav Fechner may have been the first to use the normal ogive in the mid-1860s to model stimulus intensity (Bock, 1997). Interest in normal-ogive models for analyzing binary response data was sparked in the early twentieth century by Karl Pearson's (1900) work using the univariate and bivariate normal-ogive models for estimating the biserial and tetrachoric correlations, respectively. In fact, Pearson's tetrachoric series, as we shall see later, continues to play an important role in the analysis of multidimensional binary item responses.

Using the normal-ogive model, Thomson (1919) estimated the Fechner–Müller–Urban thresholds using a method that is remarkably similar to the maximum-likelihood procedure introduced by Fisher in 1925, one of the procedures that is currently in use in the estimation of an examinee's ability. In the testing context, Richardson (1936) and Ferguson (1942) used the normal-ogive model for scaling difficulties of dichotomously scored items. Lawley (1943, 1944) was the first to formally employ the normal-ogive model to directly model binary item response data, using the maximum-likelihood procedure to estimate threshold and discrimination parameters under the assumption that the observed test score serves as a measure of the ability of the examinee; this procedure, as pointed out by Finney (1944), was already widely known in toxicology as probit regression. Tucker (1946) used the term "item curve" repeatedly to indicate the relationship between item response and ability and in doing so anticipated the current term "item characteristic curve." These early attempts at modeling binary response data culminated in the work of Lord (1952,

[*] This chapter is dedicated to the memory of Roderick P. McDonald, who was the mentor, colleague, and friend of the first author. The material in this chapter draws heavily from his writings on the subject. Errors are solely ours.

1953, 1980) who, unlike the early researchers, treated ability as a latent trait to be estimated and in doing so, laid the foundation for item response theory (IRT).

Meanwhile, on a separate front, there was some concern among factor analysts that wide variations in item difficulty induced low interitem correlations and consequently produced spurious or difficulty factors (Ferguson, 1941; Guilford, 1941; Carroll, 1945). Gibson (1960, p. 381) argued that these difficulty factors were the result of "curvilinear relations among the tests, such curvilinearities being forced by differential difficulty of the tests." McDonald (1962b, p. 398), noting that "there is a need for a factor analytic method which would allow for nonlinear, possibly nonmonotonic relations between manifest and latent variates," developed factor analytic models and methods for handling nonlinear relationships between the (observed) response and latent variables (McDonald, 1962b; 1967; 1982; 1985). In order to unify the treatment of discrete and continuous response as well as latent variables, McDonald (1962a), independently of Anderson (1959), articulated the strong and weak principles of local independence. The strong principle invokes statistical independence of the responses conditional on the set of latent variables. More formally, according to the strong principle of local independence,

$$g(\mathbf{U} \mid \boldsymbol{\theta}) = \prod_{i=1}^{I} g_i(U_i \mid \boldsymbol{\theta}), \tag{11.1}$$

where $\mathbf{U}$ is the $I \times 1$ vector of responses (e.g., the vector of responses of a subject on I items or variables), $\boldsymbol{\theta}$ is the $D \times 1$ vector of latent variables or traits, and g (.) and g_i(.) are the conditional densities of $\mathbf{U}$ and U_i given $\boldsymbol{\theta}$, respectively. The strong principle of local independence implies the weak principle, which requires only that the conditional covariances vanish, that is,

$$E\{(\mathbf{U} - E(\mathbf{U} \mid \boldsymbol{\theta})(\mathbf{U} - E(\mathbf{U} \mid \boldsymbol{\theta})'\} = \boldsymbol{\Delta}, \tag{11.2}$$

where $\boldsymbol{\Delta}$ is a diagonal matrix, E(.) is the expectation operator, and $E(\mathbf{U} \mid \boldsymbol{\theta} \equiv f(\boldsymbol{\theta}))$ is a function of the latent traits, recognizable in the context of IRT as the *item response function*. The strong principle reduces to the weak in the case where the observations are assumed to be multivariate normally distributed, as in linear factor models, since in this case the first two moments contain all the necessary information. The strong and weak principles provide the basis for assessing the adequacy of fit of a model with a specified number of latent traits.

McDonald's nonlinear factor analysis approach entails the fitting of polynomial functions of the latent traits $\boldsymbol{\theta}$ to the observed data. McDonald (1982, p. 384) classified latent trait models into (1) *strictly linear*, meaning that $f(\boldsymbol{\theta})$ is linear in both the coefficients and the latent traits $\boldsymbol{\theta}$, (2) *wide-sense linear*, meaning that $f(\boldsymbol{\theta})$ is linear in the coefficients but not linear in $\boldsymbol{\theta}$, and (3) *strictly nonlinear*, meaning that $f(\boldsymbol{\theta})$ cannot be expressed as a wide-sense linear model with a finite number of terms. An example of the first case is the linear factor model. An example of the second case is the nonlinear model described by Etezadi-Amoli and McDonald (1983) for 12 (continuous) response variables with two latent variables θ_1 and θ_2 and a response function defined as

$$E(y_i \mid \boldsymbol{\theta}) = \mu_i + \lambda_{i1}\theta_1 + \lambda_{i2}\theta_1 + \lambda_{i3}\theta_1^2 + \lambda_{i4}\theta_1\theta_2 = \mu_i + \boldsymbol{\lambda}_i'\boldsymbol{\theta}, \quad i = 1, \ldots, 12. \tag{11.3}$$

In this model, $\mathbf{y} = \boldsymbol{\mu} + \boldsymbol{\Lambda}\boldsymbol{\theta} + \mathbf{e}$ with variance–covariance matrix $\boldsymbol{\Sigma} = \boldsymbol{\Lambda}\boldsymbol{\Lambda}' + \boldsymbol{\Delta}$, where $\mathbf{y}$ is a 12×1 vector of response variables, $\boldsymbol{\theta}' = [\theta_1\, \theta_2\, \theta_1^2\, \theta_1\, \theta_2]$, and $\boldsymbol{\Lambda}$ is the 12×4 matrix of factor loadings with ith row $\boldsymbol{\lambda}_i'$. The model in Equation 11.3 and a linear factor model with four factors cannot be distinguished using the weak form of local independence, as the residual correlation/covariance matrix will be diagonal in both cases. The fitting of this model to the data is described in detail in Etezadi-Amoli and McDonald (1983).

An example of the third case is the item response model for binary responses where the response function is defined by the normal-ogive or the logistic cumulative function, and it is this case that is of interest here. Unlike the wide-sense linear model defined above, the normal-ogive or the logistic item response function cannot be expressed as polynomials with a *finite* number of terms; they can, however, be expressed as an infinite series $\sum_{k=0}^{\infty} c_k h_k(x)$ where $h_k(x)$ is a polynomial of degree k with $h_0(x) = 1$, or according to the famous theorem of Weierstrass, they can be *approximated* by a polynomial $\sum_{k=0}^{r} c_k h_k(x)$ of degree r. The coefficients, c_k, functions of the model parameters, are determined by minimizing

$$\int_R \left[f(x) - \sum_{k=0}^{\infty \text{ or } r} c_k h_k(x) \right]^2 dx.$$

Differentiating under the integral with respect to c_j and setting the derivative equal to zero, we obtain the equation

$$\int_R \left(\sum_{k=0}^{\infty \text{ or } r} c_k h_k(x) \right) h_\ell(x) dx = \int_R \left[f(x) h_\ell(x) \right] dx.$$

If the polynomials are orthonormal with respect to a weight function $g(x)$, then $\int h_k(x) h_\ell(x) g(x) dx = 0$ if $k \neq j$ and 1 if $k = j$, and coefficient c_k is

$$c_k = \int_R f(x) h_k(x) g(x) dx. \tag{11.4}$$

The proofs of these assertions along with convergence of these series have been the concern of many mathematicians (see Courant and Hilbert, 1953, Chapter 2; Hille, 1926). The expansion of a function in terms of orthogonal polynomials is not Fourier expansion in the strict sense, as the function to be expanded is not periodic. However, the term "Fourier coefficients" may be used because "the expansion considered is a generalization of the Fourier series expansion" (Courant and Hilbert, 1953, p. 51). Using the method outlined above, the normal-ogive or the logistic response function can be expressed in terms of orthogonal polynomials, and through this representation, item response models can be readily fitted to data using the method developed by McDonald (1962a, 1967). This procedure will be considered in detail in the next section.

To summarize, the conjecture by factor analysts that items varying in difficulty produced factors that are related to nonlinearities in the relationships between response and latent variables prompted the development of nonlinear factor models. While there were early attempts at formulating nonlinear factor models (Gibson, 1960), McDonald (1962b, 1967) was probably the first to develop models to handle nonlinear relationships between response variables and latent traits and provide a unified treatment of the factor analysis of continuous and binary response variables. The approach taken by McDonald using polynomials to model curvilinear response functions generalizes readily to nonlinear unidimensional as well as multidimensional item response models. Implicit in this development is the recognition that item response models are in fact nonlinear factor models. Although McDonald (1985) lamented that the "overlap between IRT and common factor theory is still not widely appreciated" (p. 127), it should be noted that the relationship between factor analysis and item response models was known to several researchers, notably Bock and Lieberman (1970), Bock and Aitkin (1981), Christoffersson (1975), and Muthén (1978). These researchers developed models to handle nonlinearities that arise when modeling relationships between binary response variables and continuous latent traits. In fact, Bock and Lieberman (1970), Christoffersson (1975), and Muthén (1978) used the term *factor analysis* for the analysis of binary item responses. The approaches of these researchers and that of McDonald for fitting unidimensional and multidimensional models to binary data have differences as well as commonalities. Descriptions of these approaches are found in the papers cited above and in van der Linden and Hambleton (1997), and Muthén and Asparouhov (Volume One, Chapter 31). In the present chapter, we shall deal only with the normal-ogive item response model as presented by McDonald (1967, 1982) in the context of nonlinear factor analysis. A description of the multidimensional logistic item response model is provided by Reckase (Volume One, Chapter 12).

11.2 Presentation of the Model

In order to facilitate the formulation of multidimensional models, we shall consider the unidimensional model first and then show that with trivial modifications, the procedure extends to the multidimensional case. The unidimensional normal-ogive item response model for person $p(=1, \ldots, P)$ and item i $(=1, \ldots, I)$ is conventionally expressed as (Lord and Novick, 1968, p. 366)

$$\Pr\{U_{pi} = 1 \mid \theta\} = \int_{-\infty}^{a_i(\theta_p - b_i)} \varphi(x)dx = P(a_i(\theta_p - b_i)),$$

where $\varphi(x)$ is the standard normal density and $P(.)$ is the cumulative normal distribution function, that is, the normal ogive. The parameterization in terms of item discrimination a_i and difficulty b_i given above, however, does not carry over to multidimensional models. For this reason, we shall write the model for item i (dropping the subscript for person) in terms of slope a_i and intercept β_i as

$$\Pr\{U_i = 1 \mid \theta\} = \int_{-\infty}^{a_i\theta + \beta_i} \varphi(x)dx = P(a_i\theta + \beta_i), \tag{11.5}$$

where $\beta_i = -a_i b_i$. The model as it stands is not identified. In order to identify it, we fix the mean and standard deviation of θ to zero and one, respectively. For more details of the normal-ogive link function, see Albert (Volume Two, Chapter 1).

As noted earlier, while the normal-ogive function $P(a_i\theta + \beta_i)$ cannot be expressed as the sum of a finite number of orthonormal polynomials, it can be *approximated* to any degree of accuracy by a finite number of polynomials in the least squares sense. The normal-ogive function can, however, be expressed as an infinite sum of suitably chosen orthogonal polynomials belonging to the weight function $g(x)$. Since the domain of θ is the real line, the natural choice of orthogonal polynomials is the Hermite–Chebychev polynomials, sometimes referred to simply as the Hermite polynomials or even as Chebychev–Hermite polynomials. (The latter order is the correct one since Chebychev published his work on the polynomials several years before Hermite.) For example, Kendall (1941) refers to the polynomials as the Hermite polynomials, while in Kendall and Stuart (1969, Volume One, p. 155), they are referred to as the Chebychev–Hermite polynomials! The Hermite–Chebychev polynomials are defined as

$$H_r(\theta) = (-1)^r exp\left(\frac{\theta^2}{2}\right)\frac{d^r}{dx^r}exp\left(-\frac{\theta^2}{2}\right). \tag{11.6}$$

The set $\{H_r(\theta)\}$ is orthogonal on the real line with respect to the weight function $g(\theta) = \left(1/\sqrt{2\pi}\right)exp(-\theta^2/2)$ that is,

$$\int_{-\infty}^{\infty} H_k(\theta)H_\ell(\theta)g(\theta)d\theta = k!\delta_{k\ell}, \tag{11.7}$$

where $\delta_{k\ell} = 1$ if $\ell = k$ and zero otherwise. It follows then that $\{H_k(\theta)/\sqrt{k!}\}$ constitutes an orthonormal set (Courant and Hilbert, 1953, p. 92; Hille, 1926). Our task is to express the normal-ogive function as

$$P[(a_i\theta + \beta_i)] = \sum_{k=0}^{\infty} c_k \frac{1}{\sqrt{k!}}H_k(\theta). \tag{11.8}$$

In expressing the function as an infinite series, it has to be shown that the series converges to the function $P(\beta_i + a_i\theta)$. The usual requirement for Fourier expansion of a function in terms of orthogonal polynomials is that the function is square integrable. Unfortunately, $P(\beta_i + a_i\theta)$ is not integrable. Hille (1926) has provided a rigorous treatment of the expansion of functions in terms of Hermite–Chebychev polynomials. He showed that if $\int_{-\infty}^{\infty} e^{-1/2x^2}\,|f(x)|\,dx$ exists, then the series is summable and converges to $f(x)$. The function $P(\beta_i + a_i\theta)$ is continuous, bounded, and positive for $-\infty < \theta < \infty$. While $P(\beta_i + a_i\theta)$ is not integrable, it is easy to see that the integral $\int_{-\infty}^{\infty} e^{-1/2\theta^2}P(\beta_i + a_i\theta)d\theta$ exists, and hence the series $\sum_{k=0}^{\infty} c_k H_k(\theta)$ converges to $P(\beta_i + a_i\theta)$. In the sequel, we shall use the infinite series representation of the normal ogive with the understanding that a finite number of terms could be used to approximate the function to any degree of accuracy.

In usual IRT modeling, it is not necessary to specify a density for θ except with Bayesian or maximum marginal-likelihood estimation procedures. In the present case, choosing the density of θ to coincide with the weight function $g(\theta)$, that is, $g(\theta) = \varphi(\theta)$, results in considerable simplification of the mathematics, and therefore, we shall assume that $\theta \sim N(0,1)$. Hence, for the representation $P(\beta_i + a_i\theta) = \sum_{k=0}^{\infty} c_k \left(1/\sqrt{k!}\right)H_k(\theta)$, we obtain, using Equations 11.4 and 11.8,

$$c_{ik} = \int_{-\infty}^{\infty} P(\beta_i + a_i\theta)\frac{1}{\sqrt{k!}}H_k(\theta)\varphi(\theta)d\theta. \tag{11.9}$$

As $H_0(x) = 1$, it can be shown after some manipulations that

$$c_{i0} = \frac{1}{2\pi}\int_{-\infty}^{\infty}\left\{\int_{-\infty}^{a_i\theta+\beta_i} exp\left(-\frac{x^2}{2}\right)dx\right\} exp\left(-\frac{\theta^2}{2}\right)d\theta = P\left\{\frac{\beta_i}{\sqrt{(1+a_i^2)}}\right\}. \quad (11.10)$$

Determining the coefficient c_{ik} $(k \neq 0)$ is tedious and best carried out using Fourier transforms. McDonald (1967, Appendix 4.4) showed that

$$c_{ik} = \varphi\left\{\frac{\beta_i}{\sqrt{(1+a_i^2)}}\right\}\frac{1}{\sqrt{k!}}\left\{\frac{a_i}{\sqrt{(1+a_i^2)}}\right\}^k H_{k-1}\left\{\frac{\beta_i}{\sqrt{(1+a_i^2)}}\right\}. \quad (11.11)$$

Combining Equations 11.9 and 11.10, we have

$$P(\beta_i + a_i\theta) = P\left\{\frac{\beta_i}{\sqrt{(1+a_i^2)}}\right\} + \varphi\left\{\frac{\beta_i}{\sqrt{(1+a_i^2)}}\right\}\sum_{k=1}^{\infty}\frac{1}{k!}\left\{\frac{a_i}{\sqrt{(1+a_i^2)}}\right\}^k H_{k-1}\left\{\frac{\beta_i}{\sqrt{(1+a_i^2)}}\right\}H_k(\theta). \quad (11.12)$$

In order to complete the modeling process, we need to relate the parameters of the response function to the moments of the distribution of the observations. Given that U_i is a Bernoulli variable, the appropriate moments are the proportions π_i of responding correctly to item i correctly, and the joint proportions π_{ij} of responding correctly to items i and j. Including all higher-order joint proportions will provide complete information, but will increase the complexity of estimation. Hence, only the first two moments are used here.

Two pertinent results for the following development are (a) $E[H_k(\theta)] = 0$ and (b) $E[H_k(\theta)H_\ell(\theta)] = k!\delta_{k\ell}$. Since $H_0(\theta) = 1$, we have from the orthogonal property of the polynomials in Equation 11.7,

$$E[H_k(\theta)] = \int_{-\infty}^{\infty} H_k(\theta)\varphi(\theta)d\theta = \int_{-\infty}^{\infty} H_0(\theta)H_k(\theta)\varphi(\theta)d\theta = 0 \quad (11.13)$$

and

$$E[H_k(\theta)H_\ell(\theta)] = \int_{-\infty}^{\infty} H_k(\theta)H_\ell(\theta)\varphi(\theta)d\theta = k!\delta_{k\ell}. \quad (11.14)$$

Now, given that U_i is 1 or 0, we obtain, using Equation 11.13,

$$\begin{aligned}\pi_i &= \Pr\{U_i\} = E(U_i) = E_\theta E(U_i \mid \theta) = E_\theta[P(\beta_i + a_i\theta)] \\ &= P\left\{\frac{\beta_i}{\sqrt{(1+a_i^2)}}\right\} + \sum_{k=1}^{\infty} c_{ik}E_\theta[\frac{1}{\sqrt{k!}}H_k(\theta)] = P\left\{\frac{\beta_i}{\sqrt{(1+a_i^2)}}\right\}.\end{aligned} \quad (11.15)$$

Furthermore, $\Pr\{U_i = 1, U_j = 1\} = \Pr\{U_i = 1 \cap U_j = 1\} = E(U_iU_j)$. Hence, using the principle of local independence and Equation 11.14, we obtain

$$\pi_{ij} = E(U_iU_j) = E_\theta E(U_iU_j \mid \theta) = E_\theta E(U_i \mid \theta)E(U_j \mid \theta) = E_\theta[P(\beta_i + a_i\theta)P(\beta_j + a_j\theta)]$$

$$= E_\theta\left(\sum_{k=0}^{r} c_{ik}\frac{1}{\sqrt{k!}}H_r(\theta)\right)\left(\sum_{m=1}^{\infty} c_{jm}\frac{1}{\sqrt{m!}}H_m(\theta)\right) = \left(\sum_{k=0}^{\infty}\sum_{m=0}^{\infty}\frac{1}{\sqrt{k!}}\frac{1}{\sqrt{m!}}c_{ik}c_{jm}E_\theta[H_k(\theta)H_m(\theta)]\right)$$

$$= P\left\{\frac{\beta_i}{\sqrt{(1+a_i^2)}}\right\}P\left\{\frac{\beta_j}{\sqrt{(1+a_j^2)}}\right\} + \left(\sum_{k=1}^{\infty} c_{ik}c_{jk}\right). \tag{11.16}$$

Equations 11.15 and 11.16 provide expressions for the first two moments in terms of the parameters a_i, β_i, $i = 1, \ldots, I$, of the model. We thus have a system of estimating equations as in linear factor analysis but with the mean and the covariances replaced by the first- and second-order proportions of responding correctly to item i and to the pair (i, j), respectively.

Multidimensional item response models are obtained by a direct extension of the models and procedures described for the unidimensional model. Let $\boldsymbol{\theta}' = [\theta_1\, \theta_2 \cdots \theta_D]$ denote the $(D \times 1)$ vector of latent traits measured by the test with I items. Furthermore, let the vector of discriminations for item i be denoted as $\boldsymbol{a}_i' = [a_{i1}\, a_{i2} \ldots a_{iD}]$ and the intercept as β_i. Then the multidimensional model corresponding to the univariate model is simply

$$\Pr\{U_i = 1 \mid \boldsymbol{\theta}\} = \int_{-\infty}^{a_i'\theta+\beta_i} \varphi(x)dx = P(\boldsymbol{a}_i'\boldsymbol{\theta} + \beta_i). \tag{11.17}$$

To identify the model, we shall assume that $\boldsymbol{\theta} \sim N(\mathbf{0}, \boldsymbol{\Phi})$ where $\text{diag}(\boldsymbol{\Phi}) = \mathbf{I}$ is the $(D \times D)$ identity matrix.

If we define $\eta_i = \mathbf{a}_i'\boldsymbol{\theta}$, then $\Pr\{U_i = 1 \mid \boldsymbol{\theta}\} = P(\eta_i + \beta_i)$, a model that bears a striking resemblance to the unidimensional model in Equation 11.6. We could then attempt to expand the normal-ogive function $P(\eta_i + \beta_i)$ in terms of the Hermite–Chebychev polynomials as in Equation 11.8. However, the weight function for the Hermite–Chebychev polynomials is the standard normal density. Since $\boldsymbol{\theta} \sim N(\mathbf{0}, \boldsymbol{\Phi})$, $\eta_i = \boldsymbol{a}_i'\boldsymbol{\theta} \sim N(0, \boldsymbol{a}_i'\boldsymbol{\Phi}\boldsymbol{a}_i)$, $\eta_i/\sigma_i \sim N(0,1)$, where $\sigma_i = \sqrt{\boldsymbol{a}_i'\boldsymbol{\Phi}\boldsymbol{a}_i}$. We now have a variable that is $N(0,1)$, and hence, Equation 11.17 can be expressed as

$$\Pr\{U_i = 1 \mid \boldsymbol{\theta}\} = P\left[\sigma_i\left(\frac{\eta_i}{\sigma_i}\right) + \beta_i\right] = \sum_{k=0}^{\infty} c_{ik}\frac{1}{\sqrt{k!}}H_k\left(\frac{\eta_i}{\sigma_i}\right). \tag{11.18}$$

The Fourier coefficients c_{ik} for the expansion in Equation 11.18 are obtained by simply replacing a_i by $\sigma_i = \sqrt{\boldsymbol{a}_i'\boldsymbol{\Phi}\boldsymbol{a}_i}$ in Equations 11.10 and 11.11. The proportions π_i are given as

$$\pi_i = P\left\{\frac{\beta_i}{\sqrt{(1+\sigma_i^2)}}\right\} \equiv P\left\{\frac{\beta_i}{\sqrt{(1+\boldsymbol{a}_i'\boldsymbol{\Phi}\boldsymbol{a}_i)}}\right\}. \tag{11.19}$$

The joint proportions π_{ij}, however, cannot be obtained from Equation 11.16 using the orthogonality property in Equation 11.14, since unlike the unidimensional case, we have the product of correlated variables η_i and η_j for item pair (i, j). To determine the expected value of the product $H_k(x)H_m(y)$, we invoke the Mehler identity (Kendall, 1941)

$$\varphi(x,y)=\varphi(x)\varphi(x)\sum_{k=0}^{\infty}\frac{\rho_{xy}^{k}}{k!}H_k(x)H_k(y), \tag{11.20}$$

where $\varphi(x, y)$ is the bivariate normal density. This celebrated identity was first discovered by Mehler in 1864 (Kendall, 1941) and later rediscovered by Pearson (1900) as the tetrachoric series. According to the great British mathematician G. H. Hardy (1932, p. 192), Einar Hille, in his communication with him, noted that Mehler's identity "has been rediscovered by almost everybody who has worked in this field." Given its importance in mathematical physics, some of the leading mathematicians of the twentieth century, including Hille (1926), Watson (1933), and Hardy, as communicated to Watson (1933), have provided rigorous proofs of this ubiquitous identity.

Multiplying both sides of Equation 11.20 by $H_k(x)H_m(y)$, and integrating, we obtain

$$E[H_m(x)H_m(y)]=\iint H_m(x)H_m(y)\varphi(x,y)dxdy$$

$$=\iint \varphi(x)\varphi(y)\sum_{k=0}^{\infty}\frac{\rho_{xy}^{k}}{k!}H_k(x)H_k(y)H_m(x)H_m(y)dxdy$$

$$=\iint \varphi(x)\varphi(y)H_0(x)H_0(y)H_m(x)H_m(y)dxdy+\sum_{k=1}^{\infty}\frac{\rho^{k}}{k!}\iint \varphi(x)\varphi(y)H_k(x)H_k(y)H_m(x)H_m(y)dxdy.$$

By virtue of the orthogonality of the polynomials, the first term is zero and the integral in the second term reduces to $k!$. Thus,

$$E[H_k(x)H_k(y)]=\rho_{xy}^{k}. \tag{11.21}$$

Alternatively, the above expression can be derived following the suggestion of Lancaster (1958). Using the generating function of the Hermite–Chebychev polynomials, namely, $exp(tx-\frac{1}{2}t^2)$ (Courant and Hilbert, 1953, p. 92), and taking the expected value of $exp(tx-\frac{1}{2}t^2)exp(uy-\frac{1}{2}u^2)$, we obtain

$$E\{e^{(tx-1/2t^2)}e^{(uy-1/2u^2)}\}=e^{(-1/2t^2-1/2u^2)}\int_{-\infty}^{\infty}\int_{-\infty}^{\infty}e^{(tx+uy)}\varphi(x,y)dxdy=e^{\rho tu}=\sum_{k=0}^{\infty}\frac{\rho^{k}}{k!}t^k u^k. \tag{11.22}$$

The result in Equation 11.22 follows since the term under the integrals is the moment-generating function of the bivariate normal, $exp\frac{1}{2}\{t^2+u^2+2\rho ut\}$. The coefficient of $(tu)^k/k!$ in the expansion of $exp(\rho tu)$ is the kth joint moment, ρ_{xy}^{k}. Thus, the joint moment can be derived using either the Mehler identity or independently of it. The importance of this alternative derivation will be made clear later. Using Equation 11.21 or 11.22 in Equation 11.16, we obtain for the joint proportion,

$$\begin{aligned}\pi_{ij} = \Pr\{U_iU_j\} &= \left(\sum_{k=0}^{\infty}\sum_{m=0}^{\infty}\frac{1}{\sqrt{k!}}\frac{1}{\sqrt{m!}}c_{ik}c_{jm}E\left[H_k\left(\frac{\eta_i}{\sigma_i}\right)H_m\left(\frac{\eta_j}{\sigma_j}\right)\right]\right)\\ &= P\left(\frac{\beta_i}{\sqrt{(1+\sigma_i^2)}}\right)P\left(\frac{\beta_j}{\sqrt{(1+\sigma_j^2)}}\right)+\left(\sum_{k=1}^{\infty}c_{ik}c_{jk}\rho_{ij}^k\right),\end{aligned} \tag{11.23}$$

where $\rho_{ij} = E[\boldsymbol{a}_i'\boldsymbol{\theta}\boldsymbol{\theta}'\boldsymbol{a}_j/\sigma_i\sigma_j] = \boldsymbol{a}_i'\boldsymbol{\Phi}\boldsymbol{a}_j/\sigma_i\sigma_j$.

The development above uses an axiomatic approach to the modeling of the probability of a correct response in the sense that any monotonically increasing function bounded in the interval [0,1] can be used. However, for relating IRT and factor analysis, it may be more meaningful to motivate the normal-ogive model from the point of view provided by Lord and Novick (1968, pp. 370–371).

We suppose that there is an underlying continuous vector of latent variables $\boldsymbol{y}$ of dimension $(I \times 1)$ such that $\boldsymbol{y} = \boldsymbol{\Lambda\theta} + \boldsymbol{\delta}$, where $\boldsymbol{\theta}$ is the $(D \times 1)$ vector of latent traits $\boldsymbol{\theta} \sim N(\mathbf{0}, \Phi)$ with $\text{diag}(\Phi) = \mathbf{I}$; $\boldsymbol{\Lambda}$ is the $(I \times D)$ matrix of factor loadings, and $\boldsymbol{\delta}$ is $(I \times 1)$ vector of unique scores, normally distributed with mean vector $\mathbf{0}$ and diagonal variance–covariance matrix Δ. It follows from standard multivariate distribution theory that the conditional distribution of Y_i given $\boldsymbol{\theta}$ is normal with mean $\boldsymbol{\lambda}_i'\boldsymbol{\theta}$ and variance $\Delta_i = 1-\boldsymbol{\lambda}_i'\boldsymbol{\Phi\lambda}_i$. We further assume that there is a threshold γ_i such that if $y_i > \gamma_i$, then the observed response $U_i = 1$, and if $y_i > \gamma_i$, $U_i = 0$. (This representation of the normal-ogive model is particularly useful for developing the Gibbs sampling approach for the estimation of parameters in item response models; see Junker et al., Volume Two, Chapter 15.) Hence,

$$\Pr(U_i = 1 \mid \boldsymbol{\theta}) = \Pr(y_i > \gamma_i \mid \boldsymbol{\theta}) = \Pr\left\{Z > \frac{\gamma_i - \boldsymbol{\lambda}_i'\boldsymbol{\theta}}{\sqrt{1-\boldsymbol{\lambda}_i'\boldsymbol{\Phi\lambda}_i}}\right\} = P\left[-\frac{\gamma_i - \boldsymbol{\lambda}_i'\boldsymbol{\theta}}{\sqrt{1-\boldsymbol{\lambda}_i'\boldsymbol{\Phi\lambda}_i}}\right]. \tag{11.24}$$

On comparing Equation 11.24 with Equation 11.17, it is immediately evident that the two formulations are the same except for a linear transformation of the parameters. In comparing the two models, we note the correspondence between the two formulations:

$$\boldsymbol{a}_i = \frac{\boldsymbol{\lambda}_i}{\sqrt{1-\boldsymbol{\lambda}_i'\boldsymbol{\Phi\lambda}_i}}, \quad \beta_i = \frac{-\gamma_i}{\sqrt{1-\boldsymbol{\lambda}_i'\boldsymbol{\Phi\lambda}_i}}. \tag{11.25}$$

To reverse this relationship, we first multiply $\boldsymbol{a}_i'$ by $\boldsymbol{\Phi a}_i$ to obtain $\boldsymbol{a}_i'\boldsymbol{\Phi a}_i = \boldsymbol{\lambda}_i'\boldsymbol{\Phi\lambda}_i/1-\boldsymbol{\lambda}_i'\boldsymbol{\Phi\lambda}_i$. It follows immediately that $\boldsymbol{\lambda}_i'\boldsymbol{\Phi\lambda}_i = \boldsymbol{a}_i'\boldsymbol{\Phi a}_i/1+\boldsymbol{a}_i'\boldsymbol{\Phi a}_i$. Hence, from Equation 11.25, we have

$$\boldsymbol{\lambda}_i = \frac{\boldsymbol{a}_i}{\sqrt{1+\boldsymbol{a}_i'\boldsymbol{\Phi a}_i}}, \quad \gamma_i = \frac{\beta_i}{\sqrt{1+\boldsymbol{a}_i'\boldsymbol{\Phi a}_i}}. \tag{11.26}$$

For the unidimensional model, the above correspondence reduces to

$$a_i = \frac{\lambda_i}{\sqrt{1-\lambda_i^2}}, \beta_i = \frac{-\gamma_i}{\sqrt{1-\lambda_i^2}}; \quad \lambda_i = \frac{a_i}{\sqrt{1+a_i^2}}, \gamma_i = -\frac{\beta_i}{\sqrt{1+a_i^2}}. \tag{11.27}$$

The relationship (11.26) is termed the delta parameterization by Muthén and Asparouhov (2002) and is useful for aligning the estimates obtained from such software as *Mplus* (Muthén and Muthén, 2010) with that of the *NOHARM* program (Fraser and McDonald, 1988).

We rewrite the multidimensional item response model in Equation 11.17 using the factor-analytic parameterization as

$$P\left[-\frac{\gamma_i - \boldsymbol{\lambda}_i'\boldsymbol{\theta}}{\sqrt{1-\boldsymbol{\lambda}_i'\boldsymbol{\Phi}\boldsymbol{\lambda}_i}}\right] = P(-\gamma_i) + \varphi(-\gamma_i)\sum_{k=1}^{\infty}\frac{1}{k!}\{\boldsymbol{\lambda}_i'\boldsymbol{\Phi}\boldsymbol{\lambda}_i\}^{k/2} H_{k-1}(-\gamma_i)H_k\left(\frac{\boldsymbol{\lambda}_i'\boldsymbol{\theta}}{\sqrt{\boldsymbol{\lambda}_i'\boldsymbol{\Phi}\boldsymbol{\lambda}_i}}\right). \tag{11.28}$$

The first- and second-order proportions π_i and π_{ij} in Equations 11.19 and 11.23, when expressed in terms of the thresholds and factor loadings, become

$$\pi_i = P(-\gamma_i), \tag{11.29}$$

$$\pi_{ij} = P(-\gamma_i)P(-\gamma_j) + \varphi(-\gamma_i)\varphi(-\gamma_j)\sum_{k=1}^{\infty}\frac{\rho_{ij}^k}{k!}H_{k-1}(-\gamma_i)H_{k-1}(-\gamma_j), \tag{11.30}$$

where $\rho_{ij} = \boldsymbol{\lambda}_i'\boldsymbol{\Phi}\boldsymbol{\lambda}_j$ is the correlation between the latent variables (y_i, y_j). The series in Equation 11.30, the tetrachoric series, is usually written using the tetrachoric function, $\tau_k(x)$, which is just the normalized Hermite–Chebychev polynomial, that is, $\tau_k(x) = H_{k-1}(x)/\sqrt{k!}$. The expression for π_{ij} simplifies considerably in the unidimensional case. Setting $\rho_{ij} = 1$ in Equation 11.30, we obtain the equivalent expression to Equation 11.16 as

$$\pi_{ij} = P(-\gamma_i)P(-\gamma_j) + \varphi(-\gamma_i)\varphi(-\gamma_j)\sum_{k=1}^{\infty}\frac{1}{k!}\tau_k(-\gamma_i)\tau_k(-\gamma_j).$$

Christoffersson's (1975) approach to factor analysis of dichotomous items uses the normal-ogive model in the form given in Equation 11.24. He arrived at Equation 11.26 using the tetrachoric series. In contrast, in the approach described above, we arrive at the same result, albeit with different but equivalent parameterizations, using the Fourier expansion of the normal-ogive item response function in Equation 11.27. While at first blush, these two approaches may appear different, they are connected through the Mehler identity in Equation 11.20, which is the Fourier expansion of the bivariate normal density in terms of Hermite–Chebychev polynomials and, expanded in this manner, resolves the bivariate density into its component univariate normal densities. Pearson's tetrachoric series is obtained by integrating the Mehler identity and hence is the Fourier expansion of the bivariate normal ogive in terms of Hermite–Chebychev polynomials. The McDonald approach uses the Fourier expansion of the univariate normal ogive in terms of the Hermite–Chebychev polynomials and the Mehler identity in evaluating the joint proportion (11.21). However, if we used the generating function approach instead of the Mehler identity, then it would seem that McDonald was yet another researcher who discovered the Mehler identity! There is, therefore, some truth in McDonald's remark that "it is pleasing to find that the same result can be obtained either as the solution to the problem of evaluating a double integral connected with the normal distribution or as a solution to the problem of approximating a strictly nonlinear function by a wide-sense linear regression function" (McDonald, 1997, p. 263).

Interpretation of the multidimensional item parameters poses a challenge. In unidimensional item response models written in the form $P(a_i(\theta - b_i))$, b_i is the point on the θ scale where $P(a_i(\theta - b_i)) = 0.5$, that is, the value of θ that is the solution of the equation $a_i(\theta - b_i) = 0$. Interpreted slightly differently, b_i is the distance of the line $\theta - b_i = 0$ from the origin. Expressed in terms of slope a_i and intercept β, this point on the θ scale is $-\beta_i/a_i$, or the shortest distance from the origin to the line $\beta_i + a_i\theta = 0$. In the multidimensional case, with the normal-ogive item response model given by $P(\beta_i + \boldsymbol{a}_i'\boldsymbol{\theta})$, $Pr\{U_i = 1|\theta\} = 0.5$ occurs at points on the plane $\beta_i + \boldsymbol{a}_i'\boldsymbol{\theta} = 0$. This equation does not yield a unique solution for $\boldsymbol{\theta}$. The normal to this plane is given by the direction vector $\boldsymbol{a}_i$, and the shortest distance from the origin to this plane along the normal is $|\beta_i|/\sqrt{\boldsymbol{a}_i'\boldsymbol{a}_i}$. Reckase (1997, p. 276; Volume One, Chapter 12), by drawing the parallel to the unidimensional model, defined item difficulty as $-\beta_i/\sqrt{\boldsymbol{a}_i'\boldsymbol{a}_i}$ and labeled $\sqrt{\boldsymbol{a}_i'\boldsymbol{a}_i}$ the multidimensional discrimination index (MDISC). The discrimination of item i on dimension d is then as defined as $\alpha_{id} = \cos^{-1}(a_{id}/\sqrt{\boldsymbol{a}_i'\boldsymbol{a}_i})$, where α_{id} is the angle of the line of steepest slope from the origin to the item response surface. This definition using Euclidean distance is appropriate when the basis for the D-dimensional latent space is orthogonal, that is, when the θs are uncorrelated. However, when they are correlated, use of the Mahalanobis distance from the origin to the plane $\beta_i + \boldsymbol{a}_i'\boldsymbol{\theta} = 0$ is more appropriate; the distance is $-\beta_i/\sqrt{1 + \boldsymbol{a}_i'\boldsymbol{\Phi}\boldsymbol{a}_i} = \gamma_i$, which is the threshold in the factor analytic formulation. Item difficulty defined as the threshold γ_i follows the classic definition of item difficulty as the transformation of the proportion of examinees responding correctly to an item. Similarly, discrimination defined as the factor loading $\lambda_i = a_i/\sqrt{1 + \boldsymbol{a}_i'\boldsymbol{\Phi}\boldsymbol{a}_i}$ agrees with the classical interpretation as the correlation between the item score and the latent trait.

11.3 Parameter Estimation

The proportions π_i and π_{ij} are functions of the parameters β_i, $\boldsymbol{a}_i'$ of the model $P(\beta_i + \boldsymbol{a}_i'\boldsymbol{\theta})$. Only these two moments are used for estimating the parameters. Needless to say, the higher-order joint proportions, π_{ijk}, $\pi_{ijk\ell}$, and so on, contain the complete information, and we can expect that using the full information available would produce more accurate estimates of the parameters. This full-information estimation is time consuming, though. But the limited-information estimation involving only the first two moments may be almost as accurate as the full-information procedure in practice, a conjecture that must be empirically verified. Meanwhile, the limited-information procedure is employed to estimate the parameters β_i, $\boldsymbol{a}_i'$ using only the first and second moments.

We define the vector $\boldsymbol{\pi}' = [\boldsymbol{\pi}_1' \, \boldsymbol{\pi}_2']$ to denote the $I(I+1)/2$ vector of parameters, where $\boldsymbol{\pi}_1' = [\pi_1 \, \pi_2 \ldots]$ is the $(I \times 1)$ vector of the first-order proportions and $\boldsymbol{\pi}_2' = [\pi_{12}\pi_{13} \ldots \pi_{ij}\pi_{I-1,I}]$ is the $(I(I-1)/2 \times 1)$ vector of second-order or joint proportions. Let $\hat{\boldsymbol{\pi}}' = [\hat{\boldsymbol{\pi}}_1' \hat{\boldsymbol{\pi}}_2']$ be the estimator of $\boldsymbol{\pi}'$. The estimates of the model parameters $\beta_i, \boldsymbol{a}_i' (i = 1,\ldots,I)$ are obtained as the values that minimize the unweighted least squares (ULS) function

$$\phi = (\hat{\boldsymbol{\pi}} - \boldsymbol{\pi}_{(r)})'(\hat{\boldsymbol{\pi}} - \boldsymbol{\pi}_{(r)}). \tag{11.31}$$

Here, $\boldsymbol{\pi}_{(r)}$ denotes the vector of the first two moments obtained by approximating the normal-ogive function using a polynomial of degree r, that is, terminating the series in Equations 11.23 and 11.30 after r terms. McDonald (1997) indicated that using the least squares criterion with the assumption that θ is random and the weak principle of local independence enables the estimation of parameters in small samples, a distinct advantage in low volume testing situations.

The estimation procedure described above has been implemented in the program *NOHARM* (normal-ogive harmonic analysis robust method). Details of this program were provided by Fraser and McDonald (1988). McDonald (1982) intimated that a polynomial of degree three is in general sufficient to approximate the normal-ogive function. This suggestion has been followed in *NOHARM*. While this approach may work well with the estimation of parameters, it may not work as well in assessing the fit of the models, as the approximation will be poor at the tails of the response function, and hence, the residuals will be determined poorly at extreme values of $\boldsymbol{\theta}$. The estimation is carried out in two stages: parameters $\gamma_i (= -\beta_i / \sqrt{1 + \boldsymbol{a}_i' \boldsymbol{\Phi} \boldsymbol{a}_i})$ are estimated first from Equation 11.29 or equivalently, Equation 11.19, as $P^{-1}(-\hat{\pi}_i)$. They are then fixed at these values, and $\boldsymbol{a}_i, \boldsymbol{\Phi}$ are estimated using Equation 11.23, minimizing the function using the conjugate gradients method. At convergence, the β_i (and γ_i) parameters are recomputed.

The *NOHARM* program is flexible in that both confirmatory and exploratory analyses can be carried out. In addition, while the c-parameters (the pseudo-chance-level parameters for multiple-choice responses) cannot be estimated, they can be fixed at values set *a priori* while estimating the item parameters β_i and $\boldsymbol{a}_i'$. To eliminate the rotational indeterminacy in the exploratory analysis, $D(D-1)/2$ constraints are imposed by specifying the $(I \times D)$ matrix of the discrimination parameters to have the upper echelon form. The correlation matrix $\boldsymbol{\Phi}$ is set to the $(D \times D)$ identity matrix. The initial solution is then rotated using the Varimax or Promax criterion.

11.3.1 Asymptotic Distribution of Estimators

A shortcoming of the harmonic analysis method of McDonald (1967, 1997) is that distributional properties of the estimators are not provided. As McDonald (1997) noted: "an obvious lacuna, following from the use of ULS, is the statistical test of significance of the model" (p. 265). However, it is possible to derive statistical tests of significance for the ULS estimators, and thereby eliminate the lacuna.

The following general theorem is applicable to the generalized least squares (GLS) estimators and specializes to the ULS estimators. In the following development, n denotes the sample size. Suppose that (a) $\boldsymbol{\pi}$ is the $\mathbf{I}(I+1)/2$ vector of population parameters π_i, π_{ij} and contains only the nonduplicate elements π_{ij}; (b) $\hat{\boldsymbol{\pi}}_n$ is the estimator of $\boldsymbol{\pi}$ based on a sample of size n such that $\sqrt{n}(\hat{\boldsymbol{\pi}}_n - \boldsymbol{\pi}) \xrightarrow{d} N(0, \boldsymbol{\Psi})$; (c) $\boldsymbol{\pi}(\boldsymbol{\vartheta})$ is a function of the structural parameter vector $\boldsymbol{\vartheta}$; (d) $\boldsymbol{\vartheta}_0$ is such that when the model fits, $\boldsymbol{\pi}(\boldsymbol{\vartheta}_0) = \boldsymbol{\pi}$; and (e) the estimates of the structural parameters are obtained as the values that minimize the criterion

$\phi_n(\boldsymbol{\vartheta}) = [\hat{\boldsymbol{\pi}}_n - \boldsymbol{\pi}(\boldsymbol{\vartheta})]' \mathbf{W}_n [\hat{\boldsymbol{\pi}}_n - \boldsymbol{\pi}(\boldsymbol{\vartheta})]$. We shall assume that the model $\boldsymbol{\pi}(\boldsymbol{\vartheta})$ is identified. For example, in the linear factor analysis case, $\boldsymbol{\pi}$ is the vector of variances and covariances (upper diagonal), $\hat{\boldsymbol{\pi}}_n$ is the vector of sample variances and covariances (upper diagonal), and ϑ is the vector of factor loadings, factor variance and covariances, and error variances.

Theorem 11.1

Assume that conditions (a) through (e) hold. Also, assume that the parameter space is compact and $\hat{\boldsymbol{\vartheta}}_n \xrightarrow{p} \boldsymbol{\vartheta}_0$, $\mathbf{W}_n$ converges to a positive definite matrix $\mathbf{W}$, and $\mathbf{G}_n = \partial\pi(\hat{\boldsymbol{\vartheta}}_n)/\partial\boldsymbol{\vartheta}$ is continuous. Then

$$\sqrt{n}(\hat{\boldsymbol{\vartheta}}_n - \boldsymbol{\vartheta}_0) \xrightarrow{d} N[0, (\mathbf{G}'\mathbf{W}\mathbf{G})^{-1}\mathbf{G}'\mathbf{W}\boldsymbol{\Psi}\mathbf{W}\mathbf{G}(\mathbf{G}'\mathbf{W}\mathbf{G})^{-1}]. \tag{11.32}$$

Proof: (Modeled after Newey and McFadden, 1994, p. 2145). By virtue of (e), $\partial\phi_n(\hat{\boldsymbol{\vartheta}})/\partial\boldsymbol{\vartheta} = 0$, that is, $\mathbf{G}_n'\mathbf{W}_n[\hat{\boldsymbol{\pi}}_n - \boldsymbol{\pi}(\hat{\boldsymbol{\vartheta}}_n)] = \mathbf{0}$. By the mean-value theorem, $\mathbf{G}_n'\mathbf{W}_n\{\hat{\boldsymbol{\pi}}_n - [\boldsymbol{\pi}(\boldsymbol{\vartheta}_0) + \mathbf{G}(\boldsymbol{\vartheta}^*)(\hat{\boldsymbol{\vartheta}}_n - \boldsymbol{\vartheta}_0)]\} = 0$, where $\boldsymbol{\vartheta}^*$ is a point between $\boldsymbol{\vartheta}_0$ and $\hat{\boldsymbol{\vartheta}}$. Solving for $(\hat{\boldsymbol{\vartheta}}_n - \boldsymbol{\vartheta}_0)$ and multiplying by $\sqrt{n}$, we obtain $\sqrt{n}(\hat{\boldsymbol{\vartheta}}_n - \boldsymbol{\vartheta}_0)] = [\mathbf{G}_n'\mathbf{W}_n\mathbf{G}_n(\boldsymbol{\vartheta}^*)]^{-1}\mathbf{G}_n'\mathbf{W}_n\sqrt{n}[\hat{\boldsymbol{\pi}}_n - \boldsymbol{\pi}(\boldsymbol{\vartheta}_0)]$. As $\hat{\boldsymbol{\vartheta}}_n \xrightarrow{p} \boldsymbol{\vartheta}_0$ and $\mathbf{W}_n \xrightarrow{p} \mathbf{W}$, the Slutsky theorem (Amemiya, 1985, p. 89; Rao, 1973, pp. 122–124) assures us that $G_n(\hat{\boldsymbol{\vartheta}})$, $G_n(\boldsymbol{\vartheta}^*) \xrightarrow{p} \mathbf{G}$. (Rao, 1973, provides the proof, but does not refer to the result as the Slutsky theorem) When the model holds, $\boldsymbol{\pi}(\boldsymbol{\vartheta}_0) = \boldsymbol{\pi}$, and hence

$$\sqrt{n}(\hat{\boldsymbol{\vartheta}}_n - \boldsymbol{\vartheta}_0) \xrightarrow{p} [\mathbf{G}'\mathbf{W}\mathbf{G}]^{-1}\mathbf{G}'\mathbf{W}\sqrt{n}[\hat{\boldsymbol{\pi}}_n - \boldsymbol{\pi}(\boldsymbol{\vartheta}_0)] = \mathbf{A}\sqrt{n}[\hat{\boldsymbol{\pi}}_n - \boldsymbol{\pi}(\boldsymbol{\vartheta}_0)].$$

From (b), $\sqrt{n}\ [\hat{\boldsymbol{\pi}}_n - \boldsymbol{\pi}(\boldsymbol{\vartheta}_0)] \xrightarrow{d} N(\mathbf{0}, \boldsymbol{\Psi})$. It follows from the Slutsky theorem or Rao (1973) that, asymptotically, $\sqrt{n}(\hat{\boldsymbol{\vartheta}}_n - \boldsymbol{\vartheta}_0)$ behaves as a linear combination of normally distributed variables with mean zero and variance–covariance matrix $\mathbf{A}\boldsymbol{\Psi}\mathbf{A}'$, that is,

$$\sqrt{n}(\hat{\boldsymbol{\vartheta}}_n - \boldsymbol{\vartheta}_0) \xrightarrow{d} N[0, (\mathbf{G}'\mathbf{W}\mathbf{G})^{-1}\mathbf{G}'\mathbf{W}\boldsymbol{\Psi}\mathbf{W}\mathbf{G}(\mathbf{G}'\mathbf{W}\mathbf{G})^{-1}].$$

For the ULS estimator, setting $\mathbf{W} = \mathbf{I}$, the result simplifies to

$$\sqrt{n}(\hat{\boldsymbol{\vartheta}}_n - \boldsymbol{\vartheta}_0) \xrightarrow{d} N[0, (\mathbf{G}'\mathbf{G})^{-1}\mathbf{G}'\boldsymbol{\Psi}\mathbf{G}(\mathbf{G}'\mathbf{G})^{-1}]. \tag{11.33}$$

To apply this theorem to the *NOHARM* estimators, we first note that the sample proportions $\hat{\pi}_i, \hat{\pi}_{ij}$ are consistent estimators of π_i, π_{ij} and condition (b) holds. The other regularity conditions are general conditions that apply to almost all estimation procedures; the requirement that the parameter space is compact ensures that the limit point $\boldsymbol{\vartheta}_0$ lies within the parameter space.

We now have to determine an estimate of $\boldsymbol{\Psi}$. Christoffersson (1975) gave explicit expressions for computing the elements of $\hat{\boldsymbol{\Psi}}$. Alternatively, we denote $\boldsymbol{\pi}' = [\boldsymbol{\pi}_1' \ \boldsymbol{\pi}_2']$ as the $I(I+1)/2$ vector of first- and second-order proportions and $\hat{\boldsymbol{\pi}}_p' = [\hat{\boldsymbol{\pi}}_{p1}' \ \hat{\boldsymbol{\pi}}_{p2}']$ as its estimate for subject p; $\hat{\boldsymbol{\pi}}_{p1}'$ is the $(I \times 1)$ vector with elements $\hat{\pi}_{pi}$, $i = 1, 2, \ldots I$, and $\hat{\boldsymbol{\pi}}_{p2}'$ is the

$\frac{1}{2}I(I-1)\times 1$ vector with elements $\hat{\pi}_{pij}$, $i=1,2,\ldots,I-1$; $j=i+1,i+2,\ldots,I$. The estimate of the $\frac{1}{2}I(I+1)\times\frac{1}{2}I(I+1)$ variance–covariance matrix is then obtained in the usual manner as $\hat{\mathbf{\Psi}}=(1/P-1)\sum_{p=1}^{P}\left(\hat{\pi}_p-\bar{\hat{\pi}}\right)\left(\hat{\pi}_p-\bar{\hat{\pi}}\right)'$, where $\bar{\hat{\pi}}$ is the mean of the vectors.

The Jacobian $\mathbf{G}=(\partial\boldsymbol{\pi}/\partial\boldsymbol{\vartheta})$ needs to be computed next. In organizing the elements of $\mathbf{G}$, we shall follow the convention that if $\boldsymbol{\pi}$ is $k\times 1$ and $\boldsymbol{\vartheta}$ is $\ell\times 1$, then $\mathbf{G}$ is a $k\times\ell$ matrix. Because of its notational simplicity, the factor-analytic formulation will be used to obtain the derivatives and the asymptotic standard errors of the estimates. We partition $\boldsymbol{\vartheta}'=[\boldsymbol{\gamma}',\boldsymbol{\tau}']$, where $\boldsymbol{\tau}$ contains the free elements of the rows of $\mathbf{\Lambda}$ arranged as column vectors followed by the upper diagonal elements of $\mathbf{\Phi}$ arranged as a $\frac{1}{2}D(D-1)\times 1$ column vector to yield a stacked vector of dimension $m\times 1$. Hence, $\mathbf{B}$ is a matrix of order $\frac{1}{2}I(I+1)\times(m+I)$. The matrix elements of $\mathbf{G}$ are organized as

$$\mathbf{G}=\begin{bmatrix}\mathbf{G}_{11}=\dfrac{\partial\boldsymbol{\pi}_1}{\partial\boldsymbol{\gamma}} & \mathbf{G}_{12}=\dfrac{\partial\boldsymbol{\pi}_1}{\partial\boldsymbol{\tau}}\\ \mathbf{G}_{21}=\dfrac{\partial\boldsymbol{\pi}_2}{\partial\boldsymbol{\gamma}} & \mathbf{G}_{22}=\dfrac{\partial\boldsymbol{\pi}_2}{\partial\boldsymbol{\tau}}\end{bmatrix}. \tag{11.34}$$

The dimensions of $\mathbf{G}_{11}$, $\mathbf{G}_{12}$, $\mathbf{G}_{21}$, and $\mathbf{G}_{22}$ are $I\times I$, $I\times m$, $\frac{1}{2}I(I-1)$, and $\frac{1}{2}I(I-1\times m)$, respectively. It is immediately seen from Equation 11.29 that $\mathbf{G}_{11}$ is a diagonal matrix, with diagonal elements $\varphi(-\gamma_i)$, and that $\mathbf{G}_{12}=\mathbf{O}$. The evaluation of $\mathbf{G}_{21}$ may be carried out using Equation 11.30 but, following Muthén (1978), it is more easily obtained by differentiating $\pi_{ij}=\int_{\gamma_i}^{\infty}\int_{\gamma_j}^{\infty}\varphi(x,y)dxdy$ directly. Using a change of variable, it can be seen that

$$\frac{\partial\pi_{ij}}{\partial\gamma_i}=-\varphi(\gamma_i)P\left(\frac{-\gamma_j+\rho_{ij}\gamma_i}{\sqrt{1-\rho_{ij}^2}}\right),\quad \frac{\partial\pi_{ij}}{\partial\gamma_j}=-\varphi(\gamma_j)P\left(\frac{-\gamma_i+\rho_{ij}\gamma_j}{\sqrt{1-\rho_{ij}^2}}\right),\quad \frac{\partial\pi_{ij}}{\partial\gamma_\ell}=0. \tag{11.35}$$

Let $\boldsymbol{\rho}$ denote the $\frac{1}{2}I(I-1)$ vector of correlations between items, where $\rho_{ij}=\boldsymbol{\lambda}_i'\mathbf{\Phi}\boldsymbol{\lambda}_j$. To evaluate $\mathbf{G}_{22}$, we express $\mathbf{G}_{22}=\partial\boldsymbol{\pi}_2/\partial\boldsymbol{\tau}=\partial\boldsymbol{\rho}/\partial\boldsymbol{\tau}\,\partial\boldsymbol{\pi}_2/\partial\boldsymbol{\rho}=\mathbf{G}_1\mathbf{G}_2$. To evaluate $\mathbf{G}_2$, we note that term-by-term integration of the infinite series in Equation 11.30, that is, the Mehler identity, yields, upon a little reflection,

$$\pi_{ij}=\int_{\gamma_i}^{\infty}\int_{\gamma_j}^{\infty}\varphi(x,y,\rho_{ij})dxdy=\int_0^{\rho_{ij}}\varphi(\gamma_i,\gamma_j,r_{ij})dr_{ij}, \tag{11.36}$$

a result obtained by Pearson (1900, p. 7). The result in Equation 11.36 immediately shows that $\mathbf{G}_2$ is a diagonal matrix with elements $\varphi(\gamma_i,\gamma_j,\rho_{ij})$. Alternatively, the result could be obtained by showing that $\partial^2\varphi(x,y,\rho)/\partial y\partial x=\partial\varphi(x,y,\rho)/\partial\rho$ using the characteristic function or moment generating function, and integrating both sides with respect to x, y to obtain the bivariate integral for π_{ij}. Upon interchanging the integral and the derivative, we obtain the elements of $\mathbf{G}_2$. To obtain $\mathbf{G}_1$, we need the derivatives of $\rho_{ij}(=\boldsymbol{\lambda}_i'\mathbf{\Phi}\boldsymbol{\lambda}_j)$ with respect to the D elements of $\boldsymbol{\lambda}_i,\boldsymbol{\lambda}_j$, and the derivatives with respect to the nondiagonal elements of $\mathbf{\Phi}$. The necessary derivatives are

$$\frac{\partial \rho_{ij}}{\partial \lambda_{ik}} = \{\mathbf{\Phi}\boldsymbol{\lambda}_j\}_k = \sum_{s=1}^{D} \Phi_{ks}\lambda_{js}; \quad \frac{\partial \rho_{ij}}{\partial \lambda_{jk}} = \{\mathbf{\Phi}\boldsymbol{\lambda}_i\}_k = \sum_{s=1}^{D} \Phi_{ks}\lambda_{is}; \quad \frac{\partial \rho_{ij}}{\partial \boldsymbol{\lambda}_{\ell m}} = 0;$$
$$\frac{\partial \rho_{ij}}{\partial \Phi} = \boldsymbol{\lambda}_i\boldsymbol{\lambda}_j' + \boldsymbol{\lambda}_j\boldsymbol{\lambda}_i' \Rightarrow \frac{\partial \rho_{ij}}{\partial \Phi_{rs}} = \lambda_{ir}\lambda_{js} + \lambda_{is}\lambda_{jr}, i \neq j, r \neq s, \Phi_{rr} = 1. \tag{11.37}$$

The matrix $\mathbf{G}_1$ has $I(I-1)/2$ columns corresponding to the distinct elements of ρ_{ij}, and m rows corresponding to the total number of distinct elements in $\mathbf{\Lambda}$ and $\mathbf{\Phi}$. Once $\mathbf{G}$ is assembled, Equation 11.33 can be used to obtain the variance of the estimators, and hence the standard errors.

The expression for the variance–covariance matrix in Equation 11.33 is appropriate when all parameters are estimated simultaneously. It is not strictly appropriate for the two-stage estimation procedure employed in *NOHARM*, where γ is estimated first and then held fixed when estimating factor loadings and correlations. Maydeu-Olivares (2001) has derived the asymptotic properties of the two-stage *NOHARM* estimators. The development given below differs somewhat from that given by him.

In the first stage, the minimizing function is $\mathbf{G}_{11n}'[\hat{\boldsymbol{\pi}}_{1n} - \boldsymbol{\pi}_1(\boldsymbol{\gamma})] = 0$. Proceeding as in the proof of Theorem 11.1 and noting that $\mathbf{G}_{11}$ is diagonal, we have from Equation 11.33

$$\sqrt{n}(\hat{\boldsymbol{\gamma}} - \boldsymbol{\gamma}_0) = \mathbf{G}_{11}^{-1}\sqrt{n}(\hat{\boldsymbol{\pi}}_1 - \boldsymbol{\pi}_1) \xrightarrow{d} N(0, \mathbf{G}_{11}^{-1}\mathbf{\Psi}_{11}\mathbf{G}_{11}^{-1}). \tag{11.38}$$

In the second stage, the estimating equation for $\boldsymbol{\tau}$ with $\boldsymbol{\gamma} = \hat{\boldsymbol{\gamma}}$ is $\partial\boldsymbol{\pi}_2(\boldsymbol{\tau})'/\partial\boldsymbol{\tau}[\hat{\boldsymbol{\pi}}_{2n} - \boldsymbol{\pi}_2(\hat{\boldsymbol{\tau}}_n, \hat{\boldsymbol{\gamma}})] = 0$, that is, $\mathbf{G}_{22n}'[\hat{\boldsymbol{\pi}}_{2n} - \boldsymbol{\pi}_2(\hat{\boldsymbol{\tau}}_n, \hat{\boldsymbol{\gamma}})] = 0$. Expanding $\mathbf{G}_{22n}'[\hat{\boldsymbol{\pi}}_{2n} - \boldsymbol{\pi}_2(\hat{\boldsymbol{\tau}}_n, \hat{\boldsymbol{\gamma}})]$ using a doubly multivariate Taylor series, we have

$$\mathbf{G}_{22n}'[\hat{\boldsymbol{\pi}}_{2n} - \boldsymbol{\pi}_2(\hat{\boldsymbol{\tau}}_n, \hat{\boldsymbol{\gamma}})] = \mathbf{G}_{22n}'[\hat{\boldsymbol{\pi}}_{2n} - \boldsymbol{\pi}_2(\boldsymbol{\tau}_0, \boldsymbol{\gamma}_0) - \mathbf{G}_{22n}(\hat{\boldsymbol{\tau}}_n - \boldsymbol{\tau}_0) - \mathbf{G}_{21n}(\hat{\boldsymbol{\gamma}} - \boldsymbol{\gamma}_0)] = 0. \tag{11.39}$$

Using Equation 11.38, solving for $(\hat{\boldsymbol{\tau}}_n - \boldsymbol{\tau}_0)$, collecting terms, and simplifying,

$$\begin{aligned}\mathbf{G}_{22n}'\mathbf{G}_{22n}(\hat{\boldsymbol{\tau}}_n - \boldsymbol{\tau}_0) &= -\mathbf{G}_{22n}'\mathbf{G}_{21n}\mathbf{G}_{11n}^{-1}(\hat{\boldsymbol{\pi}}_{1n} - \boldsymbol{\pi}_1(\boldsymbol{\gamma}_0)) + \mathbf{G}_{22n}'[\hat{\boldsymbol{\pi}}_{2n} - \boldsymbol{\pi}_2(\boldsymbol{\tau}_0, \boldsymbol{\gamma}_0)] \\ &= \mathbf{G}_{22n}'[-\mathbf{G}_{21n}\mathbf{G}_{11n}^{-1} : \mathbf{I}](\hat{\boldsymbol{\pi}}_n - \boldsymbol{\pi}(\boldsymbol{\gamma}_0)) = \mathbf{G}_{22n}'\mathbf{M}_n(\hat{\boldsymbol{\pi}}_n - \boldsymbol{\pi}(\boldsymbol{\tau}_0, \boldsymbol{\gamma}_0)),\end{aligned} \tag{11.40}$$

whence,

$$\sqrt{n}(\hat{\boldsymbol{\tau}}_n - \boldsymbol{\tau}_0) \xrightarrow{d} [\mathbf{G}_{22}'\mathbf{G}_{22}]^{-1}\mathbf{G}_{22n}'\mathbf{M} * N(0, \mathbf{\Psi}),$$

by Rao (1973) or the Slutsky theorem. Here, $\mathbf{M} = [-\mathbf{G}_{21n}\mathbf{G}_{11n}^{-1} : \mathbf{I}\ ]$ is the $\frac{1}{2}I(I-1) \times I$ partitioned matrix, $\mathbf{I}$ is the identity matrix of order $\frac{1}{2}I(I-1)$, and $(\hat{\boldsymbol{\pi}}_n - \boldsymbol{\pi})$ is obtained by stacking the vectors $[\hat{\boldsymbol{\pi}}_{1n} - \boldsymbol{\pi}_1(\boldsymbol{\gamma}_0)]$ and $[\hat{\boldsymbol{\pi}}_{2n} - \boldsymbol{\pi}_2(\boldsymbol{\tau}_0, \boldsymbol{\gamma}_0)]$ one below the other. The distribution of the second-stage estimator $\hat{\boldsymbol{\tau}}_{\gamma=\hat{\gamma}}$ is, therefore,

$$\sqrt{n}(\hat{\boldsymbol{\tau}}_{\gamma=\hat{\gamma}} - \boldsymbol{\tau}_0) \xrightarrow{d} N[0, (\mathbf{G}_{22}'\mathbf{G}_{22})^{-1}\mathbf{G}_{22n}'\mathbf{M}\mathbf{\Psi}\mathbf{M}'\mathbf{G}_{22}(\mathbf{G}_{22}'\mathbf{G}_{22})^{-1}]. \tag{11.41}$$

11.4 Goodness of Fit

Goodness-of-fit measures may be obtained by examining the second-stage residuals. McDonald (1997) suggested that a rough indication of fit is that the average residual is of the order $\sqrt{n^{-1}}$. Alternatively, we can use the asymptotic theory provided above to obtain the distribution of the residuals. Denote the residual for the second-stage model as $\hat{e} = [\hat{\boldsymbol{\pi}}_2 - \boldsymbol{\pi}_2(\hat{\boldsymbol{\tau}}, \hat{\boldsymbol{\gamma}})]$. Since $\mathbf{G}_{22}^{\prime}[\hat{\boldsymbol{\pi}}_2 - \boldsymbol{\pi}_2(\hat{\boldsymbol{\tau}}, \hat{\boldsymbol{\gamma}})] = 0$, we can find the distribution of $\hat{e} = [\hat{\boldsymbol{\pi}}_2 - \boldsymbol{\pi}_2(\hat{\boldsymbol{\tau}}, \hat{\boldsymbol{\gamma}})]$ by multiplying it by the orthogonal complement of $\mathbf{G}_{22}$ denoted as $\mathbf{G}_{22}^{c}$ such that $\mathbf{G}_{22}^{c}\mathbf{G}_{22} = \mathbf{O}$. It is easily verified that $\mathbf{G}_{22}^{c} = \mathbf{I} - \mathbf{G}_{22}(\mathbf{G}_{22}^{\prime}\mathbf{G}_{22})^{-1}\mathbf{G}_{22}^{\prime}$ is the orthogonal complement of $\mathbf{G}_{22}$. The matrix $\mathbf{G}_{22}^{c}$ is idempotent and, needless to say, a symmetric matrix of rank $r = \frac{1}{2}I(I-1) - m$; such an orthogonal complement plays an important role in linear models for obtaining the distribution of the residuals. Using Equation 11.39, left multiplying $[\hat{\boldsymbol{\pi}}_2 - \boldsymbol{\pi}_2(\hat{\boldsymbol{\tau}}, \hat{\boldsymbol{\gamma}})]$ by $\mathbf{G}_{22}^{c} = \mathbf{I} - \mathbf{G}_{22}(\mathbf{G}_{22}^{\prime}\mathbf{G}_{22})^{-1}\mathbf{G}_{22}^{\prime}$, expanding using Taylor series, and simplifying using the results $\mathbf{G}_{22n}^{c}\mathbf{G}_{22n} = \mathbf{O}$ and $\sqrt{n}(\hat{\boldsymbol{\gamma}} - \boldsymbol{\gamma}_0) = \mathbf{G}_{21n}\mathbf{G}_{11n}^{-1}\sqrt{n}(\hat{\boldsymbol{\pi}}_1 - \boldsymbol{\pi}_1)$, we have

$$
\begin{aligned}
\mathbf{G}_{22n}^{c}\sqrt{n}[\hat{\boldsymbol{\pi}}_{2n} - \boldsymbol{\pi}_2(\hat{\boldsymbol{\tau}}_n, \hat{\boldsymbol{\gamma}}_n)] &= \mathbf{G}_{22n}^{c}\sqrt{n}[\hat{\boldsymbol{\pi}}_{2n} - \boldsymbol{\pi}_2(\boldsymbol{\tau}_0, \boldsymbol{\gamma}_0) - \mathbf{G}_{21n}\mathbf{G}_{11n}^{-1}(\hat{\boldsymbol{\pi}}_1 - \boldsymbol{\gamma}_0)] \\
&= \mathbf{G}_{22n}^{c}\sqrt{n}[-\mathbf{G}_{21n}\mathbf{G}_{11n}^{-1} : \mathbf{I}][\hat{\boldsymbol{\pi}} - \boldsymbol{\pi}(\boldsymbol{\tau}_0, \boldsymbol{\gamma}_0)] \\
&= \mathbf{G}_{22n}^{c}\mathbf{M}\sqrt{n}[\hat{\boldsymbol{\pi}} - \boldsymbol{\pi}(\boldsymbol{\tau}_0, \boldsymbol{\gamma}_0)].
\end{aligned}
$$

It follows then that

$$
\sqrt{n}\hat{\mathbf{e}} = \sqrt{n}\,[\hat{\boldsymbol{\pi}}_{2n} - \boldsymbol{\pi}_2(\hat{\boldsymbol{\tau}}_n, \hat{\boldsymbol{\lambda}}_n)] \xrightarrow{d} N(0, \mathbf{G}_{22}^{c}\mathbf{M}\boldsymbol{\Psi}\mathbf{M}^{\prime}\mathbf{G}_{22}^{c}). \tag{11.42}
$$

The square roots of the diagonal elements of $\mathbf{G}_{22n}^{c}\mathbf{M}\boldsymbol{\Psi}\mathbf{M}^{\prime}\mathbf{G}_{22n}^{c}$ are the standard errors of the residuals. The standardized residuals follow the standard normal distribution (asymptotically) and hence provide an indication of misfit. Furthermore, since $\mathbf{G}_{22}^{c}$ is idempotent,

$$
\sqrt{n}\mathbf{G}_{22n}^{c}\hat{\mathbf{e}}_n = \sqrt{n}\,\mathbf{G}_{22n}^{c}[\hat{\boldsymbol{\pi}}_{2n} - \boldsymbol{\pi}_2(\hat{\boldsymbol{\tau}}_n, \hat{\boldsymbol{\gamma}}_n)] \xrightarrow{d} N(0, \mathbf{G}_{22}^{c}\mathbf{M}\boldsymbol{\Psi}\mathbf{M}^{\prime}\mathbf{G}_{22}^{c}), \tag{11.43}
$$

and hence the statistic $T = (\sqrt{n}\mathbf{G}_{22n}^{c}\hat{\mathbf{e}}_n)^{\prime}(\sqrt{n}\mathbf{G}_{22n}^{c}\hat{\mathbf{e}}_n) = n\hat{\mathbf{e}}_n^{\prime}\mathbf{G}_{22n}^{c}\hat{\mathbf{e}}_n$ provides an overall test of fit of the model. Although, according to Equation 11.43, $\hat{\mathbf{e}}$ is asymptotically multivariate normal, the quadratic form $n\hat{\mathbf{e}}_n^{\prime}\mathbf{G}_{22n}^{c}\hat{\mathbf{e}}_n$ does not yield a simple chi-square distribution. Based on the result obtained by Box (1954), Satorra (1989) and Satorra and Bentler (1994) provided the following two scaled test statistics that are approximately distributed as chi-square variates:

$$
(a)\;\; T_1 = r c_1^{-1} T \sim \chi_r^2\,; \qquad (b)\;\; T_2 = c_2^{-1} T \sim \chi_c^2, \tag{11.44}
$$

where, in terms of the of the distribution of the residuals in Equation 11.42,

$$
c_1 = Tr(\mathbf{G}_{22}^{c}\mathbf{M}\boldsymbol{\Psi}\mathbf{M}^{\prime}), \quad c_2 = \frac{Tr[(\mathbf{G}_{22}^{c}\mathbf{M}\boldsymbol{\Psi}\mathbf{M}^{\prime})^2]}{Tr(\mathbf{G}_{22}^{c}\mathbf{M}\boldsymbol{\Psi}\mathbf{M}^{\prime})}, \quad c = \frac{Tr[(\mathbf{G}_{22}^{c}\mathbf{M}\boldsymbol{\Psi}\mathbf{M}^{\prime})]^2}{Tr(\mathbf{G}_{22}^{c}\mathbf{M}\boldsymbol{\Psi}\mathbf{M}^{\prime})}. \tag{11.45}
$$

Maydeu-Olivares (2001) used these test statistics to assess the fit of the *NOHARM* model. Albeit for models different from that considered here, Yuan and Bentler (2010) compared the test statistics (*a*) and (*b*) in detail and concluded that the latter performed better than the former in most cases. Maydeu-Olivares (2001), in his examples however, found little difference between the two statistics when fitting normal-ogive models.

11.5 Empirical Example

We illustrate the *NOHARM* procedure using simulated data. Parameters for the simulation were taken from those obtained in Example 7 in the *TESTFACT* manual (Du Toit, 2003). The data were the responses of 1000 students to a 32-item science assessment covering biology, chemistry, and physics. As reported in the manual, a bifactor model was fitted to the data, with a general factor and specific factors for each of the subject areas. Thus, a four-dimensional solution was obtained. We selected 20 of the 32 items for the present simulation, and generated data using the parameter estimates for these items. Additionally, we fitted a unidimensional model to the data and used the resulting parameter estimates to simulate unidimensional data for the 20 items. For both models, we generated 1000 replications of samples of size 200, 500, and 1000, and calculated the RMSE, bias, standard deviation, and average theoretical standard error in Equation 11.41 over replications for all item parameters. We also calculated the goodness-of-fit statistics in Equation 11.44. We provide *Mplus* and *TESTFACT* results for comparison. *Mplus* uses a GLS estimator by default (Muthén and Muthén, Volume Three, Chapter 28), and *TESTFACT* uses a full-information estimation procedure. In computing the residuals for the *NOHARM* procedure, we evaluated the bivariate integrals directly instead of using the series given in Equation 11.30. The direct evaluation of the bivariate integral produced more accurate results than using the series in Equation 11.30 to more than 10 terms.

Accuracy results are reported in Table 11.1. For the unidimensional model, the root mean square error (RMSE), bias, and average standard errors for *NOHARM* were almost identical to those obtained by *Mplus* and *TESTFACT*. Overall, the *NOHARM* results were slightly better than those of the other procedures. For the bifactor model, *Mplus* failed to converge in 539 of the 1000 replications for sample sizes of 200, and 166 replications for sample sizes of 500. Overall, the *NOHARM* results were almost as good as those of the full-information *TESTFACT* procedure.

The theoretical standard errors obtained for *NOHARM* estimates compared well with the empirical standard errors provided by the standard deviation of estimates over replications and agreed well with those produced by *Mplus* when it converged. These results are not reported in the interests of space.

The mean-adjusted chi-square goodness-of-fit statistics for *NOHARM*, reported in Table 11.2, appeared to have the expected distribution. For the unidimensional model, the expected mean and standard deviation were 170 and 18.44, respectively. The means over replications ranged from 169.42 to 170.66 and the standard deviations ranged from 18.64 to 19.42. For the bifactor model, the expected mean and standard deviation were 150 and 17.32, respectively. The means over replications ranged from 145.99 to 151.82 and the standard deviations ranged from 18.57 to 19.65. These results were better than those obtained for the Satorra–Bentler chi-square fit statistics reported in *Mplus*.

TABLE 11.1

Average Accuracy Indices across 20 Items for Different Models and Sample Sizes

		NOHARM			*Mplus*			*TESTFACT*		
N	**Parameter**	**RMSE**	**Bias**	**SD**	**RMSE**	**Bias**	**SD**	**RMSE**	**Bias**	**SD**
Unidimensional Model										
200	τ	0.091	0.003	0.090	0.091	0.003	0.090	0.092	−0.004	0.091
	λ	0.093	−0.018	0.090	0.094	−0.008	0.093	0.092	−0.025	0.089
500	τ	0.058	0.007	0.056	0.058	0.007	0.056	0.062	−0.008	0.060
	λ	0.061	−0.019	0.058	0.061	−0.016	0.058	0.063	−0.026	0.057
1000	τ	0.040	0.003	0.040	0.040	0.003	0.040	0.041	−0.003	0.040
	λ	0.041	−0.008	0.040	0.041	−0.007	0.040	0.042	−0.016	0.039
Bifactor Model										
200[a]	τ	0.096	−0.033	0.082	0.096	−0.033	0.082	0.096	−0.033	0.082
	λ	0.144	−0.006	0.141	0.150	−0.005	0.144	0.137	−0.004	0.135
500[b]	τ	0.057	0.010	0.052	0.057	0.010	0.052	0.057	0.011	0.052
	λ	0.089	−0.006	0.088	0.092	−0.007	0.090	0.085	−0.006	0.084
1000	τ	0.048	0.030	0.037	0.048	0.030	0.037	0.048	0.030	0.037
	λ	0.061	0.000	0.060	0.062	0.000	0.060	0.058	0.000	0.057

[a] 539 of 1000 replications failed for *Mplus*; 16 replications failed for *TESTFACT*.
[b] 166 of 1000 replications failed for *Mplus*; 1 replication failed for *TESTFACT*.

TABLE 11.2

Mean and Standard Deviations of Model Chi-Square Fit Statistics

			NOHARM		*Mplus*	
Model	**df**	**N**	**Mean**	**SD**	**Mean**	**SD**
Unidimensional	170	200	170.58	19.42	180.07	16.23
		500	170.66	18.64	174.76	16.09
		1000	169.42	19.09	171.51	17.51
Bifactor	150	200[a]	145.99	18.57	165.14	19.16
		500[b]	148.70	19.30	161.83	20.82
		1000	151.82	19.65	159.04	21.76

[a] 539 of 1000 replications failed for *Mplus*.
[b] 166 of 1000 replications failed for *Mplus*.

11.6 Conclusion

The nonlinear factor analysis approach of McDonald coupled with the strong and weak principles of local independence provide a general framework that unifies linear and nonlinear models, showing the relationship between the traditions of factor analysis and IRT. The clear advantage of this unified treatment is that it leads naturally to the conceptualization and operationalization of multidimensional item response models. The harmonic-analysis approach is deeply rooted in the traditions of mathematics and mathematical physics. A major weakness, however, with the *NOHARM* approach was that neither standard errors for parameter estimates nor goodness-of-fit procedures were available. The work of Maydeu-Olivares (2001) as described above has remedied this situation. One area that needs further development is the extension of *NOHARM* for handling ordinal data. A further refinement that is worthy of exploration is to examine the improvement that may be offered by the inclusion of joint proportions for triples and quadruples of items. This extension may provide a compromise between a full-information procedure and a limited-information procedure that only uses the first two moments.

Although multidimensional item response models and estimation procedures have become available (Bock and Lieberman, 1970; Bock and Aitkin, 1981; Gibbons and Hedeker, 1992; Muthén and Muthén, 2010), the *NOHARM* approach has remained popular with researchers for fitting multidimensional item response models. The last few years, however, have seen a flurry of activity in this field. Bayesian approaches using the Markov chain Monte Carlo (MCMC) procedures have been made available by Muthén and Muthén (2010) and implemented in *Mplus* and by Edwards (2010). Cai (2010a,b,c) introduced a hybrid MCMC–Robbins Monroe procedure that has been implemented in *IRTPRO* (2011). Despite these advances, as shown in the examples, *NOHARM* performs well in small as well as in large samples, and almost always converges quickly to acceptable solutions. The rapidity of convergence and ability to handle large or small number of items together with the availability of standard errors of estimates and goodness-of-fit statistics make *NOHARM* a viable approach for fitting multidimensional item response models. To sum, in the words commonly attributed to Hippocrates, "first do *NOHARM*"!

References

Anderson, T. W. 1959. Some scaling models and estimation procedures in the latent class model. In: O. Grenander (Ed.), *Probability and Statistics* (*The Harold Cramer Volume*.) New York: Wiley.

Amemiya 1985. *A Course in Econometrics*. Cambridge, MA: Harvard University Press.

Bock, R. D. 1997. A brief history of item response theory. *Educational Measurement: Issues and Practice*, 16, 21–33.

Bock, R. D., and Lieberman, M. 1970. Fitting a response model for *n* dichotomously scored items. *Psychometrika*, 35, 179–197.

Bock, R. D., and Aitkin, M. 1981. Marginal maximum likelihood estimation of item parameters: Application of an EM algorithm. *Psychometrika*, 46, 443–459.

Box, G. E. P. 1954. Some theorems on quadratic forms applied in the study of analysis of variance problems: I. Effect of inequality of variance in the one-way classification. *Annals of Mathematical Statistics*, 16, 769–771.

Cai, L. 2010a. A two-tier full information item factor analysis by a Metropolis–Hastings Robbins–Monro algorithm. *Psychometrika*, 75, 581–612.

Cai, L. 2010b. High dimensional exploratory item factor analysis by a Metropolis–Hastings Robbins–Monro algorithm. *Psychometrika*, 75, 33–57.

Cai, L. 2010c. Metropolis–Hastings Robbins–Monro algorithm for confirmatory item factor analysis. *Journal of Educational and Behavioral Statistics*, 35, 307–335.

Carroll, J. B. 1945. The effect of difficulty and chance success on correlation between items or between tests. *Psychometrika*, 10, 1–19.

Christoffersson, A. 1975. Factor analysis of dichotomized variables. *Psychometrika*, 40, 5–32.

Courant, R. and Hilbert, D. 1953. *Methods of Mathematical Physics* (Volume I). New York: Interscience Publishers.

Du Toit, M. (Ed.) 2003. *IRT from SSI: BILOG-MG, MULTILOG, PARSCALE, TESTFACT.* Lincolnwood, IL: Scientific Software International.

Edwards, M. C. 2010. A Markov chain Monte Carlo approach to confirmatory item factor analysis. *Psychometrika, 75*, 474–487.
Etezadi-Amoli, J., and McDonald, R. P. 1983. A second generation nonlinear factor analysis. *Psychometrika*, 48, 315–342.
Finney, D. J. 1944. The application of probit analysis to the results of mental tests. *Psychometrika*, 9, 31–39.
Fraser, C., and McDonald, R. P. 1988. *NOHARM*: Least squares item factor analysis. *Multivariate Behavioral Research*, 23, 267–269.
Ferguson, G. A. 1941. Factorial interpretation of test difficulty. *Psychometrika*, 6, 323–329.
Ferguson, G. A. 1942. Item selection by the constant process. *Psychometrika*, 7, 19–29.
Gibbons R. D., and Hedeker, D. 1992. Full information item bi-factor analysis. *Psychometrika*, 57, 423–436.
Gibson, W. A. 1960. Non-linear factors in two dimensions. *Psychometrika*, 25, 381–392.
Guilford, J. P. 1941. The difficulty of a test and its factor composition. *Psychometrika*, 6, 67–77.
Hardy, G. H. 1932. Addendum: Summation of a series of polynomials of Laguerre. *Journal of the London Mathematical Society*, 7, 192.
Hille, E. 1926. A class of reciprocal functions. *Annals of Mathematics*, 27, 427–464.
IRTPRO [Computer Software] 2011. Lincolnwood, IL: Scientific Software International.
Kendall, M. G. 1941. Relations connected with the tetrachoric series and its generalization. *Biometrika*, 32, 196–198.
Kendall, M. G., and Stuart, A. 1969. *Advanced Theory of Statistics* (Volume 1). London: Charles Griffin and Co.
Lancaster, H. O. 1958. Structure of bivariate distributions. *The Annals of Mathematical Statistics*, 29, 719–736.
Lawley, D. N. 1943. On problems connected with item selection and test construction. *Proceedings of the Royal Society of Edinburgh*, 61, 273–287.
Lawley, D. N. 1944. The factorial invariance of multiple item tests. *Proceedings of the Royal Society of Edinburgh*, 62-A, 74–82.
Lord, F. M. 1952. A theory of test scores. *Psychometric Monographs*, No. 7.
Lord, F. M. 1953. An application of confidence intervals and of maximum likelihood to the estimation of an examinee's ability. *Psychometrika*, 18, 57–75.
Lord, F. M. 1980. *Applications of Item Response Theory to Practical Testing Problems*. Hillsdale, NJ: Erlbaum.
Lord, F. M., and Novick, M. 1968. *Statistical Theories of Mental Test Scores*. Reading, MA: Addison Wesley.
Maydeu-Olivares, A. 2001. Multidimensional item response theory modeling of binary data: Large sample properties of *NOHARM* estimates. *Journal of Educational and Behavioral Statistics*, 26, 51–71.
McDonald, R. P. 1962a. A note on the derivation of the general latent class model. *Psychometrika*, 27, 203–206.
McDonald, R. P. 1962b. A general approach to nonlinear factor analysis. *Psychometrika*, 27, 397–415.
McDonald, R. P. 1967. Nonlinear factor analysis. *Psychometric Monographs*, No. 15.
McDonald, R. P. 1982. Linear vs. nonlinear models in latent trait theory. *Applied Psychological Measurement*, 6, 379–396.
McDonald, R. P. 1985. Unidimensional and multidimensional models for item response theory. In D. J. Weiss (Ed.), *Proceedings of the 1982 Item Response and Computerized Adaptive Testing Conference* (pp. 120–148). Minneapolis: University of Minnesota.
McDonald, R. P. 1997. Normal-ogive multidimensional model. In W. J. van der Linden and R. K. Hambleton (Eds.), *Handbook of Modern Item Response Theory* (pp. 257–269). New York: Springer.
Muthén, B. 1978. Contributions to factor analysis of dichotomous variables. *Psychometrika*, 43, 551–560.
Muthén, L. K., and Muthén, B. 2010. *Mplus User's Guide*. Los Angeles, CA: Muthén and Muthén.

Muthén, B., and Asparouhov, T. 2002. Latent variable analysis with categorical outcomes: multiple groups growth modeling in *Mplus*. *Mplus Web Notes, Number 4.*

Newey, W. K., and McFadden, D. L. 1994. Large sample estimation and hypothesis testing. In: R. F. Engle and D. L. McFadden (Eds.) *Handbook of Econometrics* (Volume IV). Amsterdam: Elsevier Science B.V.

Pearson, K. 1900. On the correlation of characters not quantitatively measurable. *Royal Society Philosophical Transactions*, Series A, 195, 1–47.

Rao, C. R. 1973. *Linear Statistical Inference and Its Applications.* New York: Wiley.

Reckase, M. D. 1997. A linear logistic multidimensional model for dichotomous item response data. In: W. J. van der Linden and R. K. Hambleton (Eds.), *Handbook of Modern Item Response Theory* (pp. 271–286). New York: Springer.

Richardson, M. W. 1936. The relation between the difficulty and the differential validity of a test. *Psychometrika*, 1, 33–49.

Satorra, A. 1989. Alternative test criteria in covariance structure analysis: A unified approach. *Psychometrika*, 54, 131–151.

Satorra, A., and Bentler, P. M. 1994. Corrections to test statistics and standard errors in covariance structure analysis. In: A. Von Eye and C. C. Clogg (Eds.), *Latent Variable Analysis. Applications for Developmental Research* (pp. 399–419). Thousand Oaks, CA: Sage.

Thomson, G. H. 1919. A direct deduction of the constant process used in the method of right and wrong. *Psychological Review*, 26, 454–466.

Tucker, L. R. 1946. Maximum validity of a test with equivalent items. *Psychometrika*, 11, 1–13.

van der Linden, W. J., and Hambleton, R. K. 1997. *Handbook of Modern Item Response Theory.* New York: Springer.

Watson, G. N. 1933. Notes on generating functions of polynomials: (2) Hermite polynomials. *Journal of the London Mathematical Society*, 8, 194–199.

Yuan K. H., and Bentler, P. M. 2010. Two simple approximations to the distributions of quadratic forms. *British Journal of Mathematical and Statistical Psychology*, 63, 273–291.

12

Logistic Multidimensional Models

Mark D. Reckase

CONTENTS

12.1 Introduction 189
12.2 Presentation of the Models 190
12.3 Parameter Estimation 203
12.4 Model Fit 204
12.5 Empirical Example 205
12.6 Discussion 207
References 208

12.1 Introduction

Multidimensional item response theory (MIRT) is not a theory in the usual scientific sense of the use of the word, but rather, it is a general class of theories. These theories are about the relationship between the probability of a response to a task on a test and the location of an examinee in a construct space that represents the skills and knowledge that are the targets of measurement for the test. The tasks on the test are usually called test items. Included in the development of the general class of theories are implicit assumptions that people vary in many ways related to the constructs that are the target of the test, and the test items are sensitive to differences on multiple constructs. An important aspect of MIRT is that certain mathematical functions can be used to approximate the relationship between the probabilities of the scores on a test item and the locations of individuals in the construct space (Albert, Volume Two, Chapter 1). This chapter focuses on special cases of MIRT models that are based on one common form for the mathematical function, the cumulative logistic function. This particular functional form is used to relate the probability of the item score for a test item to the location of an individual in the latent space. If this function is found to fit well a matrix of item responses obtained from administering a test to a group of individuals, then there are a number of useful results that can be obtained from the use of this function as a mathematical model. This chapter describes several forms of the MIRT models based on the cumulative logistic function and shows how they can be used to provide useful information about the locations of examinees in the multidimensional latent space.

MIRT models based on the cumulative logistic function can be considered either as generalizations of unidimensional item response theory (UIRT) models or as special cases of factor analysis or similar latent variable models. There are a number of variations of what are called logistic UIRT models based on the work of Birnbaum (1968) who suggested the use of the logistic mathematical form to model item response data. The forms of the MIRT models presented here are generalizations of the logistic UIRT models that were developed by replacing the scalar examinee and discrimination parameters of those models with vectors. There is an extensive literature on logistic UIRT models, including several excellent books (e.g., de Ayala, 2009; Hambleton and Swaminathan, 1985; Lord, 1980; Nering and Ostini, 2010); for a chapter in this handbook, see van der Linden (Volume One, Chapter 2). The models discussed in this literature make the assumption that the goal of the measurement task is to locate the examinee on a single latent continuum. This is usually called the unidimensionality assumption.

MIRT models can also be conceptualized as special cases of factor analysis or structural equation modeling using unstandardized observed variables when doing the analysis (Raykov and Marcoulides, 2011), or as the general case with these other methods as special cases based on standardized variables (McDonald, 1999). In either case, there is a strong connection between factor analysis and MIRT models. Horst (1965) set the basic groundwork for MIRT models by suggesting performing factor analysis on the raw data matrix rather than on correlation or covariance matrices. McDonald (1967) suggested using a nonlinear function in the factor analysis context to deal with the problems of analyzing dichotomous item score data. Christoffersson (1975), Muthen (1978), and Bock and Aitkin (1981) completed the connection between factor analytic procedures and the MIRT approach. This approach involves analyzing the characteristics of test items and using those characteristics to estimate the locations of examinees in the multidimensional construct space. A multilevel perspective is offered by Cai (Volume One, Chapter 25).

12.2 Presentation of the Models

Because of the complexity of real test items and the multiple skills and knowledge that examinees bring to bear when responding to the items, the mathematical model of the relationship between the probability of response to the test items and the location of an examinee in the construct space uses a vector to represent the location of the examinee, $\boldsymbol{\theta}_p = (\theta_{p1}, \theta_{p2}, \ldots, \theta_{pD})$, where p is the index for the examinee and D is the number of coordinate axes (dimensions) needed to describe the capabilities that come to bear when responding to the test items. The elements of the vector provide coordinates for the location of the examinee in a multidimensional space. Differences among examinees are indicated by differences in locations in the space. Two assumptions that are often unstated, but that are critical to the use of the elements of the $\boldsymbol{\theta}_p$-vector, are that the elements of the vector represent points in a Euclidean space and the points are represented by coordinates in a Cartesian coordinate system. The latter point indicates that the elements of the vector represent coordinates on orthogonal axes. These two assumptions allow the use of the standard distance formula for determining the distance between the location of two examinees, p and r,

$$d_{pr} = \sqrt{\sum_{d=1}^{D} (\theta_{pd} - \theta_{rd})^2} \tag{12.1}$$

where d_{pr} is the distance between the two points and the other symbols are previously defined.

Without further assumptions or constraints on the process for estimating the elements of the $\boldsymbol{\theta}_p$-vector, the solution is indeterminate. No origin or units have been defined for the space and any order preserving transformation of the locations of individuals in the space yields coordinates of locations that are equally good as coordinates based on any other such transformation. To define the specific characteristics of the space of point estimates of examinees' locations, typically called defining the metric of the space, three other constraints are usually used. Two of these are that the mean and variance of estimates of the coordinates for a sample of examinees are set to zero and one. These values set the origin and unit for the estimates of the elements of the $\boldsymbol{\theta}_p$-vector. There are other approaches to setting the origin and unit of measurement for the coordinates in the space such as fixing values of item parameters or setting the means of item parameters to specific values. These methods will be discussed after describing the full mathematical forms of the models.

The third constraint that is needed to specify the metric of the space is the form of the relationship between the probabilities of the item responses for a test item and the locations of examinees in the D-dimensional coordinate space. Through the specification of the form of the relationship, the specific characteristics of the metric for the space are defined.

The possible forms for the relationship between the probabilities of a correct response to an item and the locations of examinees in the space are limited only by the imagination of the developer. Many alternative mathematical forms have been proposed. However, at least for MIRT models for achievement or aptitude test items, the form of acceptable models is often limited by an additional constraint. That constraint is that the relationship between probability of correct response and location in the space be monotonically increasing. That is, for an increase in the values of any of the coordinates, the probability of correct response is assumed to increase.

MIRT models are typically models for the relationship between the probability of an item score for one item and the location of one examinee in the coordinate space. That is, the basic model only contains parameters for the characteristics of one item and the location for one examinee. The item parameters are assumed to be a fixed characteristic of the test item, but the probability of the item responses changes with the location of the examinee. In any application, however, the model is used for an entire set of items. Another assumption is then included, which is that the probability of response to one item conditional on the examinee's location is independent of the probability of the response to another item conditional on the same location. The assumption is expressed mathematically as

$$\Pr\{u_1, u_2, \ldots, u_I; \boldsymbol{\theta}_p\} = \prod_{i=1}^{I} \Pr\{u_i; \boldsymbol{\theta}_p\} \tag{12.2}$$

where u_i is the response to item i and the other symbols are as previously defined. This assumption is important for the estimation of the parameters and it is assumed to hold only if $\boldsymbol{\theta}_p$ contains a sufficient number of dimensions to represent the constructs needed to respond to all of the test items. If too few dimensions are included in $\boldsymbol{\theta}_p$, the assumption of local independence will not likely hold. This provides a way of testing if a sufficient number of dimensions has been included for the modeling of the scores on the items.

Although many different types of MIRT models have been proposed, two basic forms appear most frequently in the research literature and in applications of the models. The first of these two models is usually called the *compensatory* MIRT model because an examinee whose location in the space has a high value for one coordinate can balance out a low value for another coordinate to yield the same probability of the item score as an examinee who is located at a place represented by coordinates of roughly equal magnitude. The logistic form of a MIRT model with this compensatory property is given by

$$\Pr\{u_i = 1; \boldsymbol{\theta}_p\} = \frac{e^{\boldsymbol{a}_i'\boldsymbol{\theta}_p + d_i}}{1 + e^{\boldsymbol{a}_i'\boldsymbol{\theta}_p + d_i}} \tag{12.3}$$

where e is the mathematical constant, 2.718 …, that is the base of the natural logarithms, $\boldsymbol{a}_i$ is a vector of item parameters that indicate the rate the probability of a correct response changes as the location on the corresponding coordinate dimension changes, and d_i is a scalar item parameter that is related to the difficulty of the test question. More intuitive meanings for these parameters will be provided later.

The exponent in Equation 12.3 is of the form

$$a_{i1}\theta_{p1} + a_{i2}\theta_{p2} + \cdots + a_{iD}\theta_{pD} + d_i \tag{12.4}$$

From this representation of the exponent, it is clear that if it is set equal to a particular value, k, there here are many combinations of θ-values that will lead to the same sum, even when the a- and d-parameters are held constant. It is this property of the exponent of the model that gives it the label of "compensatory." Reducing the value of one coordinate can be balanced by raising the value of another coordinate in the examinee's $\boldsymbol{\theta}_p$-vector.

The general form of the relationship between the probability of a correct response, $U_i = 1$, and the location of an examinee in the $\boldsymbol{\theta}_p$-space can be seen from a plot of the model when $D = 2$. The plot for the case when $a_{i1} = 1.5$, $a_{i2} = 0.5$, and $d_i = 0.7$ is presented in Figure 12.1.

Several important features of the compensatory MIRT model can be noted from the graphs in Figure 12.1. First, the item response surface increases more quickly along the θ_1-dimension then along the θ_2-dimension. This is a result of the difference in the a-parameters for the item. Also, note from the contour plot that the equiprobable contours are straight

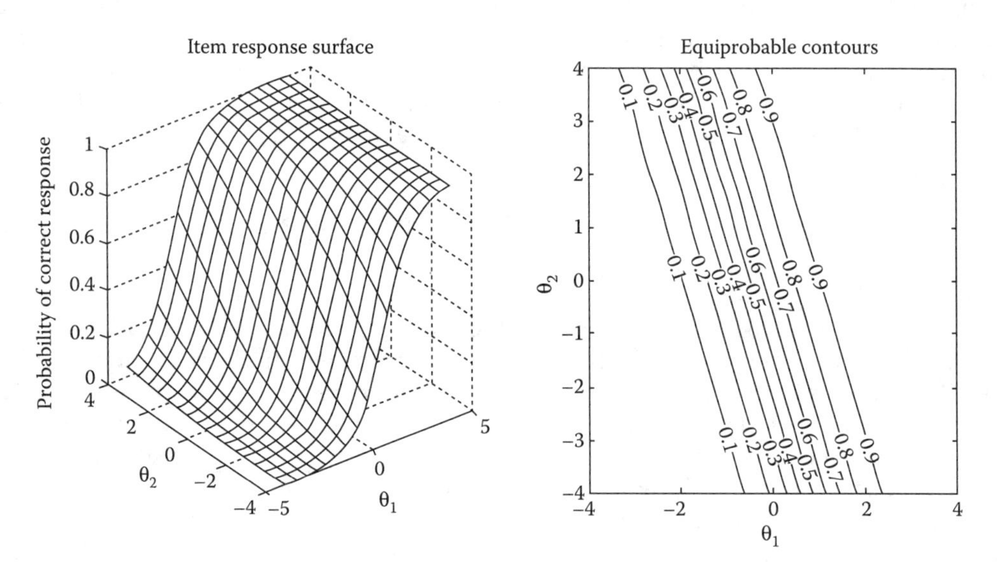

FIGURE 12.1
Graphical representations of the compensatory model—item response surface and equiprobable contours for an item with $a_{i1} = 1.5$, $a_{i2} = 0.5$, and $d_i = 0.7$.

lines. This is the result of the expression that is the exponent of e in the model equation. It is the expression for a line in D-dimensional space. For the example given here, the exponent of e is of the following form:

$$1.5\theta_{p1} + 0.5\theta_{p2} + 0.7 = k \quad (12.5)$$

If k is equal to 0, then the probability of correct response is 0.5 and any combination of θs that result in a value of the expression being 0 define the coordinates of the equiprobable contour of 0.5. Different equiprobable contours are specified by changing the value of k. The orientation of the equiprobable contours is dependent on the a-parameters for the item and the location of the lines is also dependent on the d-parameter.

Two variations of the model given in Equation 12.3 are often used. One variation includes a parameter that indicates the probability of a correct response for examinees who are very low on all of the θ coordinates. This model is specifically designed for use with multiple-choice items. Such items have a nonzero probability of correct response even when examinees taking the items have very little skill or knowledge because it is possible to guess at the answer. The parameter is usually labeled c and the form of the model is given in Equation 12.6.

$$\Pr\{u_i = 1; \theta_p\} = c_i + (1 - c_i)\frac{e^{\mathbf{a}_i'\theta_p + d_i}}{1 + e^{\mathbf{a}_i'\theta_p + d_i}} \quad (12.6)$$

For this model, even if the factor on the right (which is the same as that in Equation 12.3) is very close to 0, the probability of correct response will still be c_i or slightly greater. The value of c_i is a lower asymptote for the probability of correct response for the item. The c-parameter is a scalar rather than a vector because the process used to respond to items when an examinee has limited skills or knowledge is not expected to be related to any specific ability. Although the actual psychological processes examinees use for responding to items when they have limited ability are not known, they are assumed to be related to guessing. However, the c-parameter estimates are often lower than the chance guessing level for an item. Therefore, the parameter is often called the pseudo-guessing parameter.

The second variation of the model given in Equation 12.3 is derived from a family of models that have observable sufficient statistics for the item and examinee parameters. These models were developed based on the work of Rasch (1960, 1961; Volume One, Chapter 3). A typical characteristic of the Rasch family of models is that all of the a-parameters are assumed to have the same value, usually one. However, if all of the a-parameters in the model presented in Equation 12.3 are assumed to be equal, the power of e simplifies to

$$a_i(\theta_{p1} + \theta_{p2} + \cdots + \theta_{pD}) + d_i \quad (12.7)$$

If a_i is the same for all items, it is only possible to estimate the sum of the elements of the $\boldsymbol{\theta}$-vector and not the individual elements. Therefore, this approach to creating a simplified version of the model in Equation 12.3 is not viable.

The approach that has been used is an extension of a general formulation of a model with observable sufficient statistics provide by Rasch (1961). The model can be applied to a wide variety of testing situations, but only the multidimensional version for dichotomous items is presented here. The full version of the model is presented in Adams et al. (1997)

as well as Adams et al. (Volume One, Chapter 32). The model using the same symbols as defined above is given by

$$\Pr\{u_i = 1; \boldsymbol{\theta}_p\} = \frac{e^{\mathbf{a}_i'\boldsymbol{\theta}_p + \mathbf{w}_i'\mathbf{d}}}{1 + e^{\mathbf{a}_i'\boldsymbol{\theta}_p + \mathbf{w}_i'\mathbf{d}}} \tag{12.8}$$

where $\mathbf{a}_i$ is a $D \times 1$ vector of scoring weights that are not estimated, but are specified in advance by the psychometrician using the model; $\mathbf{d}$ is a D-dimensional vector of item characteristics that is constant across all items; and $\mathbf{w}_i$ is a $D \times 1$ set of weights that are set in advance that indicate the way the item characteristics influence the items. In many cases, the elements of $\mathbf{a}_i$ and $\mathbf{w}_i$ are 0 s or 1 s and the vectors serve to select the examinee dimensions and item characteristics that apply for a specific item. It is the fact that the $\mathbf{a}_i$ and $\mathbf{w}_i$ vectors are set in advance that result in the observable sufficient statistics property of these models.

The second of the two basic forms of the MIRT models is often called the *noncompensatory* model, although it is more correctly labeled as the *partially compensatory* model because it does not totally remove the compensation effect present in the model in Equation 12.3. The partially compensatory nature of the model is described in more detail below. The form of this model was first proposed by Sympson (1978) as an alternative to the model given in Equation 12.6. The mathematical form of the model is given in Equation 12.9.

$$\Pr\{u_i = 1; \boldsymbol{\theta}_p\} = c_i + (1 - c_i)\prod_{d=1}^{D} \frac{e^{a_{id}(\theta_{pd} - b_{id})}}{1 + e^{a_{id}(\theta_{pd} - b_{id})}} \tag{12.9}$$

where d is the index for coordinate dimensions and b_{id} is a parameter that indicates the difficulty of performing the tasks related to dimension d. The item response surface and contour plot for this model are presented in Figure 12.2 for the two-dimensional case with

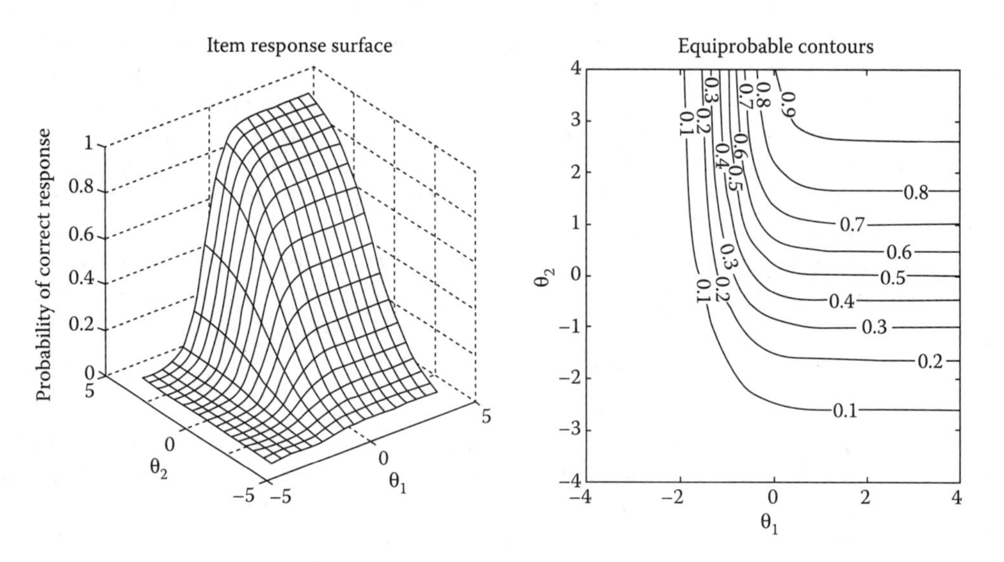

FIGURE 12.2
Graphical representation of the partially compensatory model—item response surface and equiprobable contours for an item with $a_{i1} = 1.5$, $a_{i2} = 0.5$, $b_{i1} = -1$, $b_{i2} = 0$, and $c_i = 0$.

$a_{i1} = 1.5$, $a_{i2} = 0.5$, $b_{i1} = -1$, and $b_{i2} = 0$. The c-parameter for the figure has been set to 0 so that it could be more easily compared to Figure 12.1.

The partially compensatory nature of this model is shown by the curvature of the surface. The overall probability of correct response for an item based on this model can never be higher than the lowest value in the product. Suppose that an examinee has a very low value for one coordinate yielding a value of 0.25 for that term in the product in Equation 12.9. No matter how high the values are for the other coordinates, the overall probability cannot exceed $c + (1 - c)*0.25$ because the maximum for each of the other terms in the product is 1.00. However, for the compensatory model, if one coordinate is very low, the sum in the exponent of Equation 12.6 can still become large enough to yield a very high probability of correct response. Thus, for the partially compensatory model, an increase in one or more of the coordinates can increase the overall probability of correct response somewhat, but only up to the limit set by the lowest coordinate value used in the product.

If the factors of the product are represented in this two-dimensional case as p_1 and p_2, it is easy to see that the curves that define the equiprobable contours are hyperbolas in terms of these probability values, $k = p_1p_2$, where k is the value for any one of the contours, and p_1 and p_2 are the terms in the product in Equation 12.9. The actual curves plotted in Figure 12.2 are not hyperbolas because they are mapped through the logistic functions to the $\boldsymbol{\theta}$ coordinate system. However, like the hyperbola, the curves asymptote to a value as the elements of $\boldsymbol{\theta}$ approach infinity.

The model in Equation 12.9 can be simplified by setting c_i to zero. Then the model parallels the compensatory model given in Equation 12.3. A further simplification of this model can be obtained by setting all of the a_{id} equal to 1. This model is given by

$$\Pr\{u_i = 1; \boldsymbol{\theta}_p\} = \prod_{d=1}^{D} \frac{e^{(\theta_{pd} - b_{id})}}{1 + e^{(\theta_{pd} - b_{id})}} \tag{12.10}$$

Whitely (1980) suggested this model as a way to describe the interactions among the various cognitive components that enter into the performance on an item (see also Volume One, Chapter 14). In this case, these terms in the product follow the Rasch model form with only a difficulty parameter for each dimension.

These two forms of MIRT models have quite different properties. For the compensatory model, as any θ_p approaches infinity, the probability of correct response approaches 1.0, even if the other elements of the $\boldsymbol{\theta}$-vector are very small negative values. For the partially compensatory model, the lowest value of the $\boldsymbol{\theta}$-vector controls the value of the probability of correct response. If all other values of the $\boldsymbol{\theta}$-vector approach infinity except the lowest value, the probability of correct response will still be very small—one value in the product in Equation 12.10 will be very small and all of the other values will be near 1.0. The product is still a small number.

Another difference between the two models is the influence of the number of dimensions in the model on the scaling of the coordinate space. Suppose that the c_i- and d_i-parameters in Equation 12.6 are all zero. Also, suppose that an examinee has elements of the $\boldsymbol{\theta}$-vector that are all zero as well. The probability of correct response in that case is 0.5. Adding another dimension with a $\theta_j = 0$ does not change the probability at all because the exponent remains zero. However, consider the same case for Equation 12.10 with the b-parameters equal to zero. For one dimension, the probability of correct response is 0.5, for two dimensions 0.25, for three dimensions 0.125, etc. In general, the probability of correct response for a constant probability in each component of the model, p, will be p^D, where D is the number of dimensions. If the observed proportion of correct responses for 100 examinees at the same point in the $\boldsymbol{\theta}$-space is 0.5, the values of the parameters needed to give that value will change depending on the number of dimensions. That is, the estimates of $\boldsymbol{\theta}$ for a two-dimensional solution will be quite different in magnitude than estimates of $\boldsymbol{\theta}$ for a three-dimensional solution. This complicates the interpretation of the elements of the $\boldsymbol{\theta}$-vector for the partially compensatory model. The range of values that is considered high or low changes depending on the number of dimensions used in the model.

If the MIRT models provide a reasonable representation of the data obtained from the interaction of examinees with a test item, then the characteristics of the test item are described by the surface defined by the model and they are summarized by its parameters. For example, if the compensatory model given in Equation 12.6 represents the relationship between locations in the $\boldsymbol{\theta}$-space and the proportion of correct responses to an item for a sample of examinees, then the characteristics of the item can be summarized by the **a**-vector, and the c- and d-parameters.

Unfortunately, the model parameters do not have intuitive meaning. To help in the interpretation of the MIRT models, several statistics have been derived from the parameters to help describe the workings of the models. These statistics indicate the combination of skills best measured by the item, the difficulty of the item, and the usefulness of the item for differentiating between examinees at different points in the $\boldsymbol{\theta}$-space.

Generally, the goal of measurement is to differentiate among objects using a particular characteristic of the objects. In this case, the objects are examinees who are located at different places in the $\boldsymbol{\theta}$-space. Their locations are represented by vectors of coordinates (e.g., $\boldsymbol{\theta}_p$, $\boldsymbol{\theta}_r$). Test items are used to gain information about the differences in location of the examinees. Assuming the MIRT model describes the functioning of the test items, the probability computed from the model gives the likelihood of the item response given the examinee's location in the $\boldsymbol{\theta}$-space. The 0.5-equiprobable contour is an important feature of an item because the principle of maximum likelihood indicates that if an examinee gets a correct response to an item, they are estimated to be located on the side of the 0.5 contour that has the higher probability of a correct response. If there is an incorrect response to the item, the examinee is estimated to be on the side of the 0.5 contour that has the lower probabilities of correct response.

Another important characteristic of a test item is how well it differentiates between two examinees located at different points in the $\boldsymbol{\theta}$-space. In the context of MIRT models, differentiation is indicated by a difference in the probability of correct response to the item. If the probability of correct response to the item for the locations of two examinees is the same, the item provides no information about whether the examinees are at the same point or different points. However, if the difference in probability of correct response is large, then examinees are located at different points in the $\boldsymbol{\theta}$-space. Differences in probability of correct response for an item are largest where the slope of the item response surface is greatest, and when points in the space differ in a way that is perpendicular to the equiprobable contours for the item response surface.

These uses of a test item and the characteristics of the item response surface suggest three descriptive measures for an item: (1) the direction of steepest slope for an item; (2) the distance of the point of steepest slope from the origin of the space; and (3) the magnitude of the slope at the point of steepest slope. These three descriptive measures can easily be derived from the mathematical form of the compensatory item response surface, but not for the noncompensatory surface. Measures of direction and distance need to be taken from a particular location. For convenience, the location that is used for these measures is the origin of the space—the $\boldsymbol{\theta}$-vector with all zero elements. The details of the derivations of these item characteristics are given in Reckase (2009, Chapter 5).

The direction of steepest slope from the origin of the $\boldsymbol{\theta}$-space is given by the expression in Equation 12.11 in terms of direction cosines with each of the coordinate axes.

$$\cos\alpha_{id} = \frac{a_{id}}{\sqrt{\sum_{k=1}^{D} a_{ik}^2}} \tag{12.11}$$

where α_{id} is the angle between coordinate axis d and the direction of best measurement (steepest slope of the surface) for item i, and a_{id} is the dth element of the **a**-parameter vector. This result gives the angles with each of the axes with the line from the origin that is perpendicular to the lines of equal probability for the test item response surface. This direction indicates the particular direction between two points in the space that the test item provides the most information about differences in the points.

Another useful piece of information is the distance from the origin to the point of steepest slope in the direction defined by Equation 12.11. This distance is given by

$$Dist_i = \frac{-d_i}{\sqrt{\sum_{k=1}^{D} a_{ik}^2}} \tag{12.12}$$

$Dist_i$ is often called the multidimensional difficulty, or MDIFF, of the item because it is on a scale that is analogous to that used in UIRT models. In those models, the exponent of e is of the form $a(\theta - b) = a\theta - ab$. The ab term is analogous to the d-parameter in the MIRT model, with the sign reversed. To get the equivalent of the b-parameter, the last term is divided by the discrimination parameter, a.

The slope at the point of steepest slope in the direction specified by Equation 12.11 is

$$Slope = \frac{1}{4}\sqrt{\sum_{d=1}^{D} a_{id}^2} \tag{12.13}$$

This value without the 1/4 is often referred to as the multidimensional discrimination, or MDISC for the item. Note that MDISC is in the denominator of Equations 12.11 and 12.12. The results in Equations 12.11 through 12.13 also apply to the model in Equation 12.9 with $c_i > 0$. Changing the c-parameter only raises the item response surface, but does not change the location of the equiprobable contour that has the steepest slope. However, the probability value for that contour is $(c_i + 1)/2$ rather than 0.5.

The values given in Equations 12.11 through 12.13 provide a convenient way to represent items when two or three dimensions are used to model the item response data. Each item can be represented as a vector that points in the direction of increasing probability for the item response surface. The base of the vector can be placed on the equiprobable contour that is the line of steepest slope. The base is a distance MDIFF away from the origin. The length of the vector is a scaled value of MDISC. The direction the vector is pointing is given by Equation 12.11. An example of this vector representation of an item is given on the contour plot of an item response surface in Figure 12.3. Further discussion of the development of these measures is given in Reckase (1985) and Reckase and McKinley (1991).

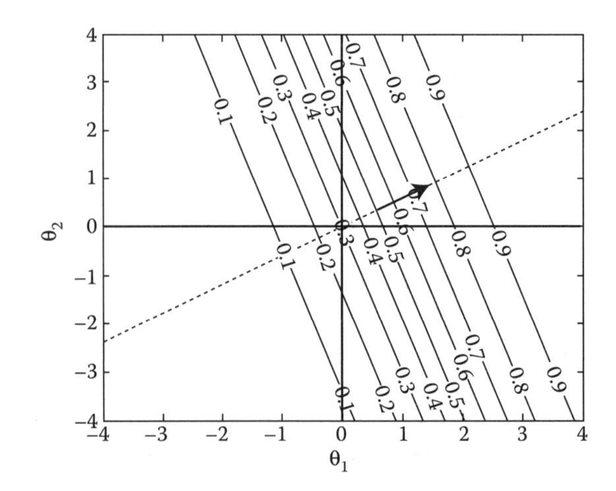

FIGURE 12.3
Item vector representing MDIFF, MDISC, and the direction of best measurement of an item ($a_{i1} = 1.2$, $a_{i2} = 0.4$, $d_i = -0.84$).

The vector plot for the item with the parameters specified in the legend has an MDIFF value of 0.66, and MDISC value of 1.26, and angles with the two axes of 18° and 72°, respectively. The use of item vectors allows many items to be graphically represented at the same time. It is also possible to show items in three dimensions, something that is not possible with response surfaces or contour plots. Figure 12.4 shows items from a science test using three dimensions. This graph shows that the items measure in different directions as the

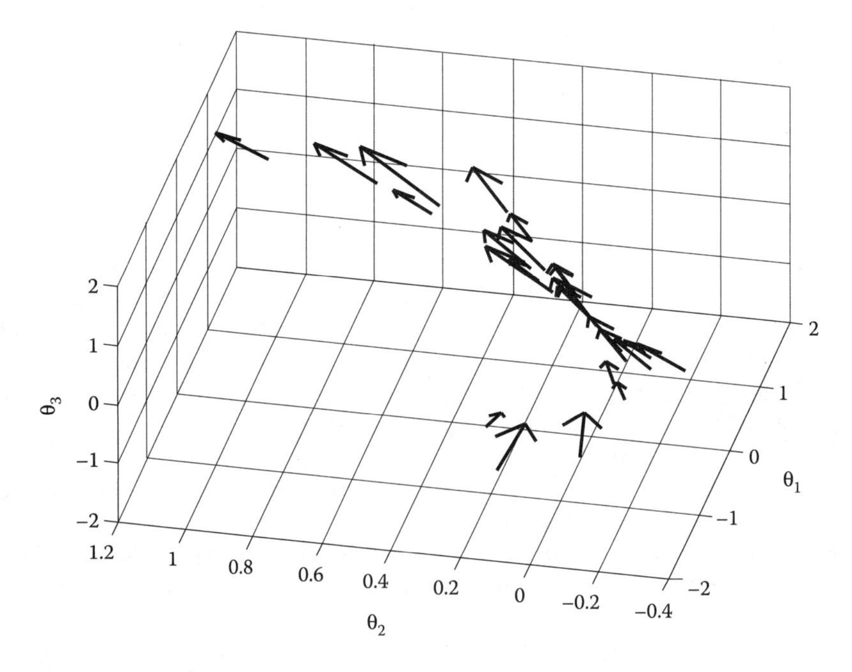

FIGURE 12.4
Vector plot of items in three dimensions.

MDIFF values for the items change. Relatively easy items measure along θ_1, while more difficult items measure along θ_2 and θ_3.

Another way of representing the way that an item functions is its capability to differentiate between two points in the space. This capability is usually labeled as the *information* provided about the location of the examinee in the multidimensional space by the response to the test item. Fisher's information is generally defined by the expected curvature of a likelihood function, more precisely as minus the expected value of its second derivative (Chang and Ying, Volume Two, Chapter 7). For the case of a scalar θ and dichotomous responses, Lord (1980) showed that its expression is equal to

$$I(\theta) = \frac{[\partial P\{\theta\}/\partial\theta]^2}{P\{\theta\}[1-P\{\theta\}]} \tag{12.14}$$

When $\boldsymbol{\theta}$ is a vector rather than a scalar, the information about the difference between a $\boldsymbol{\theta}$-point and a nearby point depends on the direction between the two points because the slope of the surface depends on the direction. Therefore, the numerator of Equation 12.14 is changed to include the directional derivative in angle $\boldsymbol{\alpha}$, where the elements of $\boldsymbol{\alpha}$ are the angles with each coordinate axis, rather than the standard derivative

$$I_{\alpha}(\boldsymbol{\theta}) = \frac{[\nabla_{\alpha} P\{\boldsymbol{\theta}\}]^2}{P\{\boldsymbol{\theta}\}[1-P\{\boldsymbol{\theta}\}]} \tag{12.15}$$

where the directional derivative is defined as follows:

$$\nabla_{\alpha} P\{\boldsymbol{\theta}\} = \frac{\partial P\{\boldsymbol{\theta}\}}{\partial\theta_1}\cos\alpha_1 + \frac{\partial P\{\boldsymbol{\theta}\}}{\partial\theta_2}\cos\alpha_2 + \cdots + \frac{\partial P\{\boldsymbol{\theta}\}}{\partial\theta_D}\cos\alpha_D \tag{12.16}$$

This definition of information is a special case of the more general definition of Fisher information where the information matrix is pre and post multiplied by the vector of direction cosines to give the information in a particular direction. See Chang et al. (Volume Two, Chapter 7, Equation 7.11) for additional details.

Figure 12.5 shows the information for an item with the same parameters as used in Figure 12.3 for directions that are 0°, 45°, and 90° from the θ_1 axis. The information

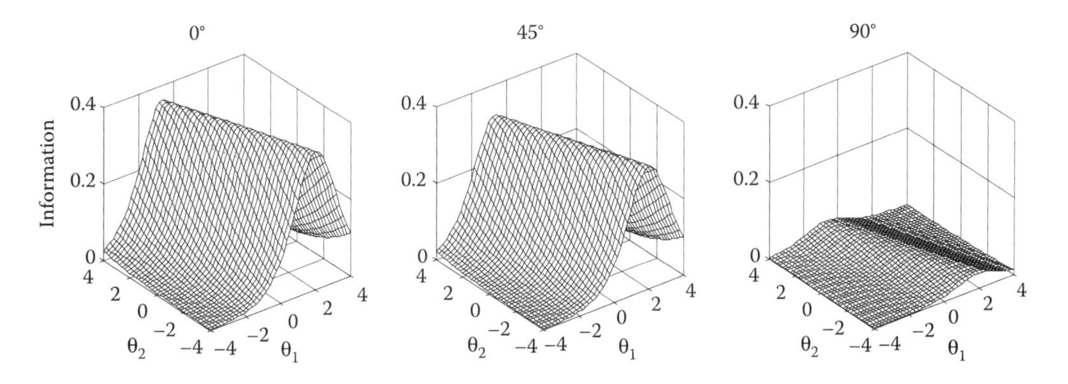

FIGURE 12.5
Information functions for an item in three different directions with the θ_1 axis.

TABLE 12.1

Item Parameters for a 20-Item Test

Item Number	a_{i1}	a_{i2}	d_i	Item Number	a_{i1}	a_{i2}	d_i
1	1.81	0.86	1.46	11	0.24	1.14	−0.95
2	1.22	0.02	0.17	12	0.51	1.21	−1.00
3	1.57	0.36	0.67	13	0.76	0.59	−0.96
4	0.71	0.53	0.44	14	0.01	1.94	−1.92
5	0.86	0.19	0.10	15	0.39	1.77	−1.57
6	1.72	0.18	0.44	16	0.76	0.99	−1.36
7	1.86	0.29	0.38	17	0.49	1.10	−0.81
8	1.33	0.34	0.69	18	0.29	1.10	−0.99
9	1.19	1.57	0.17	19	0.48	1.00	−1.56
10	2.00	0.00	0.38	20	0.42	0.75	−1.61

functions have the same orientation over the $\boldsymbol{\theta}$-plane, but the heights differ. The information functions have the highest points over the 0.5-contour of the surface shown in Figure 12.3. The angle of steepest slope for the surface is 18° which is closest to the 0° direction in the figure. Therefore, the information surface has the highest values in the leftmost panel of Figure 12.5. The 90° direction has substantial lower information values than the others because that direction is almost parallel to the equiprobable contours for the surface.

Item response theory (IRT) in general provides a number of useful ways to describe the characteristics of a test by combining the characteristics of items. For example, because the probability of correct response is the same as the expected score on an item, and because the expected value of a sum is the sum of the expected values, a test characteristic surface can be computed by summing the item characteristic surfaces. The item parameters for 20 items that are represented in a two-dimensional coordinate system are given in Table 12.1. For each of the items on this test, the item characteristic surface was computed. These 20 surfaces were summed to form the test characteristic surface. That test characteristic surface is shown in Figure 12.6.

The height of the surface represents the estimated (classical test theory) true score in the number-correct metric for the test. For examinees who are high on both θ coordinates, the estimated true score is near the maximum summed score on the test of 20. Those low on both θ coordinates have estimated true scores near zero. The surface does not have linear equal score contours because of the combination of multiple items that are measuring best in different directions. Overall, the surface shows how number-correct scores are expected to increase with increases in the coordinate dimensions.

The relationship between item information and the information for the full test is similar to the relationship between the item characteristic surface and the test characteristic surface—the test information is the sum of the item information values in a particular direction. Figure 12.7 shows the test information for the set of items given in Table 12.1 in the 0°, 45°, and 90° directions from the θ_1 axis. The different test information functions show that the form of the information surfaces is not the same for the different directions. The set of items provides more information near the origin of the space in a 45° direction than the other two directions. There are other more subtle differences as well such as having information higher along the θ_2 dimension when measuring in the 90° direction.

The amount of information provided by a test is related to the direction in the space that the test best differentiates between nearby points. The test shown in Figure 12.7 is very

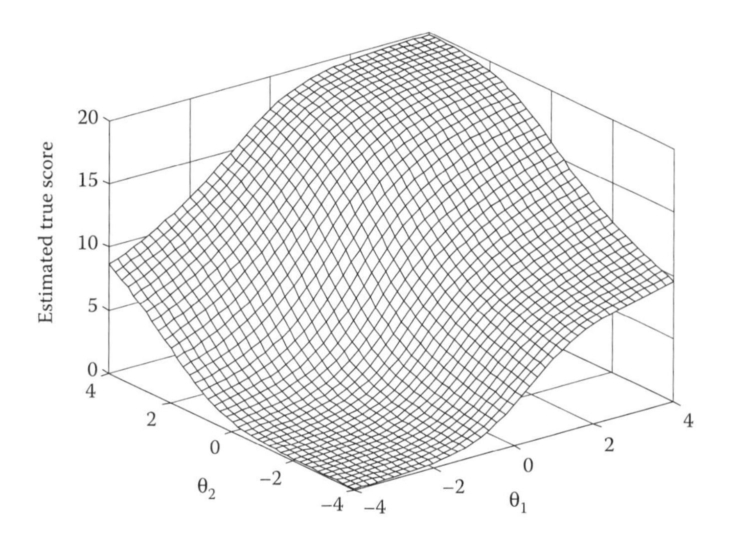

FIGURE 12.6
Test characteristic surface for items in Table 12.1.

good at distinguishing between examinee locations between 0 and 1 on the θ_1 dimension in a direction parallel to that dimension, but is better at distinguishing examinees located between 2 and 3 along the θ_2 dimension in the direction parallel to that dimension.

The single direction best measured by this set of items was derived by Wang (1985). She showed that the direction could be determined from the eigenvector of the **a′a** matrix that corresponds to the largest eigenvalue of that matrix. The values of the eigenvector are the equivalent of the elements of the **a**-parameter vector for an item and can be used to find the direction of measurement for a unidimensional calibration of the test. Wang (1985) labeled the line defined by this eigenvector the *reference composite* for the test. In the case of the set of items in Table 12.1, the reference composite is a line through the origin that has

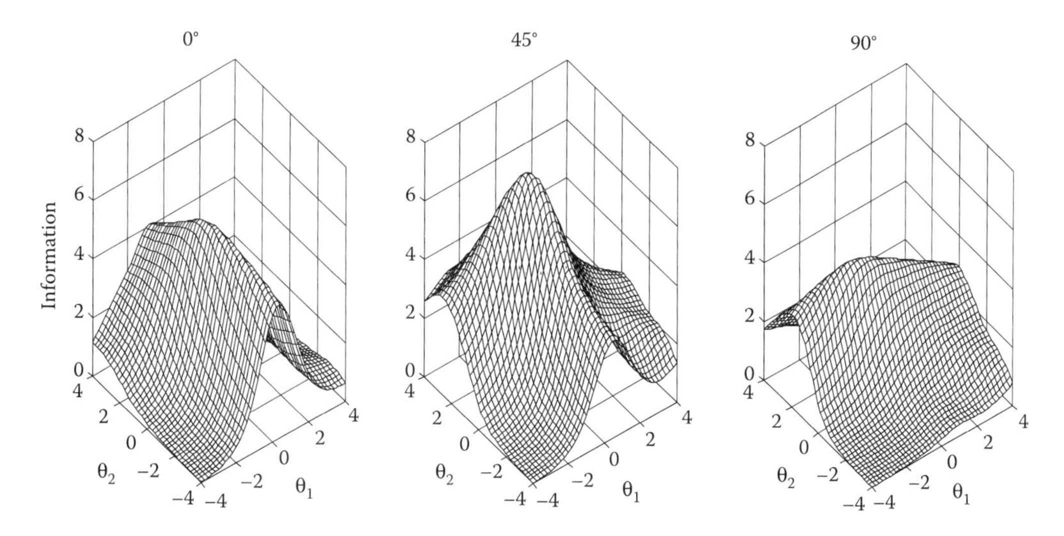

FIGURE 12.7
Test information functions in 0°, 45°, and 90° directions from the θ_1 axis.

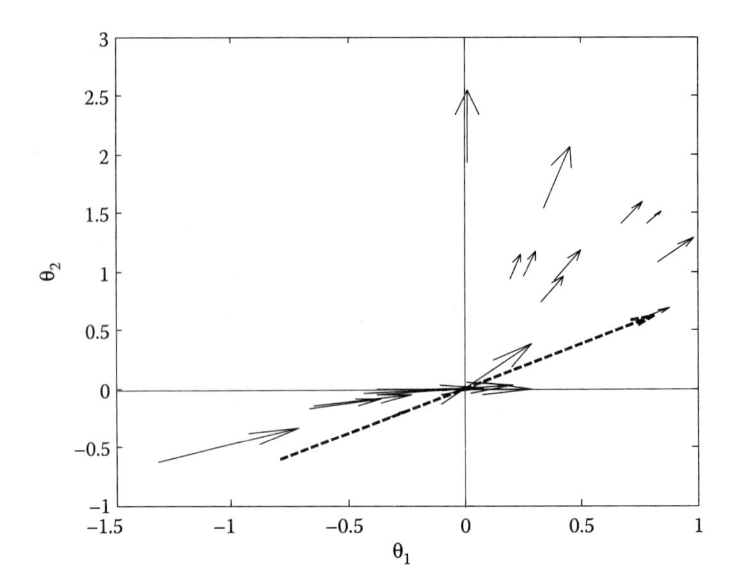

FIGURE 12.8
Item vectors and reference composite vector for the items in Table 12.1.

a 37° angle with the θ_1 axis. The reference composite is shown by the long dashed vector in Figure 12.8.

The concept of a reference composite can be used with sets of items as well as the full test. For example, if the first 10 items in the test are sensitive to differences on one content area, and the second 10 are sensitive to differences on a second content area, reference composites can be determined for each set of items and the directions of the unidimensional scale for those two item sets can be determined. Those two reference composites are shown in Figure 12.9 by the vectors drawn with thick lines. Note that the item vectors for

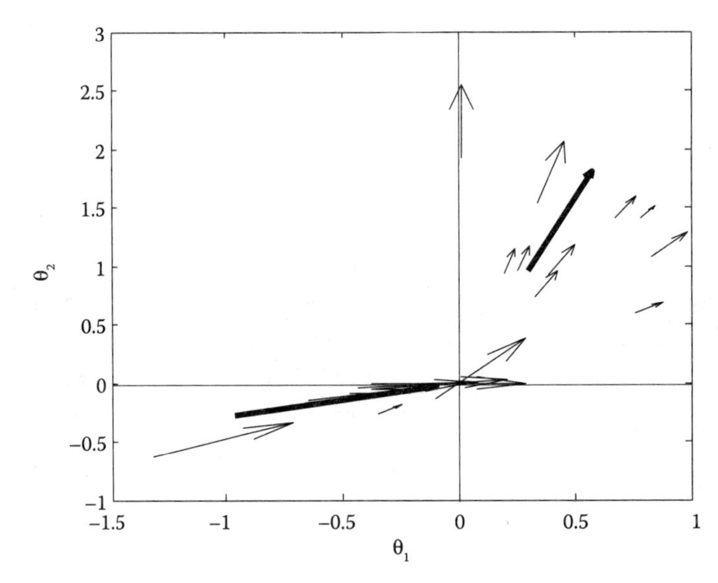

FIGURE 12.9
Reference composites for the first 10 and second 10 items for the items in Table 12.1.

the first 10 items, the easier items, are much more tightly grouped around their reference composite than those for the second set of 10 items, the harder items. When the item vectors for a set of items are all pointing in the same direction, that set of items can be represented using a unidimensional model. The first 10 items meet that condition.

The second 10 items are much more widely dispersed in their directions of best measurement. Although the reference composite for those items indicates the direction of measurement for a unidimensional analysis of them, its set of item response data would not be very well fit by a UIRT model.

12.3 Parameter Estimation

The most frequently applied estimation procedures for MIRT were implemented in two computer programs, NOHARM and TESTFACT. NOHARM (Fraser, 1988) is based on the methodology developed by McDonald (1997; Volume One, Chapter 11) using the normal ogive model. The method programmed in NOHARM uses a four-term polynomial approximation to the normal ogive model. The approximation is used to predict the joint frequency of scores to item pairs using least squares polynomial regression. The item response data from the test is tallied to provide the data in the cells of a two-by-two table of frequencies of response combinations for each pair of items. Such a table is Table 12.2. In the table, n_{ijuv} gives the frequency of the response pair u, v, where u and v can take on the values of 0 and 1 for items i and j. The program uses ordinary least squares to estimate the parameters a and d that minimizes

$$\sum_{i,j,u,v} (n_{ijuv} - \hat{n}_{ijuv})^2 \tag{12.17}$$

where $\hat{n}_{ijuv}$ is the value predicted from the polynomial equation. McDonald (1997) and others have found this procedure to be computationally fast and robust. It has been used to estimate parameters for up to 50 dimensions. However, this approach is only used to estimate the item parameters. The $\boldsymbol{\theta}$-vectors are not estimated in NOHARM, but are assumed to be distributed as multivariate normal with a specified variance–covariance matrix. The program allows fixing parameters to prespecified values.

The other commonly used program for the estimation of parameters is TESTFACT (du Toit, 2003, Chapter 5). This program uses methodology developed by Bock et al. (1988). Like the methodology used in NOHARM, this program assumes the normal ogive formulation of the MIRT model. However, rather than predicting the elements of the 2×2 table of frequencies, TESTFACT determines the parameter estimates that maximize the

TABLE 12.2

Frequency of Item Score Pairs for Two Items

		Item j	
		0	1
Item i	0	n_{ij00}	n_{ij01}
	1	n_{ij10}	n_{ij11}

probabilities of the observed frequencies of the full response pattern on all of the items for the examinees. Because the full response pattern is used, this methodology has been labeled "full information factor analysis." The maximization is done assuming a multivariate normal distribution of proficiencies. The marginal probabilities of the response data used in the estimation are given by

$$P\{\mathbf{U} = \mathbf{U}_s\} = \int_{\boldsymbol{\theta}} L_s(\boldsymbol{\theta}) g(\boldsymbol{\theta}) d\boldsymbol{\theta} \tag{12.18}$$

where $\mathbf{U}_s$ is a particular binary response pattern for a set of items, $L_s(\boldsymbol{\theta})$ is the likelihood of response string at a particular point in the $\boldsymbol{\theta}$-space, and $g(\boldsymbol{\theta})$ is the multivariate normal density for the proficiency values. This integral is approximated using m-fold Gauss–Hermite quadrature. Because estimation is done assuming a known marginal distribution of proficiencies, this estimation procedure is called marginal maximum likelihood (MML). The actual implementation of the estimation procedure uses the expectation maximization (EM) algorithm (Dempster et al., 1977). The details of the process are given in Bock et al. (1988). A new version of this estimation method has been included in the IRTPRO software (Cai, Volume Three, Chapter 24; Cai et al., 2009).

In the last few years, there have been a number of applications of Markov chain Monte Carlo (MCMC) methods to the estimation of MIRT model parameters (for details, see Junker et al., Volume Two, Chapter 15). These methods have been applied to both the compensatory (Bolt and Lall, 2003; Chalmers, 2012; Yao, 2003) and partially compensatory MIRT models (Babcock, 2009). Bolt and Lall (2003) used the Metropolis–Hastings algorithm to estimate the parameters of both the compensatory and partially compensatory logistic models presented in Equations 12.8 and 12.10. Generally, the procedure was able to recover the model parameters in a simulation study, although the recovery was better for the compensatory than the partially compensatory model. A negative aspect of these methods was that estimation took approximately 40 hours for 31 items and 3000 examinees. This should not be taken as too serious a problem because computers are already substantially faster than those used on 2002. Yao (2003) has developed another version of an MCMC-based estimation program.

There are several other programs available for MIRT analysis at various websites, but the capabilities of these programs have not been checked. Users will need to do their own parameter recovery studies to determine how well these programs function.

12.4 Model Fit

The methods for checking the fit of the MIRT models to data are directly related to the method for estimating the model parameters. Methods based on MML such as TESTFACT and IRTPRO provide a chi-squared statistic based on the log-likelihood of the data given the model and the estimated parameters. These chi-squared statistics invariably reject the null hypothesis of model fit because of their strong power to detect small deviations from the model. However, the difference in the chi-squared values is useful for checking the number of dimensions needed to model the data matrix. Shilling and Bock (2005) suggest this approach for selecting the number of dimensions for an analysis.

The NOHARM software provides an option to give a residual matrix that shows the difference between the observed and predicted product moments. Newer versions of the software also provide some of the statistics commonly used in structural equation modeling such as the root mean square error (RMSE) and Akaike information criterion (AIC) as fit measures. These do not provide formal statistical tests of fit but there are frequently used heuristics that are applied to these statistics (Cohen and Cho, Volume Two, Chapter 18; Wells and Hambleton, Volume Two, Chapter 20).

The MCMC methods check model fit using the range of diagnostics provided by those methods (Sinharay, Volume Two, Chapter 19; see Yao, 2003, for a summary of model fit checks based on the BMIRT software).

12.5 Empirical Example

In the area of teacher preparation, there is substantial literature (see Silverman and Thompson [2008] for example) that suggest that good teachers need three types of cognitive skills and knowledge to properly assist students to learn the required subject matter. The three types are (1) knowledge of the materials that their students are going to learn; (2) more advanced knowledge of the subject matter that allows the teacher to put student learning in the proper context; and (3) knowledge of the subject matter specifically focused on teaching. This last type of knowledge includes understanding of common student misconceptions and ways of presenting content to novice learners of the material.

One way of testing this view of the knowledge requirements of teachers is to develop a cognitive test of the subject matter knowledge that is built to assess the three types of knowledge and then check to determine if the three types are distinguishable. McCrory et al. (2012) provides a description of such a research project in the area of secondary mathematics—specifically the teaching of algebra concepts. In that research project, two forms of a test were produced to assess the teacher's knowledge and skills of the three hypothesized types and MIRT analyses were used to check the hypothesis and to refine the instruments and the reporting process.

The tests consisted of two, 20-item forms with five common items. There are 35 items in total. The common items were included so that IRT calibrations of the forms could be linked. The forms were administered to approximately 1100 individuals. The analysis sample was a mix of teachers and teacher candidates. Of the 1100, 841 were used in validation studies and that sample is the focus of this example.

The first step in the analysis process was to determine the number of dimensions to use in the MIRT model. The test plan suggested three or more dimensions because along with the basic theoretical constructs, process skills were being assessed. A parallel analysis procedure applied to the data suggested that two dimensions would be sufficient to explain the relationships in the data matrix. However, Reckase and Hirsch (1991) found that it was better to overestimate the number of dimensions then underestimate it. Underestimation results in constructs being projected into the same place in the latent space, sometimes causing different constructs to be combined into one. To avoid this problem, an analysis was run on the data specifying six dimensions using the TESTFACT software. TESTFACT does not estimate c-parameters, so these were first estimated using BILOG and then they were fixed as parameters in TESTFACT.

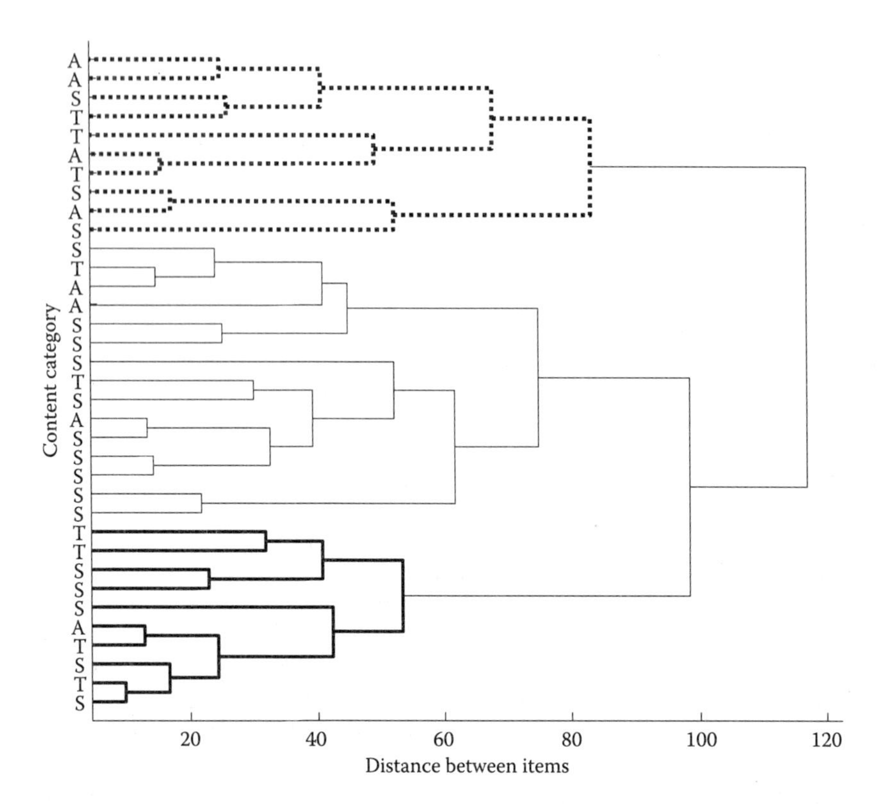

FIGURE 12.10
Dendrogram showing three cluster structure for the test of teacher skills and knowledge.

A cluster analysis was then performed on the angles between the directions of best measurement for the items in the six-dimensional space. If the angle between the directions of best measurement of two items is 0 degrees, the items are measuring the same composite of skills and knowledge. Errors in estimation of the angles of best measurement make it difficult to draw strong conclusions, but the clusters should be suggestive of the constructs assessed by the tests. The developers were also asked to identify the items that were the best exemplars of each of the types of knowledge so that they could be used to identify the clusters and the directions in the space that correspond to the target constructs.

The dendrogram from the cluster analysis is shown in Figure 12.10. The vertical axis gives the items on the test with the classification of the items into school mathematics (S), advanced mathematics (A), and knowledge of mathematics for teaching (T). As the dendrogram indicates, there are three main clusters of items and the content categorizations of the items from the item writers have a weak connection between the clusters and the intended constructs. The top cluster has four of the seven advanced mathematics items. The bottom cluster has four of the nine "teaching" items, and the middle cluster is predominantly school mathematics. Given that the sample size was somewhat small for this type of analysis resulting in fairly large estimation error in the a-parameter estimates, this result was felt to give moderate support to the three-dimensional structure for the test and the basic definition of the constructs. It also indicates that there is quite a bit of overlap between mathematics for teaching and school-level mathematics, and advanced mathematics could be hypothesized as an upward extension of school mathematics.

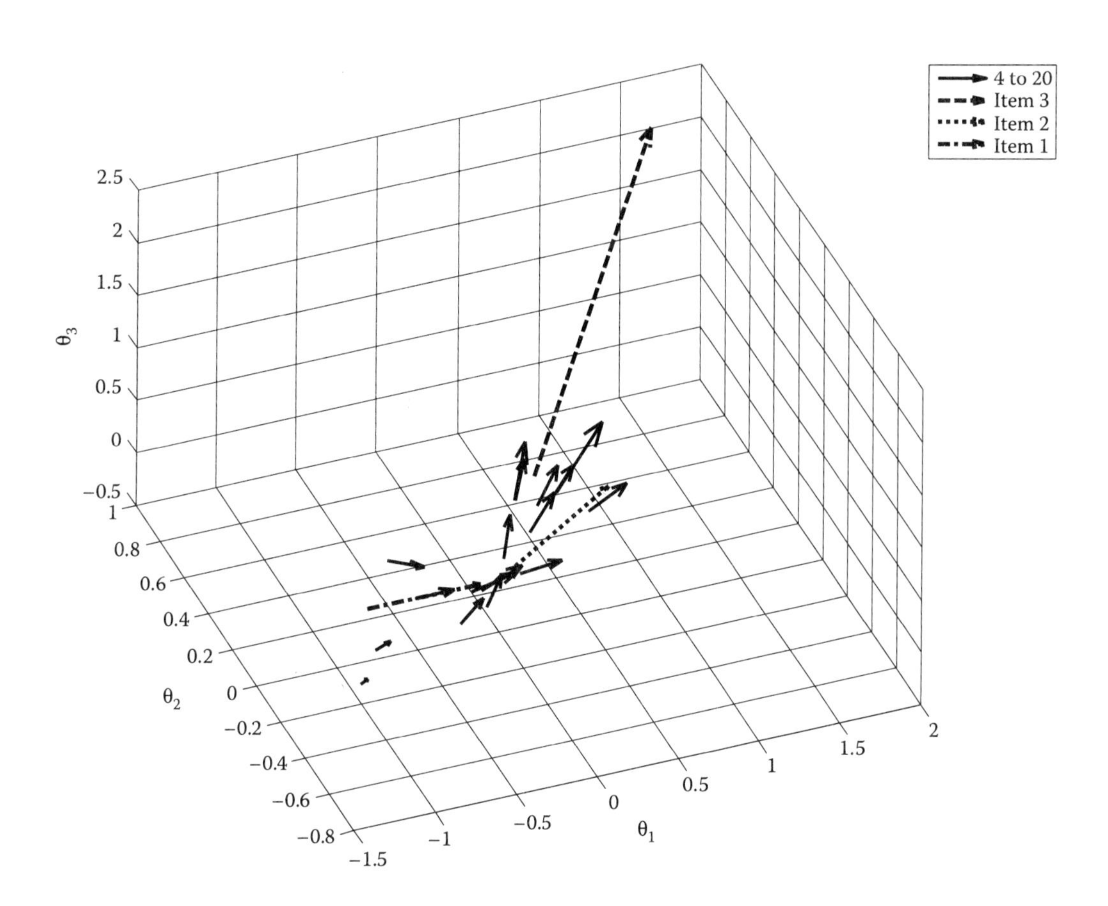

FIGURE 12.11
Three-dimensional item vector representation with exemplar items.

Because the dendrogram gave support for three constructs, the data were calibrated a second time using the NOHARM software with the three exemplar items used to anchor the orientation of the coordinate axes. NOHARM sets the direction of the first axis to correspond to the direction of best measurement of the first item in the data set, then the second axis is set to have the directions of best measurement of the first two items in the same plane, and so on for successive items. Because three dimensions were used for that analysis, the results can be shown graphically using the concept of item vectors. The item vectors are shown in Figure 12.11 with the exemplar items shown with broken and dotted lines. If the exemplar items can be considered as defining the reference constructs for the three constructs in the test, the locations of examinees in the space can be projected onto those reference composites as a way of estimating subscores.

12.6 Discussion

MIRT is a very active area of research, so the information provided in this chapter may already become dated as it was written. The use of these models show promise as methods for identifying and estimating subscore, for understanding the structure of constructs that are the focus of tests, and computerized adaptive methods for measuring multiple constructs. There is also more work being done on estimation procedures and more in-depth understanding of how the models work. It is likely that major applications of these models will become practical in the next several years.

References

Adams, R. J., Wilson, M., and Wang, W. 1997. The multidimensional random coefficients multinomial logit model. *Applied Psychological Measurement, 21*, 1–23.

Babcock, B. G. E. 2009. Estimating a noncompensatory IRT model using a modified Metropolis algorithm. Unpublished doctoral dissertation, University of Minnesota, Minneapolis, MN.

Birnbaum, A. 1968. Some latent trait models and their use in inferring an examinee's ability. In F. M. Lord and M. R. Novick, *Statistical Theories of Mental Test Scores.* Addison-Wesley, Reading, MA.

Bock, R. D. and Aitkin, M. 1981. Marginal maximum likelihood estimation of item parameters. *Psychometrika, 46*, 443–459.

Bock, R. D., Gibbons, R., and Muraki, E. 1988. Full information item factor analysis. *Applied Psychological Measurement, 12*, 261–280.

Bolt, D. M. and Lall, V. F. 2003. Estimation of compensatory and noncompensatory multidimensional item response models using Markov Chain Monte Carlo. *Applied Psychological Measurement, 27*, 395–414.

Cai, L., du Toit, S. H. C., and Thissen, D. 2009. *IRTPRO: Flexible, Multidimensional, Multiple Categorical IRT Modeling* [Computer software]. SSI International, Chicago.

Chalmers, R. P. 2012. MIRT: A multidimensional item response theory package for the R environment. *Journal of Statistical Software, 48*(6), 1–29.

Christoffersson, A. 1975. Factor analysis of dichotomized variables. *Pyschometrika, 40*, 5–32.

de Ayala, R. J. 2009. *The Theory and Practice of Item Response Theory.* The Guilford Press, New York.

Dempster, A. P., Laird, N. M., and Rubin, D. B. 1977. Maximum likelihood estimation from incomplete data via the EM algorithm (with discussion). *Journal of the Royal Statistical Society, Series B, 39*, 1–38.

du Toit, M. 2003. *IRT from SSI: BILOG-MG, MULTILOG, PARSCALE, TESTFACT.* Scientific Software International, Lincolnwood, IL.

Fraser, C. 1998. NOHARM: A FORTRAN program for fitting unidimensional and multidimensional normal ogive models in latent trait theory. The University of New England, Center for Behavioral Studies, Armidale, Australia.

Hambleton, R. K. and Swaminathan, H. 1985. *Item Response Theory: Principles and Applications.* Kluwer, Boston.

Horst, P. 1965. *Factor Analysis of Data Matrices.* Holt, Rinehart and Winston, New York.

Lord, F. M. 1980. *Application of Item Response Theory to Practical Testing Problems.* Lawrence Erlbaum Associates, Hillsdale, NJ.

McCrory, R., Floden, B., Ferrini-Mundy, J., Reckase, M. D., and Senk, S. L. 2012. Knowledge of algebra for teaching: A framework of knowledge and practices. *Journal for Research in Mathematics Education, 43*, 584–615.

McDonald, R. P. 1967. Nonlinear factor analysis. *Psychometric Monographs*, No. 15.

McDonald, R. P. 1997. Normal-ogive multidimensional model. In W. J. van der Linden and R. K. Hambleton (eds.). *Handbook of Modern Item Response Theory.* New York: Springer.

McDonald, R. P. 1999. *Test Theory: A Unified Treatment.* Lawrence Erlbaum Associates, Mahwah, NJ.

Muthen, B. 1978. Contributions to factor analysis of dichotomous variables. *Psychometrika, 43*, 551–560.

Nering, M. L. and Ostini, R. 2010. *Handbook of Polytomous Item Response Theory Models.* Routledge, New York.
Rasch, G. 1960. *Probabilistic Models for Some Intelligence and Attainment Tests.* Danish Institute for Educational Research, Copenhagen.
Rasch, G. 1961. On general laws and the meaning of measurement in psychology. *Proceedings of the Fourth Berkeley Symposium on Mathematical Statistics and Probability, 4,* 321–334.
Raykov, T. and Marcoulides, G. A. 2011. *Introduction to Psychometric Theory.* Routledge, New York.
Reckase, M. D. 1985. The difficulty of test items that measure more than one ability. *Applied Psychological Measurement, 9,* 401–412.
Reckase, M. D. and McKinley, R. L. 1991. The discriminating power of items that measure more than one dimension. *Applied Psychological Measurement, 15,* 361–373.
Reckase, M. D. and Hirsch, T. M. 1991. Interpretation of number-correct scores when the true number of dimensions assessed by a test is greater than two. Paper presented at that annual meeting of the National Council on Measurement in Education, Chicago, IL.
Schilling, S. and Bock, R. D. 2005. High-dimensional maximum marginal likelihood item factor analysis by adaptive quadrature. *Psychometrika, 70,* 533–555.
Silverman, J. and Thompson, P. W. 2008. Toward a framework for the development of mathematical knowledge for teaching. *Journal of Mathematics Teacher Education, 11,* 499–511.
Sympson, J. B. 1978. A model for testing with multidimensional items. *Proceedings of the 1977 Computerized Adaptive Testing Conference.* University of Minnesota, Minneapolis.
Wang, M. 1985. *Fitting a Unidimensional Model to Multidimensional Item Response Data: The Effect of Latent Space Misspecification on the Application of IRT* (Research Report MW: 6-24-85). University of Iowa, Iowa City, IA.
Whitely, S. E. 1980. Multicomponent latent trait models for ability tests. *Psychometrika, 45,* 479–494.
Yao, L. 2003. *BMIRT: Bayesian Multivariate Item Response Theory.* CTB/McGraw-Hill, Monterey, CA.

13

Linear Logistic Models

Rianne Janssen

CONTENTS

13.1 Introduction 211
13.2 Presentation of the Models 213
13.2.1 LLTM 213
13.2.2 Incorporating Random Item Variation 214
13.2.3 Estimating the Item Characteristics from Subtask Data 214
13.2.4 Including Local Dependencies between Items 215
13.2.5 Incorporating Individual Differences in the Effect of Item Features 215
13.2.6 Incorporating Group Differences in the Effects of Item Features 216
13.3 Parameter Estimation 216
13.4 Model Fit 217
13.5 Empirical Example 217
13.5.1 Model and Data Set 217
13.5.2 Method 219
13.5.2.1 Bayesian Model Estimation 219
13.5.2.2 Model Checking 219
13.5.3 Results 219
13.5.3.1 Parameter Estimates 219
13.5.3.2 Fit of the Model to the Data 220
13.5.3.3 Fit of the Model for the Item Parameters 220
13.5.4 Discussion of the Results 220
13.6 Discussion 221
References 221

13.1 Introduction

This chapter deals with response models for dichotomous items that specify item difficulty parameters as a linear function of a set of parameters for the features of the item. Since Fischer's (1973, 1983) seminal work on the linear logistic test model (LLTM), these explanatory item response models (Volume One, Chapter 33) have always been present at the psychometric forefront. Several models have been developed that extend the basic LLTM. Besides, an empirical line of research has been established dealing with the effects of item features on item difficulty. For instance, both developments have led to research on automated item generation (Volume One, Chapter 26), in which the item features found to influence item difficulty are integrated in a set of item-generating rules.

Fischer (1973) proposed the LLTM because for some tests, item difficulty can be conceived of as a function of certain cognitive operations involved in the solution process, as was shown by Scheiblechner (1972). In the LLTM, each cognitive operation is characterized by a so-called basic parameter η_k, referring to the difficulty of executing it. Besides, each test item is characterized by the number of times (q_{ik}) the cognitive operation is needed to solve the item. Consequently, the item difficulty of the test item is assumed to be equal to a sum over all K cognitive operations of the basic parameters weighted by the frequency of their occurrence in the item. Later applications of the LLTM did not stick to modeling cognitive operations but generalized the approach to deal with content features that can also account for differences in item difficulty, such as the familiarity of the vocabulary in a reading comprehension test. For these approaches, q_{ik} is no longer restricted to represent frequencies of cognitive operations. As summarized in Table 13.1, the LLTM and related models have been applied in a variety of domains, including mathematics, science, reasoning, reading comprehension, personality, and emotions.

Apart from modeling item difficulty as a function of substantive item features, the LLTM can also be used to investigate formal item characteristics or test administration effects, such as item-position effects (Debeer and Janssen, 2012; Fischer, 1995b; Hahne, 2008; Hohensinn et al., 2008; Kubinger, 2008, 2009), effects of different response formats (Kubinger and Draxler, 2006), or rater effects (Draney and Wilson, 2008). In each of these applications, the LLTM is applied to a generalized or virtual item (e.g., item × condition combination, or item × rater combination), which is assumed to consist of two components: one for the original item and one for the condition or rater. In fact, Linacre (1989) followed up on this idea in his facets model, which in addition to the two main facets considered in the Rasch model (item difficulty and subject ability) allows for context or situation facets

TABLE 13.1

Examples of Applications of the LLTM and Related Models

Domain	Subdomain	Examples of Publications
Mathematics	Calculus	Fischer (1973)
	Mathematical problem solving	Daniel and Embretson (2010)
Science	Elementary mechanics	Spada and May (1982)
	Statistical word problems	Holling et al. (2008)
		Kubinger (1979, 1980)
Reasoning	Figural matrices	Embretson (1985, 1998)
		Fischer and Formann (1982)
		Freund et al. (2008)
		Hornke and Habon (1986)
	Latin square task	Zeuch et al. (2011)
	Spatial ability	Ivie and Embretson (2010)
		Smith et al. (1992)
Reading	Reading comprehension	Embretson and Wetzel (1987)
		Gorin (2005); Gorin and Embretson (2006)
		Sonnleitner (2008)
	Document literacy	Sheehan and Mislevy (1990)
	Phoneme segmentation	Bouwmeester et al. (2011)
Personality	Self-reported hostility	Van Steelandt (1999)
	Verbal aggression	De Boeck and Wilson (2004)
Emotions	Feeling guilty	Smits and De Boeck (2003)

(or factors), such as different contexts or time points when the items were administered, or different judges evaluating the responses to the items. As in the LLTM, these additional facets may also be regarded as a decomposition of the original single Rasch item difficulty parameter.

13.2 Presentation of the Models

We first present the LLTM in more detail and then discuss several extensions of it. These extensions lead to two different types of models. The first type extends the LLTM on the item side. A first example is to allow for random variation in item difficulty in addition to the item characteristics. In a second example, the item characteristics do not have to be known *a priori*, but are estimated from subtask data. A final example of the extension of the LLTM on the item side takes into account local item dependencies when modeling the effect of item features. The second type of extensions of the LLTM is on the person side, allowing for individual or group differences in the effect of the item features. Parallel to the development of the LLTM described here, Fischer (1995a) also developed his linear logistic model with relaxed assumptions (LLRA); for an application of this type of model to the measurement of change, see Fischer (Volume Three, Chapter 19).

13.2.1 LLTM

The LLTM assumes the Rasch model to hold for the response of examinee p to item i (Volume One, Chapter 3). However, it replaces the difficulty parameter b_i by a model for it. Taking the logit of the success probability, the LLTM can be written as

$$logit[\Pr\{U_{pi}=1\}]=\theta_p-\sum_{k=1}^{K}\eta_k q_{ik} \tag{13.1}$$

where q_{ik} denotes an *a priori* weight for item i on the kth item feature and η_k refer to the weight of the item feature k. The latter were called the basic parameters of the LLTM by Fischer (1983, 1995b). An item-design matrix Q can be defined to contain the weights of all items on each item characteristic. Commonly, Q contains a constant for the first item feature, which serves as a normalization constant comparable to the intercept in a regression model. For the LLTM to be identified, either the mean ability of the examinees or the normalization constant is set to zero.

Embretson (1999) combined the two-parameter logistic (2PL) model with the LLTM. In her so-called 2PL-constrained model, both the item difficulty and discrimination parameters are decomposed into a weighted sum of item properties. More formally,

$$logit[\Pr\{U_{pi}=1\}]=\left(\sum_{k=1}^{K}\tau_k q_{ik}\right)\left(\theta_p-\sum_{k=1}^{K}\eta_k q_{ik}\right) \tag{13.2}$$

where τ_k refers to the weight of the item characteristic k for its impact on the item discrimination.

13.2.2 Incorporating Random Item Variation

The LLTM assumes that the item difficulty parameters are perfectly recovered by the item features design matrix Q. In practice, this is a very stringent assumption; likelihood ratio tests of the LLTM against the unconstrained Rasch model almost invariably lead to a significant deviance. For such cases, Fischer (1995b) suggested that "... if the formal model of item difficulty specified by an LLTM for a given item universe is at least approximately true, it should be possible to predict item difficulty for new items of the same universe in new samples of examinees" (p. 148). In line with this suggestion, Janssen et al. (2004; see also De Boeck, 2008; Mislevy, 1988) developed a random-effects version of the LLTM, called the LLTM-R. In this model, item difficulty is permitted not to be fully determined by the linear structure basic parameters. This change is formalized by the inclusion of a random error term ε_i for each item:

$$logit[\Pr\{U_{pi}=1\}]=\theta_p-\sum_{k=1}^{K}\eta_k q_{ik}+\varepsilon_i \tag{13.3}$$

where $\varepsilon_i \sim \mathcal{N}(0,\sigma^2)$ and $\mathcal{N}$ denotes the normal distribution. The smaller the error variance, the better the item difficulty is determined by the item features.

The LLTM-R can be interpreted in two interrelated ways. In the *random-error* interpretation of the model, σ^2 is considered to be residual variance, representing random variation in the item difficulties of items with the same item design; that is, the item design can be considered as approximately true. Consequently, it becomes possible to infer the amount of variance in the difficulty parameters explained by the item design (see the empirical example below). In the *random-sampling* interpretation of the LLTM-R, the items are considered to be exchangeable members of a population of items defined by the item-design vector q_i. As can be inferred from Equation 13.3, this item population is modeled as a normal distribution with mean $\sum_{k=1}^{K}\eta_k q_{ik}$ and variance σ^2. Under this interpretation, the LLTM-R allows to predict item difficulty for new items of the same item universe, and, hence, can serve as a hierarchical model for automated item generation based on the design (Volume One, Chapter 26).

13.2.3 Estimating the Item Characteristics from Subtask Data

In the original LLTM, the elements in the Q matrix are assumed to be fixed constants based on information external to the data. Butter et al. (1998) proposed a version of the LLTM, in which the elements in the Q matrix are parameters that are estimated from the data. Hence, its restrictions on the item difficulty become internal to the data. Therefore, their model is known MIRID, which is an acronym for model with internal restrictions on item difficulty. MIRID is a multicomponent response model (Volume One, Chapter 14) that considers each item as a composite of K subtasks. For example, for the composite task of finding a synonym for a given word, one can think of two components that can be measured with a subtask: a generation subtask ("give all words you think of when trying to find a synonym for ...") and an evaluation subtask ("which of the following words is a true synonym for ..."). For each item i of subtask k, the Rasch model holds with item difficulty parameter β_{ik}. For each item of the composite task, the difficulty parameter is an additive combination of the subtask difficulties, or formally:

$$logit[\Pr\{U_{pi0}=1\}]=\theta_p-\sum_{k=1}^{K}\sigma_k\beta_{ik}+\tau \quad (13.4)$$

where τ is a normalization constant.

Butter et al. (1998) applied the MIRID to a test consisting of the spelling task of plural Dutch nouns, which implies writing the correct number of consonants (Subtask 1) and vowels (Subtask 2). Smits and De Boeck (2003) developed an extension of the MIRID that allowed the subtasks to have different but fixed discrimination values, like in the one-parameter logistic model (OPLM) (Verhelst and Glas, 1995). They applied the OPLM–MIRID to self-report measures of guilt, which treat guilt reactions to certain situations (composite item) as functions of the interpretation of the situation (appraisals in terms of responsibility and norm violation) and the subjective reactions to these them (negative self-evaluation, ruminative worrying, and action tendencies related to restitution).

13.2.4 Including Local Dependencies between Items

Ip et al. (2009) extended the LLTM to account for local dependencies (LID) among the item responses, which may tend to arise when items cluster within a test. For example, in a reading test, an item cluster comprises items that are related to a common reading passage. Apart from modeling item difficulty when local dependencies are present, the LID–LLTM also allows for examinee covariates to model differences between subgroups of them.

13.2.5 Incorporating Individual Differences in the Effect of Item Features

In the LLTM, item features are treated as sources of item difficulty; however, item features can also be seen as sources of individual differences. For example, when an item has both a verbal and a numerical aspect, one can expect that varying these aspects across items will differentially affect the difficulty of the items for examinees that are low versus high on numerical and verbal ability. Hence, the assumption of the LLTM of equal effects of the item features for all examinees might be too stringent in some situations. Therefore, Rijmen and De Boeck (2002) developed the random weights linear logistic test model (RWLLTM). The model divides the item features in two groups: features with fixed effects (for which η_k is equal for all examinees) and features with random effects (for which η_{kp} is specific for every examinee p). Formally, the RWLLTM states that

$$logit[\Pr\{U_{pi}=1\}]=\theta_p-\sum_{k=1}^{K_1}\eta_k q_{ik}-\sum_{k=K_1+1}^{K}\eta_{kp}q_{ik} \quad (13.5)$$

where K_1 is the number of fixed effects and $K_2=K-K_1$ is the number of random effects. Note that in the LLTM-R, the effects are random across items with the same pattern of q_{ik} in the item design matrix, whereas in the RWLLTM, the effects are random across examinees. Because of the introduction of random weights, the RWLLTM becomes a multidimensional model.

Rijmen and De Boeck (2002) applied the RWLLTM to a test of deductive reasoning and showed that the weights of one item feature (indicating whether the item formulated a disjunctive rather than a conditional problem) turned out to be a major source of individual differences, much larger than the individual differences in the general ability θ_p. Debeer and Janssen (2012) applied the RWLLMT to investigate individual differences in item-position effects. They showed that, in a low-stakes assessment, there were individual differences in a fatigue effect. Hence, the fact that items tend to become more difficult toward the end of the test did not affect every respondent in the same way.

13.2.6 Incorporating Group Differences in the Effects of Item Features

In the RWLLTM, the effect of an item feature may differ across examinees. Meulders and Xie (2004) and Xie and Wilson (2008) proposed an extension of the LLTM where the effects of item features may differ across groups of examinees. They called the phenomenon differential facet functioning (DFF), where a facet refers to a group of items that share a common feature. In case there are two groups of respondents that are dummy coded, the model for the reference group is given in Equation 13.1, while the logit expression for the focal group becomes

$$logit[\Pr\{U_{pi}=1\}]=\theta_p-\sum_{k=1}^{K}\eta_k q_{ik}-\sum_{k=1}^{K}\gamma_k q_{ik} \tag{13.6}$$

where γ_k is the group × feature interaction effect for the kth item feature.

Meulders and Xie (2004) indicated that DFF can be of interest for two reasons. First, DFF may be a parsimonious way to summarize and explain differential item functioning (DIF) of several items that have a specific item feature in common (for DIF, see Gamerman et al., Volume Three, Chapter 4). Hence, instead of including DIF-parameters for every single item, one could restrict the corresponding DIF-parameters to be equal, which is formally equivalent to including an interaction between the item feature and the group indicator. Second, modeling DFF is useful to investigate both the main and interaction effects of item features. In fact, Tanzer et al. (1995) used the LLTM in this way to check the cross-cultural stability of the item-complexity factors in a spatial ability test.

Note that the RWLLTM and DFF can also be combined. If γ_k in Equation 13.6 is allowed to be a random, then within-group differences in the effect of an item feature can be investigated as well.

13.3 Parameter Estimation

The LLTM and several of its related models are cases of generalized mixed models (GLMM). Consequently, the proposed models can be estimated using general statistical packages (Rijmen et al., 2003; De Boeck and Wilson, 2004). For example, the lmer function from the lme4 package (Bates et al., 2011) of R (R Development Core Team, 2011) provides a very flexible tool for analyzing GLLM (De Boeck et al., 2011). An overview of other statistical packages is given by Tuerlinckx et al. (2004). Of course, the LLTM and related models can also be estimated using Bayesian methods (Johnson and Sinharay, Volume Two, Chapter 13), for instance, with Markov chain Monte Carlo methods to sample from the posterior distributions of the parameters (Junker et al., Volume Two, Chapter 15).

13.4 Model Fit

Testing the fit of the LLTM requires two independent steps (Fischer, 1995b): (a) testing the fit of the Rasch model (e.g., Glas, Volume Two, Chapter 17) and (b) subsequent testing of restrictions on item difficulty. For the latter test, a likelihood ratio test is possible, comparing the likelihood of the LLTM (L_{LLTM}) relative to the one for the simple Rasch model (L_{RASCH}); more specifically,

$$-2\ln\left(\frac{L_{Rasch}}{L_{LLTM}}\right) \sim \chi^2_{I-K} \tag{13.7}$$

with the degrees of freedom equal to the number of estimated item parameter minus the number of estimated weight parameters in the LLTM. However, such likelihood ratio tests typically tend to be significant and lead to the rejection of the LLTM. Consequently, a simple graphical heuristic is often used to evaluate the relative goodness of fit of the LLTM and the Rasch model by plotting the Rasch difficulties against the estimated item difficulties according to the LLTM (Fischer, 1995b). In fact, this is equivalent to the LLTM-R approach. As will be illustrated empirically below, Bayesian estimation allows for tailor-made goodness-of-fit statistics for the LLTM-R, often leading to a more detailed insight of the cause of misfit (Sinharay, Volume Two, Chapter 19).

When testing the goodness of fit of nested models against each other, the likelihood ratio test can be used. When dealing with additional random effects, as in a comparison between the RWLLTM with the LLTM, mixtures of chi-square distributions can be used to tackle certain boundary problems (Verbeke and Molenberghs, 2000, p. 64–76). For nonnested models (e.g., when comparing different Q matrices), fit can be compared on the basis of such goodness-of-fit measure as Akaike's information criterion (AIC) (Akaike, 1977) or the Bayesian information criterion (BIC) (Schwarz, 1978), as is explained by Cohen and Cho (Volume Two, Chapter 18).

Because the models within the proposed framework are GLMM or nonlinear mixed models (NLMM), the significance tests of such parameters as the value of η_k in the LLTM can be tested using Wald tests. Medina-Días (1993) showed how a Lagrange multiplier test can be used to assess the appropriateness of single elements of Q.

13.5 Empirical Example

13.5.1 Model and Data Set

Vansteelandt (1999) formalized a probabilistic version of the competency-demand hypothesis by Wright and Mischel (1987). The hypothesis states that an examinee will fail to produce a correct response whenever the demands of a situation (represented by the parameter γ_s) exceed the competency of the examinee (represented by the ability parameter θ_p). As the degree of activation needed to evoke a response may vary, Vansteelandt (1999) extended the competency-demand hypothesis with a threshold δ_r for each particular failure r. The extended competency-demand hypothesis then states that the probability of examinee p producing a specific incorrect response r in situation s (i.e., $U_{psr} = 0$)

is a function of how much the difference between demand and competency exceeds the threshold δ_r. More formally,

$$logit[\Pr\{U_{psr}=0\}]=(\gamma_s-\theta_p)-\delta_r=\beta_{sr}-\theta_p \tag{13.8}$$

As Equation 13.8 reveals, the hypothesis corresponds to an LLTM with difficulty parameter β_{sr} for situation–response combination being a linear function of the situation and response parameters. The reversal of the signs for β_{sr} and θ_p relative to Equation 13.1 is due to the fact that Equation 13.8 models a failure rather than a correct response.

Vansteelandt (1999) collected a data set consisting of the self-report measures on hostile behavior in frustrating situations (a similar, but different, data set was analyzed in De Boeck and Wilson, 2004). Displaying a hostile reaction was taken to be a failure. The demands of the situation refer to the amount of frustration in the situation, with high values of γ_s indicating high frustration. The competency of the examinee refers to his or her frustration tolerance, with high values of θ_p referring to examinees who show a hostile reaction only in highly frustrating situations. Vansteelandt (1999) presented 56 items to 316 first-year students in psychology. The items followed a 14×4 situation-by-response design. The 14 frustrating situations are presented in the left-hand column of Table 13.2. The four failure responses were "to become tense," "to feel irritated," "to curse," and "wanting to strike something or someone."

According to Vansteelandt (1999), the Rasch model (with $14\times4=56$ difficulty parameters) fitted the data with a nonsignificant R_{1c} statistic (Glas and Verhelst, 1995): R_{1c} $(165)=190.24$, $p=0.09$. A likelihood ratio test of the LLTM against the Rasch model was significant ($\chi^2=1182.93$, $df=39$, $p<0.001$). However, despite the statistical rejection of the LLTM, the 14 situations and 4 response types explained 87% of the variance in the estimated Rasch item difficulties.

In this chapter, we apply the LLTM-R to these data in order to show its flexibility relative to the LLTM. Adapting Equation 13.8 to an LLTM-R model, it becomes

TABLE 13.2

PM of the γ_s of the 14 Situations

Situation	γ_s
You are trying to study and there is incessant noise	0.95
You have just found out that someone has told lies about you	0.79
Your instructor unfairly accuses you of cheating on an examination	0.69
You are driving to a party and suddenly your car has a flat tire	0.66
You are waiting at the bus stop and the bus fails to stop for you	0.63
You arranged to meet someone and (s)he does not show up	0.56
Someone has opened your personal mail	0.53
Someone has lost an important book of yours	0.32
You are talking to someone and (s)he does not answer you	0[a]
You accidentally bang your shins against a park bench	−0.07
You are carrying a cup of coffee to the table and someone bumps into you	−0.13
You are in a restaurant and have been waiting for a long time to be served	−0.14
You woke up early to attend a special class at 8 a.m. and the instructor does not show up	−0.52
You are very tired and have just gone to bed, when you are awakened by the arrival of friends	−0.63

[a] This situation served as the reference category of the dummy coding of the 14 situations.

$$logit[\Pr\{U_{psr}=0\}]=\beta^*_{sr}-\theta_p \tag{13.9}$$

where β_{sr} in Equation 13.8 equals $\beta^*_{sr}+\varepsilon_i$, which is equivalent to $\gamma_s+\delta_r+\varepsilon_i$. The LLTM-R is estimated within a Bayesian estimation framework using Markov chain Monte Carlo methods.

13.5.2 Method

13.5.2.1 Bayesian Model Estimation

For the estimation of LLTM-R to this data set, the item design was represented using dummy variables. The model was estimated using a data-augmented Gibbs sampler (Albert, 1992). As priors for η, θ_p, and σ^2, an informative normal, standard normal, and a scaled inverse χ^2 ($\nu=1$, $s=0.01$) were used, respectively. Five chains of 2000 iterations each were run. The first half of each chain was reserved for burn-in. The posterior mean (PM) and posterior standard deviation (PSD) of each parameter were estimated by calculating the corresponding statistic pooling the remaining 5000 Gibbs iterations. Gelman and Rubin's (1992) measure indicated convergence for all parameters.

13.5.2.2 Model Checking

First, the fit of the LLTM-R to the data was investigated using the technique of posterior predictive checks (PPC) (Gelman et al., 1995, 2004; Gelman and Meng, 1996). The basic idea of a PPC is that "if the model fits, then replicated data generated under the model should look similar to observed data" (Gelman et al., 1995, p. 165). The fit of a model was assessed using a χ^2-statistic as a discrepancy measure (Janssen et al., 2000). For each draw from the posterior, eight adjacent and nonoverlapping ability groups were formed on the basis of the sorted vector of the θ_p from that draw. The statistic compared the observed and expected number of responses for each item in the different ability groups calculated at each Gibbs iteration for both the observed and the replicated data. The degree of deviance of the model was expressed by a Bayesian p-value, estimated as one minus the proportion of draws from the posterior for which the χ^2-statistic was higher for the observed than the replicated data. Small p-values indicated bad fit.

Second, the goodness of fit of the model for the item parameters was investigated, both across all items and for each item separately. As a global measure of fit, the squared correlation between β_{sr} and β^*_{sr} was calculated for each draw from the posterior as an indicator of the proportion of explained variance. At the item level, the residuals $\beta_{sr}-\beta^*_{sr}$ were calculated for each draw from the posterior and summarized as 90% and 95% central posterior intervals. Items for which the interval did not contain the value zero were considered as deviant. In the present application, deviant items are taken to indicate that their difficulty was not well captured by the main effects of the situation and the response expressed by the item, and, hence, interaction between the situation and response was likely.

13.5.3 Results

13.5.3.1 Parameter Estimates

Table 13.2 presents the PMs of the γ_s parameter for each of the 14 situations. Examples of very frustrating situations were those with incessant noise, a flat tire, and being unfairly accused of cheating. Low frustrating situations were being awakened after you have just gone to bed or having to attend an early class at 8 a.m. that was cancelled. The PM of the thresholds of the four responses were –0.85, 0 (reference category for the dummy codes of the responses), 0.26 and 1.60, for "to feel irritated," "to become tense," "to curse," and "wanting to strike something or someone," respectively. Hence, in general, the two anger-in responses (feeling irritated and becoming tense) are displayed more easily than the two anger-out responses (cursing, wanting to strike).

13.5.3.2 Fit of the Model to the Data

The Bayesian p-value of the χ^2 statistic summed over items was 0.08, indicating that the Rasch model fits the data. The Bayesian p-value of only one item was smaller than 0.05. This item was about the cancelled class at 8 a.m. and the response of cursing. Its misfit was due to the fact that the item discriminated less well than predicted by the model. The lower discrimination for this item became clear when comparing a plot of the proportion of correct responses for the eight ability groups (as defined for the χ^2-statistic and drawn for 20 randomly selected Gibbs iterations) with the plot of the item characteristic curve of the item based on the PM of β_{sr}.

13.5.3.3 Fit of the Model for the Item Parameters

The situation-by-response design for the items explained on the average 82% of the variance of the β_{sr} according to the PM of the squared correlation between β_{sr} and β^*_{sr}. This percentage is a bit lower than the 87% reported by Vansteelandt (1999) for the regression of the estimated Rasch difficulties on the design matrix Q. The difference is due to the fact that the measure of explained variance defined for the LLTM-R also takes the uncertainty of the parameter estimates into account.

The 95% and 90% central posterior intervals for $\beta_{sr} - \beta^*_{sr}$ did not include the value zero for seven out of the 56 items. These seven items referred to three different situations. In the situation of banging your shins against a park bench, the difficulty β_{sr} of three of the four responses was not well predicted by the main situation and response effects (as summarized by the β^*_{sr} parameters). Cursing had the lowest threshold but becoming tense and feeling irritated had a higher threshold relative to what the item design predicted. In the situation of talking to someone who does not respond, the threshold for feeling irritated was lower than predicted, while the cursing response had a higher threshold than predicted. The same two responses showed deviant residuals in the same direction for the situation of being unfairly accused of cheating. In this situation, the responses of feeling irritated and becoming tense had similar threshold values. In contrast with the first item, the latter involved social situations where cursing is less likely but feeling irritated is more likely. The reverse was noticed for the item about the park bench, where in addition, feeling tense also became less likely.

13.5.4 Discussion of the Results

The empirical examples showed the LLTM-R as an alternative to separate estimation of a Rasch model and an LLTM. First, as for the relative fit of the models to the data, it was shown that the PPC measure based on the χ^2-statistic for the LLTM-R gave comparable results as a χ^2-based goodness-of-fit measure for the Rasch model. Hence, when applying the LLTM-R, one can detect whether the Rasch model holds, both for the item set as a whole and for each item separately. One should note, however, that in the LLTM-R, the β_{sr} parameters in

Equation 13.8 are partially influenced by the β^*_{sr} parameters. Hence, strictly taken, the proposed PPC goodness-of-fit measure is not a goodness-of-fit test of the Rasch model.

Second, with respect to the additional modeling of the item parameters, it was shown that the LLTM-R can give information beyond what can be derived from a joint application of the Rasch model and the LLTM. Whereas a likelihood ratio test only shows that there is a global (and possibly small) discrepancy between the LLTM and the Rasch model, the residuals $\beta_{sr} - \beta^*_{sr}$ for the LLTM-R also enable us to detect discrepancies at the item level. These residuals can be very helpful in finding out what changes in the item-design matrix Q are needed. Additionally, descriptive measures of the global goodness of fit of the model for the item parameters obtained from a regression analysis of the estimated item difficulty parameters of a Rasch model on Q can be obtained directly from the LLTM-R. Further, as it takes the uncertainty about the parameter into account, the LLTM-R gives a more conservative estimate of the proportion of explained variance. Summarizing, with the LLTM-R, one can both estimate and test the quality of the restrictions on item difficulty imposed by the Q matrix, all within one model.

13.6 Discussion

A common feature of the linear logistic models discussed in this chapter is that they model response data by taking the item features of the items into account. Incorporation of an item design matrix in the models of the LLTM family can serve different purposes with respect to establishing the validity of the test. First, the item design matrix may be helpful in showing the content validity of the test. Developing a set of items covering a carefully constructed item design with respect to their cognitive determinants reduces the risk of missing important subdomains of the domain tested. Second, examination of the relationship between item features and their difficulties can be regarded as an investigation of the construct validity of the test. More specifically, showing the effects of the variables in the Q matrix on the item response probabilities confirms the construct representation of the test (Embretson, 1983, 1985), that is, it identifies the theoretical mechanisms underlying the responses to the items. Finally, the item features of the Q matrix can be helpful with score interpretation, as they may give a meaningful interpretation to different positions on the measurement scale. Similarly, as Hartig et al. (2012) showed, they can also be used to define cut-off points between different proficiency levels and, in doing so, help to classify students into different proficiency levels.

Q matrices are helpful in cognitive diagnosis, as their elements may be seen not only as attributes of the items but also as skills of the examinees. But, of course, since all models belonging to the LLTM family assume continuous proficiency constructs, they differ clearly from cognitive diagnosis models in this respect (Wang and Chang, Volume Three, Chapter 16).

References

Akaike, H. 1977. On entropy maximization principle. In: P. R. Krishnaiah (Ed.), *Applications of Statistics* (pp. 27–41). Amsterdam: North-Holland.

Albert, J. H. 1992. Bayesian estimation of normal ogive item response curves using Gibbs sampling. *Journal of Educational Statistics*, 17, 251–269.

Bates, D., Maechler, M., and Bolker, B. 2011. *lme4: Linear Mixed Effects Models Using S4 Classes.* Retrieved July 17, 2012 from http://cran.r-project.org/web/packages/lme4.

Bouwmeester, S., van Rijen, E. H. M., and Sijtsma, K. 2011. Understanding phoneme segmentation performance by analyzing abilities and word properties. *European Journal of Psychological Assessment*, 27, 95–102.

Butter, R., De Boeck, P., and Verhelst, N. 1998. An item response model with internal restrictions on item difficulty. *Psychometrika*, 6, 47–63.

Daniel, R. C., and Embretson, S. E. 2010. Designing cognitive complexity in mathematical problem-solving items. *Applied Psychological Measurement*, 34, 348–364.

Debeer, D., and Janssen, R. 2012. *Modeling Item-Position Effects within an IRT Framework.* Paper presented at the Annual Meeting of the National Council on Measurement in Education (NCME), Vancouver.

De Boeck, P. 2008. Random item IRT models. *Psychometrika*, 73, 533–559.

De Boeck, P., Bakker, M., Zwitser, R., Nivard, M., Hofman, A., Tuerlinckx, F., and Partchev, I. 2011. The estimation of item response models with the lmer function from the lme4 package in R. *Journal of Statistical Software*, 39, 1–28.

De Boeck, P., and Wilson, M. 2004. *Explanatory Item Response Models: A Generalized Linear and Nonlinear Approach.* New York: Springer.

Draney, K., and Wilson, M. 2008. A LLTM approach to the examination of teachers' ratings of classroom assessment tasks. *Psychology Science Quarterly*, 50, 417–432.

Embretson, S. E. 1983. Construct validity: Construct representation versus nomothetic span. *Psychological Bulletin*, 93, 179–197.

Embretson, S. E. 1985. *Test Design: Developments in Psychology and Psychometrics.* New York: Academic Press.

Embretson, S. E. 1998. A cognitive design system approach to generating valid tests: Application to abstract reasoning. *Psychological Methods*, 3, 380–396.

Embretson, S. E. 1999. Generating items during testing: Psychometric issues and models. *Psychometrika*, 64, 407–433.

Embretson, S. E., and Wetzel, C. D. 1987. Component latent trait models for paragraph comprehension. *Applied Psychological Measurement*, 11, 175–193.

Fischer, G. H. 1973. The linear logistic test model as an instrument in educational research. *Acta Psychologica*, 37, 359–374.

Fischer, G. H. 1983. Logistic latent trait models with linear constraints. *Psychometrika*, 48, 3–26.

Fischer, G. H. 1995a. Linear logistic models for change. In: G. H. Fischer and I. W. Molenaar (Eds.) *Rasch Models: Foundations, Recent Developments, and Applications* (pp. 157–180). New York: Springer.

Fischer, G. H. 1995b. The linear logistic test model. In: G. H. Fischer and I. W. Molenaar (Eds.), *Rasch Models: Foundations, Recent Developments, and Applications* (pp. 131–155). New York: Springer.

Fischer, G. H., and Formann, A. K. 1982. Some applications of logistic latent trait models with linear constraints on the parameters. *Applied Psychological Measurement*, 4, 397–416.

Freund, Ph. A., Hofer, S., and Holling, H. 2008. Explaining and controlling for the psychometric properties of computer-generated figural matrix items. *Applied Psychological Measurement*, 32, 195 – 210.

Gelman, A., Carlin, J. B., Stern, H. S., and Rubin, D. B. 1995. *Bayesian Data Analysis.* New York: Chapman and Hall.

Gelman, A., Carlin, J. B., Stern, H. S., and Rubin, D. B. 2004. *Bayesian Data Analysis* (2nd ed.). New York: Chapman and Hall.

Gelman, A., and Meng, X. L. 1996. Model checking and model improvement. In: W. R. Gilks, S. Richardson, and D. J. Spiegelhalter (Eds.), *Markov Chain Monte Carlo in Practice* (pp. 189–201). New York: Chapman and Hall.

Gelman, A., and Rubin, D. B. 1992. Inference from iterative simulation using multiple sequences (with discussion). *Statistical Science*, 7, 457–511.

Glas, C. A. W., and Verhelst, N. D. 1995. Testing the Rasch model. In: G. H. Fischer and I. W. Molenaar (Eds.), *Rasch Models: Foundations, Recent Developments and Applications* (pp. 69–95). New York: Springer.

Gorin, J. S. 2005. Manipulating processing difficulty of reading comprehension questions: The feasibility of verbal item generation. *Journal of Educational Measurement*, 42, 351–373.

Gorin, J. S., and Embretson, S. E. 2006. Item difficulty modeling of paragraph comprehension items. *Applied Psychological Measurement*, 30, 394–411.

Hahne, J. 2008. Analyzing position effects within reasoning items using the LLTM for structurally incomplete data. *Psychology Science Quarterly*, 50, 379–390.

Hartig, J., Frey, A., Nold, G., and Klieme, E. 2012. An application of explanatory item response modeling for model-based proficiency scaling. *Educational and Psychological Measurement*, 72, 665–686.

Hohensinn, C., Kubinger, K. D., Reif, M., Holocher-Ertl, S., Khorramdel, L., and Frebort, M. 2008. Examining item-position effects in large-scale assessment using the linear logistic test model. *Psychology Science Quarterly*, 50(3), 391–402.

Holling, H., Blank, H., Kuchenbäcker, K., and Kuhn, J.-T. 2008. Rule-based item design of statistical word problems: A review and first implementation. *Psychology Science Quarterly*, 50, 363–378.

Hornke, L. F., and Habon, M. W. 1986. Rule-based item bank construction and evaluation within the linear logistic framework. *Applied Psychological Measurement*, 10, 369–380.

Ip, E. H., Smits, D. J. M., and De Boeck, P. 2009. Locally dependent linear logistic test model with person covariates. *Applied Psychological Measurement*, 33, 555–569.

Ivie, J. L. and Embretson, S. E. 2010. Cognitive process modeling of spatial ability: The assembling objects task. *Intelligence*, 38, 324–335.

Janssen, R., Schepers, J., and Peres, D. 2004. Models with item and item group predictors. In: P. De Boeck and M. Wilson (Eds.), *Explanatory Item Response Models: A Generalized Linear and Nonlinear Approach* (pp. 189–212). New York: Springer.

Janssen, R., Tuerlinckx, F., Meulders, M., and De Boeck, P. 2000. A hierarchical IRT model for criterion-referenced measurement. *Journal of Educational and Behavioral Statistics*, 25, 285–306.

Kubinger, K. D. 1979. Das Problemlöseverhalten bei der statistischen Auswertung psychologischer Experimente: Ein Beispiel hochschuldidaktischer Forschung [Problem solving behavior in the case of statistical analyses of psychological experiments: An example of research on universities didactics]. *Zeitschrift für Experimentelle und Angewandte Psychologie*, 26, 467–495 (In German).

Kubinger, K. D. 1980. Die Bestimmung der Effektivität universitärer Lehre unter Verwendung des Linearen Logistischen Testmodells von Fischer: Neue Ergebnisse [The evaluation of effectiveness of university lecturing with the help of the linear logistic test model by Fischer: New results]. *Archiv für Psychologie*, 133, 69–79 (In German).

Kubinger, K. D. 2008. On the revival of the Rasch model-based LLTM: From constructing tests using item generating rules to measuring item administration effects. *Psychology Science Quarterly*, 50, 311–327.

Kubinger, K. D. 2009. Applications of the linear logistic test model in psychometric research. *Educational and Psychological Measurement*, 69, 232–244.

Kubinger, K. D., and Draxler, C. 2006. A comparison of the Rasch model and constrained item response theory models for pertinent psychological test data. In: M. von Davier and C. H. Carstensen (Eds.), *Multivariate and Mixture Distribution Rasch Models: Extensions and Applications* (pp. 295–312). New York: Springer.

Linacre, J. M. 1989. *Multi-Facet Rasch Measurement*. Chicago: MESA Press.

Medina-Días, M. 1993. Analysis of cognitive structure. *Applied Psychological Measurement*, 17, 117–130.

Meulders, M., and Xie, Y. 2004. Person-by-item predictors. In: P. De Boeck and M. Wilson (Eds.), *Explanatory Item Response Models: A Generalized Linear and Nonlinear Approach* (pp. 213–240). New York: Springer.

Mislevy, R. J. 1988. Exploiting auxiliary information about items in the estimation of Rasch item difficulty parameters. *Applied Psychological Measurement*, 12, 725–737.

Rijmen, F., and De Boeck, P. 2002. The random weights linear logistic test model. *Applied Psychological Measurement*, 26, 271–285.

Rijmen, F., Tuerlinckx, F., De Boeck, P., and Kuppens, P. 2003. A nonlinear mixed model framework for item response theory. *Psychological Methods*, 8, 185–205.

Scheiblechner, H. H. 1972. Das Lernen und Lösen komplexer Denkaufgaben [Learning and solving complex cognitive tasks]. *Zeitschrift für Experimentelle und Angewandte Psychologie*, 19, 476–506 (In German).

Schwarz, G. 1978. Estimating the dimension of a model. *The Annals of Statistics*, 6, 461–464.

Sheehan, K., and Mislevy, R. J. 1990. Integrating cognitive and psychometric models to measure document literacy. *Journal of Educational Measurement*, 27, 255–272.

Smith, R. M., Kramer, G. A., and Kubiak, A. T. 1992. Components of difficulty in spatial ability test items. In: M. Wilson (Ed.), *Objective Measurement: Theory into Practice, Volume I* (pp. 157–174). Norwood, NJ: Ablex.

Smits, D. J. M., and De Boeck, P. 2003. A componential IRT model for guilt. *Multivariate Behavioral Research*, 38, 161–188.

Smits, D. J. M., De Boeck, P., and Verhelst, N. 2003. Estimation of the MIRID: A program and a SAS-based approach. *Behavior Research Methods, Instruments, and Computers*, 35, 537–549.

Sonnleitner, Ph. 2008. Using the LLTM to evaluate an item-generating system for reading comprehension. *Psychology Science Quarterly*, 50, 345–362.

Spada, H., and May, R. 1982. The linear logistic test model and its application in educational research. In: D. Spearritt (Ed.), *The Improvement of Measurement in Education and Psychology* (pp. 67–84). Hawthorn, Victoria: The Australian Council for Educational Research.

Tanzer, N. K., Glittler, G., and Ellis, B. B. 1995. Cross-cultural validation of item complexity in a LLTM-calibrated spatial ability test. *European Journal of Psychological Assessment*, 11, 170–183.

Tuerlinckx, F., Rijmen, F., Molenberghs, G., Verbeke, G., Briggs, D., Van den Noortgate, W., Meulders, M., and De Boeck, P. 2004. Estimation and software. In: P. De Boeck, and M. Wilson (Eds.), *Explanatory Item Response Models: A Generalized Linear and Nonlinear Approach* (pp. 343–373). New York: Springer.

Vansteelandt, K. 1999. A formal model for the competency-demand hypothesis. *European Journal of Personality*, 13, 429–442.

Verbeke, G., and Molenberghs, G. 2000. *Linear Mixed Models for Longitudinal Data*. New York: Springer.

Verhelst, N. D., and Glas, C. A. W. 1995. One parameter logistic model. In: G. H. Fischer and I. W. Molenaar (Eds.), *Rasch Models: Foundations, Recent Developments and Applications* (pp. 215–238). New York: Springer.

Wright, J. C., and Mischel, W. 1987. A conditional approach to dispositional constructs: The local predictability of social behavior. *Journal of Personality and Social Psychology*, 55, 454–469.

Xie, Y., and Wilson, M. 2008. Investigating DIF and extensions using an LLTM approach and also an individual differences approach: an international testing context. *Psychology Science Quarterly*, 50, 403–416.

Zeuch, N., Holling, H., and Kuhn, J.-T. 2011. Analysis of the Latin square task with linear logistic test models. *Learning and Individual Differences*, 21, 629–632.

14

Multicomponent Models

Susan E. Embretson

CONTENTS

14.1 Introduction 225
  14.1.1 History of the Models 226
14.2 Presentation of the Models 228
  14.2.1 Multicomponent Models for Subtask Data 228
  14.2.2 Multicomponent Models for Varying Components 230
14.3 Parameter Estimation 231
14.4 Model Fit 232
14.5 Empirical Examples 234
  14.5.1 Standards-Based Mathematical Skills 234
  14.5.2 Cognitive Components in Mathematical Achievement Items 236
14.6 Discussion 239
Acknowledgment 241
References 241

14.1 Introduction

Identifying the sources of cognitive complexity on ability and achievement test items has been an active research area; for a review, see Bejar (2010). Several advantages result from developing an empirically plausible cognitive theory for a test. First, the results provide evidence for construct validity, especially the response process aspect, as required by the *Standards for Educational and Psychological Tests* (AERA/NCME/APA, 1999). Second, as noted by several researchers (e.g., Henson and Douglas, 2005), results from cognitive processing research are relevant to test design. Identifying the levels and sources of cognitive complexity in each item has implications for both selecting items and designing items. Third, the increasing attractiveness of automatic item generation (see Bejar, 2010; Gierl and Haladyna, 2012) depends on the predictability of the psychometric properties of the generated items. Cognitive models of item difficulty and other psychometric properties can provide the necessary predictions. Finally, assessments of individual differences in cognitive skills are useful for more specific interpretations of test scores. Diagnostic assessment, when applied in the context of appropriate psychometric models (Hensen et al., 2009; von Davier, 2008), provides important supplementary information beyond overall test scores (for reviews, see Leighton and Gierl, 2007; Rupp et al., 2010).

Multicomponent models are multidimensional item response theory (MIRT) models that interface theories of cognitive complexity with item psychometric properties and person measurement. Multicomponent models differ from traditional MIRT models (Volume One, Chapters 11 and 12) in several ways. First, multicomponent models are confirmatory models. Component involvement in items is specified from a theory or conceptual framework. Traditional MIRT models, in contrast, are not often used as confirmatory models. Second, item success in multicomponent models depends on a noncompensatory combination of traits, rather than a compensatory combination as in traditional MIRT. The noncompensatory feature of multicomponent models reflects the postulated relationships between processing components; namely, successful completion of all components is required for item solution. As noted by Stout (2007), this assumption appears more consistent with cognitive theories for complex tasks. Third, separate item difficulties are estimated for each dimension in multicomponent models. This feature is also consistent with cognitive research because tasks are designed to test theories by varying the difficulty of underlying processes. In contrast, traditional MIRT models typically contain a single difficulty estimate for each item.

Applications of multicomponent models have the several advantages described above for research on cognitive complexity: Obtaining empirically plausible multicomponent models provides validity evidence as well as parameter estimates useful for selecting and designing items and the predictions of item difficulty. Further, individual differences in the latent dimensions of multicomponent models can enhance the interpretation of test scores, including diagnostic assessments from some multicomponent models.

Three multicomponent models will be presented in this chapter: (a) the multicomponent latent trait model (MLTM) (Whitely, 1980); (b) the general component latent trait model (GLTM) (Embretson, 1984); and (c) the multicomponent latent trait model for diagnosis (MLTM-D) (Embretson and Yang, 2013). Prior to presenting the models, a brief overview of their history and features will be considered.

14.1.1 History of the Models

The early MLTM and GLTM were proposed for application to complex ability test items. Several early studies on cognitive components in test items had identified two or more processing components on a variety of test items, including analogies (Sternberg, 1977), spatial tasks (Pellegrino et al., 1985), and mathematical problems (Mayer et al., 1984).

MLTM was formulated to estimate both item and person differences on these underlying components, using subtasks developed to represent components as specified in a theory. Since the early research findings indicated that even relatively narrow tasks, such as verbal analogies or spatial items, involved outcomes from multiple components, a noncompensatory model was required. Further, from this perspective, item responses are inherently multidimensional if persons and items vary on the components. Although unidimensional models can be fit to multidimensional tasks (i.e., by eliminating items), the resulting dimension may be a confounded composite. Thus, one motivating feature of MLTM was to identify person and item differences to determine the nature of the composite trait. The estimated component complexity from MLTM could be used as evidence for response process aspect of validity. For example, assessing the extent to which performance on an analogy test represented inductive and deductive reasoning, evidence on components in the test items would be relevant (Embretson, 1997). Further, if the balance of component processes was not deemed optimal, items that reflected predominantly the less desired component impact could be identified and removed from the item bank.

GLTM was developed to link differences in component difficulty to specific stimulus features of the items. Stimulus features are typically manipulated in experimental cognitive studies to control the difficulty of specific processes or components. Thus, GLTM permits a more complete linkage to cognitive theory than the original MLTM. Analogous to the linear logistic test model (LLTM) (Fischer, 1973; Volume One, Chapter 13), GLTM includes an option to link scored stimulus features of items to their difficulty for each component. Such features are contained in Q_k matrices of order $i \times m$ for each component k, where $i = 1, \ldots, I$ represents the items and $m = 1, \ldots, M$ their scored features. When the relationships between the scored features and component complexity are substantial, they have implications for item design. For example, the strength of verbal encoding in mathematical items can be manipulated by changing the levels of relevant variables, such as syntactic or propositional complexity, vocabulary level, and the proportion of the item content devoted to describing an application context.

For both MLTM and GLTM, it was assumed that the same subtasks were involved in all items to permit estimation by the methods that were available during their development. Consider the mathematical achievement item and two subtasks that are presented in Table 14.1. The total task is shown in the upper section. According to one theory of mathematical problem solving (Daniel and Embretson, 2010; Embretson and Daniel, 2008), the total task involves several processing stages. In Table 14.1, the total task is divided into two subtasks. The first subtask involves translating the problem into a representation and integrating the results to find an appropriate equation. No solution planning or execution is involved in this subtask. The second subtask involves solution planning and execution because the appropriate equation is given. For this item, MLTM would give the probability of solving the full task as the product of the probability of correctly completing each subtask. GLTM would model the full task probably in the same manner, except that a weighted combination of stimulus features in the items would replace item difficulty.

TABLE 14.1

Total Task and Two Subtasks for a Mathematical Word Problem

Total Task

Joni is going to run a 5000-meter race that is split into two unequal segments. The second segment of the race is 2000 meters shorter than the first segment of the race. What is the length of the first segment (m)?

A. m = 3400

B. m = 2500

C. m = 3000

D. X m = 1500

Subtask 1: Translation and Integration

Joni is going to run a 5000-meter race that is split into two unequal segments. The second segment of the race is 2000 meters shorter than the first segment of the race. Which equation could be used to find the length of the first segment (m)?

A. m + m + 2000 = 5000

B. m + m = 5000

C. m − 2000 = 5000

D. X m + (m − 2000) = 5000

Subtask 2: Solution Planning and Solution Execution

Find the value of m, given that m + (m − 2000) = 5000?

A. m = 3400

B. m = 2500

C. m = 3000

D. X m = 1500

However, existing tests rarely contain both full items and their subtasks, as shown in Table 14.1. Although various methods to circumvent this assumption were attempted, including data augmentation (Embretson and Yang, 2006), constraints on components (Embretson and McCollam, 2000), and Monte Carlo methods (Bolt and Lall, 2003), these methods had limited success.

More recently, a generalization of the earlier models, the MLTM-D, was developed to extend the scope of application for multicomponent models. MLTM-D was developed to be applicable to existing tests with items that vary in subtask involvement. For example, complex tests of quantitative reasoning, such as the *Graduate Record Examination*, or broad tests of mathematical achievement, such as state accountability tests, contain heterogeneous collections of items that vary substantially in cognitive complexity. In such tests, for example, items as shown in Table 14.1 may appear as separate items with varying surface content (i.e., persons, mode of travel, distances, etc.). That is, some items contain only equations that must be solved, while other items contain only verbal descriptions of relationships for which an equation must be constructed but not solved. In the context of a theory of mathematical problem solving (Embretson and Daniel, 2008), the varying involvement of components in items can be scored. Furthermore, if the theory specifies item substantive features that impact component difficulty, items can be further scored for relevant features. For example, the complexity of the problem translation stage in Embretson and Daniel's (2008) theory is impacted by both language complexity and mathematical symbolic representation in an item, which can be scored for each item. Thus, with MLTM-D, the heterogeneity of the items provides a method to estimate the impact of components and their relevant substantive features on item responses. With MLTM-D, applications to existing tests, with no special construction of subtasks, are feasible.

MLTM-D also has potential for diagnostic assessment. Obtaining diagnostic assessments about skill mastery is a very active area in educational measurement (Stout, 2007). MLTM-D, unlike other diagnostic models, is especially appropriate for broad educational achievement tests, such as high-stakes state accountability tests, licensure tests, and certification tests. Such tests are typically constructed with hierarchically organized blueprints that specify broad categories of performance, with more specific combinations of skills nested within the various blueprint categories. While the major goal of high-stakes achievement tests is providing an overall summary of proficiency, for many such tests, a more refined diagnosis of skills is possible, given application of an appropriate psychometric model.

14.2 Presentation of the Models

Two different types of multicomponent models will be presented separately: (a) multicomponent models for items with subtask data, which includes both the MLTM and GLTM, and (b) multicomponent models for heterogeneous items with varying component involvement, which includes the MLTM-D.

14.2.1 Multicomponent Models for Subtask Data

Let U_{pik} represent the response of examinee $p = 1, \ldots, P$ to subtask $k = 1, \ldots, K$ on item i and U_{piT} represent the response to the total intact item, T. The MLTM specifies the relationship of successful execution of the K subtasks to the probability of solving the total task as the product of the probabilities of subtask completion as follows:

$$\Pr\{U_{piT}=1\mid\boldsymbol{\theta}_p,\boldsymbol{\beta}_i\}=\tau_i+(\varsigma_i-\tau_i)\prod_{k=1}^{K}\Pr\{U_{pik}=1\mid\theta_{pk},\beta_{ik}\}, \tag{14.1}$$

where $\boldsymbol{\theta}_p$ is the vector of component traits for examinee p, $\boldsymbol{\beta}_i$ is the vector of component difficulties for item i, ς_i is the probability of applying the component outcomes to solving item i, and τ_i is the probability of solving item i by guessing.

If responses to the processing components, U_{pik}, are governed by a simple Rasch model, the full MLTM can be given as

$$\Pr\{U_{piT}=1\mid\boldsymbol{\theta}_p,\boldsymbol{\beta}_i\}=\tau_i+(\varsigma_i-\tau_i)\prod_{k=1}^{K}\frac{\exp(\theta_{pk}-\beta_{ik})}{1+\exp(\theta_{pk}-\beta_{ik})}. \tag{14.2}$$

Equation 14.2 is not identified and need additional restrictions. The most direct restriction is that both the total item response and the responses to the subtasks representing its components be observable (and thus known). As Equation 14.2 incorporates the Rasch model, the scale of each component should have a mean set to a fixed value.

The relative impact of the subtasks on the total item has interesting implications for item design. Consider the response functions for two subtasks in Figure 14.1. The relationship of the probability of solving the total item to the trait for one component depends on the logit of the other component $\xi_{pik}=\theta_{pk}-\beta_{ik}$. As shown in Figure 14.1, θ_{p2} is most strongly related to $\Pr\{U_{iT}=1\}$ as ξ_{pi1} increases. Thus, items will measure θ_{p2} most strongly when β_{i1} is low. Conversely, items will measure θ_{p1} most strongly when β_{i2} is low. The implications of these relationships for item validity will be considered in the section with empirical examples.

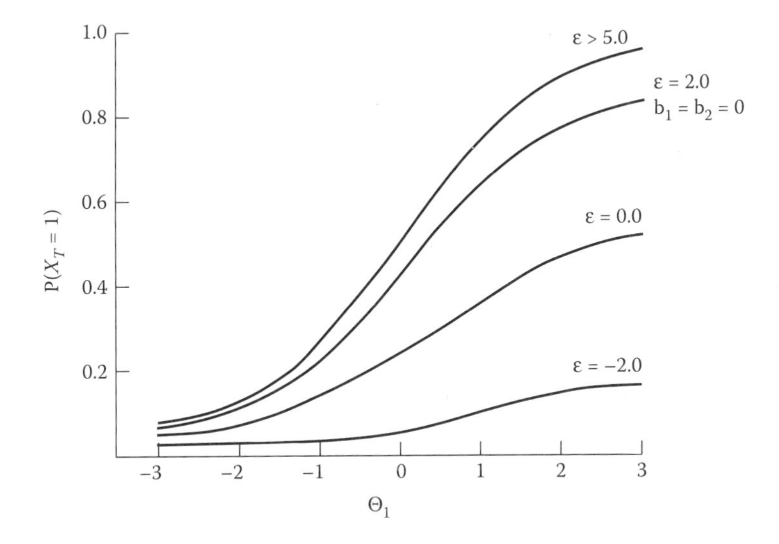

FIGURE 14.1
Regression of total item response probability on one component ability as a function of the logit of a second component.

For the GLTM, component item difficulties are replaced by a model of item difficulties from scored substantive features of the items. Let q_{im_k} be the score for item i in substantive feature m within component k, η_{m_k} be the weight of the feature m on the difficulty of component k, and η_{0k} be an intercept. Then, GLTM can be expressed as follows:

$$\Pr\{U_{piT}=1\,|\,\boldsymbol{\theta}_p,\boldsymbol{\beta}_i\}=\tau_i+(\varsigma_i-\tau_i)\prod_{k=1}^{K}\frac{\exp\left(\theta_{pk}-\sum_m q_{im_k}\eta_{m_k}+\eta_{0k}\right)}{1+\exp\left(\theta_{pk}-\sum_m q_{im_k}\eta_{m_k}+\eta_{0k}\right)}. \tag{14.3}$$

The relative strength of the stimulus features in GLTM, as indicated by η_{m_k}, has further implications for test design. That is, the stimulus features can be manipulated in the item to change the relative strength of component m in the item.

14.2.2 Multicomponent Models for Varying Components

For the MLTM-D, it is assumed that a single binary response ($U_{pi}=0, 1$) is associated with each item and that items vary in involvement of the m components. Still, responses to the components are unobservable. To provide estimates for components and for stimulus features within components, the MLTM-D involves two sets of scored variables on the items, c_{ik} is a binary variable (0,1) for the involvement of component k on item i, and q_{im_k} is the score for substantive feature m on item i that is relevant to component k with no restriction on range. At the task probability level, the MLTM-D gives the probability that examinee p solves item i as a function of the probabilities of the underlying and unobserved responses to the component, $\Pr\{U_{pik}=1\}$, that are postulated for involvement in item i as follows:

$$P\{U_{pi}=1\,|\,\boldsymbol{\theta}_p,\boldsymbol{\beta}_i\}=\prod_k[P(U_{pik}=1\,|\,\theta_{pk},\beta_{ik})]^{c_{ik}}. \tag{14.4}$$

Thus, if $c_{ik}=0$, when component k is not involved in item i, then $\Pr\{U_{pik}=1\}$. Additional restrictions must be imposed to make the model identifiable, as discussed below.

To relate stimulus features to item difficulty, the full MLTM-D gives the probability that examinee p solves item i as

$$\Pr\{U_{pi}=1\,|\,\boldsymbol{\theta}_p,\boldsymbol{\beta}_i\}=\prod_{k=1}^{K}\left[\frac{\exp\left(\theta_{pk}-\sum_m q_{im_k}\eta_{m_{lk}}+\eta_{0k}\right)}{1+\exp\left(\theta_{pk}-\sum_m q_{im_k}\eta_{m_k}+\eta_{0k}\right)}\right]^{c_{ik}}. \tag{14.5}$$

Equation 14.5 can be generalized to incorporate the additional parameters of the MLTM and GLTM to represent guessing and strategy application as

$$\Pr\{U_{pi}=1\,|\,\boldsymbol{\theta}_p,\boldsymbol{\beta}_i\}=\tau_i+(\varsigma_i-\tau_i)\prod_{k=1}^{K}\left[\frac{\exp\left(\theta_{pk}-\sum_m q_{im_k}\eta_{m(k)}+\eta_{0k}\right)}{1+\exp\left(\theta_{pk}-\sum_m q_{im_k}\eta_{m(k)}+\eta_{0k}\right)}\right]^{c_{ik}}. \tag{14.6}$$

This model in Equation 14.6 is included here only for completeness; actual parameter estimation and application awaits future research.

Model identification requires constraints both between and within components of the MLTM-D. Between components, because MLTM-D is multidimensional, model identification requires restrictions on the structure of matrix C with elements c_{bk}, where b represents blocks of items with identical c_{ik}. Within components, model identification requires restrictions on the structure of the M_k attributes in Q as well as restrictions to identify the scales of the latent components.

To illustrate a sufficient set of restrictions between components, consider the special case of items differing in the involvement of components, as given by c_{bk}, but not in difficulty within components or blocks. Then, the probability that an item in block b is solved by examinee p, $\ln P^*_{pb}$, is

$$\ln P^*_{pb} = \sum_k c_{bk} \ln P_{pbk}, \tag{14.7}$$

where $\ln P_{pbk}$ is the log probability that component k on an item in block b is responded to correctly by examinee p.

Define P^B as the matrix with elements $\ln P^*_{pb}$ and P^C as the matrix with elements $\ln P_{pbk}$ as defined above. Then, it can be shown that

$$P^B = CP^C. \tag{14.8}$$

If $B = K$, that is, the number of blocks equals the number of components, and C is of full rank, then

$$P^C = C^{-1}P^B. \tag{14.9}$$

That is, Equation 14.9 presents a just identified case where the log probabilities of the components may be obtained as linear combinations of the log probabilities in the blocks. If $B > K$, Embretson and Yang (2013) show that model identification requires a submatrix of C to be of full rank.

Within components, MLTM-D, MLTM, and GLTM are formulated as Rasch family models. Thus, implicitly, item discrimination is fixed to 1 for all items. Thus, if $\boldsymbol{\theta} \sim MVN(\boldsymbol{\mu}, \boldsymbol{\Sigma})$, we may set $\boldsymbol{\mu} = \mathbf{0}$ and estimate $\boldsymbol{\Sigma}$. Alternatively, the models can be formulated within components with a common item discrimination parameter. In this case, the model is identified by setting $\text{diag}(\boldsymbol{\Sigma}) = \mathbf{1}$ and $\mu = 0$. The latter method is often applied to ease computational burden in marginal maximum likelihood (MML) estimation. The parameter estimates can be linearly transformed into the Rasch model formulations above. The full hierarchical model in Equation 14.5 is identified when Q consists of K_m independent column vectors.

14.3 Parameter Estimation

The primary estimation method for multicomponent models is MML estimation. By using a general nonlinear mixed modeling software package, the models shown above may be specified and estimated. The approach is preferred over currently available Bayesian algorithms due to the shorter computational time, as shown below.

Embretson and Yang (2013) show that the MLTM-D may be efficiently estimated by MML with multidimensional integration. To elaborate, probability of $\mathbf{u}_p = \{U_{p1}, U_{p2}, \ldots, U_{pI}\}$ given $\boldsymbol{\theta}_p = \{\theta_{p1}, \theta_{p2}, \ldots, \theta_{pK}\}$ may be expressed as

$$\Pr\{\mathbf{u}_p \mid \boldsymbol{\theta}_p\} = \prod_{i=1}^{n} P_{pi}^{u_{pi}} (1 - P_{pi})^{1-u_{pi}}. \tag{14.10}$$

Then, the probability for a particular response pattern $\mathbf{u}_p$ for a randomly selected person p is given as

$$P\{\mathbf{u}_p\} = \int_{-\infty}^{\infty} \cdots \int_{-\infty}^{\infty} \int_{-\infty}^{\infty} P(\mathbf{u}_p \mid \boldsymbol{\theta}) g(\boldsymbol{\theta}) d\boldsymbol{\theta} = \int_{\theta} P(\mathbf{u}_p \mid \boldsymbol{\theta}) g(\boldsymbol{\theta}) d\boldsymbol{\theta} \tag{14.11}$$

and $g(\theta)$ is the probability density function of $\boldsymbol{\theta}$, assuming that $\boldsymbol{\theta} \sim MVN(\mathbf{0}, \Sigma)$.

For convenience in multidimensional integration, assume that diag $\Sigma = 1$. Then, let $\mathbf{U}$ be the item response matrix for a sample of examinees, $p = 1, \ldots, P$. The likelihood equation for parameter η_{mk} in Equation 14.5 is

$$\frac{\partial l}{\partial \eta_{mk}} = \int_{\boldsymbol{\theta}} \left[\sum_{i=1}^{I} \left(\frac{\tilde{r}_i - \tilde{N} P_{pi}}{P_{pi}(1 - P_{pi})} c_{ik} \prod_{h=1, h \neq k}^{K} P_{ijh}^{c_{ih}} (-q_{imk}) P_{ijm} (1 - P_{ijm}) \right) \right] g(\boldsymbol{\theta}) d\boldsymbol{\theta} \tag{14.12}$$

with

$$\tilde{N} = \left[\frac{\sum_{j=1}^{N} L(\mathbf{u}_p \mid \boldsymbol{\theta})}{P(\mathbf{u}_p)} \right], \quad \tilde{r}_{iT} = \left[\frac{\sum_{j=1}^{N} \mathbf{u}_{ip} L(\mathbf{u}_p \mid \boldsymbol{\theta})}{P(\mathbf{u}_p)} \right].$$

M-fold Hermite–Gauss quadrature can be applied to approximate the multiple integrals to simultaneously estimate the system of likelihood equations for the model, such as shown in Equation 14.12. Current technology usually does not permit Hermite–Gauss quadrature for more than six dimensions. Also, it should be noted that MLTM-D parameters are often estimated as one-parameter models within components in the normal metric to reduce computational burden and then transformed to the Rasch model formulations as in the equations above.

14.4 Model Fit

Several approaches to evaluating the fit of multicomponent models are available, including those that use (1) overall model indices, (2) item fit indices, and (3) graphic displays.

Overall model comparisons are useful for multicomponent models when evaluating alternative specifications of theoretical variables and/or for comparisons of models specified at different levels of complexity. For the MLTM-D with MML estimation, model comparisons based on likelihood are possible. These comparisons include nested model comparisons, information indices, and ratio fit indices. The Akaike information criterion (AIC) can be applied to compare models that are not nested. For example, models with different specifications of component inclusion in C or different Q can be compared for magnitude of AIC (or alternatives such as the Bayesian information criterion [BIC]), with the smaller values indicating better models (Cohen and Cho, Volume Two, Chapter 18). If the models to be compared have a parameter structure for model nested within the other, then comparisons with using likelihood-ratio tests are possible. That is, if L_1 and L_2 are the likelihoods of two models, with the former nested under the latter, the likelihood-ratio statistic is

$$G^2 = -2(\ln L_1 - \ln L_2), \tag{14.13}$$

which has an asymptotic χ^2 distribution with degrees of freedom equal to the difference in the number of parameters between the two models.

However, statistical tests tend to reject any model with increase in sample size, a fit index may be useful to guide the interpretations of model quality. Embretson (1997) presented a fit index for multicomponent models that utilized comparisons to both a null model (in which all items are postulated to be equally difficult) and a saturated model (in which separate difficulties are estimated for each item). Let L_0 and L_s be the likelihoods for these two models. The fit index is

$$\Delta^2 = (\ln L_0 - \ln L_m)/(\ln L_0 - \ln L_s). \tag{14.14}$$

Roughly, the denominator of Equation 14.14 indicates the amount of information that could be modeled, since it compares the saturated model with the null model. The numerator compares the information in the proposed model with the null model. The Δ^2 index is similar in magnitude and meaning to a squared multiple correlation coefficient between the target model estimates and the saturated model. Similar fit indices are routinely used to evaluate model quality in structural equation modeling (e.g., Bentler and Bonnett, 1980).

Overall goodness of fit for a multicomponent model can be evaluated by either a Pearson or likelihood-ratio statistic. The formulation of these test statistics depends on whether the data consist of the full item plus a subtask for each component, as in the original MLTM, or if component involvement varies between items, as feasible with the MLTM-D.

For data consisting of responses to a full item and its K subtasks, there are 2^{K+1} possible response patterns r for each item. Their distribution is multinomial. To obtain sufficient numbers of observations, persons must be grouped by ability estimates, jointly on the basis of K component abilities. The number of intervals for each component depends on the ability distributions and the sample size. In general, for the goodness-of-fit statistic to be distributed as χ^2, no fewer than five observations should be expected for any group. The Pearson goodness-of-fit statistic for the items is

$$G_P^2 = \Sigma_g \Sigma_r (f_{gir} - \Pr\{r_{gi}\})^2 / N_{gi} \Pr\{r_{gi}\}, \tag{14.15}$$

where $\Pr\{r_{gi}\}$ is the probability of pattern r in score group g for item i, f_{gir} its observed frequency, and N_{gi} the number of persons in score group g on item i. Thus defined, G_P^2 is assumed to be approximately distributed as Pearson χ^2.

For data in which component involvement varies across items, as accommodated by the MLTM-D, a different approach in formulating a likelihood-ratio test is required. The items have binary responses and vary in the relevant dimensions. Thus, an overall predicted item probability from the model for each examinee on each item, $\Pr\{U_{pi}\}$, is obtained. Pooling the examinees into fixed intervals g, the likelihood-ratio statistic for an item may be written as follows:

$$G_L^2 = 2\Sigma_g r_{gi} \ln[(r_{gi}/N_{gi})/\Pr\{U_{gi}\}]. \tag{14.16}$$

G_P^2 can be evaluated using the χ^2 distribution with *df* equal to the number of categories in an item.

Graphic methods, in which f_{gir} is plotted against $N_{gi}\Pr\{r_{gi}\}$, can also be useful in identifying the sources of misfit for a particular item.

14.5 Empirical Examples

To illustrate the application of multicomponent models, the same test was analyzed from two different conceptual perspectives. That is, two separate theories were applied to define components for a high-stakes test of mathematical achievement in middle school. The test contained a heterogeneous collection of items constructed to measure specified skills at Grade 8 according to a standards-based blueprint. Overall proficiency on the test had important consequences for students, teachers, and schools.

The two theoretical frameworks that were applied to illustrate applications of multicomponent models are (a) the hierarchical blueprint that defines specific skills for this standards-based test of achievement and (b) a cognitive theory of mathematical problem solving. As shown below, results from the two frameworks have different uses and implications.

14.5.1 Standards-Based Mathematical Skills

The standards-based framework provides evidence relevant to the content aspect of validity. Empirical support for the psychometric equivalence of items within the same narrow category provides additional evidence that the specified skills have been isolated. However, another potential use of this framework, in conjunction with MLTM-D, is for diagnosing skills for remediation. The example summarized below is elaborated more completely elsewhere (Embretson and Yang, 2013).

The test consisted of 86 mathematical achievement items that were organized in a hierarchy of skill specifications. At the highest level are the standards (number, algebra, geometry, and data), which define global areas in mathematics. Nested within the standards are a total of 10 benchmarks. Finally, nested with benchmarks are 25 indicators that define the most precise level of skills. The indicators are operationalized by two or more items. Although the items were developed to measure a single skill, a panel of mathematics experts and educators found that more than half the items involve more than one indicator (Poggio and Embretson, 2011) and approximately 25% of the items involve skills specified for another standard. The scores of the panel defined a C matrix for involvement of the four standards and Q matrices for the attributes (i.e., the specific indicators) represented in an item. For the current study, the Q matrix for the number component was available.

Prior to applying the multicomponent models, the one-parameter logistic (1PL) model was fit to the data, with the parameters estimated by MML. The fit of this model ($-2 \ln L = 244{,}312$; AIC = 244,486) with 86 item difficulties and a constant discrimination provides a comparison for evaluating the impact of multidimensional estimates in the MLTM-D. Likelihood-ratio tests indicated that two items did not fit the model ($p < 0.01$).

For the MLTM-D, parameters were estimated with MML using a nonlinear mixed modeling algorithm as described above. For all models, persons are specified random variables, $\boldsymbol{\theta} \sim MVN(\mathbf{0}, \Sigma)$, with the off-diagonal elements of Σ left free to permit correlated latent dimensions, and the $\text{diag}\Sigma = \mathbf{1}$ to set the scale of measurement. For convenience in multidimensional integration, the parameters of the MLTM-D were estimated in the normal metric with a constant discrimination parameter, α_k, for each component, k. Estimates can be subsequently rescaled for consistency with the MLTM-D as in Equation 14.5, where parameters within components are specified for Rasch models. With four separate components as defined by item involvement of the four standards, the specifications resulted in all models, including estimates for six covariances between the person dimensions and four constant item discrimination parameters, for a total of 10 parameters.

Three different MLTM-D models were specified. The full component model consisted of 136 estimates for component item difficulty as defined in C, plus the covariances and constant discriminations, for a total of 146 parameters. The results ($-2 \ln L = 242{,}945$; AIC = 243,237) indicated that this model fit significantly better than the 1PL comparison model ($\chi^2 = 1367$, df = 59, $p < 0.001$), and the AIC ($-2 \ln L = 244{,}312$; AIC = 244,486) was lower as well. The multidimensional specifications in MLTM-D over the traditional unidimensional model are supported.

The null model was specified as a comparison model to test the attribute specifications for the number component. In this case, the null model included 106 parameters, consisting of a constant item difficulty for the number component but 95 separately estimated item difficulties for the other components, plus the 10 covariances and constant discriminations. The resulting log likelihoods of the null model ($-2 \ln L = 255{,}439$; AIC = 255,651) were relatively low. However, the null model was the comparison model for the fit statistics, and the lower model likelihoods merely indicate that the item difficulties vary in the number component.

The restricted component model had a specified Q matrix for the number component and the C matrix specifications of item difficulty for the other components, for a total of 118 parameters, including covariances and constant discriminations. This model ($2 \ln L = 243{,}826$; AIC = 244,062) fit significantly better than the null model ($\chi^2 = 11{,}613$, df = 12, $p < 0.001$). While the fit for the restricted model was significantly worse than for the saturated model ($\chi^2 = 881$, df = 28, $p < 0.001$), the fit index ($\Delta = 0.964$) indicated relatively strong prediction of item parameters with the number component predicted by attributes.

Item fit was also evaluated using the G_L^2 statistic described above for the full component model. Fit was adequate for all but 2 of the 86 items, with the probability of $G_L^2 < 0.01$. These two items also did not fit well in the 1PL model. Overall fit was also evaluating by correlating the observed proportions correct with the MLTM-D expectation across all categories and items. Since the correlation was high ($r = 0.927$), overall model fit appears to be adequate.

Diagnostic potential for the MLTM-D dimensions was assessed in several ways (see Embretson and Yang, 2013, for more detailed analyses). First, while the correlations between the dimensions were moderately high ($0.620 < r < 0.732$), as expected for cognitive measures, the empirical reliabilities of expected a posterior (EAP) estimated person scores on the dimensions were higher ($0.783 < r_{tt} < 0.876$). Second, the fit of MLTM-D was compared to diagnostic classification models, the deterministic-input noisy-AND (DINA) model and the loglinear cognitive diagnosis model (LCDM). Again, the fit of MLTM-D was better, as indicated by AIC. Third, individual score levels on the four components varied significantly, especially near the proficiency cutline of the test, thus, permitting differential diagnosis at the component level. Finally, scores were mapped to the specific indicators of the standards for the number component through common scale measurement of attributes and persons. Thus, a more detailed description of skill mastery was feasible for this component.

Taken together, these results had several implications. First, the content aspect of validity was supported. Performance on the four standards can be distinguished because the multicomponent model did fit better than a unidimensional model. Second, for the number component, content validity was further supported by the strong prediction obtained from the indicators within this standard. That is, items categorized in the same indicator had similar difficulties. Third, the potential for diagnosing more specific skills was supported. That is, individual differences in the standards can be reliably distinguished, which has implications for possible remedial efforts to improve performance.

14.5.2 Cognitive Components in Mathematical Achievement Items

As noted by Leighton and Gierl (2007), the thinking processes that are applied by examinees to solve test items are typically assumed, rather than studied empirically. Research concerning these skills is particularly appropriate for elaborating the response process aspect of validity. Further, establishing a cognitive framework for a test also has implications for test design. Gorin (2006, 2007) describes several ways in which such a framework can impact item content.

Items on the test were scored using the cognitive variables from Embretson and Daniel's (2008) theory of mathematical problem solving that postulates several processing stages. In Embretson and Daniel (2008), the theory was supported by item difficulty modeling using LLTM applied to items from the Quantitative section on the *Graduate Record Examination*. To apply the theory to multicomponent modeling of a middle school test of mathematical achievement, adaptation was required. For the current test, two stages were scored for both component involvement, C, and the variables that impact component difficulty, Q. Consider again, the items shown in Table 14.1. Items similar to both the full task and subtasks appear as separate items with different surface content on this heterogeneous achievement test. Thus, the involvement of the integration and solution execution components could be scored for C.

The Q matrices were prepared by scoring items according to definitions of the cognitive complexity variables shown in Table 14.2. The variables that impacted the difficulty of translating the problem were used as covariates within the two stages. Within components, the integration stage variables, Q had eight variables associated with the stage, while the solution execution variables, Q, included five variables. Descriptive statistics on the item scores are shown in Table 14.3.

Estimates for a series of multicomponent models were obtained with MML estimation, using the same algorithm as described for the standards-based framework for components

TABLE 14.2

Definition of Variables in Revised Cognitive Model

Attribute	Definition
Translation and Encoding	
Mathematical encoding	Number of mathematical terms in the stem and all response options, including numerals, variables (e.g., x, y, m), axis labels, comparators (e.g., <, >)
Contextual encoding	Number of words, excluding variables, in the stem and all answer options
Diagram encoding	Indicator of the presence of a diagram, graph, or other figure, excluding tables, in the stem or answer options
Integration	
Semantic knowledge	Item requires knowledge of specific mathematical terms
Recall or access equations	Indicator of whether the examinee must recall known equations (e.g., formula for slope of a line, the Pythagorean theorem, probability equations, cost/quantity) in order to answer the item
Generate equations or plausible values	Indicator for equation source of generating or deriving equations to fit the context
Translate diagram	Indicator for mathematical representation or equation to be acquired from a diagram or table
Visualization	Indicator for whether examinee must draw or otherwise visualize a diagram or figure to understand or answer the item
Diagram/table aid	The response options include a table, chart, or diagram that represents the problem stated in the stem
Representation aid	The distractors present a representation of the problem that aids integration
Solution Execution	
Operations	Operations to isolate unknown required
Procedural knowledge	The maximum procedural knowledge necessary in solving the item
Number of procedures	The total number of procedures necessary for solving the item (e.g., if two different fractions are involved in an equation to be solved, number of procedures would be 2)
Number of computations	The total number of computations necessary for solving the item; including computations necessary in evaluating answer options and in stem
Number of subgoals	The total number of substeps necessary for answering an item (e.g., finding a slope for the equation of a line)

above. Three models were fit to the data. First, the null model with two components had only five parameters; one covariance, two constant difficulties, and two constant discriminations. With so few parameters, overall model likelihoods were not high ($-2 \ln L = 285{,}599$, AIC = 285,609) for this comparison model. Second, the full component model, with 134 parameters, the log likelihood was substantially higher ($-2 \ln L = 242{,}185$, AIC = 242,452) and significantly different from the null model ($\chi^2 = 43{,}414$, df = 129, $p < 0.001$). Further, the log likelihood of the two-component model was also significantly smaller than the comparison 1PL model ($\chi^2 = 2127$, df = 49, $p < 0.001$) and the AIC ($-2 \ln L = 244{,}312$; AIC = 244,486) is lower. However, the correlation between dimensions ($r = 0.845$) was high. For the restricted model, with the parameter estimates for the cognitive variables for a total of 24 parameters

TABLE 14.3

Descriptive Statistics and MLTM-D Estimates for Cognitive Variables

	Descriptive Statistics		Model Estimates		
	Mean	Standard Deviation	Coefficient η_{m_k}	Standard Error	*t*-Test
Overall Components					
Translation and Integration	0.88	0.321	−1.6140	0.03466	N/A
Solution Planning and Execution	0.64	0.480	−1.7957	0.02333	N/A
Translation and Integration					
Mathematical encoding	19.77	13.582	0.032	0.001	31.51
Contextual encoding	34.63	18.799	0.009	<0.001	19.65
Diagram encoding	0.29	0.454	0.509	0.031	16.51
Semantic knowledge	0.16	0.369	2.123	0.037	57.87
Translate diagram	0.27	0.443	0.047	0.028	1.68
Translate word equations	0.05	0.211	−1.513	0.074	−20.35
Recall or access equations	0.30	0.459	0.537	0.026	20.33
Generate equations	0.28	0.449	0.719	0.027	26.30
Visualization	0.08	0.273	0.765	0.031	22.23
Representation aid	0.30	0.459	−0.510	0.033	−15.63
Diagram/table aid	0.09	0.290	0.412	0.035	11.73
Solution Planning and Execution					
Operations	0.23	0.422	0.751	0.033	23.06
Number of subgoals	0.14	0.532	0.410	0.019	21.48
Procedural knowledge level	1.80	1.964	0.269	0.009	29.91
Number of procedures	1.08	0.955	−0.614	0.026	−23.16
Number of computations	1.94	2.656	−0.088	0.006	−14.29

($-2 \ln L = 269{,}716$, AIC = 269,764), the log likelihoods were significantly lower than for the full model ($\chi^2 = 27{,}531$, df = 110, $p < 0.001$). However, the likelihood-ratio statistic, which is comparable in magnitude to a multiple correlation coefficient as described above, showed moderate prediction of item difficulty within the components from the cognitive model ($\Delta = 0.604$).

Table 14.3 presents parameter estimates for the cognitive model and the null model. For the null model, Table 14.3 shows that the constant difficulty estimates for the components were very similar, indicating comparable difficulty levels. Further, the item discrimination for the integration component ($\alpha_1 = 0.445$, $\sigma_{\alpha 1} = 0.006$) was somewhat lower than for the solution execution component ($\alpha_2 = 0.689$, $\sigma_{\alpha 2} = 0.010$). For the saturated model, not shown in Table 14.3, the constant item discriminations were generally higher, and the same difference between the integration component ($\alpha_1 = 0.508$, $\sigma_{\alpha 1} = 0.006$) and the solution execution component ($\alpha_2 = 0.780$, $\sigma_{\alpha 2} = 0.012$) was observed. Thus, in summary, although the components had comparable difficulties, items on the solution execution component were somewhat more discriminating.

The results for the *t*-tests in Table 14.3 show that most cognitive complexity variables had significant impact on item difficulty ($p < 0.001$). The translation variables, which are included for both components, had significant impact on item difficulty. Mathematical encoding and diagram encoding, which are mathematically relevant variables, were significant; however, context encoding, which is a nonmathematical source of complexity, also had significant impact on item difficulty. For the integration component, the highest *t* value was for semantic knowledge, for which definitions for specific mathematical terms are required to represent the problem. Other variables with high positive impact on the integration component were generate equations, recall or access equations, and visualization, all of which increased item difficulty. In contrast, providing representation aids (usually equations) in the response options and translating equations from words were associated with decreased item difficulty. For the solution execution component, item difficulty increased when items required special operations to isolate the unknowns or had subgoals or a high procedural knowledge level. In contrast, items with more computations and more computational procedures were easier.

Taken together, these results have several implications. First, evidence relevant to the response process aspect of construct validity was obtained, that is, the separation of two underlying processing stages, with different variables that impact their difficulty, was supported. Thus, the current test represented a composite of these sources of item complexity. Second, the results can guide further item selection to represent the desired balance of these processes. Third, the results are also useful in guiding item revision. For example, the impact of verbal abilities and skills could be deemphasized by reducing the verbal context in items and by reducing the role of semantic knowledge. Fourth, although the results provide enhanced meaning to scores through supporting the response process aspect of validity, individual differences in the components cannot be reliably distinguished due to the high correlation between dimensions.

14.6 Discussion

In summary, multicomponent models can link item performance to underlying components and attributes, as defined by a cognitive theory or conceptual framework. Generalizations of the original MLTM (Whitely, 1980), GLTM (Embretson, 1984), and MLTM-D (Embretson and Yang, 2013) were developed to accommodate additional aspects of cognitive or conceptual complexity and to increase the scope of applications. The GLTM increased linkage to cognitive theory by including parameters to represent the impact of stimulus features on component difficulty. These parameter estimates can provide useful information about how to redesign items for specific levels and sources of component difficulty.

The MLTM-D increased applicability to existing tests because it includes varying component structures *between* items. That is, unlike MLTM and GLTM for which obtaining adequate estimates of component parameters required developing subtasks *within* items, the MLTM-D parameters can be estimated through scores that reflect component involvement. Further, the MLTM-D also has potential for diagnosis of skill proficiency patterns from heterogeneous achievement tests that are based on well-specified blueprints. The scope of diagnostic applications of multicomponent models is broad, since many high-stakes achievement tests, licensure tests, and certification tests are based on blueprints that precisely define skills. The MLTM-D may also be applied to other types of tests, particularly ability tests, if appropriate structures can be defined.

The details of the models were presented, along with appropriate estimation techniques. MML is currently the estimation procedure recommended for multicomponent models, due to its relative speed and derived standard errors. However, MML requires multidimensional integration which is not very successful with more than four or five dimensions when the current algorithms are applied. Thus, identifying and applying other estimation procedures may be a topic for future research to extend applications of MLTM-D.

The goodness-of-fit indices presented included both model comparisons and item fit. Model comparisons are useful in evaluating model quality if appropriate alternative models are estimated. Likelihood-ratio tests are appropriate for nested models, in which the parameters of one model are a subset of another model. Otherwise, information indices such as AIC are useful. Empirical examples with the MLTM-D specified according to two different theoretical frameworks were applied to demonstrate the goodness-of-fit tests. Interesting comparison models included a unidimensional model and models with differing degrees of parameter specification. Furthermore, a likelihood-ratio test also was demonstrated in conjunction with model comparisons. Item fit, on the other hand, was evaluated using traditional likelihood-ratio tests, reformulated for application to multicomponent models.

The advantages of applying multicomponent models were demonstrated in the two empirical examples, a cognitive model framework and a standards-based conceptual framework, that were applied with the MLTM-D to the same test of mathematical achievement. The cognitive model framework results were relevant to the response process aspect of validity. Two underlying components in problem solving, integration and solution execution, were supported. As pointed out by Leighton and Gierl (2007), elaborating the cognitive processes in item solving supports inferences about examinees' thinking skills. Further, results from the cognitive framework were also relevant to item design. As suggested by Gorin (2007), estimating the impact of substantive variables in items can provide guidance to item developers about how to change the level and source of cognitive complexity. Finally, the results provided moderate prediction of item difficulty, as would be needed in effective automatic item generation. The level of prediction that was obtained from the cognitive variables was similar to the levels that have been regarded by some researchers as useful for items with little or no tryout (Mislevy et al., 1993)

In contrast, the results from the standards-based conceptual framework had different implications. Since a test blueprint specifies test content in this framework, the results are relevant to the content aspect of validity, supported by finding that a single index for items involving the same narrow skill categories is highly predictive of item difficulty. Furthermore, the results indicated potential for diagnostic assessment, especially at the highest level in the blueprint. Many items in the test required skills from more than one standard to be applied for their solution. The dimensions associated with the standards were reliably distinguished, thus permitting differential diagnosis of competency at this level. As pointed out by Stout (2007), traditional MIRT does not have this potential because compensatory dimensions would be appropriate for items that could be solved by alternative strategies but not for items that require two or more components for solution.

Given the increasing importance of diagnostic assessment from high-stakes tests (Rupp et al., 2010), future research should be devoted to applying multicomponent models in that context. The primary advantage of multicomponent models, in contrast to diagnostic classification models (Hensen et al., 2009; von Davier, 2008), is that they are applicable to heterogeneous tests, especially those that are based on hierarchical blueprints. In fact, the identification of the multidimensional components depends on heterogeneity to separate the dimensions. In contrast, the diagnostic classification models are best applied to more narrow tests. However, the application of multicomponent models in the diagnostic context will require supporting research, including developing more efficient estimation algorithms, especially to estimate models with more parameters such as guessing and slipping, and developing conceptual frameworks that more adequately explain the sources and levels of item complexity.

Acknowledgment

The work described in this chapter is partially supported by Institute of Educational Sciences Grant R305A100234 *An Adaptive Testing System for Diagnosing Sources of Mathematics Difficulties*, Susan Embretson, PI.

References

American Educational Research Association. 1999. *Standards for Educational and Psychological Testing*. Washington, DC: American Educational Research Association, American Psychological Association, National Council on Measurement in Education.

Bejar, I. 2010. Recent developments and prospects in item generation. In S. E. Embretson (Ed.), *Measuring Psychological Constructs: Advances in Model-Based Approaches* (pp. 201–226). Washington, DC: American Psychological Association Books.

Bentler, P. M. and Bonnett, D. G. 1980. Significance tests and goodness of fit in the analysis of covariance structures. *Psychological Bulletin, 88*, 588–606.

Bolt, D. M. and Lall, V. F. 2003. Estimation of compensatory and noncompensatory multidimensional item response models using Markov chain Monte Carlo. *Applied Psychological Measurement, 27*, 395–414.

Daniel, R. C. and Embretson, S. E. 2010. Designing cognitive complexity in mathematical problem solving items. *Applied Psychological Measurement, 34*, 348–64.

Embretson, S. E. 1984. A general multicomponent latent trait model for response processes. *Psychometrika, 49*, 175–186.

Embretson, S. E. 1997. Multicomponent latent trait models. In W. J. van der Linden and R. Hambleton (Eds.), *Handbook of Modern Item Response Theory* (pp. 305–322). New York: Springer.

Embretson, S. E. and Daniel, R. C. 2008. Understanding and quantifying cognitive complexity level in mathematical problem solving items. *Psychology Science, 50*, 328–344.

Embretson, S. E. and McCollam, K. M. 2000. A multicomponent Rasch model for covert processes. In M. Wilson and G. Engelhard (Eds.), *Objective Measurement V* (pp. 203–213). Norwood, NJ: Ablex.

Embretson, S. E. and Yang, X. 2006. Multicomponent latent trait models for complex tasks. *Journal of Applied Measurement, 7*, 540–557.

Embretson, S. E. and Yang, X. 2013. A multicomponent latent trait model for diagnosis. *Psychometrika, 78*, 14–36.

Fischer, G. H. 1973. The linear logistic test model as an instrument in educational research. *Acta Psychologica, 37*, 359–374.

Gierl, M. and Haladyna, T. 2012. *Automatic Item Generation*. New York: Taylor and Francis: Rutledge.

Gorin, J. S. 2006. Item design with cognition in mind. *Educational Measurement: Issues and Practice, 25*, 21–35.

Gorin, J. S. 2007. Test construction and diagnostic testing. In J. P. Leighton and M. J. Gierl (Eds.), *Cognitive Diagnostic Assessment in Education: Theory and Practice* (pp. 173–201). New York: Cambridge University Press.

Henson, R. and Douglas, J. 2005. Test construction for cognitive diagnosis. *Applied Psychological Measurement, 29*, 262–277.

Hensen, R. A., Templin, J. L., and Willse, J. T. 2009. Defining a family of cognitive diagnosis models using log-linear models with latent variables. *Psychometrika, 74*, 1919–210.

Leighton, J. P. and Gierl, M. J. (Eds.) 2007 *Cognitive Diagnostic Assessment in Education: Theory and Practice*. New York: Cambridge University Press.

Mayer, R. E., Larkin, J., and Kadane, J. B. 1984. A cognitive analysis of mathematical problem solving ability. In R. Sternberg (Ed.), *Advances in the Psychology of Human Intelligence (Volume 2)* (pp. 231–273). New York: Psychology Press.

Mislevy, R. J., Sheehan, K. M., and Wingersky, M. 1993. How to equate tests with little or no data? *Journal of Educational Measurement, 30*, 55–76.

Pellegrino, J. W., Mumaw, R., and Shute, V. 1985. Analyses of spatial aptitude and expertise. In S. Embretson (Ed.), *Test Design: Developments in Psychology and Psychometrics* (pp. 45–76). New York: Academic Press.

Poggio, J. and Embretson, S. E. 2011. *Multiple Indicators and Standards in Complex Mathematical Achievement Test Items for Middle School.* Report IES1002A-2011 for Institute of Educational Sciences Grant R305A100234. Cognitive Measurement Laboratory, Georgia Institute of Technology: Atlanta, Georgia.

Rupp, A., Templin, J., and Henson, R. 2010. *Diagnostic Assessment: Theory, Methods, and Applications.* New York: Guilford.

Sternberg, R. J. 1977. *Intelligence, Information-Processing and Analogical Reasoning: The Componential Analysis of Human Abilities.* Hillsdale, NJ: Erlbaum.

Stout, W. 2007. Skills diagnosis using IRT-based continuous latent trait models. *Journal of Educational Measurement, 44*, 313–324.

von Davier, M. 2008. A general diagnostic model applied to language testing data. *British Journal of Mathematical and Statistical Psychology, 61*, 287–307.

Whitely, S. E. 1980. Multicomponent latent trait models for ability tests. *Psychometrika, 45*, 479–494.

Section IV

Models for Response Times

15

Poisson and Gamma Models for Reading Speed and Error

Margo G. H. Jansen

CONTENTS

15.1 Introduction.. 245
15.2 Rasch Poisson Counts Model.. 246
15.2.1 Derivation of the Model.. 246
15.3 Parameter Estimation.. 247
15.3.1 Joint Maximum Likelihood Estimation.. 247
15.3.2 Maximum Marginal Likelihood Estimation 248
15.4 Rasch Model for Speed 249
15.4.1 Derivation of the Model.. 249
15.4.2 Parameter Estimation: MML.. 250
15.4.3 Model Extensions.. 251
15.4.4 Model Fit 252
15.4.5 Examples 253
15.5 Discussion.. 257
References.. 258

15.1 Introduction

Most of the currently used item response models are designed for testing situations, where test takers are presented with multiple, written items that are usually (not necessarily) scored dichotomously. A particularly well-known model for these kinds of test data is the one-parameter logistic (1-PL) or Rasch model (Volume One, Chapter 3). In such a model, the chance of a test taker answering an item correctly is a function of a person ability parameter and an item difficulty parameter as in 1-PL. Not all tests follow this item format and some of them require another type of modeling.

In his monograph *Probabilistic Models for Some Intelligence and Attainment Tests,* Rasch (1980) has proposed two Poisson process models that are useful for analyzing results from certain specialized tests, in this test of reading ability. Both models can be referred to as Poisson process models because of the basic similarity of the underlying assumptions. The first model Rasch proposed can be used in testing situations where an examinee is presented with a series of highly similar items, or when the same task is given repeatedly, and the test score is the number of errors or unsuccessful attempts. Spray (1997) coined, in the context of measuring psychomotor skills, the term multiple-attempt single-item (MASI) tests, as opposed to the more usual single-attempt multiple-items tests in the cognitive domain. Assuming small and independent error probabilities, Rasch derived a Poisson distribution for the error counts in these types of tests.

The evaluation of another aspect of reading ability, namely reading speed, was the focus of Rasch's second Poisson process model. In educational measurement, the focus is usually on how well a test taker can perform on a task, and not on the time it takes to complete it. Among the exceptions is early reading. Oral reading tests are usually given in a continuous format, where the test takers are instructed to read a set of single words or a reading passage as rapidly as possible without making errors. Several studies have shown that speed and accuracy in reading are related to speed and accuracy in word retrieval as demonstrated in various types of naming tests. Suppose that the test is administered using a time limit. Since examinees differ in speed, they will tend to complete a different number of items (words) in any fixed time period and the test score will be the total number of items completed. Obviously, it is also possible to administer the same test without a time limit, allowing the test takers to complete the test. The event of a test taker completing a number of items in a given time t exceeds some given number m that is equivalent to the event of the time T required to complete m items being less than t. The test score to record and analyze under unlimited time conditions will therefore be the response time. On the basis of similar assumptions, as before, Rasch derived a gamma distribution for the response time, and a Poisson distribution for the number of items completed.

Psychometric models for item response times have been presented among others, Fischer and Kisser (1981), Maris (1993), Teurlinckx, Molenaar, and van der Maas (Volume One, Chapter 17), and van der Linden (Volume One, Chapter 16). Among the examples of item response models where the response times are taken into account is a model developed by Thissen (1983). In Thissen's model, a two-parameter logistic item response model is combined with a linear model for the logarithm of the response times, including a normally distributed error term. Some research has also been done to develop variants of Rasch's 1-PL item response model taking both speed and power into account (Thissen, 1983; Van Breukelen, 1989; Verhelst et al., 1997; van der Linden, 2006). Mainly these theoretical efforts laid the foundation for applied work (van der Linden, 2008).

15.2 Rasch Poisson Counts Model

15.2.1 Derivation of the Model

As already noted, the Rasch Poisson counts model (RPCM) is a unidimensional response trait model for tests where the same item or task is given repeatedly, and the test score is the number of (un)successful attempts. The Poisson distribution can be derived as the limit of a sum of independent, identically distributed Bernoulli variables (Rasch, 1980; Masters and Wright, 1984). As pointed out by Lord and Novick (1968), the derivation can equally well be based on the more general case of unequal but small probabilities. The probability of a specific observed count y can be expressed as

$$p(y;\lambda) = \frac{(\lambda)^y}{y!} e^{-\lambda} \tag{15.1}$$

where λ is known as the intensity or rate parameter, and $y! = y \times (y-1) \times \cdots \times 2 \times 1$.

The expectation and variance of such a Poisson-distributed variable are respectively

$$E(\lambda) = Var(Y) = \lambda$$

Now, let p denote the examinees and let i denote the (nested within test takers) repeated observations. The latter might refer to sets of tests measuring the same trait, and also to longitudinal test administration design where the same measure is obtained on different occasions (Jansen, 1995). Rasch assumed a multiplicative decomposition for λ. Adding n_i allows for differences between the tests in length. So

$$\lambda_{pi} = n_i \times \tau_i \times \theta_p \tag{15.2}$$

for a specific examinee p and test i. The θ's refer to the test taker's ability and the τ's refer to the test difficulty parameters. In Rasch's original model, both the ability and the test parameters were assumed to be fixed. It is reasonable to think of the set of examinees as a sample from some population and therefore to introduce a common underlying distribution for the ability parameters. This model allows for a phenomenon, often observed in count data, which is known as overdispersion or extra-Poisson variation (Lawless, 1987; Thall and Vail, 1990; Ogasawara, 1996). The mixed Rasch model can be further extended by imposing a structure on the item difficulties and by incorporating manifest, categorical or continuous, and explanatory variables (Jansen, 1995a,b; Ogasawara, 1996; Hung, 2012).

Since the person parameters are nonnegative quantities, the gamma distribution, being conjugate to the Poisson, is a convenient choice. Using other, nonconjugate distributions, such as the lognormal distribution leads to less-tractable results. Assuming that the θ_i's are independently and identically gamma distributed with index parameter c and scale b, the distribution can be written as follows:

$$f(\theta; c, b) = \frac{1}{b^c}\frac{1}{\Gamma(c)}\theta^{c-1}e^{-(\theta/b)} \tag{15.3}$$

where $\Gamma(.)$ is the gamma function. When the argument of $\Gamma(.)$ is a positive integer, the following expression applies: $\Gamma(c) = (c - 1)!$ Alternatively, the gamma distribution can be parameterized in terms of a shape parameter c and an inverse- scale parameter, $s = b^{-1}$, also known as a rate parameter. The first two moments of the distribution are simple functions of these shape and rate parameters

$$\mathrm{E}(\theta) = \frac{c}{s}$$

$$\mathrm{Var}(\theta) = \frac{c}{s^2}$$

15.3 Parameter Estimation

15.3.1 Joint Maximum Likelihood Estimation

We assume that all pupils $p = 1, \ldots, P$ have taken the same $i = 1, \ldots, I$ tasks. Assuming local independence, the likelihood of the data based on Equation 15.1 can be written as

$$L = \prod_p \prod_i \frac{\lambda_{pi}^{y_{pi}}}{y_{pi}!} \exp(-\lambda_{pi}) \tag{15.4}$$

Joint maximum likelihood (JML) estimates of the parameters in Equation 15.2 can be derived by taking the logarithm of Equation 15.4 and obtaining the derivatives with respect to the parameters

$$\ell = \sum_{pi}[y_{pi}\ln(\lambda_{pi}) - \lambda_{pi}] - \sum_{pi}\ln(y_{pi}!) \tag{15.5}$$

$$\frac{\partial \ell}{\partial \tau_i} = \frac{y_{+i}}{\tau_i} - n_i \sum_p \theta_p \tag{15.6}$$

$$\frac{\partial \ell}{\partial \theta_p} = \frac{y_{p+}}{\theta_p} - \sum_i n_i \tau_i \tag{15.7}$$

For identification purposes, we need one restriction on the parameters. Choosing $\Sigma_j n_j \tau_j$ to be equal to an arbitrary constant C, we obtain the following JML estimators from Equations 15.6 and 15.7:

$$\hat{\theta}_p = \frac{y_{p+}}{C} \tag{15.8}$$

$$\hat{\tau}_i = \frac{y_{+i}}{n_i \Sigma_p \hat{\theta}_p} = \frac{C y_{+i}}{n_i \Sigma_p y_{p+}} \tag{15.9}$$

Note that these equations can be solved explicitly.

15.3.2 Maximum Marginal Likelihood Estimation

For given ($\boldsymbol{\tau}$, c, b), the joint density of ($\mathbf{y}$, $\boldsymbol{\theta}$) is given by

$$f(y, \boldsymbol{\theta}; \boldsymbol{\tau}, c, b) = p(y \mid \boldsymbol{\theta}; \boldsymbol{\tau}) g(\boldsymbol{\theta}; c, b) \tag{15.10}$$

from which the marginal likelihood can be obtained by integration with respect to $\boldsymbol{\theta}$. The marginal likelihood can be written in the following convenient way:

$$L = \prod_p \frac{\Gamma(\Sigma_i y_{pi} + c)(c/b)^c}{(\Sigma_i y_{pi})!\Gamma(c)} \left[\frac{(c/b)}{\Sigma_i n_i \tau_i + (c/b)}\right]^{\Sigma_i y_{pi}} \left[\frac{1}{\Sigma_i n_i \tau_i + (c/b)}\right]^c$$
$$\times \prod_p \left(\sum_j y_{pi}\right)! \prod_i \frac{(n_i \tau_i / (\Sigma_i n_i \tau_i))^{y_{pi}}}{(y_{pi})!} \tag{15.11}$$

The first main factor is a product of negative binomial likelihoods for the row sums y_{p+} ($y_{p+} = \Sigma_i y_{pi}$) depending on the distribution of the ability parameters and the test parameters through the sums $\Sigma_i n_i \tau_i$, and the second factor is a product of multinomial likelihoods for the vectors $(y_{p1}, \ldots, y_{pI})$ given row sum y_{p+}, which depends only on the parameter vector τ. Maximum marginal likelihood (MML) estimators for the model parameters are derived by taking the logarithm of the expression in Equation 15.10, obtaining the partial derivatives with respect to the parameters, and setting the resulting expressions equal to zero.

Maximum conditional and marginal likelihood procedures lead to identical estimation equations for the test parameters τ. MML estimators of model parameters c and s are derived by taking the logarithm of the expression in Equation 15.9, obtaining the partial derivatives with respect to the parameters, and setting the resulting expressions equal to zero.

$$\frac{c}{s} = \frac{1}{P} \sum_{p}^{P} \frac{(y_{p+} + c)}{(\Sigma_j n_j \tau_j + s)} \tag{15.12}$$

$$\psi(c) - \ln(s) = \frac{1}{P} \sum_{p}^{P} \psi(y_{p+} + c) - \ln(\Sigma_i n_i \tau_i + s) \tag{15.13}$$

Approximate standard errors for the estimates can be obtained by inverting the observed information matrix.

The posterior distribution of the θ's is a gamma distribution with parameters $(y_{p+} + c)$ and $(\Sigma n_i \tau_i + s)$. So, writing C for $\Sigma n_i \tau_i$, their posterior means are

$$\varepsilon(\theta \mid y) = \frac{y + c}{C + s} \tag{15.14}$$

and their standard deviation are

$$sd(\theta \mid y) = \frac{\sqrt{y + c}}{C + s} \tag{15.15}$$

It is possible to adapt the model to the case of not all tasks given to all test takers. However, by allowing for missing data, we lose a very attractive characteristic of the model for complete data: Estimation of the parameters of the latent distribution, c and b, independently of the test parameters τ, is then no longer possible (Jansen, 1995a,b).

In many situations, the assumption of fixed test parameters is too restrictive. A tractable and reasonably flexible way to describe variation over test takers is to assume that the test taker's vectors of occasion parameters $(\tau_{p1}, \ldots, \tau_{pI})$ are sampled from a Dirichlet distribution (Nelson, 1984; van Duijn, 1993; Jansen, 1995).

15.4 Rasch Model for Speed

15.4.1 Derivation of the Model

The focus of Rasch's second Poisson process model is response speed. Rasch developed the model originally for tests aimed at measuring reading speed, where test takers had to read texts aloud. The words in the text were the basic units. Rasch assumed that the probability that a test taker completing a word in any small time interval depends on the length of the interval, and that as the interval length goes to zero, the probability of more than one completion occurring in the interval is negligible. The probability of completing y readings in m time periods can then be approximated by a Poisson distribution. Now, by writing mp as λt in the limit, with λ as the average number of words read per time unit, the Poisson distribution for the number of items y completed in total test time t is

$$p(y;t) = \frac{(\lambda t)^y}{y!} e^{-\lambda t} \tag{15.16}$$

If we instruct test takers to complete the entire test as fast as possible without making errors, instead of imposing a time limit, then, the total amount of time that takes the test takers to do so is the response variable to record. The probability model for the response rate can be directly obtained from Equation 15.1. Namely, if the Poisson distribution holds for the number of items completed, then, the between-response times are negative exponentially distributed with scale parameter λ, and the time t it takes to complete m items is gamma distributed with a known index parameter m and scale (rate) parameter λ. The probability distribution for the total response time of a test taker on an m-item test can be written as

$$f(t;\lambda) = \frac{\lambda^m}{\Gamma(m)} t^{m-1} e^{-\lambda t}. \tag{15.17}$$

For the rate parameter, we use the same type of multiplicative decomposition as Rasch proposed for the counts model

$$\lambda_{pi} = \theta_p \times \tau_i$$

where θ_p refers to the ability and τ_i refers to the test "difficulty." The larger θ_p and τ_i, the shorter the expected test response time. A drawback of the factorization is the lack of simple sufficient statistics. Rasch was therefore unable to deal with the parameter estimation problem in a completely satisfactory way. Here, too, we assume the θ_i's to be gamma-distributed random variates:

$$f(\theta;c,s) = \frac{s^c}{\Gamma(c)} \theta^{c-1} e^{-s\theta}. \tag{15.18}$$

MML methods can be used to obtain model parameter estimates.

15.4.2 Parameter Estimation: MML

For given ($\boldsymbol{\tau}$, c, s), upon some arrangement, the joint density of ($\mathbf{t}$, $\boldsymbol{\theta}$) can be written as

$$f(\boldsymbol{t},\boldsymbol{\theta};\boldsymbol{\tau},c,s) = p(\boldsymbol{t}\,|\boldsymbol{\theta};\boldsymbol{\tau}) g(\boldsymbol{\theta};c,s) = \prod_p \frac{(s)^c \theta_p^{\Sigma m_i + c - 1} \exp(-(s + \Sigma_i \tau_i t_{pi})\theta_p)}{\Gamma(c)} \prod_i \frac{\tau_i^{m_i} t_{pi}^{m_i - 1}}{(m_i - 1)!} \tag{15.19}$$

The marginal likelihood function, which in this particular case is a closed expression, is obtained by integrating with respect to θ. It can be written as

$$L_m = \prod_p \frac{s^c \Gamma(c+\Sigma_i m_i)\prod_i\left(\tau_i^{m_i} t_{pi}^{m_i-1}\right)}{\Gamma(c)(s+\Sigma_i \tau_i t_{pi})^{c+\Sigma_i m_i}\prod_i (m_i-1)!} \tag{15.20}$$

where Γ(.) denotes the earlier gamma function. From Equation 15.16, MML estimators can be derived for the test difficulties and the parameters of the latent ability distribution can be derived by taking derivatives with respect to the parameters and setting the resulting expressions equal to zero:

$$\frac{c}{s} = \frac{1}{P}\sum_p^P \frac{(m_+ + c)}{(\Sigma_i \tau_i t_{pi} + s)} \tag{15.21}$$

$$\psi(c) - \ln(s) = \frac{1}{P}\sum_P^P \psi(m_+ + c) - \ln(\Sigma_i^I \tau_i t_{pi} + s) \tag{15.22}$$

$$\frac{m_i}{\tau_i} = \frac{1}{P}\sum_P^P \frac{(m_+ + c)}{(\Sigma_i \tau_i t_{pi} + s)} t_{pi} \tag{15.23}$$

Equations 15.21 through 15.23 can be solved using a Newton–Raphson or expectation–maximization (EM) algorithm (Aitkin, Volume Two, Chapter 12). Approximate standard errors for the estimates can be obtained by inverting the observed information matrix (Jansen, 1997a).

It is easy to show that the joint posterior density of the θ_p's is a product of independent gamma distributions, with parameters $(c + \Sigma m_i)$ and $(s + \Sigma \tau_i t_{pi})$

$$f(\theta \mid t) = L_m^{-1} p(t \mid \theta; \tau) g(\theta; c, s) \tag{15.24}$$

where L_m is the marginal likelihood function. Note that the posterior distribution depends on the observed data through the weighted sums $\Sigma \tau_i t_{pi}$. For given $\tau_1, \ldots, \tau_I$, c, and s, the posterior means could be used as estimators for the ability parameters θ_p $(p = 1, \ldots, P)$. Substituting the MML estimators gives

$$\mathcal{E}(\theta \mid t) = \frac{\Sigma_i m_i + c}{\Sigma_i \tau_i t_{pi} + s} \tag{15.25}$$

and the standard deviations

$$sd(\theta \mid t) = \frac{\sqrt{\Sigma_i m_i + c}}{\Sigma_i \tau_i t_{pi} + s} \tag{15.26}$$

15.4.3 Model Extensions

Both the RPCM and gamma–gamma model for response speed have been extended to incorporate explanatory variables. When the test takers are cross-classified according to between-test taker factors, group differences can be modeled by allowing the parameters index and scale parameters (and therefore the mean and variance) of the person parameter distribution to vary over groups (Jansen and van Duijn, 1992). Another approach, which we will take here, enables us to incorporate manifest, categorical, as well as continuous, explanatory variables for both the test and ability parameters in the models. The following reparameterization for the test difficulties was proposed by Jansen (1995a):

$$\ln(\tau_i) = z_i^T \boldsymbol{\eta} \tag{15.27}$$

Likewise, variation in the test taker parameters can be explained by the following log-linear model:

$$\ln(\theta_p) = x_p^T \boldsymbol{\beta} + \ln(\zeta) \tag{15.28}$$

where x_p is a set of predictors on test taker level (Jansen, 1997b; Jansen, 2003; Hung, 2012). The error term $ln\zeta$ has a mean equal to $(\psi(c) - \ln(c))$ and variance $\psi'(c)$. For large values of c, the mean goes to zero.

It can be shown that, for large c, $\ln\zeta$ will be approximately normally distributed, with mean equal to zero and variance $(c - 0.5)^{-1}$. Therefore, in practice, it might be difficult to discriminate between a gamma- multiplicative model for θ_p, and a normal linear model for $\ln(\theta_p)$.

15.4.4 Model Fit

Like other item response theory (IRT) models, likelihood-ratio statistics have been proposed for assessing model fit, both for the Rasch Poisson model and gamma model for speed.

The Rasch model for speed tests implies a restricted variance–covariance matrix: It can be shown that the correlations between the weighted relative response rates S, given the total weighted response times, have a specific pattern depending on the test length m (Jansen, 1997). Defining the relative response time of test taker p on i as

$$s_{pj} = \frac{\tau_i t_{pi}}{\Sigma_i \tau_i t_{pi}}$$

the expected correlation between S_k and S_l can be written as

$$r_{kl} = -\sqrt{\frac{m_k m_l}{(\Sigma_h m_h - m_k)(\Sigma_h m_h - m_l)}} \tag{15.29}$$

Alternative procedures for checking the model, based on the principle of Lagrange multiplier tests, were discussed by Jansen and Glas (2001) and Jansen (2007); see also Glas (Volume Two, Chapter 17). This type of test, which has proved to be extremely useful with several other latent trait models, is for a specific model against a generalization obtained by relaxing a basic assumption. Suppose the special model has parameters $\boldsymbol{\Phi}_1$, while its generalization is obtained by adding extra parameters $\boldsymbol{\Phi}_2$ to represent a possible violation. The special model returns by setting $\boldsymbol{\Phi}_2$ equal to zero. The LM-statistic is defined as

$$LM = h(\mathbf{\Phi}_2)' H^{-1} h(\mathbf{\Phi}_2) \tag{15.30}$$

where

$$H = \Sigma_{22} - \Sigma_{21}(\Sigma_{11})^{-1}\Sigma_{12} \tag{15.31}$$

and

$$\underset{pq}{\Sigma} = \underset{h}{\Sigma} h_n(\mathbf{\Phi}p) h_n(\mathbf{\Phi}p)' \tag{15.32}$$

for p = 1, 2 and q = 1, 2. The $h(\mathbf{\Phi}_p)$'s stand for the first-order derivatives with respect to the parameters $\mathbf{\Phi}_p$. The statistic is approximately chi-square distributed with degrees of freedom equal to the number of parameters in $\mathbf{\Phi}_2$. An attractive property of the LM statistic is that it can be calculated from the estimates of only $\mathbf{\Phi}_1$. This is an obvious advantage since estimating parameters under a general model can become complicated (Glas and Suarez Falcon, 2003). An LM test for differential test functioning was described by Jansen and Glas and a test for the violation of local independence using the same approach was described by Jansen (Jansen and Glas, 2001; Jansen, 2007). These tests proved to perform well in simulation studies.

15.4.5 Examples

In our first example, we analyze data from a larger study of the development of arithmetic and language skills among (Dutch) primary school pupils. One of the measurement instruments was the Brus test, which is a time-limit word-reading test. Its score consists of the number of words that an examinee can read within 1 min. The Brus test was administered at approximately 3-month intervals during the second, third, fourth, and fifth grades. From these 12 administrations, we only use the first six, since some pupils were not further tested after they had reached a cutting score of 80. Our sample consisted of 153 test takers (including incomplete cases). Among the background variables available were gender, socioeconomic background, and home language. We restricted our analysis to the test takers with complete data. Table 15.1 presents some relevant descriptive statistics, such as the means and standard deviations of the reading scores per occasion.

For the estimation of the model parameters, we used a Markov chain Monte Carlo (MCMC) method (Spiegelhalter et al., 2003) implemented in WinBUGS. To do so, we expanded our model with priors for the fixed effects as well as hyperparameters for their random effects. The following specifications were used:

```
Model:
  y[i,j] ~ Poisson(λ[i,j])
      with λ[i,j] = τ[j]* θ[i], i=1,..,N j=1,..K
where θ is the random component
      θ[i]~ gamma(α, β)
A non-informative prior for the fixed effects
      τ[j] ~ unif(0,1)
and for the hyper-parameters
      α, β ~ Gamma(0.001, 0.001)
```

TABLE 15.1

Means and Standard Deviations of the Number of Words Read on the Brus Test (N = 154)

Occasion	Mean	Std Dev	N	# Miss.
1	27.1	14.52	147	7
2	35.7	15.82	148	6
3	45.6	16.07	148	6
4	49.4	16.79	153	1
5	55.3	17.01	153	1
6	61.1	16.46	153	1

The results were obtained from an analysis of 10,000 WinBUGS simulations, after a burn-in of 5000. Two different sets of starting values were used. Convergence was monitored by the inspection of history plots and the use of Gelman–Rubin statistics.

Table 15.2 gives the estimates of the occasion parameters based on the unrestricted model, which can be viewed as a model for the proportional distribution of the total score over the occasions as well as the estimates of the parameters of the gamma distribution. To be considered as accurate, the Monte Carlo error for each parameter of interest (Column 4) should have been smaller than 5% of the standard deviation (Column 3) (Spiegelhalter et al., 2003).

The current estimates were nearly identical to those reported in Jansen (1995a), where maximum likelihood methods were used. Next, a linear growth model with gender as the background variable (1 = boy; 0 = girl) was fitted to the data. The occasions were assumed to be equally spaced (time = 0, 1, 2, 3, 4, 5).

```
Model:
  y[i,j] ~ Poisson(λ[i,j])
       with log(λ[i,j]) <- betanul + beta*sex[i]+
            gamma*time[j]+ log(ξ[i])
with the following non-informative priors
            betanul ~ dnorm(0.0,1.0E-6)
            beta    ~ dnorm(0.0,1.0E-6)
            gamma   ~ dnorm(0.0,1.0E-6)
```

The results are summarized in Table 15.3. The weight of the linear time component was positive and significant. The negative value of the regression weight for gender (beta = –0.13) was significant, but relatively small. The Monte Carlo error for this parameter (Column 4),

TABLE 15.2

Parameter Estimates

Node	Mean	sd	MC Error	2.5%	Median	97.5%	Start	Sample
Alpha	8.553	1.045	0.06516	6.694	8.503	10.7	5001	10,000
Beta	0.031	0.004887	2.429E–4	0.024	0.03074	0.03891	5001	10,000
Tau[1]	0.099	0.001508	2.053E–5	0.096	0.09923	0.1023	5001	10,000
Tau[2]	0.129	0.001644	2.435E–5	0.1254	0.1287	0.1319	5001	10,000
Tau[3]	0.165	0.001849	2.421E–5	0.1608	0.1645	0.1681	5001	10,000
Tau[4]	0.181	0.001949	2.203E–5	0.1776	0.1815	0.1851	5001	10,000
Tau[5]	0.203	0.001988	2.319E–5	0.1991	0.203	0.2068	5001	10,000
Tau[6]	0.223	0.002077	3.605E–5	0.2191	0.2232	0.2273	5001	10,000

TABLE 15.3

Parameter Estimates of the Model with Linear Growth and Sex as Explanatory Variables

Node	Mean	sd	MC Error	2.5%	Median	97.5%	Start	Sample
Betanul	3.43	0.04174	0.002375	3.346	3.431	3.512	5001	20,000
Beta	–0.13	0.06229	0.003584	–0.2472	–0.13	–0.0093	5001	20,000
Gamma	0.1496	0.00299	5.09E–5	0.1439	0.1496	0.1555	5001	20,000
Sigma	0.3599	0.02249	2.261E–4	0.3189	0.359	0.4068	5001	20,000

which should have been smaller than 5% of the standard deviation (sd = 0.063), was on the larger side; so, the estimate was presumably not very accurate; see also Figure 15.1. A negative weight means that on average, the boys read slightly slower than the girls. These results were in line with the literature; see, among others, Majeres (1999). Parameter sigma represents the standard deviation of $\ln(\theta)$.

In the second example, we considered an application of the extended Rasch model for speed to rapid automated naming (RAN) tasks. It had already been shown that, for various types of naming tests, speed and accuracy of reading are related to speed and accuracy of word retrieval. Naming speed tests use a variety of visual stimuli, such as alphanumeric symbols, colors, and pictures, given in different formats.

In our study, naming speed was measured by several tasks, but in this example, we used only four different tasks based on alphanumerical stimuli (van den Bos et al., 2003): two letter naming and two number-naming tasks. In the naming tasks, five letters (s, e, g, r, k and d, o, a, s, p) or numbers (2, 4, 5, 8, 9 and 2, 3, 6, 7, 8) were repeated 10 times in random order. The test takers were instructed to read the stimuli aloud, as fast as possible, without making errors. The tasks were administered individually in a fixed order: letters 1, numbers 1, letters 2, and numbers 2. The pupils were timed after the first 50 items; in the analysis, we therefore assumed a test length of 50.

The sample consisted of Grade 2 (n = 43), Grade 4 (n = 30), and Grade 6 (n = 38) typical primary school pupils, and approximately half of them were boys. Gender (1 = boy, 0 = girl) and grade (second, fourth, or sixth grade) were used as explanatory variables for the ability parameters.

Table 15.4 shows the means and the standard deviations for the complete cases (N = 107). Note that the first letter-naming task was more difficult and had a much-larger standard deviation than the other three tasks.

The model parameters were estimated using MCMC methods (Spiegelhalter et al., 2003) implemented in WinBUGS. To do so, the model was expanded with priors for its fixed effects and hyperparameters of the random effects.

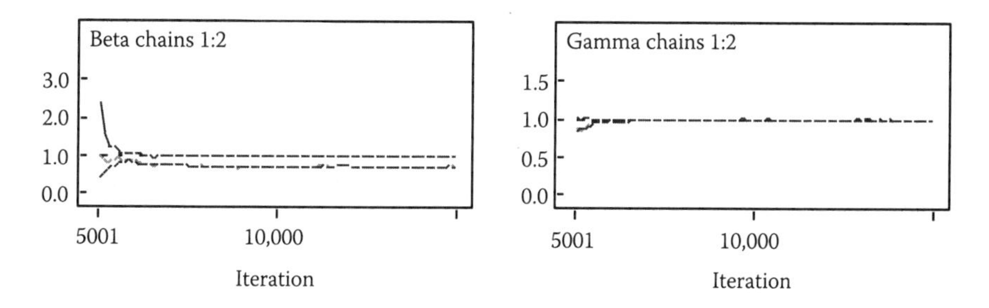

FIGURE 15.1
Gellman–Rubin statistics Brus test.

TABLE 15.4

Means and Standard Deviations of the Response Times on the Rapid Naming Tasks

	Boys			Girls					
Task Name	**Mean**	**Std Dev**	**N**	**Mean**	**Std Dev**	**N**			
Numbers 1	29.3	7.06	52	27.8	7.87	55			
Numbers 2	30.3	7.75	52	27.6	7.33	55			
Letters 1	37.5	17.14	52	31.8	16.06	55			
Letters 2	29.9	10.18	52	28.2	9.79	55			
	Grade 2			**Grade 4**			**Grade 6**		
Task Name	**Mean**	**Std Dev**	**N**	**Mean**	**Std Dev**	**N**	**Mean**	**Std Dev**	**N**
Numbers 1	34.9	6.75	40	26.6	4.27	29	23.2	4.74	38
Numbers 2	35.4	6.74	40	26.9	4.22	29	23.6	5.25	38
Letters 1	49.5	19.30	40	28.9	6.12	29	23.2	4.70	38
Letters 2	38.4	10.30	40	27.8	5.31	29	22.4	4.93	38

```
Model:
      t[i,j] ~ Gamma(m[j], λ[i,j])
          with λ[i,j] =τ[j]* θ[i], i=1,..,N j=1,..K
Systematic component: 3 dummy variables
      log(λ[i,j])= betanul+beta.1*X1[i] +beta.2 *X2[i]
      + beta.3*X3[i]+ log(τ[j]) +log(θ[i])
With
      θ[i] ~ Gamma(c,c)
and with the following (non-informative) priors for the fixed effects and
hyper-parameters
      τ[j] ~ Gamma(0.001, 0.001)
      c ~ Gamma(0.001, 0.001)
and
      betanul, beta.1, beta.2 and beta.3 ~ N(0,0.001)
```

The results of the analysis were based on 20,000 WinBUGS simulations, obtained after a burn-in of 5000. Two different sets of starting values were used. Convergence was monitored by inspecting history plots and using the Gelman–Rubin statistics. Table 15.5 shows the estimates of the parameters of interest and the regression weights of the explanatory variables (Figure 15.2).

The negative regression weight for sex (beta.1 = −0.1226) was significant, indicating that boys were on average slower than girls. The weights for the dummy-coded grade variable

TABLE 15.5

Parameter Estimates

Node	Mean	sd	MC Error	2.5%	Median	97.5%	Start	Sample
C	30.99	5.014	0.04609	22.16	30.67	41.74	5001	40,000
betanul	2.25	0.03884	0.001946	2.173	2.249	2.327	5001	40,000
beta.1	−0.1226	0.03749	0.001556	−0.1929	−0.1244	−0.04675	5001	40,000
beta.2	−0.5146	0.0452	0.002008	−0.6029	−0.5153	−0.4213	5001	40,000
beta.3	−0.1878	0.05111	0.002243	−0.2896	−0.1872	−0.08715	5001	40,000

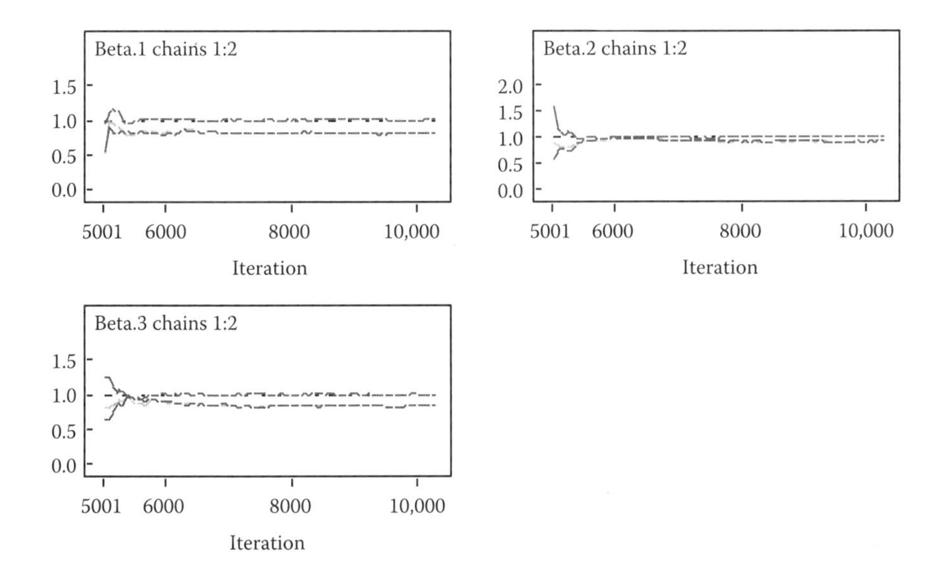

FIGURE 15.2
Gelman–Rubin statistics rapid naming tasks.

were also significant. Grade-2 pupils demonstrated a lower naming speed than Grade-4 pupils, who, in turn, had a lower speed than Grade-6 pupils.

15.5 Discussion

Both Rasch models are basically two-level models with fixed and random effects, where the random effects are assumed to be gamma distributed. Clearly, other possibilities of modeling positive random variates than gamma distribution exist, the lognormal distribution being an obvious alternative candidate. Given the plethora of suitable computer programs available now, WinBUGS being one of them, the computational tractability of alternative choices has become less of a problem.

Obviously, the applicability of the RPCM is not limited to count data observed in achievement testing. Several authors have described applications to the measurement of psychomotor skills (Safrit et al., 1992; Spray, 1997; Zhu and Safrit, 2003). In an educational context, we might also consider application to such count data as frequency of school absenteeism, dropouts, etc. (Jansen, 1995a).

Response times are considered as relevant in a wide variety of psychological and educational measurement situations. They can be seen as (achievement) measures in their own right, and also as explanatory variables for other types of test behavior (van der Maas and Jansen, 2003; van der Linden, 2008). Studies with varying types of naming tests have shown that speed and accuracy in reading are related to speed and accuracy in word retrieval. And there is a vast amount of literature on the connection between naming speed and other aspects of reading and language acquisition ability. For instance, naming speed tests appear to consistently distinguish between good and poor readers and are related to specific reading deficiencies such as dyslexia (van den Bos et al., 2003).

References

Fischer, G. H. and Kisser, R. 1981. Notes on the exponential latency model and an empirical application. In H. Wainer and S. Messick (Eds.), *Principles of Modern Psychological Measurement*. Hillsdale, NJ: Lawrence Erlbaum.

Glas, C. A. W. and Suarez Falcon, J. C. 2003. A comparison of item-fit statistics for the three-parameter logistic model. *Applied Psychological Measurement*, 27, 87–106.

Hung, L.-F. 2012. A negative binomial regression model for accuracy tests. *Applied Psychological Measurement*, 36, 88–103.

Jansen, M. G. H. 1995a. Applications of Rasch's Poisson counts model to longitudinal count data. In J. Rost and R. Langeheine (Eds.), *Applications of Latent Trait and Latent Class Models in the Social Sciences* (pp. 380–389). Münster, Germany: Waxmann.

Jansen, M. G. H. 1995b. The Rasch Poisson counts model for incomplete data: An application of the EM algorithm. *Applied Psychological Measurement*, 19, 291–302.

Jansen, M. G. H. 1997a. The Rasch model for speed tests and some extensions with applications to incomplete designs. *Journal of Educational and Behavioral Statistics*, 22, 125–140.

Jansen, M. G. H. 1997b. Rasch's model for reading speed with manifest explanatory variables. *Psychometrika*, 62, 393–409.

Jansen, M. G. H. 2003. Estimating the parameters of a structural model for the latent traits in Rasch's model for speed tests. *Applied Psychological Measurement*, 27, 138–151.

Jansen, M. G. H. 2007. Testing for local dependence in Rasch's multiplicative gamma model for speed tests. *Journal of Educational and Behavioral Statistics*, 32, 24–38.

Jansen, M. G. H. and Glas, C. A. W. 2001. Statistical tests for differential test functioning in Rasch's model for speed tests. In A. Boomsma, M. A. J. van Duijn, and T. A. B. Snijders (Eds.), *Essays on Item Response Theory* (pp. 149–162). New York: Springer.

Jansen, M. G. M. and van Duijn, M. A. J. 1992. Extensions of Rasch's multiplicative Poison model. *Psychometric*, 57, 405–419.

Lawless, J. F. 1987. Negative binomial and mixed Poisson regression. *Canadian Journal of Statistics*, 15, 209–225.

Lord, F. M. and Novick, M. R. 1968. *Statistical Theories of Mental Test Scores*. Reading, MA: Addison-Wesley.

Majeres, R. L. 1999. Sex differences in phonological processes: Speeded matching and word reading. *Memory and Cognition*, 27, 246–253.

Masters, G. N. and Wright, B. D. 1984. The essential process in a family of measurement models. *Psychometrika*, 49, 529–544.

Maris, E. 1993. Additive and multiplicative models for gamma distributed random variables, and their application as psychometric models for response times. *Psychometrika*, 3, 445–469.

Nelson, J. F. 1984. The Dirichlet–Gamma–Poisson model of repeated events. *Sociological Methods and Research*, 12, 347–373.

Ogasawara, H. 1996. Rasch's multiplicative Poisson model with covariates. *Psychometrika*, 61, 73–92.

Rasch, G. 1980. *Probabilistic Models for Some Intelligence and Attainment Tests*. Chicago: The University of Chicago Press. (Original work published in 1960.)

Safrit, M., Zhu, W., Costa, M., and Zhang, L. 1992. The difficulty of sit-up tests: An empirical investigation. *Research Quarterly for Exercise and Sport*, 63, 277–283.

Spiegelhalter, D. J., Thomas, A., Best, N., and Lunn, D. 2003. *WinBUGS Manual [Version 1.4]*. Cambridge, England: MRC Biostatistics Unit.

Spray, J. A. 1997. Multiple-attempt, single-item response models. In W. J. van der Linden and R. K. Hambleton (Eds.), *Handbook of Modern Item Response Theory* (pp. 207–223). New York: Springer.

Thall, P. F. and Vail, S. C. 1990. Some covariance models for longitudinal count data with overdispersion. *Biometrics*, 46(3), 657–671.

Thissen, D. 1983. Timed testing: An approach using item response theory. In D. J. Weiss (Ed.), *New Horizons in Testing: Latent Trait Theory and Computerized Adaptive Testing* (pp. 179–203). New York: Academic Press.

Van Breukelen. 1989. *Concentration, Speed and Precision in Mental Tests*. Unpublished doctoral dissertation. University of Nijmegen. The Netherlands.

van den Bos, K. P., Zijlstra, B. J. H., and van den Broeck, W. 2003. Specific relations between alphanumeric-naming speed and reading speeds of monosyllabic and multisyllabic words. *Applied Psycholinguistics*, 24, 407–430.

van der Linden, W. J. 2006. A lognormal model for response times on test items. *Journal of Educational and Behavioral Statistics*, 31, 181–204.

van der Linden, W. J. 2008. Using response times for item selection in adaptive testing. *Journal of Educational and Behavioral Statistics*, 33, 5–20.

van der Maas, H. L. J. and Jansen, R. J. 2003. What response times tell of children's behavior on the balance scale task. *Journal of Experimental Child Psychology*, 85, 141–187.

van Duijn, M. A. J. 1993. *Mixed Models for Repeated Count Data*. Leiden, Netherlands: DSWO Press.

van Duijn, M. A. J. and Jansen, M. G. H. 1995. Modeling repeated count data: Some extensions of the Rasch Poisson counts model. *Journal of Educational and Behavioral Statistics*, 20, 241–258.

Verhelst, N., Verstralen, H., and Jansen, M. G. H. 1997. Models for time-limit tests. In W. J. van der Linden and R. K. Hambleton (Eds.), *Handbook of Modern Item Response Theory* (pp. 169–186). New York: Springer.

Zhu, W. and Safrit, M. 2003. The calibration of a sit-ups task using the Rasch Poisson counts model. *Canadian Journal of Applied Physiology*, 18, 207–219.

16

Lognormal Response-Time Model

Wim J. van der Linden

CONTENTS

16.1 Introduction 261
16.2 Presentation of the Model 264
16.2.1 Assumptions 264
16.2.2 Formal Model 266
16.2.3 Parameter Interpretation 267
16.2.4 Parameter Identifiability and Linking 268
16.2.5 Moments of RT Distributions 270
16.2.6 Relationships with Other Models 271
16.3 Parameter Estimation 272
16.3.1 Estimating Both Item and Person Parameters 272
16.3.2 Estimating the Speed Parameters 273
16.4 Model Fit 274
16.5 Empirical Example 276
16.6 Discussion 278
References 280

16.1 Introduction

Raw responses to the items in an ability test typically have the physical form of bubbles on an answer sheet, clicks with a computer mouse, or a few lines of text. Each of these responses has two important aspects: (i) its degree of correctness and (ii) the time used to produce it. The former is usually recorded on a numerical scale, typically a 0–1 scale for incorrect–correct responses or a more-refined scale for the evaluation of polytomous items. Until recently, response times (RTs) on items could be recorded only for individually proctored tests. But since the arrival of computers and handheld electronic devices in testing, they have become automatically available for any type of test.

It may be tempting to view the correctness of a response and the time required to produce it as the result of two highly dependent processes, or even one single process. In fact, this is exactly what two different traditions of descriptive analyses of responses and RTs seem to suggest. The first tradition is that of item analysis in educational and psychological testing, where predominantly negative correlations between the correctness of responses and RTs have been found for tests ranging from adaptive matrices (e.g., Hornke, 2000) to medical-licensing exams (e.g., Swanson et al., 2001). It seems easy to explain the sign of these correlations: Test takers with lower abilities struggle for a longer time to find answers to the items, and when they find one, it is likely to be wrong. On the other hand, more able test takers usually find correct answers faster.

The other tradition is that of psychological research on speed–accuracy trade-offs (SATs), typically conducted as experiments with subjects replicating a standardized task multiple times under conditions inducing different levels of speed. These experiments invariably result in positive correlations between the proportions of correct responses and the average RT (for a review, see Luce, 1986, Section 6.5). The standard explanation of these results seems convincing as well: Subjects who work faster do so at the expense of their accuracy and consequently produce fewer correct responses.

To add to the confusion, consider the case of the RT patterns of different test takers responding to the same items. Table 16.1 shows the RTs by two test takers on the first 10 items in a 65-item cognitive ability test. As the test takers were arbitrarily selected from a much larger set, it seems safe to assume that they worked entirely independently. Nevertheless, the full patterns with the RTs for the two students correlated to 0.89! In fact, positive correlations can be found for nearly any pair of students on any kind of ability test.

How is it possible for these three different types of correlations to be so conflicting with each other, or with reality? The answer is: aggregation of data across hidden (or latent) covariates. For instance, for the RT patterns in Table 16.1, the pertinent covariate is the labor intensity of the items in the test. The solution for some of the test items just required more labor than for others. As a consequence, even when they work completely independently, the expected RTs of any pair of test takers simply go up and down jointly with the amount of labor required by the items. An example of a factor that contributes to the difference in labor between the items in an arithmetic test is the length of the computations. The longer-addition item in Figure 16.1 obviously demands more time than the shorter item (even though the latter may be more difficult because of the presence of three-digit numbers). In several other data sets analyzed by this author, the expected RTs required by the items easily differed by a factor of more than 5; for the data set for a large-scale adaptive test analyzed in van der Linden and Guo (2008, Table 16.1), they even ranged from some 14 to 812 s per item.

The same type of spuriousness due to hidden covariates explains the difference between the negative and positive correlations in the two earlier traditions of item analysis and

TABLE 16.1

RTs by Two Test Takers on the First 10 Items in an Ability Test

	Test Taker	
Item	$p = 1$	$p = 2$
1	22	26
2	19	38
3	40	101
4	43	57
5	27	37
6	21	27
7	45	116
8	23	44
9	14	10
10	47	117
r	0.89	

Item 1	Item 2
286	73
155	25
——— +	52
	93
	18
	41
	——— +

FIGURE 16.1
Two addition items of unequal length differing in the amount of cognitive labor they require.

SAT research. Typical covariates associated with test items are their difficulty and time intensity. If we aggregate responses and RTs across items, depending on the sign of the correlation between these covariates, spurious positive or negative correlations between the responses and RTs may be observed. The same holds for aggregation across test takers who vary in their ability and speed of work. If we aggregate both across items and test takers, the correlations between their covariates interact and, depending on their signs and relative strengths, the results are difficult to predict without any further formal analysis.

An effective solution to the problem of spurious correlation is to explicitly model the relationship between the observed variables and all covariates. If the model does fit the empirical data satisfactorily, we are able to neutralize the effects of the covariates by conditioning on them. This principle, which underlies all latent variable modeling, has found applications in diverse areas such as physics and psychometrics; for a discussion of some of the parallels between them, see van der Linden (2011a).

Several attempts at modeling RTs have been made. The main differences between them exist in how they treat the relationship between the responses and RTs on the items. One category consists of attempts to incorporate the observed RTs as an explanatory variable in the model for the responses. For instance, Roskam's (1987, 1997) model,

$$\Pr\{U_{pi}=1\mid\theta_p\}=\frac{\exp[(\theta_p+\ln t_{pi}-b_i)]}{1+\exp[(\theta_p+\ln t_{pi}-b_i)]},\tag{16.1}$$

which basically was an extension of the well-known Rasch model (Volume One, Chapter 3), was motivated by the success of psychological research on the SAT. The presence of $\theta_j+\ln t_{ij}$ (instead of just θ_j) relative to the difficulty parameter b_i in its parameter structure represents the assumption that more time spent by the test taker on the item should have the same effect on his probability of success as a higher ability in the regular Rasch model. Other models in this category (e.g., Verhelst et al., 1997; Wang and Hanson, 2005) used RT parameters instead of the observed RTs but were motivated similarly.

The models in the second category are the other way around; they incorporate response parameters in a model for the distribution of the RTs. A well-known example is Thissen's (1983) model, which is the product of a normal model for the logtimes

$$\ln T_{pi}\sim N(\mu+\tau_p+\beta_i-\rho(a_i\theta_p-b_i),\sigma^2)\tag{16.2}$$

with a regular 2PL or 3PL model for the responses. Parameters τ_p and β_i in Equation 16.2 are slowness parameters for test taker p and item i, respectively, $a_i\theta_p-b_i$ is the parameter structure of the 2PL or 3PL model assumed to hold for the responses, and ρ serves as a slope parameter in the regression of this structure on the expected RT. The minus sign between $\mu + \tau_p + \beta_i$ and $\rho(a_i\theta_p - b_i)$ represents a trade-off between the probability of success on the items and the slowness of the response, with ρ controlling the strength of the trade-off. Other models in this category were designed to mimic the same effect (e.g., Gaviria, 2005). Ferrando and Lorenzo-Seva (2007) proposed a variation that was claimed to be more appropriate for personality tests.

The third category consists of completely distinct models for RTs. One of the first was Rasch's (1960) model for reading speed, which postulates a density for the time required to read an item of m words equal to

$$p(t_{pi} \mid \xi_p, \delta_i) \equiv \frac{(\xi_p / \delta_i)^m}{\Gamma(m)} t^{m-1} e^{-\xi_p t/\delta_i}, \tag{16.3}$$

where ξ_p is the ability of the test taker, δ_i is the difficulty of the text, and $\Gamma(m) \equiv (m-1)!$ is the gamma function. The density is a standard gamma density with its intensity parameter modeled as $\lambda_{pi} \equiv \xi_p/\delta_i$; that is, the number of words expected to be read in a given time unit is assumed to be a function of the test taker's speed and the difficulty of the text only. More details on this model and its counterpart in the form of a Poisson model for reading errors can be found in Jansen (Volume One, Chapter 15). Similar models in this category are the gamma models by Maris (1993), an exponential model with an additive version of the basic parameterization in Equation 16.3 by Oosterloo (1975) and Scheiblechner (1979), and a Weibull model by Tatsuoka and Tatsuoka (1980). The lognormal model in this chapter also belongs to this category.

A category of models not reviewed here are the reaction-time models used in mathematical psychology. These models have been developed mainly to explain the stochastic processes leading to reaction times observed in psychological experiments. As nearly all of them assume a single standardized task replicated by groups of subjects considered as exchangeable, they miss a fundamental feature shared by all item response theory (IRT) models—separation of person and item effects. These reaction-time models are therefore less relevant to educational and psychological testing, which needs test taker parameters to adjust for their effects when calibrating the items and, conversely, item parameters to adjust the scores of test takers. An exception is the version of the diffusion model with item and person parameters developed by Tuerlinckx, Molenaar, and van der Maas (Volume One, Chapter 17).

16.2 Presentation of the Model

16.2.1 Assumptions

One of the most careless practices in the educational and psychological literature is the equating of the time on an item or task with the test taker's speed. For instance, nearly all the literature on the SAT (e.g., Luce, 1986, Chapter 6) measures speed as the average time recorded for its experimental subjects (as well as accuracy simply as the proportion of correct responses). However, time and speed are definitely distinct variables. If they were not, it would be unnecessary for our cars to have a speedometer in addition to a clock.

The first assumption underlying the RT model in this chapter is just a definition of speed. We follow the general practice of defining speed as the rate of change of a substantive measure with respect to time. A prime example of this format is the definition of the average speed of a moving object in physics:

$$\text{Average speed} = \frac{\text{Distance traveled}}{\text{Time elapsed}}. \tag{16.4}$$

Two other examples are speed of inflation as the amount of inflation over time in economics and speed of the spread of a bacterial infection measured as the increase in the number of patients infected over time in epidemiology. In this context, it seems natural to define the speed by a test taker on an item as the amount of cognitive labor performed on it over time (van der Linden, 2009a).

Generally, cognitive labor cannot be measured directly. Even for the two simple addition items in Figure 16.1, although their length clearly is a factor with an impact on the amount of labor they require, it would be wrong to equate the two; other factors do have an impact as well, for instance, the combinations of the numbers that have to be added at each step. However, although cognitive labor cannot be observed directly, we do observe its effect on the RT and have the option to introduce a latent item parameter for it in our model. We will refer to this parameter as a time-intensity parameter, just to remind ourselves that it represents the amount of labor required by the items indirectly, through its effect on the observed time spent on them.

The second assumption is that of constancy of speed during testing. This assumption has both a between- and within-item aspect. The assumption of constant speed between items not only has the technical advantage of avoiding overparameterization (i.e., the introduction of a separate speed parameter for each item) but is also consistent with the assumption of constant ability that underlies all the mainstream IRT models used in educational and psychological testing. The history of a satisfactory fit of these models would have been impossible when ability did vary substantially during testing. Given the strong evidence on the existence of SATs in performance tasks collected in experimental psychology, it seems inconsistent to entertain the possibility of constant ability but varying speed during testing.

Strictly speaking, an assumption of constant speed within items is not required. The only thing typically observed is the total time spent on each of them, and whether we should consider the speed parameter to represent the test taker's average or instantaneous speed on each item is beyond possible verification.

Of course, test takers will always vary somewhat in their speed during real-world testing. But rather than complicating the RT model by trying to allow for such changes, it is more practical to assume constant speed and use the actual RTs to check on the seriousness of the violations of the assumption. An example of this type of residual analysis is offered later in this chapter. As for possible larger systematic trends in speed, for instance, in the form of an increase in speed toward the end of the test due to a tight time limit, they typically have nothing to do with the intended ability measured by the test and should be avoided when designing the test. In fact, the RT model introduced in this chapter can be used to assemble test forms or set time limits on them to guarantee an acceptable level of speededness; the methodology required to do so is reviewed in van der Linden (Volume Three, Chapter 13).

The third assumption is conditional independence of RTs given the test taker's speed. The assumption is entirely analogous to that of conditional (or "local") independence adopted throughout IRT. It may seem to be at odds with the (marginal) correlations found between the RTs on items in the earlier descriptive studies. But, as already argued, these studies necessarily aggregate data across test takers and/or items and tend to create spurious results. To model their results, we need to take the step from conditional, single-level modeling of fixed effects in this chapter to the hierarchical modeling in van der Linden and Fox (Volume One, Chapter 29).

16.2.2 Formal Model

The final assumption is that of a family of densities for the RT distributions of the test takers on the items. Let τ_p^* denote the speed of test takers $p = 1, \ldots, P$ on items $i = 1, \ldots, I$, which are assumed to have parameters β_i^* for their time intensities. Entirely analogous to Equation 16.4, we define speed as

$$\tau_p^* \equiv \frac{\beta_i^*}{t_{pi}}, \tag{16.5}$$

where t_{pi} is the observed RT by test taker p on item i. Equivalently,

$$t_{pi} = \frac{\tau_p^*}{\beta_i^*}, \tag{16.6}$$

which more explicitly demonstrates the assumed separation of the observed RT into an unknown test taker and item effect.

As RT distributions tend to be positively skewed, it is customary to work with logtimes, which leads to

$$\ln t_{pi} = \tau_p - \beta_i. \tag{16.7}$$

Acknowledging the random nature of RTs, the addition of a normally distributed random term gives us

$$\ln T_{pi} = \tau_p - \beta_i + \epsilon_i, \epsilon_i \sim N(0, \alpha_i^{-2}). \tag{16.8}$$

The result is a normal distribution of ln T_{pi} with mean $\beta_i - \tau_p$ and variance α_i^{-2}. We refer to the a_i parameter as the item discrimination parameter in the model.

The same distribution can be presented by the lognormal density

$$f(t_{pi}; \tau_p, \alpha_i, \beta_i) = \frac{\alpha_i}{t_{pi}\sqrt{2\pi}} \exp\left\{-\frac{1}{2}[\alpha_i(\ln t_{pi} - (\beta_i - \tau_p))]^2\right\} \tag{16.9}$$

with $\tau_p \in \mathbb{R}$, $\beta_i \in \mathbb{R}$, and $\alpha_i > 0$.

A third representation is

$$T_{pi} = \exp\left(\beta_i - \tau_p + \frac{Z}{\alpha_i}\right), Z \sim N(0,1), \tag{16.10}$$

which reveals that Equation 16.9 actually is the density of an exponential rather than the log of a normally distributed variable. The use of "lognormal" for the name of this distribution is thus a misnomer. But since it has been accepted without much practical confusion, we follow the tradition. As demonstrated throughout the rest of this chapter, transformations of random variables with an assumed normal distribution have obvious advantages. Other choices than the log of RTs are possible, though. If the interest is in flexible curve fitting and less in the applications reviewed later in this chapter, the class of Box–Cox transformations could be considered (Klein Entink et al., 2009).

16.2.3 Parameter Interpretation

We need to check if the behavior of each of the model parameters is consistent with the interpretation for it claimed so far. As a starting point, observe that Equations 16.8 through 16.10 do not have the standard parameterization of a lognormal distribution (Volume Two, Chapter 3, Equation 16.30), which would have implied the choice of

$$f(\ln t_{pi};\mu_{pi},\sigma_{pi}^2) \equiv \frac{1}{\sigma_{pi}\sqrt{2\pi}}\exp\left\{-\frac{1}{2}\left(\frac{\ln t_{pi}-\mu_{pi}}{\sigma_{pi}}\right)^2\right\} \tag{16.11}$$

with μ_{pi} and σ_{pi}^2 parameters representing the mean and variance of the logtime by the test taker on the item. The current model differs because of our assumption of $\mu_{pi} \equiv \beta_i - \tau_p$ as a function of a separate item and test taker effects and the choice of $\sigma_{pi}^2 \equiv \alpha_i^{-2}$ as a function of only the items.

The interpretation of τ_p as a speed parameter was primarily motivated by the agreement between its definition in Equation 16.5 and the notion of speed generally adopted throughout science. But it is also consistent with the behavior of the model. For test takers with a higher value of τ_p, the model returns a lower-expected RT on each item. Likewise, a higher value of β_i results in a higher-expected RT for every test taker. Both are exactly what we would expect from parameters introduced to represent the opposite effects on RTs by the test taker's speed and the laboriousness of the items.

The interpretation of α_i as a discrimination parameter can be motivated by its contribution to the precision of the estimate of the test taker's speed produced by the test. The key quantity is the standard error of estimation for the speed parameter. Consider a test from an item bank calibrated with enough data to treat all item parameters as known. The Fisher information in the logtimes on any τ parameter for an item with parameters α_i and β_i is

$$\begin{aligned}
\mathbf{I}_i(\tau) &\equiv \mathcal{E}\left(\frac{\partial}{\partial\tau_p}\ln f(\ln T;\tau,\alpha_i,\beta_i)\right)^2 \\
&= \mathcal{E}\left(-a_i^2(\ln T-(\beta_i-\tau)\right)^2 \\
&= \alpha_i^4\mathcal{E}(\ln T-(\beta_i-\tau))^2 \\
&= \alpha_i^2.
\end{aligned} \tag{16.12}$$

Hence, because of conditional independence, for an n-item test, the information is $\sum_{i=1}^{n} \alpha_i^2$, and the maximum-likelihood estimation (MLE) of speed parameter τ has an asymptotic standard error

$$\mathrm{SE}(\hat{\tau} \mid \alpha_1, \ldots, \alpha_n) = \left(\sum_{i=1}^{n} \alpha_i^2 \right)^{-1/2}. \tag{16.13}$$

The discrimination parameter is thus the only parameter in the model that controls how much each item contributes to the accuracy of the estimation of τ. Remarkably, the true value of the τ parameter does not play any role. Neither do the time-intensity parameters of the items. It thus holds that no matter the speed at which a test taker operates, it can be estimated equally well with more or less laborious items—a conclusion with a potentially large practical value.

The reason we have opted for α_i instead of its reciprocal in Equation 16.11 is to obtain three model parameters with a similar interpretation as the θ_p, a_i, and b_i parameters in the 2PL and 3PL response models with their well-established history in educational and psychological testing. The θ_p and b_i parameters in these two models have an effect on the success probability of the item compared to that of τ_p and β_i on the RT. As for the discrimination parameters, entirely analogous to Equation 16.12, the 2PL and 3PL models have a_i^2 as the main factor in their expressions for the item and test information functions (Volume Three, Chapter 9, Equations 16.2 and 16.3).

The similarity of interpretation does *not* extend to the formal structure of these response models, though. The lognormal model specifies a density directly for the distribution of the RTs T_{pi}, whereas the two response models are for the success parameters in the Bernoulli distributions of response variables U_{pi}. Consequently, the former has t_{pi} in its argument but the latter do not contain u_{pi}. Nevertheless, to enhance the similarities between the two types of models, it has been suggested to include an additional discrimination parameter in the lognormal model, adopting

$$\alpha_i(\ln t_{pi} - \phi_i(\beta_i - \tau_p)) \tag{16.14}$$

as the core structure of it (e.g., Fox, 2010, Chapter 8). Although the product of $\phi_i(\beta_i - \tau_p)$ does remind us of the parameter structure of a 2PL or 3PL model, the addition of ϕ_i is unnecessary. In fact, the adoption of two parameters in one model with an entirely similar impact on the distribution of the RT seems an obvious case of overparameterization, which is likely to result in poor identifiability of both.

When test takers guess on an item, their behavior does not necessarily imply anything for their RTs—the guess may have been immediate or the final result of a long uphill battle. Therefore, the RT model does not require any "guessing parameter" to constrain the RTs from below (beyond their natural lower bound at zero).

16.2.4 Parameter Identifiability and Linking

Models with latent parameters generally lack identifiability and it may be difficult to find restrictions that make them identifiable (Volume Two, Chapter 8). However, the presence of the manifest variable $\ln t_{pi}$ in the current RT model makes the job much easier. As a starting point, observe that if we had chosen Equation 16.11 with parameters μ_{pi} and σ^2_{pi} as our density of $\ln T_{pi}$, the model would have been fully identified (provided we have enough test takers and items). Also, the substitution of $\sigma_{pi} = \alpha_i^{-1}$ for all p, which for $\alpha_i > 0$ amounts to a monotone transformation, would not have led to any loss of identifiability. The possible lack of identifiability of Equation 16.9 can thus only be a consequence of its further parameterization of μ_{pi} as $\beta_i - \tau_p$. Indeed, for any value of ε, the transformations $\beta_i - \varepsilon$ and $\tau_p - \varepsilon$ yield identical RT distributions.

A convenient restriction to establish full identifiability of the model is setting the τ_p parameters for the test takers equal to zero; that is

$$\mu_\tau \equiv 0. \tag{16.15}$$

The choice implies speed parameters with values that are deviations from their average. As for the time-intensity parameters, averaging μ_{pi} across all test takers and items gives an overall mean equal to $\mu \equiv \mu_\beta - \mu_\tau$. Hence, the choice of Equation 16.15 also implies that

$$\mu_\beta = \mu, \tag{16.16}$$

which implies that the individual time-intensity parameters can now be viewed as deviations from the expected logtime across all test takers and items.

As an alternative to Equation 16.15, we could set the sum of the β_i parameters equal to zero. This choice results in time intensity and speed parameters that are deviations from their average and the expected logtime for all test takers and items, respectively.

The necessity to introduce an additional identifiability restriction creates a linking problem for parameters estimated from different samples. For samples that include different test takers and/or items, Equation 16.15 implies true parameter values that are deviations from different overall means. Linking functions that adjust for these differences are

$$\tau_{p_2} = \tau_{p_1} + v \tag{16.17}$$

and

$$\beta_{i_2} = \beta_{i_1} + v, \tag{16.18}$$

where the extra subscripts have been added to indicate two separate calibration studies.

For a common-item design, the unknown linking parameter v is equal to

$$v = \mu_{\beta_1} - \mu_{\beta_2} \tag{16.19}$$

with the averages taken across the common items in the two calibrations. Similarly, for a common-person design linking, parameter v is equal to the difference between its average speed parameters for the two test forms.

Obviously, in practice, v has to be calculated from an estimated item or person parameters. Standard errors of linking due to parameter estimation error for a variety of linking designs were presented in van der Linden (2010). In fact, since this reference, it has become clear that a statistically more accurate form of linking is possible using precision-weighted averaging rather than plain averages as in Equation 16.19. For a more general treatment of the linking problem in IRT models, including the use of this improved-type estimator and its optimization through the application of optimal design principles, see van der Linden and Barrett (Volume Three, Chapter 2).

16.2.5 Moments of RT Distributions

For the standard family of lognormal densities, with parameters μ and σ^2 for the mean and variance of the natural log of the variate (Volume Two, Chapter 3), the kth moment about the origin is equal to

$$m_k \equiv \mathcal{E}(X^k) = \exp(k\mu + k^2\sigma^2/2) \tag{16.20}$$

(Kotz and Johnson, 1985, 134–136; note that one of the squares is accidentally missing in this reference). From Equation 16.20, the first two moments for the current RT model can be derived as

$$\mathcal{E}(T_{pi}) = \exp(-\tau_p)\exp(\beta_i + \alpha_i^{-2}/2), \tag{16.21}$$

$$\mathcal{E}(T_{pi}^2) = \exp(-2\tau_p)\exp(2\beta_i + 2\alpha_i^{-2}) \tag{16.22}$$

(van der Linden, 2011b,c). Consequently, its second cumulant can be written as

$$\mathcal{E}[T_{pi} - \mathcal{E}(T_{pi})]^2 = \exp(-2\tau_p)\exp(2\beta_i + \alpha_i^{-2})[\exp(\alpha_i^{-2}) - 1]. \tag{16.23}$$

Observe how these expressions factor into separate components for the test takers and the items, a feature that suggests the following reparameterization:

$$q_i \equiv \exp(\beta_i + \alpha_i^{-2}/2), \tag{16.24}$$

$$r_i \equiv \exp(2\beta_i + \alpha_i^{-2})[\exp(\alpha_i^{-2}) - 1] \tag{16.25}$$

without any change of the speed parameters. Two remarkably simple expressions are resulting for the mean and variance of the RT of a test taker on an item:

$$\text{mean}(T_{pi}) = \exp(-\tau_p)q_i, \tag{16.26}$$

$$\text{var}(T_{pi}) = \exp(-2\tau_p)r_i. \tag{16.27}$$

Because of conditional independence, the distribution of the total time of a test taker on a test of n items has mean

$$\text{mean}(T_p) = \exp(-\tau_p)\sum_{i=1}^{n} q_i \tag{16.28}$$

and variance

$$\mathrm{var}(T_p) = \exp(-2\tau)\sum_{i=1}^{n} r_i. \tag{16.29}$$

Although the exact shape of the total-time distribution of a test taker is not known to be lognormal, it can be approximated surprisingly accurately by a standard lognormal with μ and σ^2 matching the mean and variance in Equations 16.28 and 16.29, though. Further treatment of the approximation is given in van der Linden (Volume Two, Chapter 6).

16.2.6 Relationships with Other Models

As pointed out by Finger and Chee (2009), the lognormal model in Equation 16.8 can be reparameterized as a one-factor model with item-dependent intercepts

$$\ln t_{pi} \equiv \upsilon_i + \lambda_i \xi_p + \varepsilon_i. \tag{16.30}$$

That is, under its standard assumptions of $\varepsilon_i \sim N(0,\psi_i^2)$, $\mathcal{E}(\xi_p) \equiv 0$, and $\mathcal{E}(\xi_p \varepsilon_i) \equiv 0$, substitution of $\psi_i \equiv \alpha_i^{-1}$, and setting $\lambda_i \equiv -1$ for all i, the time-intensity parameter β_i corresponds to intercept υ_i and speed parameters τ_p with the factor score ξ_p in Equation 16.30. A more general treatment of the correspondence between IRT models for continuous response variables and factor analysis models is offered by Mellenbergh (Volume One, Chapter 10).

For the analysis of possible dependencies between the RTs of pairs of test takers without any confounding due to effects of the latent covariates associated with the items or test takers, a bivariate version of the model is available. Let p and q denote two different test takers with RTs that fit the lognormal model. Their joint RTs (T_{pi}, T_{qi}) are then distributed with density

$$\begin{aligned} &f(t_{pi}, t_{qi}; \tau_p, \tau_q, \alpha_i, \beta_i, \rho_{pq}) \\ &\equiv \frac{\alpha_i^2}{t_{pi} t_{qi} 2\pi\sqrt{1-\rho_{pq}^2}} \exp\left\{\frac{-1}{2(1-\rho_{pq}^2)}(\psi_{pi}^2 - 2\rho_{pq}\psi_{pi}\psi_{qi} + \psi_{qi}^2)\right\}, \end{aligned} \tag{16.31}$$

where

$$\psi_{pi} \equiv \alpha_i[\ln t_{pi} - (\beta_i - \tau_p)] \tag{16.32}$$

is the residual RT after adjustment for the item and person effects and ρ_{pq} is the correlation between ψ_{pi} and ψ_{qi} (van der Linden, 2009b).

This bivariate version of the model can be used, for instance, for the detection of collusion between test takers during testing. Because all regular item and person effects have been removed, no matter the properties of the items, independently working test takers should have $\rho_{pq} = 0$. Deviations from this null value could thus point at a potentially suspicious behavior (Volume Three, Chapter 13).

16.3 Parameter Estimation

Our primary estimation method is Bayesian with Gibbs sampling of the joint posterior distribution of all model parameters (Fox, 2010, Chapter 4; Junker, Patz, and Vanhoudnos, Volume Two, Chapter 15). It is convenient to use the version of the model in Equation 16.8; that is, estimate the parameters from the logtimes. The sampler proposed in van der Linden (2006) alternates conveniently between slight modifications of the two standard cases of (i) normal data with known variances and a conjugate normal prior for the unknown speed parameters and (ii) normal data with a conjugate normal-gamma gamma prior for the unknown time-intensity and discrimination parameters (e.g., Gelman et al., 2014, Sections 2.5 and 3.3). The two modifications are: First, because of our choice of $\alpha_i \equiv \sigma^{-1}$, the inverse-$\chi^2$ in the second case has to be replaced by a gamma distribution. Second, when sampling the item parameters, the data need to be adjusted for the speed parameters, and vice versa.

16.3.1 Estimating Both Item and Person Parameters

The common prior distribution for the speed parameters τ_p is

$$\tau_p \sim N(\mu_\tau, \sigma_\tau^2), \tag{16.33}$$

while the one for the item parameters (α_i, β_i) is specified to be a normal gamma:

$$\beta_i \mid \alpha_i \sim N(\mu_\beta, (\alpha_i^2 \kappa)^{-1}); \tag{16.34}$$

$$\alpha_i^2 \sim G(\nu/2, \nu/(2\lambda)). \tag{16.35}$$

Thus, μ_τ is our prior guess of the value of the speed parameters with σ_τ^2 expressing the strength of our evidence for it. Likewise, μ_β is our guess of the value of β_i given α_i, now with $\alpha_i^2\kappa$ representing the strength of our evidence, while λ represents our prior guess of α_i^2, this time with ν expressing the strength of our evidence. The identifiability restriction in Equation 16.15 allows us to set the prior means equal to $\mu_\tau = 0$ and $\mu_\beta = \overline{\ln t}$, where $\overline{\ln t}$ is the average observed logtime across all test takers and items in the sample. Our choices for σ_τ^2 and κ should reflect the expected variability of the speed and time-intensity parameters. Similarly, it makes sense to set λ relative to the observed variance of the logtimes.

The Gibbs sampler alternates between the blocks of speed parameters $\tau_1, \ldots, \tau_P$, and item parameters $(\alpha_1, \beta_1), \ldots, (\alpha_I, \beta_I)$. At step k, the conditional posterior distributions that need to be sampled for these blocks have the following two forms:

1. When sampling each $\tau_p^{(k)}$, the item parameters have fixed values of $\alpha_i^{(k-1)}$ and $\beta_i^{(k-1)}$. Using the latter to redefine the data on each test taker as $\beta_i^{(k-1)} - \ln t_{pi}$, $i = 1, \ldots, I$, the posterior distribution of τ_p is the one for normal data with known variance $(\alpha_i^{(k-1)})^{-2}$ but an unknown mean τ_p with a conjugate normal prior. Consequently, assuming a prior mean $\mu_\tau \equiv 0$, $\tau_j^{(k)}$ has to be drawn from a normal distribution with mean

$$\frac{\sum_{i=1}^{I} (\alpha_i^{(k-1)})^2 (\beta_i^{(k-1)} - \ln t_{pi})}{\sigma_\tau^{-2} + \sum_{i=1}^{n} (\alpha_i^{(k-1)})^2} \tag{16.36}$$

and variance

$$\left(\sigma_\tau^{-2}+\sum_{i=1}^{I}(\alpha_i^{(k-1)})^2\right)^{-1}. \tag{16.37}$$

2. When sampling each $(\alpha_i^{(k)},\beta_i^{(k)})$, the speed parameters have fixed values of $\tau_j^{(k)}$. Using them to redefine the data on each item as $\ln t_{pi}+\tau_p^{(k)}$, $p=1,\ldots,P$, the posterior distribution of (α_i, β_i) is now the one for normal data with an unknown mean β_i and unknown variance α_i^{-2}. Given their conjugate prior distribution, the values for these parameters have to be drawn from a normal-gamma posterior. That is, assuming a prior mean $\mu_\beta \equiv \overline{\ln t}$, $\alpha_i^{(k)}$ is drawn from a gamma distribution $G(\phi/2, \omega/2)$ with parameters

$$\phi=\nu+P; \tag{16.38}$$

$$\omega=\nu\lambda^{-1}+\sum_{p=1}^{P}(\ln t_{pi}-\overline{\ln t_i}+\tau_p^{(k)})^2+\frac{\kappa P\left[\overline{\ln t_i}-\overline{\ln t}\right]^2}{\kappa+P}, \tag{16.39}$$

while $\beta_i^{(k)}$ is drawn from a normal distribution with mean

$$\frac{\kappa\overline{\ln t}+\sum_{p=1}^{P}(\ln t_{pi}+\tau_p^{(k)})}{\kappa+P} \tag{16.40}$$

and variance

$$(\kappa+P)^{-1}. \tag{16.41}$$

As a frequentist alternative to this Bayesian method, the item and person parameters can be estimated by confirmatory factor analysis of the logtimes with an appropriate back transformation of the discrimination parameters (Section 16.2.6), for instance, using *MPlus* (Volume Three, Chapter 28). A comparative study with empirical RTs by Finger and Chee (2009) showed correlations between the factor analysis and Gibbs sampler estimates uniformly greater than 0.99 for all parameters.

16.3.2 Estimating the Speed Parameters

When speed parameters τ_p have to be reported for examinees on a test with items already calibrated under the RT model, an attractive option is to sample their posterior distributions using an adjusted version of the above Gibbs sampler. The average of the sampled values for each test taker is their expected a posterior (EAP) estimate, which is the standard deviation of their posterior standard error.

The adjusted sampler has the following two alternating steps:

1. Drawing the τ_p parameters from the normal posterior distribution with the mean and variance in Equations 16.36 and 16.37.
2. Resampling of vectors of draws from the stationary posterior distributions of the α_i and β_i parameters for the items in the test saved during their calibration.

The estimation is extremely fast because only the draws for the τ_p parameters need to converge; the draws for the item parameters are already from extremely narrow posterior distributions centered at their true values. The vectors with the saved draws for the item parameter do not need to contain more than 1000 or so well-spaced draws by the stationary Gibbs sampler during their calibration. For a proof of the convergence of this type of Gibbs sampler, see van der Linden and Ren (2015).

Again, as a frequentist alternative to this Bayesian method, if the items have been calibrated with enough precision to treat the remaining uncertainty as negligible, the speed parameters can be estimated using the following MLE with point estimates substituted for the item parameters:

$$\hat{\tau}_p = \frac{\sum_{i=1}^{n} \hat{\alpha}_i^2(\hat{\beta}_i - \ln t_{pi})}{\sum_{i=1}^{n} \hat{\alpha}_i^2}. \tag{16.42}$$

As the expression reveals, the speed of an individual test taker is then estimated as the precision-weighted average of $\hat{\beta}_i - \ln t_{pi}$ (i.e., the logtime on each item adjusted for its estimated time intensity). Returning to our earlier discussion of the interpretation of the α_i parameters as a discrimination parameter, observe how they serve as precision weights in this estimator. The same could already have been observed for the similar role played by these parameters in the expressions for the posterior mean of τ_p in Equations 16.36 and 16.37.

16.4 Model Fit

It is convenient to combine these two Gibbs samplers with posterior predictive checks of the fit of the model to the items and test takers. As we have one observation per combination of test taker and item, an obvious choice is marginal checks with aggregation of the results across test takers or items to evaluate their fit more specifically.

Let $\widetilde{\ln t}_{pi}$ denote the predicted logtime for test taker p on item i. The (lower-tail) posterior predictive p-values of the observed logtimes are defined as

$$\begin{aligned}\pi_{pi} &\equiv \Pr\{\widetilde{\ln t}_{pi} < \ln t_{pi}\} \\ &= F_{\widetilde{\ln T}_{pi}}(\ln t_{pi}) \\ &= \int\int\int \Phi(\ln t_{pi}; \tau_p, \alpha_i, \beta_i) f(\tau_p, \alpha_i, \beta_i \mid \mathbf{t}) d\tau_p\, d\alpha_i\, d\beta_i,\end{aligned} \tag{16.43}$$

where $\Phi(\ln t_{pi}; \tau_p, \alpha_i, \beta_i)$ is the cdf for the normal distribution of $\ln T_{pi}$ in Equation 16.8 and $f(\tau_p, \alpha_i, \beta_i | \mathbf{t})$ is the posterior density of its parameters given the observed data $\mathbf{t} \equiv (\mathbf{t}_{ij})$.

With increasing sample size and test length, the posterior density in Equation 16.43 converges to that of a degenerate distribution at the true parameter values, and consequently $F_{\widetilde{\ln T_{pi}}}(\ln t_{pi}) \to \Phi(\ln t_{pi}; \tau_p, \alpha_i, \beta_i)$. The probability integral transform theorem (e.g., Casella and Berger, 2002, Section 2.1) tells us that $F_{\widetilde{\ln T_{pi}}}(X)$ has the $U(0,1)$ distribution. Combining the two observations, it follows that each π_{pi}-value has the same asymptotic uniform distribution on [0,1]. For posterior distributions with noticeable variance, the predictive distribution of $\widetilde{\ln t_{pi}}$ is still wider than that of $\ln T_{pi}$. Depending on the true value of π_{pi}, the observed distribution will then tend to be concentrated more toward a point somewhere in the middle of its scale.

Upon stationarity of the Gibbs sampler, for $k = 1, \ldots, K$ additional draws from the posterior distribution, each of these π_{pi}-values can be approximated as

$$\pi_{pi} \approx K^{-1} \sum_{k=1}^{K} \Phi(\ln t_{pi}; \tau_p^{(k)}, \alpha_i^{(k)}, \beta_i^{(k)}). \tag{16.44}$$

The proposed tool for checking on the fit of the model are plots of the cumulative distributions of these Bayesian π_{pi}-values across items or test takers. The use of cumulative distributions allows us to visually establish the lack of uniformity as deviations from the identity line. However, as just discussed, due to less than full convergence of the posterior distributions, the curves may be somewhat lower than the identity line at the lower end of the scale with the corresponding compensation at the upper end.

Lagrange multiplier (LM) tests (or score tests) of the assumption of conditional independence of RTs have been presented for the two cases of marginal MLE of all model parameters (Glas and van der Linden, 2010) and MLE of the speed parameters for known item parameters (van der Linden and Glas, 2010). Both statistical tests are based on the assumption of the bivariate lognormal distribution in Equation 16.31, this time posited for the case of the RTs of a single test taker on pairs of items i and i'. More specifically, they evaluate $H_0: \rho_{ii'} = 0$ against $H_1: \rho_{ii'} \neq 0$. LM tests are known to be uniformly most powerful.

When the items are taken from a bank gone through a process of item calibration careful enough to allow treatment of their parameters as known, the use of the second LM test has the advantage of a test statistic in a closed form

$$LM(\rho_{ii'}) = \frac{\left(\sum_{p=1}^{P} \hat{\psi}_{pi} \hat{\psi}_{pi'}\right)^2}{\sum_{p=1}^{P} \left[\hat{\psi}_{pi}^2 + \hat{\psi}_{pi'}^2 - 1 - \left((\alpha_i \hat{\psi}_{pi} + \alpha_{i'} \hat{\psi}_{pi'})^2 \Big/ \sum_{i=1}^{n} \alpha_i^2\right)\right]}, \tag{16.45}$$

where

$$\hat{\psi}_{pi} = \alpha_i [\ln t_{pi} - (\beta_i - \hat{\tau}_p)] \tag{16.46}$$

is the estimate of the standardized residual RT for test taker p on item i in Equation 16.32, $\hat{\psi}_{pi'}$ is defined analogously for item i', and $\hat{\tau}_p$ is the MLE in Equation 16.42. Under H_0, the statistic has a χ^2 distribution with one degree of freedom.

Observe that, since the model posits a normal distribution of each ln T_{pi}, their standardized versions will have $N(0,1)$ as an asymptotic distribution. The fact has been used by van der Linden et al. (2007) to check observed distributions of residuals for changes in speed during test administration. The procedure is illustrated in the empirical example in the next section.

16.5 Empirical Example

The data set is from a computerized high-stakes test existing of the RTs of 396 test takers on 65 multiple-choice items. The lognormal model was fitted using the Gibbs sampler in Equations 16.33 through 16.41 with a burn-in of 1500 iterations and 4500 additional iterations for the calculation of the EAP estimates of the parameters. These numbers have been shown to suffice in extensive explorations of the model, provided the following implementation choices are made: As an identifiability restriction, we used $\mu_\tau = 0$ in Equation 16.15, which was implemented simply by recentering the draws for all τ_p parameters after each iteration. All prior distributions were set to have parameter values according to our suggestions above. As variance of the common prior distribution of the τ_p parameters, $\sigma_\tau^2 = 100$ was chosen. Because of the identifiability restriction, we were able to use Equation 16.16 and set the prior mean of the β_i parameters equal to $\mu_\beta = \overline{\ln t} = 3.88$. In addition, we set $\kappa = 1$. As for the common prior distribution for the α_i parameters, $\lambda = \text{var}(\ln t) = 0.68$ was used in combination with the choice of $\upsilon = 1$. The choice of values for σ_τ^2, κ, and υ were all low informative. The Gibbs sampler was initialized using their prior mean as starting value for each of its parameters.

The estimates of the β_i parameters ranged from 2.75 to 5.32 with an average of 3.89. Observe that, for test takers operating at the average speed of $\tau = 0$, the endpoints of this range correspond to the expected RTs of exp(2.75) = 15.64 and exp(5.32) = 204.38 s, respectively. The estimates of the τ_p parameters ranged from −0.43 to 0.70 around their mean of zero with a standard deviation equal to 0.23. The variation was much smaller than for the time-intensity parameters of the items—a result typical of high-stakes tests with a well-chosen time limit. Finally, all estimates of the α_i parameters ranged from 1.33 to 2.55 with a mean of 1.96.

The fit of the model to the dataset was evaluated using the tools described earlier. Figure 16.2 shows the plot of the cumulative distribution of the Bayesian π-values in Equation 16.44 across all test takers and items in the set. The curve shows a minor deviation from the identity line at the lowest π-values due to the remaining posterior variance of the parameter estimates discussed earlier; otherwise, its shape is entirely according to expectation. It is important to note, however, that this result only represents a necessary condition for the overall model fit. It is still possible for the individual items or test takers to have opposite types of misfit canceling out at this higher level of aggregation.

Figure 16.3 shows the same cumulative curves for each of the 65 items. As each of them is based on some 2.5% of the data only, the results are less stable. However, some of the variation may also reflect a true misfit of some of the items. For the three items with the lowest curves at the lower end of the scale, there certainly was some misfit. For instance, each of them had an observed proportion of π-values of some 0.07 lower than expected at the nominal value of 0.30.

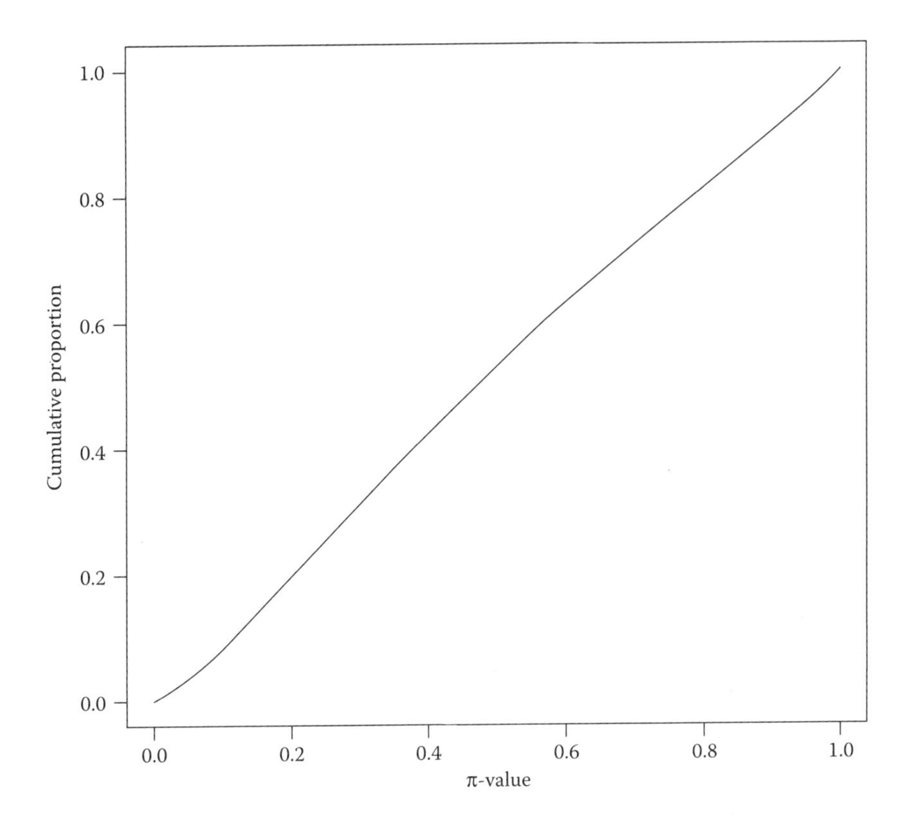

FIGURE 16.2
Cumulative distribution of Bayesian π-values across all 65 items and 396 test takers in the dataset.

Obviously, the same type of plot with the curves for the 396 test takers in Figure 16.4 shows even more variation as each of the curves is now based on the observed RTs on only 65 items. Nevertheless, the plot is still useful to identify test takers with aberrant behavior. For instance, the outlying curve at the higher end of the scale was for a test taker whose observed proportion of π-values below 0.75 was equal to 0.48, a result clearly below expectation. Inspection of his record revealed some six items RTs in the range from 0 to 10 s. Apparently, this test taker had inspected only these items quickly entering random responses to move to the next item.

As for the assumption of constancy of speed, it is possible to check on systematic violations by plotting the mean-standardized residual logtimes on the items in Equation 16.46 as a function of their position in the test. The results in Figure 16.5 reveal generally small mean residuals, varying about their expected value of zero in the range from −0.05 to 0.05. The slight tendency for the residuals to be more negative toward the end of the test may point at a corresponding increase in speed. However, as a mean-standardized residual logtime of 0.05 corresponds to just exp(0.05) = 1.1 s, the observed violations hardly have any practical value. The standard deviations of the residuals were close to their expected value of 1. Their values ranged from 0.94 to 1.02, with a mean equal to 0.99.

As a check on the assumption of local independence, we calculated the correlations between the residual RTs in Equation 16.32 on the pairs of adjacent items in the test beginning with the first item. The results are presented in Table 16.2. As expected, the correlations

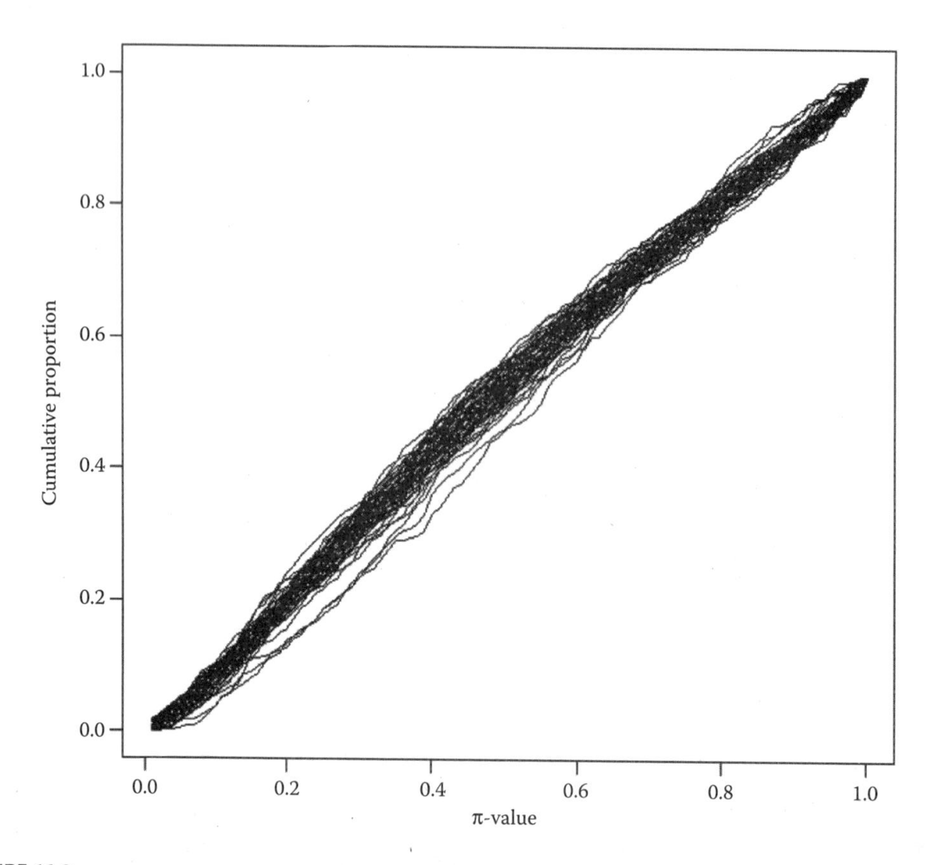

FIGURE 16.3
Cumulative distribution of Bayesian π-values for each of the 65 items across all 396 test takers in the dataset.

varied around zero and, with a few exceptions, were generally low. We also calculated the LM test in Equation 16.45, which returned all correlations with an absolute value greater than 0.10 as significant at $\alpha = 0.05$. Of course, these results were greatly influenced by the high power of the test for a sample size of $P = 369$ test takers. Violations of the assumption of local independence of this size do not necessarily have a practical meaning. Generally, they lead to standard errors for the speed parameters based on a somewhat lower-than-anticipated effective number of observations but do not imply any unnecessary bias or even inconsistency of the parameter estimates. The only violation with a noticeable impact on the standard error might be Item Pair (45,46). If its correlation of 0.21 is deemed too high, replacement of one of its items in the test would solve the problem.

16.6 Discussion

We began this chapter by observing that every response to a test item has both an aspect of correctness and time. During its first century, test theory exclusively focused on responses scored for their correctness. It had to do so because, except for an individually proctored

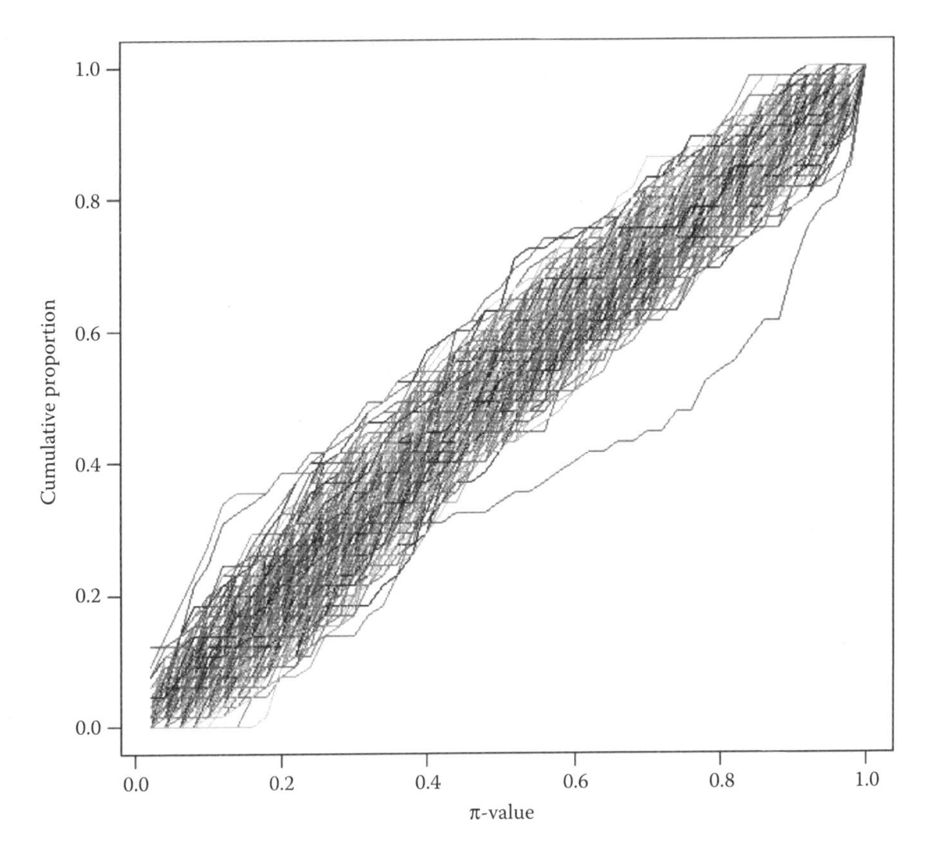

FIGURE 16.4
Cumulative distribution of Bayesian π-values for each of the 396 test takers across all 65 items in the test.

test, accurate recording of the RTs at the level of the individual test takers and items was practically infeasible. The arrival of computers and handheld electronic devices in test administration has removed this limitation. Consequently, test theorists will have to rethink their basic notions and applications. Examples already present in the literature focus on the notion of test speededness (e.g., Bolt et al., 2002; Goegebeur et al., 2008; Lu and Sireci, 2007; Schmitt et al., 2010; Volume Three, Chapter 12), the role of motivation in test performance (e.g., Wise and DeMars, 2009; Wise and Kong, 2005), and the nature of parameter drift (e.g., Li and Shen, 2010; Wollack et al., 2003). More in-depth psychological analyses of the role time limits and the time spent on test items can be found in Goldhammer and Kroehne (2014), Goldhammer et al. (2014), and Ranger and Kuhn (2015). Besides, it has become possible to improve such practices as test item calibration (van der Linden et al., 2010) and test accommodations (e.g., Stretch and Osborne, 2005). Adaptive testing has already profited from the use of RTs. Fan et al. (2012), in the spirit of Woodbury and Novick's (1968) pioneering work on the assembly of fixed forms, worked out the details of maximization of its information about a test takers' ability parameter for a given time interval while van der Linden (2008) showed how RTs can be used as collateral information to optimize item selection and ability estimation in adaptive testing. More complete reviews of all these changes and new promises of RT analyses can be found in Lee and Chen (2011) and van der Linden (2011a).

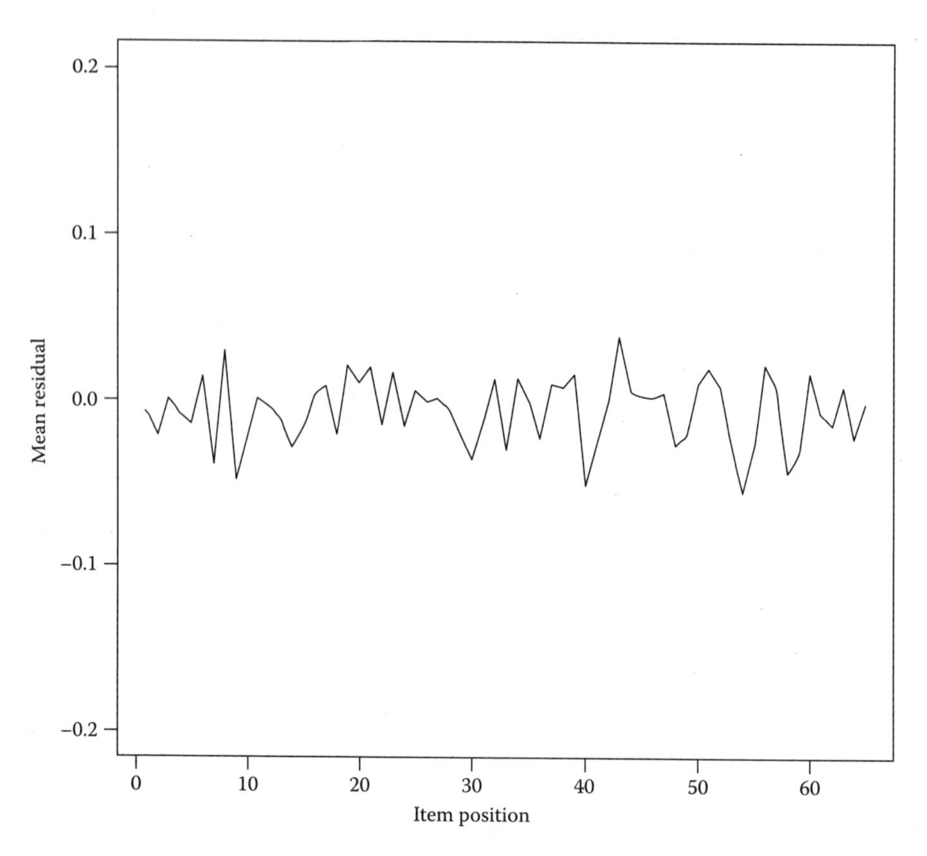

FIGURE 16.5
Mean residual logtimes on the 65 items as a function of their position in the test.

TABLE 16.2
Correlations between Residual RTs on the Items

Items	ρ	Items	ρ	Items	ρ	Items	ρ
1,2	0.00	17,18	−0.01	33,34	0.06	49,50	−0.07
3,4	0.03	19,20	0.06	35,36	−0.12	51,52	0.00
5,6	0.00	21,22	−0.02	37,38	−0.12	53,54	0.00
7,8	0.15	23,24	0.01	39,40	−0.05	55,56	−0.06
9,10	−0.11	25,26	−0.03	41,42	0.03	57,58	0.07
11,12	0.04	27,28	−0.02	43,44	−0.15	59,60	−0.07
13,14	0.01	29,30	0.09	45,56	0.21	61,62	−0.04
15,16	0.08	31,32	−0.02	47,48	0.08	63,64	−0.10

References

Bolt, D. M., Cohen, A. S., and Wollack, J. A. 2002. Item parameter estimation under conditions of test speededness: Application of a mixture Rasch model with ordinal constraints. *Journal of Educational Measurement, 39*, 331–348.

Casella, G. and Berger, R. L. 2002. *Statistical Inference* (2nd ed.). Pacific Grove, CA: Duxbury.

Fan, Z., Wang, C., Chang, H.-H., and Douglas, J. 2012. Utilizing response time distributions for item selection in CAT. *Journal of Educational and Behavioral Statistics, 37*, 655–670.

Ferrando, P. J. and Lorenzo-Seva, U. 2007. An item-response model incorporating response time data in binary personality items. *Applied Psychological Measurement, 31*, 525–543.

Finger, M. S. and Chee, C. S. 2009. Response-time model estimation via confirmatory factor analysis. Paper presented at the *Annual Meeting of the National Council on Measurement in Education*, San Diego, CA, April.

Fox, J.-P. 2010. *Bayesian Item Response Modeling*. New York: Springer.

Gaviria, J.-L. 2005. Increase in precision when estimating parameters in computer assisted testing using response times. *Quality and Quantity, 39*, 45–69.

Gelman, A., Carlin, J. B., Stern, H., Dunson, D. B., Vehtari, A., and Rubin, D. B. 2014. *Bayesian Data Analysis* (3rd ed.). Boca Raton, FL: Chapman and Hall/CRC.

Glas, C. A. W. and van der Linden, W. J. 2010. Marginal likelihood inference for a model for item responses and response times. *British Journal of Mathematical and Statistical Psychology, 63*, 603–626.

Goegebeur, Y., De Boeck, P., Wollack, J. A., and Cohen, A. S. 2008. A speeded response model with gradual process change. *Psychometrika, 73*, 65–87.

Goldhammer, F. and Kroehne, U. 2014. Controlling individuals' time spent on task in speeded performance measures: Experimental time limits, posterior time limits, and response-time modeling. *Applied Psychological Measurement, 38*, 255–267.

Goldhammer, F., Naumann, J., Stelter, A., Toth, K., Roelle, H., and Klieme, E. 2014. The time on task effect in reading and problem solving is moderated by task difficulty and skill: Insights from a computer-based large scale assessment. *Journal of Educational Psychology, 106*, 608–626.

Hornke, L. F. 2000. Item response times in computerized adaptive testing. *Psicológica, 21*, 175–189.

Klein Entink, R. H., van der Linden, W. J., and Fox, J.-P. 2009. A Box–Cox normal model for response times. *British Journal of Mathematical and Statistical Psychology, 62*, 621–640.

Kotz, S. and Johnson, N. L. 1985. *Encyclopedia of Statistical Science* (Volume 5). New York: Wiley.

Lee, Y.-H. and Chen, H. 2011. A review of recent response-time analyses in educational testing. *Psychological Test and Assessment Modeling, 53*, 359–379.

Li, F. and Shen, L. 2010. Detecting item parameter drift by item response and item response time in a computer-based exam. Paper presented at the *Annual Meeting of the National Council on Measurement in Education*, Denver, CO, April 29–May 3.

Lu, Y. and Sireci, S. G. 2007. Validity issues in test speededness. *Educational Measurement: Issues and Practice, 26*(4), 29–37.

Luce, R. D. 1986. *Response Times: Their Roles in Inferring Elementary Mental Organization*. Oxford, UK: Oxford University Press.

Maris, E. 1993. Additive and multiplicative models for gamma distributed variables, and their application as psychometric models for response times. *Psychometrika, 58*, 445–469.

Oosterloo, S. J. 1975. *Modellen voor reactie-tijden* [Models for reaction times]. Unpublished master's thesis, Faculty of Psychology, University of Groningen, The Netherlands.

Ranger, J. and Kuhn, J.-T. 2015. Modeling information accumulation in psychological tests using item response times. *Journal of Educational and Behavioral Statistics, 40*, 274–306.

Rasch, G. 1960. *Probabilistic Models for Some Intelligence and Attainment Tests*. Chicago: University of Chicago Press.

Roskam, E. E. 1987. Toward a psychometric theory of intelligence. In E. E. Roskam and R. Suck (Eds.). *Progress in Mathematical Psychology* (pp. 151–171). Amsterdam: North-Holland.

Roskam, E. E. 1997. Models for speed and time-limit tests. In W. J. van der Linden and R. K. Hambleton (Eds.). *Handbook of Modern Item Response Theory* (pp. 187–208). New York: Springer.

Scheiblechner, H. 1979. Specific objective stochastic latency mechanisms. *Journal of Mathematical Psychology, 19*, 18–38.

Schmitt, T. A., Sass, D. A., Sullivan, J. R., and Walker, C. M. 2010. A Monte Carlo simulation investigating the validity and reliability of ability estimation in item response theory with speeded computer adaptive tests. *International Journal of Testing, 10*, 230–261.

Stretch, L. S. and Osborne, J. W. 2005. Extended time test accommodation: Directions for future research and practice. *Practical Assessment, Research and Evaluation, 10,* 1–8.

Swanson, D. B., Case, S. E., Ripkey, D. R., Clauser, B. E., and Holtman, M. C. 2001. Relationships among item characteristics, examinee characteristics, and response times on USMLE, Step 1. *Academic Medicine, 76,* 114–116.

Tatsuoka, K. K. and Tatsuoka, M. M. 1980. A model for incorporating response-time data in scoring achievement tests. In D. J. Weiss (Ed.). *Proceedings of the 1979 Computerized Adaptive Testing Conference* (pp. 236–256). Minneapolis: University of Minnesota, Department of Psychology, Psychometric Methods Program.

Thissen, D. 1983. Timed testing: An approach using item response theory. In D. J. Weiss (Ed.). *New Horizons in Testing: Latent Trait Test Theory and Computerized Adaptive Testing* (pp. 179–203). New York: Academic Press.

van der Linden, W. J. 2006. A lognormal model for response times on test items. *Journal of Educational and Behavioral Statistics, 31,* 181–204.

van der Linden, W. J. 2008. Using response times for item selection in adaptive testing. *Journal of Educational and Behavioral Statistics, 33,* 5–20.

van der Linden, W. J. 2009a. Conceptual issues in response-time modeling. *Journal of Educational Measurement, 46,* 247–272.

van der Linden, W. J. 2009b. A bivariate lognormal model response-time model for the detection of collusion between test takers. *Journal of Educational and Behavioral Statistics, 34,* 378–394.

van der Linden, W. J. 2010. Linking response-time parameters onto a common scale. *Journal of Educational Measurement, 47,* 92–114.

van der Linden, W. J. 2011a. Modeling response times with latent variables: Principles and applications. *Psychological Test and Assessment Modeling, 53,* 334–358.

van der Linden, W. J. 2011b. Setting time limits on tests. *Applied Psychological Measurement, 35,* 183–199.

van der Linden, W. J. 2011c. Test design and speededness. *Journal of Educational Measurement, 48,* 44–60.

van der Linden, W. J., Breithaupt, K., Chuah, S. C., and Zhang, Y. 2007. Detecting differential speededness in multistage testing. *Journal of Educational Measurement, 44,* 117–130.

van der Linden, W. J. and Glas, C. A. W. 2010. Statistical tests of conditional independence between responses and/or response times on test items. *Psychometrika, 75,* 120–139.

van der Linden, W. J. and Guo, F. 2008. Bayesian procedures for identifying aberrant response-time patterns in adaptive testing. *Psychometrika, 73,* 365–384.

van der Linden, W. J., and Klein Entink, R. H., and Fox, J.-P. 2010. Item parameter estimation with response times as collateral information. *Applied Psychological Measurement, 34,* 327–347.

van der Linden, W. J., and Ren, H. 2015. Optimal Bayesian adaptive design for test-item calibration. *Psychometrika, 80,* 263–288.

Verhelst, N. D., Verstralen, H. H. F. M., and Jansen, M. G. 1997. A logistic model for time-limit tests. In W. J. van der Linden and R. K. Hambleton (Eds.). *Handbook of Modern Item Response Theory* (pp. 169–185). New York: Springer.

Wang, T. and Hanson, B. A. 2005. Development and calibration of an item response model that incorporates response time. *Applied Psychological Measurement, 29,* 323–339.

Wise, S. L. and DeMars, C. E. 2009. An application of item-response time: The effort-moderated IRT model. *Journal of Educational Measurement, 43,* 19–38.

Wise, S. L. and Kong, C. 2005. Response time effort: A new measure of examinee motivation in computer-based tests. *Applied Measurement in Education, 18,* 163–183.

Wollack, J. A., Cohen, A. S., and Wells, C. S. 2003. A method for maintaining scale stability in the presence of test speededness. *Journal of Educational Measurement, 40,* 307–330.

Woodbury, M. A. and Novick, M. R. 1968. Maximizing the validity of a test battery as a function of relative test length for a fixed total testing time. *Journal of Mathematical Psychology, 5,* 242–259.

17

Diffusion-Based Response-Time Models

Francis Tuerlinckx, Dylan Molenaar, and Han L. J. van der Maas

CONTENTS

17.1 Introduction 283
17.2 Models 284
17.2.1 The D-Diffusion Model 287
17.2.2 The Q-Diffusion Model 289
17.2.3 Relations to Other Models and Old Ideas 289
17.3 Parameter Estimation 291
17.4 Model Fit 292
17.5 Empirical Example 292
17.6 Discussion 297
References 299

17.1 Introduction

A large part of cognitive psychological research operates under the assumption that the time it takes to complete a task (e.g., taking a decision) carries information on how the task is solved. The earliest use of response time (RT) in scientific psychological research can be traced back to the late nineteenth century (e.g., Donders, 1869/1969). In fact, since then, RT has probably been the most important dependent variable used in cognitive psychology.

Apart from experimental manipulation, cognitive psychology has a rich tradition of mathematical modeling of RT (Luce, 1986). Particularly, successful is the class of sequential sampling models (e.g., Luce, 1986), in which information required for a particular response is assumed to accumulate over time (and of which the diffusion model is an example). Sequential sampling models in general, and the diffusion model in particular, have been able to give a valid account both of the accuracy of the responses and the RT across many different types of experiments (Wagenmakers, 2009).

However, unlike cognitive psychology, test theory has not (yet) reserved a front seat in research on RT. Indeed, the major focus has been on the response itself, as observed by Goldstein and Wood (1989, p. 139), who keenly comment: "It is still the same scenario first projected in 1942: person with ability x encounters an item, which has difficulty and discrimination, and the outcome is a binary response (or, perhaps, graded)." One reason for the neglect of RT is that most Item Response Theory (IRT) models have been developed for application to pure power tests (as opposed to speed tests). However, as noted by many authors (e.g., van der Linden and Hambleton, 1997; Verhelst et al., 1997; van Breukelen, 2005; Volume One, Chapter 29), the concept of a pure power test is an illusion in every testing situation, there is always some time pressure. It is remarkable that this simple fact has been largely ignored for a long time by most psychometricians (for some early exceptions, see, e.g., Thurstone, 1937).

However, the tide is turning (see, e.g., Klein Entink et al., 2009; van der Linden, 2009; van der Maas et al., 2011; Ranger and Kuhn, 2012). Without any doubt, technological advances have fueled this change, because the recording of RT has become relatively cheap in computerized testing. But there is also a deeper reason. A growing group of test theorists is embracing the idea that the time to respond reveals something about the ability that is measured. As a result, the past two decades have shown a surge of studies on the joint modeling response correctness and RTs.

Broadly, we distinguish three conceptual perspectives to deal with RT and item responses jointly in testing. First, RTs can be considered to be collateral information. The prime interest is still in the item response model, but by using RTs the item response parameters (e.g., the person's ability θ) can be estimated more efficiently (see, e.g., Gaviria, 2005; van Breukelen, 2005; van der Linden, 2007, 2009). As a result, an important goal is to increase the information about θ by making use of RT. However, other applications are also possible, such as cheating detection and the diagnosis of test speededness (e.g., van der Linden, 2011).

The second approach to jointly modeling RT and responses is normative in nature. In one variant, a scoring rule is employed, assuming that people will change their behavior to comply with the scoring rule. Although less often applied in psychometric modeling, scoring rules are popular in video games and sports such as show jumping (for a recent theoretical contribution, see Maris and van der Maas, 2012). In another variant of the normative approach, the requirement of Rasch's specific objectivity is imposed on RT models. Examples of the latter include for instance Scheiblechner (1979) and Vorberg and Schwarz (1990).

A third and final approach, and one that is the focus of this chapter, is the process approach, where an underlying process model is assumed to give simultaneous rise to responses and RT. The research in this chapter builds on earlier contributions to this approach by Tuerlinckx and De Boeck (2005) and van der Maas et al. (2011). The same approach is linked to the ideas on validity of Borsboom et al. (2003, 2004), who argue more generally that a claim of validity requires a process model of how the ability that is assumed to be measured causes the item responses.

In the following, we derive a general class of IRT models that incorporate the underlying response process explicitly. Our focus will be mainly on two important instances of these models that are useful for applications to personality and ability measurement. We present estimation and model fit techniques for these models and apply them to a data set from personality research.

17.2 Models

We use one of the most successful models from mathematical psychology—the diffusion model (Ratcliff, 1978; Ratcliff and Rouder, 1998)—as the basic measurement model for the administration of an item to a person (called a single trial).

The diffusion model assumes that noisy information is sampled and accumulated (or integrated) across time to reach a decision. This stochastic information accumulation takes place in continuous time.

In order to explain the mechanics of the diffusion model, consider the example of a person given an item from an extraversion personality test. A possible item could be: "Do you enjoy organizing parties?" where the person has to decide between the possible responses "yes" and "no." In the remainder of this chapter, we will restrict our attention to the same case of items with two response alternatives.

Figure 17.1 shows a graphical representation of the simplest version of the diffusion process for a single trial. The total RT T can be split into the decision time D and a residual time t_{er}. The decision time D is the time needed for the stochastic information accumulation process (i.e., the sample path in Figure 17.1) to reach one of two decision boundaries. Each decision boundary is associated with a response. In Figure 17.1, the two boundaries are associated with the responses "yes" and "no."

The model has four basic parameters (and one constant), with each parameter representing a specific aspect of the response process. (Note that we consider a single person responding to a single item and hence we will not distinguish between item and person parameters at this point. This distinction will be made below.) First, there is the drift rate, v ($v \in \mathbb{R}$), which is the average rate of information accumulation toward the upper boundary. If drift rate is negative, there is tendency for the process to drift off to the lower boundary. The drift rate represents the quality of the information present during the decision making. If the absolute value $|v|$ is large, the person has enough information available for this item to make the decision. For our example, this means that the person can easily respond with "yes" or "no" to the question "Do you enjoy organizing parties?" because much information has been accumulated for one of the two responses. Second, there is the distance between the response boundaries, called boundary separation, which is represented by parameter α (with $\alpha > 0$). The larger the boundary separation, the more information the person needs before being able to respond to the item. Therefore, boundary separation α quantifies response caution. A third key parameter represents the starting point of the information accumulation process, z, which, by definition, is located between the lower and upper boundary; hence $0 < z < \alpha$. If z deviates from $\alpha/2$, there is a prior bias toward one of the response options. In this chapter, we will assume that the process always starts exactly halfway between the two boundaries and that there is thus no bias. A fourth parameter is the nondecision or residual time t_{er} (where "e" stands for encoding and "r" for response; $t_{er} > 0$). The information accumulation process only accounts for the decision time, but the total RT T also consists of other components, such

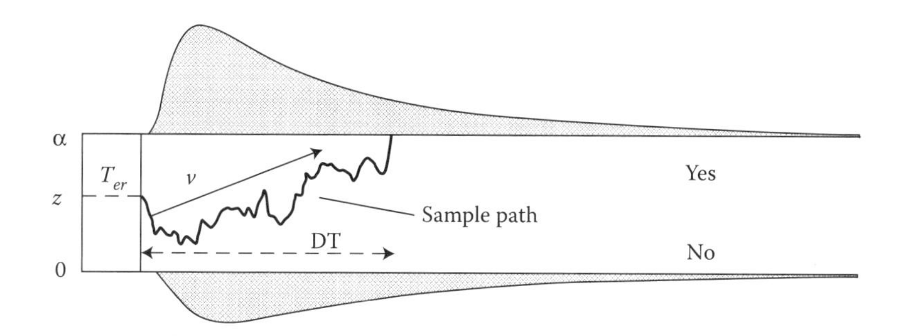

FIGURE 17.1
Graphical representation of the diffusion model for two-choice decisions. The jagged gray line represents a single sample path resulting in a "yes" response. The ensemble of sample paths gives rise to the RT densities at the top and bottom boundary. The area of the densities corresponds to the probability of giving the corresponding response.

as perception time (e.g., reading the question) and motor response execution time (e.g., pressing the response key).

Finally, as the process of integration of information is noisy, the variability of the information accumulation process can be quantified by the variance σ^2 (with $\sigma^2 > 0$). However, commonly, σ^2 is not treated as a free parameter but set to 1 for identification purposes (when treated as a free parameter, another diffusion process with exactly the same properties can be found by multiplying α, v, z, and σ with the same constant).

In the mathematical psychological literature, several other versions of the diffusion model have been proposed. However, in this chapter, we will only deal with the simplest version of the model containing the four free parameters listed above. This version appears to serve the modeling item responses well.

The diffusion model defines a bivariate distribution for the pair of random variables (U, T), where U is the chosen response and T is the RT. The RT T is a positive continuous variable, consisting of the decision time D and the residual time t_{er}. The binary variable U is defined to be 1 (yes response) and 0 (no response). In the literature on stochastic processes (e.g., Cox and Miller, 1965), the joint density of (U, T) for the process described above can be found described as

$$p_{U,T}(u,t;\alpha,\nu,t_{er},z) = \frac{\pi}{\alpha^2}\exp\left((\alpha u - z)v - \frac{v^2}{2}(t-t_{er})\right)$$
$$\times \sum_{k=1}^{\infty} k\sin\left(\frac{\pi k(\alpha u - 2uz + z)}{\alpha}\right)\exp\left(-\frac{1}{2}\frac{\pi^2 k^2}{\alpha^2}(t-t_{er})\right). \tag{17.1}$$

From Equation 17.1, the marginal probability of the event $U = 1$ is (with $\alpha > 0$)

$$\Pr\{U=1;\alpha,\nu,z\} = \frac{1-\exp(-2zv)}{1-\exp(-2\alpha v)}.$$

Under the assumption of an unbiased diffusion process (i.e., = $\alpha/2$), this probability simplifies to

$$\Pr\{U=1;\alpha,\nu\} = \frac{\exp(\alpha v)}{1+\exp(\alpha v)}, \tag{17.2}$$

which is the common logistic form. It is also helpful to have an expression for the mean decision time

$$E(T;\alpha,\nu,t_{er}) = \begin{cases} t_{er} + \dfrac{\alpha}{2v}\dfrac{1-\exp(-\alpha v)}{1+\exp(-\alpha v)} \approx t_{er} + \dfrac{\alpha}{2|v|} & \text{if } v \neq 0, \\ t_{er} + \dfrac{1}{4}\alpha^2 & \text{if } v = 0, \end{cases} \tag{17.3}$$

where the approximation of the first equation is valid when $|\alpha v|$ is large enough (this approximation should not be used for computational purposes, but it is helpful to give insight into the mechanics of the diffusion model).

From Equations 17.2 and 17.3, some important relations between the diffusion model parameters and the observable data can be inferred. First, larger and smaller drift rates give rise to choice probabilities closer to one and zero, respectively, as well as faster expected RTs. This observation is a consequence of the fact that if the amount of information is higher, the decision will be more certain and quicker. Second, larger boundary separation lead both to more extreme choice probabilities (depending on the sign of the drift rate) and a greater expected RT. This observation is consistent with the interpretation of α as response caution: more cautious responding leads to more compatible choices (e.g., positive drift rate and saying "yes") but also takes longer. Note that both drift and boundary separation influence accuracy and speed. Third, the parameter t_{er} only influences the speed and not the accuracy. It represents a minimum RT, below which no responses are possible.

The model described thus far can be considered as the response model for a single encounter of a person and an item. It is now extended to the traditional measurement context of P persons responding to I items. Consequently, we will need to make the parameters of the diffusion model person and item dependent. In principle, this can be done for all parameters, but we will focus on the most important ones: drift rate v_{pi} and boundary separation α_{pi}. Both parameters will now be treated as a function of a person and an item parameter: $v_{pi} = f(\theta_p, b_i)$ and $\alpha_{pi} = g(\gamma_p, a_i)$. The nature of these person and item parameters will be explained below. Two choices have been proposed for the functional forms of f and g in the literature and will be discussed in the next sections.

17.2.1 The D-Diffusion Model

Tuerlinckx and De Boeck (2005) and van der Maas et al. (2011) propose to take the drift rate as the simple difference between the person and item parameters: $v_{pi} = f(\theta_p, b_i) = \theta_p - b_i$ (with $\theta_p, b_i \in \mathbb{R}$). For boundary separation, van der Maas et al. (2011) proposed a ratio function: $\alpha_{pi} = g(\gamma_p, a_i) = \gamma_p/a_i$, with $\gamma_p, a_i > 0$. Because of the additive decomposition of the drift rate, this model will be called the D-diffusion model ("D" denoting difference). The person parameter θ_p is the speed of information processing for person p and the item parameter b_i operates as an item threshold. The other pair of parameters, γ_p and a_i, is related to the response caution: parameter γ_p is under control of the person and represents the general level of cautiousness of the person. On the other hand, the a_i is a time pressure parameter induced by the context; for example, it may be influenced by the instructions at the start of the test (e.g., strong emphasis on speed or not) or the position of the item in the test (e.g., more toward the end where the time pressure increases).

Substituting these equations into Equation 17.2 (adding person and item indices) gives

$$\Pr\{U_{pi} = 1; a_i, \gamma_p, \theta_p, b_i\} = \frac{\exp((\gamma_p / a_i)(\theta_p - b_i))}{1 + \exp((\gamma_p / a_i)(\theta_p - b_i))} \tag{17.4}$$

and

$$\begin{aligned} E(T_{pi}; a_i, \gamma_p, \theta_p, b_i, t_{er}) &\approx t_{er} + \frac{\gamma_p}{2a_i|\theta_p - b_i|} \quad \text{if } \theta_p \neq b_i, \\ E(T_{pi}; a_i, \gamma_p, \theta_p, b_i, t_{er}) &= t_{er} + \frac{\gamma_p^2}{4a_i^2} \quad \text{if } \theta_p = b_i. \end{aligned} \tag{17.5}$$

The D-diffusion model assumes that both persons and items can be ordered on an information continuum. As an example, take the trait neuroticism as measured by items as "Do you enjoy organizing parties?" with possible answers "yes" and "no." Items indicative of this trait are on the continuum from introvert to extravert (i.e., locations of the b_i's). Persons also ordered on this continuum (i.e., locations of the θ_p's). If a person dominates the item ($\theta_p > b_i$), the probability of responding "yes" is larger than 0.5 and vice versa. When $\theta_p = b_i$, the choice probability $\Pr\{U_{pi} = 1; a_i, \gamma_p, \theta_p, b_i\}$ is 0.5. It has been found that respondents are less confident in such a case (Hanley, 1963).

Interestingly, the D-diffusion model predicts the mean RT as an inverse function of the absolute difference $|\theta_p - b_i|$ (holding everything else constant). This inverse relation means that the closer a person is to the item, the larger the mean RT. In fact, this phenomenon has been called the distance-difficulty hypothesis (Ferrando and Lorenzo-Seva, 2007) or the inverted-U RT effect (Kuiper, 1981), a well-known finding in the literature on RTs in personality assessment (see, e.g., Fekken and Holden, 1992). The distance-difficulty effect for the D-diffusion model is illustrated in Figure 17.2.

The person-specific response caution parameter γ_p controls the so-called speed-accuracy trade-off (SAT) (e.g., Luce, 1986; van Breukelen, 2005): by being less cautious, people can respond quicker, but at the expense of making more errors. Also, the item time pressure a_i influences the SAT: items with different a_i's lead to responses generated under different SAT conditions. It is important to realize that it only makes sense to compare SAT differences within a person-item combination (comparisons across persons or items are determined by many other factors).

The D-diffusion model generates a number of interesting predictions for personality and attitude items, but is less suited for the analysis of ability test data for two reasons: first, the distance-difficulty hypothesis is unrealistic for ability test items because the model implies that very able and very unable people generate the fastest (and in fact, equally fast) responses.

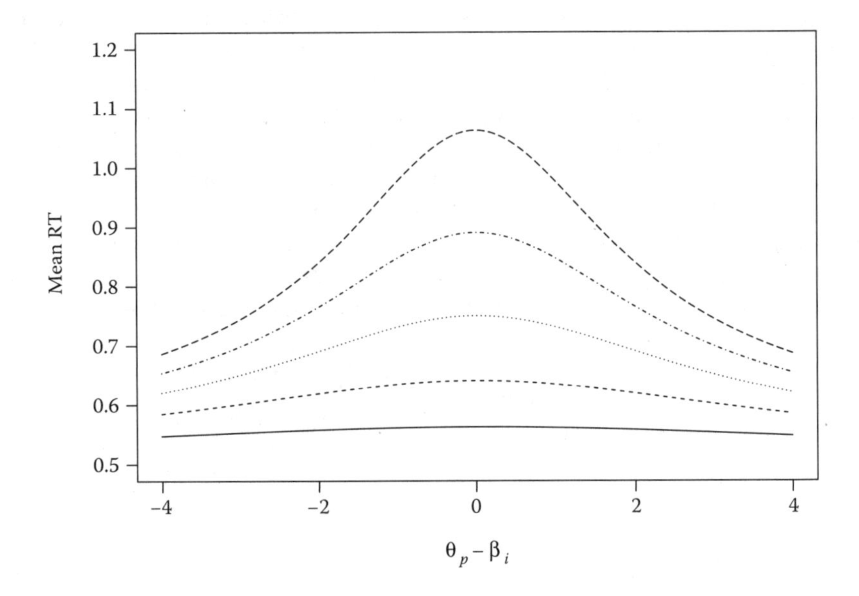

FIGURE 17.2
Graphical illustration of the distance-difficulty hypothesis as predicted by the D-diffusion model. The mean RT as a function of $\theta_p - b_i$ for different values of γ_p/a_i (from bottom to top, the values of γ_p/a_i are 0.5, 0.75, 1, 1.25, and 1.5, respectively).

Second, increasing the speed (lowering a_i) will lower the probability of a correct response for the less able persons (with $\theta_p < b_i$), but increase the probability of a correct response for the more able persons (with $\theta_p > b_i$), which seems very unlikely for ability tests. Therefore, we introduce another version of the diffusion model.

17.2.2 The Q-Diffusion Model

In the Q-diffusion model (van der Maas, 2011), the decomposition function for boundary separation is the same as in the D-diffusion model: $\alpha_{pi} = g(\gamma_p, a_i) = \gamma_p/a_i$ with $\gamma_p, a_i > 0$. But for the drift rate, a ratio function is proposed as well ("Q" in Q-diffusion stands for quotient): $v_{pi} = f(\theta_p, b_i) = \theta_p / b_i$ with $\theta_p, b_i > 0$. The probability of a correct response in the Q-diffusion model then becomes

$$\Pr\{U_{pi} = 1; a_i, \gamma_p, \theta_p, b_i\} = \frac{\exp((\gamma_p/a_i)(\theta_p/b_i))}{1+\exp((\gamma_p/a_i)(\theta_p/b_i))}. \tag{17.6}$$

The mean RT is

$$\begin{aligned} E(T_{pi}; a_i, \gamma_p, \theta_p, b_i, t_{er}) &\approx t_{er} + \frac{\gamma_p b_i}{2a_i\theta_p} \quad \text{if } \theta_p, b_i > 0, \\ E(T_{pi}; a_i, \gamma_p, \theta_p, b_i, t_{er}) &= t_{er} + \frac{\gamma_p^2}{4a_i^2} \quad \text{otherwise.} \end{aligned} \tag{17.7}$$

Like the D-diffusion model, parameters γ_p and a_i control the SAT. If a person increases his/her γ_p, both the probability correct and mean RT will go up.

The Q-diffusion model has a number of attractive properties in the case of ability testing. First, the key ability parameter θ_p and the item difficulty b_i in the Q-diffusion model are positive. As argued by van der Maas et al. (2011), this restriction to positive values is conceptually attractive. At worst, someone's ability is zero (e.g., a toddler answering multiple-choice items on quantum mechanics). In the Q-diffusion model, this event would yield response accuracy at chance level. Thus, below-chance responding is excluded from the model by restricting parameters to positive values instead of introducing an additional guessing parameter, as in the 3PL model. Interestingly, this setup gives the Q-diffusion model ratio scale properties (for further discussion, see van der Maas et al., 2011). In Figure 17.3, we show several item characteristic curves (ICCs) for the Q-diffusion model.

Note that the version of the Q-diffusion model presented here is only applicable to two-choice problems because if α_{pi} or v_{pi} goes to zero, the probability of a correct response goes to 0.5. For a modification of the Q-diffusion model to deal with general multiple-choice items and open-ended questions, we refer to van der Maas et al. (2011).

17.2.3 Relations to Other Models and Old Ideas

For the following reduced version of the D-diffusion model, the response probabilities are the same as for the 2PL model: if all test takers have the same level of response caution (and thus operate at the same SAT level), then $\gamma_p = \gamma$ in Equation 17.4. As a consequence, the logit of $\Pr\{U_{pi} = 1; a_i, \gamma_p, \theta_p, b_i\}$ becomes $\gamma / a_i(\theta_p - b_i) = a_i^*(\theta_p - b_i)$, with a_i^* as the discrimination parameter.

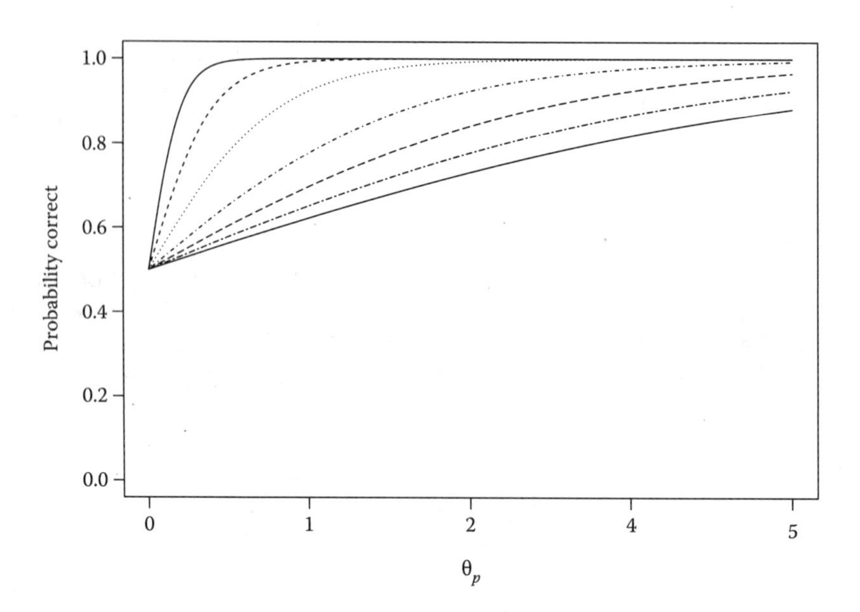

FIGURE 17.3
ICCs for the probability correct of the Q-diffusion model with two response options. In the figure, $\gamma_p/a_i = 1$ and the item component of drift rate, b_i, takes values 0.1, 0.2, 0.4, 0.8, 1.2, 1.6, and 2.0 (ICCs from left to right).

There exists a close relationship between the probability of a correct response in the Q-diffusion model and the quotient model (QM) test model in Ramsay (1989). In fact, if we write $\theta_p^* = \gamma_p\theta_p$ and $b_i^* = a_i b_i$, one obtains Ramsay's QM test model.

Without any reduction, neither of these two diffusion models give response probabilities exactly equivalent to the 2PL model. Suppose that responses are generated by the process underlying the D-diffusion model. Furthermore, assume that the positive person parameters γ_p are rewritten as mean-deviation scores γ_p^* and mean M_γ. Then, we may reformulate the logit of Equation 17.4 as

$$\begin{aligned} \text{logit}(\Pr\{U_{pi} = 1; a_i, \gamma_p, \theta_p, b_i\}) &= \frac{\gamma_p}{a_i}(\theta_p - b_i) = \frac{(\gamma_p^* + M_\gamma)}{a_i}(\theta_p - b_i) \\ &= -\frac{M_\gamma}{a_i}b_i + \frac{M_\gamma}{a_i}\theta_p - \frac{b_i}{a_i}\gamma_p^* + \frac{1}{a_i}\gamma_p^*\theta_p, \end{aligned} \tag{17.8}$$

which can be considered as a special case of a generalized latent variable (GLV) model with two interacting zero-centered latent variables (see Rizopoulos and Moustaki, 2008). This is an important result in that it shows that if D-diffusion is the data-generating process, then the response data measure two latent variables with an interaction between them. Only if it can be assumed that people operate at the same SAT level, then the 2PL model is an appropriate model for the responses. This analysis also shows that when a test follows a 2PL model but is administered under a reasonable amount of time pressure, person ability (which is the target of the measurement) is likely to be contaminated by response caution.

For the Q-diffusion model, a similar derivation holds. In this case, both θ_p and γ_p are positive latent traits that can be written as $\theta_p^* + M_\theta$ and $\gamma_p^* + M_\gamma$, respectively. After some algebra, we again find a GLV structure with two interacting latent variables:

$$\text{logit}(\Pr\{U_{pi}=1;a_i,\gamma_p,\theta_p,b_i\})=\frac{\gamma_p}{a_i}\frac{\theta_p}{b_i}=\frac{(\gamma_p^*+M_\gamma)}{a_i}\frac{(\theta_p^*+M_\theta)}{b_i}=\frac{M_\gamma M_\theta}{a_ib_i}+\frac{M_\gamma}{a_ib_i}\theta_p^*+\frac{M_\theta}{a_ib_i}\gamma_p^*+\frac{1}{a_ib_i}\gamma_p^*\theta_p^*.$$

Apart from the link to the 2PL model, we may consider relations to other models as well. An interesting joint model for responses and RTs is the hierarchical model by van der Linden (2007, 2009). The response model in the hierarchical model of van der Linden is the 2PL model and the relations with D- and Q-diffusion are already discussed. The RT model is a lognormal distribution (independent of the response model, given the person and item parameters) such that $(\log(T_{pi})) = \lambda_i(\xi_i - \tau_p)$ (for an introduction, see Volume One, Chapter 16). The parameter τ_p refers to the person's speed and ξ_i to the item's time intensity (or labor); λ_i is a discrimination parameter at the RT level. Van der Maas et al. (2011) show that for the Q-diffusion model (setting $t_{er} = 0$), it holds that $\xi_i \approx \log(b_i/a_i)$, $\lambda_i = 1$, and $\tau_p \approx \log(\theta_p/\gamma_p)$. For the D-diffusion model, there are no known equivalence relations with the RT part of van der Linden's model.

Both for the D- and Q-diffusion model, the concept of response caution has somewhat unexpected but important implications for another research area in IRT. In model-based approaches to person misfit, the logistic person response function plays a major role (e.g., Conijn et al., 2012). In the person response function, the discrimination parameter is person specific. Therefore, the D-diffusion model can be seen as a 2PL model with a person and item discrimination parameter (Strandmark and Linn, 1987). Likewise, the Q-diffusion model is a QM test model (Ramsay, 1989) with a person discrimination parameter. In our empirical example below, we will show how the D-diffusion model can be used to investigate person fit.

17.3 Parameter Estimation

Estimation in the D- and Q-diffusion is carried out through marginal maximum likelihood (Glas, Volume Two, Chapter 11). Before fitting the diffusion models, we recommend applying the 2PL model and the GLV models to the response data.

First, assuming local independence within a person (given the person parameters), the diffusion densities for the I items are multiplied. Next, this joint density is integrated over the person parameters (i.e., γ_p and θ_p), which are now considered to be random effects. Because, for convenience, we assume normally distributed random effects, the parameters that are strictly positive (e.g., γ_p for the D-diffusion model) are transformed to the real line and the integration is carried out with respect to the transformed parameter (e.g., $\tilde{\gamma}_p = \log(\gamma_p)$). Likewise, the strictly positive item parameters (e.g., a_i for the D-diffusion model) undergo a logarithmic transformation. Upon multiplying the P person contributions to the likelihood, we have the likelihood for the full D- or Q-diffusion model. Because some of the integrals over the random effects do not have an analytical solution, they are approximated by a numerical approximation (Laplace or Gauss–Hermite). Note that in the D-diffusion model, for a normal density, the integral over θ_p has a closed-form solution (for details, see Tuerlinckx, 2004). Thus, for this model only the integral over γ_p has to be approximated numerically. Next, a standard optimization routine is used to maximize the log-likelihood, with starting points are provided by the 2PL model parameters.

17.4 Model Fit

In order to evaluate the fit of the model, a mixed strategy should be pursued. First, we test qualitative predictions from the D- and Q-diffusion model. For the D-diffusion model, a prime candidate is the distance-difficulty hypothesis, which can be tested by fitting the response model to the binary response data (independently of the RT data) and checking the inverted U-shape of mean RTs as a function of the drift rate. Second, we compare the model with several competing models by using the traditional information criteria (Akaike information criterion [AIC] and Bayesian information criterion [BIC]). Third, we evaluate the absolute fit of the model to the response data using limited-information goodness-of-fit statistics (Maydeu-Olivares and Joe, 2005). Finally, both for the response accuracies and RTs, we inspect plots of the observed data against model predictions.

17.5 Empirical Example

The D-diffusion model was applied to a subset of the data from Ferrando and Lorenzo-Seva (2007), who administered two 11-item tests measuring extraversion and neuroticism (for more information on the items, see Ferrando and Lorenzo-Seva, 2007). Both sets of items were administered to a single block of 750 persons in an intermixed mode, but the data were analyzed separately. No speed instruction was given to the subjects. For reasons of computation time, a random sample of 400 subjects was analyzed here.

For each person, both the responses and RTs were recorded. The responses were coded as 1 (endorsing the item which is an indication of a higher level of extraversion or neuroticism) or 0 (not endorsing the item, indicative of a lower level of extraversion or neuroticism). There were no missing responses, but for the diffusion model analysis, we analyzed only the RTs between 1.1 and 15 s (about 97% of the data). The shorter RTs had a suspicious mode around 0.65 s and the RTs longer than 15 s were clearly outliers with a potentially excessive impact on the data. The results reported from our analyses are for extraversion; however, the results for neuroticism were generally the same.

A set of item response models was fitted to the response data to the extraversion items. The relative fit statistics are given in Table 17.1. AIC and BIC do not give a similar conclusion: AIC supports a nonadditive GLV model (model 4, i.e., a two-dimensional IRT model with an interaction between the latent traits), while BIC supports the 2PL model. Nevertheless, we concluded that there was evidence for two interacting latent traits.

TABLE 17.1

AIC and BIC Values for Four Item Response Models Fitted to Binary Responses to Extraversion Items

	Model	AIC	BIC
1	Rasch model	4898	4946
2	2PL model	4846	4934
3	Additive GLV model	4816	4948
4	Nonadditive GLV model	4805	4981

Note: These models were fitted to the complete matrix of item responses.

This finding is consistent with the D-diffusion model shown in Equation 17.8. The same conclusion held for neuroticism.

From the GLV model with two interacting traits in Equation 17.8, an approximate estimate of β_i was derived. Hence, we were able to link the absolute difference between the person and item parameter (i.e., $|\hat{\theta}_p - \hat{\beta}_i|$) to T_{pi}, which is the observed RT for person p and item i. As shown in Figure 17.4, for every item, T_{pi} was regressed on $|\hat{\theta}_p - \hat{\beta}_i|$ using a nonparametric regression procedure. As can be seen from Figure 17.4, the data supported the distance-difficulty hypothesis. Note that this test was quite strong, since the nonadditive GLV model estimates were not influenced by the RTs.

Fitting a series of diffusion models to the response data and the RTs resulted in the fit statistics reported in Table 17.2 (where we have considered also models with an item specific t_{er_i}). Both AIC and BIC pointed to the full D-diffusion model as the best fitting model. As for the parameter estimates, the standard deviation of the logarithm of the

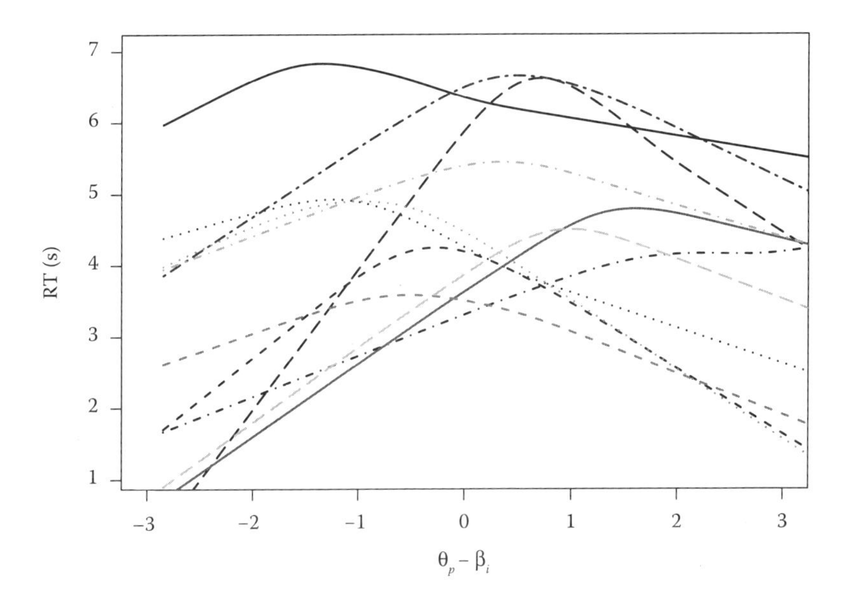

FIGURE 17.4
Graphical illustration of the distance-difficulty effect in the data. Each line corresponds to an item. See the main text for more explanation.

TABLE 17.2
AIC and BIC Values for Four D-Diffusion Models Fitted Both to Binary Response Data and RTs on Extraversion Items

	Model	AIC	BIC
1	D-diffusion (full model: a_i, a_p, t_{eri})	20,747	20,887
2	D-diffusion (a_i, a_p, $T_{eri} = t_{er}$)	21,338	21,438
3	D-diffusion (a_i, $a_p = a$, t_{eri})	21,407	21,629
4	D-diffusion ($a_i = a'$, a_p, t_{eri})	21,095	21,195
5	D-diffusion ($a_p = a$, $a_i = a'$, t_{eri})	21,664	21,823
6	D-diffusion ($a_p = a$, $a_i = a'$, $t_{eri} = t_{er}$)	22,170	22,259

Note: Only the data with RTs between 1.1 and 15 s were selected.

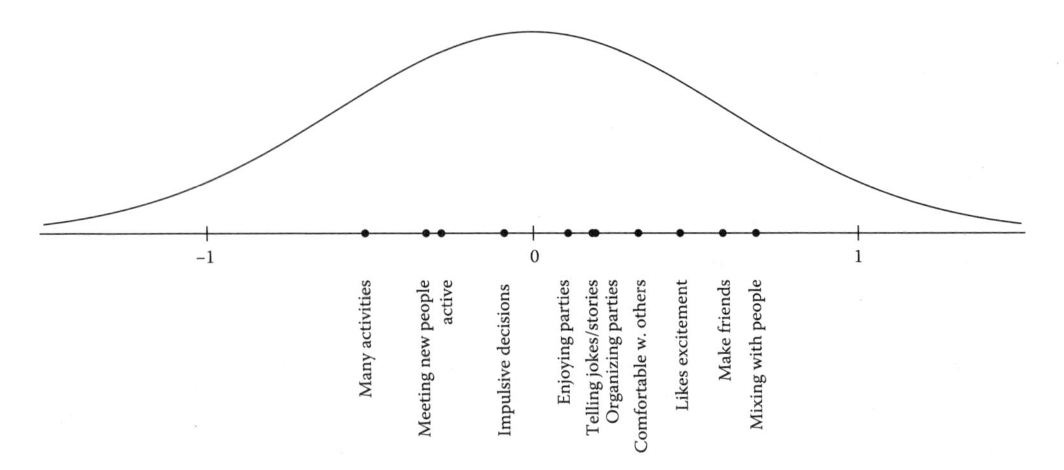

FIGURE 17.5
Graphical display of the drift components β_i for the 11 extraversion items and the estimated population distribution for θ_p, the person component to the drift.

response caution parameter (i.e., $\tilde{\gamma}_p = \log(\gamma_p)$) was estimated to be 0.27 ($SE = 0.01$). The estimated standard deviation of the person components of the drift rate (i.e., θ_p) was 0.60 ($SE = 0.04$). The estimated item drift rates ranged from −0.51 ($SE = 0.05$) ("Do you often take on more activities than you have time for?") to 0.69 ($SE = 0.06$) ("Do you like mixing with people?"). A graphical representation of the item drift components and the estimated population distribution of person drift components can be found in Figure 17.5.

The results for the neuroticism data (not shown here) were very similar. Figure 17.6 shows the relation between the person estimates for extraversion and neuroticism (both for θ_p and γ_p). The weak correlation between the θ_p's for the same persons was to be expected

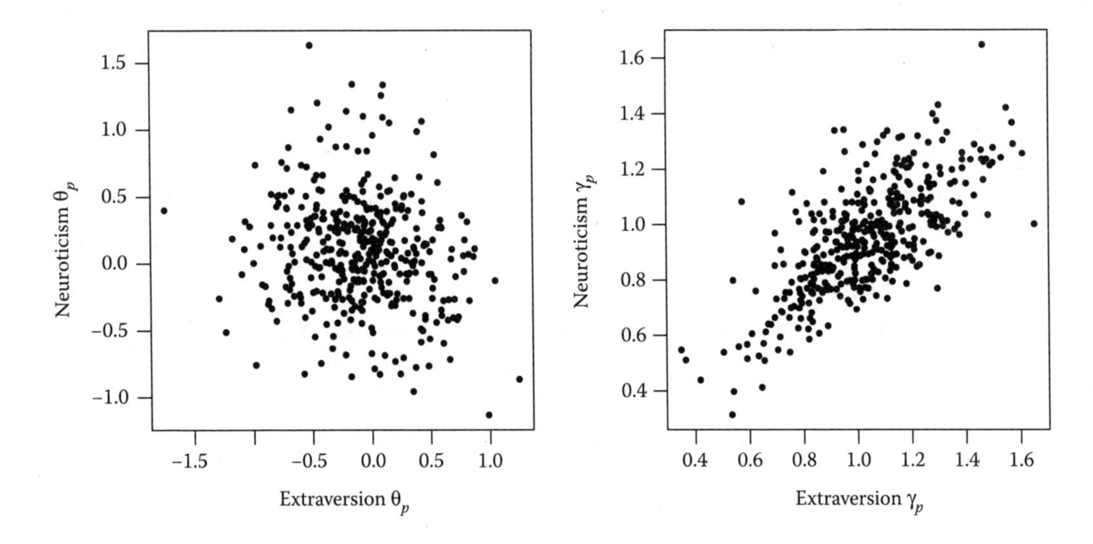

FIGURE 17.6
Relation between the person estimates for extraversion and neuroticism. In the left panel, the relation between the drift rate components are shown (the correlation is −0.16) and in the right panel the response cautions (the correlation is 0.71).

because the two scales were constructed to be independent. On the other hand, there was a strong positive correlation between the γ_p's from the same persons, which is consistent with the interpretation of γ_p as a stable response caution parameter.

As explained earlier, one of the major strengths of the diffusion model is the natural way in which it accounts for the SAT. Figure 17.7 shows the model-implied SAT functions for 40 randomly selected persons (in gray), given a standard item with $\beta_i = 0$, $a_i = 1$, and $t_{eri} = 1.2$. As can be seen, the SAT functions appeared to fan out from the guessing point (t_{eri}, 0.5) to the right. As the mean RT further increased, the SAT functions stabilized. From the left to the right, the SAT function increased ($\theta_p > 0$), remained constant ($\theta_p = 0$), or decreased ($\theta_p < 0$). This diverse behavior is specific for personality and attitude tests: given more time to deliberate, a person will choose a response consistent with the sign of $\theta_p - \beta_i$ (note that such behavior is not to be expected with ability tests). The black dots on the gray lines indicate the estimated γ_p of the persons. Figure 17.7 shows a curvilinear relation between θ_p and γ_p.

In addition to the qualitative tests (Figures 17.4 and 17.6), we also performed a limited-information goodness-of-fit test based on the two-way associations in the accuracy data (Maydeu-Olivares and Joe, 2005). Specifically, we compared all observed two-way tables with the predicted tables using the M_2 statistic, which is approximately χ^2-distributed. For the extraversion data, the test was significant ($M_2 = 189.57$, $df = 42$, $p < 0.001$), indicating some misfit. However, as can be seen in Figures 17.8 and 17.9, a graphical model check for the response data and the RTs did not indicate any large misfit. Figure 17.8 shows the expected marginal ICC according to the D-diffusion model (with γ_p integrated out) with the observed proportion "yes" within 10 subgroups relatively homogenous on θ_p. As can be seen from this figure, the fit of the model was acceptable for all

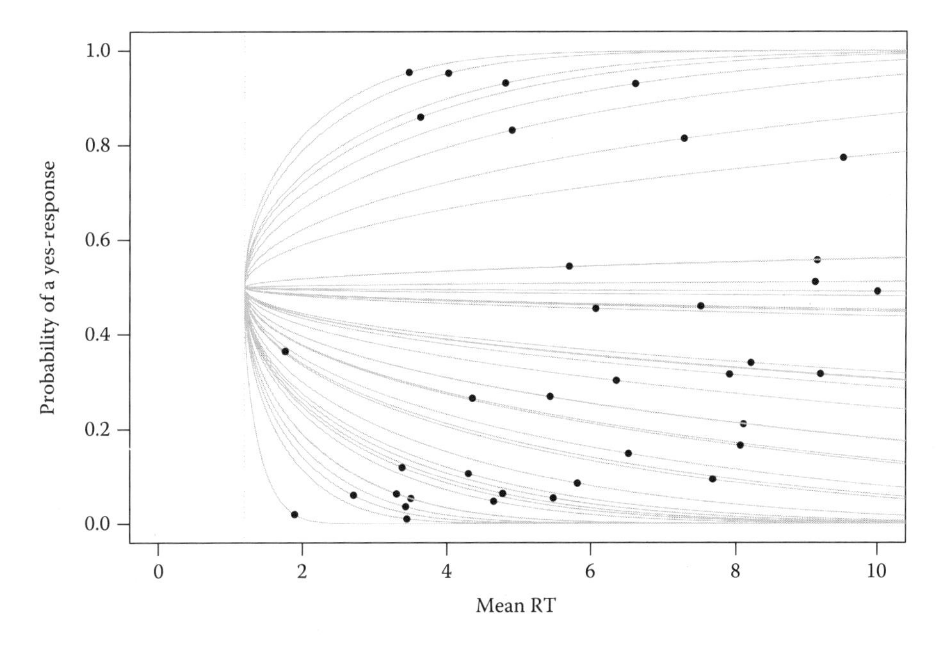

FIGURE 17.7
Graphical presentation of the model-implied SAT functions for 40 randomly selected individuals (the solid gray lines) for a standard item with $\beta_i = 0$ and $a_i = e^{-1.5}$. The lines trace out the relation between the probability of a yes response (Equation 17.4) and the mean RT (Equation 17.5) for a varying α_p. The black dot indicates the actual SAT point. The minimum RT is 1.2 s, and is indicated by the dotted vertical line.

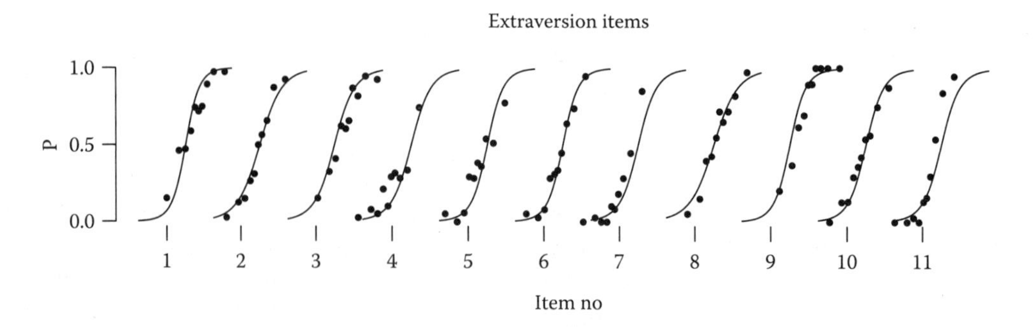

FIGURE 17.8
Comparison of the predicted and observed choice probabilities for the 11 extraversion items. The solid lines show the probability of choosing response "yes" as a function of θ_p (for each item, the relevant part of the θ_p scale is chosen). The dots are the observed choice probabilities for 10 binned categories of θ_p. The person specific γ_p is integrated out.

items but items 7 and 11. In Figure 17.9, the fit of the RTs is visualized by means of the deciles of a QQ-plot of the observed and expected RTs. Generally, the fit of all items was acceptable. However, for large RTs (right tail), some misfit was apparent. We return to this point in the discussion below. Overall, we concluded that the D-diffusion model fitted the data acceptably.

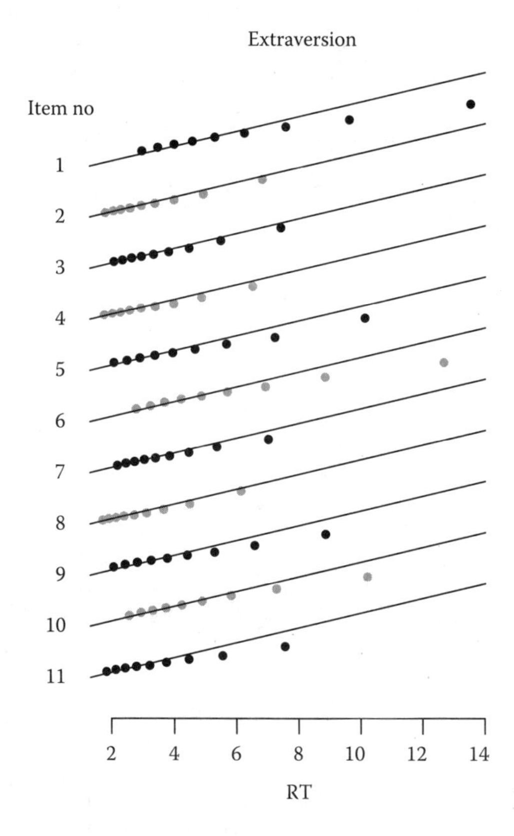

FIGURE 17.9
QQ plots (using deciles) for the 11 items based on the marginal RT distribution (i.e., θ_p and γ_p are integrated).

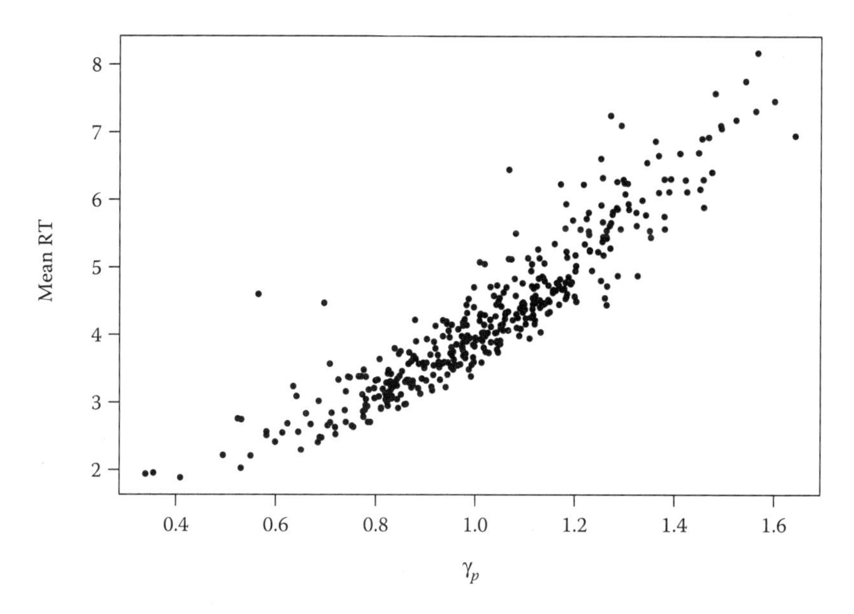

FIGURE 17.10
Relation between the estimates of γ_p, the person-specific response caution, and the mean RT for person p.

In Figure 17.10, the relationship between the estimates of the person-specific response caution parameters γ_p and the mean RTs for the persons are depicted. The figure shows a strong, almost linear relation between them. Because γ_p is considered as the person discrimination parameter in the response model, and therefore indicative of person misfit, Figure 17.10 illustrates how one can measure and subsequently use RTs to investigate person fit. Persons that are overly cautious have a relatively large γ_p and therefore a large mean RT (they ponder longer over their answers). The response patterns of such respondents tend to follow a Guttman scale. On the other hand, people that are careless responders (small γ_p) have a low mean RT (i.e., they arrive at their response quickly). Their response patterns are essentially random (i.e., not much influenced by the underlying latent variable θ_p).

17.6 Discussion

In this chapter, we linked a powerful model from cognitive psychology—the diffusion model—to one of the main item response models. Two variants were discussed, the D- and Q-diffusion models. For both models, the SAT is governed by the person's response caution and the time pressure imposed on the item. In the D-diffusion model, the speed of information processing (i.e., drift rate) for a person-item combination is an additive function of a person and an item part. We expect this model to be especially suited for the analysis of personality and attitude items. An attractive feature of the D-diffusion model is its inherent inverted U-shape relation between the person-item combination drift rate and the mean RT: if a person is close to the item, a larger RT is expected. The Q-diffusion model seems better suited for ability tests. In this model, both the person's ability and the difficulty of the item are assumed to be positive.

Our application of the D-diffusion model to the data set for the extraversion items yielded a generally acceptable fit, excepts for some misfit at the right tails of the RT distributions. There are many possible reasons for this misfit, however, the fact that no speed instruction was given seems to be the most important one. As a result of the lack of instruction, the respondents may have engaged in other activities, giving rise to longer RTs than predicted by the diffusion model.

In addition to the potential fit of the diffusion model to observed accuracies and RT distributions for a limited number of parameters, it has a number of promising, more substantive features. First, the model gives a natural interpretation of the item discrimination parameter. This parameter has always been somewhat enigmatic: it quantifies how well the latent variable is measured by the item, but also measures the nonlinear correlation between the item and the latent variable. However, these interpretations are tautological. In the diffusion model, on the other hand, the item discrimination is directly linked to the time pressure for a given item. Second, the setting of the response caution is a matter of strategy. Each test taker wants to attain a certain goal (e.g., getting the highest possible score within the time limit), and the response caution can be adjusted to reach an optimal level of performance given the goal. An optimality analysis similar to the one carried out by Bogacz et al. (2006) could reveal the setting of the response caution in a test situation.

A possible criticism of the application of diffusion models to response data on psychological tests is that their tasks are generally more complex than perceptual decisions used in experimental psychology and the recorded RTs tend to be much longer. Typically, the diffusion model is applied to much simpler tasks with single decision processes. For the personality items discussed in this chapter, it is not unreasonable to assume that this holds. However, for many other test items or questions, the response process is without doubt more complicated. But in such cases, the diffusion process may serve as a reasonable first approximation.

As illustrated, the D-diffusion model (but the Q-diffusion model as well) opens an interesting perspective on person fit. Viewing person misfit from this perspective has two main advantages. First, the person discrimination parameter obtains a psychological interpretation in terms of response caution—a step from description to explanation with practical implications. According to the D-diffusion model, low person discrimination leads to fast responses, because low response caution implies low boundary separation. The other way around, for subjects with very high person discrimination parameters, we expect slow responses. This point brings us to the second advantage. Fitting the 2PL model with item and person discrimination parameters meets all kinds of statistical problems (Conijn et al., 2012). But because of the additional information from the RTs, we are able estimate both the item and person discrimination parameters in the D- or Q-diffusion model.

Of course, our approach is model based and consequently, these advantages only apply when the underlying model is correct. Besides variation in response caution, other factors may very well cause person misfit, for instance, when a multidimensional instead of a scalar parameter θ_p drives the decision process. Applying the unidimensional drift rate D-diffusion model might not work in such a case.

Notwithstanding the potential of the model, there are still a few issues that need to be addressed regarding diffusion IRT models. For example, a very important issue is the application of the Q-diffusion model to regular multiple-choice and open-ended questions that needs to be further investigated, building upon the partial solution put forward by van der Maas et al. (2011).

References

Bogacz, R., Brown, E., Moehlis, J., Holmes, P., and Cohen, J. D. 2006. The physics of optimal decision making: A formal analysis of models of performance in two-alternative forced choice tasks. *Psychological Review*, 113, 700–765.

Borsboom, D., Mellenbergh, G. J., and van Heerden, J. 2003. The theoretical status of latent variables. *Psychological Review*, 110, 203–219.

Borsboom, D., Mellenbergh, G. J., and van Heerden, J. 2004. The concept of validity. *Psychological Review*, 111, 1061–1071.

Conijn, J. M., Emons, W. H. M., van Assen, M. A. L. M., and Sijtsma, K. 2011. On the usefulness of a multilevel logistic regression approach to person-fit analysis. *Multivariate Behavioral Research*, 46, 365–388.

Cox, D. R., and Miller, H. D. 1965. *The Theory of Stochastic Processes*. London: Methuen.

Donders, F. C. 1869/1969. On the speed of mental processes. *Acta Psychologica*, 30, 412–431.

Fekken, G. C., and Holden, R. R. 1992. Response latency evidence for viewing personality traits as schema indicators. *Journal of Research in Personality*, 26, 103–120.

Ferrando, P. J., and Lorenzo-Seva, U. 2007. An item response theory model for incorporating response time data in binary personality items. *Applied Psychological Measurement*, 31, 525–543.

Gaviria, J.-L. 2005. Increase in precision when estimating parameters in computer assisted testing using response time. *Quality and Quantity*, 39, 45–69.

Goldstein, H., and Wood, R. 1989. Five decades of item response modelling. *British Journal of Mathematical and Statistical Psychology*, 42, 139–167.

Hanley, C. 1963. The "difficulty" of a personality inventory item. *Educational and Psychological Measurement*, 22, 577–584.

Klein Entink, R. H., Kuhn, J. T., Hornke, L. F., and Fox, J.-P. 2009. Evaluating cognitive theory: A joint modeling approach using responses and response times. *Psychological Methods*, 14, 54–75.

Kuiper, N. A. 1981. Convergent evidence for the self as a prototype: The "inverted-U RT effect" for self and other judgments. *Personality and Social Psychology Bulletin*, 7, 438–443.

Luce, R. D. 1986. *Response Times*. New York: Oxford University Press.

Maris, G., and van der Maas, H. L. J. 2012. Speed-accuracy response models: Scoring rules based on response time and accuracy. *Psychometrika*, 77, 615–633.

Maydeu-Olivares, A., and Joe, H. 2005. Limited and full information estimation and testing in $2n$ contingency tables: A unified framework. *Journal of the American Statistical Association*, 100, 1009–1020.

Ramsay, J. O. 1989. A comparison of three simple test theory models. *Psychometrika*, 54, 487–499.

Ranger, J., and Kuhn, J.-G. 2012. A flexible latent trait model for response times in tests. *Psychometrika*, 77, 31–47.

Ratcliff, R. 1978. A theory of memory retrieval. *Psychological Review*, 85, 59–108.

Ratcliff, R., and Rouder, J. 1998. Modeling response times for two-choice decisions. *Psychological Science*, 9, 347–356.

Rizopoulos, D., and Moustaki, I. 2008. Generalized latent variable models with non-linear effects. *British Journal of Mathematical and Statistical Psychology*, 61, 415–438.

Scheiblechner, H. 1979. Specific objective stochastic latency mechanisms. *Journal of Mathematical Psychology*, 19, 18–38.

Strandmark, N. L., and Linn, R. L. 1987. A generalized logistic item response model parameterizing test score inappropriateness. *Applied Psychological Measurement*, 11, 355–370.

Thurstone, L. L. 1937. Ability, motivation, and speed. *Psychometrika*, 2, 249–254.

Tuerlinckx, F. 2004. The efficient computation of the distribution function of the diffusion process. *Behavior Research Methods, Instruments, and Computers*, 36, 702–716.

Tuerlinckx, F., and De Boeck, P. 2005. Two interpretations of the discrimination parameter. *Psychometrika*, 70, 629–650.

van Breukelen, G. J. P. 2005. Psychometric modeling of response speed and accuracy with mixed and conditional regression. *Psychometrika*, 70, 359–376.
van der Linden, W. J. 2007. A hierarchical framework for modeling speed and accuracy on test items. *Psychometrika*, 72, 287–308.
van der Linden, W. J. 2009. Conceptual issues in response-time modeling. *Journal of Educational Measurement*, 46, 247–272.
van der Linden, W. J. 2011. Modeling response times with latent variables: Principles and applications. *Psychological Test and Assessment Modeling*, 53, 334–358.
van der Linden, W. J. and Hambleton, R. K. 1997. *Handbook of Modern Item Response Theory*. New York: Springer.
van der Maas, H. L. J., Molenaar, D., Maris, G., Kievit, R. A., and Borsboom, D. 2011. Cognitive psychology meets psychometric theory: On the relation between process models for decision making and latent variable models for individual differences. *Psychological Review*, 118, 339–356.
Verhelst, N. D., Verstralen, H. H., and Jansen, M. G. H. 1997. A logistic model for time-limit tests. In van der Linden, W. J. and Hambleton, R. K. (Eds), *Handbook of Modern Item Response Theory* (pp. 169–185). New York: Springer.
Vorberg, D., and Schwarz, W. 1990. Rasch-representable reaction time distributions. *Psychometrika*, 55, 617–632.
Wagenmakers, E.-J. 2009. Methodological and empirical developments for the Ratcliff diffusion model of response times and accuracy. *European Journal of Cognitive Psychology*, 21, 641–671.

Section V

Nonparametric Models

18

Mokken Models

Klaas Sijtsma and Ivo W. Molenaar*

CONTENTS

18.1 Introduction 303
18.2 Presentation of the Models 304
18.3 Parameter Estimation 308
18.4 Model Fit 310
 18.4.1 Conditional Association and Local Independence 310
 18.4.2 Manifest Monotonicity and Latent Monotonicity 311
 18.4.3 Manifest and Latent Invariant Item Ordering 312
 18.4.4 Item Selection 313
 18.4.5 Person Fit and Reliability 314
18.5 Empirical Example 315
18.6 Discussion 319
References 319

18.1 Introduction

Mokken (1971) titled his book *A Theory and Procedure of Scale Analysis.* He proposed two models that were based on assumptions that nowadays locate them in the family of item response theory (IRT) models. The novelty of Mokken's approach was that item response functions were not parametrically defined but were subjected only to order restrictions. For example, each item response function is assumed to be nondecreasing in the latent variable. The use of only order restrictions led later authors to speak of nonparametric IRT models (Sijtsma and Molenaar, 2002). The Mokken models are less restrictive than several of the alternative IRT models based on parametric response functions, such as the one-, two-, and three-parameter logistic models and their normal-ogive analogs.

Mokken (1971, p. 115) noticed that the use of parametric item response functions requires a profound knowledge of the items and the examinees. He assumed that knowledge at this level can only be obtained when the researcher has access to a large set of items and large numbers of examinees and is in a position to set up a continuous research program that enables the accumulation of knowledge relevant to the measurement of the attribute of interest. Mokken (1971, p. 116) also assumed that this would be possible for ability and achievement testing but less so for measurement of attributes in the other areas of social research. In these areas, such knowledge has not yet accumulated to a sufficient degree.

* The authors are grateful to the researchers from the CoRPS research center, Tilburg University, for allowing them to analyze the data used in this chapter.

Moreover, measurement of attributes such as the attitude toward religion (sociology) and elections (political science), is often confined to a specific object of investigation rather than a broad ability represented by a large item domain. Using larger numbers of items with the same persons may also be limited by respondent fatigue and allowed questionnaire length. We notice that today, the same practical limitations apply to medical and health research, in which the use of long inventories for anxiety and depression measurement is prohibited by patients' physical condition and restrictions imposed by medical–ethical committees.

On the basis of these considerations, Mokken motivated the use of order restrictions as consistent with the relatively low knowledge level in many social science areas and with the practical conditions on measurement existing in them. To keep the connection with parametric IRT models that were believed to be representative of higher knowledge levels, Mokken proposed to define nonparametric models, of which the current parametric models were mathematical special cases. In addition, he proposed to derive properties from his models that were distribution-free; that is, true irrespective of the population distribution of the latent variable across persons.

This was 1971 but how are things today? IRT has invaded almost all areas in which multi-item tests and questionnaires are used for attribute measurement, probably most pervasively in education and psychology. When Mokken wrote about social research, he referred to sociology and political science, but since then, attribute measurement and IRT use have also become popular in marketing, medical and health research, and policy and opinion research. In all these areas, it is common to restrain instruments to a limited number of items, either due to time limits and other constraints or a limited theoretical basis for obtaining large numbers of items, while sample sizes vary greatly. The same holds for measurement of personality traits, common in many areas of psychology. Sometimes, personality inventories are large—for example, the NEO-PPI (Costa and McCrae, 1992) for measuring the big five contains 240 items—but regularly practical or theoretical limitations lead to the use of short inventories.

Parametric IRT models have proven to be useful, in particular, in educational measurement. Here, testing tends to be on a large scale (involving large sets of items and many examinees), is repeated periodically, and often at high stakes (has important consequences for the individual). Consequently, the parameters for different sets of items used for measuring the same knowledge domain have to be linked (van der Linden and Barrett, Volume Three, Chapter 2) and this is where parametric IRT models prove to be particularly useful. Further, parametric IRT models are used in all areas in which nonparametric IRT models are also used; that is, when a test or questionnaire has only one version and measurement is limited to one administration. One reason for using a nonparametric IRT model could be the often substantively weak theoretical basis of measurement, so that the use of a measurement model based on less-restrictive assumptions would be more realistic. Another reason could be the increased availability of adequate goodness-of-fit methods developed by researchers working in the Mokken tradition (e.g., Van Schuur, 2011). Ramsay (1991; also, Volume One, Chapter 20), Stout (2002), and Karabatsos (Volume One, Chapter 19) propose nonparametric IRT approaches related to those in this chapter.

18.2 Presentation of the Models

A set of I items indexed as $i = 1, \ldots, I$ is administered to P persons indexed as $p = 1, \ldots, P$. The final test or questionnaire contains n items. This means that not all I items in the initial item set need to be included in the final measurement instrument. Item scores are ordered, integer valued, as with rating-scale scoring, and they are denoted by $a = 0, \ldots, A$. For dichotomously scored items, for example, incorrect–correct scoring, $a = 0, 1$. The random variable for the score of person p on item i is denoted as $U_{pi} = a$.

Mokken (1971, Chapter 4) proposed two nonparametric IRT models for dichotomous items, the monotone homogeneity model and the double-monotonicity model. The latter is a special case of the former. The models have three assumptions in common. Let θ denote the latent variable common in IRT models. The first assumption is local independence, which says that the multivariate conditional distribution of n items is the product of the n univariate conditional distributions. Let Pr denote a probability and let $\Pr(U_i = 1|\theta)$ denote the item response function. Further, let $\mathbf{U} = (U_1, \ldots, U_n)$ with realization $\mathbf{u} = (a_1, \ldots, a_n)$; then

$$\Pr(\mathbf{U} = \mathbf{u} \mid \theta) = \prod_{i=1}^{n} \Pr(U_i = 1 \mid \theta)^{a_i} [1 - \Pr(U_i = 1 \mid \theta)]^{(1-a_i)}.$$

A consequence of local independence is that all conditional covariances between items are zero; that is, for any pair of items, i and j, local independence implies that

$$Cov(U_i, U_j \mid \theta) = 0.$$

The second assumption of the monotone homogeneity model is unidimensionality; that is, only one latent variable suffices to obtain local independence. In other IRT models with multiple latent variables (Volume One, Chapters 11 through 14), we would have to condition on the whole set of latent variables to obtain local independence.

The third assumption is monotonicity; for any pair of values, $\theta_p < \theta_q$, monotonicity means that

$$\Pr(U_i = 1 \mid \theta_p) \le \Pr(U_i = 1 \mid \theta_q), \quad \text{for all } i = 1, \ldots, n.$$

Thus, $\Pr(U_i = 1|\theta)$ is nondecreasing in the latent variable θ.

The three assumptions of local independence, unidimensionality, and monotonicity together constitute the monotone homogeneity model. Figure 18.1a shows four response functions that are monotone and depend on the latent variable θ, and can have any shape provided they do not decrease anywhere along the scale of θ. The monotone homogeneity model encompasses any IRT model that is locally independent, unidimensional, and has monotone, parametric item response functions. Examples are the one-, two-, and three-parameter logistic models and their normal-ogive counterparts (for an introduction, see Volume One, Chapter 2). A difference with these specialized parametric models is that the likelihood of the monotone homogeneity model does not enable the numerical estimation of the latent variable θ for persons. Since the model is defined only by order restrictions, it does not contain any latent item parameters; hence, such parameters cannot be estimated either.

The monotone homogeneity model is a practically useful IRT model because the three assumptions imply the stochastic ordering of the latent variable θ by the total score, $X = \Sigma_{i=1}^{n} U_i$. That is, for any pair of values, $0 \le X_p < X_q \le n$, and for any value t of θ, stochastic ordering means that

$$\Pr(\theta > t \mid X_p) \le \Pr(\theta > t \mid X_q).$$

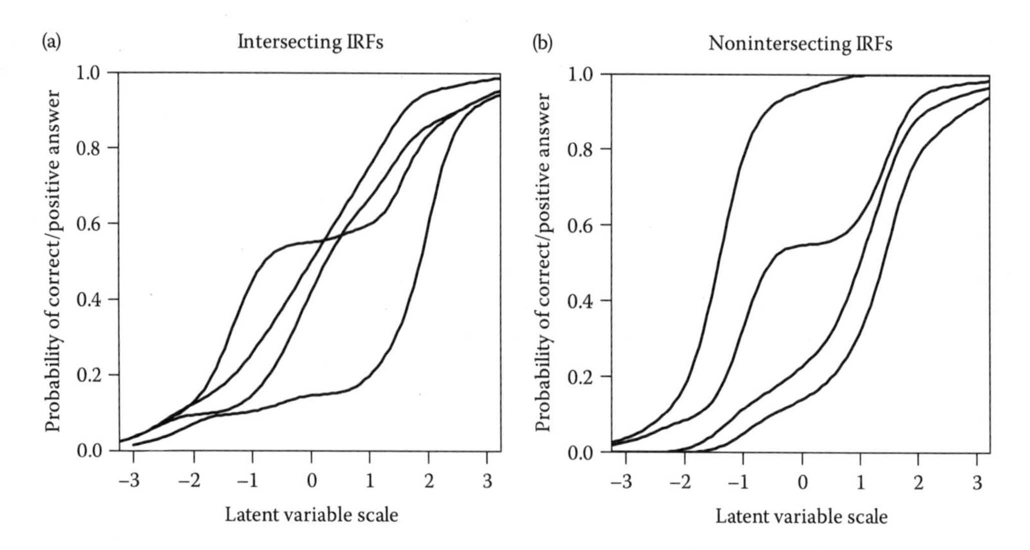

FIGURE 18.1
Examples of monotone, intersecting item response functions ((a) monotone homogeneity model) and monotone item response functions that are nonintersecting ((b) double-monotonicity model).

The stochastic ordering property implies that

$$E(\theta \mid X_p) \leq E(\theta \mid X_q).$$

This inequality is easy to understand: Compared to the subgroup with the lower total score X_p, in the group with the higher total score X_q, the expected score on the latent variable is at least as high. Hence, even though the monotone homogeneity model does not enable the numerical estimation of any latent variable, it implies an ordering on such a variable by means of the total score. Thus, if the monotone homogeneity model fits the data, persons can be ordered by means of their total scores and this ordering estimates their ordering on an assumed common latent variable, θ. This constitutes an ordinal scale for person measurement.

The double-monotonicity model has one additional assumption relative to the monotone homogeneity model, in that it assumes that item response functions do not intersect. For any pair of items, *i* and *j*, if we know that there is a value θ_0 for which $\Pr(U_i = 1 \mid \theta_0) < \Pr(U_j = 1 \mid \theta_0)$, then, nonintersection means that for all other θ values

$$\Pr(U_i = 1 \mid \theta) \leq \Pr(U_j = 1 \mid \theta). \tag{18.1}$$

Sijtsma and Junker (1996) defined that *n* response functions have an invariant item ordering if they do not intersect, with the possibility of ties for certain θ values or in certain θ intervals. Figure 18.1b shows four functions that do not intersect. The one-parameter logistic model, also known as the Rasch model, is a special case of the double-monotonicity model.

If a set of items have an invariant item ordering, the same item is the easiest or the most popular for each person taking the test, another item is the second easiest or the second most popular, and so on, until the last and most difficult or least-popular item, which is also the same item for every person. Figure 18.1b shows that for all θ values, the items have the same ordering.

An invariant item ordering is important in several situations. In sociological measurement and personality measurement, items often represent different intensity levels with respect to the underlying attribute, and it is assumed that the ordering of the items is the same for all persons tested. For example, in testing religiosity, a statement like "I attend mass on Christmas Eve" can be assumed to represent a lower level than "I pray before every meal," and in testing the personality trait of introversion, one may assume that the statement "I like to be alone now and then" represents a lower level of introversion than "I feel uneasy in the company of other people." In intelligence testing, the items are often presented to children in the order from the easiest to the most difficult, not only to avoid discouraging of children by starting with difficult items but also because starting and stopping rules dependent on age group are often based on item ordering by increasing difficulty level. Each of the examples assumes an invariant item ordering. Whether this is a reasonable assumption must be investigated in the data. Even though several measurement applications assume an invariant item ordering, test constructors are often unaware of it and thus do not investigate whether it is justified to assume this ordering.

Molenaar (1997) generalized the two Mokken models to the case of items with more than two ordered scores, for example, rating-scale items. Such items are called polytomous items. With polytomous items, item category response functions are defined by $\Pr(U_i \geq a \mid \theta)$, with $\Pr(U_i \geq 0 \mid \theta) = 1$. For the monotone homogeneity model, for each item, its A category functions are assumed to be monotone nondecreasing, whereas the functions of different items can intersect, and in practice often do so. The double-monotonicity model in addition assumes that category response functions of different items do not intersect. The Mokken models for polytomous items do not imply the stochastic ordering of the latent variable θ by the total score X (Hemker et al., 1997), but, based on extensive simulation studies, Van der Ark (2005) showed that stochastic ordering of the latent variable θ by the total score X holds approximately for the monotone homogeneity model for polytomous items. In addition, Van der Ark and Bergsma (2010) proved that the monotone homogeneity model for polytomous items implies a weak stochastic ordering; that is, for any fixed value x of the total score X, they proved that

$$\Pr(\theta > t \mid X < x) \leq \Pr(\theta > t \mid X \geq x), \quad \text{for all } t \text{ and } 0 < x \leq nA.$$

Weak stochastic ordering says that a subgroup with total scores in excess of a cut score x_c has a higher mean value of θ than any subgroup scoring below the cut score.

Owing to space limitations, this chapter focuses on the case of dichotomous items. For polytomous-item tests, assessing the fit of the monotone homogeneity model to real data by investigating the monotonicity of nA category response functions typically provides many detailed results showing multiple nonmonotonicities. The consequences of their presence for person ordering are unclear. The problem of multiple detailed results and sampling violations as well as systematic but small violations also occur in the assessment of nonintersection of the category response functions of different items in the context of the double-monotonicity model. Moreover, this model does not imply an invariant item ordering at the level of polytomous items (Sijtsma and Hemker, 1998). An alternative, valid definition is the following: Given it holds for θ_0 that $E(U_i|\theta_0) < E(U_j|\theta_0)$, then, for all other θ values, it holds that

$$E(U_i \mid \theta) \leq E(U_j \mid \theta).$$

Equation 18.1 is a special case of this definition for dichotomous items. Ligtvoet et al. (2010) proposed to assess model fit for polytomous items at the higher aggregation level of $E(U_i|\theta)$, which summarizes information about monotonicity and invariant item ordering at the item level rather than their categories. They considered this to be a useful approach because the items, rather than the constituent parts described by their category response functions are the units that test constructors and practitioners work with. We return to this approach with the empirical example in Section 18.5.

18.3 Parameter Estimation

As already noted, since Mokken models do not have parametric item response functions but define order restrictions on item response functions, the likelihood of the models is a product of not fully specified response probabilities from which no latent person parameters and latent item location and item slope parameters typical of parametric IRT models can be solved. Alternatively, the unknown person parameter in the two Mokken models can be taken to be the expected total score, $E(X_p)$, defined across hypothetical, independent replications of the same test administered to the same person. This definition of an expected total score is the same as in classical test theory. As the total score depends on the items used in the test and on the test length, n, the precision with which the expected total score is estimated depends on the particular items used and the number of items used.

Item difficulty (in the context of ability measurement) or item popularity (in the context of trait and attitude measurement) is defined as the proportion of correct (ability) or positive (trait, attitude) responses in the population of interest; that is,

$$\pi_i = \int \Pr(U_i = 1 \mid \theta) dG(\theta), \quad \text{for all } i,$$

where $G(\theta)$ denotes the cumulative distribution function of the latent variable θ in this population. Proportion π_i is estimated by the fraction of 1 scores in the sample, and is denoted as p_i. It may be noted that, because in the monotone homogeneity model the response functions intersect, the π values are only informative about the ordering of the items at the population level but not their conditional ordering given fixed θ values. In the double-monotonicity model, if for some θ_0 the inequality $\Pr(U_i = 1|\theta_0) < \Pr(U_j = 1|\theta_0)$ holds strictly but for all other θ values $\Pr(U_i = 1|\theta) - \Pr(U_j = 1|\theta) \leq 0$ holds, then it is true that

$$\pi_i - \pi_j = \int [\Pr(U_i = 1 \mid \theta) - \Pr(U_j = 1 \mid \theta)] dG(\theta) < 0,$$

because all inequalities between the brackets are either 0 or negative. Hence, once it has been established that the double-monotonicity model fits the data, the invariant ordering of items is reflected by $\pi_i < \pi_j$, which is estimated by the ordering of the sample fractions, p.

Finally, we introduce the scalability coefficients H_i for individual items and H for the set of n items. The scalability coefficients can be written in different ways, each allowing a different interpretation. One fruitful interpretation of the item coefficient H_i is that it is an index for the slope of a response function. Similarly, the overall coefficient H can be interpreted as a weighted mean slope for the n response functions in the test. Let $Cov(U_i, U_j) = Cov_{ij}$ denote the interitem covariance and $Cov(U_i, U_j)^{(\max)} = Cov_{ij}^{(\max)}$ the maximum interitem covariance possible given the π_i and π_j values; then, for items i and j, the overall scalability coefficient is defined as $H_{ij} = Cov_{ij}/Cov_{ij}^{(\max)}$.

Let $R_{(i)} = \sum_{j \neq i} U_i$ be the total score on $n-1$ items excluding item i, also called the rest score. For item i, the scalability coefficient is defined as

$$H_i = \frac{\sum_{j \neq i} Cov_{ij}}{\sum_{j \neq i} Cov_{ij}^{(\max)}} = \frac{Cov(U_i, R_{(i)})}{Cov(U_i, R_{(i)})^{(\max)}}.$$

Hence, item scalability coefficient H_i expresses the strength of the association between item i and the total score on the other $n-1$ items in the test. As any total score on a subset of items from the test for which the monotone homogeneity model holds is an estimator of the latent variable θ, coefficient H_i expresses the strength of the association between item i and estimator $R_{(i)}$ of the same latent variable. Given that response functions are monotone, a higher H_i value expresses a stronger association and a better discrimination between lower- and higher-scale values. Hence, such items should contribute better to a reliable person ordering by means of total score X.

Coefficient H is the weighted average of the n H_i coefficients. Let $\pi_{ij}^{(0)} = \pi_i(1-\pi_j)$ if $\pi_i \leq \pi_j$ and let $\pi_{ij}^{(0)} = (1-\pi_i)\pi_j$ if $\pi_i > \pi_j$; then, coefficient H can be written as

$$H = \frac{\sum_{i=1}^{n} \sum_{j \neq i} \pi_{ij}^{(0)} H_i}{\sum_{i=1}^{n} \sum_{j \neq i} \pi_{ij}^{(0)}}.$$

Tests having monotone response functions and a high H value have a high weighted mean H_i value and allow a reliable person ordering. It can be proven that, given the monotone homogeneity model, $0 \leq H_{ij}, H_i, H \leq 1$. On the basis of practical experience, the following rules of thumb are used:

$H < 0.3$ implies that the items are unscalable

$0.3 \leq H < 0.4$, the scale is weak

$0.4 \leq H < 0.5$, the scale is medium

$H \geq 0.5$, the scale is strong

Van der Ark et al. (2008) used marginal models to estimate standard errors for scalability coefficients that were estimated from a sample of persons, and Kuijpers et al. (2013) extended the results to polytomous items and long tests.

18.4 Model Fit

18.4.1 Conditional Association and Local Independence

Holland and Rosenbaum (1986) proved that, for a general class of nonparametric IRT models that includes the two Mokken models as special cases, conditional association holds. Let the set of n-item random variables in $\mathbf{U}$ be divided into two mutually exclusive subsets. The variables from the two subsets are collected in vectors $\mathbf{Y}$ and $\mathbf{Z}$, such that $\mathbf{U} = (\mathbf{Y}, \mathbf{Z})$. Further, f_1 and f_2 are two nondecreasing functions defined on the scores in $\mathbf{Y}$ and h is a function defined on those in $\mathbf{Z}$. The monotone homogeneity model implies conditional association

$$Cov\ [f_1(\mathbf{Y}), f_2(\mathbf{Y}) \mid h(\mathbf{Z}) = \mathbf{z}] \geq 0.$$

A simple case of conditional association is nonnegative interitem covariances; that is, for $f_1(\mathbf{Y}) = U_i$ and $f_2(\mathbf{Y}) = U_j$, and ignoring conditioning on $h(\mathbf{Z})$ (so that we consider the association in the whole population), conditional association implies that all interitem covariances are nonnegative (and so are their product–moment correlations, $\rho(U_i,U_j) = \rho_{ij}$); that is

$$Cov(U_i, U_j) \geq 0, \quad \text{for all } i,\ j,\ i \neq j \tag{18.2}$$

(Mokken, 1971, p. 120, lemma 1.1.1).

In fact, Mokken (1971, p. 184) used Equation 18.2 as part of his definition of a scale:

Definition

A set of n items constitute a scale at level c—henceforth, a Mokken scale—if

1. $\rho_{ij} > 0, \quad$ for all i, j, and $i \neq j$,

and, for some user-specified lower-bound value $0 < c < 1$,

2. $H_i \geq c, \quad$ for all i.

It can be shown that, for any set of items, because always $\min(H_i) \leq H \leq \max(H_i)$, then, if the definition applies, the second condition implies that $H \geq c$. The choice of c provides the researcher with the opportunity to define what he or she considers to be an acceptable lower bound for item discrimination reflected by H_i and reliable person ordering reflected by H.

Two different implications of conditional association are the following. First, for any item k let $h(\mathbf{Z}) = U_k$. Then, conditional association implies that

$$Cov(U_i,\ U_j \mid U_k = a_k) \geq 0, \quad \text{for all } i,\ j,\ i \neq j, \quad \text{and} \quad k \neq i,\ j. \tag{18.3}$$

More precision can be obtained by defining the rest score as the total score on $n-2$ items except for items i and j; that is, $h(\mathbf{Z}) = S_{(ij)} = \sum_{k \neq i,j} U_k$. It then holds that

$$Cov(U_i,\ U_j \mid S_{(ij)} = s) \geq 0, \quad \text{for all } i,\ j,\ i \neq j, \quad \text{and} \quad s = 0, \ldots, n-2. \tag{18.4}$$

These and other observable consequences of conditional association provide many possibilities for assessing the fit of the monotone homogeneity model to data. However, there are two obvious drawbacks. First, there are numerous possibilities to choose functions f_1, f_2, and h, and it is difficult to say when one has inspected enough of them to make additional choices superfluous. Second, most observable consequences imply a large number of covariances (in many cases: thousands) that must be checked. It is then difficult to decide on how many negative signs can be tolerated and how one can control sampling error before rejecting the monotone homogeneity model as the population model that generated the data.

Straat et al. (2016) proposed a procedure that uses Equations 18.2 through 18.4 to assess whether a set of I items complies with the assumption of local independence. The procedure identifies the items that are most frequently involved in local dependencies with others from the set. It may be noted that if only one item pair were mutually locally dependent, then, the removal of either of them implies local independence between all the remaining items. To maintain a relatively large, locally independent subset of the original item set, the procedure removes as few items as possible. Because this method still is in an experimental phase, it was not used in the empirical example in Section 18.5. Van Abswoude et al. (2004; also Stout, 2002) discussed and compared different cluster methods for assessing both local independence and unidimensionality of item sets in the context of nonparametric IRT.

18.4.2 Manifest Monotonicity and Latent Monotonicity

For the rest score $R_{(i)}$, Junker (1993) proved that locally independent, unidimensional, monotone IRT models imply the property of manifest monotonicity

$$\Pr(U_i = 1 \mid R_{(i)} = r) \text{ is nondecreasing,} \quad \text{for } r = 0, \ldots, n-1, \text{ and for all } i. \tag{18.5}$$

As the result may be unexpected, we remind that, when conditioning on the total score X, the monotone homogeneity model does not imply a nondecreasing sequence of probabilities; that is, $\Pr(U_i = 1|X = x)$ is not necessarily nondecreasing in the total score x (Junker and Sijtsma, 2000).

A simple nonparametric regression method to assess monotonicity is binning (Fox, 1997). The method requires the calculation of both the fractions from the total sample that have the rest score r, denoted as p_r, and the fractions that have the same rest score r together with a correct response on item i, denoted as p_{ri}. Then, by division, $p_{i|r} = p_{ri}/p_r$.

According to the property of manifest monotonicity defined in Equation 18.5, a sample result like $p_{i|r} > p_{i|(r+w)}$ (w is a positive integer) represents a violation of monotonicity, which can be tested for significance using an approximation to the hypergeometric distribution (Molenaar and Sijtsma, 2000, p. 72). Small violations, for example, $0 < p_{i|r} - p_{i|(r+w)} < 0.03$, are quite common for real test data and are expected not to distort the person ordering seriously; they can safely be ignored when testing for significance. In samples smaller than, say, 300 observations, and tests consisting of, say, more than 10 items, rest-score groups in the tails of their distribution may be small and uninformative about manifest monotonicity. They may be pooled with adjacent rest-score groups until a sufficiently large combined group has been assembled and the sample proportions yield more precise estimates.

The computer program *Mokken Scale analysis for Polytomous items* (MSP; Molenaar and Sijtsma, 2000) and R package mokken (Van der Ark, 2012) contain binning to assess monotonicity. Ramsay (1991; Volume One, Chapter 20) suggested kernel smoothing and spline regression for estimating item response functions and assessing monotonicity. The program *Testgraf98* (Ramsay, 2000) implements kernel smoothing to estimate item response functions using the following procedure. First, total scores are replaced by rank scores, randomly ordering examinees with the same total score within their total-score group so that P different rank scores emerge. Second, the rank scores are grouped as $P+1$ quantiles covering areas of size $1/(P+1)$ under the standard normal distribution. Third, kernel smoothing is applied to the transformed rank scores; that is, each score is replaced by a weighted average of all scores in its neighborhood using a kernel function to specify the weights. A common choice of a kernel function is a normal density function that assigns the largest weight to each score considered at that moment and smaller weights to scores in their neighborhood. For more details, see Ramsay (2000); *Testgraf98* is downloadable from http://www.psych.mcgill.ca/faculty/ramsay/ramsay.html.

18.4.3 Manifest and Latent Invariant Item Ordering

Ligtvoet et al. (2010) proposed the following procedure for assessing invariant item ordering: Suppose a set of response functions has an invariant item ordering. The closer the functions are to each other, the less precise the information about their possible lack of intersection available in the data. As a result, the item ordering is determined with less precision than when the response functions are further apart. The H^T coefficient is defined on the transposed data matrix with items in the rows and persons in the columns, and H^T is computed like the H coefficient but with persons playing the roles of the items. Ligtvoet et al. (2010) argued that H^T can be interpreted as an index for the distance between nonintersecting response functions and hence for the accuracy of the ordering of the item response functions: higher H^T values suggest a more accurate ordering.

The procedure by Ligtvoet et al. (2010) is as follows: The I response functions are estimated to assess whether or not they intersect. If they do not intersect, H^T is estimated and expresses the degree to which the data allow an accurate item ordering. In more detail, the two-step procedure is as follows: Equation 18.1 implies that

$$\Pr(U_i = 1 \mid S_{(ij)} = s) \le \Pr(U_j = 1 \mid S_{(ij)} = s), \quad \text{for } s = 0, \ldots, n-2 \tag{18.6}$$

(Ligtvoet et al. (2011) and Tijmstra et al. (2011) showed that Equation 18.6 also holds when conditioning on the total score X). Equation 18.6 is evaluated by using $p_{i|s} = p_{si}/p_s$ and $p_{j|s} = p_{sj}/p_s$. If one finds in the sample that for item difficulties or popularities $p_i < p_j$, then this ordering is taken as the overall ordering of the items and the relation between $p_{i|s}$ and $p_{j|s}$ is reevaluated for each value of the rest score s. Suppose that, for some value of s, one finds the reverse sample ordering, $p_{i|s} > p_{j|s}$. Then, a hypergeometric test is used to test the null hypothesis of equality, $\pi_{i|s} = \pi_{j|s}$, which is the borderline case consistent with the expected item ordering, against the alternative that $\pi_{i|s} > \pi_{j|s}$, which violates the expected ordering. Rejection of the null hypothesis implies an intersection of the response functions. As with the earlier evaluation of the monotonicity assumption, it may be necessary to pool adjacent rest-score groups when they are small and ignore sample reversals that are smaller than a value chosen by the researcher, for example, 0.03.

For each of the I items, one counts of how often it is involved in significant intersections. If there are no violations whatsoever, the I items have an invariant item ordering; otherwise, the item with the largest count is removed, and the evaluation is repeated for the remaining $I-1$ items; and so on, until a subset of n items remains that can be assumed to have an invariant item ordering. The second step of the procedure entails the computation of coefficient H^T for the item subset. The R package mokken (Van der Ark, 2012) contains a procedure for finding an invariantly ordered item set as an automated algorithm. The researcher can manipulate the pooling of rest-score groups by choosing a minimum size for the pooled group; he or she can also try different minimum numbers of reversals of fractions that determine when statistical testing is pursued. Sijtsma and Molenaar (2002, Chapter 6) discussed other methods for investigating invariant item ordering that are included in the program MSP (Molenaar and Sijtsma, 2000).

18.4.4 Item Selection

Mokken proposed an automatic, bottom-up item-selection procedure that aims at selecting as many items as possible fitting a common scale for a choice of a lower bound c. The procedure works as follows:

Step 1: From a set of I items, the procedure selects the pair that has the highest, positive, significant H_{ij} value, or the researcher defines a subset of at least two items that he or she considers to be the core of the scale. Item selection is identical for both options.

Step 2: The next item k to be selected has (a) positive correlations with the already-selected items, (b) an H_k value in excess of c, and (c) produces the highest H value among all items satisfying the conditions in (a) and (b) with the already-selected items.

Step 3: The procedure repeats Step 2 selecting items from the remaining, unselected items until no items satisfying the conditions in (2a) and (2b) are left.

If any items remain in the pool, the procedure starts selecting items into a second scale following the same procedure, then a third scale, and so on, until no items have remained unselected or none of the remaining items satisfy Step 1. Items left are unscalable.

The procedure proceeds selecting one item at a time, and once an item is selected, it cannot be excluded from the current scale, nor is it available for selection in another scale. Consequently, the end result may not be the maximum number of items possible per scale satisfying the definition of a Mokken scale. Straat et al. (2013) proposed a genetic algorithm that considers many item clusters in an effort to find the global maximum. The authors found that the genetic algorithm performs better than Mokken's procedure, especially when, at the moment of their selection, for some items, the H_i value just exceeded the lower bound c.

Item selection based on positive interitem correlations and positive H_i values does not automatically produce scales that are consistent with the monotone homogeneity model, though. As positive correlations are a necessary condition for the monotone homogeneity model and positive, preferably high, positive H_i values are to be preferred, the selected scales are expected not to contain gross violations of the monotone homogeneity model assumptions. Nevertheless, for each selected scale, it is recommended to use the conditional associations in Equations 18.2 through 18.4 to identify possibly locally dependent items and the manifest monotonicity requirement in Equation 18.5 to identify possibly non-monotone item response functions.

In addition, contrary to what is sometimes believed, the definition of a Mokken scale does not imply an invariant item ordering (Sijtsma et al., 2011). Researchers wishing to know whether an item set has an invariant item ordering are advised to use the manifest invariant item ordering property in Equation 18.6 to evaluate their item response functions.

MSP and the R package `mokken` include the automated item-selection procedure. The R package `mokken` includes the genetic algorithm.

18.4.5 Person Fit and Reliability

Just like the H coefficient for a set of n items can be specialized to an item coefficient H_i, coefficient H^T for a population of P persons can be specialized to a person coefficient H_p^T. The person coefficient expresses the degree to which the pattern of item scores produced by person p matches the order of the items implied by the total set of item scores produced by the $P-1$ other persons. Karabatsos (2003) found that, out of 36 available person-fit statistics quantifying how well a person's pattern of item scores matches the item ordering or the predictions made by an IRT model, person coefficient H_p^T was one of the five statistics that was the most successful at detecting aberrant patterns. Persons producing such aberrant patterns may be diagnosed to find the cause of their aberrant performance. Alternatively, their item scores may be classified as outliers and removed from the data set (Meijer and Sijtsma, 2001).

Finally, we define the reliability of the total score X in the context of the double-monotonicity model (Mokken, 1971, pp. 142–147). Suppose item i is administered twice to the same population under exactly the same conditions; the observed responses are the result of independent replications of item i. Let π_{ii} be the joint proportion of a correct response on the two independent replications. Obviously, in practice, items are administered only once and even if they could be administered twice, the two sets of responses would be dependent, for example, due to memory effects. Hence, proportion π_{ii} cannot be estimated directly from the data. However, the reliability of the total score X can be written as

$$\rho_{XX'} = 1 - \frac{\sum_{i=1}^{n}(\pi_i - \pi_{ii})}{\sigma_X^2}. \tag{18.7}$$

It may be noted that proportions π_i $(i = 1, \ldots, n)$ and total-score variance σ_X^2 can be estimated from the sample data. Invoking the assumption of local independence, the joint proportion π_{ii} $(i = 1, \ldots, n)$ can be written as

$$\pi_{ii} = \int \Pr(X_i = 1 \mid \theta)\Pr(X_i = 1 \mid \theta)dG(\theta). \tag{18.8}$$

It is proposed to replace one of the probabilities in Equation 18.8 by the weighted average of two neighboring response functions, indexed as $i-1$ and $i+1$, such that, for a choice of weights f, g, and h,

$$\widetilde{\Pr}(X_i = 1 \mid \theta) = f + g\Pr(U_{i-1} = 1 \mid \theta) + h\Pr(U_{i+1} = 1 \mid \theta).$$

Integration then yields

$$\tilde{\pi}_{ii} = f\pi_i + g\pi_{i-1,i} + h\pi_{i,i+1}.$$

This equation contains proportions that can be estimated from sample data and inserted in Equation 18.7 to obtain a reliability estimate. Of course, different choices of f, g, and h

produce different reliability estimates; see Sijtsma and Molenaar (1987). Van der Ark et al. (2011) discuss bias and accuracy of this reliability method.

18.5 Empirical Example

In this chapter, we focused on the theory for dichotomous items but we discuss a polytomous-item empirical example to better illustrate the potential of the Mokken models. The *Hospital Anxiety and Depression Scale* (HADS; Zigmond and Snaith, 1983; also, see Emons et al., 2012) is a frequently used questionnaire in health research. It contains 14 items, seven of which measure anxiety using four-point rating scales. Table 18.1 shows the item texts. We analyzed a sample that consisted of 3111 adults representing the Dutch population. Quota sampling was used to ensure that all age groups between 20 and 80 years and gender groups were equally represented. The purpose of this example was to decide whether the seven anxiety items form a scale consistent with the monotone homogeneity model and had an invariant item ordering. To evaluate the monotonicity requirement, we investigated whether the response functions $E(U_i|\theta)$ were monotone, and to evaluate the item ordering, we investigated whether the pairs $E(U_i|\theta)$ and $E(U_j|\theta)$ intersected.

All correlations between the seven items were positive. Table 18.2 shows that the H_i coefficients ranged from 0.320 to 0.449 and that, for the entire set of items, $H = 0.391$ (standard error $SE = 0.010$). Following the earlier rules of thumb, the scale was weak. The *SE* showed that we could not rule out that in the population the scale might have medium strength but based on this sample, the odds are in favor of a weak scale. The automated item-selection procedure with a lower bound $c = 0.3$ selected the seven items in the same scale. For $c = 0.4$, Items 1, 2, 3, 5, and 7 were selected into the same medium scale ($H = 0.496$) but Items 4 and 6, both measuring restlessness, were unscalable (i.e., their item scalability values relative to the five scale items were below 0.4: $H_4 = 0.326$ [$SE = 0.015$] and $H_6 = 0.300$ [$SE = 0.016$]; results not tabulated). For $c = 0.5$, Items 1 and 5 formed one scale ($H_{1,5} = 0.500$). The same holds for Items 2 and 3 ($H_{2,3} = 0.622$), but the remaining three items were unscalable

TABLE 18.1

Anxiety Items from the HADS

Item#	Item Contents
1	*I feel tense or "wound up"* Most of the time—A lot of the time—From time to time, occasionally—Not at all
2	*I get a sort of frightened feeling as if something awful is about to happen* Very definitely and quite badly—Yes, but not too badly—A little, but it does not worry me—Not at all
3	*Worrying thoughts go through my mind* A great deal of the time—A lot of the time—From time to time, but not too often—Only occasionally
4	*I can sit at ease and feel relaxed* Definitely—Usually—Not often—Not at all
5	*I get a sort of frightened feeling like "butterflies" in the stomach* Not at all—Occasionally—Quite often—Very often
6	*I feel restless as if I have to be on the move* Very much indeed—Quite a lot—Not very much—Not at all
7	*I get sudden feelings of panic* Very often indeed—Quite often—Not very often—Not at all

TABLE 18.2

Scalability Coefficients for Two Lower Bounds c

		No Selection/$c = 0.3$		$c = 0.4$	
Item#	Item Mean	H_i	SE	H_i	SE
1	0.89	0.439	0.014	0.488	0.016
2	0.53	0.423	0.013	0.513	0.014
3	0.86	0.449	0.012	0.543	0.013
4	0.75	0.341	0.014		
5	0.44	0.397	0.014	0.462	0.016
6	0.96	0.320	0.014		
7	0.44	0.399	0.015	0.464	0.017
Scale H		0.391	0.010	0.496	0.013

(no further details in Table 18.2). The genetic algorithm produced exactly the same selection results for the same three lower-bound values.

Let $G_{(i)}$ denote a pooled combination of adjacent rest-score groups with $R_{(i)} = r$ such that the size of the group exceeds a given lower bound. For each of the seven items, Figure 18.2 shows the three category response functions as well as the item response function. The category response functions show several violations of monotonicity that do not cause the corresponding item response function to decrease. Therefore, we focus on the latter. Item response functions that do show local decreases are those of Item 1 (one decrease smaller than 0.03), Item 2 (one decrease equal to 0.09), and Item 6 (two decreases; one smaller than 0.03 and the other equal to 0.05). More specifically, the second graph in Figure 18.2 shows $\Pr(U_2 \geq a \mid G_{(2)})$ for Item 2 ($a = 1, 2, 3$), along with its sample item response function defined as the mean of its category response functions, that is, $(\bar{U}_2 \mid G_{(2)}) / 3$. The item response function shows a decrease between groups $G_{(2)} = 11$, representing $R_{(2)} = 10$ ($n = 68$), and $G_{(2)} = 12$, representing $R_{(2)} = 11$ ($n = 35$), where the mean score on Item 2 decreases from 0.58 to 0.49. The next group $G_{(2)} = 13$ pools persons with rest scores $R_{(2)}$ running from 12 to 18 ($n = 63$). It has a mean item score equal to 0.70, which is higher than 0.58 and thus consistent with the assumption of monotonicity. The difference $0.58 - 0.49 = 0.09$ exceeds the minimum of 0.03, below which any differences are ignored. Program `mokken` provides statistical tests only for category response functions, but we consider 0.09 as large enough to be an important violation of monotonicity.

The results for our evaluation of invariant item ordering were as follows. Given the overall ordering of the items by their sample means, we found one significant violation against this ordering (for items 1 and 3). The violation occurred at the high end of the scale. Thus, the observed item response functions seemed to be invariantly ordered but the corresponding $H^T = 0.19$ suggested that the ordering was inaccurate. Ligtvoet et al. (2010) tentatively suggest that $H^T < 0.3$ is too low to be interpretable.

The conclusion from the example was that the seven anxiety items in the HADS constitute a scale on which individuals can be ordered by means of their total scores X. Table 18.3 provides the frequency distribution of the total score X. Given a score range of 21, the mean-scale score of 4.9 is low, which is understandable given that the sample was from a normal population, which was not expected to be excessively anxious. The positive skewness of 0.9 suggests that high-scale values were relatively rare, again which was to be expected in a normal population. Hence, the standard deviation of 3.4 refers to a large spread of anxiety scores on the lower half of the scale. The kurtosis was 0.8, suggesting a flat distribution

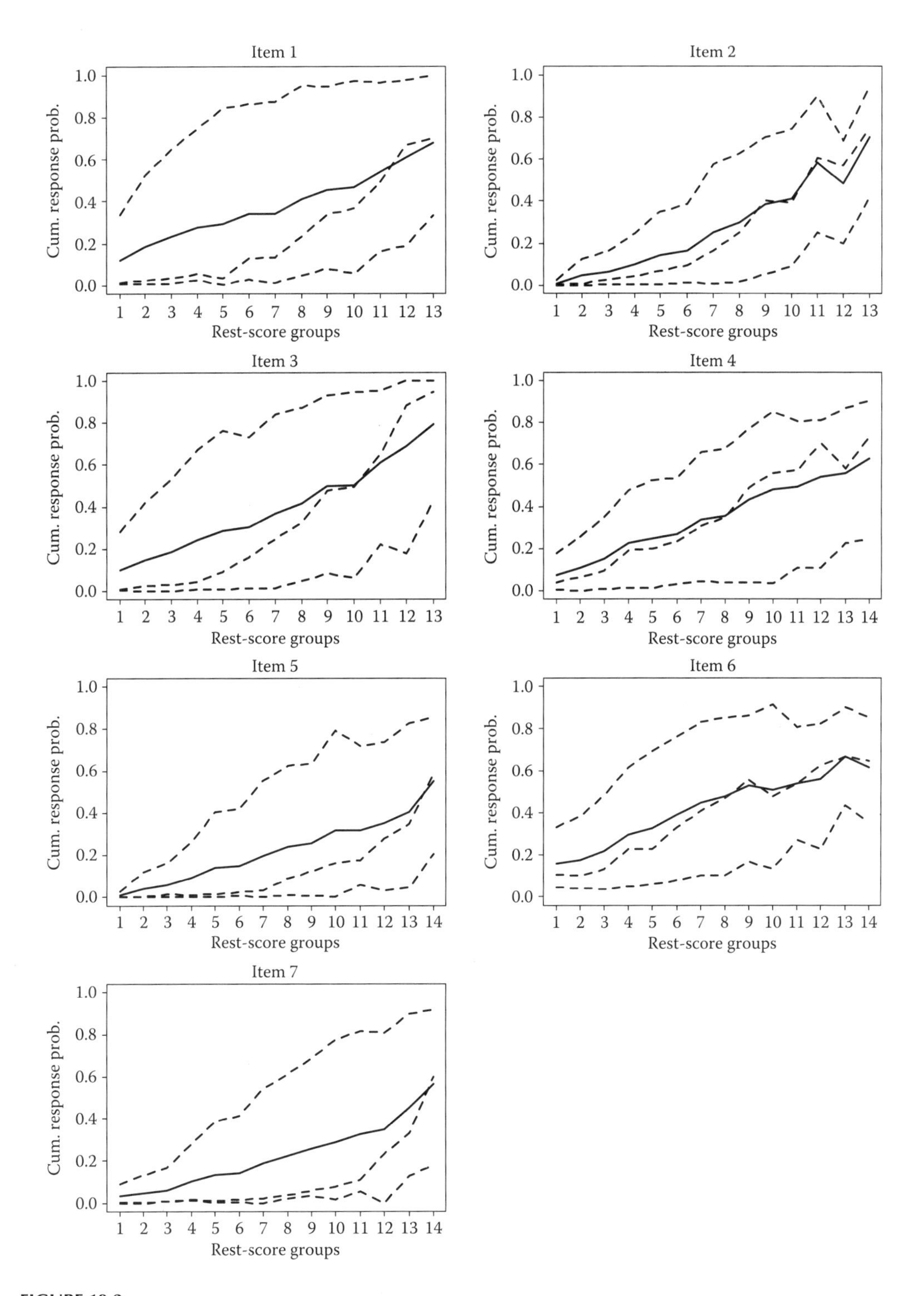

FIGURE 18.2
For each of the seven HADS anxiety items, a panel shows the three category response functions (dashed lines) and the item response function (solid line). "Rest-score groups" can also be combinations of adjacent rest-score groups.

TABLE 18.3

Frequency (*Fr*) Distribution of the Total Score *X* ($N = 3111$)

X	0	1	2	3	4	5	6	7	8	9	10	11	12	13	14	15	16	17	18	19	20	21
Fr	200	302	364	373	364	370	274	233	181	129	112	72	37	40	18	18	11	9	1	0	2	1

Mean = 4.87
Standard deviation = 3.42
Skewness = 0.88
Kurtosis = 0.77

relative to, for example, a standard normal distribution (for which the kurtosis equals 3). The reliability of the total score based on all seven items was 0.78. This value was relatively high given the small number of items. To assess whether a particular total score significantly exceeds a cut score representing a minimally acceptable level of anxiety, the standard error of measurement should be taken into account. This topic, however, is beyond the scope of this chapter.

18.6 Discussion

Mokken scales are constructed in almost all areas in which tests or questionnaires are used to measure attributes on which individuals show variation in scores. The use of the simple total score rather than a hypothetical latent variable undoubtedly has contributed to the popularity of Mokken-scale analysis. Another feature is its automated item-selection procedure, which only requires a researcher to push a button to produce one or more scales. The downside of this feature, however, is that researchers may be satisfied with the outcome without reporting an investigation of the monotonicity and invariant item ordering requirement, which are properties not automatically guaranteed by item selection based only on scalability coefficients. Although the automated item-selection procedure tends to select items with monotone item response functions (which are not necessarily ordered invariantly), one would wish that researchers be more careful about this aspect of scaling.

Mokken-scale analysis, as well as nonparametric IRT in general, has proven to be useful to study general measurement properties that are also relevant for parametric IRT models, which are special cases of nonparametric models. Examples supporting this claim are from research into ordinal measurement properties (Hemker et al., 1997; Van der Ark and Bergsma, 2010), monotonicity (Junker and Sijtsma, 2000), and invariant item ordering (Ligtvoet et al., 2011; Sijtsma and Junker, 1996; Tijmstra et al., 2011). The current research on Mokken models mainly focuses on the use of marginal models for estimating standard errors of interesting new statistics, on improvements of Mokken's automated item-selection procedure, and person-fit analysis for models that involve explanatory variables.

References

Costa, P. T. Jr., and McCrae, R. R. 1992. *Revised NEO Personality Inventory (NEO PI-RTM) and NEO Five-Factor Inventory (NEO-FFI) Professional Manual.* Odessa, FL: Psychological Assessment Resources.

Emons, W. H. M., Sijtsma, K., and Pedersen, S. S. 2012. Dimensionality of the Hospital Anxiety and Depression Scale (HADS) in cardiac patients: Comparison of Mokken scale analysis and factor analysis. *Assessment*, 19, 337–353.

Fox, J. 1997. *Applied Regression Analysis, Linear Models, and Related Methods.* Thousand Oaks, CA: Sage.

Hemker, B. T., Sijtsma, K., Molenaar, I. W., and Junker, B. W. 1997. Stochastic ordering using the latent trait and the sum score in polytomous IRT models. *Psychometrika*, 62, 331–347.

Holland, P. W., and Rosenbaum, P. R. 1986. Conditional association and unidimensionality in monotone latent variable models. *The Annals of Statistics*, 14, 1523–1543.

Junker, B. W. 1993. Conditional association, essential independence, and monotone unidimensional item response models. *The Annals of Statistics*, 21, 1359–1378.

Junker, B. W., and Sijtsma, K. 2000. Latent and manifest monotonicity in item response models. *Applied Psychological Measurement*, 24, 65–81.

Karabatsos, G. 2003. Comparing the aberrant response detection performance of thirty-six person-fit statistics. *Applied Measurement in Education*, 16, 277–298.

Kuijpers, R. E., Van der Ark, L. A., and Croon, M. A. 2013. Standard errors and confidence intervals for scalability coefficients in Mokken scale analysis using marginal models. *Sociological Methodology*, 43, 42–69.

Ligtvoet, R., Van der Ark, L. A., Bergsma, W. P., and Sijtsma, K. 2011. Polytomous latent scales for the investigation of the ordering of items. *Psychometrika*, 76, 200–216.

Ligtvoet, R., Van der Ark, L. A., Te Marvelde, J. M., and Sijtsma, K. 2010. Investigating an invariant item ordering for polytomously scored items. *Educational and Psychological Measurement*, 70, 578–595.

Meijer, R. R., and Sijtsma, K. 2001. Methodology review: Evaluating person fit. *Applied Psychological Measurement*, 25, 107–135.

Mokken, R. J. 1971. *A Theory and Procedure of Scale Analysis*. The Hague: Mouton/Berlin: De Gruyter.

Molenaar, I. W. 1997. Nonparametric models for polytomous items. In W. J. van der Linden and R. K. Hambleton (eds.), *Handbook of Modern Item Response Theory* (pp. 369–380). New York: Springer.

Molenaar, I. W., and Sijtsma, K. 2000. *MSP5 for Windows. A Program for Mokken Scale Analysis for Polytomous Items*. Groningen: iecProGAMMA.

Ramsay, J. C. 1991. Kernel smoothing approaches to nonparametric item characteristic curve estimation. *Psychometrika*, 56, 611–630.

Ramsay, J. O. 2000. *TestGraf. A Program for the Graphical Analysis of Multiple Choice Test and Questionnaire Data*. Montreal, Canada: McGill University.

Sijtsma, K., and Hemker, B. T. 1998. Nonparametric polytomous IRT models for invariant item ordering, with results for parametric models. *Psychometrika*, 63, 183–200.

Sijtsma, K., and Junker, B. W. 1996. A survey of theory and methods of invariant item ordering. *British Journal of Mathematical and Statistical Psychology*, 49, 79–105.

Sijtsma, K., Meijer, R. R., and Van der Ark, L. A. 2011. Mokken scale analysis as time goes by: An update for scaling practitioners. *Personality and Individual Differences*, 50, 31–37.

Sijtsma, K., and Molenaar, I. W. 1987. Reliability of test scores in nonparametric item response theory. *Psychometrika*, 52, 79–97.

Sijtsma, K., and Molenaar, I. W. 2002. *Introduction to Nonparametric Item Response Theory*. Thousand Oaks, CA: Sage.

Stout, W. F. 2002. Psychometrics: From practice to theory and back. *Psychometrika*, 67, 485–518.

Straat, J. H., Van der Ark, L. A., and Sijtsma, K. 2013. Comparing optimization algorithms for item selection in Mokken scale analysis. *Journal of Classification*, 30, 75–99.

Straat, J. H., Van der Ark, L. A., and Sijtsma, K. 2016. Using conditional association to identify locally independent item sets. Submitted for publication.

Tijmstra, J., Hessen, D. J., Van der Heijden, P. G. M., and Sijtsma, K. 2011. Invariant ordering of item-total regressions. *Psychometrika*, 76, 217–227.

Van Abswoude, A. A. H., Van der Ark, L. A., and Sijtsma, K. 2004. A comparative study of test dimensionality assessment procedures under nonparametric IRT models. *Applied Psychological Measurement*, 28, 3–24.

Van der Ark, L. A. 2005. Stochastic ordering of the latent trait by the sum score under various polytomous IRT models. *Psychometrika*, 70, 283–304.

Van der Ark, L. A. 2012. New developments in Mokken scale analysis in R. *Journal of Statistical Software*, 48, 1–27.

Van der Ark, L. A., and Bergsma, W. P. 2010. A note on stochastic ordering of the latent trait using the sum of polytomous item scores. *Psychometrika*, 75, 272–279.

Van der Ark, L. A., Croon, M. A., and Sijtsma, K. 2008. Mokken scale analysis for dichotomous items using marginal models. *Psychometrika*, 73, 183–208.

Van der Ark, L. A., Van der Palm, D. W., and Sijtsma, K. 2011. A latent class approach to estimating test-score reliability. *Applied Psychological Measurement*, 35, 380–392.
Van Schuur, W. H. 2011. *Ordinal Item Response Theory. Mokken Scale Analysis.* Thousand Oaks, CA: Sage.
Zigmond, A. S., and Snaith, R. P. 1983. The Hospital Anxiety and Depression Scale. *Acta Psychiatrica Scandinavica*, 67, 361–370.

19

*Bayesian Nonparametric Response Models**

George Karabatsos

CONTENTS

19.1 Introduction 323
19.2 Mixture IRT and BNPs 325
19.3 Presentation of the Model 326
19.4 Parameter Estimation 330
19.5 Model Fit 331
19.6 Empirical Example 332
19.7 Discussion 333
References 334

19.1 Introduction

Given a set of data, consisting of a person's individual responses to items of a test, an item response theory (IRT) model aims to infer each person's ability on the test, and to infer the test item parameters. In typical applications of an IRT model, each item response is categorized into one of two or more categories. For example, each item response may be scored as either correct (1) or incorrect (0). From this perspective, a categorical regression model, which includes person ability parameters and item difficulty parameters, provides an interpretable approach to inferring from item response data. One basic example is the Rasch (1960) model. This model can be characterized as a logistic regression model, having the dichotomous item score as the dependent variable. The predictors (covariates) of this model include N person-indicator (0,1) variables, corresponding to regression coefficients that define the person ability parameters; and include I item-indicator (0,–1) variables, corresponding to coefficients that define the item difficulty parameters.

In many item response data sets, there are observable and unobservable covariates that influence the item responses, in addition to the person and item factors. If the additional covariates are not fully accounted for in the given IRT model, then the estimates of person ability and item difficulty parameters can become noticeably biased. Such biases can be (at least) partially alleviated by including the other, observable covariates into the IRT (regression) model, as control variables. However, for most data collection protocols, it is not possible to collect data on all the covariates that help to determine the item responses (e.g., due to time, financial, or ethical constraints). Then, the unobserved covariates, which influence the item responses, can bias the estimates of the ability and item parameters in an IRT model that does not account for these covariates.

* This material is based upon work supported by National Science Foundation grant SES-1156372 from the Program in Methodology, Measurement, and Statistics. The author gives thanks to Wim J. van der Linden and Brian Junker for feedback on this work.

A flexible mixture IRT model can provide robust estimates of person ability parameters and item difficulty parameters, by accounting for any additional unobserved latent covariates that influence the item responses. Modeling flexibility can be maximized through the use of a Bayesian nonparametric (BNP) modeling approach.

In this chapter, we present a BNP approach to infinite-mixture IRT modeling, based on the general BNP regression model introduced by Karabatsos and Walker (2012a). We then illustrate this model, called the BNP-IRT model, through the analysis of real item response data. The analysis was conducted using menu-driven (point-and-click) software, developed by the author (Karabatsos 2014a,b).

In the next section, we give a brief overview of the concepts of mixture IRT modeling and BNP infinite-mixture modeling. Then in Section 19.3, we introduce our basic BNP-IRT model. This is a regression model consisting of person ability and item difficulty parameters, constructed via the appropriate specification of person and item-indicator predictor variables, as mentioned above. While the basic model assumes dichotomous item scores and unidimensional person ability, our model can be easily extended to handle polytomous responses (with item response categories not necessarily ordered), extra person-level and/or item-level covariates, and/or multidimensional person ability parameters. In Section 19.4, we describe the Markov chain Monte Carlo (MCMC) methods that can be used to estimate the posterior distribution of the model parameters. (This is a highly technical section which can be skipped when reading this chapter.) In Section 19.5, we describe methods for evaluating the fit of our BNP-IRT model. Section 19.6 provides an empirical illustration of the BNP-IRT model through the analysis of polytomous response data. The data were obtained from an administration of a questionnaire that was designed to measure teacher preparation. Section 19.7 ends with a brief overview of how to use the menu-driven software to perform data analysis using the BNP-IRT model. That section also includes a brief discussion of how to extend the BNP-IRT model for cognitive IRT.

The remained of this chapter makes use of the following notational conventions. Let $\mathbf{U} = (U_1,\ldots,U_i,\ldots,U_I)^\top$ denote a random vector for the scores on a test with I items. A realized value of the item response vector is denoted by $\boldsymbol{u} = (u_1,\ldots,u_i,\ldots,u_I)^\top$. We assume that each item $i = 1, \ldots, I$ has $m_i + 1$ possible discrete-valued scores, indexed by $u = 0, 1, \ldots, m_i$.

We use lower cases to denote a probability mass function (pmf) of a value u discrete random variable (or vector, $\mathbf{u}$) or a probability density function (pdf) of a value u of a continuous random variable (or $\mathbf{u}$), such as $f(u)$ or $f(\mathbf{u})$, respectively. The given pmf (or pdf) $f(u)$ corresponds to a cumulative distribution function (cdf), denoted by upper case $F(u)$, which gives the probability that the random variable U does not exceed u. $F(u)$ is sometimes more simply referred to as the distribution function. Thus, for example, $\mathrm{N}(\mu,\sigma^2)$, $\mathrm{U}(0,b)$, $\mathrm{IG}(a,b)$, and $\mathrm{Be}(a,b)$ (or cdfs $\mathrm{N}(\cdot|\mu,\sigma^2)$, $\mathrm{U}(\cdot|0,b)$, $\mathrm{IG}(\cdot|a,b)$, and $\mathrm{Be}(\cdot|a,b)$, respectively), denote the univariate normal, uniform, inverse-gamma, and beta distribution functions, respectively. They correspond to pdfs $\mathrm{n}(\cdot|\mu,\sigma^2)$, $\mathrm{u}(\cdot|0,b)$, $\mathrm{ig}(\cdot|a,b)$, and $\mathrm{be}(\cdot|a,b)$, with mean and variance parameters (μ,σ^2), minimum and maximum parameters $(0,b)$, shape and rate parameters (a,b), and shape parameters (a,b), respectively. Also, if $\boldsymbol{\beta}$ is a realized value of a K-dimensional random vector, then $\mathrm{N}(\boldsymbol{\beta}|\mathbf{0},\mathbf{V})$ denotes the cdf of the multivariate (K-variate) normal distribution with mean vector of zeros $\mathbf{0}$ and $K \times K$ variance–covariance matrix $\mathbf{V}$, distribution function $\mathrm{n}(\mathbf{0},\mathbf{V})$, and corresponding to pdf $\mathrm{n}(\boldsymbol{\beta}|\mathbf{0},\mathbf{V})$. The pmf or pdf of u given values of one or more variables $\mathbf{x}$ is written as $f(u|\mathbf{x})$ (with corresponding cdf $F(u|\mathbf{x})$); given a vector of parameter values $\boldsymbol{\zeta}$ is written as $f(u|\boldsymbol{\zeta})$ (with corresponding cdf $F(u|\boldsymbol{\zeta})$), and conditionally on variables and given parameters is written as $f(u|\mathbf{x};\boldsymbol{\zeta})$ (with corresponding cdf $F(u|\mathbf{x};\boldsymbol{\zeta})$). Also, $\sim$ means "distributed as," $\sim_{ind}$ means "independently distributed," and $\sim_{iid}$ means "independently and identically distributed." For example, $U \sim F$, $U \sim_{iid} F(u)$, $U \sim F(u|\mathbf{x};\boldsymbol{\zeta})$,

$U \sim F(u|\zeta)$, $U \sim_{iid} F(\zeta)$, $\boldsymbol{\beta} \sim \mathrm{N}(\mathbf{0},\mathbf{V})$, or $\sigma^2 \sim \mathrm{IG}(a,b)$. The preceding notation may replace U by $\mathbf{U}$, replace F by f, replace N by n, and/or replace IG by ig.

19.2 Mixture IRT and BNPs

For any given vector of item response data $u = (u_1,\ldots,u_i,\ldots,u_I)^{\mathrm{T}}$, a discrete-mixture IRT model admits the general form

$$f_{G_x}(\mathbf{u}\,|\,\mathbf{x}) = \int f(\mathbf{u}\,|\,\mathbf{x};\boldsymbol{\beta},\boldsymbol{\Psi}(\mathbf{x}))dG_{\mathbf{x}}(\Psi) = \sum_{j=1}^{J} f(\mathbf{u}\,|\,\mathbf{x};\boldsymbol{\beta},\boldsymbol{\Psi}_j(\mathbf{x}))\omega_j(\mathbf{x}) \tag{19.1}$$

conditionally on any given value of a vector of any covariates $\mathbf{x}$. In this expression, $f(u|\mathbf{x};\boldsymbol{\beta},\Psi(\mathbf{x}))$ is the kernel of the mixture and $G_{\mathbf{x}}$ is a mixture distribution that may (or may not) depend on the same covariates.

Also, as shown in Equation 19.1, this pmf is based on a mixture of J pmfs $f(u|\mathbf{x};\,\boldsymbol{\beta},\,\Psi_j(\mathbf{x}))$, $j = 1, \ldots, J$. Here, $\boldsymbol{\beta}$ is a vector of (any available) fixed parameters that are not subject to the mixture, the $\boldsymbol{\Psi}_j(\mathbf{x})$, $j = 1, \ldots, J$ are random parameters that are subject to the mixture that may be covariate dependent, and J is the number of mixture components. In addition, $\omega_j(\mathbf{x})$, $j = 1, \ldots, J$ are mixture weights that sum to one for every given covariate value $\mathbf{x} \in \mathcal{X}$. The mixture model (19.1) is called a discrete (continuous) mixture model if $G_{\mathbf{x}}$ is discrete (continuous); it is called a finite (infinite) mixture model if J is finite (infinite).

A simple example is given by the finite mixture Rasch model for dichotomous item scores (Rost, 1990, 1991; Volume One, Chapter 23), which assumes that

$$f(u\,|\,\mathbf{x};\boldsymbol{\beta},\boldsymbol{\Psi}_j(\mathbf{x})) = \prod_{i=1}^{I} \frac{\exp(\theta_j - \beta_{ij})^{u_i}}{1+\exp(\theta_j - \beta_{ij})} \tag{19.2}$$

with a finite number of J components and mixture weights that are not covariate dependent (i.e., $\omega_j(\mathbf{x}) = \omega_j$, $j = 1, \ldots, J < \infty$). The ordinary Rasch (1960) model for dichotomous item scores is the special case of the model defined by Equations 19.1 and 19.2 for $J = 1$.

An infinite-mixture model is given by Equation 19.1 for $J = \infty$. A general BNP infinite-mixture IRT model assumes that the mixture distribution has the general form

$$G_{\mathbf{x}}(\cdot) = \sum_{j=1}^{\infty} \omega_j(\mathbf{x})\delta_{\boldsymbol{\Psi}_j(\mathbf{x})}(\cdot), \tag{19.3}$$

where $\delta_{\boldsymbol{\Psi}}(\cdot)$ denotes a degenerate distribution with support Ψ. This Bayesian model is completed by the specification of a prior distribution on $\{\Psi_j(\mathbf{x})\}_j = 1, 2, \ldots, \{\omega_j(\mathbf{x})\}_j = 1, 2, \ldots$, and $\boldsymbol{\beta}$ with large supports.

A common example is a Dirichlet process (DP) mixed IRT model, which assumes that the mixing distribution is not covariate dependent (i.e., $G_{\mathbf{x}}(\cdot) = G(\cdot)$), along with a random mixing distribution $G(\cdot)$ constructed as $G(\cdot) = \sum_{j=1}^{\infty} \omega_j \delta_{\boldsymbol{\Psi}_j}(\cdot)$ where $\omega_j = \upsilon_j \prod_{k=1}^{j-1}(1-\upsilon_k)$ for random draws $\upsilon_j \sim_{iid} \text{Be}(1,\alpha)$ and $\boldsymbol{\Psi}_j \sim_{iid} G_0$, for $j = 1, 2, \ldots$. Here, G is a DP, denoted $G \sim \text{DP}(\alpha,G_0)$, with baseline parameter G_0 and precision parameter α (Sethuraman, 1994). The DP (α,G_0) has mean (expectation) $\mathbb{E}[G(\cdot)] = G_0(\cdot)$ and variance $\mathbb{V}[G(\cdot)] = G_0(\cdot)\{1-G_0(\cdot)\}/(\alpha+1)$ (Ferguson, 1973).

An important generalization of the DP prior includes the Pitman–Yor (Poisson–Dirichlet) prior (Ishwaran and James, 2001), which assumes that $\upsilon_j \sim_{iid} \text{Be}(\alpha_{1j},\alpha_{2j})$, for $j = 1, 2, \ldots$, for some $\alpha_{1j} = 1 - \alpha_1$ and $\alpha_{2j} = \alpha_2 + j\alpha_1$ with $0 \le \alpha_1 < 1$ and $\alpha_2 > -\alpha_1$. The special case defined by $\alpha_1 = 0$ and $\alpha_2 = \alpha$ results in the $\text{DP}(\alpha G_0)$.

Another important generalization of the DP is given by the dependent Dirichlet process (DDP) (MacEachern, 1999, 2000, 2001), which provides a model for the covariate-dependent random distribution, denoted $G_{\mathbf{x}}$. The DDP model assumes that $G_{\mathbf{x}} \sim \text{DP}(\alpha_{\mathbf{x}},G_{0\mathbf{x}})$, marginally for each $\mathbf{x}$. Specifically, the DDP defines a covariate-dependent random distribution $G_{\mathbf{x}}$ of the form given in Equation 19.3, and incorporates this dependence either through covariate-dependent atoms $\boldsymbol{\Psi}_j(\mathbf{x})$, a covariate-dependent baseline $G_{0\mathbf{x}}$, and/or covariate-dependent stick-breaking weights of the form $\omega_j(\mathbf{x}) = \upsilon_j(\mathbf{x})\prod_{k=1}^{j-1}(1-\upsilon_k(\mathbf{x}))$, for $j = 1, 2, \ldots$. For example, the ANOVA-linear DDP (De Iorio et al., 2004), denoted $G_{\mathbf{x}} \sim$ ANOVA-DDP $(\alpha,G_0,\mathbf{x})$, constructs a dependent random distribution $G_{\mathbf{x}}(\cdot) = \sum_{j=1}^{\infty}\omega_j\delta_{\mathbf{x}^{\mathrm{T}}\boldsymbol{\beta}_j}(\cdot)$, via covariate-dependent atoms $\mathbf{x}^{\mathrm{T}}\boldsymbol{\beta}_j$, along with $\boldsymbol{\beta} \sim G$ and $G \sim \text{DP}(\alpha,G_0(\boldsymbol{\beta}))$.

Many examples of DP-mixture and DDP mixture IRT models can be found in the literature (Duncan and MacEachern, 2008; Farina et al., 2009; Karabatsos and Walker, 2012a,b; Miyazaki and Hoshino, 2009; San Martín et al., 2011, 2013; Qin, 1998).

19.3 Presentation of the Model

The BNP-IRT model is a special case of the BNP regression (infinite-mixture) model introduced by Karabatsos and Walker (2012a). These authors demonstrated that the model tended to have better predictive performance relative to DP-mixed and DDP mixed regression models. As will be shown, the BNP-IRT model is suitable for dichotomous or polytomous item responses.

First, we present the basic BNP-IRT model for dichotomous item responses. Let $\mathcal{D} = \{(u_{pi}, \mathbf{x}_{pi})_{i=1}^{I}\}_{p=1}^{P}$ denote a set of item-response data, including dichotomous responses $u_{pi} \in \{0,1\}$. Also, $\mathbf{x}_{pi}$ denotes a covariate vector that describes person $p = 1, \ldots, P$ and item $i = 1, \ldots, I$.

The basic BNP-IRT model is defined as

$$f(\mathcal{D}\,|\,\mathbf{X};\boldsymbol{\zeta}) = \prod_{p=1}^{P}\prod_{i=1}^{I} f(u_{pi}\,|\,\mathbf{x}_{pi};\boldsymbol{\zeta}), \tag{19.4}$$

$$f(u_{pi}\,|\,\mathbf{x}_{pi};\boldsymbol{\zeta}) = P(U_{pi}=1\,|\,\mathbf{x}_{pi};\boldsymbol{\zeta})^{u_{pi}}[1-P(U_{pi}=1\,|\,\mathbf{x}_{pi};\boldsymbol{\zeta})]^{1-u_{pi}}, \tag{19.5}$$

$$\Pr(U=1\,|\,\mathbf{x};\boldsymbol{\zeta}) = 1-F^{*}(0\,|\,\mathbf{x};\boldsymbol{\zeta}) = \int_{0}^{\infty} f(u_{pi}^{*}\,|\,\mathbf{x}_{pi};\boldsymbol{\zeta})du^{*}, \tag{19.6}$$

$$= \int_0^\infty \sum_{j=-\infty}^{\infty} \mathrm{n}(u^* \mid \mu_j + \mathbf{x}_{pi}^\top \boldsymbol{\beta}, \sigma^2)\omega_j(\mathbf{x}_{pi}; \boldsymbol{\beta}_\omega, \sigma_\omega) du^*, \tag{19.7}$$

$$\omega_j(\mathbf{x}; \boldsymbol{\beta}_\omega, \sigma_\omega) = \Phi\left(\frac{j - \mathbf{x}^\top \boldsymbol{\beta}_\omega}{\sigma_\omega}\right) - \Phi\left(\frac{j - 1 - \mathbf{x}^\top \boldsymbol{\beta}_\omega}{\sigma_\omega}\right), \tag{19.8}$$

$$(\mu_j, \sigma_\mu^2) \sim \mathrm{N}(\mu_j \mid 0, \sigma_\mu^2)\mathrm{U}(\sigma_\mu \mid 0, b_{\sigma\mu}), \tag{19.9}$$

$$(\boldsymbol{\beta}, \boldsymbol{\beta}_\omega) \sim \mathrm{N}(\boldsymbol{\beta} \mid \mathbf{0}, \sigma^2 \mathrm{v}\,\mathrm{diag}(\infty, \mathbf{J}_{NI}^\top))\mathrm{N}(\boldsymbol{\beta}_\omega \mid \mathbf{0}, \sigma_\omega^2 \mathrm{v}_\omega \mathbf{I}_{NI+1}), \tag{19.10}$$

$$(\sigma^2, \sigma_\omega^2) \sim \mathrm{IG}(\sigma^2 \mid \mathrm{a}_0/2, \mathrm{a}_0/2)\mathrm{IG}(\sigma_\omega^2 \mid \mathrm{a}_\omega/2, \mathrm{a}_\omega/2). \tag{19.11}$$

Under the model, the data likelihood is given by Equations 19.4 through 19.8 given parameters $\boldsymbol{\zeta} = (\boldsymbol{\mu}, \sigma_\mu^2, \boldsymbol{\beta}, \boldsymbol{\beta}_\omega, \sigma^2, \sigma_\omega)$ with $\boldsymbol{\mu} = (\mu_j)_{j=-\infty}^{\infty}$. By default, the model assumes that $\mathbf{x}_{pi}$ is a binary indicator vector with $NI + 1$ rows, having constant (19.1) in the first entry, a "1" in entry $p + 1$ to indicate person p and "–1" in entry $i + (p + 1)$ to indicate item i. Specifically, each vector $\mathbf{x}_{pi}$ is defined by

$$\mathbf{x}_{pi} = (1, \mathbf{1}(p = 1), \ldots, \mathbf{1}(p = N), -\mathbf{1}(i = 1), \ldots, -\mathbf{1}(i = I))^\top,$$

where $\mathbf{1}(\cdot)$ denotes the indicator (0,1) function. Then, in terms of the coefficient vector $\boldsymbol{\beta} = (\beta_0, \beta_1, \ldots, \beta_{PI})$, each coefficient $\beta_{p+1} = \theta_p$ represents the ability of person $p = 1, \ldots, P$. Likewise, each coefficient β_{i+p+1} represents the difficulty of item $i = 1, \ldots, I$. The covariate-dependent mixture weights $\omega_j(\mathbf{x})$ in Equation 19.8 are specified by a cumulative ordered probits regression, based on the choice of a standard normal cdf for $\Phi(\cdot)$ with latent mean $\mathbf{x}^\top \boldsymbol{\beta}_\omega$ and variance σ_ω^2, for the "ordinal categories" $j = 0, \pm 1, \pm 2, \ldots$, where coefficient vector $\boldsymbol{\beta}_\omega$ contains additional person parameters and item parameters.

As shown in Equations 19.9 through 19.11, the Bayesian model parameters $\boldsymbol{\zeta}$ have joint prior density

$$\begin{aligned}\pi(\boldsymbol{\zeta}) = \prod_{j=-\infty}^{\infty} &\mathrm{n}(\mu_j \mid 0, \sigma_\mu^2)\mathrm{u}(\sigma_\mu \mid 0, b_{\sigma\mu})\mathrm{n}(\boldsymbol{\beta} \mid \mathbf{0}, \sigma^2 \mathrm{diag}(\infty, \mathrm{v}\mathbf{J}_{NI}^\top)) \\ &\times \mathrm{n}(\boldsymbol{\beta}_\omega \mid \mathbf{0}, \sigma_\omega^2 \mathrm{v}_\omega \mathbf{I}_{NI+1})\mathrm{ig}(\sigma^2 \mid a_0/2, a_0/2)\mathrm{ig}(\sigma_\omega^2 \mid a_\omega/2, a_\omega/2),\end{aligned} \tag{19.12}$$

where $\mathbf{J}_{NI}^\top$ denotes the vector of NI ones and $\mathbf{I}_{NI+1}$ is the identity matrix of dimension $NI + 1$. As shown in Equation 19.12, the full specification of their prior density relies on the choice of the parameters $(b_{\sigma\mu}, \mathrm{v}, a_0, \mathrm{v}_w, a_w)$. In Section 19.6, where we illustrate the BNP-IRT model through the analysis of a real item response data set, we suggest some useful default choices for these prior parameters.

As shown by the model equations in Equations 19.4 through 19.7, the item response function $\Pr(U = 1 \mid \mathbf{x}; \boldsymbol{\zeta})$ is modeled by a covariate($\mathbf{x}$)-dependent location mixture of normal distributions for the latent variables u_{pi}^*. The random locations μ_j of this mixture corresponds to mixture weights $\omega_j(\mathbf{x})$, $j = 0, \pm1, \pm2, \ldots$. Conditionally on a covariate vector, $\mathbf{x}_{pi}$ and model parameters, the latent mean and variance of the mixture can be written as

$$\mathbb{E}[U_{pi}^* \mid \mathbf{x}_{pi};\boldsymbol{\beta},\boldsymbol{\beta}_\omega,\sigma^2,\sigma_\omega] = \mu_{pi}^* = \sum_{j=-\infty}^{\infty}(\mu_j + \mathbf{x}_{pi}^\top\boldsymbol{\beta})\omega_j(\mathbf{x}_{pi};\boldsymbol{\beta}_\omega,\sigma_\omega),$$

$$\mathbb{V}[U_{pi}^* \mid \mathbf{x}_{pi};\boldsymbol{\beta},\boldsymbol{\beta}_\omega,\sigma^2,\sigma_\omega] = \sum_{j=-\infty}^{\infty}\{[(\mu_j + \mathbf{x}_{pi}^\top\boldsymbol{\beta}) - \mu_{pi}^*]^2 + \sigma^2\}\omega_j(\mathbf{x}_{pi};\boldsymbol{\beta}_\omega,\sigma_\omega),$$

respectively (Marron and Wand, 1992).

The BNP-IRT model can be viewed as an extension of the DP-mixed binary logistic generalized linear model (Mukhopadhyay and Gelfand, 1997). In terms of the responses u, the extension can be written as

$$f(u \mid \mathbf{x}) = \sum_{j=1}^{\infty}\frac{\exp(\mu_j + \mathbf{x}^\top\boldsymbol{\beta})^u}{1+\exp(\mu_j + \mathbf{x}^\top\boldsymbol{\beta})}\omega_j,$$

$$\omega_j = \upsilon_j\prod_{k=1}^{j-1}(1-\upsilon_k),$$

$$\upsilon_j \sim \mathrm{Be}(1,\alpha),\ j = 1, 2, \ldots,$$

$$\mu_j \sim \mathrm{N}(0,\sigma_\mu^2),\ j = 1, 2, \ldots,$$

$$\boldsymbol{\beta} \sim \mathrm{N}(0,\Sigma_\beta).$$

This model thus defines a mixture of logistic cdfs for the inverse link function, with weights ω_j that are not covariate dependent. In contrast, as shown in Equations 19.6 and 19.7, the BNP-IRT model in Equations 19.4 through 19.11 is based on a mixture of normal cdfs for the inverse link function. The BNP-IRT model is more flexible than the DP model, because the former uses covariate-dependent mixture weights, as shown in Equation 19.8.

In other words, if $\mu_j = 0$ for all j, then the BNP-IRT model reduces to the Rasch IRT model with "normal-ogive" response functions; all items are assumed to have common slope (discrimination) parameter that is proportional to $1/\sigma$. Nonzero values of μ_j, along with the covariate-dependent mixture weights $\omega_j(\mathbf{x};\boldsymbol{\beta}_\omega,\sigma_\omega)$, for $j = 0, \pm1, \pm2, \ldots$, allows the BNP-IRT model to shift the location of each response function across persons and items. Value of $\mu_j > 0$ ($\mu_j < 0$) shifts the response function to the left (right). The BNP-IRT model allows for this shifting in a flexible manner, accounting for any outlying responses (relative to a normal-ogive Rasch model). This feature enables inferences of person and item parameters from the BNP-IRT that model are robust against such outliers.

According to Bayes' theorem, a set of data $\mathcal{D}$ updates of the prior probability density $\pi(\boldsymbol{\zeta})$ in Equation 19.12 leads to posterior probability density

$$\pi(\boldsymbol{\zeta} \mid \mathcal{D}) = \frac{f(\mathcal{D} \mid \mathbf{X};\boldsymbol{\zeta})\pi(\boldsymbol{\zeta})}{\int f(\mathcal{D} \mid \mathbf{X};\boldsymbol{\zeta})\pi(\boldsymbol{\zeta})d\boldsymbol{\zeta}}.$$

Also, conditionally on $(\mathbf{x}_{pi}, \mathcal{D})$, the posterior predictive pmf and the posterior expectation ($\mathbb{E}$) and variance ($\mathbb{V}$) of the item response U_{pi} are given by

$$f(u_{pi} \mid \mathbf{x}_{pi}, \mathcal{D}) = \int f(u_{pi} \mid \mathbf{x}_{pi}; \boldsymbol{\zeta}) \pi(\boldsymbol{\zeta} \mid \mathcal{D}) \, d\boldsymbol{\zeta}, \tag{19.13}$$

$$\mathbb{E}[U_{pi} \mid \mathbf{x}_{pi}, \mathcal{D}] = f(U_{pi} = 1 \mid \mathbf{x}_{pi}, \mathcal{D}) = f(1 \mid \mathbf{x}_{pi}, \mathcal{D}), \tag{19.14}$$

$$\mathbb{V}[U_{pi} \mid \mathbf{x}_{pi}, \mathcal{D}] = f(1 \mid \mathbf{x}_{pi}, \mathcal{D})[1 - f(1 \mid \mathbf{x}_{pi}, \mathcal{D})], \tag{19.15}$$

respectively.

It is straightforward to extend the BNP-IRT regression model to other types of response data by making appropriate choices of covariate vector $\mathbf{x}$ (corresponding to coefficients β, β_ω). Such extensions are described as follows:

1. Suppose that for each item $i = 1, \ldots, I$, the responses are each scored in more than two categories, say $m_i + 1$ nominal or ordinal categories denoted as $u' = 0, 1, \ldots, m_i$, with $u' = 0$ the reference category. Then, the model can be extended to handle such polytomous item responses using the Begg and Gray (1984) method. Specifically, the model would assume the response to be defined by $u_{pi} = \mathbf{1}(u'_{pi} > 0)$ each covariate vector $\mathbf{x}_{pi}$ to be defined by a binary indicator vector

$$\mathbf{x}_{pi} = (1, \mathbf{1}(p=1), \ldots, \mathbf{1}(p=N), \mathbf{1}(i=1)\mathbf{1}(u'_{pi}=1), \ldots, \mathbf{1}(i=I)\mathbf{1}(u'_{pi}=1), \ldots,$$
$$\mathbf{1}(i=1)\mathbf{1}(u'_{pi}=m_i), \ldots, \mathbf{1}(i=I)\mathbf{1}(u'_{pi}=m_i))^{\top}.$$

   Then in terms of coefficient vector $\boldsymbol{\beta} = (\beta_0, \beta_1, \ldots, \beta_{1+p+m^*I})$, coefficient $\beta_{1+p} = \theta_p$, $p = 1, \ldots, P$ represents the latent ability of person p and the coefficient $\beta_{1+p+(u-1)I+i}$ represents the latent difficulty of item $i = 1, \ldots, I$ and category $u = 1, \ldots, m^*$, where $m^* = \max_i m_i$.

2. If the data have additional covariates $(x_1, \ldots, x_q)$ which describe either the persons (e.g., socioeconomic status), test items (e.g., item type), or type of response (e.g., response time), associated with each person p and item i, then these covariates can be added as the last q elements to each of the covariate vectors $\mathbf{x}_{pi}$, such that $\mathbf{x}_{pi} = (\ldots, x_{1i}, \ldots, x_{qi})^{\top}$, $p = 1, \ldots, P$, and $i = 1, \ldots, I$. Then, specific elements of coefficient vector $\boldsymbol{\beta}$, namely the elements β_k, $k = \dim(\boldsymbol{\beta}) - q + 1, \ldots, \dim(\boldsymbol{\beta})$, would represent the associations of the q covariates with the responses.

3. Similarly, suppose that given test consists of measuring one or more of $D \leq I$ measurement dimensions. Then, we can extend the model to represent such multidimensional items, by including D binary (0,1) covariates into the covariate vectors $\mathbf{x}_{pi}$, $p = 1, \ldots, P$ and $i = 1, \ldots, I$, such that the first set of elements of $\mathbf{x}_{pi}$ defined by

$$\mathbf{x}_{pi} = (1, \mathbf{1}(p=1)\mathbf{1}(d_i=1), \ldots, \mathbf{1}(p=N)\mathbf{1}(d_i=1), \ldots, \mathbf{1}(p=1)\mathbf{1}(d_i=D), \ldots,$$
$$\mathbf{1}(p=N)\mathbf{1}(d_i=D), \ldots)^{\top},$$

   where $d_i \in \{1, \ldots, D\}$ denotes the measurement dimension of item i. Then, specific elements of the coefficient vector $\boldsymbol{\beta}$, namely the elements β_k, for $k = 2, \ldots, ND + 1$, indicate each person's ability on dimension $d = 1, \ldots, D$.

19.4 Parameter Estimation

By using latent-variable Gibbs sampling methods for Bayesian infinite-mixture models (Kalli et al., 2011), it is possible to conduct exact MCMC sampling from the posterior distribution of the BNP-IRT model parameters. More specifically, introducing latent variables $(\underline{u}_{pi}, z_{pi} \in \mathbb{Z}, u^*_{pi} \in \mathbb{R})_{N\times I}$ and a fixed decreasing function such as $\xi_l = \exp(-l)$, the conditional likelihood of the BNP-IRT model can be written as

$$\prod_{p=1}^{P}\prod_{i=1}^{I}\mathbf{1}(0 < \underline{u}_{pi} < \xi_{|z_{pi}|})\xi^{-1}_{|z_{pi}|}\mathrm{n}(u^*_{\mathrm{pi}} \mid \mu_{z_{\mathrm{pi}}} + \mathbf{x}^{\mathrm{T}}_{pi}\boldsymbol{\beta}, \sigma^2)\omega_{z_{\mathrm{pi}}}(\mathbf{x}^{\mathrm{T}}_{\mathrm{i}}\boldsymbol{\beta}_{\omega}, \sigma_{\omega}). \quad (19.16)$$

For each (p,i), after marginalizing over the latent variables in Equation 19.16, we obtain the original model likelihood $f(u_{pi}|\mathbf{x}_{pi}; \boldsymbol{\zeta})$ in Equation 19.4. Importantly, conditionally on the latent variables, the infinite-dimensional BNP-IRT model can be treated as a finite-dimensional model, which then makes the task of MCMC sampling feasible (of course, even a computer cannot handle an infinite number of parameters). Given all variables, save the latent variables $(z_i)_{i=1}^n$, the choice of each z_i has finite maximum value $\pm N_{\max}$, where $N_{\max} = \max_p[\max_i\{\max_j \mathbb{I}(\underline{u}_{pi} < \xi_j)|j|\}]$.

Then, standard MCMC methods can be used to sample the full conditional posterior distributions of each latent variable and model parameter repeatedly for a sufficiently large number of times, S. If the prior $\pi(\boldsymbol{\zeta})$ is proper (Robert and Casella, 2004, Section 10.4.3), then, for $S \to \infty$, this sampling process constructs a discrete-time Harris ergodic Markov chain

$$\{((\underline{u}^{(s)}_{pi}),(z^{(s)}_{pi}),(z^{*(s)}_{pi}),\boldsymbol{\zeta}^{(s)}) = ((\underline{u}_{pi}),(z_{pi}),\boldsymbol{\mu},\sigma_{\mu},\boldsymbol{\beta},\sigma^2,\boldsymbol{\beta}_{\omega},\sigma_{\omega})^{(s)}\}_{s=1}^{S},$$

which, upon after marginalizing out all the latent variables $(\underline{u}^{(s)}_{pi}),(z^{(s)}_{pi}),(z^{*(s)}_{pi})$, has the posterior distribution $\Pi(\boldsymbol{\zeta}|\mathcal{D}_n)$ as its stationary distribution (for definitions, see Nummelin, 1984; Meyn and Tweedie, 1993; Roberts and Rosenthal, 2004). (The next paragraph provides more details about the latent variables, $z^{*(s)}_{pi}$.)

The full conditional posterior distribution are as follows: the one of $\underline{u}_{pi}$ is $\mathrm{u}(\underline{u}_{pi} \mid 0, \xi_{|z_{pi}|})$; u^*_{pi} has a truncated normal distribution; the one of z_{pi} is a multinomial distribution independently for $p = 1, \ldots, P$ and $i = 1, \ldots, I$; the full conditional distribution of μ_j is a normal distribution (sampled using a Metropolis–Hastings algorithm), independently for $j = -N_{\max}, \ldots, N_{\max}$; σ_{μ} can be sampled using a slice sampling algorithm involving a stepping-out procedure (Neal, 2003); the one $\boldsymbol{\beta}$ is multivariate normal distribution; and the full conditional posterior distribution of σ^2 is inverse gamma. Also, upon sampling of truncated normal latent variables z^*_{pi} that have full conditional densities proportional to $\mathrm{n}(z^*_{pi}|\mathbf{x}^{\mathrm{T}}_i\boldsymbol{\beta}_{\omega}, \sigma_{\omega})\mathbf{1}(z_{pi} - 1 < z^*_{pi} < z_{pi})$, independently for $p = 1, \ldots, P$ and $i = 1, \ldots, I$, the full conditional posterior distribution of $\boldsymbol{\beta}_{\omega}$ is multivariate normal distribution and the one of σ^2_{ω} is inverse-gamma distribution. For further details of the MCMC algorithm, see Karabatsos and Walker (2012a).

In practice, obviously, only an MCMC chain based on a finite number S can be generated. The convergence of finite MCMC chains to samples from posterior distributions can be assessed using the following two procedures (Geyer, 2011): (i) viewing univariate trace plots of the model parameters to evaluate MCMC mixing (Robert and Casella, 2004) and (ii) conducting a batch-mean (or subsampling) analysis of the finite chain, which would provide 95% Monte Carlo confidence intervals (MCCIs) of all the posterior mean and quantile estimates of the model parameters (Flegal and Jones, 2011). Convergence can be confirmed both by trace plots that look stable and "hairy" and 95% MCCIs that, for all practical purposes, are sufficiently small. If convergence is not attained for the current choice of S samples of an MCMC chain, additional MCMC samples should be generated until convergence is obtained.

19.5 Model Fit

The fit of the BNP-IRT model to a set of item response data, $\mathcal{D}$, can be assessed on the basis of its posterior predictive pmf, defined in Equation 19.13.

More specifically, the fit to a given response u_{pi} can be assessed by its standardized response residual

$$r_{pi} = \frac{u_{pi} - \mathbb{E}[U_{pi} \mid \mathbf{x}_{pi}, \mathcal{D}]}{\sqrt{\mathbb{V}_n[U_{pi} \mid \mathbf{x}_{pi}]}}.$$

Response u_{pi} can be judged to be an outlier when $|r_{pi}|$ is greater than two or three.

A global measure of the predictive fit of a regression model, indexed by $\underline{m} \in \{1, \ldots, M\}$, is provided by the mean-squared predictive error criterion

$$D(\underline{m}) = \sum_{p=1}^{P} \sum_{i=1}^{I} \{u_{pi} - \mathbb{E}[U_{pi} \mid \mathbf{x}_{pi}, \mathcal{D}]\}^2 + \sum_{p=1}^{P} \sum_{i=1}^{I} \mathbb{V}_n[U_{pi} \mid \mathbf{x}_{pi}, \underline{m}].$$

(Laud and Ibrahim, 1995; Gelfand and Ghosh, 1998). The first term of $D(\underline{m})$ measures the goodness-of-fit (Gof $(\underline{m})$) of the model to the data, while its second term is a penalty ($P(\underline{m})$) for model complexity. Among a set of $\underline{m} = 1, \ldots, M$ that is compared, the model with the highest predictive accuracy for the data set $\mathcal{D}$ is identified as the one with the smallest value of $D(\underline{m})$.

The proportion of variance explained by the regression model is given by the R-squared (R^2) statistic

$$R^2 = 1 - \frac{\sum_{p=1}^{P} \sum_{i=1}^{I} \{u_{pi} - \mathbb{E}[U_{pi} \mid \mathbf{x}_{pi}, \mathcal{D}]\}^2}{\sum_{p=1}^{P} \sum_{i=1}^{I} \{u_{pi} - \overline{u}\}^2},$$

where $\overline{u} = (1/PI) \sum_{p=1}^{P} \sum_{i=1}^{I} u_{pi}$.

The standardized residuals r_{pi}, the $D(\underline{m})$ criterion, and R^2 can each be estimated as a simple by-product of an MCMC algorithm.

19.6 Empirical Example

Using the BNP-IRT model, we analyzed a set of polytomous response data obtained from the 2006 *Progress in International Reading Literacy Study*. A total of $N = 244$ fourth-grade U.S. teachers rated their own teaching preparation level in a 10-item questionnaire ($I = 10$). Each item was scored on a scale ranging from zero to two.

For this questionnaire, the latent person ability was assumed to represent the level of teaching preparation. The 10 items addressed the following areas: education level (named CERTIFICATE), English LANGUAGE, LITERATURE, teaching reading (PEDAGOGY), PSYCHOLOGY, REMEDIAL reading, THEORY of reading, children's language development (LANGDEV), special education (SPED), and second language (SECLANG) learning. The CERTIFICATE item was scored on a scale of 0 = bachelor's, 1 = master's, and 2 = doctoral, while the other nine questionnaire items were each scored on a scale consisting of 0 = not at all, 1 = overview or introduction to topic, and 2 = area of emphasis. Each of the 10 items described a type of training for literacy teachers, as prescribed by the *National Research Council* (2010).

We considered three additional covariates for the BNP-IRT model, namely AGE level (scored in nine ordinal categories), FEMALE status, and Miss:FEMALE, an indicator (0,1) of missing value for FEMALE status. Overall, 2419 of the total possible 2440 item were observed. Three of the 244 teachers had missing values for FEMALE, which were imputed using information from the observed values of all the variables mentioned above.

Given that each of the 10 items, item was scored on a polytomous scale (three categories), and that we were interested in additional covariates over and beyond the person-indicator and item-indicator covariates, we analyzed the data using the BNP-IRT model, using extensions nos 1 and 2 of the basic BNP-IRT model in Section 19.3 above. Also, the parameters of the prior pdf (5) of the model were chosen as $(b_{\sigma\mu}, v, a_0, v_w, a_w) = (1,10,1000,1,0.01)$.

To estimate the posterior distribution of the BNP-IRT model parameters, we ran the MCMC sampling algorithm in Section 19.4 for 62,000 iterations. We used 12,000 MCMC samples for posterior inference, retaining every fifth sample beyond the first 2000 iterations (burn-in) to obtain (pseudo-) independence between them. Trace plots for the univariate parameters displayed adequate mixing (i.e., exploration of the posterior distribution), and a batch-mean (subsampling) analysis of the 12,000 MCMC samples revealed 95% MCCIs of the posterior mean and quantile estimates (reported below) that typically had half-widths less than 0.2. If desired, smaller half-widths could have been obtained by generating additional MCMC samples.

For the BNP-IRT model, the standardized response residuals ranged from −0.21 to 0.20, meaning that the model had no outliers (i.e., all the absolute standardized residuals were well below two). Globally, the model fit analyses yielded criterion value $D(m) = 2.76$ (with $\mathrm{Gof}(m) = 0.03$ and Penalty $P(m) = 2.73$) for the 2419 responses in the data set. Also, the BNP-IRT model attained an R-squared of one.

The estimated posterior means of the person ability parameters were found to be distributed with mean 0.00, standard deviation 0.46, minimum −0.66, and maximum 3.68 for the 244 persons. Figure 19.1 presents a box plot of the marginal posterior distributions (full range, interquartile range, and median), for all the remaining parameters, including the item-difficulty parameters and the slope coefficients of the covariates AGE, FEMALE, and Miss:FEMALE. Parameter labels such as CERTIFICATE(1) and CERTIFICATE(2) refer to the difficulty of the CERTIFICATE item, with respect to its rating categories 1 and 2, respectively. The most difficult item was REMEDIAL(2) (with posterior median difficulty

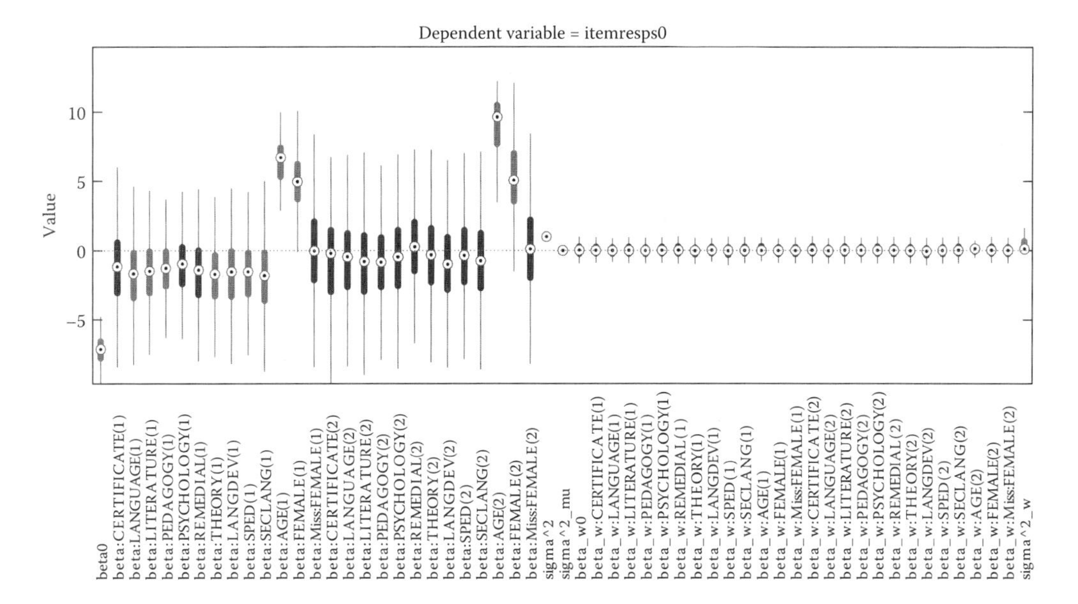

FIGURE 19.1
For the BNP-IRT model, a boxplot of the marginal posterior distributions of the item, covariate, and prior parameters. For each of these model parameters, the boxplot presents the range, interquartile range, and median.

of 0.27) and the easiest item was SECLANG(1) (posterior median difficulty –1.81). Also, the covariates AGE and FEMALE were each found to have a significant positive association with the rating response, since they had coefficients with 75% posterior intervals that excluded zero (this type of interpretation of significance was justified by Li and Lin, 2010). The box plot also presents the marginal posterior distributions for all the item and covariate parameters in $\boldsymbol{\beta}_{\omega}$, the mixture weights, and the variance parameters σ_{μ}, σ^2, and σ_{ω}^2.

19.7 Discussion

In this chapter, we proposed and illustrated a practical and yet flexible BNP-IRT model, which can provide robust estimates of person ability and item difficulty parameters. We demonstrated the suitability of the model through the analysis of real polytomous item response data. The model showed excellent predictive performance for the data, with no item response outliers.

For the BNP-IRT model, a user-friendly and menu-driven software, entitled: "Bayesian Regression: Nonparametric and Parametric Models" is freely downloadable from the authors website (see Karabatsos, 2014, 2015, 2016). The BNP-IRT model can be easily specified by clicking the menu options "Specify new model" and "Binary infinite homoscedastic probits regression model." Afterwards, the response variable, covariates, and prior parameters can be selected by the user. Then, to run for data analysis, the user can click the "Run posterior analysis" button to start the MCMC sampling algorithm in Section 19.4 for a chosen number of iterations. Upon completion of the MCMC run, the software automatically opens a text output file containing the results, which includes summaries of the posterior distribution of the model obtained from the MCMC samples. The software also allows the user to check for MCMC convergence through menu options that can be clicked to construct trace plots or run a batch-mean analysis that produces 95% MCCIs of the posterior estimates of the model parameters. Other menu options allow the user to construct plots (e.g., box plots) and text with the estimated marginal posterior distributions of the model parameters or residual plots and text reports the fit of the BNP-IRT model in greater detail.

Currently, the software provides a choice of 59 statistical models, including a large number of BNP regression models. The choice allows the user to specify DP-mixture (or more generally, stick-breaking-mixture) IRT models, with the mixing done either on the intercept parameter or the entire vector of regression coefficient parameters.

An interesting extension of the BNP-IRT model would involve specifying the kernel of the mixture by a cognitive model. For example, one may consider the multinomial processing tree model (e.g., Batchelder and Riefer, 1999) with parameters that describe the latent processes underlying the responses. Such an extension would provide a flexible, infinite-mixture of cognitive models that allows cognitive parameters to vary flexibly as a function of (infinitely-many) covariate-dependent mixture weights.

References

Batchelder, W. and Riefer, D. 1999. Theoretical and empirical review of multinomial processing tree modeling. *Psychonomic Bulletin and Review, 6*, 57–86.

Begg, C. and Gray, R. 1984. Calculation of polychotomous logistic regression parameters using individualized regressions. *Biometrika, 71*, 11–18.

DeIorio, M., Müller, P., Rosner, G., and MacEachern, S. 2004. An ANOVA model for dependent random measures. *Journal of the American Statistical Association, 99*, 205–215.

Duncan, K., and MacEachern, S. 2008. Nonparametric Bayesian modelling for item response. *Statistical Modelling, 8*, 41–66.

Farina, P., Quintana, F., Martin, E. S., and Jara, A. 2009. A dependent semiparametric Rasch model for the analysis of Chilean educational data (Unpublished manuscript).

Ferguson, T. 1973. A Bayesian analysis of some nonparametric problems. *Annals of Statistics, 1*, 209–230.

Flegal, J. and Jones, G. 2011. Implementing Markov chain Monte Carlo: Estimating with confidence. In S. Brooks, A. Gelman, G. Jones, and X. Meng (Eds), *Handbook of Markov Chain Monte Carlo* (p. 175–197). Boca Raton, Florida: CRC.

Gelfand, A. and Ghosh, J. 1998. Model choice: A minimum posterior predictive loss approach. *Biometrika, 85*, 1–11.

Geyer, C. 2011. Introduction to MCMC. In S. Brooks, A. Gelman, G. Jones, and X. Meng (Eds), *Handbook of Markov Chain Monte Carlo* (p. 3–48). Boca Raton, Florida: CRC.

Ishwaran, H., and James, L. 2001. Gibbs sampling methods for stick-breaking priors. *Journal of the American Statistical Association, 96*, 161–173.

Kalli, M., Griffin, J., and Walker, S. 2011. Slice sampling mixture models. *Statistics and Computing, 21*, 93–105.

Karabatsos, G. 2014. Bayesian regression: Nonparametric and parametric models. Software users manual, University of Illinois-Chicago, http://www.uic.edu/georgek/HomePage/BayesSoftware.html.

Karabatsos, G. 2015. A menu-driven software package for Bayesian regression analysis. *ISBA Bulletin*, 22(4), 13–16.

Karabatsos, G. 2016. A menu-driven software package of Bayesian nonparametric (and parametric) mixed models for regression analysis and density estimation. *Behavior Research Methods.* Submitted for publication.

Karabatsos, G. and Walker, S. 2012a. Adaptive-modal Bayesian nonparametric regression. *Electronic Journal of Statistics, 6,* 2038–2068.

Karabatsos, G. and Walker, S. 2012b. Bayesian nonparametric mixed random utility models. *Computational Statistics and Data Analysis, 56(6),* 1714–1722.

Laud, P., and Ibrahim, J. 1995. Predictive model selection. *Journal of the Royal Statistical Society, Series B, 57,* 247–262.

Li, Q. and Lin, N. 2010. The Bayesian elastic net. *Bayesian Analysis, 5,* 151–170.

MacEachern, S. 1999. Dependent nonparametric processes. *Proceedings of the Bayesian Statistical Sciences Section of the American Statistical Association,* Alexandria, VA, 50–55.

MacEachern, S. 2000. Dependent Dirichlet processes, Technical Report, Department of Statistics, Columbus, OH: The Ohio State University.

MacEachern, S. 2001. Decision theoretic aspects of dependent nonparametric processes. In E. George (Ed.), *Bayesian Methods with Applications to Science, Policy and Official Statistics* (pp. 551–560). Creta, Greece: International Society for Bayesian Analysis.

Marron, J. and Wand, M. 1992. Exact mean integrated squared error. *The Annals of Statistics, 20,* 712–736.

Meyn, S. and Tweedie, R. 1993. *Markov Chains and Stochastic Stability.* London: Springer-Verlag.

Miyazaki, K. and Hoshino, T. 2009. A Bayesian semiparametric item response model with Dirichlet process priors. *Psychometrika, 74,* 375–393.

Mukhopadhyay, S. and Gelfand, A. 1997. Dirichlet process mixed generalized linear models. *Journal of the American Statistical Association, 92,* 633–639.

National Research Council. 2010. *Preparing Teachers: Building Evidence for Sound Policy.* Washington, DC: National Academies Press.

Neal, R. 2003. Slice sampling (with discussion). *Annals of Statistics, 31,* 705–767.

Nummelin, E. 1984. *General Irreducible Markov Chains and Non-Negative Operators.* London: Cambridge University Press.

Qin, L. 1998. Nonparametric Bayesian models for item response data, PhD thesis, Unpublished doctoral dissertation, Columbus, OH: The Ohio State University.

Rasch, G. 1960. *Probabilistic Models for Some Intelligence and Achievement Tests.* Copenhagen, Denmark: Danish Institute for Educational Research (Expanded edition, 1980, Chicago, Illinois: University of Chicago Press).

Robert, C. and Casella, G. 2004. *Monte Carlo Statistical Methods* (2nd edn). New York: Springer.

Roberts, G. and Rosenthal, J. 2004b. General state space Markov chains and MCMC algorithms. *Probability Surveys, 1,* 20–71.

Rost, J. 1990. Rasch models in latent classes: An integration of two approaches to item analysis. *Applied Psychological Measurement, 14,* 271–282.

Rost, J. 1991. A logistic mixture distribution model for polychotomous item responses. *British Journal of Mathematical and Statistical Psychology, 44,* 75–92.

San Martín, E., Jara, A., Rolin, J.-M., and Mouchart, M. 2011. On the Bayesian nonparametric generalization of IRT-type models. *Psychometrika, 76,* 385–409.

San Martín, E., Rolin, J.-M., and Castro, L. M. 2013. Identification of the 1PL model with guessing parameter: Parametric and semi-parametric results. *Psychometrika, 78,* 341–379.

Sethuraman, J. 1994. A constructive definition of Dirichlet priors. *Statistica Sinica, 4,* 639–650.

20

Functional Approaches to Modeling Response Data

James O. Ramsay

CONTENTS

20.1 Introduction 337
20.2 Modeling Item Response Manifolds 338
20.2.1 Performance-Indexing Systems 341
20.2.1.1 Thurstone Tradition 341
20.2.2 Uniform Measure or Rank 342
20.2.2.1 Arc-Length Metric 342
20.2.2.2 Metrics for [0, n] 343
20.3 Estimating Performance Manifolds $\mathcal{P}$ 344
20.3.1 Improving Sum Score by Weighting 344
20.3.2 A Functional Test Analysis Algorithm 346
20.4 Plotting Performance Manifolds 348
20.5 Conclusion 349
Acknowledgment 350
References 350

20.1 Introduction

Statistical applications often involve the estimation of smooth curves, and the item response function (IRF) $P_i(\theta)$, where $i = 1, \ldots, n$, the probability of success on item i that is conditional on value θ, is an iconic example. Functional data analysis as articulated in Ramsay and Silverman (2005) and Ramsay et al. (2009) aims to allow the data analyst as much flexibility as is required for this task, while at the same time providing tools to ensure that the estimated curves satisfy whatever smoothness and shape constraints the application requires. Chapter 9 of Ramsay and Silverman (2002) illustrated these principles in a test analysis situation.

This chapter reviews item response theory (IRT) from a functional perspective, and argues that IRFs, as well as other models making use of unobserved latent variables, are in fact modeling data using a manifold. The space curve defined by a one-dimensional item response model is such an object, and the next section discusses its fundamental invariance properties. In particular, both the domain over which θ is defined and its distribution over this domain can be chosen arbitrarily.

The chapter goes on from this somewhat abstract point to show how these two choices can help test takers and interpreters to more easily understand IRT by seeing how it is applied, without any loss of generality, within a context that is more familiar, namely performance estimates θ distributed over the interval [0, n]. A simple method for estimating IRFs is outlined that is easy to turn into a code, fast, and remarkably efficient from a statistical

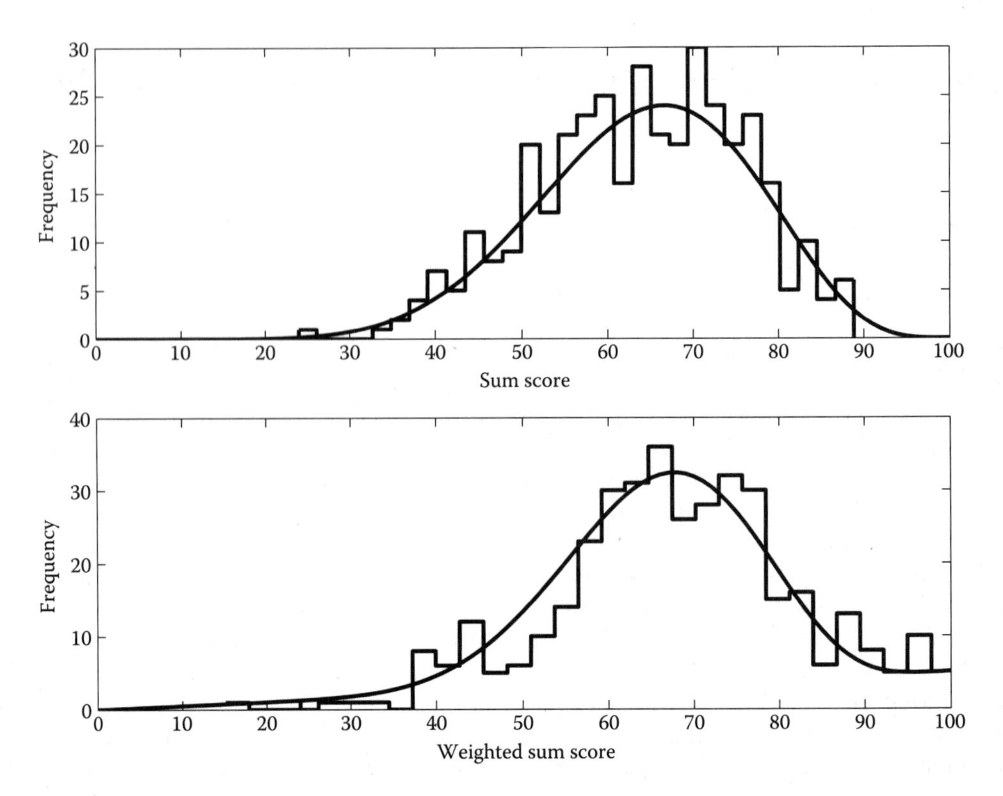

FIGURE 20.1
In the top panel, the step change line is a histogram of the frequency of binned sum score values for the introductory psychology test containing 100 items and given to 379 examinees. The smooth curve is the TSβ density fit to these data and rescaled to overlap the histogram. The bottom panel plots these results for the weighted sum score estimates defined in Equation 20.7.

perspective. These IRFs are then used to estimate weighted sum score estimates of ability that are more robust than sum scores with respect to the presence of faulty test items.

This chapter focuses on the application of IRT to course evaluation in teaching institutions, and to this end, the methods in this chapter will be illustrated by the analysis of data from an introductory psychology exam given at McGill University. Figure 20.1 displays a histogram of the sum scores X_p, $p = 1, \ldots, N$ distributed over 31 bins for an examination containing 100 items, each with five options, and given to $N = 379$ examinees. The minimum and maximum sum scores were 24 and 91, respectively. The smooth curve is a rescaled tilted scaled beta (TSβ) density function that is fit to these data and that is defined by four parameters chosen to capture the main shape characteristics of the distribution of sum scores. This distribution will be discussed in Section 20.2.2.2.

20.2 Modeling Item Response Manifolds

Most of the functional aspects of IRT are captured in Figure 20.2, which displays the IRFs for three selected items from the introductory psychology exam displayed in Figure 20.1. Item 3 was a typical easy item, Item 4 was of intermediate difficulty, and Item 29 was one of the few really challenging items in the exam.

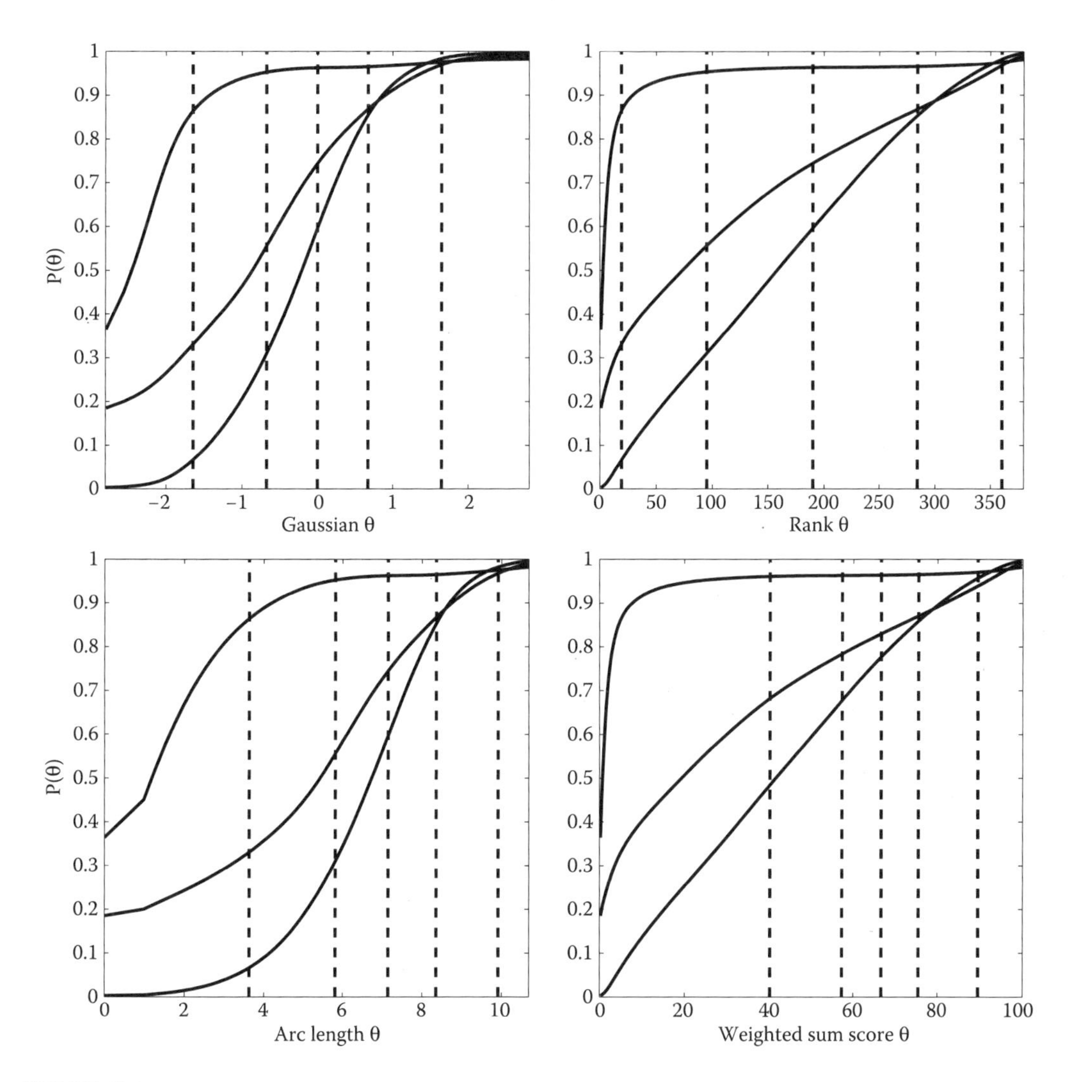

FIGURE 20.2
The upper-left panel shows the IRFs $P_i(\theta)$ for Items 3, 4, and 29 in the introductory psychology test, assuming that θ has a standard normal distribution in the population of examinees. The upper-right panel displays the curve defined over the ranks of the examinees. The lower-left panel shows the same curves defined over the arc length along the performance space curve. The lower-right panel shows these curves plotted over the TSβ distribution fit to the weighted sum score data. The vertical dashed lines in each panel indicate the 5%, 25%, 50%, 75%, and 95% quantiles.

The notation $P_i(\theta)$ suggests that P_i is a function. But in IRT, θ is not observed; and, consequently, the domain of P is not like time, space, wavelength, and other continua that are directly measurable. Instead, θ plays the role of an *index*, each value of which is associated with its own unique probability value.

As with indices in general, we can therefore contemplate any transformation $h(\theta)$ that is one to one and therefore also uniquely identifies the value of a probability. That is, for any such h, there exists an alternative but equivalent function P^* such that

$$P^*[h(\theta)] = P(\theta) \quad \text{for all } \theta \tag{20.1}$$

Since transformation h is one to one or bijective, there exists an inverse transformation denoted by h^{-1} such that $\theta = h^{-1}[h(\theta)]$ for all θ. Defining $\phi = h(\theta)$, we see that

$$P^*(\phi) = P[h^{-1}(\phi)] = P(\theta) \tag{20.2}$$

or more compactly that $P^* \circ h = P$ and $P^* = P \circ h^{-1}$. In other words, $P_i(\theta)$ denotes an *equivalent class* of functions, each member of which corresponds to a bijective transformation of the baseline argument θ.

We may require other additional properties of h, such as being strictly increasing, being continuous, or being smooth in the sense of being differentiable up to some order. Each of these additional constraints are imposed when we conjecture that they capture something about the learning process that we and our community of collaborators consider as given.

Any realization of a unidimensional item response model will, as a consequence, involve two choices: The interval over which θ is defined, and the probability distribution $\pi(\theta)$ over this interval, which defines a metric for purposes of integration and differentiation of P with respect to θ. Both these choices will impact the visual appearance of IRFs and how they are interpreted. The panels of Figure 20.2 plot success probabilities as functions of four choices of θ indices, each having a credible motivation, but nevertheless offering quite different perspectives on how these three items reflect performance on the test.

Before discussing the merits of these four representations, we consider what is in common among these models. Figure 20.3 hides index θ by plotting the three sets of P-values

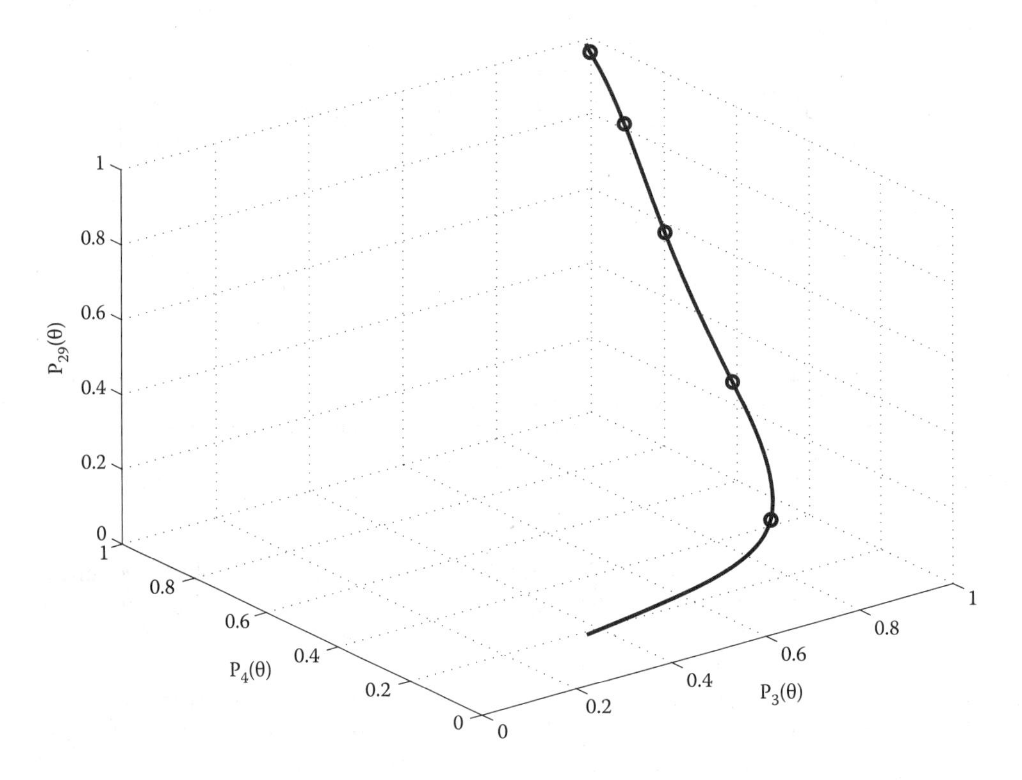

FIGURE 20.3
Values of $P_3(\theta)$, $P_4(\theta)$, and $P_{29}(\theta)$ for the introductory psychology test plotted as a space curve. The circles indicate the position of the 5%, 25%, 50%, 75%, and 95% quantiles of the TSβ density fit to the data.

as points in a three-dimensional cube. Now, we see what is invariant with respect to transformation h: the space curve along which all examinees are assumed to be positioned, and which we denote by $\mathcal{P}$. The trajectory $\mathcal{P}$ does not depend on how we choose the indexing system for charting locations along the curve. The triplets of 0–1 values associated with each examinee's data are located at the $2^3 = 8$ vertices of this cube, and the three coordinates $P_3(\theta)$, $P_4(\theta)$, and $P_{29}(\theta)$ are the probabilities of success for these items for all examinees located at or near any specific point θ on the curve.

A time-like interpretation of the curve suggests that examinees first learn what Item 3 tests, and then make a steady progress toward mastering the content of Items 4 and 29, starting from a platform of about 0.4 for Item 4 but close to 0 for Item 29.

More generally, performance on an n-item test is modeled as a space curve $\mathcal{P}$ in a closed n-dimensional hypercube that we might call the *performance space* for the test (Ramsay, 1996; Rossi et al., 2002). The curve itself is the *performance manifold*, and any point on this manifold corresponds to a specific level of performance. When the responses to items are multinomial with k_i categories, the probability of a specific response combination for each item is represented as a point in a simplex of dimension $k - 1$, and the resulting total response structure is a *simplicial hypercube* of dimension n, each "edge" of which is a simplex.

Indexing systems such as θ define *charts*, using the terminology of Boothby (1975), which lay out a navigational system within a manifold, in much the same way that longitude charts position on the equator; and, when combined with latitude, charts the entire globe. The manifold is the model for the data in much the same sense that the k-dimensional subspace identified by principal components analysis, called a *Grassmann mainfold*, is the model for the p-dimensional multivariate data.

20.2.1 Performance-Indexing Systems

We now consider the four index choices displayed in Figure 20.2, where three selected IRFs (3, easy; 4, medium; and 29, hard) are plotted against four different choices of interval and metric for θ. The values of the curves were estimated by the methodology used in TestGraf (Ramsay, 1991), in which examinees are ranked in terms of sum scores perturbed by a small random deviate, and the value of θ then allocated to each examinee being the quantile of the distribution defining the plot and corresponding to the examinee's rank.

20.2.1.1 Thurstone Tradition

The indexing system most often used by psychometricians is displayed in the upper-left panel of Figure 20.2, and was introduced as a choice model by Thurstone (1927). The interval $(-\infty, \infty)$ and Gaussian measure were handy for expressing ogival shapes in terms of a small number of parameters, such as the single-item difficulty parameter in the Rasch model, the difficulty/discriminability parameters in the two-parameter logistic (2PL) model, and the mixture of the uniform and 2PL model in the three-parameter logistic (3PL) model.

There are two downsides of the Gaussian metric. The first is the difficulty of explaining the meaning of a θ-value in this metric to anyone not possessing a high level of statistical literacy. In contrast, the magnitudes that constitute most of our experiential framework incline us to conjecture a fixed lower-bound zero on performance, a result achieved by an examinee that allocates the exam period to a nice nap. Moreover, and especially for multiple-choice tests, there can be an achievable upper limit of complete mastery as measured by the test, a value often assigned as the value of 100%.

The second problem is rather more subtle but even more devastating from a marketing perspective. This is the widespread claim made both explicitly and implicitly through terms such as "psychometrics" and "educational measurement" that IRT provides a metric for performance that has the additive properties of scientific measurement. That is, the test theory literature often loses sight of the infinite transformability of θ, including the obvious fact that there is no justification for assuming that performances for different examinee populations will conform to a common distribution.

20.2.2 Uniform Measure or Rank

The upper-right panel of Figure 20.2 plots the success probability against the student's rank in the class, which is a legitimate indexing system; and it is at once clear that only the bottom 10 students have any trouble with Item 3, although we wonder why a failure rate of around 5% persists over almost all of the rest of the class. Only the top 50 or so get Item 4 right with a probability of 0.4 or higher, and the bottom half of the class fail to get Item 29 half of the time or more.

Whether we employ the interval $[0, P]$ containing ranks or the unit interval $[0, 1]$ is of no consequence, the uniform distribution violates the intuition that performance in anything has a few impossibly clumsy persons, which is a much-larger collection of amateurs ranging from passable to high-end, and a small group of olympic-class athletes. That is, the difference between percentiles 45 and 59 seems rather less interesting than that between 85 and 99.

20.2.2.1 Arc-Length Metric

The lower-left panel of Figure 20.2 displays an indexing system that is uniquely defined by a space curve and, in addition, actually does have metric properties. Arc length

$$s(\theta) = \int_0^\theta \sqrt{\sum_i^n \left(\frac{dP_i}{dz}\right)^2}\, dz \tag{20.3}$$

where $z \in [0, n]$ is a variable of integration, is invariant with respect to transformations of θ, and measures how far within performance space $\mathcal{P}$ one has to go to achieve a target performance s. A real-world unit of performance can be the energy, time, effort, money, or other resources required to advance by a fixed amount Δs along the curve. Moreover, the length of the tangent vector $(dP_1/ds, \ldots, dP_n/ds)$ with respect to arc length is one everywhere, and the angles of this vector with respect to the coordinates taken at any point on the curve are useful measures of item-wise progress toward mastery.

Arc length also has a natural interpretation. Since all examinees are assumed to traverse on the same space curve to achieve mastery, and since the curve is defined in terms of coordinates representing n aspects of test performance, one's position on the curve is a measure of total test performance, and the length of the curve from this point to its terminus is the remaining performance improvement required to reach the level of mastery of the best examinees. The unit of arc length can be any positive value, so that if we normalize the curve to have a total arc length of 100%, we can speak of "performance mastery."

An important geometrical feature of the space curve in Figure 20.3 is its anchoring at its upper end in the $(1, \ldots, 1)$ vertex, but there is no obvious meaning to the lower end. Perhaps, the dummies in next year's class will extend the curve in the general direction of the complete failure corner (0, ..., 0), but it seems unlikely that any examinee will get there, unless by intent. This suggests that the arc length along the curve might be better started at the mastery vertex, thereby measuring what effort remains for achieving complete mastery. However, the distribution of arc length does not conform to our intuitions concerning the higher frequency of mediocrity.

20.2.2.2 Metrics for [0, n]

What is missing from these three displays is a metric that could be explained to a student who has just taken an exam in history of the Roman empire, where a knowledge of the Gaussian distribution cannot be assumed. What students do know well is the experience of taking multiple-choice examinations in large courses, or university entrance examinations such as the Scholastic Aptitude Test of Educational Testing Service. By comparing notes with other examinees, they are reasonably aware of the chances of falling into either the failing or exceptional performance range, and know, too, that such exams vary considerably in average scores from one course and occasion to another. These exams are almost surely marked using the sum score X_p.

Ramsay and Wiberg (2016) proposed the TSβ distribution with density

$$p(S|\,\alpha,\beta,h_0,h_n)=\frac{h_0(1-S/n)+h_1(S/n)+(S/n)^{\alpha-1}(1-S/n)^{\beta-1}}{n(h_0+h_n)/2+B(\alpha,\beta)} \tag{20.4}$$

where $B(\alpha, \beta)$ is the beta function defined by parameters α and β. The TSβ density has the shape of the beta density, but is scaled to be over $[0, n]$ and departs from beta shape in having the potential for positive end-point heights h_0 and h_n. The reality of large sets of testing data is that there are almost always some examinees who will either fail every item or pass all items.

Maximum-likelihood estimation for the TSβ distribution proceeds smoothly provided α and β are greater than 2; otherwise, the gradient is unbounded as θ approaches its boundaries. Consequently, the log likelihood is optimized with respect to the transformed parameters $\alpha^* = \log(\alpha-2), \beta^* = \log(\beta-2), h_0^* = \log(h_0)$, and $h_n^* = \log(h_n)$ and obtains rapid convergence using the quasi-Newton algorithm. Only score values that are not at the boundaries are used in the estimation.

Parameters α and β were estimated at 9.1 and 5.1, respectively, indicating a strongly peaked and right-shifted density, as is apparent in the top panel of Figure 20.1. The two end-point height parameters were estimated as virtually zero, but this might not happen for much of the data that we have analyzed. In fact, we will see that this test actually has items that penalize the brightest students, and therefore keeps sum scores from reaching 100 items.

The lower-left panel of Figure 20.2 shows how the three IRFs look like when plotted with respect to the TSβ density fit to the weighted sum scores defined in Section 20.3. The vertical quantile lines show that the lower 40% of the plotting interval is empty of all but 5% of the weighted sum scores. Over the central 90% of the sum scores, the curves are remarkably straight, and therefore easy to describe in terms of the average slope and level.

We conclude that all four displays are useful in at least some sense. In fact, the lack of identifiability of IRFs with respect to their domain and measure is actually a valuable source of versatility, enabling a range of displays, each of which will be considered informative by some class of observers.

20.3 Estimating Performance Manifolds $\mathcal{P}$

Ramsay (1991) developed an approach and a program TestGraf that was algorithmic in nature rather than requiring the optimization of a fitting criterion or the use of Bayesian Markov chain Monte Carlo (MCMC) sampling. This method was remarkable for its computational efficiency; but it also turned out to be surprisingly accurate when applied to simulated data (Lei et al., 2004), and was shown by Douglas (2001) to asymptotically identify the true IRFs. In this section, a new version of the TestGraf strategy that replaces kernel smoothing by more powerful spline-smoothing methods is proposed for the simultaneous estimation of IRFs and weighted sum score ability estimates.

20.3.1 Improving Sum Score by Weighting

We now turn to the essential question of how we use the data in vector $\mathbf{u}_p$ for examinee p to estimate θ_p. Any estimation strategy must produce results that are invariant with respect to differentiable strictly monotonic transformations $\theta^* = h(\theta)$ so as to preserve the order of two examinees who have taken an arbitrarily long test and are therefore separated by an arbitrarily small amount $d\theta$.

For example, the sum score satisfies this criterion since its expectation $\Sigma_i U_{pi}$ is a linear function of the values of the n IRFs and these values, as we have seen, are invariant in this way. However, the sum score as an estimated performance measure is flawed in the sense that it treats all items symmetrically for all values of θ when in fact items tend to vary widely in their information value at all ability levels.

Maximum-likelihood estimates of θ conditional on values of $P_i(\theta)$ and $W_i(\theta)$ also satisfy this invariant principle. From the log likelihood

$$-\log L(\theta | \mathbf{W}, \mathbf{u}) = \sum_{i=1}^{n} [U_{pi} W_i(\theta) - \log[1 + \exp W_i(\theta)] \tag{20.5}$$

where $\mathbf{W}$ is the n-vector containing the log-odds functions W_i and $\mathbf{u}$ is the n binar data vector, we have the maximum-likelihood estimate $\hat{\theta}$ that solves the stationary equation

$$\sum_{i}^{n} [U_i - P_i(\theta)] \frac{\partial W_i}{\partial \theta} = 0 \tag{20.6}$$

By the chain rule

$$\frac{\partial W_i}{\partial \theta^*} = \frac{\partial W_i}{\partial \theta} \frac{dh}{d\theta}$$

and, since by strict monotonicity $dh/d\theta$ is positive, invariance follows. This estimate has the virtue that it weighs items by the slopes of the W-functions, so that items that have near-zero slopes are virtually ignored (see, e.g., Figure 20.5), and getting an item "right" at a point where the slope is negative can actually decrease the θ estimate.

The weighting of the residuals by the value of the derivative of the log-odds ratio in this equation suggests the use of the weighted average estimating equation implicit in θ

$$\hat{\theta} = n\left[\sum_i U_i \frac{\partial W_i}{\partial \hat{\theta}}\right] \Bigg/ \left[\sum_i \frac{\partial W_i}{\partial \hat{\theta}}\right] \tag{20.7}$$

That is, if we define θ in this way, we see again by the chain rule that its expectation is also invariant; with respect to *h*, we achieve the desired *W*-slope weighting, and at the same time turn the weighted sum into a weighted average that keeps its expected value within [0, *n*].

These values of θ converge rapidly for most examinees when this equation is iterated starting with the sum score values. Figure 20.4 displays the difference between the sum scores and their iterated weighted counterparts after 10 iterations of Equation 20.7. The result is a substantial improvement in the quality of the ability estimate, and especially so for examinees at the extremes. The better examinees have increases equivalent to as much as eight items, allowing the best to come close to 100% performance. This is due to the virtual elimination of the influence of the six poor items on their estimates. On the other hand, the hapless low-end examinees experience decreases by as much as 10 items. The median of the iterated values is 1.5 items higher than that of the sum scores.

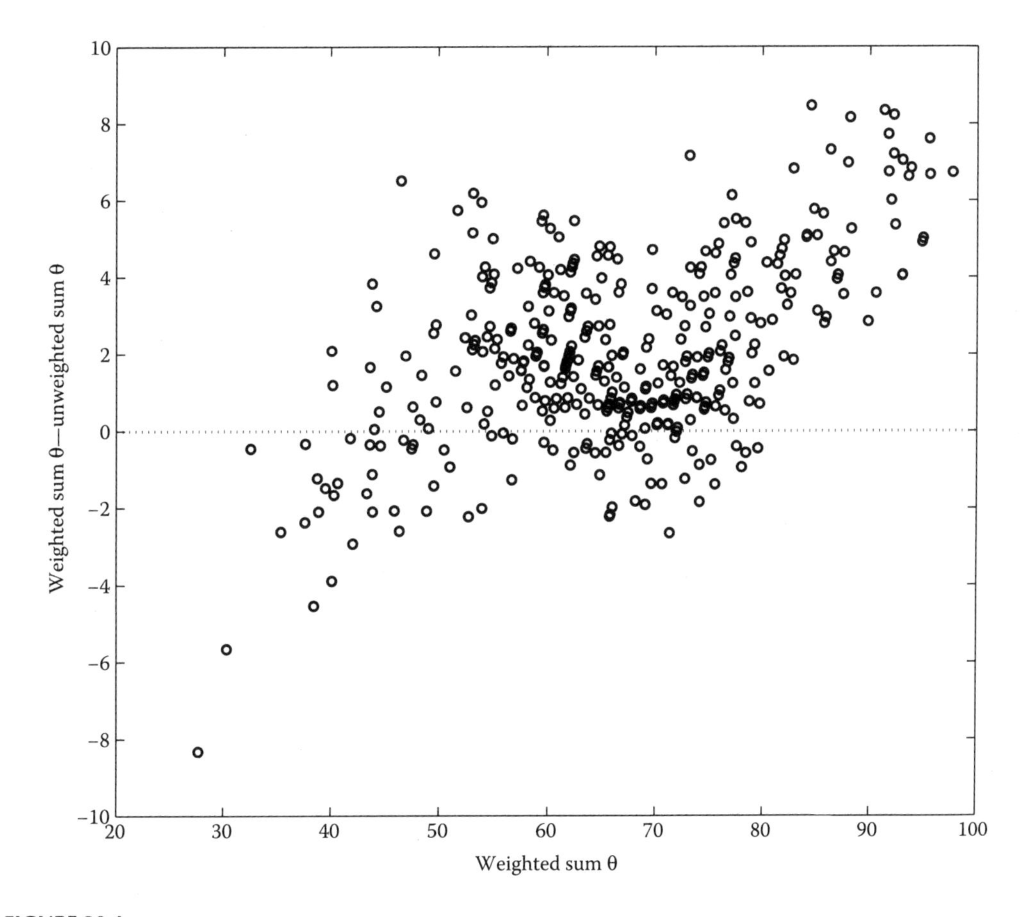

FIGURE 20.4
Difference between the weighted sum scores and the corresponding sum score estimates in Equation 20.7 of θ are plotted against the weighted estimates.

Figure 20.1 displays in its bottom panel the histogram of the weighted sum score estimates satisfying Equation 20.7 along with a rescaled plot of the TSβ density function fit to these values. A comparison with the top panel shows that the variance of the performance estimates has increased, and the positive tail now has a positive probability estimated for a score of $n = 100$ correct items.

20.3.2 A Functional Test Analysis Algorithm

A density is selected, which may be one of the four described here or some other, such as a more fine-grained nonparametric representation of the sum score distribution. The estimation algorithm alternates between the smoothing of binned binary data to estimate the log-odds functions W_i and the application of Equation 20.7 as an updater of previous ability estimates $\hat{\theta}_p$. The number K of bins chosen will depend on the number N of examinees and the desired resolution of the W_i's. Eleven bins were used for the introductory psychology data.

The steps are as follows:

1. *Initialization*: The initial steps are:
   a. *Initializing* θ_p: Unweighted sum scores were the initial values for θ_p in the analysis of the introductory psychology data, but any other initial values may be used depending on the choice of interval and density.
   b. *Initializing the density*: In this chapter, the TSβ density over $[0, n]$ was initialized to the distribution of unweighted sum scores.
2. *Iterative updates*: At each iteration $\nu = 1, \ldots$ the following steps are taken:
   a. *Converting ranks into quantiles*: The normalized rank values $\tilde{u}_p = p/(N+1)$, are converted into values $\hat{\theta}_p^{\nu-1}$ of the variate having the selected distribution by using its quantile function values $Q^{\nu-1}(u)$ for the distribution estimated in the previous iteration.
   b. *Computing bin boundaries*: The values $Q^{\nu-1}[(0{:}\ 2K)/(2K)]$ are calculated if the domain is closed; or $Q^{\nu-1}[(1{:}\ 2K)/(2K-1)]$ otherwise plus suitably extreme endpoints, where K is the number of planned bins. The odd-numbered quantiles are bin centers and the even-numbered ones are bin boundaries. The numbers of rank–quantiles falling into each bin will not vary a great deal.
   c. *Log-odd transforms of within-bin averages*: The P quantile values are assigned to the sorted previous ability estimates, with ties being broken, if they occur, by adding small random deviates. Within-bin averages of the data are approximate probabilities, and, adjusting a little for values of 0 and 1 if they occur, the log-odds-transformed bin averages are used as estimates of the log-odds functions $W_i(\theta)$ in Equations 20.5 through 20.7 evaluated at bin centers.
   d. *Spline smoothing of log-odds values*: A spline basis, typically of order four, is set up with knots located at bin centers, and a spline-smoothing function such as `smooth.basis` in the R package `fda` is used to smooth the within averages in Step 2c over the bin centers. This smooth should penalize roughness, and a second-derivative penalty with a penalty parameter of 100 works well in general.

e. *Ability updates*: The values $\hat{\theta}_p^{\nu-1}$ from Step 1a are updated by the application of Equation 20.7 using the log-odds functions $W_i(\theta)$ from Step 2c. Values outside the boundaries, if they occur, are replaced by boundary values. A convergence criterion is also computed and used to terminate iterations if appropriate.

f. *Update of the density*: If desired, as was the case for the analysis reported here, the density function is reestimated for the updated ability values.

This process will only require a few seconds when even tens of thousands of exam records are involved. Ramsay and Wiberg (2016) report that when maximum-likelihood estimation method uses these values as starting values, the improvement in root-mean-squared error tends to be about 5%–7%, suggesting that in many if not most applications, the computational overhead involved will scarcely be justified.

Figure 20.5 displays all 100 IRFs for the introductory psychology exam estimated in this way using a smoothing parameter value of 1000. Six problem items are highlighted, and these explain why the highest sum score was 91, and also why the flexibility of functional

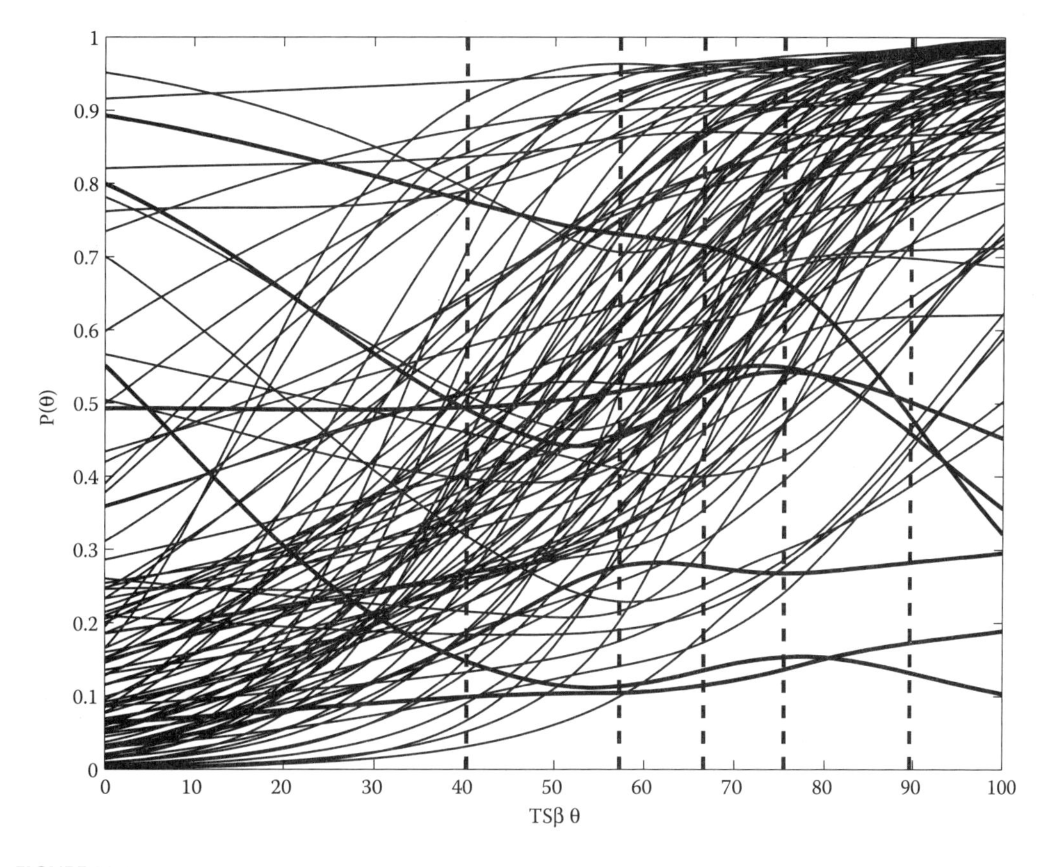

FIGURE 20.5
100 IRFs for the introductory psychology test estimated by spline smoothing of log-odds-transformed binned data over the quantiles of the TSβ density in Figure 20.2. The six heavy lines show IRFs that are inconsistent with the other 94 IRFs and which prevented the best students from achieving the highest scores.

methods is essential in the rather messy world of postsecondary testing. The inclusion of the five quantile locations conveys some information about the distribution of sum scores and supports the interpretation of curve shapes.

20.4 Plotting Performance Manifolds

Imagining what plotting all the $P_i(\theta)$ values against each other looks like in 100-dimensional space is beyond most of us. The space curve for the introductory psychology exam, fortunately, does not actually require all these dimensions. A singular-value decomposition of the 1001-by-100 matrix **P** containing 100 IRFs evaluated over a mesh of 1001 points, followed by a rotation of these curve values into their three-dimensional principal-component subspace captures 99.1% of the shape variation in $\mathcal{P}$. Figure 20.6 displays this projection into the three-dimensional principal-component subspace, and shows that most of the movement is in the principal component I–II plane, accounting for 94.1% shape variation. But there is an excursion into the third principal component for examinees of intermediate performance.

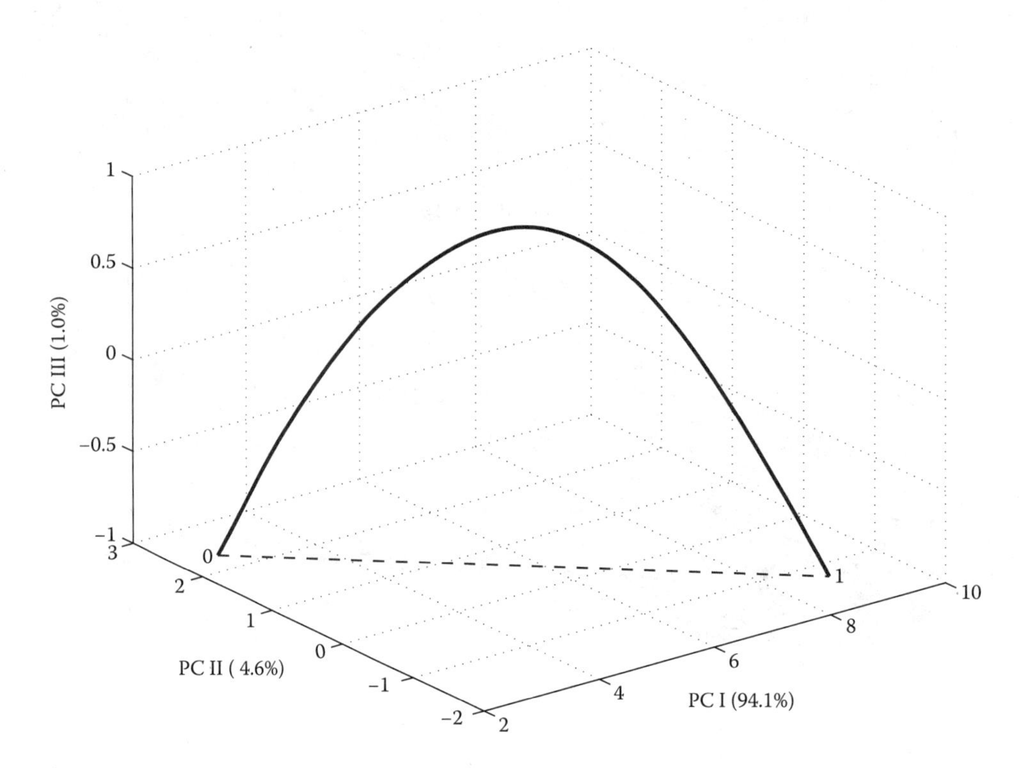

FIGURE 20.6
Item response or performance space curve projected into its first three principal components of variation, which accommodate 99.1% of its shape. The asterisk indicates the highest-performance end of the curve and the circle indicates the lowest end.

20.5 Conclusion

By some accounts, modern test theory is now in its eighth decade. What has been its impact on educational assessment? Certainly, the IRF has proven itself to be a great aid to test and scale developers working in the context of a large professional assessment organization as well as those blessed with a relatively advanced knowledge of statistics and test theory. However, even the largest testing agencies fall back on reporting results based on sum score estimates, and sum scores continue to be used everywhere as performance measures in teaching institutions and applications of psychological scales. One looks for an explanation of why IRT has not yet penetrated through the practice of assessment more deeply.

This chapter proposes that an answer may involve the lack of understanding the nature of the IRT model in general, and the role of the latent trait θ in particular. This in turn may have prevented the use of index topologies and measures that would be far easier to explain to a test taker than would a signed number θ occurring anywhere on the real line. It is suggested that the sum score distribution at least defines a familiar statistical space that we can use to our advantage in selling IRT to the general public. Weighted sum scores taking values on $[0, n]$ such as Equation 20.7 should be easy to sell to the public not only for their better statistical efficiency, but also because of their robustness with respect to faulty items.

In this chapter, the TSβ distribution is proposed as a replacement for the Gaussian distribution for the purposes of providing a useful graphical caricature of the distribution of performance. Of course, its role is essentially cosmetic since the distribution of θ is in principle unidentifiable in IRT, and therefore may be chosen without any regard for data. But there are a host of pedagogical and other reasons on why examinees and instructors would want to know how specific performance levels compare to those obtained by others to whom the test is administered, and also those for larger populations. The TSβ distribution, or perhaps an extension of it or some other low-dimensional summary of distribution, therefore plays an essential role.

A great deal of the literature on IRT is preoccupied with high-end estimation methods employing computationally intensive maximum likelihood and Bayesian procedures; but much testing takes place in relatively messy environments where test design procedures are less than optimal, and where results are required quickly and with a minimum of fuss. Functional data-analysis methodology can be used to estimate IRFs rapidly and with more-than-enough accuracy for most testing situations. The TestGraf approach described in the previous sections of this chapter and the algorithm outlined here are easy to program, very fast, and, with the addition of first-class statistical graphics, seem to be the kind of testing tool that should in principle be available with minimal cost and effort to any test designer.

The chapter has focused on binary data, but there is useful information in the distractor multiple-choice options assigned to 0. The extension of the algorithm to the estimation of position in the simplicial hypercube is a simple matter, as it was for TestGraf, and promises to provide further improvement in the quality of θ-estimation.

A good plot such as Figure 20.6 highlights an essential structure by hiding an inessential detail. It can also highlight what might be missing. Why should we imagine that all students follow the same trajectory? Psychology has important roots in other sciences such as biology, and the course contained considerable material on this and other disciplines such as neurophysiology, genetics, and even mathematics. If we were to analyze the data for the biology majors in the class, would we not see a somewhat different trajectory? Or the same for a segmentation by arts, science, and other faculties? Or perhaps even for gender? In other words, $P_i(\theta)$ can benefit from being extended to include covariate effects, such as $P(\theta,\beta'\mathbf{z}_i)$.

Can we in the psychometric community mobilize around improving the penetration of our valuable technology into the academic workplace? Terms such as "educational measurement" may be a slight exaggeration from a technical perspective, but they capture well what we are about, and deserve a better public image than they seem to have.

Acknowledgment

Preparation of this chapter has been supported by a grant from the Natural Science and Engineering Research Council of Canada to J. O. Ramsay.

References

Boothby, W. M. 1975. *An Introduction to Differentiable Manifolds and Riemannian Geometry.* New York: Academic Press.

Douglas, J. 2001. Asymptotic identifiability of nonparametric item response models. *Psychometrika, 66,* 531–540.

Lei, P.-W., Dunbar, S. B., and Kolen, M. J. 2004. Comparison of parametric and nonparametric approaches to item analysis for multiple-choice tests. *Educational and Behavioral Measurement, 64,* 565–587.

Ramsay, J. O. 1991. Kernel smoothing approaches to nonparametric item characteristic curve estimation. *Psychometrika, 56,* 611–630.

Ramsay, J. O. 1996. A geometrical approach to item response theory. *Behaviormetrika, 23,* 3–17.

Ramsay, J. O. and Silverman, B. W. 2002. *Applied Functional Data Analysis.* New York: Springer.

Ramsay, J. O. and Silverman, B. W. 2005. *Functional Data Analysis.* New York: Springer.

Ramsay, J. O., Hooker, G., and Graves, S. 2009. *Functional Data Analysis with R and MATLAB.* New York: Springer.

Ramsay, J. O. and Wiberg, M. 2016. *Nonparametric Estimation of Item Characteristic Curves Using Parameter Cascading.* Unpublished manuscript.

Rossi, N., Wang, X., and Ramsay, J. O. 2002. Nonparametric item response function estimates with the EM algorithm. *Journal of the Educational and Behavioral Sciences, 27,* 291–317.

Thurstone, L. L. 1927. A law of comparative judgement. *Psychological Review, 34,* 273–286.

Section VI

Models for Nonmonotone Items

21

Hyperbolic Cosine Model for Unfolding Responses

David Andrich

CONTENTS

21.1 Introduction....353
21.2 Presentation of the Model....353
  21.2.1 Latitude of Acceptance....357
21.3 Parameter Estimation....358
  21.3.1 Solution Algorithm....359
  21.3.2 Initial Estimates....360
  21.3.3 Inconsistency of Parameter Estimates and a Constraint on Parameter ρ_i....360
21.4 Goodness of Fit....361
21.5 Example....362
21.6 Discussion....365
References....365

21.1 Introduction

The two main mechanisms for characterizing dichotomous responses of persons to items on a single dimension are the *cumulative* and the *unfolding*. In the former, the probability of a positive response is a monotonic function of the location of the relevant parameters on the continuum; in the latter, it is single-peaked. This chapter presents a unidimensional IRT model for unfolding, referred to as the hyperbolic cosine model (HCM). Table 21.1 shows a deterministic unfolding response pattern and, for purposes of contrast, the deterministic cumulative response patterns, for five items ordered on the continuum. It is evident that although the total score recovers the pattern of responses in the latter (Guttman, 1950), this is not the case with the unfolding patterns. In the former, if a person responds positively to an item, then the person will respond positively to all items to the left in the order; in the latter, the person responds positively only to items close to the person's location on the continuum, and negatively at greater distances from the person's location on both ends of the continuum.

21.2 Presentation of the Model

Although introduced by Thurstone (1927, 1928) for the measurement of attitude, the study of unfolding is associated with Coombs (1964) who worked within a deterministic framework, and when more than four or so items are involved, the analysis is extremely complex.

TABLE 21.1

Deterministic Unfolding and Cumulative and Response Patterns to Five Statements

Statements						Statements					
1	2	3	4	5	Sum	1	2	3	4	5	Sum
0	0	0	0	0	0	0	0	0	0	0	0
1	0	0	0	0	1	1	0	0	0	0	1
1	1	0	0	0	2	1	1	0	0	0	2
0	1	1	0	0	2	1	1	1	0	0	3
0	0	1	1	0	2	1	1	1	1	0	4
0	0	0	0	1	1	1	1	1	1	1	5
Illustrative unfolding response patterns for five ordered statements						Cumulative response patterns for five ordered statements					

Michell (1994) has provided a rationale for constructing such items of different intensity on a continuum, referred to originally by Thurstone as the items' *affective* values. Probabilistic models have been introduced subsequently (e.g., Andrich, 1988; Davison, 1977; Post, 1992; van Schuur, 1984, 1989). In this chapter, a rationale that links the dichotomous responses of persons to items directly to the unfolding mechanism through a graded response structure is used to construct the probabilistic HCM for dichotomous responses of the unfolding kind (Andrich and Luo, 1993).

In addition to the measurement of attitude, unfolding models have been applied to the study of development along a continuum, for example, in psychological development (Coombs and Smith, 1973), development in learning goals (Volet and Chalmers, 1992), social development (Leik and Matthews, 1968), general preference studies (Coombs and Avrunin, 1977), and political science (van Blokland-Vogelesang, 1991; van Schuur, 1987). The model studied here is developed from first principles using a concrete example in attitude measurement.

Consider a dichotomous response of agree or disagree of a person to the statement "I think capital punishment is necessary but I wish it were not." This statement appears in the example later in the chapter, and reflects an ambivalent attitude to capital punishment. The person's location on the continuum is referred to as the person's *ideal point*. If the person's ideal point is close to the location of the statement, then the person will tend to agree to the statement; on the other hand, if a person's location is far from that of the statement—either very much *for* or very much *against* capital punishment, then the probability of agree will decrease and that of disagree will increase correspondingly. This gives the single-peaked response function (RF) for the agree response and a complementary RF for the disagree response, both shown in Figure 21.1. Figure 21.1 shows in dotted lines two other response functions, which are part of the construction of HCM described below.

In developing the model, it is instructive to consider first the two sources of the disagree response. Let the locations of person p and statement i on a continuum be θ_p and δ_i, respectively. Then, formally, if $\theta_p \ll \delta_i$ or $\theta_p \gg \delta_i$, the probability of a disagree response tends to 1.0. This reveals that the disagree response occurs for *two* latent reasons, one because the person has a much stronger attitude *against* capital punishment, the other because the person has a much stronger attitude *for* capital punishment, than reflected by the statement. The RF with broken lines in Figure 21.1 shows the resolution of a single disagree response into these two constituent components. This resolution shows that there are three possible latent responses that correspond to the two possible manifest responses of agree/disagree: (i) disagree because persons consider themselves below the location of the statement (disagree below—DB); (ii) agree because the persons consider themselves close to the location of the statement (agree close—AC); and (iii) disagree because the persons consider themselves above the location of the statement (disagree above—DA). Further, the probabilities of the two disagree responses have a monotonic decreasing and increasing shape, respectively. This means that, as shown in Figure 21.1, the three responses take the form of the responses to three graded responses. Accordingly, a model for graded responses can be applied to this structure.

Because it is the simplest of models for graded response, and because it can be expressed efficiently and simply in the case of three categories, the rating scale parameterization (Andrich, 1978; Andersen, 1977; Wright and Masters, 1982) of the Rasch (1961) model is used. The parameterization can take a number of forms, all equivalent to

$$P\{Y_{pi}=y;\theta_p,\delta_i,\boldsymbol{\tau}_i\}=\gamma_{pi}^{-1}\exp\left[-\sum_{a=0}^{y}\tau_{ai}+y(\theta_p-\delta_i)\right], \tag{21.1}$$

where $y=0,1,2,\ldots,m$ are scores of $m+1$ successive categories, τ_{ai}, $a=1,2,\ldots,m$ are m thresholds on the continuum dividing the categories with $\tau_{0i}\equiv 0$ for convenience, $\gamma_{pi}=\sum_{y=0}^{m}\exp\left[-\sum_{a=0}^{y}\tau_{ai}+y(\theta_p-\delta_i)\right]$ is the normalizing factor which ensures that the sum of the probabilities is 1.0 and without loss of generality, $\sum_{a=0}^{m}\tau_{ai}=0$.

The correspondence between the three responses in Figure 21.1 and the random variable Y in Equation 21.1, in which case $m=2$, is as follows: $y=0\leftrightarrow DB; y=1\leftrightarrow AC; y=2\leftrightarrow DA$. It is evident that the RFs in Figure 21.1 operate so that when θ_p is close to δ_i, the probability of the score of 1 (AC) is greater than the probability of either 0 (DB) or 2 (DA). This is as required in the unfolding mechanism.

We now take two steps to obtain the final version of the model. First, define $\lambda_i=(\tau_{2i}-\tau_{1i})/2$. Then the location parameter δ_i is in the middle of the two thresholds, and λ_i is the distance

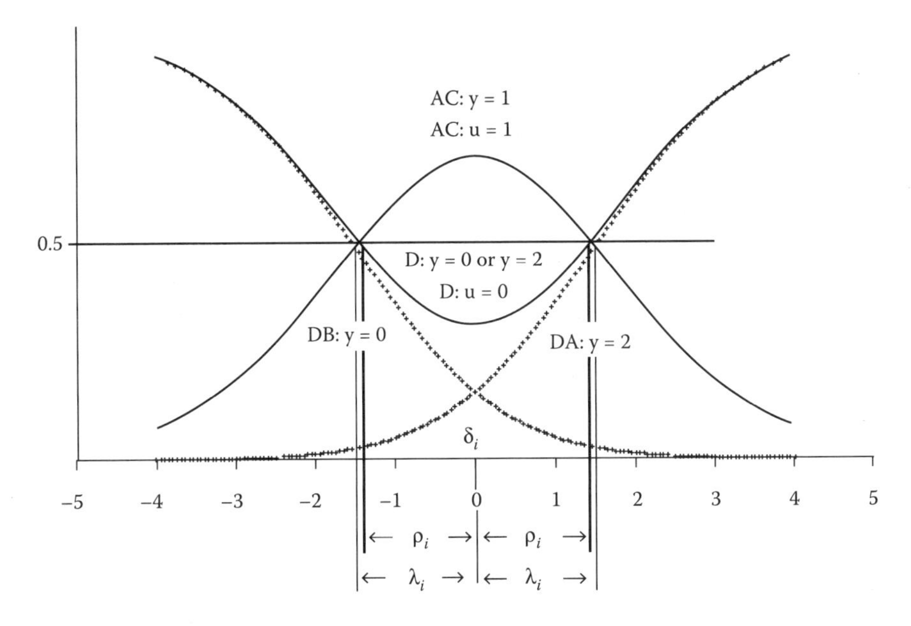

FIGURE 21.1
Response functions of the agree close (AC) and the disagree (D) responses and the resolution of the disagree response into its constituent components: disagree below (DB) and disagree above (DA) shown in dotted lines.

from δ_i to each of the thresholds (Andrich, 1982). The parameters δ_i and λ_i are also shown in Figure 21.1. It is now instructive to write the probability of each RF explicitly in terms of parameter λ_i:

$$P\{Y_{pi} = y; \theta_p, \delta_i, \lambda_i\} = \gamma_{pi}^{-1}, \tag{21.2}$$

$$P\{Y_{pi} = y; \theta_p, \delta_i, \lambda_i\} = \gamma_{pi}^{-1} \exp[\lambda_i + (\theta_p - \delta_i)], \tag{21.3}$$

$$P\{Y_{pi} = y; \theta_p, \delta_i, \lambda_i\} = \gamma_{pi}^{-1} \exp 2(\theta_p - \delta_i), \tag{21.4}$$

where $\gamma_{pi} = 1 + \exp[\lambda_i + (\theta_p - \delta_i)] + \exp[2(\theta_p - \delta_i)]$. Because the thresholds define the points where the probability of the DA and DB responses becomes greater than the AC response, the probabilities of the pairs AC and DB, and AC and DA, are equal at $\delta_i \pm \lambda_i$. Although the model characterizes the three implied responses of a person to a statement, there are, nevertheless, only two manifest responses, and, in particular, only one manifest disagree response. That is, $y = 0$ and $y = 2$ are not distinguished in the data.

In order to make the model correspond to the data, define a new random variable $U_{pi} = u$, where $u = 0$ when $y = 0$ or $y = 2$ and $u = 1$ when $y = 1$. Then

$$P\{u = 0\} = P\{y = 0\} + P\{y = 2\}, \tag{21.5}$$

$$P\{u = 1\} = P\{y = 1\}. \tag{21.6}$$

Inserting the explicit expressions of Equations 21.2, 21.3, 21.4 into Equations 21.5, 21.6 gives

$$P\{U_{pi} = 0; \theta_p, \delta_i, \lambda_i\} = \gamma_{pi}^{-1}[1 + \exp 2(\theta_p - \delta_i)], \tag{21.7}$$

$$P\{U_{pi} = 1; \theta_p, \delta_i, \lambda_i\} = \gamma_{pi}^{-1} \exp[\lambda_i + (\theta_p - \delta_i)]. \tag{21.8}$$

Because U_{pi} is dichotomous, it is efficient to focus on only one of Equation 21.7 or 21.8. Focusing on Equation 21.8 and inserting the normalizing factor explicitly gives

$$P\{U_{pi} = 1; \theta_p, \delta_i, \lambda_i\} = \frac{\exp[\lambda_i + (\theta_p - \delta_i)]}{1 + \exp[\lambda_i + (\theta_p - \delta_i)] + \exp 2(\theta_p - \delta_i)}. \tag{21.9}$$

Multiplying the numerator and denominator by $\exp[-(\theta_p - \delta_i)]$ and simplifying gives

$$P\{U_{pi} = 1; \theta_p, \delta_i, \lambda_i\} = \frac{\exp \lambda_i}{\exp \lambda_i + \exp[-(\theta_p - \delta_i)] + \exp[(\theta_p - \delta_i)]}. \tag{21.10}$$

Recognizing that the hyperbolic cosine $\cosh(a) = [\exp(-a) + \exp(a)]/2$ gives

$$P\{U_{pi} = 0 : \theta_p, \delta_i, \lambda_i\} = \gamma_{ni}^{-1} 2 \cos h(\theta_p - \delta_i) \tag{21.11}$$

and

$$P\{U_{pi}=1;\theta_p,\delta_i,\lambda_i\}=\gamma_{pi}^{-1}\exp\lambda_i, \tag{21.12}$$

where now $\gamma_{pi}=\exp\lambda_i+2\cosh(\theta_p-\delta_i)$ is the normalizing factor. It is obvious now why the model of Equations 21.11 and 21.12 is referred to as the HCM.

Equations 21.11 and 21.12 are relatively simple and, as shown in Figure 21.1, the parameters are readily interpreted. However, by considering the points at which $P\{u_{ij}=0\}=P\{u_{ij}=1\}$, Equations 21.11 and 21.12 can be simplified considerably, making the model not only more elegant but also providing a tangible interpretation of a new parameter (Luo, 1998). Equating Equations 21.11 and 21.12 gives

$$2\cosh(\theta_p-\delta_i)=\exp(\lambda_i). \tag{21.13}$$

Now, let $|\theta_p-\delta_i|=\rho_i$ be the distance on the continuum that characterizes the point of equal probability: $P\{u_{ij}=0\}=P\{u_{ij}=1\}$. Then,

$$2\cosh(\rho_i)=\exp(\lambda_i) \tag{21.14}$$

and, on cancelation of the factor 2 in the numerator and denominator, Equations 21.11 and 21.12 become

$$P\{u_{ij}=0\}=(\cosh(\theta_p-\delta_i))/\gamma_{pi} \tag{21.15}$$

and

$$P\{u_{ij}=1\}=(\cosh\rho_i)/\gamma_{pi}, \tag{21.16}$$

where $\gamma_{pi}=\cosh\rho_i+\cosh(\theta_p-\delta_i)$ is the new normalizing factor of the two equations which ensures that the sum of their probabilities is 1.0. Equations 21.15 and 21.16 can be written as the single expression

$$P\{U_{pi}=u;\theta_p,\delta_i,\rho_i\}=[(\cosh\rho_i)^u(\cosh(\theta_p-\delta_i))^{1-u}]/\gamma_{pi}. \tag{21.17}$$

21.2.1 Latitude of Acceptance

The parameter ρ_i is the range within which the agree response has a greater probability than the disagree response. This range is known as the *latitude of acceptance*, which is an important concept in attitude measurement (Coombs, 1964, Hovland et al., 1957). The concept can be characterized as a person or an item parameter. If it is characterized as a person parameter, it can be difficult to estimate from a relatively few items for each person. However, Luo, Andrich, and Styles (1998) showed it could be formalized as a group parameter and estimated successfully. In this chapter, the latitude of acceptance remains as an item parameter.

Equation 21.17 is extremely elegant—it has the same function "cosh" of the two relevant parameters, $(\theta_p-\delta_i)$, the distance between the location θ_p of the person and the location δ_i

of the item. Luo (1998) abstracted this structure of Equation 21.17 to provide the following general probabilistic formulation:

$$P\{U_{pi}=1;\theta_p,\delta_i,\rho_i\}=\Psi(\rho)]/[\Psi(\rho)+\Psi(\theta_p-\delta_i)], \tag{21.18}$$

where $\Psi(.)$ is called the operational function which is (i) nonnegative ($\Psi(t)>0, -\infty<t<\infty$), (ii) monotonic in the positive domain ($\Psi(t_1)>\Psi(t_2), t_1>t_2\geq 0$), and (iii) symmetric about the origin ($\Psi(-t)=\Psi(t), -\infty<t<\infty$). Again, the latitude of acceptance parameter can be subscripted as a person, person-group, or item parameter. In addition to $\Psi(.)=\cosh(.)$, the square function $\Psi(.)=(.)^2$ also satisfies the above three conditions. Luo (1998) showed that many of the then-known models, for example, the simple square logistic model (Andrich 1988), De Sarbo–Hoffman model (DeSarbo and Hoffman, 1986), and PARELLA model (Hoijtink, 1991) could be reexpressed in the form of Equation 21.18, making explicit in these models the implied, specific latitude of acceptance parameter.

21.3 Parameter Estimation

Although HCM is constructed from a Rasch model for graded responses, by combining two categories, the distinguishing feature of sufficient statistics in the Rasch models for the person and statement parameters (Andersen, 1977) is destroyed. Therefore, methods of estimation that involve conditioning on simple sufficient statistics are not available. But other methods, including the joint maximum likelihood (JML) and the EM algorithm, can be applied. The procedure described here is the simplest of these—the JML. It has the advantage that no assumptions need be made regarding the distribution of the person parameters. This is shown to be an advantage with the example below.

From Equation 21.17, the likelihood of the matrix of responses for P persons and I items is given by

$$\begin{aligned} L &= \prod_{p=1}^{P}\prod_{i=1}^{I}\frac{[\cosh(\rho_i)]^{u}[\cosh(\theta_p-\delta_i)]^{1-u}}{\cosh(\rho_i)+\cosh(\theta_p-\delta_i)} \\ &= \frac{\prod_{i=1}^{I}[\cosh(\rho_i)]^{s_i}\prod_{p=1}^{P}\prod_{i=1}^{I}[\cosh(\theta_p-\delta_i)]^{I-u}}{\prod_{p=1}^{P}\prod_{i=1}^{I}\cosh(\rho_i)+\cosh(\theta_p-\delta_i)}, \end{aligned} \tag{21.19}$$

where $s_i=\sum_{p=1}^{P}u_{pi}$. According to the factorization theorem (Lehmann, 1983, p. 39), s_i is a sufficient statistic for ρ_i provided the other parameters are known. However, there is no jointly sufficient statistic for the parameters $\{\theta_p,\delta_i,\rho_i\}$.

Thus, the log likelihood is

$$\begin{aligned} Log\ L &= \sum_{i=1}^{I}s_i\log(\cosh(\rho_i))+\sum_{p=1}^{P}\sum_{i=1}^{I}(1-u_{pi})\log(\cosh(\theta_p-\delta_i)) \\ &\quad -\sum_{p=1}^{P}\sum_{i=1}^{I}\log[\cosh(\rho_i)+\cosh(\theta_p-\delta_i)]. \end{aligned} \tag{21.20}$$

Differentiating Equation 21.20 partially with respect to each of the parameters and equating to zero gives the JML equations. On considerable but standard simplification, these are

$$\varphi(\rho_i) = s_i - \sum_{p=1}^{P} \pi_{pi} = 0, \quad i = 1,2,\dots,I; \tag{21.21}$$

$$\varphi(\delta_i) = \sum_{p=1}^{P} \tanh(\theta_p - \delta_i)(u_{pi} - \pi_{pi}) = 0, \quad i = 1,2,\dots,I; \tag{21.22}$$

$$\varphi(\theta_p) = -\sum_{i=1}^{i} \tanh(\theta_p - \delta_i)(u_{pi} - \pi_{pi}) = 0, \quad p = 1,2,\dots,p \tag{21.23}$$

and

$$\sum_{i=1}^{I} \hat{\delta}_i = 0, \tag{21.24}$$

where $\pi_{pi} \equiv P\{U_{pi} = 1 \mid \theta_p, \delta_i, \rho_i\}$ and Equation 21.18 is an arbitrary but necessary constraint.

21.3.1 Solution Algorithm

The Newton–Raphson algorithm is efficient in reaching an iterative solution to Equations 21.21 through 21.24. The complete multivariate algorithm is not feasible because as P increases, the matrix of second derivatives becomes too large to invert. The alternative algorithm commonly used in such situations is to take the statement parameters as fixed while the person parameters are improved individually to the required convergence criterion; the person parameters are then fixed, and the statement parameters are improved individually until the whole set of estimates converges to a specified criterion. For the $(q+1)$th iteration, the algorithm takes the form

$$\hat{\rho}_i^{(q+1)} = \hat{\rho}_i^{(q)} - [\psi_{\rho\rho(i)}]^{-1}\varphi(\rho_i), \quad i = 1,2,\dots,I; \tag{21.25}$$

$$\hat{\delta}_i^{(q+1)} = \hat{\delta}_i^{(q)} - [\psi_{\delta\delta(i)}]^{-1}\varphi(\delta_i), \quad i = 1,2,\dots,I; \tag{21.26}$$

$$\hat{\theta}_p^{(q+1)} = \hat{\theta}_p^{(q)} - [\psi_{\theta\theta(p)}]^{-1}\varphi(\theta_p), \quad i = 1,2,\dots,I; \tag{21.27}$$

where $\psi_{\rho\rho(i)} = E[\partial^2 \log L/\partial\rho_i^2]$, $\psi_{\delta\delta(i)} = E[\partial^2 \log L/\partial\delta_i^2]$, $\psi_{\theta\theta(p)} = E[\partial^2 \log L/\partial\theta_p^2]$. Further, the respective information functions are $I(\rho_i) = -[\psi_{\rho\rho(i)}]^{-1}$; $I(\delta_i) = -[\psi_{\delta\delta(i)}]^{-1}$; $\theta_p = -[\psi_{\theta\theta(p)}]^{-1}$, the reciprocals of which give asymptotic estimates of the respective standard errors, $\hat{\sigma}_\rho, \hat{\sigma}_\delta, \hat{\sigma}_\theta$.

21.3.2 Initial Estimates

The initial values for ρ_i are set to $\hat{\rho}_i^{(0)} = 0$ and to obtain initial estimates for $\hat{\delta}_i^{(0)}$, all initial values for θ_p are set to $\theta_p^{(0)} = 0$. This gives $\hat{\delta}_i^{(0)} = \cosh^{-1}((P - s_i)/s_i)$, where, as earlier, $s_i = \sum_{p=1}^{P} u_{pi}$ is the total score for an item. Because $\cosh(\delta_i) \geq 1$, if for some i $(P - s_i)/s_i \leq 1$, the minimum of these values for the items is taken and

$$\hat{\delta}_i^{(0)} = \pm\cosh^{-1}\left(\alpha + \frac{P - s_i}{s_i}\right), \tag{21.28}$$

where $\alpha = 1 - \min(\alpha + (P - s_i)/s_i)$.

The unfolding structure implies that statements at genuinely different locations can have the same total score statistic s_i This difference can be taken into account by assigning a negative sign to the initial estimates of approximately half of the statements and a positive sign to the other half. There are two ways of choosing the statements that receive a negative sign, one empirical and the other theoretical, and of course, both can be used together. First, if data have an unfolding structure, then the traditional factor loadings on the first factor of a factor analysis unfolds into the positive and negative loadings (Davison, 1977). Luo (2000) elaborates on this result in assigning a positive or a negative sign to each item. The sign according to the above analysis should be checked for consistency with the theory governing the construction of the variable and the operationalization of the statements.

The initial estimates for $\hat{\theta}_p$ are obtained from the initial estimates of the statement locations according to $\hat{\theta}_p^{(0)} = [\sum_{i=1}^{I} u_{pi}\hat{\delta}_i^{(0)}]/I_p$, where I_p is the number of statements for which person p has a score $u_{pi} = 1$. This is the average values of the statements to which each person agreed to and is the value that Thurstone (1927, 1928) used to calculate a person's location given an estimate of the statement locations using an independent method, usually that of paired comparisons.

21.3.3 Inconsistency of Parameter Estimates and a Constraint on Parameter ρ_i

Two further related issues need to be addressed in solving Equations 21.21 through 21.24. First, Luo (2000) shows that in principle there can be a trade-off between the estimates of the two statement parameters ρ_i and δ_i resulting in inflated estimates. Second, because person parameters are involved as *incidental* parameters, the estimates are inconsistent in the sense that as the number of persons increases, the estimates of the item parameters do not converge to the correct values (Neyman and Scott, 1948). This inconsistency is evident also in JML estimation in the Rasch and IRT models (Andersen, 1973; Wright and Douglas, 1977). However, as indicated earlier, no assumptions regarding the distribution of person parameters need to be made, and this is particularly desirable in attitude measurement, where, for example, attitudes may be polarized and the empirical study of the shape of this distribution is important.

To control the trade-off between the estimates of ρ_i and δ_i, a common latitude of acceptance parameter for all items, ρ, is estimated by simply constraining Equation 21.21 to have the same estimate for all items. It is a kind of average of the latitude of ρ_i among the items. To control the inconsistency, and from simulation studies, the correction factor

$$\hat{\rho}_c = \frac{\hat{\rho}(I-2)}{I} \tag{21.29}$$

is imposed. Then the estimates $\hat{\rho}_i$ from Equation 21.21 are constrained so that

$$\left(\sum_{i=1}^{I} \hat{\rho}_i\right) I = \hat{\rho}_c. \tag{21.30}$$

The constraint and correction factor of Equations 21.29 and 21.30 imposed on the estimate of ρ_i controls the consistency in the estimates of both the θ_p and δ_i parameters. However, because generally there are few items relative to the number of persons, checks need to be made that the estimate of each θ_p from Equation 21.23 arises from a global maximum and not either a local maximum or a local minimum. With modern computer speeds, this check is readily implemented (Luo et al., 1998).

21.4 Goodness of Fit

In considering tests of fit between the data and a model, the perspective taken in this chapter is that the model is an explicit rendition of a theory, and that the data collection is in turn governed by the model. Therefore, to the degree that the data accord with the model, to that degree they confirm both the theory and its operationalization in the data collection, and vice versa. In the application of IRT, the model chosen is expected to summarize the responses with respect to some substantive variable in which the statements reflect differences in degree on the latent continuum. Thus, following Thurstone (1928), the aim should be to construct statements at different locations on the continuum, and the relative order of the statements becomes a hypothesis about the data. This theoretical ordering, taken as a hypothesis, is especially significant when the model that reflects the response process is single-peaked because intrinsic to the response structure and the single-peaked response function, there are always two person locations that give the same probability of a positive response.

The hypothesis that the responses take a single-peaked form can be checked by estimating the locations of the persons and the statements, dividing the persons into class intervals, and checking if the proportions of persons responding positively across class intervals takes the single-peaked form specified by the model. This is a common general test of fit, and can be formalized as

$$\chi^2 \approx \sum_{i=1}^{I}\sum_{g=1}^{G} \frac{\left(\sum_{p\in g} u_{pi} - \sum_{p\in g} E[u_{pi}]\right)^2}{\sum_{p\in g} V[u_{pi}]}, \tag{21.31}$$

where $g = 1,2, \ldots, G$, are the class intervals. Intuitively, as the number of persons and statements increases, this statistic should approximately be the χ^2 distribution on $(G-1)(I-1)$

degrees of freedom. The power of this test of fit is governed by the relative locations of the persons and the statements, where the greater the variation among these, the greater the power, and of course the number of persons where the greater the number of persons, again the greater the power.

Complementary to this formal procedure, a telling and simpler check is to display the responses of the persons to the statements when both are ordered according to their estimated, affective values. Then the empirical order of the statement should accord with the theoretical ordering, and the matrix of responses should show the parallelogram form shown in Table 21.1. The study of the fit between the data and model using the combination of these approaches, the theoretical and the statistical, is illustrated in the example.

21.5 Example

The example involves the measurement of an attitude to capital punishment using a set of eight statements originally constructed by Thurstone (1927, 1928), and subsequently studied again by Wohlwill (1963) and Andrich (1988). The data involve the responses of a small sample of 41 persons in a class in educational measurement. The example with the estimate of the parameter λ_i of Equations 21.11 and 21.12, rather than the latitude of acceptance parameter ρ_i of Equation 21.17 as in this chapter, was shown in Andrich (1995).

Table 21.2 shows the statements, their estimated affective values, and standard errors under the HCM model in which first, all statements are assumed to have the same latitude

TABLE 21.2

Scale Values of Statements about Capital Punishment from Direct Responses

	Statement	$\chi^2, 2df$ (p)	Equal $\hat{\rho}$: $\hat{\delta}$ (σ_δ)	Equal $\hat{\rho}$: $\hat{\rho}$ (σ_ρ)	Estimated $\hat{\rho}$: $\hat{\delta}$ (σ_δ)	Estimated $\hat{\rho}$: $\hat{\rho}$ (σ_ρ)
1.	Capital punishment is one of the most hideous practices of our time	1.147 (0.61)	−6.89 (0.83)	4.36 (0.78)	−6.41 (0.82)	4.57 (0.77)
2.	Capital punishment is not an effective deterrent to crime	1.10 (1.00)	−5.65 (0.80)	4.36 (0.76)	−5.70 (0.79)	5.66 (0.76)
3.	The state cannot teach the sacredness of human life by destroying it	6.03 (0.05)	−3.00 (0.46)	4.36 (0.46)	−3.68 (0.49)	4.16 (0.49)
4.	I don't believe in capital punishment but I am not sure it isn't necessary	1.10 (0.61)	−1.57 (0.41)	4.36 (0.41)	−1.45 (0.41)	4.07 (0.41)
5.	I think capital punishment is necessary but I wish it were not	2.90 (0.22)	2.58 (0.61)	4.36 (0.61)	2.71 (0.60)	4.77 (0.56)
6.	Until we find a more civilized way to prevent crime, we must have capital punishment	1.50 (0.61)	4.07 (0.66)	4.36 (0.66)	4.54 (0.72)	4.42 (0.68)
7.	Capital punishment is justified because it does act as a deterrent to crime	0.99 (0.60)	4.45 (0.65)	4.36 (0.65)	4.69 (0.67)	3.32 (0.63)
8.	Capital punishment gives the criminal what he deserves	1.99 (0.37)	6.02 (0.54)	4.36 (0.54)	5.30 (0.55)	4.57 (0.50)

TABLE 21.3

Distribution of Attitude Scale Values with a Constant Latitude of Acceptance $\hat{\rho} = 4.36$

Estimate $\hat{\theta}$	SE $\hat{\sigma}_{\hat{\theta}}$	Score	Response Pattern								Frequency
−6.62	1.45	3	1	1	1	0	0	0	0	0	5
−4.50	1.55	4	1	1	1	1	0	0	0	0	7
−1.73	0.98	2	0	0	1	1	0	0	0	0	1
−1.73	0.98	4	0	1	1	1	1	0	0	0	1
−0.03	0.91	4	0	0	1	1	1	0	1	0	1
0.81	0.94	5	0	0	1	1	1	1	1	0	2
0.81	0.94	5	0	1	0	1	1	1	0	1	1
1.72	1.00	4	0	0	1	0	1	1	1	0	1
1.73	1.00	6	0	0	1	1	1	1	1	1	2
1.73	1.00	4	0	0	0	1	1	1	1	0	1
1.73	1.01	4	0	0	1	0	1	1	0	1	1
2.86	1.22	5	0	0	1	0	1	1	1	1	1
2.86	1.22	5	0	0	0	1	1	1	1	1	3
2.87	1.22	3	0	0	0	0	1	1	1	0	2
4.91	1.92	4	0	0	0	1	0	1	1	1	1
4.97	1.92	4	0	0	0	0	1	1	1	1	10
7.23	1.28	3	0	0	0	0	0	1	1	1	1

of acceptance parameter ρ_i, and second, where ρ_i is free to vary among statements. Table 21.3 shows the frequencies of all observed patterns, the estimated attitudes, and their standard errors. In Table 21.2, the statements are ordered according to their estimated affective values, and in Table 21.3, the persons are likewise ordered according to the estimate of their attitudes. This ordering helps appreciate the definition of the variable and is a check on the fit. Although it may be considered that the sample is small, it has the advantage that the analyses can be studied very closely, and in addition, it shows that the model can be applied successfully even in the case of small samples. In the current case, there is already strong theoretical and empirical evidence that the statements do work as a scale and so the analysis takes a confirmatory role.

Table 21.2 shows that the ordering of the statements begins with an attitude that is strongly against capital punishment, through an ambivalent attitude, to one that is strongly for capital punishment. (In the questionnaire administered, the ordering was not the one shown in Table 21.2.) In addition, the likelihood ratio test, that the latitude of acceptance is not significantly different among the items, gives, $Log\,L(\hat{\rho}_i) = -58.067$, $Log\,L(H_0 : \hat{\rho}_i = \hat{\rho}) = -61.696$, $\chi^2 = 7.258$, $df = 6$, $p = 0.339$, confirming that the latitude of acceptance across statements can be taken as equivalent. This result is further confirmed by the closeness of the estimates of the affective values of statements when equal and unequal latitude of acceptance values $\hat{\rho}_i$ are assumed as shown in Table 21.2. The interpretation of the affective values of the statements on the continuum is made substantially easier when the latitude of acceptance parameter is equivalent across statements.

The ordering of the persons and statements in Table 21.3 shows the required feature of 1s in a parallelogram around the diagonal. Because of the evidence that the latitude of acceptance parameter was equivalent across items, only the estimates based on a single estimate of $\hat{\rho}$ are shown. Figure 21.2 shows the distribution of the persons, and it is evident it is bimodal with more people against capital punishment than for it. Nevertheless, there

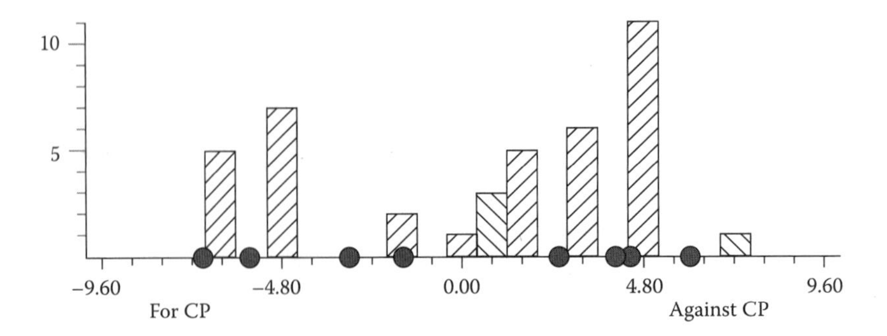

FIGURE 21.2
Distribution of persons and the affective values of the statements.

are some people who have an ambivalent attitude. It is clear that assuming a distribution, for example, the typical normal distribution, might have been erroneous in this example.

Even without any further checks on the fit, it should be apparent that the data are consistent with the model, and therefore confirm the hypothesis of the ordering of the statements and their usefulness for measuring an attitude to capital punishment. However, for completeness, Table 21.2 also shows the χ^2 test of fit when the persons are placed into three class intervals for a common latitude of acceptance parameter estimate. Figure 21.2 shows the ICC and the observed means for the ambivalent Statement 4, *I don't believe in capital punishment but I am not sure it is not necessary.*

The global test of fit according to Equation 21.31, in which the sample is divided into three class intervals, has a value $\chi^2 = 16.04$, $df = 14$, $p = 0.32$, which confirms this impression. In addition to the distribution of the persons, Figure 21.3 shows the locations of the statements on the continuum. Given the evidence that the data conform to the HCM with equal latitude of acceptance $\hat{\rho}$ for all items, this figure summarizes the effort to simultaneously locate statements and persons on an attitude continuum when direct responses of persons to statements subscribe to the unfolding mechanism.

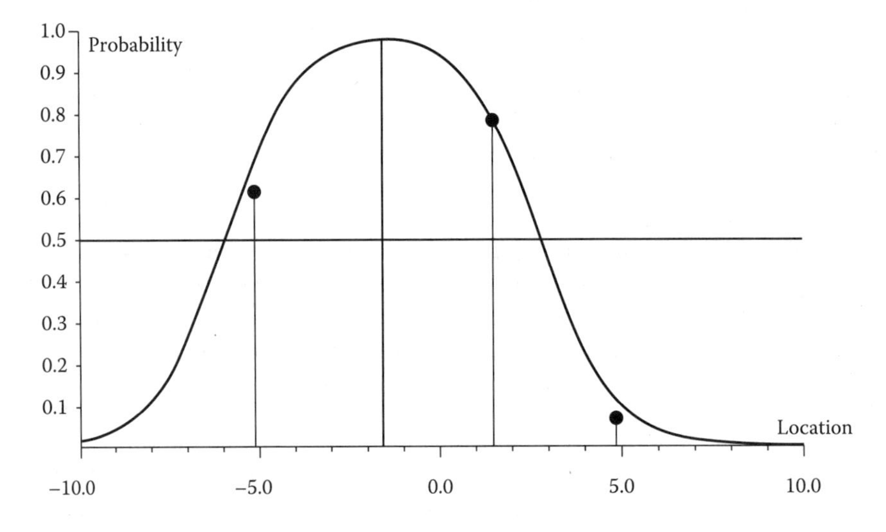

FIGURE 21.3
Item characteristic curve for Item 4 (location −1.57) with observed means of responses in three class intervals.

21.6 Discussion

The original data involved 50 persons, but the questionnaires of two persons had no responses, and of three persons had incomplete responses. Although the models can be operationalized to cater for missing data, it was decided to use only complete data. It was surmised that, because the class had a number of students with non-English-speaking backgrounds, the persons who did not complete their papers might have been from such backgrounds.

Lack of familiarity with the language of response is relevant in all assessment, but it has a special role in the measurement of attitude where the statements often deliberately take the form of a cliché, catch phrase, or proverb, and it is not expected that a person ponders at length to make a response. Indeed, it is required to have a relatively quick and spontaneous affective, rather than a reasoned cognitive, response, and this generally requires fluency in the vernacular.

The responses of the remaining 45 students with complete direct responses were then analyzed according to the model. The theoretical ordering was violated quite noticeably in that the statement *I don't believe in capital punishment, but I am not sure it is not necessary* was estimated to have a stronger attitude against capital punishment than the statement *Capital punishment is one of the most hideous practices of our time.* A close examination of the data showed that the responses of four persons had glaring anomalies and they seemed to agree to statements randomly. It was surmised that these four persons may also had had trouble understanding these statements or were otherwise distracted. Therefore, they too were removed, leaving the sample shown in Table 21.3. Although there is no space to describe the example in detail, responses from the same persons to the same statements were obtained according to a pairwise preference design and analyzed according to a model derived from the HCM. When the responses of these four persons were also eliminated from the analysis of the pairwise preference data, the test of fit between the data and the model again improved (Andrich, 1995). This improvement in fit confirmed that these four persons had responses whose validity was doubtful in both sets of data.

The point of this discussion is that four persons in a sample of 45 were able to produce responses that provided an ordering of statements that grossly violated the theoretical ordering of two statements, and that, with a strong theoretical perspective, these four persons could be identified. Ideally, these persons, if they were known, would be interviewed regarding their responses. Then, either it would be confirmed that they had trouble with the language, or if they did not, that they perhaps read the statements in some way different from the intended reading. It is by such a process of closer study that anomalies are disclosed, and in this case, possible ambiguities in statements are understood.

References

Andersen, E. B. 1973. Conditional inference for multiple choice questionnaires. *British Journal of Mathematical and Statistical Psychology*, 26, 31–44.

Andersen, E. B. 1977. Sufficient statistics and latent trait models. *Psychometrika*, 42, 69–81.

Andrich, D. 1978. A rating formulation for ordered response categories. *Psychometrika*, 43, 357–374.

Andrich, D. 1982. An extension of the Rasch model for ratings providing both location and dispersion parameters. *Psychometrika*, 47, 105–115.

Andrich, D. 1988. The application of an unfolding model of the PIRT type for the measurement of attitude. *Applied Psychological Measurement*, 12, 33–51.

Andrich, D. 1995. Hyperbolic cosine latent trait models for unfolding direct responses and pairwise preferences. *Applied Psychological Measurement*, 19, 269–290.

Andrich, D. and Luo, G. 1993. A hyperbolic cosine latent trait model for unfolding dichotomous single-stimulus responses. *Applied Psychological Measurement*, 17, 253–276.

Coombs, C. H. 1964. *A Theory of Data*. New York: Wiley.

Coombs, C. H. and Avrunin, C. S. 1977. Single-peaked functions and the theory of preference. *Psychological Review*, 84, 216–230.

Coombs, C. H. and Smith, J. E. K. 1973. On the detection of structure in attitudes and developmental process *Psychological Review*, 80, 337–351.

Davison, M. 1977. On a metric, unidimensional unfolding model for attitudinal and developmental data. *Psychometrika*, 42, 523–548.

DeSarbo, W. A. and Hoffman, D. L. 1986. Simple and weighted unfolding threshold models for the spatial representation of binary choice data. *Applied Psychological Measurement*, 10, 247–264.

Guttman, L. 1950. The problem of attitude and opinion measurement. In: S. A. Stouffer, L. Guttman, E. A. Suchman, P. F. Lazarsfeld, S. A. Star, and J. A. Clausen (Eds.), *Measurement and Prediction* (pp. 46–59). New York: Wiley.

Hovland, C. I., Harvey, O. J., and Sherif, M. 1957. Assimilation and contrast effects in reactions to communication and attitude change. *Journal of Abnormal and Social Psychology*, 55, 244–252.

Hoijtink, H. 1991. The measurement of latent traits by proximity items. *Applied Psychological Measurement*, 15, 153–159.

Lehmann, E. L. 1983. *Theory of Point Estimation*. New York: Wiley.

Leik, R. K. and Matthews, M. 1968. A scale for developmental processes. *American Sociological Review*, 33, 62075.

Luo, G. 1998. A general formulation of unidimensional unfolding and pairwise preference models: Making explicit the latitude of acceptance. *Journal of Mathematical Psychology*, 42, 400–417.

Luo, G. 2000. The JML estimation procedure of the HCM for single stimulus responses. *Applied Psychological Measurement*, 24, 33–49.

Luo, G., Andrich, D., and Styles, I. M. 1998. The JML estimation of the generalised unfolding model incorporating the latitude of acceptance parameter. *Australian Journal of Psychology*, 50, 187–198.

Michell, J. 1994. Measuring dimensions of belief by unidimensional unfolding. *Journal of Mathematical Psychology*, 38, 244–273.

Neyman, J. and Scott, E. L. 1948. Consistent estimates based on partially consistent observations. *Econometrika*, 16, 1–32.

Post, W. J. 1992. *Nonparametric Unfolding Models: A Latent Structure Approach*. Leiden: DSWO Press.

Rasch, G. 1961. On general laws and the meaning of measurement in psychology. In J. Neyman (ed.), *Proceedings of the Fourth Berkeley Symposium on Mathematical Statistics and Probability IV* (pp. 321–334). Berkeley, CA: University of California Press.

Thurstone, L. L. 1927. A law of comparative judgement. *Psychological Review*, 34, 278–286.

Thurstone, L. L. 1928. Attitudes can be measured. *American Journal of Sociology*, 33, 529–554.

van Blokland-Vogelesang, R. 1991. *Unfolding and Group Consensus Ranking for Individual Preferences*. Leiden: DSWPO Press.

van Schuur, W. H. 1984. *Structure in Political Beliefs: A New Model for Stochastic Unfolding with Application to European Party Activists*. Amsterdam: CT Press.

van Schuur, W. H. 1987. Constraint in European party activists' sympathy scores for interest groups: The left-right dimension as dominant structuring principle. *European Journal of Political Research*, 15, 347–362.

van Schuur, W. H. 1989. Unfolding German political parties: A description and application of multiple unidimensional unfolding. In: G. de Soete, H. Ferger, and K. W. Klauser (Eds.), *New Developments in Psychological Choice Modelling* (pp. 259–277). Amsterdam: North Holland.

Volet, S. E. and Chalmers, D. 1992. Investigation of qualitative differences in university students' learning goals, based on an unfolding model of stage development. *British Journal of Educational Psychology*, 62, 17–34.

Wohlwill, J. F. 1963. The measurement of scalability for noncumulative items. *Educational Psychological Measurement*, 23, 543–555.

Wright, B. D. and Douglas, G. A. 1977. Conditional versus unconditional procedures for sample-free item analysis. *Educational and Psychological Measurement*, 37, 47–60.

Wright, B. D. and Masters, G. N. 1982. *Rating Scale Analysis: Rasch Measurement*. Chicago: MESA Press.

22

Generalized Graded Unfolding Model

James S. Roberts

CONTENTS

22.1 Introduction 369
22.2 Presentation of the Model 370
22.3 Parameter Estimation 375
22.4 Model Fit 380
22.5 Empirical Example 382
22.6 Discussion 386
Appendix 22A 386
References 388

22.1 Introduction

The notion of "unfolding models" for item responses dates back to at least the early attitude measurement work of Thurstone (Thurstone, 1928, 1931, 1932; Thurstone and Chave, 1929) where he assumed that a "relevant" attitude statement attracted endorsements from those individuals whose opinion was matched well by the statement content. Thurstone represented the locations of both persons and items on an underlying unidimensional attitude continuum in which a person was located on the basis of his or her attitude and an item was located according to its content. Thurstone's idea gave rise to single-peaked, empirical item response functions in which persons were more likely to endorse attitude statements to the extent that they were located at nearby positions on the underlying continuum. It was an implicit model for item responses that followed from a proximity-based response process (Roberts et al., 1999).

Since the early work of Thurstone, unfolding response models have been formalized in many ways. Coombs (1950) developed the first formal unfolding model for preference ranks that led simultaneously to a joint representation of persons and items. Persons with conformable preference ranks were hypothesized to fall along given regions of the latent continuum that led to distinct rank orders in a deterministic fashion. Bennett and Hays (1960) generalized this deterministic approach to a multidimensional context. Other researchers followed with alternative multidimensional methods in the form of nonmetric multidimensional unfolding (Kruskal and Carroll, 1969) and metric multidimensional unfolding given known stimulus coordinates (Carroll, 1972). These approaches continued to analyze preference ranks rather than direct single-stimulus ratings. Eventually, Andrich (1988) proposed the first metric item response theory (IRT) model for unfolding single-stimulus preference (i.e., disagree–agree) judgments. Soon after, there were several metric IRT unfolding models available to psychometricians (Volume One, Chapter 21; Andrich and Luo, 1993; Hoijtink, 1990; Roberts and Laughlin, 1996) along with specialized software to estimate model parameters and calculate diagnostics.

During the last two decades, a series of partially nested IRT models have been developed for unfolding direct ratings to single-stimulus responses (Roberts, 1995; Roberts and Laughlin, 1996; Roberts et al., 2000). The most flexible of these models is referred to as the generalized graded unfolding model (GGUM). In the text that follow, I refer to the set of models obtained by intuitively constraining the GGUM as the GGUM family of models.

The GGUM family of models has been applied in the analysis of self-report data to better measure attitude (Roberts and Laughlin, 1996; Roberts et al., 2000), personality (Stark et al., 2006; Chernyshenko et al., 2007; Weekers and Meijer, 2008), and clinically oriented constructs (Noël, 1999). In this chapter, the focus will be on modeling graded responses to typical statements on a Thurstone or Likert-style attitude questionnaire. The primary difference between these two questionnaire types is that during the development of the latter questionnaire, more moderate items that exhibit nonmonotonic item response functions tend to be eliminated in favor of those closer to monotonicity. Nonetheless, the model is still appropriate in either case presuming that item responses follow from a proximity-based response process.

22.2 Presentation of the Model

The GGUM presumes that when a subject encounters a questionnaire item, the subject evaluates how closely the item content matches his or her own opinion about the attitude object. The subject is expected to agree with the item to the extent that this match is high and disagree with the item to the extent that the match is low. From a psychometric perspective, the GGUM jointly represents persons and items on a unidimensional latent continuum where item locations correspond to the valence of the item content and person locations indicate the opinion of the individual. Higher item scores are indicative of higher levels of agreement and are expected to the extent that the person is located closer to a given item. Hence, the GGUM is a model for responses arising from a proximity-based response process (Roberts et al., 1999).

To define the model, one must first consider the notion of proximity as the subject presumably experiences it (i.e., the latent response) and how this experience translates into an observed response. Suppose that a subject is presented with an attitude item and a 4-point response scale where 1 = *strongly disagree*, 2 = *disagree*, 3 = *agree*, and 4 = *strongly agree*. The GGUM presumes that a given subject might use any of these response alternatives for either of two reasons. For example, if the subject is located far below the position of the item on the continuum, then he or she will most likely "Strongly Disagree from Below" the item. In contrast, if the subject were located far above the item, then the subject will most likely "Strongly Disagree from Above" the item. Note that in either case, the subject is relatively far from the item, and hence the item content does not match the individual's opinion. The subject is not asked to indicate which of these two reasons leads to the strong disagreement, but instead indicates only that he or she strongly disagrees with the item. The reason for the strong disagreement is not observable given the response scale, and thus, it is referred to as a "latent response." In contrast, an actual item response is referred to as an "observed response."

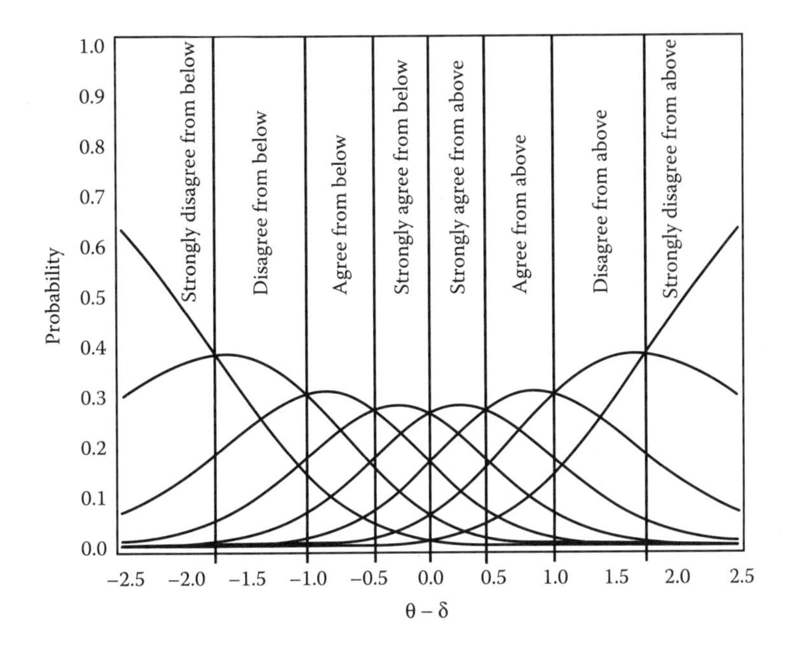

FIGURE 22.1
Latent response category probability curves for a hypothetical four-category item under the GGUM. Latent response category thresholds occur at points where successive probability curves intersect. The dominant latent response is labeled within each segment of the latent continuum.

As shown in Figure 22.1, the GGUM posits two latent responses for each observed response. Moreover, these latent responses are presumed to follow from a dominance-based response process (Coombs, 1964), and thus, are expected to be consistent with a cumulative IRT model. The generalized partial credit model (Volume One, Chapter 8) is used to represent these latent responses as it is one of the more flexible parametric IRT models for graded responses. The parametric form of this model implemented in the development of the GGUM is

$$f_{Y_{pi}}(m;\theta_p,a_i,b_i,d_{ik}) = \frac{\exp\left(a_i\left[(m-1)(\theta_p-b_i)-\sum_{k=1}^{m} d_{ik}\right]\right)}{\sum_{w=1}^{M_i}\exp\left(a_i\left[(w-1)(\theta_p-b_i)-\sum_{k=1}^{w} d_{ik}\right]\right)} \tag{22.1}$$

subject to the constraint that

$$\sum_{m=1}^{M_i} d_{ik} = 0, \tag{22.2}$$

where

Y_{pi} is a latent response from the pth person to the ith item with $m = 1, \ldots, M_i$

$m = 1$ represents the strongest level of disagreement from below item i, and $m = M_i$ represents the strongest level of disagreement from above item i as illustrated in Figure 22.1

θ_p is the location of the pth individual on the latent continuum
b_i is the location of the ith item on the latent continuum
a_i is the discrimination of the ith item
d_{ik} is the relative location of the kth latent response category threshold for the ith item

The term "thresholds" was originally used by Andrich (1978) to describe analogous locations for observed response categories in the rating scale model, and that tradition has continued in the historical development of the GGUM. The value of d_{i1} is arbitrarily defined as zero in this chapter, but any other value could be used without affecting the resulting probabilities (Muraki, 1992).

The latent response category probability curves developed with the generalized partial credit model are given for a hypothetical item in Figure 22.1. The signed distances, relative to b_i, where successive probability curves cross are the latent response category thresholds, d_{i2} through d_{iM_i}. These signed distances are ordered in Figure 22.1, but this is often not the case with real data. Disordinal latent response category thresholds are indicative of dominated latent response categories that are never most likely regardless of the proximity of an individual to an item. Nonetheless, the generalized partial credit model guarantees that the modes of the latent response probability curves are ordered from smallest to largest, and this feature enables the formation of an unfolding model based on these probability curves.

Because the latent responses are mutually exclusive and exhaustive, one can sum the category probability curves for the two latent responses in Figure 22.1 associated with each observed response to obtain the probability of that observed response. For example, one could sum the latent response probability curves for "Strongly Disagree from Below" and "Strongly Disagree from Above" to yield a probability curve for the observed response *"strongly disagree."* Similarly, the latent response probability curves could be summed for "Disagree from Below" and "Disagree from Above" to give the probability of the observed response *"disagree,"* etc. This summation yields the category probability function for the GGUM:

$$f_{U_{pi}}(a;\theta_p,a_i,b_i,d_{ik}) = f_{Y_{pi}}(a;\theta_p,a_i,b_i,d_{ik}) + f_{Y_{pi}}((M_i - a);\theta_p,a_i,b_i,d_{ik})$$

$$= \frac{\exp\left(a_i\left[(a-1)(\theta_p - b_i) - \sum_{k=1}^{a} d_{ik}\right]\right) + \exp\left(a_i\left[(M_i - a)(\theta_p - b_i) - \sum_{k=1}^{M_i-a+1} d_{ik}\right]\right)}{\sum_{w=1}^{M_i}\left[\exp\left(a_i\left[(w-1)(\theta_p - b_i) - \sum_{k=1}^{w} d_{ik}\right]\right)\right]}, \quad (22.3)$$

where

U_{pi} is an observed response from the pth person to the ith item with $a = 1, \ldots, A_i$;
$a = 1$ represents the strongest level of disagreement and $a = A_i$ represents the strongest level of agreement for the ith item; and
$M_i = 2A_i$.

Because the observed responses are indicative of proximities, the response to a single item does not provide information about the sign of $(\theta_p - b_i)$. Consequently, the values of the d_{ik} parameters cannot be determined from the item response data without further restrictions. The GGUM assumes that d_{ik} parameters are symmetric about b_i. This assumption results in the following identities $d_{i1} = d_{i(A_i+1)} = 0$, and $d_{ia} = -d_{i(M_i-a+2)}$ for $a \neq 1$. Additionally, this assumption makes the following summated terms in Equation 22.3 equivalent:

$$\sum_{k=1}^{a} d_{ik} = \sum_{k=1}^{M_i-a+1} d_{ik}, \tag{22.4}$$

which, in turn, allows the category probability function for the GGUM to be written entirely in terms of observed responses:

$$f_{U_{pi}}(a;\theta_p,a_i,b_i,d_{ik}) = \frac{\exp\left(a_i\left[(a-1)(\theta_p-b_i)-\sum_{k=1}^{a} d_{ik}\right]\right)+\exp\left(a_i\left[(M_i-a)(\theta_p-b_i)-\sum_{k=1}^{a} d_{ik}\right]\right)}{\sum_{w=1}^{A_i}\left[\exp\left(a_i\left[(w-1)(\theta_p-b_i)-\sum_{k=1}^{w} d_{ik}\right]\right)\right]+\exp\left(a_i\left[(M_i-w)(\theta_p-b_i)-\sum_{k=1}^{w} d_{ik}\right]\right)}. \tag{22.5}$$

Figure 22.2 illustrates the GGUM category probability curves associated with the same hypothetical item portrayed in Figure 22.1. The probability of an observed response category in Figure 22.2 was derived by summing the probability of the two associated latent response categories (e.g., "Strongly Disagree From Below" and "Strongly Disagree From Above" were summed to obtain *strongly disagree*, etc.). This summing process is evident in the numerator and denominator of Equation 22.5 where two exponential functions are summed together to develop each observed response category probability, and these two

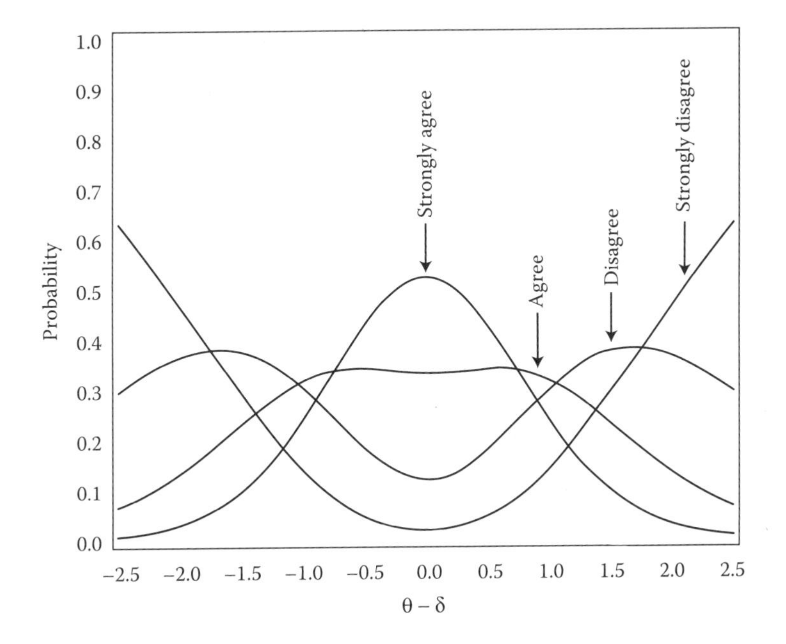

FIGURE 22.2
Category probability curves for a hypothetical four-category item under the GGUM. Each curve corresponds to a given observed response.

functions are associated with the two latent response categories corresponding to a given observed response.

An item characteristic curve (ICC) from the GGUM is derived by taking the expected value of an observed response across categories:

$$\sum_{a=1}^{A_i} a f_{U_{pi}}(a;\theta_p,a_i,b_i,d_{ik}). \tag{22.6}$$

The corresponding item characteristic curve for the hypothetical item from Figure 22.2 is shown in Figure 22.3. The item characteristic curve is single peaked and symmetric about b_i. This single-peaked nature is a natural consequence of the model and will be evident regardless of the values of item parameters as long as a_i is not equal to zero (Donoghue, 1999), in which case, the item characteristic curve would be a horizontal line.

As implied by Figures 22.1 and 22.2, the parameters of the GGUM are defined at the latent response level. The θ_p, b_i, and d_{ik} parameters all relate to locations on the latent continuum with respect to the probability of latent responses in Figure 22.1, and the a_i parameters indicate how much category probability functions vary as a function of θ_p for a fixed set of b_i and d_{ik} parameters. The θ_p and b_i parameters represent the same locations on the latent continuum when considering the probability of an observed response (Figure 22.2), but the d_{ik} and a_i parameters do not retain their respective interpretations at this level. Nonetheless, these latter quantities still affect the observed response category probability and item characteristic curves through their impact on the latent response category probabilities. Figure 22.4 illustrates the effect of increasing the a_i parameter on the ICCs for a hypothetical GGUM item with four response categories and equally spaced latent response

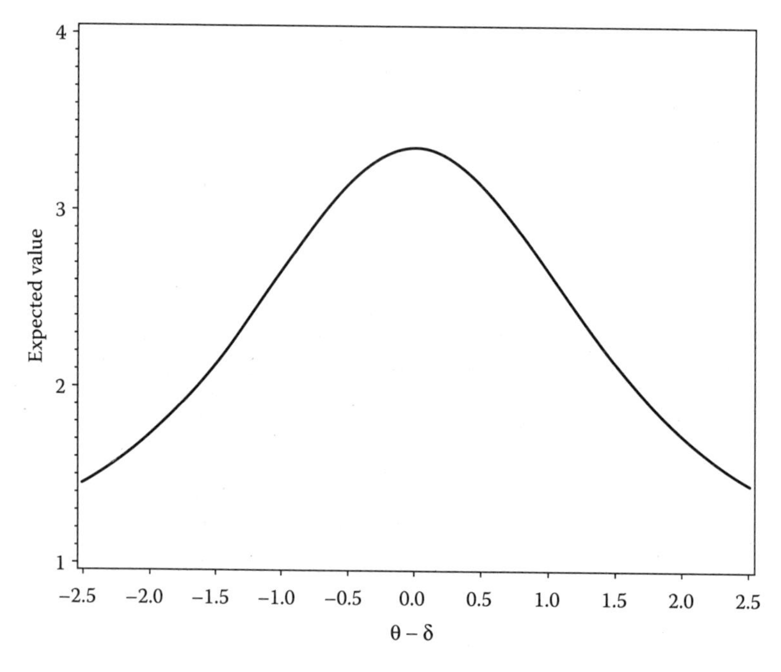

FIGURE 22.3
The item characteristic curve (i.e., expected value curve) for a hypothetical four-category item under the GGUM.

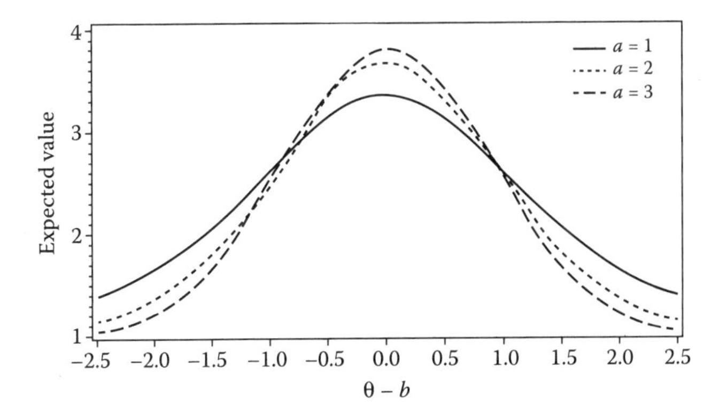

FIGURE 22.4
Item characteristic curves for a hypothetical four-category item with discrimination parameters that are varied from 1 to 3.

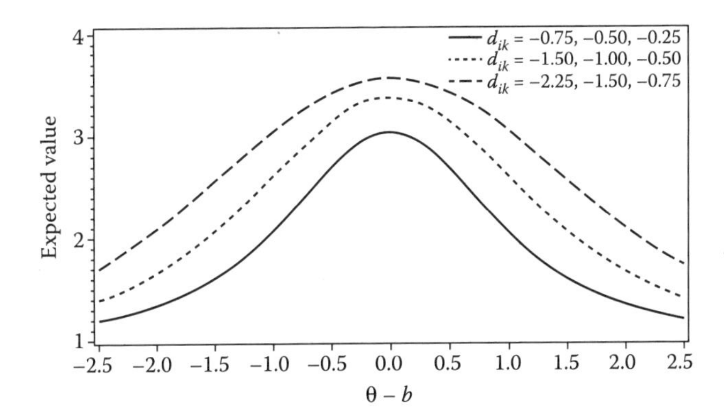

FIGURE 22.5
Item characteristic curves for a hypothetical four-category item with interthreshold distances that are varied from 0.25 to 0.75.

category thresholds located at –1.5, –1, and –0.5. (Recall that $d_{i1} = 0$.) As the a_i parameter increases from 1 to 3, the ICCs become more peaked, and the maximum expected value attained when $(\theta_p - b_i) = 0$ increases. Figure 22.5 illustrates an alternative scenario in which the latent response category thresholds are varied so that the interthreshold distance (ID) between d_{i2}, d_{i3}, and d_{i4} values varies from 0.25, to 0.5, to 0.75 while the a_i value is fixed at 1. The ICCs for the hypothetical item, again, take on higher maximum values as the ID increases, but the curves become somewhat less peaked in this case.

22.3 Parameter Estimation

Item parameters within the GGUM family of models have been estimated with a variety of methods that include joint maximum likelihood (Roberts and Laughlin, 1996), marginal maximum likelihood (MML; Roberts et al., 2000), marginal maximum *a posteriori* (MMAP;

Roberts and Thompson, 2011), and Bayesian estimation with Markov chain Monte Carlo (MCMC) sampling from the posterior distribution (de la Torre et al., 2006; Roberts and Thompson, 2011). With respect to the most general model, MMAP estimation seems generally preferable. It leads to the smallest amount of estimation error in contexts with four or fewer response categories and it is substantially faster than MCMC—its closest competitor (Roberts and Thompson, 2011). However, when five or more response categories are considered, then MML, MMAP, and MCMC produce item parameter estimates of similar accuracy. In each of these cases, person parameters have been estimated with an expected *a posteriori* method (Bock and Mislevy, 1982).

Estimation of the GGUM family of models is most widely accomplished with the GGUM2004 freeware program (Roberts et al., 2006). This visually based program will estimate the full model in Equation 22.5 as well as some systematically constrained versions of the model using an MML estimation algorithm. This most frequently used means of obtaining GGUM parameter estimates, with results similar in accuracy to those produced by MMAP and MCMC when the number of response categories is greater than four, will be elaborated on here.

Let $\boldsymbol{u}_p$ be the response vector for person p to the set of I items. Let u_{pi} refer to the ith element of $\boldsymbol{u}_p$. Under the assumption of local independence, the likelihood of observing $\boldsymbol{u}_p$ given θ_p is equal to

$$L(\theta_p;\boldsymbol{u}_p)=\prod_{i=1}^{I} f_{U_{pi}}(u_{pi};\theta_p,a_i,b_i,d_{ik}\}. \tag{22.7}$$

If all subjects are independently sampled from an identical population with a distribution denoted as $g(\theta)$, then the marginal likelihood of observing $\mathbf{U}_p = \mathbf{u}_p$ is equal to

$$L_p=\int_{\theta} L(\theta_p;\boldsymbol{u}_p)g(\theta)d\theta. \tag{22.8}$$

Let P denote the number of subjects in the sample. Under the assumption of independence between respondents, the marginal likelihood of the sample of response vectors encountered across subjects is equal to

$$L=\prod_{p=1}^{P} L_p. \tag{22.9}$$

The likelihood equations for a_i, b_i, and d_{ik} are obtained by calculating the first-order partial derivatives of the logarithm of Equation 22.9 with respect to each parameter and then setting these derivatives equal to 0. The values that solve these equations are the marginal maximum likelihood estimates of a_i, b_i, and d_{ik}. Roberts et al. (2000) described an expectation-maximization (EM) algorithm to accomplish this based on the work of Bock and Aitkin (1981) and Muraki (1992); for a general introduction, see Aitkin (Volume Two, Chapter 12). Briefly, the EM algorithm proceeds in two stages. During the expectation stage, the values of θ_p are treated as missing data, and the expected frequencies of these parameters are derived from their respective posterior distributions. With these expectations in hand, the item parameters are then estimated in a maximization stage that is divided into two steps. In the first step, the estimates of d_{ik} are calculated by maximizing the logarithm of the marginal likelihood, and this is followed by the second step in which estimates of a_i and b_i are derived using an analogous method. Splitting the maximization stage into these two steps leads to a general algorithm that is suitable for the entire GGUM family of models in which d_{ik} and a_i parameters are constrained across items in systematic ways. Given that the frequencies calculated in the expectation stage are dependent upon the current values of item parameter estimates, the expectation and maximization stages iterate back and forth until there is little change in the item parameter estimates.

Both stages of the EM algorithm involve integration over $g(\theta)$, and this is accomplished using numerical integration (i.e., quadrature). The GGUM2004 program implements a rectangular quadrature method in which the latent continuum is approximated by F equally spaced, discrete points denoted as v_f. The approximate density of $g(\theta)$ at quadrature point v_f is denoted $z(v_f)$, which is scaled so that $\sum_{f=1}^{F} z(v_f) = 1$. In the expectation stage, one calculates the expected number of subjects at quadrature point v_f who respond to item i using category a. These expected frequencies are given as

$$\bar{r}_{iaf} = \sum_{p=1}^{P} \frac{h_{pia} L(v_f; \boldsymbol{u}_p) z(v_f)}{\tilde{P}_p}, \tag{22.10}$$

where

h_{pia} is a dummy variable equal to 1 when $u_{pi} = a$; otherwise, $h_{pia} = 0$;

$L(v_f; \boldsymbol{u}_p)$ is the likelihood of response vector $\boldsymbol{u}_p$ from Equation 22.7 with θ_p set equal to v_f, and

$$\tilde{P}_p = \sum_{f=1}^{F} L(v_f; \boldsymbol{u}_p) z(v_f). \tag{22.11}$$

Once these expectations, $\bar{r}_{iaf}$, are calculated, they are treated as known values, and the log marginal likelihood is maximized with respect to the item parameters using a Fisher scoring algorithm. As mentioned earlier, the maximization is performed first with respect to d_{ik} parameters and then with respect to a_i and b_i parameters.

The Fisher scoring algorithm updates the d_{ik} parameters in an iterative fashion, and the interim values for these parameters on the qth iteration are given as follows:

$$\begin{bmatrix} d_{i2} \\ d_{i3} \\ \vdots \\ d_{iA_i} \end{bmatrix}_q = \begin{bmatrix} d_{i2} \\ d_{i3} \\ \vdots \\ d_{iA_i} \end{bmatrix}_{q-1} + \left[\tilde{\boldsymbol{I}}_{d_{ik}}\right]_{q-1}^{-1} \begin{bmatrix} \dfrac{\partial \ln(L)}{\partial d_{i2}} \\ \dfrac{\partial \ln(L)}{\partial d_{i3}} \\ \vdots \\ \dfrac{\partial \ln(L)}{\partial d_{iA_i}} \end{bmatrix}_{q-1}, \tag{22.12}$$

where $\tilde{\boldsymbol{I}}_{d_{ik}}$ is the information matrix for the unconstrained d_{ik} parameters, and $\partial \ln(L)/\partial d_{i2}$, $\partial \ln(L)/\partial d_{i3} \ldots, \partial \ln(L)/\partial d_{iA_i}$ are derivatives of the log marginal likelihood. The general form of these derivatives is given by

$$\frac{\partial \ln(L)}{\partial d_{ik}} = \left(\sum_{f=1}^{F} \sum_{a=1}^{A_i} \frac{\bar{r}_{iaf}}{f_{U_{fi}}(a;v_f,a_i,b_i,d_{ik})} \frac{\partial f_{U_{fi}}(a;v_f,a_i,b_i,d_{ik})}{\partial d_{ik}} \right) \quad \text{for } k \geq 2. \tag{22.13}$$

The last term in Equation 22.13 involves a parameter-specific derivative that is defined in the appendix. Matrix $\tilde{\boldsymbol{I}}_{d_{ik}}$ has the following structure:

$$\tilde{\boldsymbol{I}}_{d_{ik}} = \begin{bmatrix} \tilde{I}_{d_{i2}d_{i2}} & \tilde{I}_{d_{i2}d_{i3}} & \cdots & \tilde{I}_{d_{i2}d_{iA_i}} \\ \tilde{I}_{d_{i3}d_{i2}} & \tilde{I}_{d_{i3}d_{i3}} & \cdots & \tilde{I}_{d_{i3}d_{iA_i}} \\ \vdots & \vdots & \vdots & \vdots \\ \tilde{I}_{d_{iA_i}d_{i2}} & \tilde{I}_{d_{iA_i}d_{i3}} & \cdots & \tilde{I}_{d_{iA_i}d_{iA_i}} \end{bmatrix} \tag{22.14}$$

with entries equal to

$$\tilde{I}_{d_{ik}d_{ik'}} = \sum_{f=1}^{F} \bar{n}_{if} \sum_{a=1}^{A_i} \frac{1}{f_{U_{fi}}(a;v_f,a_i,b_i,d_{ik})} \frac{\partial f_{U_{fi}}(a;v_f,a_i,b_i,d_{ik})}{\partial d_{ik}} \frac{\partial f_{U_{fi}}(a;v_f,a_i,b_i,d_{ik})}{\partial d_{ik'}}. \tag{22.15}$$

In Equation 22.15, both d_{ik} and $d_{ik'}$ refer to one of the two through A_i latent response category thresholds and $\bar{n}_{if}$ is the expected number of respondents at quadrature point v_f who respond to item i (i.e., $\bar{n}_{if} = \sum_{a=1}^{A_i} \bar{r}_{iaf}$).

The updates continue until the change in d_{ik} parameters is very small from one iteration to the next (e.g., 0.001) or a predetermined maximum number of iterations has been reached (e.g., 30). After the d_{ik} parameters are updated, the values of a_i and b_i that maximize the current log marginal likelihood are found using the Fisher scoring algorithm:

$$\begin{bmatrix} a_i \\ b_i \end{bmatrix}_q = \begin{bmatrix} a_i \\ b_i \end{bmatrix}_{q-1} + \left[\tilde{\boldsymbol{I}}_{a_i b_i} \right]_{q-1}^{-1} \begin{bmatrix} \dfrac{\partial \ln(L)}{\partial a_i} \\ \dfrac{\partial \ln(L)}{\partial b_i} \end{bmatrix}_{q-1}, \tag{22.16}$$

where $\tilde{\boldsymbol{I}}_{a_i b_i}$ is the information matrix for the a_i and b_i parameters, and $\partial \ln(L)/\partial a_i$ and $\partial \ln(L)/\partial b_i$ are derivatives of the interim log marginal likelihood. The general form of these derivatives is

$$\frac{\partial \ln(L)}{\partial \phi_i} = \left(\sum_{f=1}^{F} \sum_{a=1}^{A_i} \frac{\bar{r}_{iaf}}{f_{U_{fi}}(a;v_f,a_i,b_i,d_{ik})} \frac{\partial f_{U_{fi}}(a;v_f,a_i,b_i,d_{ik})}{\partial \phi_i} \right), \tag{22.17}$$

where ϕ_i represents either a_i or b_i depending on the context. The final term in Equation 22.17 is defined in the appendix for each of these parameters. Matrix $\tilde{\boldsymbol{I}}_{a_i b_i}$ is defined as

$$\tilde{\boldsymbol{I}}_{a_i b_i} = \begin{bmatrix} \tilde{I}_{a_i a_i} & \tilde{I}_{a_i b_i} \\ \tilde{I}_{b_i a_i} & \tilde{I}_{b_i b_i} \end{bmatrix} \tag{22.18}$$

with entries equal to

$$\tilde{I}_{\phi_i \phi_i'} = \sum_{f=1}^{F} \bar{n}_{if} \sum_{a=1}^{A_i} \frac{1}{f_{U_{fi}}(a; v_f, a_i, b_i, d_{ik})} \frac{\partial f_{U_{fi}}(a; v_f, a_i, b_i, d_{ik})}{\partial \phi_i} \frac{\partial f_{U_{fi}}(a; v_f, a_i, b_i, d_{ik})}{\partial \phi_i'}, \tag{22.19}$$

where ϕ_i and ϕ_i' each refer to one of the two item parameters. The a_i and b_i parameters are updated until there is little change across successive iterations or until a predetermined maximum number of iterations is reached. This constitutes the end of a single maximization (i.e., inner) cycle. The maximization process is repeated until there is little change in any of the item parameters (i.e., d_{ik}, a_i, or b_i) between successive inner cycles or until some maximum number of inner cycles has been reached (e.g., ten). Once the inner cycles have completed, the maximization step ends.

Because the estimates of $\bar{r}_{iaf}$ and the item parameters are mutually dependent on each other, the expectation and maximization stages of the EM algorithm must be repeated until there is little change in the item parameter estimates from one repetition to the next. At this point, the algorithm has numerically converged, and the derivatives of the marginal log likelihood with respect to the item parameters are all approximately zero. Recovery simulations have shown that estimates from this procedure are generally accurate with five or more response categories and a variety of item/person parameter configurations (Roberts et al., 2002; Roberts and Thompson, 2011). However, this may not always be the case, and researchers should examine the model parameters and associated fit carefully to ascertain whether any item parameters seem illogical or misrepresent the item responses.

Calculation of the log marginal likelihood function and the associated item parameter derivatives requires the specification of a distribution for the θ_p parameters [i.e., $g(\theta)$]. A fixed mean and variance for the distribution provide a location and scale for the latent continuum. When combined with the symmetry constraint imposed on d_{ik} parameters, these constraints yield a mathematically identified model. Roberts et al. (2002) have conducted simulation studies which suggest that using an N(0,1) distribution for $g(\theta)$ generally leads to very reasonable parameter estimates even when the true θ distribution deviates substantially from normality.

As mentioned earlier, the GGUM2004 program will estimate parameters for a number of other models in the GGUM family by either constraining the a_i parameters to be equal to 1 across items or by constraining d_{ik} in various ways (Roberts and Shim, 2008). For example, d_{ik} may be constrained to be equal across items such that $d_{ik} = d_k$. The two-step maximization strategy for d_{ik} followed by a_i and b_i allows for all of the models in the GGUM family to be estimated using the same logic. Namely, for all models, the interim log marginal likelihood is maximized first with respect to the latent response category thresholds followed by the remaining item parameters (i.e., the item location and, if requested, item discrimination parameters).

After the MML procedure has completed, one can treat the item parameter estimates as known quantities and estimate the person parameters. Past research with the GGUM has shown that an expected *a posteriori* (EAP) procedure generally provides accurate estimates of θ_p parameters (Roberts et al., 2002; Roberts and Thompson, 2011). EAP estimates are advantageous because they exist for all response patterns and tend to have an average mean-square error smaller than that for most other estimators (Bock and Mislevy, 1982). The EAP estimate is given by

$$\hat{\theta}_p = \frac{\sum_{f=1}^{F} v_f \, L(v_f; \boldsymbol{u}_p) \, z(v_f)}{\sum_{f=1}^{F} L(v_f; \boldsymbol{u}_p) z(v_f)}. \tag{22.20}$$

Like the item parameter estimates obtained with the MML procedure, EAP estimates of θ_p are generally robust in cases where a normal distribution for θ_p is used, but the true distribution is not normal. However, some sensitivity to the θ_p distribution may occur for deviations in its tails because there is little information about the tails in the response data (Roberts et al., 2002).

22.4 Model Fit

GGUM fit statistics can be useful when considered along with other information such as graphical information about how items function in a given calibration, and how well the item estimates correspond with the item content. When these statistics corroborate what the user can see with various graphically based fit indices, then a stronger case can be made to reject a given item due to misfitting responses.

Prior to fitting the GGUM, a principal component analysis of the graded responses should be conducted, and the results should be compared to those expected from item responses that conform to the GGUM. Specifically, there should be two dominant components (van Schuur and Kiers, 1994), and a plot of the pattern coefficients should form a simplex-like shape with the extreme ends folded inward (Polak et al., 2009; Roberts and Laughlin, 1996). If this condition does not hold, then either the unfolding model is not appropriate or the data are multidimensional. A practical rule of thumb is that the variance of any item explained by the first two components should be greater than 0.30. Items with smaller communalities should generally be discarded because less than 30% of their variance is reflected by the hypothesized unidimensional construct (Roberts and Laughlin, 1996). Note that this does not in any way guarantee that an item is unidimensional. It simply eliminates items that exhibit little chance of unidimensionality (and thus, relevance) within the context of an unfolding model.

Perhaps the most intuitive and basic diagnostic of GGUM performance with a given set of data is simply to order statements according to their b_i values and then ascertain whether the content of each statement makes sense with the associated item position. Statement content should flow from very negative, moderately negative, neutral, moderately positive, and very positive expressions with respect to the attitude object. Although there may be gaps on the continuum where there are no items due to a lack of statements that express the corresponding sentiment, there should not be noticeable gaps when suitable items have been calibrated. For example, if there was a gap near the middle of the continuum where one might expect to find neutral items that were calibrated, then that is reason for concern. In such cases, the particular unfolding model used may be too constrained or a cumulative model may be more appropriate.

The GGUM2004 software provides a number of graphs to explore item and model fit. At the item level, one can rank order individuals according to their estimated θ_p values and form homogeneous clusters of approximately equal size. One can then plot the mean estimated θ_p value in each cluster against both the average observed and average expected item response for that cluster. Model fit is indicated to the extent that the expected and observed averages are similar. However, this graph provides additional information about how the item is functioning. For example, the d_{ik} and a_i parameters both determine the overall discrimination of an item (i.e., the steepness of its item characteristic curve), and these displays help ascertain this item feature. Items that show little signs of agreement at any point on the continuum are irrelevant items and should be discarded.

A similar graph can be produced at the model level by rank ordering the $\theta_p - b_i$ difference associated with each item response. Homogeneous groups of equal size can be formed and then the mean observed and the mean expected response for each group can be plotted against the average $\theta_p - b_i$ value. Ideally, the average observed responses should be closely mimicked by the corresponding mean expected responses. Additionally, this graph should exhibit higher levels of agreement (i.e., higher item scores) to the extent that the absolute value of $\theta_p - b_i$ is small. However, it should not be used to diagnose performance in lieu of the item fit graphs described above because it is possible to achieve the desired result with items that fit a cumulative model rather than an unfolding model.

With respect to statistics, the GGUM2004 program calculates a variety of statistics to ascertain which items or respondents do not fit the model. Some of the more useful ones are the localized infit and outfit t-statistics (Roberts and Shim, 2008). These are infit and outfit t-statistics (Wright and Masters, 1982) that have been rescaled so that their sum of squares across items is equal to I in the case of item fit, or alternatively, the sum of squares across persons is equal to P in the case of person fit. These localized t-statistics provide a relative, rather than absolute, index of item or person fit in the calibrated data set. They help identify the most aberrant items or persons given the realized degree of misfit inherent in the data set under analysis (i.e., the "localized" context). In particular, these localized t-statistics can be useful for identifying unconscientious respondents in situations where there is little incentive or motivation to attend to a questionnaire.

The most researched item fit statistics that have been used with the GGUM are the family of ${}_{S}X_i^2$ statistics described by Roberts (2008). These statistics are generalizations of Orlando and Thissen's (2000) item fit chi-square technique. They exhibit good Type I error and reasonable power under several logical violations of the estimating model. The GGUM2004 software calculates four statistics in this family. The most commonly recommended of these is defined as

$$ {}_{S}X_i^2 = \sum_{k=(I+A-1)}^{(IA-A+1)} \sum_{a=1}^{A} N_k \frac{(O_{ika} - E_{ika})^2}{E_{ika}}, \tag{22.21} $$

where

N_k is the number of respondents with an observed test score (i.e., summated score) equal to k

O_{ika} is the observed proportion of respondents with an observed test score equal to k who select response category a for item i

E_{ika} is the proportion of respondents with an observed test score equal to k who are expected to select response category a for item i

Calculation of E_{ika} follows a recursive algorithm developed by Thissen and colleagues (1995) and detailed in Roberts (2008). Variants of ${}_SX_i^2$ include versions that do not include the studied item in the calculation of the observed test score and those that only include items with approximately monotonic item characteristics in the score calculation. Generally speaking, the ${}_SX_i^2$ statistic is recommended in those situations where it has calculable degrees of freedom. In other situations, one of its variants can be used (Roberts, 2008).

As mentioned earlier, the distributions of fit statistics that have been applied to the GGUM have not been rigorously established. Moreover, these statistics test a null hypothesis of perfect fit of the model to a set of item responses which, in practice, is very unlikely to occur. Therefore, it seems reasonable to simply use these indices as indicators of relative misfit rather than absolute fit. One way in which this can be accomplished is by calculating the value of ${}_SX_i^2$ divided by its degrees of freedom and using this as an index of model fit. Items can be ordered from the worst to the best fitting items, and those items with an index greater than some rational cutoff can be selected for further examination. A chi-square per degree of freedom ratio greater than 3 has some precedence as a misfit criterion in the IRT literature (Drasgow et al., 1995; Chernyshenko et al., 2007), but whatever cutoff is ultimately used, the degree of misfit should be corroborated with other means such as graphical procedures.

22.5 Empirical Example

The following example is based on responses from 837 Georgia Institute of Technology undergraduate subjects to 21 questionnaire items that were designed to measure attitude toward abortion. Ten of the questionnaire items were previously investigated by Roberts et al. (2000) and represented pro-life, moderately pro-life, neutral, moderately pro-choice, and pro-choice opinions, with two items portraying each of these attitude positions. The remaining 11 items were newly constructed to represent the two moderate portions of the attitude continuum. Each item had six response categories—*strongly disagree, disagree, slightly disagree, slightly agree, agree,* and *strongly agree.* Items were administered randomly to each subject via computer.

Item responses were analyzed with the most general model in the GGUM family (i.e., Equation 22.5) using the GGUM2004 computer program. An initial analysis revealed that many of the newly constructed items had very low levels of strong agreement. To simplify the analysis, the *strongly agree* and *agree* categories were collapsed so that each item had five response categories. After rerunning the analysis, nine subjects with extreme values for localized infit or outfit t-statistics (i.e., beyond ±2.576) were found, and all had unusual response patterns that suggested a lack of attention to the task. These individuals were discarded from further analyses.

An analysis of the remaining 828 subjects was performed, and the estimated item parameters and associated item statistics were examined along with their statement content. The items were ordered in a manner that was consistent with their content, and their parameter estimates seemed quite reasonable. Generally speaking, the statements in the moderately pro-life portion of the continuum exhibited less fit than statements in other portions of the latent continuum. Three of these items had item characteristic curves that showed noticeably less discrimination than other items. They also had the highest ${}_{S}X^2/df$ ratios which ranged from 2.90 to 3.55. These items were deleted from the analysis, and the remaining 18 items were recalibrated. The resulting estimates and associated statement content are shown in Table 22.1. As shown in the table, the estimates were consistent with the sentiment expressed by each item, and these items formed five distinct clusters representing the desired areas of the latent continuum. All items exhibited reasonable fit, both graphically and with respect to ${}_{S}X^2/df$, and all had adequate discrimination.

Figure 22.6 illustrates the expected and observed responses for 10 representative items with the respondents clustered into 20 groups that were homogeneous with respect to their attitude estimates. As shown in the figure, items located in the pro-life and pro-choice portions of the continuum had item characteristic curves that were located in the farthest regions of the latent continuum and were approximately monotone in shape. Moderately

TABLE 22.1

GGUM Item Parameter Estimates for 18 Abortion Attitude Items

Statement	b_i	a_i	d_{i2}	d_{i3}	d_{i4}	d_{i5}
1. Abortion is inhumane	−2.6	1.8	−3.3	−2.9	−2.9	−2.6
2. Abortion is unacceptable under any circumstances	−2.4	1.5	−2.0	−2.3	−1.4	−2.5
3. Abortion is basically immoral except when the woman's physical health is in danger	−1.1	1.8	−2.2	−1.3	−1.2	−1.0
4. Abortion should be illegal except in extreme cases involving incest or rape	−1.0	1.6	−2.0	−1.1	−1.0	−0.9
5. Abortions should not normally be performed unless there is medical evidence that the baby will die before one year of age	−0.8	1.0	−1.8	−0.9	−0.7	−0.3
6. Abortion is generally unacceptable except when the child will never be able to live outside a medical institution	−0.8	1.2	−1.7	−1.1	−0.5	−0.2
7. I cannot whole-heartedly support either side of the abortion debate	0.0	2.0	−1.2	−0.5	−0.8	−0.4
8. My feelings about abortion are very mixed	0.0	1.7	−1.3	−0.6	−0.8	−0.4
9. Abortions should be allowed only if both biological parents agree to it	0.4	1.0	−1.0	−0.6	−0.8	−0.4
10. Abortions should usually be permissible, but other alternatives must be explored first	0.6	2.6	−1.7	−1.1	−1.2	−0.7
11. Abortion should be a woman's choice, but should never be used simply due to its convenience	0.6	1.6	−2.1	−1.2	−1.5	−1.4
12. Abortion is acceptable in most cases, but it should not be supported with tax dollars	0.7	1.3	−1.5	−0.8	−0.9	−0.6
13. Abortions should generally be allowed, but only when the woman obtains counseling beforehand	0.7	2.2	−1.6	−0.9	−1.1	−0.5
14. Abortion should generally be legal, but should never be used as a conventional method of birth control	0.8	2.8	−1.9	−1.3	−1.3	−1.2
15. Abortion should usually be legal except when it is performed simply to control the gender balance in a family	0.8	1.1	−0.9	−0.4	−1.2	−1.1
16. Abortions should generally be legal unless the woman is mentally incapable of making a decision to undergo the procedure	0.8	1.7	−1.9	−0.9	−1.1	−0.3
17. Society has no right to limit a woman's access to abortion	2.2	1.9	−3.3	−2.6	−2.5	−2.4
18. Abortion should be legal under any circumstances	2.3	1.4	−2.0	−2.2	−2.0	−2.2

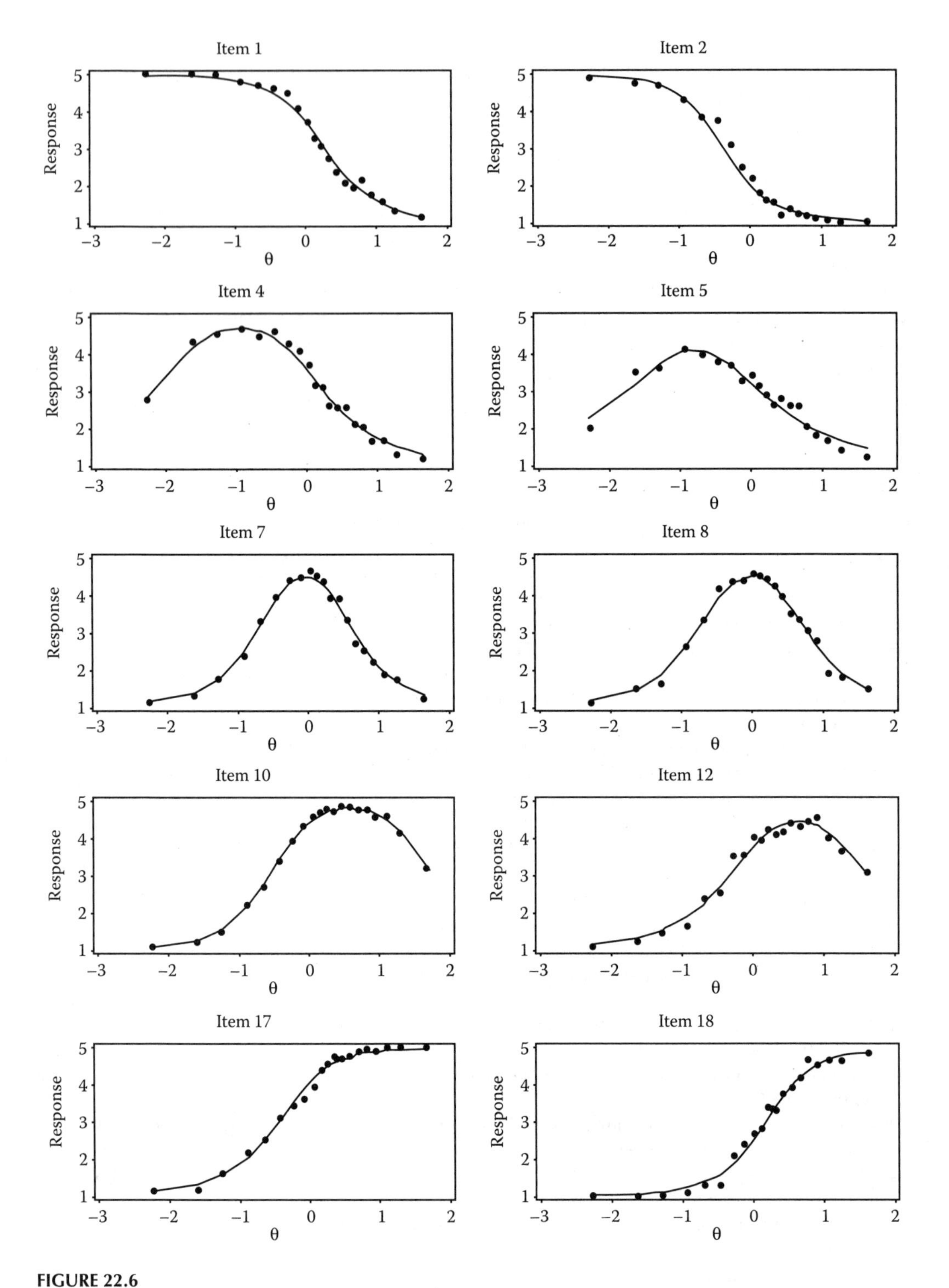

FIGURE 22.6
Mean observed responses (dots) versus mean expected responses (solid line) as a function of mean attitude estimate. Means are calculated for 20 respondent groups with homogeneous attitude estimates.

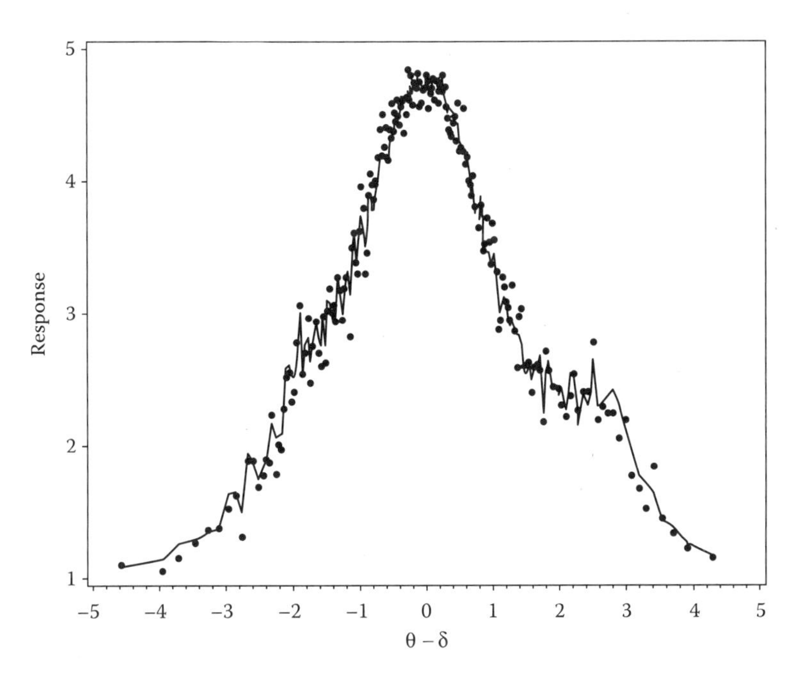

FIGURE 22.7
Mean observed item response (dots) versus mean expected item response (solid line) as a function of mean $\theta_p - b_i$ value. Means were calculated for 199 groups of data points with relatively homogeneous values of $\theta_p - b_i$.

pro-life and moderately pro-choice items were located in less extreme positions on the continuum and showed significant folding (i.e., nonmonotonicity) in the associated extreme regions of the continuum. Items with neutral content were located in the middle portion of the latent continuum and exhibited somewhat bell-shaped item characteristic curves. In each of these cases, the fit of the model to the observed responses was reasonably good. Indeed, the average absolute discrepancy between the observed and expected responses across the 10 panels of Figure 22.6 was 0.117, which is good for a 5-point response scale. In short, the item characteristics reflected exactly what would be expected from a proximity-based response process.

The distributions of the θ_p estimates was negatively skewed (skewness = −0.63) with a mean of 0.02, a standard deviation of 0.99, and a kurtosis of 0.05. The median θ_p estimate was equal to 0.17 and fell between the statements "My feelings about abortion are very mixed" ($b_i = 0.01$) and "Abortions should be allowed only if both biological parents agree to it" ($b_i = 0.43$). Thus, the median respondent exhibited a very slight pro-choice orientation.

Figure 22.7 illustrates the general fit of the GGUM to all item responses. In this plot, the estimated difference between $\theta_p - b_i$ is calculated for each response and these differences are then rank ordered and clustered into homogeneous groups of approximately equal size ($n \approx 75$). The average $\theta_p - b_i$ for each cluster is portrayed on the horizontal axis whereas the average observed and average expected item responses for the cluster are given on the vertical axis. The plot shows that the GGUM was able to order items and respondents such that more agreement was typically expected to the extent that $\theta_p - b_i$ approached zero. Additionally, there was generally good correspondence between observed and expected responses with an average absolute discrepancy of 0.087 on a 5-point scale.

22.6 Discussion

The GGUM model has been used successfully to measure a variety of constructs in the attitude and personality domains. The model is potentially applicable whenever unidimensional responses follow from a proximity-based response process. Its parameters have been estimated using a variety of methods, and each of these methods seems to work well when the number of response categories is five or more. In cases with fewer response categories, those methods that incorporate prior distributions for item parameters (i.e., MMAP and MCMC) appear to work best. A freely available computer program can be used to estimate parameters for all of the models in the GGUM family. This makes the model readily usable by the psychometric community.

Future research with the GGUM will focus on expanding it to model preference and self-to-stimulus similarity judgments. It is unlikely that these sorts of responses are unidimensional, and thus, a multidimensional version of the GGUM is necessary. Initial investigations of a multidimensional extension of the model using MCMC parameter estimation have been quite promising (Roberts and Shim, 2010), but further work is necessary to make parameter estimation more computationally efficient and user friendly.

In addition to the multidimensional extension mentioned above, other researchers have generalized the basic GGUM model in various ways. Cui (2008) developed a longitudinal version of the GGUM using a latent difference approach first explored by Embretson (1991) in the context of the Rasch model. This longitudinal GGUM could be quite useful for those researchers studying attitude change. Wu and Wang (2010) created a multidimensional version of the GGUM for multiple unidimensional constructs which are allowed to correlate. This extension should be useful to those individuals who desire to use the GGUM to measure multiple personality or attitude dimensions with a battery of items that possesses a simple structure. Usami (2011) developed a version of the GGUM that is explanatory in nature. His model explains θ_p as a function of both latent factors and observed variables using a structural equation approach. Additionally, Roberts, Rost, and Macready (2010) developed a version of the GGUM in which mixtures of respondents have alternative latent response category thresholds. This, in turn, leads to different "latitudes of acceptance" or ranges of statements that individuals are expected to endorse to some extent. Each of these extensions makes the GGUM more pertinent to particular research questions and helps ensure a bright future for item response theory approaches to unfolding.

Appendix 22A

Let

$$y = \exp\left\{ a_i \left[(a-1)(v_f - b_i) - \sum_{k=1}^{A_i} j_{ak} d_{ik} \right] \right\}, \tag{22A.1}$$

$$\tilde{y} = \exp\left\{ a_i \left[(w-1)(v_f - b_i) - \sum_{k=1}^{A_i} j_{wk} d_{ik} \right] \right\}, \tag{22A.2}$$

$$t = \exp\left\{a_i\left[(M_i - a)(v_f - b_i) - \sum_{k=1}^{A_i} j_{ak} d_{ik}\right]\right\}, \quad (22A.3)$$

$$\tilde{t} = \exp\left\{a_i\left[(M_i - w)(v_f - b_i) - \sum_{k=1}^{A_i} j_{wk} d_{ik}\right]\right\}, \quad (22A.4)$$

$$g = \sum_{w=1}^{A_i} (\tilde{y} + \tilde{t}), \quad (22A.5)$$

$$q = \sum_{k=1}^{A_i} j_{ak} d_{ik}, \quad (22A.6)$$

$$\tilde{q} = \sum_{k=1}^{A_i} j_{wk} d_{ik}, \quad (22A.7)$$

and

$$e = v_f - b_i. \quad (22A.8)$$

In the foregoing equations, j_{ak} and j_{wk} are dummy variables set equal to 1 whenever $k \le a$ or $k \le w$, respectively; otherwise, they are set equal to 0. For these definitions, the probability of a respondent located at quadrature point v_f endorsing category a of item i can be denoted as

$$f_{U_{fi}}(a; v_f, a_i, b_i, d_{ik}) = \frac{\exp(a_i[(a-1)(v_f - b_i) - \sum_{k=1}^{A_i} j_{ak}\, d_{ik}]) + \exp(a_i[(M_i - a)(v_f - b_i) - \sum_{k=1}^{A_i} j_{ak}\, d_{ik}])}{\sum_{w=1}^{A_i}[\exp(a_i[(w-1)(v_f - b_i) - \sum_{k=1}^{A_i} j_{wk}\, d_{ik}]) + \exp(a_i[(M_i - w)(v_f - b_i) - \sum_{k=1}^{A_i} j_{wk}\, d_{ik}])]}. \quad (22A.9)$$

Then the derivatives of Equation A.9 with respect to each of the GGUM item parameters are equal to

$$\frac{\partial f_{U_{fi}}(a; v_f, a_i, b_i, d_{ik})}{\partial a_i} = \frac{\{y[(a-1)e - q] + t[(M_i - a)e - q]\}g}{g^2} - \frac{(y+t)\left\{\sum_{w=1}^{A_i} \{\tilde{y}[(w-1)e - \tilde{q}] + \tilde{t}[(M_i - w)e - \tilde{q}]\}\right\}}{g^2}, \quad (22A.10)$$

$$\frac{\partial f_{U_{fi}}(a;\, v_f, a_i, b_i, d_{ik})}{\partial b_i} = \frac{\{y[-a_i(a-1)] + t[-a_i(M_i - a)]\}g}{g^2} - \frac{(y+t)\sum_{w=1}^{A_i}\{\tilde{y}[-a_i(w-1)] + \tilde{t}[-a_i(M_i - w)]\}}{g^2}, \tag{22A.11}$$

$$\frac{\partial f_{U_{fi}}(a;\, v_f, a_i, b_i, d_{ik})}{\partial d_{ik}} = \frac{(y+t)\left\{g(-a_i\, j_{ak}) - \sum_{w=1}^{A_i}(\tilde{y}+\tilde{t})(-a_i\, j_{wk})\right\}}{g^2}. \tag{22A.12}$$

References

Andrich, D. 1978. A rating formulation for ordered response categories. *Psychometrika*, 43, 561–573.

Andrich, D. 1988. The application of an unfolding model of the PIRT type to the measurement of attitude. *Applied Psychological Measurement*, 12, 33–51.

Andrich, D. 1996. A general hyperbolic cosine latent trait model for unfolding polytomous responses: Reconciling Thurstone and Likert methodologies. *British Journal of Mathematical and Statistical Psychology*, 49, 347–365.

Andrich, D. and Luo, G. 1993. A hyperbolic cosine latent trait model for unfolding dichotomous single-stimulus responses. *Applied Psychological Measurement*, 17, 253–276.

Bennett, J. F. and Hays, W. L. 1960. Multidimensional unfolding: Determining the dimensionality of ranked preference data. *Psychometrika*, 25, 27–43.

Bock, R. D. and Aitkin, M. 1981. Marginal maximum likelihood estimation of item parameters: Application of an EM algorithm. *Psychometrika*, 46, 443–459.

Bock, R. D. and Mislevy, R. J. 1982. Adaptive EAP estimation of ability in a microcomputer environment. *Applied Psychological Measurement*, 6, 431–444.

Carroll, J. D. 1972. Individual differences and multidimensional scaling. In: R. N. Shepard, A. K. Romney, and S. B. Nerlove (Eds.), *Multidimensional Scaling: Theory and Applications in the Behavioral Sciences* (Volume One, pp. 105–155). New York: Seminar Press.

Chernyshenko, O. S., Stark, S. Drasgow, F., and Roberts, B. W. 2007. Constructing personality scales under the assumptions of an ideal point response process: Toward increasing the flexibility of personality measures. *Psychological Assessment*, 19, 88–106.

Coombs, C. H. 1950. Psychological scaling without a unit of measurement. *Psychological Review*, 57, 145–158.

Coombs, C. H. 1964. *A Theory of Data*. Ann Arbor: Mathesis Press.

Cui, W. 2008. *The Multidimensional Generalized Graded Unfolding Model for Assessment of Change across Repeated Measures* (Doctoral dissertation). Retrieved from ProQuest Dissertations and Theses Database.

de la Torre, J., Stark, S., and Chernyshenko, O. S. 2006. Markov chain Monte Carlo estimation of item parameters for the generalized graded unfolding model. *Applied Psychological Measurement*, 30, 216–232.

Donoghue, J. R. 1999. *Establishing Two Important Properties of Two IRT-Based Models for Unfolding Data.* Unpublished manuscript.

Drasgow, F., Levine, M. V., Tsien, S., Williams, B. A., and Mead, A. D. 1995. Fitting polytomous item response theory models to multiple-choice tests. *Applied Psychological Measurement*, 19, 143–165.

Embretson, S. E. 1991. A multidimensional latent trait model for measuring learning and change. *Psychometrika*, 56, 495–515.

Hoijtink, H. 1990. A latent trait model for dichotomous choice data. *Psychometrika*, 55, 641–656.

Kruskal, J. B. and Carroll, J. D. 1969. Geometrical models and badness-of-fit functions. In: P. R. Krishnaiah (Ed.), *Multivariate Analysis* (Volume 2, pp. 639–671). New York: Academic Press.

Muraki, E. 1992. A generalized partial credit model: Application of an EM algorithm. *Applied Psychological Measurement*, 16, 159–176.

Noel, Y. 1999. Recovering unimodal latent patterns of change by unfolding analysis: Application to smoking cessation. *Psychological Methods*, 4, 173–191.

Orlando, M. and Thissen, D. 2000. Likelihood-based item-fit indices for dichotomous item response theory models. *Applied Psychological Measurement*, 24, 50–64.

Polak, M., Heiser, W. J., and de Rooij, M. 2009. Two types of single-peaked data: Correspondence analysis as an alternative to principal components analysis. *Computational Statistics and Data Analysis*, 53, 3117–3128.

Roberts, J. S. 1995. Item response theory approaches to attitude measurement. (Doctoral dissertation, University of South Carolina, Columbia, 1995). *Dissertation Abstracts International*, 56, 7089B.

Roberts, J. S. 2008. Modified likelihood-based item fit statistics for the generalized graded unfolding model. *Applied Psychological Measurement*, 32, 407–423.

Roberts, J. S., Donoghue, J. R., and Laughlin, J. E. 2000. A general item response theory model for unfolding unidimensional polytomous responses. *Applied Psychological Measurement*, 24, 3–32.

Roberts, J. S., Donoghue, J. R., and Laughlin, J. E. 2002. Characteristics of MML/EAP parameter estimates in the generalized graded unfolding model. *Applied Psychological Measurement*, 26, 192–207.

Roberts, J. S., Fang, H., Cui, W., and Wang, Y. 2006. GGUM2004: A Windows-based program to estimate parameters in the generalized graded unfolding model. *Applied Psychological Measurement*, 30, 64–65.

Roberts, J. S. and Laughlin, J. E. 1996. A unidimensional item response model for unfolding responses from a graded disagree-agree scale. *Applied Psychological Measurement*, 20, 231–255.

Roberts, J. S., Laughlin, J. E., and Wedell, D. H. 1999. Validity issues in the Likert and Thurstone approaches to attitude measurement. *Educational and Psychological Measurement*, 59, 211–233.

Roberts, J. S., Rost, J., and Macready, G. B. 2010. MIXUM: An unfolding mixture model to explore the latitude of acceptance concept in attitude measurement. In: Embretson, S. E. (Ed.), *Measuring Psychological Constructs: Advances in Model-Based Approaches* (pp. 175–197). Washington, DC: American Psychological Association.

Roberts, J. S. and Shim, H. 2008. *GGUM2004 Technical Reference Manual, Version 1.1.* Retrieved from http://www.psychology.gatech.edu/unfolding/ggum2004.asp

Roberts, J. S. and Shim, H. 2010, July. Multidimensional unfolding with item response theory: The multidimensional generalized graded unfolding model. Paper presented at the International Meeting of the Psychometric Society, Athens, Georgia.

Roberts, J. S. and Thompson, V. M. 2011. Marginal maximum *a posteriori* item parameter estimation for the generalized graded unfolding model. *Applied Psychological Measurement*, 35, 259–279.

Stark, S., Chernyshenko, O. S., Drasgow, F., and Williams, B. A. 2006. Item responding in personality assessment: Should ideal point methods be considered for scale development and scoring? *Journal of Applied Psychology*, 91, 25–39.

Thissen, D., Pommerich, M., Billeaud, K., and Williams, V. S. L. 1995. Item response theory for scores on tests including polytomous items with ordered responses. *Applied Psychological Measurement*, 19, 39–49.

Thurstone, L. L. 1928. Attitudes can be measured. *The American Journal of Sociology*, 33, 529–553.

Thurstone, L. L. 1931. The measurement of social attitudes. *Journal of Abnormal and Social Psychology*, 26, 249–269.

Thurstone, L. L. 1932. *Motion Pictures and Attitudes of Children.* Chicago, IL: University of Chicago Press.

Thurstone, L. L. and Chave, E. J. 1929. *The Measurement of Attitude: A Psychophysical Method and Some Experiments with a Scale for Measuring Attitude toward the Church.* Chicago, IL: University of Chicago Press.

Usami, S. 2011. Generalized graded unfolding model with structural equation for subject parameters. *Japanese Psychological Research*, 53, 221–232.

van Schuur, W. H. and Kiers, H. A. L. 1994. Why factor analysis is often the incorrect model for analyzing bipolar concepts, and what model to use instead. *Applied Psychological Measurement*, 18, 97–110.

Weekers, A. and Meijer, R. 2008. Scaling response processes on personality items using unfolding and dominance models: An illustration with a Dutch dominance and unfolding personality inventory. *European Journal of Psychological Assessment*, 24, 65–77.

Wright, B. D. and Masters, G. N. 1982. *Rating Scale Analysis.* Chicago: MESA Press.

Wu, S. and Wang, W. 2010, July. Extensions of the generalized graded unfolding model. Paper presented at the International Meeting of the Psychometric Society, Athens, Georgia.

Section VII

Hierarchical Response Models

23

Logistic Mixture-Distribution Response Models

Matthias von Davier and Jürgen Rost

CONTENTS

23.1 Introduction 393
23.2 Presentation of the Models 394
23.3 Parameter Estimation 396
23.4 Model Fit 399
23.5 Empirical Example 400
23.6 Discussion 401
References 402

23.1 Introduction

Whenever multiple populations are sampled, there is potential for detecting differential item functioning, a lack of parameter invariance, or other violations of the homogeneity assumptions made in item response theory (IRT). Such findings may indicate that some items may not fit the IRT model well enough across some or all populations. If the populations are known and well defined, multiple-group IRT (Bock and Zimowski, 1997) models can be applied, and the item parameters can be estimated using equality constraints across groups while admitting "unique" item parameters where appropriate (Oliveri and von Davier, 2011).

There are quite a few cases, however, in which the membership of test takers of the relevant populations is unknown. Examples include the preference to apply different solution strategies to the test items by members of different groups (Kelderman and Macready, 1990; Rost and von Davier, 1993), the type of test preparation through which different test takers have gone, or different response tendencies such as acquiescence or the tendency to avoid or choose extreme responses in rating scales (e.g., Rost et al., 1996; Austin et al., 2006). All these examples establish what is sometimes called a *hidden structure*—a typological variable that accounts for differences among test takers that are not sufficiently well captured by mere score differences between them.

A very flexible approach to account for hidden structure is the use of discrete mixture-distribution models. While discrete mixtures of statistical models first appeared in the literature more than 120 years ago (Newcomb, 1886; Pearson, 1894), these models did not see much use until powerful computers and adequate algorithms to estimate their parameters became available (McLachlan and Peel, 2000).

In the realm of test theory, mixture models assume that the observed response data are sampled from a composite population, that is, a population consisting of a number of not directly observable subpopulations. Like multigroup IRT models, mixture-distribution

IRT models assume the existence of mutually exclusive groups, but they try to identify the underlying grouping by means of heterogeneity in the item response data. The aim of this approach is twofold (Rost, 1990; von Davier, 2009a): (a) to obtain a number of equivalency classes that represent the latent structure—and thus the typological difference between test takers and (b) to measure test takers who are identified as belonging to one of these classes with respect to what the data were originally collected for—typically a cognitive proficiency, a skill, an attitude, or a personality trait.

The remaining sections introduce mixture-distribution response models more formally. After introducing mixture Rasch models and mixture IRT models for dichotomous and polytomous data, a mixture-distribution multidimensional IRT (MIRT) model will be introduced. We will then discuss parameter estimation, model-fit assessment, and an application of a mixture MIRT model to the analysis of personality data.

23.2 Presentation of the Models

The observed data structure for mixture IRT models is the same as for many other IRT or MIRT models: let $\boldsymbol{U} = (U_1, \ldots, U_I)$ denote a vector of discrete observed variables with $U_i \in \{0, \ldots A_i\}$. The response variable associated with the ith item is denoted by U_i. If $A_i = 1$, this is a binary response and if $A_i > 1$, this is a polytomous ordinal response variable. For each test taker $p = 1, \ldots, P$, let u_{pi} denote the test taker's answer to item i.

In the simplest case, we assume that the responses are binary and there exists a unidimensional latent trait θ and a limited number of unobserved groups $c \in \{1, \ldots, G\}$. For illustration, we assume that the IRT model is chosen to be the Rasch (1960) or one-parameter logistic model (OPLM; Volume One, Chapter 3). Mixture response models based on more general IRT models will be introduced later in this chapter. The current model is the mixture Rasch model for binary data as described by Rost (1990). The model equation for a response $a \in \{0,1\}$ to item i is

$$Pr\{U_i = a; \theta, c\} = \frac{\exp(a(\theta - b_{ic}))}{1 + \exp(\theta - b_{ic})}. \tag{23.1}$$

This equation differs from the ordinary Rasch model only in that the difficulty parameter of each item is group specific. If $G = 2$, there are two parameters, b_{i1} and b_{i2}, for each item. These are parameters to be estimated separately. Whenever b_{i1} and b_{i2} are not equal, the item functions defined by Equation 23.1 for the unobserved groups 1 and 2 differ.

The mixture Rasch model, as well as other mixture IRT models, assume local independence of the responses in the same way conventional IRT models do. The assumption yields

$$Pr\{u_1, \ldots, u_I; \theta, c\} = \prod_{i=1}^{I} Pr\{u_i; \theta, c\} \tag{23.2}$$

for the probability of a response vector $(u_1, \ldots, u_I)$ given a test taker's ability θ and group membership c.

The marginal probability of a response vector is then given by

$$Pr\{u_1,\ldots,u_I\} = \sum_{c=1}^{G} \pi_c \int_{-\infty}^{+\infty} g(\theta \mid c) Pr\{u_1,\ldots,u_I;\theta,c\} d\theta. \quad (23.3)$$

Equation 23.3 is the customary representation of the marginal distribution of observed responses when assuming a discrete mixture distribution (McLachlan and Peel, 2000). Its weighted sum of conditional densities $\int_{-\infty}^{+\infty} g(\theta \mid c)Pr\{u_1,\ldots,u_I;\theta,c\}$ combines the latent group-specific components for which an IRT model with a potentially unique set of item parameters hold. The π_c parameters are called mixing proportions, or class sizes, and quantify the relative size of the unobserved groups $c \in \{1, \ldots, G\}$. The term $g(\theta|c)$ denotes the conditional density of the distribution of the latent trait θ in class c. While Rost (1990) introduced the mixture Rasch model in a conditional maximum-likelihood framework, the presentation above follows a maximum marginal likelihood approach (Mislevy and Verhelst, 1990).

As already evident in the mixture Rasch model, the latent structure in a discrete mixture IRT model is a combination of two latent variables: it combines a continuous and a discrete latent variable. In this sense, a mixture IRT model is already a multidimensional latent variable model and special cases exist where discrete mixture IRT and MIRT coincide (e.g., Rijmen and De Boeck, 2005).

A more general, multidimensional mixture IRT model can be developed on the basis of a general latent variable model (von Davier, 2005, 2008, 2009a,b, 2010). Let $\boldsymbol{\theta} = (\theta_1, \ldots, \theta_D)$ denote a continuous latent variable with $D \geq 1$ dimensions and, as before, let $c \in \{1, \ldots, G\}$ denote a discrete latent variable representing the unobserved groups.

Let Ξ denote the joint set of model parameters, including the item parameters, mixing proportions, and parameters that describe the latent trait distribution. That is, the components of Ξ are the class sizes, mixing proportions, or marginal (class) probabilities π_c. As before, it holds that $\pi_c = P(C = c) > 0$, for all c and $\sum_c \pi_c = 1$. Let $g_\Xi(\theta|c)$ denote the (multidimensional) conditional density of θ given c. For simplicity, we may assume that $\theta|c \sim N(\mu_c, \Sigma_c)$. As defined earlier, the latent trait distribution parameters μ_c, Σ_c are components of Ξ. Other, more flexible distributions are possible, though. Together with the observed responses $\boldsymbol{U}$, the complete data record is $\Delta = (\boldsymbol{U}, \boldsymbol{\theta}, C)$. That is, each test taker p is characterized by the complete data $\delta_p = (u_{p1}, \ldots, u_{pI}, \theta_{p1}, \ldots, \theta_{pD}, c_p)$, of which only the $\boldsymbol{U}$-part is observed.

The probability of an observed response u_{pi} to item i by a test taker p with latent variable $\lambda = (\theta_1 \ldots \theta_d, c) \in \boldsymbol{\Lambda}$ is given by

$$\Pr_\Xi\{U_i = u;\lambda\} = \frac{\exp(b_{iuc} + \sum_{d=1}^{D} u a_{idc} q_{id} \theta_d)}{1 + \sum_{y=1}^{m_i} \exp(b_{iyc} + \sum_{d=1}^{D} y a_{idc} q_{id} \theta_d)}. \quad (23.4)$$

Equation 23.4 defines a multidimensional mixture IRT model for dichotomous and polytomous data (von Davier, 2005, 2008). The b_{iuc} denote item threshold parameters for category u of item i in latent group c. The a_{idc} denote slope parameters of dimension d for item i in latent group c. The $q_{id} \in \{0, 1\}$ denote entries of a design matrix; they relate dimension d to item i. Because the b_{iuc} and the a_{idc} are model parameters, they are components of Ξ, whereas the q_{id} are constants. Local independence given $\boldsymbol{\lambda} = (\theta_1 \ldots \theta_d, c)$ is assumed and the marginal probability of a response vector is obtained as

$$Pr_{\Xi}\{u_{p1},\ldots,u_{pI}\} = \sum_{c=1}^{G} \pi_c \int g_{\Xi}(\theta \mid c) Pr_{\Xi}\{u_{p1},\ldots,u_{pI};\theta,c\} d\theta. \tag{23.5}$$

The estimation of the parameters in this model can be carried out using customary maximum-likelihood methods. This approach requires a likelihood function, that is, a function to be maximized with respect to the model parameters Ξ given the observed data. The likelihood function is given by

$$L(\Xi;\mathbf{u}) = \sum_{p=1}^{P}\sum_{c=1}^{G} \pi_c \int g_{\Xi}(\theta \mid c) Pr_{\Xi}\{u_{p1},\ldots,u_{pI};\theta,c\} d\theta \tag{23.6}$$

with

$$u = \begin{bmatrix} u_1 \\ \ldots \\ u_P \end{bmatrix}$$

representing the $P \times I$ response matrix. The following section describes the algorithm used to maximize the function defined in Equation 23.6.

The notation given in Equations 23.4 through 23.6 can be taken to represent the Rasch and two-parameter logistic (2PL) models (Birnbaum, 1968), the mixture Rasch model (Rost, 1990), the polytomous mixture Rasch model (von Davier and Rost, 1995), and the mixture Birnbaum model (Smit et al., 1999), as well as the mixture generalized partial credit model (von Davier and Yamamoto, 2004). This can be seen by setting $D = 1$ for the mixture Rasch and IRT model, and in addition $G = 1$ for regular (nonmixture single group) IRT models. If $G > 1$ and the group membership is unknown, mixture (M)IRT models are obtained. If the group membership is known, multiple-group (M)IRT models are obtained. If the number of groups is large, multilevel extensions of mixture IRT (von Davier, 2010) and latent class models (Vermunt, 2003) should be considered. Constraints on item parameters across groups or latent classes should be imposed in order to link the parameters across observed and latent subpopulations, respectively (von Davier and von Davier, 2007; Xu and von Davier, 2008; Oliveri and von Davier, 2011). Other models, such as a mixture multidimensional version of the OPLM (Verhelst and Glas, 1995) and the hybrid model (Yamamoto, 1989) can also be specified in this framework by applying appropriate constraints on the model parameters.

23.3 Parameter Estimation

While different estimation methods for mixture IRT models exist, the predominant ones are the expectation maximization (EM) algorithm (Dempster et al., 1977; Bock and

Aitkin, 1981; Volume Two, Chapter 12) and Markov chain Monte Carlo (MCMC) sampling from the posterior distributions of the parameters (Junker, Patz, and Vanhoudnos, Volume Two, Chapter 15).

The EM algorithm (Dempster et al., 1977) appears to be the preferred method for estimation of mixture IRT models for a variety of reasons. For one thing, the algorithm offers a straightforward, frequently used approach to the handling of incomplete data. Mixture models, just like other latent variable models, can be regarded as incomplete data problems because the observations of the latent variable are missing. Not surprisingly, some of the literature on mixture modeling focuses on variations of the EM algorithm (e.g., Celeux and Diebolt, 1985). Also, the literature on the EM algorithm tends to use mixture models as standard examples (McLachlan and Krishnan, 1997).

MCMC estimation of mixture IRT models (e.g., Bolt et al., 2001, 2002) is typically done by writing control scripts for general MCMC software such as WinBUGS (Spiegelhalter et al., 1999; Johnson, Volume Three, Chapter 21) instead of using a general-purpose programming language. We will not compare the EM algorithm and the MCMC approach on the basis of schools of thought, but only make some practical remarks: while an MCMC approach has the clear advantage of quickly prototyping a model in WinBUGS,* the approach does not compare favorably to the EM algorithm in terms of computational costs and run time. Also, the majority of the commercial and noncommercial software programs for mixture as well as multiple-group IRT continue to use the EM algorithm or an extension of it.

The EM algorithm is an iterative procedure that uses mathematical methods for numerical optimization which typically require the first and second derivatives of the model's likelihood function and alternates between two types of calculations, the E- and M-step, until a prespecified convergence criterion is reached. Typical criteria used to determine termination of the algorithm are based on changes in estimated parameters (or log-likelihood of the data, or both) between two consecutive EM cycles that do not exceed a certain threshold.

We describe the two alternating steps of the EM algorithm for mixture IRT by integrating the descriptions for the polytomous mixture Rasch model (von Davier and Rost, 1995) and the mixture generalized partial credit model (von Davier and Yamamoto, 2004), adapting them for brevity to the case of the unidimensional binary mixture Rasch model:

1. The E-step of the algorithm produces expected counts of the sufficient statistics of the model parameters given preliminary parameter estimates from the previous M-step (or starting values in the first cycle). The calculations for the binary mixture Rasch model are as follows: let $\mathbf{\Theta}^{(t)}$ denote the vector of parameters for the model obtained in cycle t. The vector includes the item parameters $(b_{11}, \ldots, b_{1I}, \ldots, b_{G1}, \ldots, b_{GI})^{(t)}$ and the parameters of the ability distribution(s), for example, $(\mu_1, \sigma_1, \ldots, \mu_G, \sigma_G)^{(t)}$, if a normally distributed ability is assumed for each class c. Let $\hat{n}_{j1c}{}^{(t+1)} \mid \mathbf{\Xi}^{(t)}$ denote an estimate of the expected number of correct responses to item i in class c given preliminary parameters $\mathbf{\Xi}^{(t)}$. Recall that the observed item score is a sufficient statistic for the item difficulty parameter in the Rasch model. Note, however, that we do not observe this quantity for each class separately, hence the need to estimate these in the E-step. Using Bayes' theorem, the probability of

* This is in part the case because no knowledge about properties such as differentiability of the likelihood function, differentiability of derivatives, or parameter redundancy of the model to be specified for MCMC estimation is required in order to author or adapt an existing script and choose commonly used priors for the model parameters.

class membership given the observed responses, $Pr_{\Xi^{(t)}}\{c \mid u_{p1},\ldots,u_{pI}\}$ is obtained for each test taker p. Then, let

$$\hat{n}_{i1c}{}^{(t+1)} \mid \boldsymbol{\Xi}^{(t)} = \sum_{p=1}^{P} 1_{\{1\}}(u_{pi}) Pr_{\Xi^{(t)}}\{c \mid u_{p1},\ldots,u_{pI}\} \tag{23.7}$$

with $1_{\{1\}}(u_{pi}) = 1$ if $u_{pi} = 1$, and $1_{\{1\}}(u_{pi}) = 0$ otherwise. Let $E_{\Xi^{(t)}}(\theta \mid u_{p1},\ldots,u_{pI},c)$ denote the posterior mean of ability in class c given the observed responses, and let $V_{\Xi^{(t)}}(\theta \mid u_{p1},\ldots,u_{pI},c)$ denote its posterior variance. These posterior moments are obtained from the quantities

$$K_{p|\Xi^{(t)}}{}^{(m)} = \int \theta^m Pr_{\Xi^{(t)}}\{\theta \mid u_{p1},\ldots,u_{pI},c\} d\theta \tag{23.8}$$

for all test takers for $m = 1$ and $m = 2$, using

$$E_{\Xi^{(t)}}(\theta \mid u_{p1},\ldots,u_{pI},c) = K^{(1)}_{p|\Xi^{(t)}} \tag{23.9}$$

as well as

$$V_{\Xi^{(t)}}(\theta \mid u_{p1},\ldots,u_{pI},c) = K^{(2)}_{p|\Xi^{(t)}} - \left[K^{(1)}_{p|\Xi^{(t)}}\right]^2. \tag{23.10}$$

2. The M-step uses these quantities to generate $\hat{\mu}_c{}^{(t+1)} \mid \boldsymbol{\Xi}^{(t)}$, an estimate of the average ability of test takers in class c, and $\hat{\sigma}_c{}^{(t+1)} \mid \boldsymbol{\Xi}^{(t)}$, an estimate of the standard deviation of ability in class c given the parameter estimate $\boldsymbol{\Xi}^{(t)}$. These estimates are obtained as

$$\hat{\mu}_c{}^{(t+1)} \mid \boldsymbol{\Xi}^{(t)} = \frac{1}{N}\sum_{i=1}^{N} E_{\Xi^{(t)}}(\theta \mid u_{p1},\ldots,u_{pI},c)$$

and

$$\hat{\sigma}_c{}^{(t+1)} \mid \boldsymbol{\Xi}^{(t)} = \sqrt{E(V_{\Xi^{(t)}}(\theta \mid u_{p1},\ldots,u_{pI},c)) + V(E_{\Xi^{(t)}}(\theta \mid u_{p1},\ldots,u_{pI},c))}.$$

The M-step also involves an update of the item parameters b_{ic} if the Rasch model is used, and other item parameters if a more general model is used. This update typically involves derivatives of the likelihood function used in Newton, Quasi-Newton, or steepest descend methods (e.g., Kosmol and Müller-Wichards, 2011). Details can be found, for example, in Rost (1990), von Davier and Rost (1995), von Davier and Yamamoto (2004), and von Davier and Carstensen (2007).

Over the past two decades, several software programs for estimating mixture IRT models have been developed. Some examples are: Hybil (Yamamoto, 1989), MIRA (Rost and von Davier, 1992), WINMIRA (von Davier, 1994), LEM (Vermunt, 1993), WINMIRA 2001 (von Davier, 2001), Latent Gold (Vermunt, Volume Three, Chapter 30; Vermunt and Magidson, 2005), and Mplus (Muthén, Volume Three, Chapter 28; Muthén and Muthén,

1998–2007). As indicated earlier, mixture IRT models can also be estimated using MCMC methods as implemented in WinBUGS (Johnson, Volume Three, Chapter 21; Spiegelhalter et al., 1999; Bolt et al., 2001). Log-linear (mixture) Rasch models can be estimated using LOGIMO (Kelderman and Steen, 1988). In addition, mixture Rasch models can be estimated with a variety of R packages (R Core Team 2012; Rusch and Mair, Volume Three, Chapter 20). Rijmen's (2006) MATLAB toolbox BNL can also be used for mixture IRT. The family of multidimensional mixture IRT models presented in Equations 23.4 through 23.6 can be estimated with the *mdltm* software (von Davier, 2005, 2008), for which a license for noncommercial use can be requested from the first author of this chapter.

23.4 Model Fit

Assessing overall model data fit for mixture IRT models is somewhat different from the case of single-population models. Typically, mixture IRT models are evaluated using methods commonly found in latent structure analysis, and the analysis is often focused on the determination of how many latent groups are needed (e.g., Langeheine et al., 1996; von Davier, 1997).

In addition to well-known information criteria, such as the Akaike information criterion (AIC; Akaike, 1973) and the Bayesian information criterion (BIC; Schwarz, 1978) that are often used to evaluate mixture IRT models (for a review, see Cohen and Cho, Volume Two, Chapter 18), a related measure referred to as log penalty (Gilula and Haberman, 1994) may also be considered. This measure has the advantage of being strongly related to the AIC, while it quantifies the expected log-likelihood per observation. This is advantageous because it transforms the measure on a scale that does not vary with sample size, N, or number of items per test taker, I. Note, however, that the final number of latent groups chosen is somewhat dependent on the IRT model assumed to hold in each of the latent classes (Alexeev et al., 2011; Aitkin and Aitkin, 2011).

In addition to an overall evaluation of the fit of mixture IRT models, more specific measures for item- and person-fit for mixture models are available (von Davier and Molenaar, 2003). More general-fit measures (Glas and Verhelst, 1995; Glas, 2007), however, can also be used to test the fit of individual response patterns to certain classes. Such more specific fit statistics should be preferred over general measures when choosing, for example, between a mixture Rasch model and a mixture 2PL model.

Overall goodness-of-fit measures in use in latent class analysis may also be used for mixture IRT models. Statistics from the power-divergence family of goodness-of-fit statistics (Cressie and Read, 1984) can be calculated because mixture IRT models predict counts for each possible response vector. Because of this feature, fit measures such as the likelihood ratio statistics and the Pearson X^2 are readily computed. One important caveat, however, is that these statistics cannot be tested by using the well-known $\mathcal{X}^2$ approximation; the regularity conditions required for the use of the $\mathcal{X}^2$ approximation are often not met. Glas (1988) along with Glas and Verhelst (1995) offered alternatives to testing the divergence of observed and expected frequencies across all possible response patterns by suggesting the use of patterns that share some criteria (e.g., patterns with the same response to some item).

As suggested by Rost and von Davier (1995), mixture IRT models yield powerful tests of population invariance of item parameters and ability distribution. Von Davier (1997) has shown that resimulation methods can be used to check how many populations are needed to achieve conditional homogeneity and appropriate model data fit. However, resimulation-based methods are a computationally costly way to assess the accuracy of the model in predicting the observed data. Langeheine et al. (1996) and von Davier (1997) studied resimulation methods with parametric bootstrap generation of the null distributions of goodness-of-fit measures for checking the number of latent groups in latent class and mixture-distribution IRT models. For Bayesian approaches to parameter estimation, posterior predictive checks, based on draws from stationary posterior distribution generated by an MCMC chain, are recommended (Sinharay, Volume Two, Chapter 9; Sinharay et al., 2006).

23.5 Empirical Example

The data for this example are from a series of studies with mixture IRT models that used a public domain personality inventory, the International Personality Item Pool (IPIP) BIG5—a five-factor questionnaire. A number of college-bound students who participated in a study on English language proficiency and its relationship to a range of noncognitive variables were asked to fill out this questionnaire. The sample used here is a combination of multinational samples of college applicants; for further details, see Stankov (2009). For the assessment of the five personality factors (Saucier and Goldberg, 1996), the 50-item representation of the Goldberg (e.g., Goldberg, 1999) markers for the big-five factor structure, available through the IPIP (n.d.), was used. A total of 50 items was rated by the subjects on a five-point Likert-type scale ranging from 1 (very inaccurate) to 5 (very accurate).

The results presented here are from an analysis of the questionnaire with five-dimensional (5D) mixture regular and generalized partial credit models. The generalized partial credit model for 50 items with five response categories involves 200 item threshold parameters as well as 50 slope parameters per class. Due to this large number of parameters, it was decided to split the data for our 2031 test takers into two 25-item forms, A and B, with five items for each of the five dimensions. Information about additional IRT and mixture MIRT analyses conducted on the same data can be found in von Davier et al. (2012).

Table 23.1 presents the results for forms A and B, comparing the regular 5D generalized partial credit model with its mixture version.

TABLE 23.1

Comparison of Models for the IPIP Personality Data in two 25-Item Forms

Form	Model	NPAR	Loglik	AIC	BIC	Logpen
A	LCA 2	202	−64,748.07	130,304.13	131,438.62	1.29721
A	LCA 3	304	−63,417.00	128,050.00	129,757.35	1.27266
A	MIRT 5	130	−63,729.50	127,979.01	128,709.12	1.27543
A	MIRT mixture	261	−61,622.42	124,288.83	125,754.68	1.23596
B	LCA 2	202	−67,341.32	135,490.64	136,625.13	1.34236
B	LCA 3	304	−65,691.23	132,598.47	134,305.82	1.31159
B	MIRT 5	130	−66,160.96	132,841.93	133,572.04	1.31747
B	MIRT mixture	261	−63,584.88	128,213.76	129,679.61	1.26887

Note: AIC—Akaike information criterion, BIC—Bayesian information criterion, logpen—log penalty (Gilula and Haberman, 1994), LCA 2 and LCA 3 latent class analysis with 2 and 3 classes, MIRT 5—multidimensional IRT, MIRT mixture—mixture-distribution MIRT model.

TABLE 23.2

Average Class-Specific Threshold Parameters for Forms A and B

Class	Size	Form A				Size	Form B			
1	0.58	3.7097	1.6722	0.3221	−2.5553	0.57	3.3739	1.2083	0.0885	−2.6706
2	0.42	1.5788	0.8088	0.6570	−0.2997	0.43	1.3029	0.3676	0.6845	−0.3487

Note that both the AIC and BIC criteria indicate better fit for the mixture MIRT than the regular MIRT model. We therefore limited our analyses to a mixture model with two mixing components. The choice involves the need to estimate about 130 parameters per class. (More than two classes would lead to a model with close to 400 parameters for 2031 test takers, and possibly to overfitting of the model.) For this two-class mixture model, observe that while the mixture model almost doubles the number of parameters needed to fit the data, the penalties applied by the AIC and BIC criteria do not outweigh the improved likelihood of the data under the mixture MIRT model.

Our results in favor of the mixture MIRT model replicate earlier findings from unidimensional mixture IRT analysis by Rost et al. (1996) and Austin et al. (2006), who found better fit for mixture than nonmixture IRT models in analyses of personality data. These authors also found that the mixture IRT model provided a correction for extreme versus moderate response styles through different threshold distances in the parameters for each of the latent classes.

Table 23.2 shows that the results for the 5D mixture IRT model. Independent analyses of both test forms showed similar differences between the threshold parameters for the two latent classes. Note that not only are the differences of thresholds consistent across forms, but also that the class sizes are also close across the two analyses of the nonoverlapping 25 item forms.

Maij-de Meij et al. (2008) also used mixture IRT models with covariates to disentangle different types based on responses to personality items, while Egberink (2010, Chapter 5) described further applications of mixture IRT models to the analysis of personality data.

23.6 Discussion

The use of mixture IRT models is quite versatile. The models have been used for the analysis of clinical data and symptom checklists (e.g., Schmidt et al., 2005; Schultz-Larsen et al., 2007), aberrant responses in organizational surveys (Eid and Zickar, 2007), strategy differences (Kelderman and Macready, 1990; Rijkes and Kelderman, 2007), and response sets in rating data (Rost et al., 1996; Austin et al., 2006). Also, attempts have been made to use it for standard setting (Jiao et al., 2011). Special cases of these models have been used for the detection of test speededness (Yamamoto and Everson, 1997; Bolt et al., 2002; Boughton and Yamamoto, 2007). The Saltus model (Wilson, 1989), a special case of mixture IRT, has been used to analyze developmental stages (Draney and Wilson, 2007; Draney et al., 2008). Mixture MIRT models can also be used to analyze learning and growth in multiple subpopulations (von Davier et al., 2011). For a more detailed discussion of all applications, refer to the review by von Davier (2009a) and the volume edited by von Davier and Carstensen (2007).

An important remaining question is whether to use a mixture model when an IRT model does not fit the data sufficiently well (Rijmen and De Boeck, 2003; De Boeck et al., 2005).

There are cases in which an increase in the number of dimensions does not pay off in expected ways, whereas the use of a mixture model may lead to deeper understanding of individual differences (von Davier et al., 2012). Obviously, alternative models are possible, but unless there is clear evidence that a single homogenous population exists, mixture models are a powerful tool to explore data not fitted easily by conventional IRT models.

Research on mixture models has shown that profound knowledge of the factors that drive item difficulty along with mixture IRT models may facilitate deeper understanding of group differences in performance on complex tasks (Rijkes and Kelderman, 2007; Embretson, 2007). Without this synthesis of content expertise and measurement knowledge, response modeling may remain just a case of fishing in the dark for an apparently fitting model. Such an approach is likely to lead to modeling choices that cannot be replicated in future application as pointed out by Haberman and von Davier (2006) and elaborated in von Davier (2009a,b). Therefore, it is strongly suggested to combine content and methodological expertise in data analyses when using a general modeling family such as the multidimensional discrete mixture models presented in this chapter.

References

Aitkin, M., and Aitkin, I. 2011. *Statistical Modeling of the National Assessment of Educational Progress*. New York: Springer.

Akaike, H. 1973. Information theory as an extension of the maximum likelihood principle. In: B. N. Petrov and F. Csaki (Eds.), *Second International Symposium on Information Theory* (pp. 267–281). Budapest, Hungary: Akademiai Kiado.

Alexeev, N., Templin, J., and Cohen, A. S. 2011. Spurious latent classes in the mixture Rasch model. *Journal of Educational Measurement*, 48, 313–332.

Austin, E. J., Deary, I. J., and Egan, V. 2006. Individual differences in response scale use: Mixed Rasch modelling of responses to NEO-FFI items. *Personality and Individual Differences*, 40, 1235–1245.

Birnbaum, A. 1968. Some latent trait models. In: F. M. Lord and M. R. Novick (Eds.), *Statistical Theories of Mental Test Scores* (pp. 397–479). Reading, Massachusetts: Addison-Wesley.

Bock, R. D., and Aitkin, M. 1981. Marginal maximum likelihood estimation of item parameters: Application of an EM algorithm. *Psychometrika*, 46, 443–445.

Bock, R. D., and Zimowski, M. F. 1997. Multiple group IRT. In: W. J. van der Linden and R. K. Hambleton (Eds.), *Handbook of Modern Item Response Theory* (pp. 433–448). New York: Springer.

Bolt, D. M., Cohen, A. S., and Wollack, J. A. 2001. A mixture item response model for multiple-choice data. *Journal of Educational and Behavioral Statistics*, 26, 381–409.

Bolt, D. M., Cohen, A. S., and Wollack, J. A. 2002. Item parameter estimation under conditions of test speededness: Application of a mixture Rasch model with ordinal constraints. *Journal of Educational Measurement*, 39, 331–348.

Boughton, K., and Yamamoto, K. 2007. A HYBRID model for test speededness. In: M. von Davier and C. H. Carstensen (Eds.), *Multivariate and Mixture Distribution Rasch Models* (pp. 147–156). New York: Springer.

Celeux, G., and Diebolt, J. 1985. The SEM algorithm: A probabilisitic teacher algorithm derived from the EM algorithm for the mixture problem. *Computational Statistics Quaterly*, 2, 73–82.

Cressie, N., and Read, T. R. C. 1984. Multinomial goodness-of-fit tests. *Journal of the Royal Statistical Society, Series B*, 46, 440–464.

De Boeck, P., Wilson, M., and Acton, G. S. 2005. A conceptual and psychometric framework for distinguishing categories and dimensions. *Psychological Review*, 112, 129–158.

Dempster, A. P., Laird, N. M., and Rubin, D. B. 1977. Maximum likelihood from incomplete data via the EM algorithm (with discussion). *Journal of the Royal Statistical Society, Series B*, 39, 1–38.

Draney, K., and Wilson, M. 2007. Application of the Saltus model to stage-like data: Some applications and current developments. In: M. von Davier and C. H. Carstensen (Eds.), *Multivariate and Mixture Distribution Rasch Models: Extensions and Applications*. New York: Springer.

Draney, K., Wilson, M., Glück, J., and Spiel, C. 2008. Mixture models in a developmental context. In: G. R. Hancock and K. M. Samuelson (Eds.), *Latent Variable Mixture Models* (pp. 199–216). Charlotte, North Carolina: Information Age Publishing.

Eid, M., and Zickar, M. 2007. Detecting response styles and faking in personality and organizational assessments by mixed Rasch models. In: M. von Davier and C. H. Carstensen (Eds.), *Multivariate and Mixture Distribution Rasch Models: Extensions and Applications* (pp. 255–270). New York: Springer.

Egberink, I. J. L. 2010. Applications of item response theory to non-cognitive data. Unpublished dissertation, University of Groningen, Groningen, The Netherlands. Retrieved from: http://dissertations.ub.rug.nl/FILES/faculties/gmw/2010/i.j.l.egberink/thesis.pdf.

Embretson, S. 2007. Mixed Rasch models for measurement in cognitive psychology. In: M. von Davier and C. H. Carstensen (Eds.), *Multivariate and Mixture Distribution Rasch Models: Extensions and Applications* (pp. 235–253). New York: Springer.

Gilula, Z., and Haberman, S. J. 1994. Models for analyzing categorical panel data. *Journal of the American Statistical Association*, 89, 645–656.

Glas, C. A. W. 1988. The derivation of some tests for the Rasch model from the multinomial distribution. *Psychometrika*, 53, 525–546.

Glas, C. A. W. 2007. Testing generalized Rasch models. In: M. von Davier and C. H. Carstensen (Eds.), *Multivariate and Mixture Distribution Rasch Models: Extensions and Applications* (pp. 37–55). New York: Springer.

Glas, C. A. W., and Verhelst, N. D. 1995. Testing the Rasch model. In: G. H. Fischer and I. W. Molenaar (Eds.), *Rasch Models: Their Foundations, Recent Developments and Applications* (pp. 69–96). New York: Springer.

Goldberg, L. R. 1999. A broad-bandwidth, public domain, personality inventory measuring the lower-level facets of several five-factor models. In: I. Mervielde, I. Deary, F. De Fruyt, and F. Ostendorf (Eds.), *Personality Psychology in Europe* (Volume 7; pp. 7–28). Tilburg, The Netherlands: Tilburg University Press.

Haberman, S. J., and von Davier, M. 2006. Some notes on models for cognitively based skills diagnosis. In: C. R. Rao and S. Sinharay (Eds.), *Handbook of Statistics: Volume 26. Psychometrics* (pp. 1031–1038). Amsterdam: Elsevier.

IPIP (n.d.). A scientific collaboratory for the development of advanced measures of personality traits and other individual differences. International Personality Item Pool. Retrieved from: http://ipip.ori.org/.

Jiao, H., Lissitz, B., Macready, G., Wang, S., and Liang, S. 2011. Exploring using the mixture Rasch model for standard setting. *Psychological Testing and Assessment Modeling*, 53, 499–522.

Kelderman, H., and Macready, G. B. 1990. The use of loglinear models for assessing differential item functioning across manifest and latent examinee groups. *Journal of Educational Measurement*, 27, 307–327.

Kelderman, H., and Steen, R. 1988. *LOGIMO: A Program for Loglinear IRT Modeling*. Enschede, The Netherlands: Department of Education, University of Twente.

Kosmol, P., and Müller-Wichards, D. 2011. *Optimization in Function Spaces: With Stability Considerations in Orlicz Spaces*. Berlin, Germany: De Gruyter.

Langeheine, R., Pannekoek, J., and van de Pol, F. 1996. Bootstrapping goodness-of-fit measures in categorical data analysis. *Sociological Methods and Research*, 24, 492–516.

Lord, F. M., and Novick, M. R. 1968. *Statistical Theories of Mental Test Scores*. Reading, Massachusetts: Addison-Wesley.

Maij-de Meij, A. M., Kelderman, H., and van der Flier, H. 2008. Fitting a mixture item response theory model to personality questionnaire data: Characterizing latent classes and investigating possibilities for improving prediction. *Applied Psychological Measurement*, 32, 611–631.

McLachlan, G., and Peel, D. 2000. *Finite Mixture Models*. New York: Wiley.

McLachlan, G. J., and Krishnan, T. 1997. *The EM Algorithm and Extensions*. New York: Wiley.

Mislevy, R. J., and Verhelst, N. D. 1990. Modeling item responses when different subjects employ different solution strategies. *Psychometrika*, 55, 195–215.

Muthén, L. K., and Muthén, B. O. 1998–2007. *Mplus User's Guide* (5th edn). Los Angeles, California: Muthén and Muthén.

Newcomb, S. 1886. A generalized theory of the combination of observations so as to obtain the best result. *American Journal of Mathematics*, 8, 343–366.

Oliveri, M. E., and von Davier, M. 2011. Investigation of model fit and score scale comparability in international assessments. *Psychological Test and Assessment Modeling*, 53, 315–333. Retrieved September 29, 2011, from: http://www.psychologie-aktuell.com/fileadmin/download/ptam/3-2011_20110927/04_Oliveri.pdf.

Pearson, K. 1894. Contributions to the theory of mathematical evolution. *Philosophical Transactions of the Royal Society of London, A*, 185, 71–110.

R Core Team 2012. R: A Language and Environment for Statistical Computing. Vienna, Austria: R Foundation for Statistical Computing. Retrieved from http://www.R-project.org/.

Rasch, G. 1960. *Probabilistic Models for Some Intelligence and Attainment Tests*. Chicago, Illinois: University of Chicago Press.

Rijkes, C. P. M., and Kelderman, H. 2007. Latent response Rasch models for strategy shifts in problem solving processes. In: M. von Davier and C. H. Carstensen (Eds.), *Multivariate and Mixture Distribution Rasch Models: Extensions and Applications* (pp. 311–328). New York: Springer.

Rijmen, F. 2006. BNL: A Matlab toolbox for Bayesian networks with logistic regression nodes [Software manual]. Available from the Matlab Central Web site: http://www.mathworks.com/matlabcentral/fileexchange/13136

Rijmen, F., and De Boeck, P. 2005. A relation between a between-item multidimensional IRT model and the mixture-Rasch model. *Psychometrika*, 70, 481–496.

Rost, J. 1990. Rasch models in latent classes: An integration of two approaches to item analysis. *Applied Psychological Measurement*, 14, 271–282.

Rost, J., Carstensen, C., and von Davier, M. 1996. Applying the mixed Rasch model to personality questionnaires. In: J. Rost and R. Langeheine (Eds.), *Applications of Latent Trait and Latent Class Models in the Social Sciences* (pp. 324–332). Münster, Germany: Waxmann. Retrieved April 23, 2012, from: http://www.ipn.uni-kiel.de/aktuell/buecher/rostbuch/ltlc.htm.

Rost, J., and von Davier, M. 1992. *MIRA: A PC Program for the Mixed Rasch Model*. Kiel, Germany: Institute for Science Education.

Rost, J., and von Davier, M. 1993. Measuring different traits in different populations with the same items. In: R. Steyer, K. F. Wender and K. F. Widaman (Eds.), *Psychometric Methodology: Proceedings of the 7th European Meeting of the Psychometric Society in Trier* (pp. 412–417). Stuttgart, Germany: Gustav Fischer Verlag.

Rost, J., and von Davier, M. 1995. Mixture distribution Rasch models. In: G. H. Fischer and I. W. Molenaar (Eds.), *Rasch Models: Foundations, Recent Developments and Applications* (pp. 257–268). New York: Springer.

Saucier, G., and Goldberg, L. R. 1996. The language of personality: Lexical perspectives on the five-factor model. In: J. S. Wiggins (Ed.), *The Five-Factor Model of Personality: Theoretical Perspectives* (pp. 21–50). New York: Guilford.

Schmidt, S., Muhlan, H., and Power, M. 2005. The EUROHIS-QOL 8-item index: Psychometric results of a cross-cultural field study. *European Journal of Public Health*, 16, 420–428.

Schultz-Larsen, K., Kreiner, S., and Lomholt, R. K. 2007. Mini-mental status examination: Mixed Rasch model item analysis derived two different cognitive dimensions of the MMSE. *Journal of Clinical Epidemiology*, 60, 268–279.

Schwarz, G. E. 1978. Estimating the dimension of a model. *Annals of Statistics*, 6, 461–464.
Sinharay, S., Johnson, M., and Stern, H. S. 2006. Posterior predictive assessment of item response theory models. *Applied Psychological Measurement*, 30, 298–321.
Smit, A., Kelderman, H., and van der Flier, H. 1999. Collateral information and mixture Rasch models. *Methods of Psychological Research (Online)*, 4(3). Retrieved April 23, 2012, from: http://www.dgps.de/fachgruppen/methoden/mpr-online/issue8/art2/mrc.pdf.
Spiegelhalter, D. J., Thomas, A., and Best, N. G. 1999. *WinBUGS Version 1.2 User Manual [Software Manual]*. Cambridge, Massachusetts: MRC Biostatistics Unit.
Stankov, L. 2009. Conservatism and cognitive ability. *Intelligence*, 37, 294–304.
Verhelst, N. D., and Glas, C. A. W. 1995. The generalized one parameter model: OPLM. In: G. H. Fischer and I. W. Molenaar (Eds.), *Rasch Models: Their Foundations, Recent Developments and Applications* (pp. 215–238). New York: Springer.
Vermunt, J. K. 1993. *LEM 0.1: Log-Linear and Event History Analysis with Missing Data Using the EM Algorithms*. Tilburg, The Netherlands: Tilburg University.
Vermunt, J. K. 2003. Hierarchical latent class models. *Sociological Methodology*, 33, 213–239.
Vermunt, J. K., and Magidson, J. 2005. *Technical Guide for Latent GOLD Choice 4.0: Basic and Advanced*. Belmont, Massachusetts: Statistical Innovations Inc.
von Davier, M. 1994. *WINMIRA (A Windows Program for Analyses with the Rasch Model, with the Latent Class Analysis and with the Mixed Rasch model) [Computer software]*. Kiel, Germany: Institute for Science Education.
von Davier, M. 1997. Bootstrapping goodness-of-fit statistics for sparse categorical data: Results of a Monte Carlo study. *Methods of Psychological Research*, 2, 29–48. Retrieved January 14, 2010, from: http://www.dgps.de/fachgruppen/methoden/mpr-online/issue3/art5/davier.pdf.
von Davier, M. 2001. WINMIRA 2001: Software for estimating Rasch models, mixed and hybrid Rasch models, and the latent class analysis [Computer software]. Retrieved from: http://www.von-davier.com.
von Davier, M. 2005. A general diagnostic model applied to language testing data. Research Report Series RR-05-16. Princeton, New Jersey: Educational Testing Service.
von Davier, M. 2008. The mixture general diagnostic model. In: G. R. Hancock and K. M. Samuelson (Eds.), *Advances in Latent Variable Mixture Models* (pp. 255–276). Charlotte, North Carolina: Information Age Publishing.
von Davier, M. 2009a. Mixture distribution item response theory, latent class analysis, and diagnostic mixture models. In: S. Embretson (Ed.), *Measuring Psychological Constructs: Advances in Model-Based Approaches* (pp. 11–34). Washington, DC: APA Press.
von Davier, M. 2009b. Some notes on the reinvention of latent structure models as diagnostic classification models. *Measurement – Interdisciplinary Research and Perspectives*, 7, 67–74.
von Davier, M. 2010. Hierarchical mixtures of diagnostic models. *Psychological Test and Assessment Modeling*, 52, 8–28. Retrieved April 26, 2012, from: http://www.psychologie-aktuell.com/fileadmin/download/ptam/1-2010/02_vonDavier.pdf.
von Davier, M., and Carstensen, C. H. (Eds.). 2007. *Multivariate and Mixture Distribution Rasch Models*. New York: Springer.
von Davier, M., and Molenaar, I. W. 2003. A person-fit index for polytomous Rasch models, latent class models, and their mixture generalizations. *Psychometrika*, 68, 213–228.
von Davier, M., Naemi, B., and Roberts, R. D. 2012. Factorial versus typological models: A comparison of methods for personality data, *Measurement: Interdisciplinary Research and Perspectives*, 10(4), 185–208.
von Davier, M., and Rost, J. 1995. Polytomous mixed Rasch models. In: G. H. Fischer and I. W. Molenaar (Eds.), *Rasch Models: Foundations, Recent Developments and Applications* (pp. 371–379). New York: Springer.
von Davier, M., and von Davier, A. A. 2007. A unified approach to IRT scale linkage and scale transformations. *Methodology*, 3, 115–124.
von Davier, M., Xu, X., and Carstensen, C. H. 2011. Measuring growth in a longitudinal large scale assessment with a general latent variable model. *Psychometrika*, 76, 318–336.

von Davier, M., and Yamamoto, K. 2004. Partially observed mixtures of IRT models: An extension of the generalized partial credit model. *Applied Psychological Measurement*, 28, 389–406.

Wilson, M. 1989. Saltus: A psychometric model for discontinuity in cognitive development. *Psychological Bulletin*, 105, 276–289.

Xu, X., and von Davier, M. 2008. Linking with the general diagnostic model. Research Report RR-08-08. Princeton, New Jersey: Educational Testing Service.

Yamamoto, K. 1989. A hybrid model of IRT and latent class models. Research Report RR-89-41. Princeton, New Jersey: Educational Testing Service.

Yamamoto, K., and Everson, H. 1997. Modeling the effects of test length and test time on parameter estimation using the HYBRID model. In: J. Rost and R. Langeheine (Eds.), *Applications of Latent Trait and Latent Class Models in the Social Sciences.* New York: Waxmann. Retrieved April 23, 2012, from: http://www.ipn.uni-kiel.de/aktuell/buecher/rostbuch/ltlc.htm.

24

Multilevel Response Models with Covariates and Multiple Groups

Jean-Paul Fox and Cees A. W. Glas

CONTENTS

24.1 Introduction 407
24.1.1 Multilevel Modeling Perspective on IRT 408
24.2 Bayesian Multilevel IRT Modeling 409
24.3 Presentation of the Models 410
24.3.1 Multilevel IRT Model 410
24.3.2 GLMM Presentation 411
24.3.3 Multiple-Group IRT Model 411
24.3.4 Mixture IRT Model 412
24.3.5 Multilevel IRT with Random Item Parameters 412
24.4 Parameter Estimation 413
24.5 Model Fit 414
24.6 Empirical Example 415
24.6.1 Data 415
24.6.2 Model Specification 415
24.6.3 Results 415
24.7 Discussion 417
References 418

24.1 Introduction

Most educational research data have a multilevel structure. For example, the data collected may be nested in students as a second level, students in classes as a third level, classes in schools as a fourth level, and so forth. Such structures require multiple levels of analysis to account for differences between observations, students, and other higher-level units. Since the typical clustering of multilevel data leads to (marginally) dependent observations, separate linear analyses based on the assumption of independent identically distributed variables at each of these levels are inappropriate.

When the nested structure of such data is ignored, aggregation bias (i.e., group-level inferences incorrectly assumed to apply to all group members), also known as the ecological fallacy, may occur. Furthermore, the estimated measurement precision will become biased. Partly in response to these technical problems, hierarchical or multilevel modeling has emerged, which is characterized by the fact that observations within each cluster are assumed to vary as a function of cluster-specific level parameters. In turn, these parameters may vary randomly across a population of clusters as a function of higher-level parameters. Multilevel modeling takes such hierarchical structures into account, adopting appropriate variance components at each level of sampling. As a result, homogeneity of the observations of, for instance, students in the same class due to their common experiences is accounted for. Also, multilevel models can be used to describe relationships between one or more dependent variables with teacher (e.g., attitudes), school (e.g., financial resources, class sizes), and student characteristics (e.g., achievements, social background).

Aitkin and Longford (1986) were the first to show the appropriateness of multilevel models for educational research and to tackle their computational problems. From then on, the idea of multilevel modeling of hierarchically structured data has received much attention, and important contributions have been made addressing technical issues such as the estimation of appropriate error structures and the testing of statistical hypotheses of within-cluster, between-cluster, and cross-level effects (e.g., Goldstein, 2003; Longford, 1993; Raudenbush and Bryk, 2002; Snijders and Bosker, 2011).

24.1.1 Multilevel Modeling Perspective on IRT

The increasing popularity of multilevel modeling has also affected item response theory (IRT) modeling in various ways. In the straightforward multilevel approach by Adams et al. (1997; Volume One, Chapter 32), the responses to test items are accepted as first-level observations, students and items as the second level, and, for instance, the population distribution of the students' parameters as the third level. Their approach reflects a multistage sampling design often used to collect educational data. As already noted, for data collected through such designs, a standard analysis relying on the assumptions of independently and identically distributed observations is inappropriate. To deal with this issue, univariate multilevel response models were developed (e.g., Bock, 1989; Raudenbush and Bryk, 1988), which were later extended to deal with multivariate response data.

A slightly different perspective was offered by the integration of latent variable measurement models into a more general multilevel model, for instance, by Muthén (1991) and Raudenbush and Sampson (1999). Their general idea was that a multilevel design can include various latent variables at different levels. Consequently, when item responses are observed, it is natural to adopt an IRT response model as a first-level model in a more general multilevel framework.

IRT models have also been approached from the perspective of generalized linear mixed modeling (GLMM) (Volume One, Chapters 30, 31, and 33). Separate attention has been given to the specific case of GLMM formulations of extensions of the Rasch model (e.g., Adams and Wilson, 1996; Adams et al., 1997; Volume One 1, Chapter 33; Kamata, 2001; Kamata and Cheong, 2007; Pastor, 2003; Rijmen et al., 2003). Nowadays, various computer programs support the statistical treatment of GLMMs, which has made the approach accessible for use in a large variety of applications.

Unlike regular latent variable modeling, modeling of response data leads to a few new requirements. First, response data are often sparse at the individual level, which makes it difficult to obtain reliable estimates of individual effects. However, the individual responses by one respondent are typically linked to many other respondents through the higher levels in the model, and by borrowing strength from them, improved estimates of individual effects can be obtained. Second, responses are often integer valued, for instance, obtained as correct–incorrect or on a five-point or seven-point scale. This lumpy nature of the responses prevents the use of standard statistical distributions and requires a special modeling approach. Third, response data may be obtained in combination with other input variables. An example is responses obtained from students together with school variables, where the objective is to make joint inferences about individual and school effects. Again, such cases with different sources of information require accounting for the uncertainty inherent in each of the sources and can be handled most efficiently in a multilevel framework.

24.2 Bayesian Multilevel IRT Modeling

Following Aitkin and Aitkin (2011), Fox (2010), and Fox and Glas (2001) among others, a complete hierarchical modeling framework can be defined by integrating item response models with survey models for population distributions. Besides item-specific differences, this hierarchical type of item response modeling can be used to account for the survey design, background of the respondents, and clusters in which respondents are located.

A Bayesian approach to multilevel IRT provides additional features. First, it supports a flexible way of incorporating prior knowledge to account for different sources of uncertainty, complex dependencies, and other sources of information at separate levels, with subsequent inferences possible at each of these different levels from the posterior distributions of their parameters. This option is one of the natural strengths of the approach, which makes it possible to handle sampling designs with complex dependency structures.

Second, although the attractiveness of the Bayesian response modeling framework was already recognized in the 1980s (e.g., Mislevy, 1986), it only became feasible with the introduction of computational methods such as computer simulation and Markov chain Monte Carlo (MCMC). The development of powerful computational simulation techniques has introduced a tremendous positive change in the applicability of Bayesian methodology. Combining multilevel IRT modeling with powerful computational simulation techniques makes practical application in educational test and survey research possible.

This chapter provides a description of multilevel IRT modeling and shows a few applications. Besides, a few developments are discussed and directions for future research are given. Although our applications are from educational research, multilevel IRT models have been applied in other fields as well. For instance, van den Berg et al. (2007) showed how a multilevel IRT model can be applied to twin studies in genetics to account for measurement error variance that otherwise would have been interpreted as environmental variance. More specifically, they demonstrated that heritability estimates can be severely biased if the analyses are simply based on sum scores. He et al. (2010) used a multilevel response model to assess geographical variation in hospital quality. Their observations were of patients receiving or not receiving the therapy, where hospital quality was measured by the success rate of the therapy. Their patients were nested in hospitals, which, in turn, were nested in geographical units. Their complete model included an IRT model that enabled them to measure the quality score of each individual hospital while accounting for differential measurement-specific weights. In addition, its higher levels addressed the hierarchical structure of their data.

24.3 Presentation of the Models

24.3.1 Multilevel IRT Model

Assume a multistage sampling design, for instance, where schools $j = 1, \ldots, J$ are sampled from a district and subsequently students are sampled within each school j. The students' abilities are assessed using a test of $i = 1, \ldots, I$ items. To simplify the notation, a balanced test design is assumed where each student $p = 1, \ldots, P$ responds to each item. Thus, let U_{pji} denote the response of student p in school j to item i.

For dichotomous items, the following two-parameter IRT model describes the probability of a correct response of student p to item i

$$P\{U_{pji} = 1; \theta_{pj}, a_i, b_i\} = \Phi(a_i(\theta_{pj} - b_i)), \tag{24.1}$$

where $\Phi(.)$ represents the cumulative normal distribution function, a_i denotes the item discrimination parameter, b_i its difficulty parameter, and the latent variable θ_{pj} represents the student's ability (Volume One, Chapter 2).

The two-parameter IRT model in Equation 24.1 defines the first or observational level of modeling. The second level explains the within-school distribution of the abilities by variables denoted as $\mathbf{x}_{pj} = (x_{0pj}, x_{1pj}, \ldots, x_{Qpj})^t$, where x_{0pj} usually equals one. The level-2 model is

$$\begin{aligned}\theta_{pj} &= \beta_{0j} + \cdots + \beta_{qj}x_{qpj} + \cdots + \beta_{Qj}x_{Qpj} + e_{pj} \\ &= \sum_{q=0}^{Q} \beta_{qj}x_{qpj} + e_{pj},\end{aligned} \tag{24.2}$$

where the errors are independently and identically distributed with mean zero and variance σ_θ^2. The regression parameters are allowed to vary across schools. Similarly, level-3 explanatory variables can be adopted, which we denote as $\mathbf{w}_{qj} = (w_{0qj}, w_{1qj}, \ldots, w_{Sqj})^t$, where w_{0qj} typically equals one. The random regression coefficients defined in Equation 24.2 are now considered to be the result of linear regression at level 3

$$\begin{aligned}\beta_{qj} &= \gamma_{q0} + \cdots + \gamma_{qs}w_{sqj} + \cdots + \gamma_{qS}w_{Sqj} + r_{qj} \\ &= \sum_{s=0}^{S} \gamma_{qs}w_{sqj} + r_{qj}\end{aligned} \tag{24.3}$$

for $q = 0, \ldots, Q$, with level-2 error terms, $\mathbf{r}_j$, assumed to be multivariate normally distributed with mean zero and covariance matrix $\mathbf{T}$. The elements of $\mathbf{T}$ are denoted as $\tau_{qq'}^2$ for q, $q' = 0, \ldots, Q$.

Thus, within each school j, the abilities are modeled as a linear function of the student characteristics $\mathbf{x}_j$ plus an error term $\mathbf{e}_j$, where the matrix with the explanatory data $\mathbf{x}_j$ is assumed to be of full rank. Further, the level-2 random regression parameters β_j are assumed to vary across schools as a function of the school predictors $\mathbf{w}_j$ plus an error term $\mathbf{u}_j$. Both these level-2 and level-3 equations can be reduced to a single equation by substituting Equation 24.3 into Equation 24.2, stacking their matrices appropriately. The result resembles the general Bayesian linear model and allows $\mathbf{x}_j$ to be of less than full rank. Furthermore, not all level-2 parameters are necessarily random effects; some of them can also be viewed as fixed effects (i.e., not varying across schools).

24.3.2 GLMM Presentation

A multilevel Rasch model can be reformulated as a generalized linear mixed effects model (e.g., Adams et al., 1997; Kamata, 2001; Rijmen et al., 2003). Consider an unconditional multilevel Rasch model that does not have any student or school predictors. Let π_{pji} denote the probability of student p in school j answering correctly to item i. A logit link function can be used to describe the relationship between the log odds of the probability π_{pji} and a linear term with the item difficulty and ability parameter. In a more general formulation, let $\eta_{pj0}, \ldots, \eta_{pjI}$ denote the parameters of this linear term. Then, the level-1 model is represented by

$$\log\left(\frac{\pi_{pji}}{1-\pi_{pji}}\right) = \eta_{pj0} + \sum_{k=1}^{I} \eta_{pjk} D_{pjk},$$

where the indicator variable D_{pjk} for student p will equal minus one when $k = i$ and zero otherwise. A constraint $\eta_{pjI} = 0$ is given to ensure that the design matrix is of full rank.

The intercept η_{pj0} can be interpreted as the ability level of student p in school j. The effects $\eta_{pj1}, \ldots, \eta_{pj(I-1)}$ are constrained to be fixed across students such that they represent item effects. The η_{pjk} represents the offset of the kth item from the Ith item.

The variation in ability across students and schools is represented by a level-2 and level-3 model, which can be summarized as

$$\eta_{pj0} = \gamma_{00} + r_{0j} + e_{pj},$$

where γ_{00} represents the average level of ability in the population. The error terms r_{0j} and e_{pj} represent the between-student and between-school variation, respectively; they are assumed to be normally distributed. Again, if necessary, the model can be extended with student and school predictors.

24.3.3 Multiple-Group IRT Model

The notion of a multilevel IRT model is closely related to that of a multiple-group model. In some studies, the interest is in more than one specific group. Although the respondents are randomly sampled from these groups, the groups themselves are not considered to be sampled from any larger population. For this case, Bock and Zimowski (1997) proposed a multiple-group IRT model that included group-specific population distributions to handle the clustering of respondents into groups. Their model allows inferences made with respect to each of the sampled groups but not to some higher-level population across all groups.

Azevedo et al. (2012) generalized the multiple-group IRT model following a Bayesian approach with item response functions allowed to be skewed on probit, logit, or log–log functions (Albert, Volume Two, Chapter 1) and a multiple-group latent variable distribution that can be represented by a normal, Student's t, skew normal, or skew Student's t distribution, or any finite mixture of normal distributions. This flexibility of choice of response functions and population distributions is obtained by parameterizing mixtures $l = 1, \ldots, L$ of different response functions based on different cumulative distribution functions $h = 1, \ldots, H$, with possibly different latent trait distributions across items and groups. For dichotomous responses, the success probability of this generalized multiple-group IRT model is

$$P\{U_{pji} = 1; \theta_{pj}, \boldsymbol{\xi}_i, \boldsymbol{\omega}\} = \sum_{l=1}^{L} \prod_{h=1}^{H} F_{lh}(\eta_{pj}, \boldsymbol{\xi}_i, \boldsymbol{\omega}) \tag{24.4}$$

$$\theta_{pj} \mid \boldsymbol{\eta}_j \sim G(\boldsymbol{\eta}),$$

where the cumulative distribution function F_{lh} has parameter $\boldsymbol{\omega}$ and $G(\boldsymbol{\eta}_j)$ represents a continuous population distribution function with parameters $\boldsymbol{\eta}_j$ for group j. The model encompasses the well-known one-, two-, and three-parameter item response models, each with a choice from the earlier- mentioned link functions.

24.3.4 Mixture IRT Model

Both multiple-group and multilevel IRT models assume response data sampled from respondents nested in manifest groups. When the respondents are clustered but the clusters cannot be observed directly, a latent-class model can be used. The approach then captures the nesting of students in latent clusters as well as the dependencies between the response vectors it has created.

Following Rost (1997), the success probability of a correct response can now be modeled as

$$P\{U_{pgi} = 1; \theta_{pg}, \boldsymbol{\xi}_i, g\} = \Phi(a_{ig}(\theta_{pg} - b_{ig}))$$

$$\theta_{pg} \mid \mu_g, \sigma_g^2 \sim N(\mu_g, \sigma_g^2), \tag{24.5}$$

where θ_{pg} is the ability of student p in latent class g, a_{ig} and b_{ig} are class-specific item discrimination and difficulty parameters, respectively, and μ_g, σ_g^2 are class-specific mean and variance parameters, respectively.

This mixture IRT model has been used for detecting differential item functioning, differential use of response strategies, and the effects of different test accommodations. Assuming measurement invariance, the mixture-modeling approach is suitable to identify unobserved clusters of students. Cho and Cohen (2010) and Vermunt (2003) used a multilevel latent-class model to identify latent classes of students that are homogeneous with respect to item response patterns, while accounting for the manifest clustering of students in schools.

24.3.5 Multilevel IRT with Random Item Parameters

Item responses are nested both within students and items. Thus far, our attention was exclusively on the clustering of students. However, the item side of the multilevel IRT model needs to be correctly specified as well to make proper inferences.

The item characteristics are specified to be normally distributed around the average item characteristics. Furthermore, the item parameters of each item are assumed to be correlated. For the IRT model specified in Equation 24.1, a multivariate normal prior density for the item parameters $i = 1, \ldots, I$ can be specified as

$$(a_i, b_i)^t \sim N(\boldsymbol{\mu}_\xi, \boldsymbol{\Sigma}_\xi)I(a_i > 0) \tag{24.6}$$

with prior hyperparameters

$$\boldsymbol{\Sigma}_\xi \sim IW(\nu, \boldsymbol{\Sigma}_0),$$

$$\boldsymbol{\mu}_\xi | \Sigma_\xi \sim N(\boldsymbol{\mu}_0, \Sigma_\xi / K_0).$$

The specification amounts to the assumption of all item response functions being invariant across populations of test takers. Alternatively, different assumptions have been made to allow for functions that do vary across such populations. Such kind of variation in item parameters has also been discussed by De Boeck and Wilson (2004; Volume One, Chapter 33) and De Jong et al. (2007). Verhagen and Fox (2013) discussed a longitudinal IRT model for ordinal response data, where items function differently over time, using prior distributions for the items. Glas et al. (Volume One, Chapter 26) discussed a different application, where items are generated by a computer algorithm. They modeled variation in item characteristics of clone items in sets of similar items, which only differ by surface characteristics.

To allow for variation in item functioning over populations, random item parameters can be defined that allow for changes in the response functions of the items over them. Let population-specific item parameters $\tilde{a}_{ij}$ and $\tilde{b}_{ij}$ be distributed as

$$\tilde{\boldsymbol{\xi}}_{ij} = (\tilde{a}_{ij}, \tilde{b}_{ij})^t \sim N((a_i, b_i)^t, \boldsymbol{\Sigma}_{\tilde{\xi}}),$$

where a_i and b_i are means across the populations $j = 1, \ldots, J$, and independent random item parameters are defined by choosing $\Sigma_{\tilde{\xi}}$ to be diagonal. However, this random item parameter specification introduces an identification problem. For each population, both its mean ability and the mean item difficulty are not identified. Therefore, we may decide to restrict the mean item difficulty to be equal across populations, so that possible score differences between populations become attributable to differences in ability as well as residual differences in item functioning. This choice of identification lets the random item parameters explain between-population residual variance. Similarly, we can restrict average item discrimination to be constant across populations. Random parameters then explain residual between-population variation in item discrimination. Of course, for both types of parameters, the preferred case is negligible fluctuations between populations, that is, "measurement invariance." However, as described by Verhagen and Fox (2013), when the fluctuations appear to be substantial, they may be explained by background differences (e.g., culture or gender differences). Besides, it also possible to allow for cross-classified differences in item characteristics, for example, when cross-national and cross-cultural response heterogeneity is present.

24.4 Parameter Estimation

A Bayesian approach to response modeling is a natural way to account for sources of uncertainty in the estimation of the parameters. Also, it leads to estimation procedures that are easy to implement (e.g., Congdon, 2001).

A fully Bayesian approach requires the specification of prior distributions for all model parameters. In Fox (2010), noninformative inverse gamma priors are specified for the variance components, an inverse Wishart prior is specified for the covariance matrix, and vague normal priors are specified for the remaining mean parameters. Subsequently, a Gibbs sampling procedure is used to sample from the joint posterior distribution of all model parameters. More specifically, following Albert (1992), an augmentation scheme is defined to sample latent continuous responses, whereupon the item parameters and multilevel model parameters can be sampled directly from the full conditional distributions given the latent responses (Fox and Glas, 2001; Fox, 2010).

Alternatively, several packages in R and WinBUGS are available to estimate the model parameters. Cho and Cohen (2010) developed WinBUGS programs to estimate a mixture multilevel IRT model. For a GLMM presentation, various programs exist to estimate the model parameters. Tuerlinckx et al. (2004) compared the performance of different programs (GLIMMIX, HLM, MLwiN, MIXOR/MIXNO, NLMixed, and SPlus) and found generally similar results for each of them.

24.5 Model Fit

Posterior predictive checks provide a natural way to check the assumptions underlying an item response model. To use them, discrepancy measures need to be defined that provide information about specific model assumptions. The extremeness of the discrepancy measures given the data is then evaluated using data generated from their posterior predictive distribution. For instance, discrepancy measures have been proposed to evaluate the assumptions of local independence and unidimensionality. Posterior predictive checks for evaluating IRT models have been proposed, among others, by Glas and Meijer (2003), Levy and Sinharay (2009), and Sinharay et al. (2006). For an overview, see Sinharay (Volume Two, Chapter 19).

Different multilevel IRT models can be compared for their goodness of fit using the deviance information criterion (DIC); for an introduction, see Cohen and Cho, Volume Two, Chapter 18. The criterion is defined as

$$\begin{aligned} DIC &= D(\hat{\mathbf{\Omega}}) + 2p_D \\ &= -2\log p(\mathbf{y} \mid \hat{\mathbf{\Omega}}) + 2p_{D,} \end{aligned}$$

where $\mathbf{\Omega}$ represents the multilevel IRT model parameters, $D(\hat{\mathbf{\Omega}})$ represents the deviance evaluated at the posterior mean $\hat{\mathbf{\Omega}}$, and p_D represents the effective number of parameters defined as the posterior mean of the deviance minus the deviance evaluated at the posterior mean of the model parameters.

When $\mathbf{\Omega} = (\boldsymbol{\xi}, \boldsymbol{\gamma}, \sigma_\theta^2, \mathbf{T})$, the likelihood becomes

$$\begin{aligned} p\{\mathbf{u}; \boldsymbol{\xi}, \boldsymbol{\gamma}, \sigma_\theta^2, \mathbf{T}\} = \prod_j \int_{\beta_j} \Bigg[\prod_{p|j} \int_{\theta_{pj}} \prod_i p(u_{pji} \mid \theta_{pj}, \boldsymbol{\xi}_i) \\ p(\theta_{pj} \mid \boldsymbol{\beta}_j, \sigma_\theta^2) d\theta_{pj} \Bigg] p(\boldsymbol{\beta}_j \mid \boldsymbol{\gamma}, \mathbf{T}) d\boldsymbol{\beta}_j, \end{aligned} \tag{24.7}$$

such that the fit of random effects is not explicitly expressed in the (marginal) likelihood.

24.6 Empirical Example

Data from the 2003 assessment in the Programme for International Student Assessment (PISA) of the Organisation for Economic Co-operation and Development (OECD) were analyzed to illustrate the multilevel IRT model. The original data and results from PISA can be found at http://pisa2003.acer.edu.au. Similar to Fox (2010), the performances of the subpopulation of Dutch students in mathematics was investigated using various selections of background variables. Besides, a random item parameter multilevel IRT model was used to investigate measurement invariance assumptions across Dutch schools.

24.6.1 Data

The performances in mathematics were measured using 84 items. Students received credit for each item they answered correctly. However, although some items were scored with partial credit, for this example, all item responses were coded as zero (incorrect) or one (correct). Each student in PISA 2003 was given a test booklet with a different combination of clusters of items, with each mathematics item that appeared in the same number of test booklets. In total, 3829 students were sampled from 150 Dutch schools. Students with less than nine responses were not included in the current analysis.

24.6.2 Model Specification

To investigate individual and school differences in student performances, the following unconditional multilevel IRT model was used to analyze the data:

$$P\{u_{pji}=1;\theta_{pj},\xi_i\}=\Phi(a_i(\theta_{pj}-b_i))$$

$$\theta_{pj}\sim N(\beta_{0j},\sigma_\theta^2)$$

$$\beta_{0j}\sim N(\gamma_{00},\tau_{00}^2),$$

where $j=1,\ldots,150$ represent the selected schools. Three levels of modeling were chosen to account for the within-student, between-student, and between-school variability. To account for item parameter variability, random item parameters were assumed. The multilevel IRT model was identified by fixing the mean and variance of the scale at zero and one, respectively.

24.6.3 Results

All model parameters were estimated using MCMC sampling from the posterior distribution, as implemented in the package mlirt.* A total of 10,000 MCMC iterations were run, where the first 1000 iterations were used as burn-in.

In Table 24.1, the parameter estimates of the unconditional multilevel IRT model are given under the label Unconditional MLIRT. On the standardized ability scale, the between-student variance was 0.43 and the between-school variability was about 0.61. The estimated intraclass correlation coefficient, which represents the percentage of variability between

* The Splus and R package mlirt are available at www.jean-paulfox.com.

TABLE 24.1

Math Performances of Dutch Students in PISA 2003: Parameter Estimates of the Multilevel IRT Models

	Unconditional MLIRT		MLIRT	
	Mean	**HPD**	**Mean**	**HPD**
Fixed Part				
Intercept	−0.04	[−0.17, 0.09]	0.02	[−0.10, 0.14]
Student Variables				
Female			−0.16	[−0.22, −0.11]
Foreign born			−0.28	[−0.38, −0.17]
Foreign language			−0.23	[−0.34, −0.12]
Index			0.15	[0.12, 0.18]
School Variables				
Mean index			0.39	[0.08, 0.70]
Random Part				
σ_θ^2	0.43	[0.40, 0.45]	0.40	[0.37, 0.42]
τ_{00}^2	0.61	[0.47, 0.76]	0.49	[0.38, 0.61]

math scores explained by the differences between the schools, was approximately 0.59. In PISA 2003, the estimated intraclass correlation coefficient varied from country to country, with many countries scoring above 0.50.

Multilevel parameter estimates can be biased when point estimates are used as a dependent variable. To prevent this from happening, the PISA 2003 results were computed using plausible values for the student's math abilities. The plausible values were random draws from the posterior distributions of the ability parameters given the response data. Their use facilitates the computation of standard errors, while allowing for the uncertainty associated with the ability estimates. Fox (2010, Chapter 6) showed that the current multilevel IRT parameter estimates and standard deviations were not significantly different from the estimates obtained using plausible values, given the 95% highest posterior density (HPD) intervals.

The between-student and between-school differences in math performance were explained using background variables. At the student level, female, place of birth (Netherlands or foreign), native versus foreign first language, and the PISA index of economic, social, and cultural status were used as explanatory variables. At the school level, the mean index of economic, social, and cultural status for each school was used to explain variability between the average school performances. In Table 24.1, the multilevel IRT model parameter estimates are given in the column labeled MLIRT. It may be concluded that male students performed slightly better than the female students, native speakers performed better than nonnative speakers with a migrant background, and students from more advantaged socioeconomic backgrounds generally performed better. Besides, the schools' average index of economic, social, and cultural status had a significant positive effect on their average score.

The multilevel IRT analysis was performed assuming invariance of the items across schools. A version of the model with random item parameters with school-specific distributions was used to investigate whether admittance of small deviations in item functioning

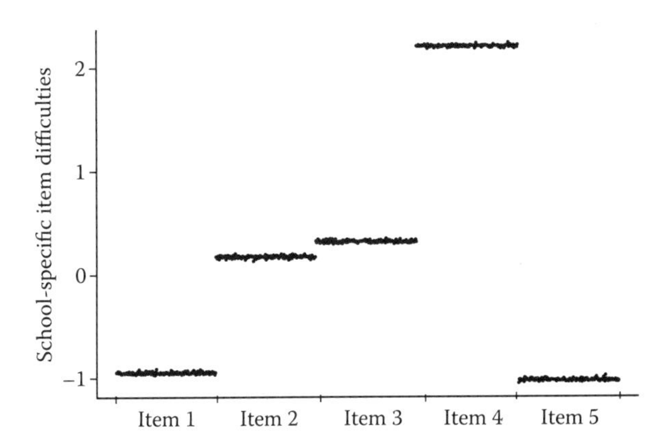

FIGURE 24.1
Random item difficulty estimates of items one to five for the 150 Dutch schools in PISA 2003.

across schools would lead to a better model fit. Although our preferred model was the measurement- invariant multilevel IRT model, this more-flexible generalization of it did capture some additional residual variance. Its estimated intraclass correlation coefficient was slightly lower than 0.43 for the previous model. However, the item parameter estimates hardly varied across schools; only small variations in item discriminations could be detected. For instance, the first five math items had average difficulty estimates equal to −0.94, 0.17, 0.32, 2.21, and −1.02. In Figure 24.1, the estimated difficulties of these five items are plotted for each school, revealing hardly any differences between schools.

24.7 Discussion

This chapter presented an overview of multilevel IRT modeling. It also reviewed extensions of multilevel models that included student and/or item population distributions. In either case, the clustering of students or items is modeled by an extra hierarchical level in the model. In addition, generalizations to different types of response data and different types of clustering were presented, and the use of a Bayesian approach with MCMC sampling from the posterior distributions of the parameters was highlighted. The same approach, via posterior predictive assessment, offers the possibility to evaluate model assumptions.

Multilevel response modeling has been shown to be useful in school effectiveness research, where typically, differences in responses and abilities within and between schools need to be explored. It also allows us to account for the nested structure of the responses and abilities.

Current multilevel models usually treat the student variable of interest as unidimensional. However, when multiple student abilities must be assumed to produce the observed responses, a multidimensional IRT model can be specified as a response model. The specification then requires a multivariate second-level model. The approach has been developed for modeling responses and response times to measure both the ability and speed of working (Volume One, Chapter 29).

References

Adams, R.J. and Wilson, M. 1996. Formulating the Rasch model as a mixed coefficients multinomial logit. In: G. Engelhard and M. Wilson (Eds.), *Objective Measurement III: Theory into Practice* (pp. 143–166). Norwood, NJ: Ablex.

Adams, R.J., Wilson, M., and Wu, M. 1997. Multilevel item response models: An approach to errors in variables regression. *Journal of Educational and Behavioral Statistics*, 22, 1 47–76.

Adams, R.J., Wilson, M.R., and Wang, W.-C. 1997. The multidimensional random coefficients multinomial logit model. *Applied Psychological Measurement*, 21, 1–23.

Aitkin, M. and Aitkin, I. 2011. *Statistical Modeling of the National Assessment of Educational Progress.* New York: Springer.

Aitkin, M. and Longford, N.T. 1986. Statistical modelling issues in school effectiveness studies. *Journal of the Royal Statistical Society*, 149, 1–43.

Albert, A.H. 1992. Bayesian estimation of normal ogive item response curves using Gibbs sampling. *Journal of Educational Statistics*, 17, 251–269.

Azevedo, C.L.N., Andrade, D.F., and Fox, J.-P. 2012. A Bayesian generalized multiple group IRT model with model-fit assessment tools. *Computational Statistics and Data Analysis*, 56, 4399–4412.

Bock, R.D. 1989. *Multilevel Analysis of Educational Data*. New York: Academic Press.

Bock, D.R. and Zimowski, M.F. 1997. Multiple group IRT. In: W.J. van der Linden and R.K. Hambleton (Eds.), *Handbook of Modern Item Response Theory* (pp. 433–448). New York: Springer-Verlag.

Congdon, P. 2001. *Bayesian Statistical Modelling.* West Sussex: John Wiley and Sons.

Cho, S.-J. and Cohen, A.S. 2010. A multilevel mixture IRT model with an application to DIF. *Journal of Educational and Behavioral Statistics*, 35, 336–370.

De Boeck, P. and Wilson, M. 2004. *Explanatory Item Response Models: A Generalized Linear and Nonlinear Approach.* New York: Springer.

De Jong, M.G., Steenkamp, J.B.E.M., and Fox, J.-P. 2007. Relaxing measurement invariance in cross-national consumer research using a hierarchical IRT model. *Journal of Consumer Research*, 34, 260–278.

Fox, J.-P. 2010. *Bayesian Item Response Modeling: Theory and Applications.* New York: Springer.

Fox, J.-P. and Glas, C.A.W. 2001. Bayesian estimation of a multilevel IRT model using Gibbs sampling. *Psychometrika*, 66, 271–288.

Glas, C.A.W. and Meijer, R.R. 2003. A Bayesian approach to person fit analysis in item response theory models. *Applied Psychological Measurement*, 27, 217–233.

Goldstein, H. 2003. *Multilevel Statistical Models.* Third edition. London: Edward Arnold.

He, Y., Wolf, R.E., and Normand, S.-L.T. 2010. Assessing geographical variations in hospital processes of care using multilevel item response models. *Health Services and Outcomes Research Methodology*, 10, 111–133.

Kamata, A. 2001. Item analysis by the hierarchical generalized linear model. *Journal of Educational Measurement*, 38, 79–93.

Kamata, A. and Cheong, Y.F. 2007. Hierarchical Rasch models. In: M. von Davier and C.H. Carstensen (Eds.), *Multivariate and Mixture Distribution Rasch Models—Extensions and Applications* (pp. 217–232). New York: Springer.

Levy, R., Mislevy, R.J., and Sinharay, S. 2009. Posterior predictive model checking for multidimensionality in item response theory. *Applied Psychological Measurement*, 33, 519–537.

Longford, N.T. 1993. *Random Coefficient Models.* Oxford: Clarendon Press.

Mislevy, R.J. 1986. Estimation of latent group effects. *Journal of the American Statistical Association*, 80, 993–997.

Muthén, B. 1991. Multilevel factor analysis of class and student achievement components. *Journal of Educational Measurement*, 28, 338–354.

Pastor, D.A. 2003. The use of multilevel item response theory modeling in applied research: An illustration. *Applied Measurement in Education*, 16, 223–243.

Raudenbush, S.W. and Bryk, A.S. 2002. *Hierarchical Linear Models: Applications and Data Analysis Methods* (2nd ed.). Newbury Park, CA: Sage.

Raudenbush, S.W. and Bryk, A.S. 1988. Methodological advances in studying effects of schools and classrooms on student learning. *Review of Research in Education*, 15, 423–476.

Raudenbush, S.W. and Sampson, R.J. 1999. "Ecometrics": Toward a science of assessing ecological settings, with application to the systematic social observation of neighborhoods. *Sociological Methodology*, 29, 1–41.

Rijmen, F., Tuerlinckx, F., De Boeck, P., and Kuppens, P. 2003. A nonlinear mixed model framework for item response theory. *Psychological Methods*, 8, 185–205.

Rost, J. 1997. Logistic mixture models. In: W.J. van der Linden and R. Hambleton (Eds.), *Handbook of Modern Item Response Theory* (pp. 449–463). New York: Springer.

Sinharay, S., Johnson, M.S., and Stern, H.S. 2006. Posterior predictive assessment of item response theory models. *Applied Psychological Measurement*, 30, 298–321.

Snijders, T.A.B. and Bosker, R.J. 2011. *Multilevel Analysis: An Introduction to Basic and Advanced Multilevel Modeling* (2nd ed.). London: Sage.

Tuerlinckx, F., Rijmen, F., Molenberghs, G., Verbeke, G., Briggs, D., Van den Noortgate, W., Meulders, M., and De Boeck, P. 2004. Estimation and software. In: P. De Boeck and M. Wilson (Eds.), *Explanatory Item Response Models: A Generalized Linear and Nonlinear Approach* (pp. 343–373). New York: Springer.

Van den Berg, S.M., Glas, C.A.W., and Boomsma, D.I. 2007. Variance decomposition using an IRT measurement model. *Behavior Genetics*, 37, 604–616.

Verhagen, J. and Fox, J.-P. 2013. Longitudinal measurement in health-related surveys. A Bayesian joint growth model for multivariate ordinal responses. *Statistics in Medicine*, 32, 2988–3005.

Vermunt, J.K. 2003. Multilevel latent class models. *Sociological Methodology*, 33, 213–239.

25

Two-Tier Item Factor Analysis Modeling

Li Cai

CONTENTS

25.1 Introduction 421
25.2 Presentation of the Model 422
25.3 Parameter Estimation 426
25.4 Model Fit 429
25.5 Empirical Example 430
25.6 Discussion 433
Acknowledgments 433
References 434

25.1 Introduction

Multidimensional item response theory (MIRT; Reckase, 2009), and in particular, full-information item factor analysis (Bock et al., 1988) has become increasingly important in educational and psychological measurement research (see a recent survey by Wirth and Edwards, 2007). In contrast to unidimensional item response theory (IRT) modeling, an item can potentially load on one or more latent dimensions (common factors) in an item factor analysis model. These common factors are also potentially correlated, in the tradition of Thurston's (1947) multiple factor analysis.

Item factor analysis can be conducted in either the exploratory (unrestricted, with rotation) or confirmatory (restricted, without rotation) flavors. Recent years have seen progress in the development of efficient full-information item factor analysis estimation algorithms, both for exploratory use (e.g., Cai, 2010a; Schilling and Bock, 2005) and for confirmatory applications (e.g., Cai, 2010b; Edwards, 2010). User-friendly and flexible software implementations (e.g., Cai, 2012; Cai et al., 2011a; Wu and Bentler, 2011) have also become available.

A particular confirmatory item factor analysis model, the item bifactor model (Gibbons and Hedeker, 1992; Gibbons et al., 2007), has attracted considerable interest among psychometricians in recent years (see, e.g., a recent review by Reise, 2012). In a prototypical bifactor model, there is one primary dimension, representing a target construct being measured, and there are S mutually orthogonal group-specific dimensions that are also orthogonal to the general dimension. All items may load on the general dimension, and at the same time, an item may load on at most one group-specific dimension. The group-specific dimensions are group factors accounting for residual dependence above and beyond the general dimension. The bifactor pattern is an example of a *hierarchical* factor solution (Holzinger and Swineford, 1937; Schmid and Leiman, 1957), which represents a somewhat different tradition in the history of factor analysis. Wainer et al.'s (2007) testlet response theory model is closely related to the bifactor model. The testlet response theory model is equivalent to a second-order item factor analysis model, but it is typically presented as a constrained version of item bifactor model with proportionality restrictions on the general and group-specific factor loadings (Glas et al., 2000; Li et al., 2006; Rijmen, 2010; Yung et al., 1999).

Cai's (2010c) two-tier item factor model represents a hybrid approach that can be understood from three distinct orientations. First, it is minimally a more general version of the bifactor hierarchical item factor model in that the number of general dimensions is not required to be equal to one and the correlations among these general dimensions may be explicitly represented and modeled. Second, it may be understood as a more general version of the Thurstonian correlated-factors MIRT model that explicitly includes an additional layer (or tier) of random effects to account for residual dependence. Finally, it is mathematically a confirmatory item factor analysis model with certain special restrictions (to be elaborated) that can facilitate efficient maximum marginal likelihood computations.

The last point is important considering the extent to which item factor analysis has been plagued by the "curse" of dimensionality. The popularity of the item bifactor model can be attributed, in no small part, to the discovery of a dimension reduction method (Gibbons and Hedeker, 1992) by which maximum marginal likelihood estimation requires at most a series of two-dimensional quadrature computations, regardless of the total number of factors in the model. Thus for the first time, high-dimensional confirmatory factor models may be fitted to item response data with reasonable numerical accuracy and stability, and, importantly, within a reasonable amount of time that is difficult to surpass even with the advent of efficient simulation-based optimization algorithms. The efficiency of dimension reduction prompted a flurry of recent activities in the technical literature (e.g., Cai et al., 2011b; Gibbons et al., 2007; Jeon et al., 2013; Rijmen, 2009; Rijmen et al., 2008), within which the computational aspects of the two-tier model is situated.

25.2 Presentation of the Model

In a two-tier model, two kinds of latent variables are considered, primary and group specific. This creates a partitioning of the vector of D latent factors $\boldsymbol{\theta} = (\theta_1, \ldots, \theta_D)$ into two mutually exclusive parts: $\boldsymbol{\theta} = (\boldsymbol{\eta}, \boldsymbol{\xi})$, where $\boldsymbol{\eta}$ is a G-dimensional vector of (potentially correlated) primary latent dimensions and $\boldsymbol{\xi}$ is an S-dimensional vector of group-specific latent dimensions that are *(at a minimum) independent conditional on the primary dimensions.* While other chapters in this handbook may refer to the individual elements of $\boldsymbol{\theta}$ directly, it is most convenient for this discussion to consider individual elements of $\boldsymbol{\eta} = (\eta_1, \ldots, \eta_G)$ and $\boldsymbol{\xi} = (\xi_1, \ldots, \xi_S)$. Most of the time, particularly for single-group analysis, the group-specific dimensions are mutually orthogonal, and also orthogonal to the primary dimensions as an identification restriction (Rijmen, 2009). In a two-tier model, an item may load on all G primary dimensions, subject to model identification (such as fixing rotational indeterminacy), and at most on 1 group-specific dimension.

Using a path diagram, Figure 25.1 shows a hypothetical two-tier model with 10 items (the rectangles) that load on $G = 2$ correlated primary dimensions, as well as $S = 4$ group-specific

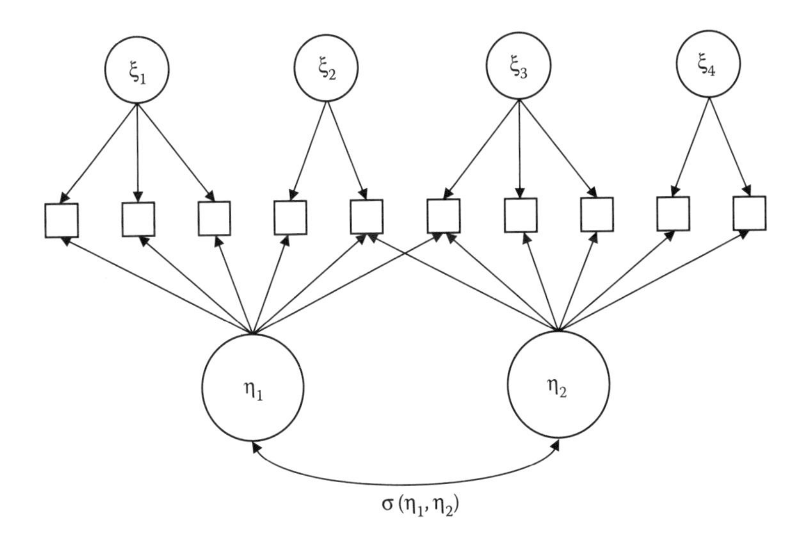

FIGURE 25.1
Two-tier model with two correlated primary dimensions and four group-specific dimensions where the factor pattern for the primary dimension is not of the independent cluster type.

dimensions that are orthogonal. In terms of factor pattern, the 10 by 6 factor pattern matrix corresponding to the model in Figure 25.1 has the following form:

$$\left(\begin{array}{cc|cccc} a & \square & a^* & \square & \square & \square \\ a & \square & a^* & \square & \square & \square \\ a & \square & a^* & \square & \square & \square \\ a & \square & \square & a^* & \square & \square \\ a & a & \square & a^* & \square & \square \\ a & a & \square & \square & a^* & \square \\ \square & a & \square & \square & a^* & \square \\ \square & a & \square & \square & a^* & \square \\ \square & a & \square & \square & \square & a^* \\ \square & a & \square & \square & \square & a^* \end{array}\right),$$

where nonempty entries indicate free parameters. The primary dimensions are shown to the left of the vertical bar. Note that the independent cluster pattern for the primary factors is by no means a requirement in general. Obviously, a two-tier model with only one primary dimension becomes a bifactor model (or a testlet model if proportionality restrictions are also imposed). On the other hand, a two-tier model without any group-specific dimension becomes a standard correlated-factors MIRT model.

The two-tier model has a number of applications in psychometric research. First, the model may be useful for calibrating longitudinal item responses. Suppose the eight items shown in Figure 25.2 in fact represent the test–retest administration of a four-item measurement instrument to the same group of individuals. The first four items are responses from Time 1 and the rest of the items are from Time 2. Even if the measurement instrument

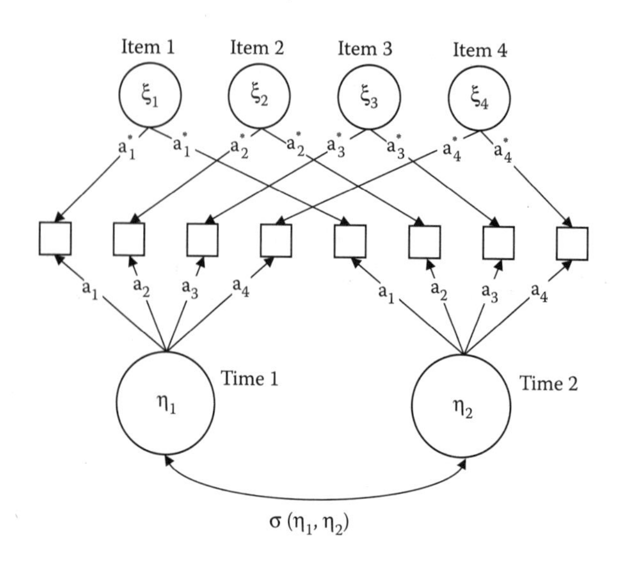

FIGURE 25.2
Two-tier model with two correlated primary dimensions and four group-specific dimensions, with potential application to longitudinal item response data.

is unidimensional at each test time point, the longitudinal item response data are inherently multidimensional. At least two occasion-specific primary latent dimensions are necessary to model potential changes in latent variable means and variances, as well as potential differences in the structure of measurement (e.g., shifts in location or discrimination of items) over time. The correlation between the primary dimensions can be understood to represent stability of the latent construct. In addition, the responses to the same item at Time 1 and at Time 2 from the same individual may be residually correlated, even after controlling for the influence of the primary dimensions. The group-specific dimensions are introduced to handle the potential residual dependence, and there are as many of these as the number of repeated items. Similar models have been considered by Hill (2006) and Vasdekis et al. (2012).

Second, the model may be useful for testlet-based educational assessments. Take the reading literacy assessment that is part of the Program for International Student Assessment as an example. The test design and specification (see, e.g., Adams and Wu, 2002) call for several potentially correlated "reading process" dimensions (the subscales) that are thought to contribute to the complex notion of reading literacy: interpretation, reflection/evaluation, and retrieval of information. In addition, the format of the assessment resembles passage-based reading tests with several questions following each reading task. This design introduces residual dependence among the items following the same task.

Figure 25.2 shows a potential model for such an assessment. The three primary dimensions correspond to the three reading process subscales. Items following the same reading task can belong to different subscales, and the total number of group-specific dimensions is equal to six, one for each task. In contrast to the longitudinal model in which the instrument is unidimensional within each time point, the two-tier model here is accounting for multidimensionality/dependence within and/or between the correlated subscales.

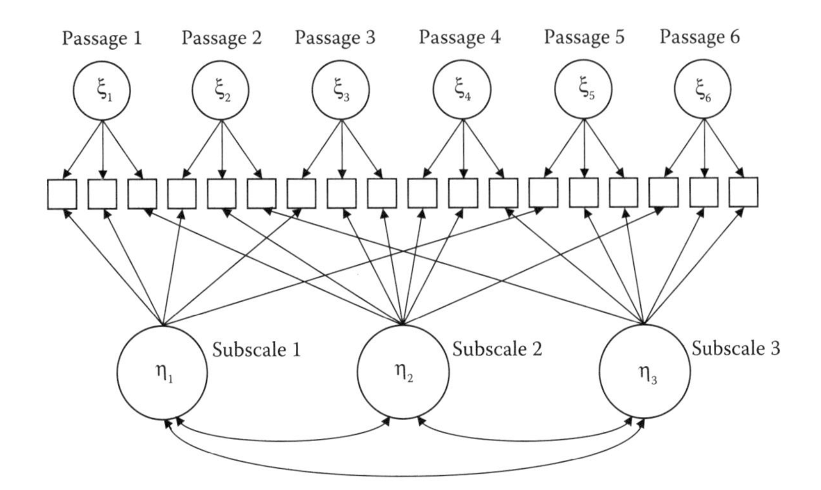

FIGURE 25.3
Two-tier model with three correlated primary dimensions and six group-specific dimensions, with potential application to testlet-based assessments.

Third, the two-tier model may be useful for test linking. The motivation comes from Thissen et al.'s (2011) novel *calibrated projection* linking method. Thissen et al. were conducting a linking study involving two patient-reported outcomes questionnaires that were designed to measure two highly similar latent constructs about the impact of asthma on patients' quality of life. A sample of respondents took both assessments. While a number of items were nearly identically phrased across the two questionnaires, the degree of overlap between the two constructs was appreciably less than perfect (at least statistically). Because the "common" items were not exactly the same across the two questionnaires, they could not function as anchor items, that is, the item parameters could not be constrained equal. A model that provided the best fit to their data resembles the model shown in Figure 25.3. The two correlated primary dimensions represent the two scales that need to be aligned. The "common" items that are highly similar are expected to share residual dependence on additional group-specific dimensions. In this case, there are five such items. The standard method of concurrent calibration is not entirely appropriate because forcing the two separate dimensions into a single shared dimension introduces bias in the standard error estimates of the scaled scores and overconfidence in the magnitude of linking error. Calibrated projection, on the other hand, uses the two-tier model as depicted. The joint estimation of the correlation between the two dimensions provides an automatic mechanism to perform projection linking. The projection linking is achieved simultaneously as the item parameters were being calibrated, hence the term calibrated projection (Figure 25.4).

Cai (2010c) contains discussions about additional applications of the two-tier model. A key recognition in all of the above referenced examples is that while the number of latent variables may appear to be large on the surface, the two-tier model structure ensures that the real dimensionality may be reduced analytically to achieve substantial efficiency and stability for parameter estimation. In a two-tier model, full-information estimation and inference requires integrating over at most $G+1$ dimensions, as opposed to $D=G+S$ dimensions. When G is equal to one, the two-tier dimension reduction becomes the well-known bifactor dimension reduction.

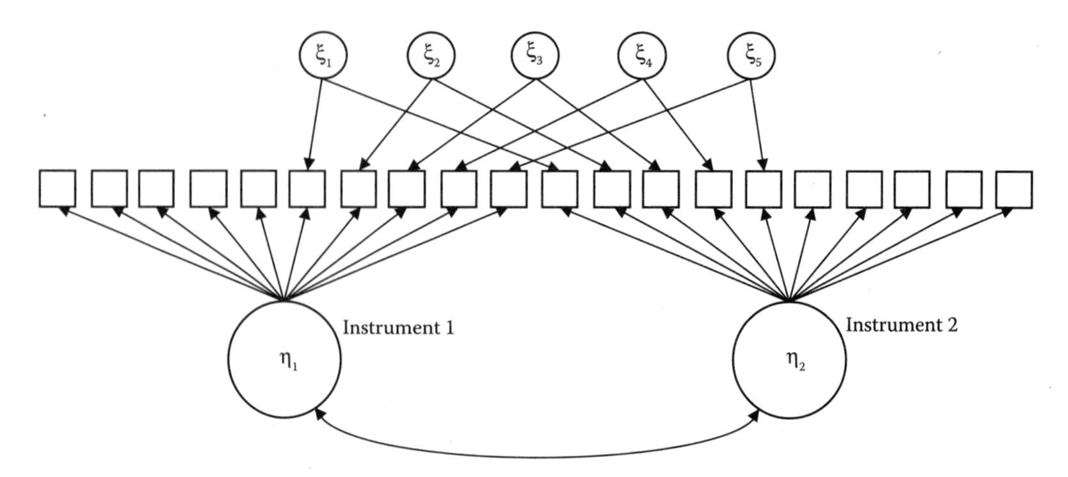

FIGURE 25.4
Two-tier model with two correlated primary dimensions and five group-specific dimensions, with potential application to calibrated projection test linking.

25.3 Parameter Estimation

Without loss of generality, let there be $i = 1, \ldots, I$ items. For item i, let the response categories be labeled $k = 0, \ldots, K_i - 1$, where K_i is the number of categories for this item. The category response probability, conditional on $\boldsymbol{\theta}$, can be written as $T_i(k|\boldsymbol{\theta})$. In a two-tier model, the category response probability can be simplified as $T_i(k|\boldsymbol{\theta}) = T_i(k|\boldsymbol{\eta},\boldsymbol{\xi}) = T_i(k|\boldsymbol{\eta},\xi_s)$, where the last equality assumes that item i loads on group-specific dimension s, where $s = 1, \ldots, S$.

In principle, any valid MIRT model may be used at the item level (see, e.g., models presented in Volume One, Chapter 12). For example, the following model is a generalization of the unidimensional two-parameter logistic model:

$$T_i(1|\boldsymbol{\eta},\xi_s) = \frac{1}{1+\exp\left(-c_i - a_{i1}\eta_1 - a_{i2}\eta_2 - \cdots - a_{iG}\eta_G - a_{is}^{*}\xi_s\right)}, \tag{25.1}$$

where the 1 category represents an endorsement or correct response. The response probability for the 0 category is $T_i(0|\boldsymbol{\eta},\xi_s) = 1 - T_i(1|\boldsymbol{\eta},\xi_s)$. The item parameters include an item intercept c_i, the G slopes for the general dimensions $a_{i1}, \ldots, a_{iG}$, and a slope for the group-specific dimension a_{is}^{*}. Similarly,

$$T_i(1|\boldsymbol{\eta},\xi_s) = g_i + \frac{1-g_i}{1+\exp\left(-c_i - a_{i1}\eta_1 - a_{i2}\eta_2 - \cdots - a_{iG}\eta_G - a_{is}^{*}\xi_s\right)} \tag{25.2}$$

is a natural extension of the unidimensional three-parameter logistic IRT model, where g_i is the pseudo-guessing parameter. For ordered polytomous responses, the multidimensional graded model is a straightforward choice. Let

$$T_i^{+}(k|\boldsymbol{\eta},\xi_s) = \frac{1}{1+\exp\left(-c_{ik} - a_{i1}\eta_1 - a_{i2}\eta_2 - \cdots - a_{iG}\eta_G - a_{is}^{*}\xi_s\right)} \tag{25.3}$$

be the cumulative response probability for categories k and above if $k = 1, \ldots, K_i - 1$. For consistency, two boundary cases are defined, $T_i^+(0|\boldsymbol{\eta}, \xi_s) = 1$ and $T_i^+(K_i|\boldsymbol{\eta}, \xi_s) = 0$, such that category k's response probability can be written as

$$T_i(k|\boldsymbol{\eta}, \xi_s) = T_i^+(k|\boldsymbol{\eta}, \xi_s) - T_i^+(k+1|\boldsymbol{\eta}, \xi_s) \tag{25.4}$$

for $k = 0, \ldots, K_i - 1$. In this graded response model, the number of item slopes is the same as in previous models, and the number of item intercepts increases to $K_i - 1$. Other polytomous MIRT models such as the nominal family (Volume One, Chapter 4; Thissen et al., 2010) may be also used but are not presented here due to space constraints.

Let u_{pi} be the observed response from person p to item i. This can be understood as the realization of a discrete random variable U_{pi} whose conditional distribution may be specified according to the following probability mass function:

$$f(u_{pi}|\boldsymbol{\eta}, \xi_s; \boldsymbol{\gamma}) = \prod_{k=0}^{K_i-1} [T_i(k|\boldsymbol{\eta}, \xi_s)]^{\chi_k(u_{pi})}, \tag{25.5}$$

where $\chi_k(u_{pi})$ is an indicator function, taking a value of 1 if and only if $u_{pi} = k$, and 0 otherwise. Equation 25.5 may be recognized as the probability mass function for a multinomial in K_i cells with trial size 1 and cell probabilities given by the T_i's. The vector $\boldsymbol{\gamma}$ is taken to be a vector of all free parameters from the two-tier model such as item slopes, intercepts, asymptote parameters, means, variances, etc. The parameters for item i are obviously components of $\boldsymbol{\gamma}$. The formulation in Equation 25.5 is flexible in that (a) constraints on item parameters (e.g., equality, proportionality, and fixing) are included without special treatment and (b) missing observations are directly handled as the indicator function excludes them from the probabilities.

Let the response pattern from person p be denoted as $\boldsymbol{u}_p = (u_{p1}, \ldots u_{pi}, \ldots u_{pI})'$. Furthermore, suppose the items can be partitioned into S mutually exclusive sets, indexed by $\mathcal{H}_1, \ldots \mathcal{H}_s, \ldots, \mathcal{H}_S$ where an item i belongs to set $\mathcal{H}_s$ (write $i \in \mathcal{H}_s$) if item i loads on group-specific dimension s. If an item does not load on any group-specific dimension, it may be conveniently grouped with the first set without any loss of generality. Assuming conditional independence of the item responses on $\boldsymbol{\theta} = (\boldsymbol{\eta}, \boldsymbol{\xi})$, the conditional probability of the response pattern may be specified as

$$f(\boldsymbol{u}_p|\boldsymbol{\theta}; \boldsymbol{\gamma}) = f(\boldsymbol{u}_p|\boldsymbol{\eta}, \boldsymbol{\xi}; \boldsymbol{\gamma}) = \prod_{s=1}^{S} \prod_{i \in \mathfrak{H}_s} f(u_{pi}|\boldsymbol{\eta}, \xi_s; \boldsymbol{\gamma}). \tag{25.6}$$

Let $h(\xi_s|\boldsymbol{\eta}; \boldsymbol{\gamma})$ be the density function of the sth group-specific dimension, conditional on $\boldsymbol{\eta}$. In most applications, a convenient choice is to let $h(\xi_s|\boldsymbol{\eta}; \boldsymbol{\gamma})$ be univariate Gaussian. Integrating out the group-specific dimensions, the probability of the response pattern conditional on the primary dimensions alone may be written as

$$\begin{aligned} f(\boldsymbol{u}_p|\boldsymbol{\eta}; \boldsymbol{\gamma}) &= \int \cdots \int \left[\prod_{s=1}^{S} \prod_{i \in \mathfrak{H}_s} f(u_{pi}|\boldsymbol{\eta}, \xi_s; \boldsymbol{\gamma}) \right] h(\xi_1|\boldsymbol{\eta}; \boldsymbol{\gamma}) \cdots h(\xi_S|\boldsymbol{\eta}; \boldsymbol{\gamma}) d\xi_1 \cdots d\xi_S \\ &= \prod_{s=1}^{S} \int \prod_{i \in \mathfrak{H}_s} f(u_{pi}|\boldsymbol{\eta}, \xi_s; \boldsymbol{\gamma}) h(\xi_s|\boldsymbol{\eta}; \boldsymbol{\gamma}) d\xi_s, \end{aligned} \tag{25.7}$$

where the integral on the first line is an S-fold multiple integral, and the two-tier restriction is utilized to reduce the integrals into a series of one-dimensional integrals on the second line. Finally, the marginal probability of the response pattern is

$$f(\boldsymbol{u}_p;\boldsymbol{\gamma}) = \int\left[\prod_{s=1}^{S}\int\prod_{i\in\mathfrak{H}_s} f(u_{pi}|\boldsymbol{\eta},\xi_s;\boldsymbol{\gamma})h(\xi_s|\boldsymbol{\eta};\boldsymbol{\gamma})d\xi_s\right]h(\boldsymbol{\eta};\boldsymbol{\gamma})d\boldsymbol{\eta}, \tag{25.8}$$

where $h(\boldsymbol{\eta};\boldsymbol{\gamma})$ is the distribution of the primary dimensions, and it is often conveniently assumed to be multivariate Gaussian. Note that in all cases, the distributions of the latent variables may depend on the vector of free parameters $\boldsymbol{\gamma}$, so that means, variances, and correlations may be estimated simultaneously with the item parameters.

The integrals in Equation 25.8 do not in general have closed-form solutions. They may, however, be numerically approximated to arbitrary precision using a set of Q-point quadrature rules as follows:

$$f(\mathbf{u}_p;\boldsymbol{\gamma}) \approx \sum_{q_1=1}^{Q}\cdots\sum_{q_G=1}^{Q}\left[\prod_{s=1}^{S}\sum_{j_s=1}^{Q}\prod_{i\in\mathfrak{H}_s} f(u_{pi}|\,Y_{q_1},\ldots,Y_{q_G},X_s;\boldsymbol{\gamma})W(X_{j_s}|\,Y_{q_1},\ldots,Y_{q_G})\right]W(Y_{q_1})\ldots W(Y_{q_G}),$$

where $Y_{q_1},\ldots,Y_{q_G}$ is a set of direct-product quadrature nodes covering the primary dimensions and X_{j_s} belongs to the quadrature nodes covering group-specific dimension s. The corresponding quadrature weights are given by the $W(\cdot)$ functions. In the case of Guassian latent variable distributions, Gauss–Hermite quadrature is a standard choice.

The marginal log-likelihood of $\boldsymbol{\gamma}$ for a sample of P respondents is therefore

$$L\left(\boldsymbol{\gamma}|\{\boldsymbol{u}_p\}_{p=1}^{P}\right) = \sum_{p=1}^{P}\log f(\boldsymbol{u}_p;\boldsymbol{\gamma}), \tag{25.9}$$

where $\{\boldsymbol{u}_p\}_{p=1}^{P}$ collects the responses from all respondents. The marginal log-likelihood can be optimized using standard algorithms such as the Bock–Aitkin expectation maximization (EM) algorithm (Aitkin, Volume Two, Chapter 12; Bock and Aitkin, 1981). Let the maximum marginal likelihood estimator be denoted as $\hat{\boldsymbol{\gamma}}$. The covariance matrix of $\hat{\boldsymbol{\gamma}}$ can be computed using the supplemented EM algorithm (Cai, 2008), which facilitates the construction of Wald tests and asymptotic confidence intervals for parameter estimates. The maximized log-likelihoods can be used for standard inferential purposes such as likelihood ratio testing of nested models. The individual scores can be computed using either maximum *a posteriori* or expected *a posteriori* methods after estimates of $\boldsymbol{\gamma}$ are found. Cai (2010c) provides detailed scoring formulas. Alternatively, if summed score based IRT scoring is desired, Cai (2011) proposed an extended Lord–Wingersky algorithm (Lord and Wingersky, 1984) for the two-tier model that also actively utilized dimension reduction to produce a summed score to IRT scaled score translation tables. As of this writing, parameter estimation and individual scoring algorithms utilizing two-tier dimension reduction are implemented in IRTPRO (Cai et al., 2011a) and flexMIRT® (Cai, 2012).

25.4 Model Fit

When all items are dichotomous and the number of items is not large, full-information model fit test statistics such as Pearson's X^2 and the likelihood ratio G^2 may be used to test the model against the general multinomial alternative. The standard asymptotic reference distribution of these goodness-of-fit statistics is central chi-square with degrees of freedom equal to the number of response pattern cross-classifications (2^I in the case of dichotomous items) minus the number of free parameters minus 1. When a test consists of perhaps more than two dozens of dichotomous items, or if it involves more than a handful of polytomous items, for example, Likert-type responses, the number of response pattern cross-classifications can very quickly exceed the number of respondents by several orders of magnitude. This is because the number of response patterns increases exponentially as the number of items increases. When the number of response pattern cross-classifications is large, the underlying contingency table becomes sparse. A consequence of the sparseness is that the asymptotic chi-squaredness of the full-information model fit test statistics breaks down. This is a well-known issue in categorical data analysis (e.g., Bartholomew and Tzamourani, 1999).

Recently, limited-information test statistics (Bartholomew and Leung, 2002; Cai et al., 2006; Maydeu-Olivares and Joe, 2005) have been suggested as successful alternatives to the full-information test statistics. These statistics typically utilize lower-order marginal residuals of the underlying response pattern cross-classification table, for example, first-order and second-order marginal residuals. Even under extreme sparseness within the full contingency table, limited-information tests can maintain their asymptotic chi-square approximations. They are also potentially more powerful than the full-information tests against model misspecification.

A particular limited-information test statistic, Maydeu-Olivares and Joe's (2005) M_2 is found to be especially effective for testing the fit of unidimensional IRT models. M_2 uses first- and second-order marginal residuals. It is asymptotically chi-squared with degrees of freedom equal to the total number of residuals used for testing minus the number of free parameters estimated. Based on M_2, indices such as the root mean square error of approximation coefficient (Browne and Cudeck, 1993) can be developed for latent variable modeling of categorical data, entirely analogous to the use of such indices in confirmatory factor analysis or structural equation modeling for continuous outcome variables.

Cai and Hansen (2013) extended the logic of limited-information goodness of fit testing to the case of hierarchical models, which include the item bifactor and two-tier models. They utilized these models' unique structure to conduct analogous analytic dimension reduction for model fit testing computations as in parameter estimation, resulting in the ability to efficiently test high-dimensional item factor models. Furthermore, they derived a test statistic based on M_2 that further reduces the marginal residuals if the categories are ordered. This new statistic is substantially better calibrated than the original M_2 statistic and is even more powerful. The M_2 statistic for two-tier models is implemented in flexMIRT (Cai, 2012).

For diagnosing the sources of model misspecification, univariate (first-order) and bivariate (second-order) residuals of the type discussed by Chen and Thissen (1997) can be useful for further decomposing a significant M_2 overall model fit chi-square. Large residuals could indicate the presence of an item or blocks of items that significantly deviate from the assumed model structure and assumptions. Using the extended Lord–Wingersky algorithm for two-tier models, Cai (2011) reasoned that item fit indices analogous to $S - X^2$ proposed by Orlando and Thissen (2000) may be computed, and the extent to which the item response functions fit the empirical item characteristics can be evaluated. These residual-based indices are routinely printed by IRTPRO (Cai et al., 2011a) and flexMIRT (Cai, 2012) in the output.

25.5 Empirical Example

The empirical data came from the Context of Adolescent Substance Use study (the Context study). The goal of the Context study was to investigate the intrapersonal and contextual factors that influence adolescent tobacco, alcohol, and drug use (Ennett et al., 2008). Adolescents enrolled in three public school systems in North Carolina were surveyed. The full longitudinal study involved several waves of data collection. For this illustration, only data from waves 1 and 2 were used.

A four-item subset of a health risk scale that measures psychological distress was used in this analysis. Both Hill (2006) and Cai (2010c) had analyzed this data set earlier, showing that these four items were sufficiently unidimensional within an occasion, and applied the two-tier model for longitudinal data (refer to Figure 25.1) to these data. This analysis extended the previous studies by showing that one could readily test longitudinal measurement invariance using the two-tier modeling framework. Overall goodness of fit test was also conducted to evaluate the absolute fit of the two-tier model. All analyses were conducted in flexMIRT with the Bock–Aitkin EM algorithm. The number of quadrature points per dimension was equal to 21, equally spaced between −5.0 and +5.0. The E-step convergence criterion was equal to 0.0001.

The item stems ware as follows: "I had trouble getting my breath"; "I got mad easily"; "I often worried about bad things happening to me"; "I was a bad person." The original responses were on the Likert response scale. For simplicity of presentation, the responses were dichotomized. The Strongly Agree, Agree, and Neither categories are recoded into an endorsement category, and the remaining two (Disagree and Strongly Disagree) indicated non-endorsement. A total of $P = 1200$ complete cases was drawn randomly from the original database. Each record contained eight items: four for the first wave and four for the second wave. A two-tier model for these items would have two primary dimensions, presenting the psychological distress latent variable at waves 1 and 2, respectively. In addition, there would be four item-specific dimensions representing residual dependence. Consequently, after applying dimension reduction, the actual dimensionality of the model was equal to three.

Table 25.1 presents the parameter estimates assuming item invariance across time points. Each item had a single intercept parameter, as shown in Equation 25.1, and these were set equal for the same item across the two waves. So were the slope parameters on the primary dimensions. For identification, the residual dependence (group-specific) slopes were also set equal. This amounted to estimating a residual correlation for the same item across the two time points. The group-specific dimensions had zero means and unit variances. They were orthogonal to the two (correlated) primary dimensions. Setting the primary dimension at wave 1 to have zero mean and unit variance, the equality restrictions enabled the estimation of the primary factor mean and variance at wave 2, as well as the covariance. In total, there were 15 free parameters.

TABLE 25.1

Parameter Estimates and Standard Errors for the Two-Tier Model Applied to Longitudinal Item Responses from the Context Study (All Items Invariant)

			Primary Slope		Group-Specific Slope			
Wave	**Item**	**Intercept**	**1**	**2**	**1**	**2**	**3**	**4**
1	1	−1.91 (0.16)	2.20 (0.19)		0.98 (0.22)			
1	2	0.37 (0.12)	2.65 (0.25)			1.37 (0.22)		
1	3	−1.13 (0.17)	3.15 (0.34)				1.49 (0.26)	
1	4	−3.28 (0.30)	3.87 (0.40)					0.45 (0.25)
2	1	−1.91 (0.16)		2.20 (0.19)	0.98 (0.22)			
2	2	0.37 (0.12)		2.65 (0.25)		1.37 (0.22)		
2	3	−1.13 (0.17)		3.15 (0.34)			1.49 (0.26)	
2	4	−3.28 (0.30)		3.87 (0.40)				0.45 (0.25)
Factor means			0.00 (—)	0.18 (0.04)	0.00 (—)	0.00 (—)	0.00 (—)	0.00 (—)
Factor covariance matrix			1.00 (—)					
			0.57 (0.06)	1.15 (0.13)				
			0.00 (—)	0.00 (—)	1.00 (—)			
			0.00 (—)	0.00 (—)	0.00 (—)	1.00 (—)		
			0.00 (—)	0.00 (—)	0.00 (—)	0.00 (—)	1.00 (—)	
			0.00 (—)	0.00 (—)	0.00 (—)	0.00 (—)	0.00 (—)	1.00 (—)

Note: The empty slope parameter entries are fixed to zero. Standard errors are shown in parentheses. Fixed factor mean and variance/covariance parameters are indicated by dashes.

One can see that the items loaded strongly on the primary dimensions. The slope estimates were all several times the magnitude of the standard error estimates. The residual dependence also appeared to be strong. The mean of the psychological distress factor at wave 2 appeared to have shifted upward about 0.18 of a standard deviation unit from wave 1. This shift was also statistically significant from zero, as indicated by the small standard error of 0.04. The variance of the psychological distress factor had not changed much (1.00 vs. 1.15 with an *SE* of 0.13). The covariance was 0.57, which can be converted into a correlation of 0.53. This indicates that just under 30% of the variability was shared across time.

The two-tier model with six-dimensions took a total of 2 min and 31 s to estimate on a PC laptop with Intel i7 CPU running 64-bit Windows 7 (this includes standard error and goodness of fit testing computations). The total number of response patterns was actually small enough in this case ($2^8 = 256$) that full-information model fit statistics could be computed and examined. The degrees of freedom were equal to $256 - 15 - 1 = 240$. The likelihood ratio G^2 statistic was equal to 264.7 ($p = 0.13$) and the Pearson X^2 is 292.9 ($p = 0.01$). At the usual 0.05 alpha level, the evidence from full-information tests was inconclusive as Pearson's statistic would reject the model and the likelihood ratio statistic would retain it. On the other hand, M_2 was equal to 46, on 21 degrees of freedom, $p = 0.001$. Previous simulations (e.g., Cai et al., 2006) suggest that in these cases the Pearson statistic and the limit-information statistic are more powerful. The highly significant M_2 indicated that the model could still be improved. On the other hand, a two-factor model that did not include the residual dependence dimensions resulted in a likelihood ratio G^2 of 328. Since the two-factor model was nested within the two-tier model, the difference in G^2 was an indicator of the plausibility of the restrictions imposed by the two-factor model on the two-tier model, that is, that residual dependence could be safely ignored. The two models differed by four parameters. The difference in G^2 was equal to 63, highly significant on four degrees of freedom. The lack of residual correlations in the two-factor model represented implausible restrictions.

The second two-tier model reported here relaxed the measurement invariance assumption imposed by the two-tier model just fitted. In the results shown in Table 25.1, the item parameters (intercepts and primary slopes) were all set equal across the same item administered at the two time points. Inspection of univariate residuals suggests that item 2 may exhibit a lack of measurement invariance. Consequently, in the second two-tier model, the item parameter invariance restrictions were relaxed for item 2. The intercepts and the primary slopes for item 2 were allowed to be freely estimated, taking occasion-specific values. Table 25.2 presents the estimates and standard errors for the second two-tier model.

As one can tell from Table 25.2, the intercepts of item 2 at waves 1 and 2 differed quite a bit. This may have contributed to the observation that the mean difference across waves dropped to 0.07 ($SE = 0.05$), which was no longer statistically significantly different from 0. There were 17 parameters in this model. The likelihood ratio G^2 statistic of this model was equal to 243.7 ($p = 0.39$) and the Pearson X^2 is 264.3 ($p = 0.11$), both on 238 degrees of freedom. The first two-tier model was nested within this model. The likelihood ratio comparison between the two models was a formal test of the measurement invariance assumption imposed by the first model. The difference in G^2 was equal to $264.7 - 243.7 = 21$, also highly

TABLE 25.2

Parameter Estimates and Standard Errors for the Two-Tier Model Applied to Longitudinal Item Responses from the Context Study (Invariance Relaxed for Item 2)

			Primary Slope		Group-Specific Slope			
Wave	**Item**	**Intercept**	**1**	**2**	**1**	**2**	**3**	**4**
1	1	−1.82 (0.15)	2.10 (0.19)		1.02 (0.21)			
1	2	0.16 (0.13)	2.68 (0.35)			1.45 (0.42)		
1	3	−0.97 (0.16)	2.98 (0.31)				1.51 (0.25)	
1	4	−3.06 (0.32)	3.63 (0.42)					0.42 (0.53)
2	1	−1.82 (0.15)		2.10 (0.19)	1.02 (0.21)			
2	2	0.91 (0.18)		2.37 (0.38)		1.45 (0.42)		
2	3	−0.97 (0.16)		2.98 (0.31)			1.51 (0.25)	
2	4	−3.06 (0.32)		3.63 (0.42)				0.42 (0.53)
Factor means			0.00 (—)	0.07 (0.05)	0.00 (—)	0.00 (—)	0.00 (—)	0.00 (—)
Factor covariance matrix			1.00 (—)					
			0.65 (0.06)	1.50 (0.21)				
			0.00 (—)	0.00 (—)	1.00 (—)			
			0.00 (—)	0.00 (—)	0.00 (—)	1.00 (—)		
			0.00 (—)	0.00 (—)	0.00 (—)	0.00 (—)	1.00 (—)	
			0.00 (—)	0.00 (—)	0.00 (—)	0.00 (—)	0.00 (—)	1.00 (—)

Note: The empty slope parameter entries are fixed to zero. Standard errors are shown in parentheses. Fixed factor mean and variance/covariance parameters are indicated by dashes.

significant on two degrees of freedom. Thus, the invariance assumption had to be rejected. On an absolute basis, the second model fitted the data. M_2 was equal to 27.3 on 19 degrees of freedom ($p = 0.10$).

25.6 Discussion

Just as with the item bifactor model, the two-tier model belongs to the general hierarchical item factor model family. The restrictions in this model seek a balance between the goal of typical assessment activities, which is to extract a small enough number of target dimensions to measure, and the reality of data analysis—that items may be locally dependent for a host of intentional or unintended reasons. The local dependence (in the forms of item doublets or triplets) has historically been a nemesis of full-information item factor analysis methods. By employing dimension reduction, the two-tier model can efficiently handle residual dependence, introduced in a number of contexts as illustrated earlier. Recent work on the two-tier modeling framework has led to the development of efficient model fit testing methods and scoring algorithms that employ dimension reduction as well.

The two-tier model has a number of limitations. First, not all residual dependence can be efficiently modeled by the single layer of orthogonal group-specific dimensions. The graphical modeling approach developed by Rijmen (2009) offers alternative derivations that can automatically simplify the marginal likelihood computations for a more general class of hierarchical item factor models. Second, the latent variables are typically assumed to be normally distributed in the two-tier model. This is for the most part driven by its mathematical simplicity. Recent work by Woods and Thissen (2006) offers new strategies to relax this restrictive assumption by adopting spline latent densities. Fully multidimensional implementations of semi-nonparametric density estimation in the two-tier model would be a highly desirable extension. Last but not the least, while the two-tier model has some degree of built-in flexibility, it is nevertheless a highly restricted confirmatory item factor model. For the factor structures that would "break" the two-tier restriction, alternative calibration and scoring methods are required. The emphasis on numerical quadrature and standard optimization methods in this chapter should not be interpreted as an unconditional preference in practice. Nor should the reader be timid in specifying more complex measurement models, as simulation-based estimation methods have matured and can handle the complexities. If the two-tier model should add anything to the repertoire of IRT researchers, the concept of dimension reduction probably ought to be highlighted as a useful general principle to adopt when deriving models and estimation algorithms.

Acknowledgments

Part of this research was supported by the Institute of Education Sciences (R305B080016 and R305D100039) and the National Institute on Drug Abuse (R01DA026943 and R01DA030466). The views expressed here belong to the author and do not reflect the views or policies of the funding agencies. The author is grateful for Dr. Susan Ennett's permission to use data from the Context project for the numerical illustrations and wishes to thank Mark Hansen for helpful feedback on an earlier draft.

References

Adams, R. and Wu, M. 2002. PISA 2000 Technical Report. Paris: Organizaton for Economic Cooperation and Development.

Bartholomew, D. J. and Leung, S. O. 2002. A goodness of fit test for sparse 2_p contingency tables. *British Journal of Mathematical and Statistical Psychology, 55*, 1–15.

Bartholomew, D. J. and Tzamourani, P. 1999. The goodness-of-fit of latent trait models in attitude measurement. *Sociological Methods and Research, 27*, 525–546.

Bock, R. D. and Aitkin, M. 1981. Marginal maximum likelihood estimation of item parameters: Application of an EM algorithm. *Psychometrika, 46*, 443–459.

Bock, R. D., Gibbons, R., and Muraki, E. 1988. Full-information item factor analysis. *Applied Psychological Measurement, 12*, 261–280.

Browne, M. W. and Cudeck, R. 1993. Alternative ways of assessing model fit. In K. Bollen and J. Long (Eds.), *Testing Structural Equation Models* (pp. 136–162). Newbury Park, California: Sage.

Cai, L. 2008. SEM of another flavour: Two new applications of the supplemented EM algorithm. *British Journal of Mathematical and Statistical Psychology, 61*, 309–329.

Cai, L. 2010a. High-dimensional exploratory item factor analysis by a Metropolis–Hastings Robbins–Monro algorithm. *Psychometrika, 75*, 33–57.

Cai, L. 2010b. Metropolis–Hastings Robbins–Monro algorithm for confirmatory item factor analysis. *Journal of Educational and Behavioral Statistics, 35*, 307–335.

Cai, L. 2010c. A two-tier full-information item factor analysis model with applications. *Psychometrika, 75*, 581–612.

Cai, L. 2011. Lord–Wingersky algorithm after 25+ Years: Version 2.0 for hierarchical item factor models. Unpublished Paper Presented at the *2011 International Meeting of the Psychometric Society*, Hong Kong, China.

Cai, L. 2012. *flexMIRT®: Flexible Multilevel Item Factor Analysis and Test Scoring* [Computer software]. Seattle, Washington: Vector Psychometric Group, LLC.

Cai, L. and Hansen, M. 2013. Limited-information goodness-of-fit testing of hierarchical item factor models. *British Journal of Mathematical and Statistical Psychology, 66*, 245–276.

Cai, L., Maydeu-Olivares, A., Coffman, D. L., and Thissen, D. 2006. Limited-information goodness-of-fit testing of item response theory models for sparse 2^P tables. *British Journal of Mathematical and Statistical Psychology, 59*, 173–194.

Cai, L., Thissen, D., and du Toit, S. H. C. 2011a. *IRTPRO: Flexible, Multidimensional, Multiple Categorical IRT Modeling* [Computer software]. Lincolnwood, Illinois: Scientific Software International, Inc.

Cai, L., Yang, J. S., and Hansen, M. 2011b. Generalized full-information item bifactor analysis. *Psychological Methods, 16*, 221–248.

Chen, W. H. and Thissen, D. 1997. Local dependence indices for item pairs using item response theory. *Journal of Educational and Behavioral Statistics, 22*, 265–289.

Edwards, M. C. 2010. A Markov chain Monte Carlo approach to confirmatory item factor analysis. *Psychometrika, 75*, 474–497.

Ennett, S. T., Foshee, V. A., Bauman, K. E., Hussong, A. M., Cai, L., Luz, H. et al. 2008. The social ecology of adolescent alcohol misuse. *Child Development, 79*, 1777–1791.

Gibbons, R. D., Bock, R. D., Hedeker, D., Weiss, D. J., Segawa, E., Bhaumik, D.K. et al. 2007. Full-information item bifactor analysis of graded response data. *Applied Psychological Measurement, 31*, 4–19.

Gibbons, R. D. and Hedeker, D. 1992. Full-information item bifactor analysis. *Psychometrika, 57*, 423–436.

Glas, C. A. W., Wainer, H., and Bradlow, E. T. 2000. Maximum marginal likelihood and expected a posteriori estimation in testlet-based adaptive testing. In W. J. van der Linden and C. A. W. Glas (Eds.), *Computerized Adaptive Testing: Theory and Practice* (pp. 271–288). Boston, Massachusetts: Kluwer Academic.

Hill, C. D. 2006. Two models for longitudinal item response data. Unpublished doctoral dissertation, Department of Psychology, University of North Carolina, Chapel Hill, North Carolina.

Holzinger, K. J. and Swineford, F. 1937. The bi-factor method. *Psychometrika, 2,* 41–54.

Jeon, M., Rijmen, F., and Rabe-Hesketh, S. 2013. Modeling differential item functioning using a generalization of the multiple-group bifactor model. *Journal of Educational and Behavioral Statistics, 38,* 32–60.

Li, Y., Bolt, D. M., and Fu, J. 2006. A comparison of alternative models for testlets. *Applied Psychological Measurement, 30,* 3–21.

Lord, F. M. and Wingersky, M. S. 1984. Comparison of IRT true-score and equipercentile observed-score "equatings." *Applied Psychological Measurement, 8,* 453–461.

Maydeu-Olivares, A., and Joe, H. 2005. Limited and full information estimation and testing in 2^n contingency tables: A unified framework. *Journal of the American Statistical Association, 100,* 1009–1020.

Orlando, M. and Thissen, D. 2000. New item fit indices for dichotomous item response theory models. *Applied Psychological Measurement, 24,* 50–64.

Reckase, M. D. 2009. *Multidimensional Item Response Theory.* New York: Springer.

Reise, S. P. 2012. The rediscovery of bifactor measurement models. *Multivariate Behavioral Research, 47,* 667–696.

Rijmen, F. 2009. Efficient full information maximum likelihood estimation for multidimensional IRT models. Technical Report No. RR-09-03. Educational Testing Service. Available at: https://www.ets.org/Media/Research/pdf/RR-09-03.pdf

Rijmen, F. 2010. Formal relations and an empirical comparison between the bi-factor, the testlet, and a second-order multidimensional IRT model. *Journal of Educational Measurement, 47,* 361–372.

Rijmen, F., Vansteelandt, K., and De Boeck, P. 2008. Latent class models for diary method data: Parameter estimation by local computations. *Psychometrika, 73,* 167–182.

Schilling, S. and Bock, R. D. 2005. High-dimensional maximum marginal likelihood item factor analysis by adaptive quadrature. *Psychometrika, 70,* 533–555.

Schmid, J. and Leiman, J. M. 1957. The development of hierarchical factor solutions. *Psychometrika, 22,* 53–61.

Thissen, D., Cai, L., and Bock, R. D. 2010. The nominal categories item response model. In M. Nering and R. Ostini (Eds.), *Handbook of Polytomous Item Response Theory Models: Developments and Applications* (pp. 43–75). New York: Taylor & Francis.

Thissen, D., Varni, J. W., Stucky, B. D., Liu, Y., Irwin, D. E., and DeWalt, D. A. 2011. Using the PedsQL™ 3.0 asthma module to obtain scores comparable with those of the PROMIS pediatric asthma impact scale (PAIS). *Quality of Life Research, 20,* 1497–1505.

Thurstone, L.L. 1947. *Multiple-Factor Analysis.* Chicago, Illinois: University of Chicago Press.

Vasdekis, V. G. S., Cagnone, S., and Moustaki, I. 2012. A composite likelihood inference in latent variable models for ordinal longitudinal responses. *Psychometrika, 77,* 425–441.

Wainer, H., Bradlow, E. T., and Wang, X. 2007. *Testlet Response Theory and Its Applications.* New York: Cambridge University Press.

Wirth, R. J. and Edwards, M. C. 2007. Item factor analysis: Current approaches and future directions. *Psychological Methods, 12,* 58–79.

Woods, C. M. and Thissen, D. 2006. Item response theory with estimation of the latent population distribution using spline-based densities. *Psychometrika, 71,* 281–301.

Wu, E. J. C. and Bentler, P. M. 2011. *EQSIRT: A User-Friendly IRT Program* [Computer software]. Encino, California: Multivariate Software, Inc.

Yung, Y. F., McLeod, L. D., and Thissen, D. 1999. On the relationship between the higher-order factor model and the hierarchical factor model. *Psychometrika, 64,* 113–128.

26

Item-Family Models

Cees A. W. Glas, Wim J. van der Linden, and Hanneke Geerlings

CONTENTS

26.1 Introduction 437
26.2 Presentation of the Models 438
26.2.1 General Model Formulation 438
26.2.2 Possible Restrictions on the Model 439
26.3 Parameter Estimation 440
26.4 Model Fit 440
26.5 Optimal Test Design 441
26.6 Empirical Example 442
26.6.1 Setup of the Study 443
26.6.2 Results 444
26.7 Discussion 446
References 447

26.1 Introduction

For a test-item pool to support a variety of applications, a necessary requirement is that the items cover its subject area for a broad range of difficulty. Given the heavy burden placed by the requirement on the resources typically available for item-pool development, the use of item cloning techniques has become attractive. For textual items, such as word problems and cloze items, item cloning may involve random selection from replacement sets for some insignificant elements of an item template (Millman and Westman, 1989) or the use of more advanced linguistic rules (Bormuth, 1970). The result of item cloning are sets of items differing in surface features only, which can be assumed to have similar psychometric properties. Such sets are referred to as item families.

Knowledge of the family structures allows us to generalize item properties across entire families. Statistical support for such generalization in the context of item response modeling is obtained by modeling item parameters as random variables. For instance, possibly upon transformation of some of them, the item parameters could be assumed to be multivariate normally distributed within each family with mean vector and covariance matrix that could be estimated from a sample from them. Each individual item parameter is then equal to its mean family parameter plus an item-specific deviation from the mean. If the deviations are small, the family parameters can be used for newly generated items from the same family; for instance, to adapt their selection to the test taker's ability estimates during adaptive testing from a pool of families.

Although the focus of this chapter is on item families generated by item cloning, the models discussed are more generally applicable to cases in which item parameters can be clustered based on prior knowledge about them. For example, Janssen et al. (2000) assumed item clustering based on the educational standards they were designed to measure. Similarly, Fox and Verhagen (2010) considered the case of clustering country-specific item parameters in cross-national assessments. As such clusters allow for differences in item behavior across countries, their use of family models with random item parameters may provide a solution to the problem of international score comparability.

26.2 Presentation of the Models

The item-family models presented in this chapter are examples of multilevel IRT models. However, they differ from the models with the multilevel structures for the ability parameter reviewed by Fox and Glas (Volume One, Chapter 24) in that our current focus is on random structures for the item parameters only.

26.2.1 General Model Formulation

At the first level, item-family models consist of a standard response model for the individual items, while their second level specifies a structural model for the families. As first-level models, Glas and van der Linden (2003) used the three-parameter logistic model for dichotomous items (van der Linden, Volume One, Chapter 2), Janssen et al. (2000) adopted the two-parameter normal-ogive model (Lord and Novick, 1968), whereas Geerlings et al. (2011) worked with a reparameterization of the two-parameter normal-ogive model suggested by Albert (1992), which allowed them to simplify estimation in a Bayesian framework. For polytomous items, Johnson and Sinharay (2005) used the generalized partial credit model (Muraki and Muraki, Volume One, Chapter 8) as their first-level model. The following sections focus on the dichotomous case. As all these models are already reviewed in other chapters of this handbook, their basic equations are not repeated here.

In addition to regular notation for the model parameters, we need an index for family membership; we use $i_f = 1, \ldots, I_f$ to indicate that item i belongs to family $f = 1, \ldots, F$ of size I_f. As already noted, to accommodate the assumption of a multivariate normal distribution as second-level model, it may be necessary to transform the item discrimination, difficulty, and guessing parameters; for instance, using

$$\boldsymbol{\xi}_{i_f} = (\alpha_{i_f} = \log a_{i_f}, b_{i_f}, \gamma_{i_f} = \text{logit } c_{i_f}) \tag{26.1}$$

The distribution of the item parameters within family f is then given by

$$\boldsymbol{\xi}_{i_f} \sim MVN(\boldsymbol{\mu}_f, \boldsymbol{\Sigma}_f) \tag{26.2}$$

where $\boldsymbol{\mu}_f = (\mu_{\alpha_f}, \mu_{b_f}, \mu_{\gamma_f})$ is the vector with the mean parameters for family f and covariance matrix

$$\boldsymbol{\Sigma}_f = \begin{bmatrix} \sigma^2_{\alpha_f} & \sigma_{\alpha b_f} & \sigma_{\alpha\gamma_f} \\ \sigma_{\alpha b_f} & \sigma^2_{b_f} & \sigma_{b\gamma_f} \\ \sigma_{\alpha\gamma_f} & \sigma_{b\gamma_f} & \sigma^2_{\gamma_f} \end{bmatrix} \quad (26.3)$$

describes its variability. Note that the item parameters are allowed to correlate within each family, which is not uncommon to occur in practice (Wingersky and Lord, 1984).

26.2.2 Possible Restrictions on the Model

For explanatory purposes, or simply for the sake of parsimony, it may be helpful to restrict the general model formulation above.

First, knowledge of the rules used to generate the item families may be used to restrict the family difficulty parameters to a sum of effect parameters:

$$\mu_{b_f} = \sum_{r=1}^{R} d_{fr}\beta_r \quad (26.4)$$

where $\boldsymbol{D} = (d_{fr})$ is a design matrix specifying which families have been constructed by rules $r = 1, \ldots, R$ with β_r representing the effect of rule r. The size of the β_r parameters can be used to assess which of them are the most important determinants of the family difficulties. Generally, the literature on rule-based item generation (e.g., Irvine, 2002) makes a distinction between two types of rules: those that have a larger fixed effect on the difficulty of the items in a family (radicals), and others that create minor random variation within families only (incidentals). The radicals can be modeled by the effect parameters β_r, whereas the incidental effects are reflected by the covariance matrix Σ_f. Although the effect of rules on family discrimination is generally difficult to predict, equations similar to Equation 26.4 can be considered for these parameters as well. Alternatively, it may be possible to impose equality restrictions on the means of the discrimination parameters of some of the families.

Second, if the within-family variability can be assumed to be rather homogeneous over families, Equation 26.2 can be replaced by

$$\xi_{i_f} \sim MVN(\boldsymbol{\mu}_f, \boldsymbol{\Sigma}) \quad (26.5)$$

where $\boldsymbol{\Sigma}$ is a covariance matrix common to all families (Geerlings et al., 2011). If, in addition, the item parameters are assumed not to correlate within each family, Equation 26.5 reduces to the model for criterion-referenced measurement by Janssen et al. (2000), who assumed a two-parameter first-level model and modeled their second level as

$$a_{i_f} \sim N(\mu_{a_f}, \sigma^2_a) \quad (26.6)$$

$$b_{i_f} \sim N(\mu_{b_f}, \sigma^2_b) \quad (26.7)$$

Finally, if the within-family variability is deemed to be ignorable altogether, the model further reduces to the identical siblings model by Sinharay et al. (2003), in which all items in the same family have identical parameters and, in fact, the model no longer has a multilevel structure.

26.3 Parameter Estimation

Although the estimation of the models in Section 26.2 in a frequentist framework is feasible (Glas and van der Linden, 2003), the Bayesian framework has been more popular (Fox and Verhagen, 2010; Geerlings et al., 2011; Janssen et al., 2000; Sinharay et al., 2003). The main reason for this difference in popularity is the possibility of Gibbs sampling, which divides the larger, complex estimation problem into smaller, simpler pieces that can be solved iteratively (Junker, Patz, and Vanhoudnos, Volume Two, Chapter 15; Sinharay and Johnson, Volume Two, Chapter 13). In each iteration of the procedure, one sample is drawn from each conditional posterior distribution of a (subset of) parameter(s) given all others. Upon convergence, the resulting samples are from the joint posterior distribution of the parameters (see, for example, Casella and George, 1992). The subsets of parameters are often chosen such that conjugate priors can be specified and the conditional posterior distributions have standard forms. For example, Glas et al. (2010) specified a normal inverse Wishart prior distribution for $(\boldsymbol{\mu}_f, \boldsymbol{\Sigma}_f)$ resulting in a normal inverse Wishart conditional posterior distribution. A standard normal distribution is often chosen as a prior for $\boldsymbol{\theta}$ to identify the model. Gibbs sampling algorithms for the different models above can be found in Geerlings (2012, Appendix A), Glas et al. (2010), and Janssen et al. (2000).

Once the family parameters $\boldsymbol{\mu} = (\boldsymbol{\mu}_f)$ and $\boldsymbol{\Sigma} = (\boldsymbol{\Sigma}_f)$ have been estimated, the expected a posteriori (EAP) estimate of the ability of test taker p, θ_p, given his or her responses $\boldsymbol{u}_p = (u_{i_f p})$, is obtained by

$$P(\theta_p \mid \boldsymbol{u}_p, \boldsymbol{\mu}, \boldsymbol{\Sigma}) = \frac{\int \theta_p p(\theta) \prod_{f=1}^{F} \prod_{i_f=1}^{I_f} p_{i_f}(\theta_p)^{u_{i_f p}} \left[1 - p_{i_f}(\theta_p)\right]^{1-u_{i_f p}} d\theta}{\int p(\theta) \prod_{f=1}^{F} \prod_{i_f=1}^{I_f} p_{i_f}(\theta_p)^{u_{i_f p}} \left[1 - p_{i_f}(\theta_p)\right]^{1-u_{i_f p}} d\theta} \tag{26.8}$$

where the probability of a correct response of test taker p on an item from family f is computed by integrating over the item parameters within the family:

$$p_{i_f}(\theta_p) = \int p(U_{i_f p} = 1; \theta_p, \boldsymbol{\xi}_{i_f}) p(\boldsymbol{\xi}_{i_f}; \boldsymbol{\mu}_f, \boldsymbol{\Sigma}_f) d\boldsymbol{\xi}_{i_f} \tag{26.9}$$

In a fully Bayesian framework, the uncertainty about the family parameters can be taken into account by averaging out the item-family parameters with respect to their posterior distribution as well (Sinharay et al., 2003).

26.4 Model Fit

Approaches to assessing the fit of the first-level models are reviewed by Glas (Volume Two, Chapter 17; Volume Three, Chapter 6), Hambleton and Wells (Volume Two, Chapter 20), and Sinharay (Volume Two, Chapter 19). However, given the interest in the family parameters in the current context, model fit analysis should not only focus on the first-level model but address the second-level family distributions as well.

Although second-level model fit has received less attention in the literature on hierarchical IRT, several developments might prove to be particularly useful here. One of them is the use of extended posterior predictive checks (EPPCs, Gelman et al., 2005; Steinbakk and Storvik, 2009), which are especially attractive because of their ease of implementation and the interpretation of their results. A posterior predictive p-value is defined as the probability of a value for a test statistic replicated under the model being more extreme than the one observed (Gelman et al., 1996; Sinharay, Volume Two, Chapter 19). An extended posterior predictive p-value (EPPP) for a test statistic is computed by additional replicated sampling of item parameters from their family distributions. That is,

$$EPPP = P\left(T(\boldsymbol{u}^{rep}, \boldsymbol{\xi}^{rep}, \boldsymbol{\mu}, \boldsymbol{\Sigma}, \boldsymbol{\theta}) > T(\boldsymbol{u}, \boldsymbol{\xi}, \boldsymbol{\mu}, \boldsymbol{\Sigma}, \boldsymbol{\theta}); \boldsymbol{u}\right) \tag{26.10}$$

where $T(.)$ is the given test statistic.

To assess whether the difficulty of family f is well predicted by its radicals, following suggestions by Sinharay and Johnson (2008), Geerlings (2012, Chapter 4) used the variance of the proportions of correct responses for the items in the family as a test statistic—a statistic indicative of within-family variation in the item difficulty parameters. When the radicals do not predict the family difficulty parameters adequately, there will be a discrepancy between their estimates and the empirical means of the item parameters. Consequently, the estimated covariance matrix will be larger and the observed variance of the proportions of correct item responses will be smaller than expected under the model.

Similarly, the variance of the point-biserial correlations for the items per family can be used to assess whether a linear structure on the family discrimination parameters analogous to Equation 26.4 would hold. Geerlings (2012, Chapter 4) applied statistics of this type to several of the models discussed previously for empirical data on families of statistical word problems. Also, similar statistics were previously used by Sinharay (2005) for posterior predictive checks on the fit of the 3PL model. However, additional study is necessary to determine their power to detect misfit.

Gilbride and Lenk (2010) proposed two fit statistics for the assumption of multivariate normality that lend themselves to application in an EPPC to assess its validity for the distribution of the item parameters within families. The first is based on the fact that, if the assumption of normality holds, the squared Mahalanobis distance $(\boldsymbol{\xi}_{i_f} - \boldsymbol{\mu}_f)^T \boldsymbol{\Sigma}_f^{-1} (\boldsymbol{\xi}_{i_f} - \boldsymbol{\mu}_f)$ follows a chi-square distribution with degrees of freedom equal to the dimension of $\boldsymbol{\Sigma}_f$. A Kolmogorov–Smirnov test statistic can be used to assess whether the replicated data are closer to the theoretical chi-square distribution than the observed data. As noted by Gilbride and Lenk (2010), the statistic is particularly sensitive to departures from the tails of the multivariate normal distribution. The second statistic is based on the observation that, under the assumption of multivariate normality, the scores on the first principal component are normal. Therefore, the correlation between the empirical and normal cumulative distribution function of these scores can be used as a test statistic. The statistic is sensitive to deviations from multivariate normality due to multimodality (Gilbride and Lenk, 2010).

26.5 Optimal Test Design

Instead of selecting items from a pregenerated item pool, tests can be constructed on the fly using the family parameters ($\boldsymbol{\mu}_f$, $\boldsymbol{\Sigma}_f$) in the two-staged adaptive procedure described by

Glas and van der Linden (2003): First, an item family is selected to be optimal at the current ability estimate. Second, an item is generated using the radicals defining the family and applying incidental features to create random variation.

Two existing item-selection criteria that have been adapted for the selection of item families are the criterion of minimum expected posterior variance and a posterior weighted information criterion (Glas and van der Linden, 2003). To select the kth family for a test, the former requires the computation of the posterior predictive probability of a response for each family f in the pool given the previous responses $\boldsymbol{u}^{(k-1)}$:

$$p\left(u_{ifp} \mid \boldsymbol{u}_p^{(k-1)}\right)=\int\left[\int p\left(u_{ifp};\theta_p,\boldsymbol{\xi}_{if}\right)p\left(\boldsymbol{\xi}_{if};\boldsymbol{\mu}_f,\boldsymbol{\Sigma}_f\right)d\boldsymbol{\xi}_{if}\right]p\left(\theta_p;\boldsymbol{u}_p^{(k-1)}\right)d\theta_p \tag{26.11}$$

These probabilities are used as weights when calculating the expected value of the posterior variance of θ_p, which selects the kth family in the test as

$$p_k=\arg\min_f\left\{\sum_{U_{ifp}=0}^{1}\mathrm{Var}(\theta_p;\boldsymbol{u}_p^{(k-1)},U_{ifp})p(U_{ifp};\boldsymbol{u}_p^{(k-1)})\right\}. \tag{26.12}$$

Family information is defined as the expected information in the response to a random item from family f on the ability parameter θ_p:

$$I_f(\theta_p)=-E_{u_{ifp}}\left[\frac{\partial^2}{\partial\theta_p^2}\ln\int p(u_{ifp};\theta_p,\boldsymbol{\xi}_{if})p(\boldsymbol{\xi}_{if};\boldsymbol{\mu}_f,\boldsymbol{\Sigma}_f)d\boldsymbol{\xi}_{if}\right] \tag{26.13}$$

The posterior weighted information criterion selects the kth family as

$$p_k=\arg\max_f\left\{\int I_f(\theta_p)p(\theta_p;\boldsymbol{u}^{(k-1)})d\theta_p\right\} \tag{26.14}$$

All integrals in Equations 26.11, 26.13, and 26.14 can be approximated by Monte Carlo integration or Gaussian quadrature. The latter was used in the simulations reported below.

Generally, the larger the uncertainty about the true item parameters within a family f (i.e., the larger $\boldsymbol{\Sigma}_f$), the less the information about the ability parameter in a response to an item (Geerlings et al., 2013). Therefore, the two criteria above tend to prefer families with smaller item-parameter variability. However, items from such families are more likely to look similar, and their use may therefore lead to easy family disclosure. In practice, such tradeoffs are mitigated by adding exposure-control constraints to the test-design process. For example, a family could be given a lower weight with increasing exposure over time.

26.6 Empirical Example

As already noted, entire item families can be calibrated using random samples of items from them. Estimates of the family parameters can then be used to optimally design a test and score its test takers. This procedure is expected to be most advantageous to adaptive testing, where a slight loss in efficiency in ability estimation due to the use of calibrated families instead of individual items is easily compensated by a slightly longer test length.

To give a quantitative impression of the loss in efficiency due to random sampling from families, an example based on simulated data is presented. Four conditions are compared:

1. Random item selection with ability estimated using the item parameters.
2. Optimal item selection based on the minimum expected posterior variance criterion with ability estimated using known first-level item parameters; The minimum expected posterior variance criterion was equal to Equation 26.12, with the model for a correct response $p(u_{ifp};\theta_p,\xi_{if})$ replaced by the three-parameter logistic model for dichotomous items (van der Linden, Volume One, Chapter 2); This condition is equivalent with standard adaptive testing.
3. Random family selection with random generation of one item from each family and ability estimated using the family means and covariance matrices.
4. Optimal family selection based on the minimum expected posterior variance criterion with random generation of one item from each family and ability estimated using the family means and covariance matrices.

26.6.1 Setup of the Study

For all conditions, the three-parameter logistic model was used as first-level response model. For Conditions 3 and 4, the model was extended with the second-level family model in Equation 26.5. In each of the 3000 replications, the mean parameters of either 100 or 400 families were drawn from a multivariate normal distribution with means (log $\mu_a = 0.0$, $\mu_b = 0.0$, logit $\mu_c = -1.5$) and between-family covariance matrix

$$\boldsymbol{\Sigma}_B = \begin{bmatrix} 0.1 & 0.0 & 0.0 \\ 0.0 & 1.5 & 0.0 \\ 0.0 & 0.0 & 0.5 \end{bmatrix}$$

The within-family covariance matrices were set equal to $r = 1.00$, 0.50, or 0.25 times the between-family covariance matrix $\boldsymbol{\Sigma}_B$. The ability parameters were either fixed at −2, −1, 0, 1, 2 or randomly selected from a standard normal distribution.

For each condition either 20 or 40 items were selected according to the following procedures: for the first two conditions, the item parameters were drawn from the between-family multivariate normal distributions defined previously, but these drawn item parameters were considered known for item selection and estimation of ability. An item was selected either randomly (Condition 1) or according to the minimum expected posterior variance criterion (Condition 2). For the last two conditions, a family was selected either randomly (Condition 3) or according to the minimum expected posterior variance criterion (Condition 4). However, to get started, the first three items in Conditions 2 and 4 were always selected randomly as well. For Conditions 3 and 4, once a family was selected, the parameters for the next item were randomly drawn from its multivariate normal distribution. All responses were generated given the true values of the parameters of the selected items and the true ability parameters of the simulated test takers. The EAP estimates of the test takers' abilities were obtained using the item parameters (Conditions 1–2) or the family parameters (Conditions 3–4).

For all four conditions, an item or family could be selected for a test taker at most once. This constraint was to avoid the case of multiple items being selected from one most informative family only. The mean absolute errors (MAEs) in the ability estimates were computed and used to compare the results for the four different conditions.

26.6.2 Results

The two figures show the results for an item pool with 100 (Figure 26.1) and 400 families (Figure 26.2). In each figure, the plots in the left (right) column correspond to a test length n of 20 (40) items while those in the first, second, and third rows correspond to within-between covariance ratios $r = 1.00$, 0.50, and 0.25, respectively. Each plot shows the MAEs

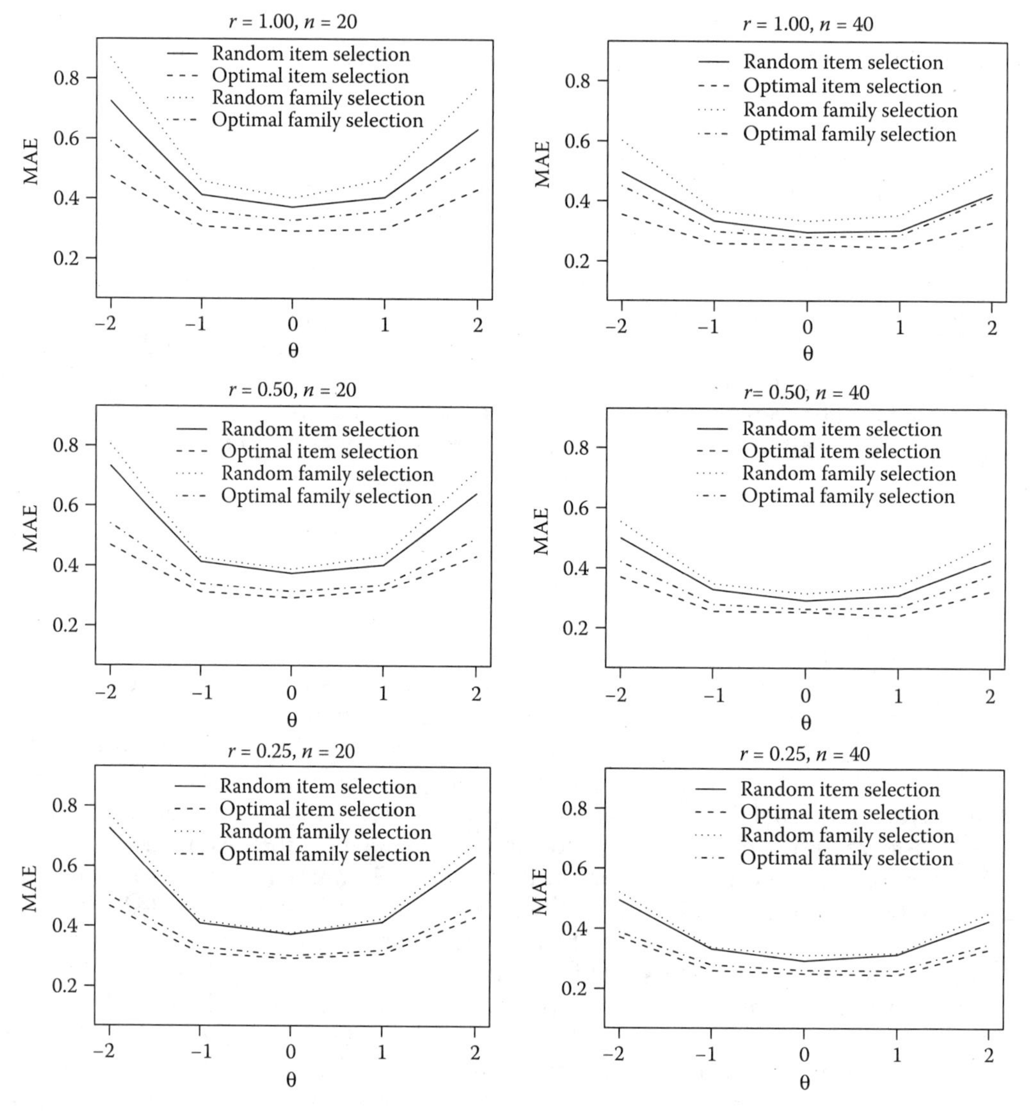

FIGURE 26.1
Mean absolute error in ability estimates for an item pool with 100 families.

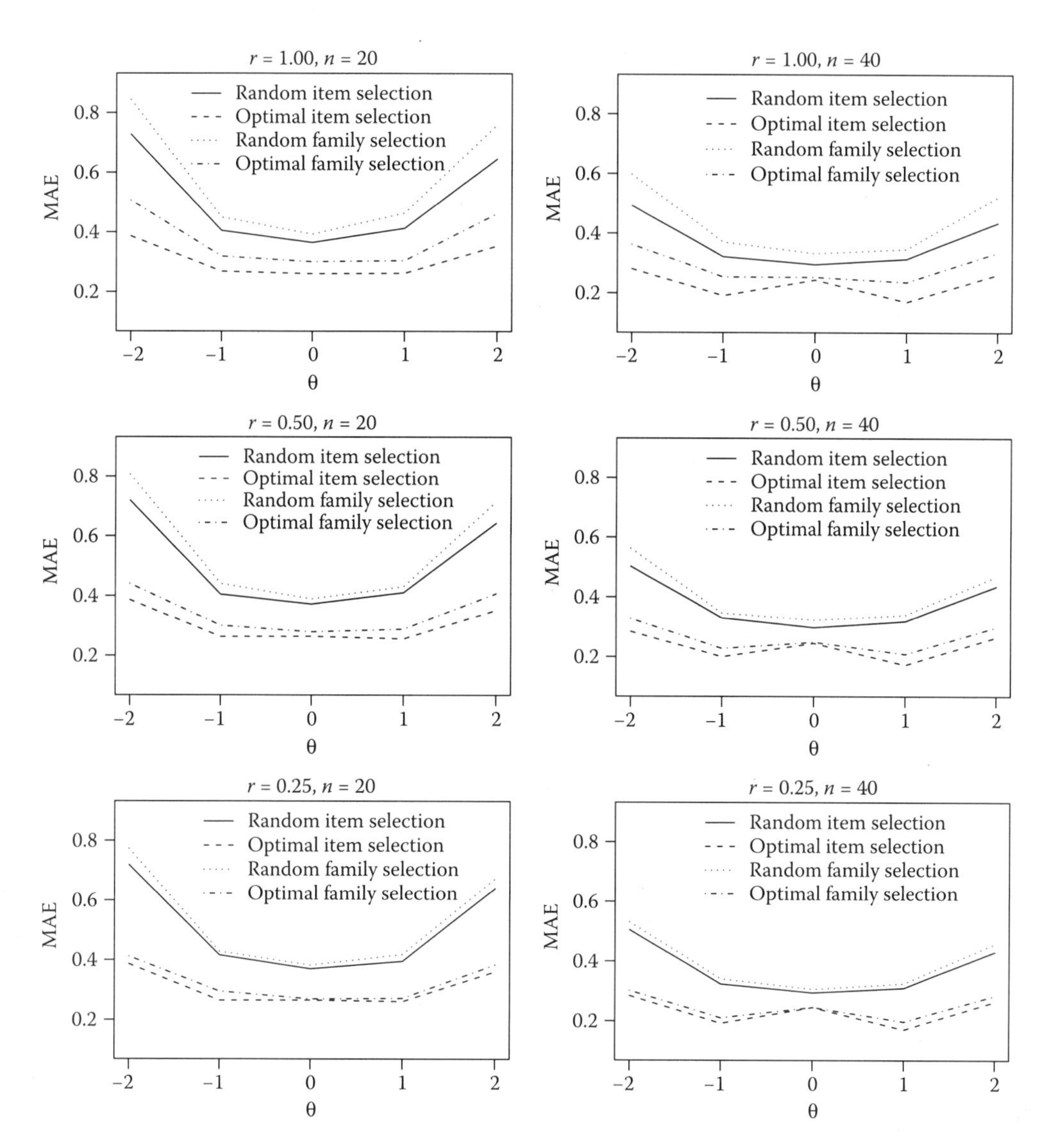

FIGURE 26.2
Mean absolute error in ability estimates for an item pool with 400 families.

over the 3000 replications for the four conditions and five true ability values used in the simulations. Table 26.1 presents the results for the case of random selection of the ability parameter from the standard normal distribution. For each condition, a larger number of items per test resulted in lower MAEs for the ability estimates. However, the trend was stronger for Conditions 1 and 3 with the random selection of the items than Conditions 2 and 4 where the items were selected according to either minimum expected posterior variance criterion.

The number of items/families in the item pool did have an effect on Conditions 2 and 4 only. As the items were optimally selected for these two conditions, the decrease in MAEs for the conditions with the larger pool of 400 families reflected the larger space of optimization created by it.

TABLE 26.1
Mean Absolute Error in Ability Estimates

			Condition			
r	*F*	*n*	1	2	3	4
1.00	100	20	0.427	0.321	0.491	0.383
		40	0.325	0.268	0.372	0.299
	400	20	0.437	0.274	0.488	0.324
		40	0.336	0.216	0.368	0.259
0.50	100	20	0.425	0.314	0.464	0.350
		40	0.332	0.259	0.364	0.289
	400	20	0.432	0.281	0.453	0.303
		40	0.330	0.213	0.359	0.245
0.25	100	20	0.437	0.324	0.437	0.339
		40	0.329	0.262	0.352	0.273
	400	20	0.426	0.278	0.446	0.302
		40	0.334	0.214	0.351	0.234

As expected, the two conditions with ability estimation based on the family parameters (Conditions 3 and 4) showed a decrease in MAE for the families with the smaller within-family item-parameter variability. Obviously, the smaller the deviations of the parameters from their family means, the smaller the error involved in using the latter for the former.

26.7 Discussion

The defining characteristic of item-family response models is their hierarchical structure with random item parameters with a different common distribution for each family of items. The main example of a family in this chapter is the set of items produced for the same choice of radicals for an automated item generator. But, as indicated earlier, other examples of item families are readily available.

The questions of how much the items in a family vary and how large the differences between different families can be expected to be are questions with empirical answers. Statistically, however, the only thing that counts is the ratio of the within-family and between-family variation. As long as the former is smaller than the latter, we gain by selecting families and then randomly sampling items from families rather than randomly sampling items from the entire pool and hence our manipulation of the ratio in the preceding empirical example.

Although we highlighted the application of family-response models to adaptive testing, application to the automated assembly of fixed test forms can be equally beneficial (Geerlings et al., 2013). We then model the test assembly problem as an instance of constraints combinatorial optimization at the level of the item families using the technique of mixed-integer programming (for a review, see van der Linden, Volume Three, Chapter 9). That is, rather than selecting the best combination from a pool of discrete items we use the optimization model to identify the best combination of item families in the pool, with subsequent generation of predetermined numbers of items from each selected family.

References

Albert, J. H. 1992. Bayesian estimation of normal ogive item response curves using Gibbs sampling. *Journal of Educational Statistics*, 17, 251–269.

Bormuth, J. R. 1970. *On the Theory of Achievement Items*. Chicago: University of Chicago Press.

Casella, G. and George, E. I. 1992. Explaining the Gibbs sampler. *The American Statistician*, 46, 167–174.

Fox, J.-P. and Verhagen, A. J. 2010. Random item effects modeling for cross-national survey data. In: E. Davidov, P. Schmidt and J. Billiet (Eds.), *Cross-Cultural Analysis: Methods and Applications* (pp. 467–488). London: Routledge Academic.

Geerlings, H. 2012. Psychometric methods for automated test design. Doctoral dissertation, University of Twente, Enschede, the Netherlands. Retrieved from http://doc.utwente.nl.

Geerlings, H., Glas, C. A. W., and van der Linden, W. J. 2011. Modeling rule-based item generation. *Psychometrika*, 76, 337–359.

Geerlings, H., van der Linden, W. J., and Glas, C. A. W. 2013. Optimal test design with rule-based item generation. *Applied Psychological Measurement*, 37, 140–161.

Gelman, A., Meng, X.-L., and Stern, H. 1996. Posterior predictive assessment of model fitness via realized discrepancies. *Statistica Sinica*, 6, 733–807.

Gelman, A., Van Mechelen, I., Verbeke, G., Heitjan, D. F., and Meulders, M. 2005. Multiple imputation for model checking: Completed-data plots with missing and latent data. *Biometrics*, 61, 74–85.

Gilbride, T. J. and Lenk, P. J. 2010. Posterior predictive model checking: An application to multivariate normal heterogeneity. *Journal of Marketing Research*, 47, 896–909.

Glas, C. A. W. and van der Linden, W. J. 2003. Computerized adaptive testing with item cloning. *Applied Psychological Measurement*, 27, 247–261.

Glas, C. A. W., van der Linden, W. J., and Geerlings, H. 2010. Estimation of the parameters in an item-cloning model for adaptive testing. In: W. J. van der Linden and C. A. W. Glas (Eds.), *Elements of Adaptive Testing* (pp. 289–314). New York: Springer.

Irvine, S. H. 2002. The foundations of item generation for mass testing. In: S. H. Irvine and P. C. Kyllonen (Eds.), *Item Generation for Test Development* (pp. 3–34). Mahwah: Lawrence Erlbaum Associates.

Janssen, R., Tuerlinckx, F., Meulders, M., and De Boeck, P. 2000. A hierarchical IRT model for criterion-referenced measurement. *Journal of Educational and Behavioral Statistics*, 25, 285–306.

Johnson, M. S. and Sinharay, S. 2005. Calibration of polytomous item families using Bayesian hierarchical modeling. *Applied Psychological Measurement*, 29, 369–400.

Lord, F. M. and Novick, M. R. 1968. *Statistical Theories of Mental Test Scores*. Reading: Addison-Wesley.

Millman, J. and Westman, R. S. 1989. Computer-assisted writing of achievement test items: Toward a future technology. *Journal of Educational Measurement*, 26, 177–190.

Sinharay, S. 2005. Assessing fit of unidimensional item response theory models using a Bayesian approach. *Journal of Educational Measurement*, 42, 375– 394.

Sinharay, S. and Johnson, M. S. 2008. Use of item models in a large-scale admissions test: A case study. *International Journal of Testing*, 8, 209–236.

Sinharay, S., Johnson, M. S., and Williamson, D. M. 2003. Calibrating item families and summarizing the results using family expected response functions. *Journal of Educational and Behavioral Statistics*, 28, 295–313.

Steinbakk, G. H. and Storvik, G. O. 2009. Posterior predictive p-values in Bayesian hierarchical models. *Scandinavian Journal of Statistics*, 36, 320–336.

Wingersky, M. S. and Lord, F. M. 1984. An investigation of methods for reducing sampling error in certain IRT procedures. *Applied Psychological Measurement*, 8, 347–364.

27

Hierarchical Rater Models[*]

Jodi M. Casabianca, Brian W. Junker, and Richard J. Patz

CONTENTS

27.1 Introduction 449
27.2 The HRM 450
27.2.1 IRT Model for Item Responses 451
27.2.2 Model for Rater Accuracy 451
27.2.3 Rater Covariates 452
27.3 Estimation 453
27.4 Assessing Model Fit 455
27.5 Example 456
27.6 Discussion 461
References 462

27.1 Introduction

Ratings of rich response formats have been a part of the assessment landscape for as long as there have been assessments. Short-answer and multiple-choice question formats largely eliminate extraneous variability in scoring, but in many areas of student and teacher assessment, rater bias, variability, and other factors affect assessment scores. Unusual rating behavior (DeCarlo, 2008; Patz et al., 2002; Wolfe and McVay, 2002), factors in raters' backgrounds (Winke et al., 2011), the circumstances of rating (Mariano and Junker, 2007), and their effects on procedures for producing and reporting assessment scores (e.g., Yen et al., 2005) continue to be of central interest. In the case of student work, the relative merits of automated machine-scoring algorithms versus human scoring continue to be of interest (Attali and Burstein, 2005; CTB/McGraw-Hill, 2012; Steedle and Elliot, 2012; J. Wang and Brown, 2012). In the case of teacher assessment, human rating continues to be the only way to assess certain aspects of professional practice (e.g., Casabianca and Junker, 2013; Casabianca et al., 2013; Farrokhi et al., 2011; Junker et al., 2006).

Many studies of rater effects—including several of those cited above—employ an item response theory (IRT) model such as the generalized partial credit model (GPCM) (Muraki, 1992; Patz and Junker, 1999a) or the facets model (Linacre, 1989). As noted by Patz et al.

* The research reported here was supported in part through a postdoctoral training program supported by the Institute of Education Sciences, U.S. Department of Education, through Grant R305B1000012 to Carnegie Mellon University. The opinions expressed are those of the authors and do not necessarily represent views of the institute or the U.S. Department of Education. The order of authorship is alphabetical.

(2002) and formally proven by Mariano (2002), however, these approaches have a fundamental flaw: as the number of raters increase—even for a single item!—the standard error of measurement for the examinee ability parameter tends to zero. This cannot be: repeatedly rating the same item response or task behavior can tell us more about the quality of that particular response, but should not reduce measurement error for the underlying latent proficiency variable to zero.

The hierarchical rater model (HRM, Patz, 1996; Patz et al., 2002) is an extension of polytomous IRT models for items with multiple ratings or scores, that corrects this flaw in IRT and facet models, by composing two measurement stages: the first stage is a "signal-detection-like" model for measuring the ideal rating of an item based on multiple raters' observed ratings; and the second stage is an IRT model relating the ideal ratings to the underlying examinee proficiency variable. Other approaches to correcting this flaw have also been proposed (Bock et al., 2002; Muckle and Karabatsos, 2009; Wilson and Hoskens, 2001); see also Donoghue and Hombo (2000) for a comparison of some of these approaches.

From the HRM, we obtain estimates of item parameters, examinee proficiency means and variance, and rater bias and variance. Since its introduction, the HRM has been extended to accommodate (i) the fitting of rater covariates (Mariano and Junker, 2007), and (ii) the fitting of rater effects beyond simple rater bias (severity) and rater variability (consistency) (DeCarlo et al., 2011).

27.2 The HRM

The HRM is composed of a three-level hierarchy. The first level models the distribution of ratings given the quality of response, the second level models the distribution of an examinee's response given their latent trait, and the third level models the distribution of the latent trait θ_p. The hierarchy is given by

$$\left.\begin{array}{lll}\theta_p & \sim & iid\,N(\mu,\sigma^2), \quad p=1,\ldots,P \\ \xi_{pi} & \sim & \text{polytomous IRT model}, \quad i=1,\ldots,I, \text{ for each } p \\ X_{pir} & \sim & \text{polytomous signal detection model}, \quad r=1,\ldots,R, \text{ for each } p,i.\end{array}\right\} \tag{27.1}$$

Here, θ_p, the latent trait for examinee $p = 1, \ldots, P$ is normally distributed with mean μ and σ^2, ξ_{pi} is the ideal rating for examinee p on item i 1, ..., I, and X_{pir} is the observed rating given by rater r for examinee p's response to item i. Note that specifying a normal distribution for the latent trait is a popular choice, but alternatives could be used instead.

This hierarchy formalizes the two-stage process discussed earlier: the examinee responds to a set of I items (with ideal ratings) and a series of R raters evaluate the responses, giving ratings conditional on the examinees' responses. This notation is for the completely crossed design where all raters score each item. Within this framework, incomplete designs are treated as missing completely at random (MCAR; Mislevy and Wu, 1996; Rubin and Little, 2002). Models for informative missingness (e.g., Glas and Pimentel, 2008; Holman and Glas, 2005) could also be directly incorporated into the HRM if needed.

27.2.1 IRT Model for Item Responses

The ideal ratings in the second level are latent variables modeled using a polytomous IRT model, such as the A-category GPCM (Muraki, 1992). The GPCM with item response ξ_{pi} and A- possible scores ($a = 1, \ldots, A$) is given by

$$P[\xi_{pi} = \xi \mid \theta_p, \alpha_i, \beta_i, \gamma_{i\xi}] = \frac{\exp\left\{\sum_{a=1}^{\xi} \alpha_i(\theta_p - \beta_i - \gamma_{ia})\right\}}{\sum_{h=0}^{A-1} \exp\left\{\sum_{a=1}^{h} \alpha_i(\theta_p - \beta_i - \gamma_{ia})\right\}}, \tag{27.2}$$

where α_i is the item discrimination, β_i is the item location, and γ_{ia} is the ath threshold parameter for item i.

For examinee p's response to item i, ideal ratings represent the quality of the response. Note that other polytomous IRT models can be used in this level, and that A, the number of response categories per item, need not be constant across items.

27.2.2 Model for Rater Accuracy

The ideal rating ξ_{pi} represents the rating that person p would get on item or task i, by a rater exhibiting no rater bias and perfect rating consistency. In the HRM, the deviations between actually observed ratings and these ideal ratings are modeled using a discrete signal-detection model that is specified to represent the quality of the response. A matrix of response probabilities defines the relationship between the observed and ideal rating probabilities such that $p_{\xi ar} = \mathrm{P}(\text{Rater } r \text{ rates } a \mid \text{ideal rating } \xi)$. Table 27.1 provides an example matrix of response probabilities with four categories ($A = 4$). The probabilities in the matrix are manipulated to represent variations in rater behavior as well as dependence of ratings on rater covariates (see the next section), and interactions between raters, items, and/or examinees.

The signal-detection model (SDM) used in the original formulation of the HRM uses a discrete unimodal distribution for each row of the matrix to give the probability of an observed rating X_{pir} given an ideal rating ξ_{pi}. The mode of this distribution is the rater bias or severity, ϕ_r, and the spread of this distribution is the rater variability or unreliability, ψ_r. Probabilities in each row of the matrix can be made proportional to a Normal density with mean $\xi + \phi_r$ and standard deviation ψ_r

$$p_{\xi ar} = P[X_{pir} = a \mid \xi_{pi} = \xi] \propto \exp\left\{-\frac{1}{2\psi_r^2}[a - (\xi + \phi_r)]^2\right\}. \tag{27.3}$$

TABLE 27.1

Rating Probability Table for the Signal-Detection Process Modeled in the HRM

Ideal Rating (ξ_{pi})	Observed Rating (a) 0	1	2	3
0	p_{11r}	p_{12r}	p_{13r}	p_{14r}
1	p_{21r}	p_{22r}	p_{23r}	p_{24r}
2	p_{31r}	p_{32r}	p_{33r}	p_{34r}
3	p_{41r}	p_{42r}	p_{43r}	p_{44r}

Estimates of the rater parameters for bias ϕ_r and variability ψ_r provide information on an individual rater r. When $\phi_r = 0$, rater r is more likely to assign a rating that matches the ideal rating $a = \xi$ (no bias). When $\phi_r < -0.5$ (less than 0 on a discrete scale), rater r is more likely to exhibit negative bias relative to the ideal rating and rate in a category lower than the ideal rating, $a < \xi$. Conversely, when $\phi_r > 0.5$ (greater than 0 on a discrete scale), rater r is more likely to exhibit positive bias relative to the ideal rating and rate in a category higher than the ideal rating, $a > \xi$. The spread parameter ψ_r indicates a rater's variability; values near 0 indicate high consistency or reliability in rating (to the rubric or scoring guidelines) and high values indicate poorer consistency in rating.

The model described above is the original model for rater accuracy (Mariano and Junker, 2007; Patz et al., 2002); however, there are possible alternatives to the signal-detection model for rater accuracy (see DeCarlo et al., 2011).

27.2.3 Rater Covariates

In practice, there may be characteristics of raters and/or of raters' ratings that influence rater bias and variability. Analysis of bias and variability effects of rater covariates may be useful for adapting features of the rating design such as rater training; nonzero covariate effects may reveal areas in which attention is needed. Here, we introduce the basic notation and framework of Mariano and Junker (2007) for the HRM with rater covariates.

Rater covariates may be fixed for all ratings from the same rater (e.g., gender, race, and hours of rater training) or they may differ over ratings from the same rater (e.g., the time to complete the rating or scoring mode). Consider a rating process with S rater covariates, $Z = \{Z_1, Z_2, \ldots, Z_S\}$, which are either quantitative or C-categorical where $Z_s + \cdots + Z_{s+C} = 1$ (Mariano and Junker, 2007). To incorporate rater covariates into the HRM, we consider a design matrix that includes observations grouped by every unique combination of rater and rater covariates (Rater × Covariate combination) instead of R groups of observations (one per rater). Each combination is referred to as a *pseudorater* (by analogy with Fischer, 1973). Let V be the number of pseudoraters ($v = 1, \ldots, V$), where V is at a minimum, equal to R, if rater covariates are fixed for all items within rater r's ratings to all I items, or $V > R$ if some or all rater covariates differ over ratings.

To determine the effect of each pseudorater in terms of bias and variability for the V Rater × Covariate combinations, we adapt the signal-detection model component of the HRM by replacing the bias ϕ_r and variability parameters ψ_r^2 in the original model given in Equation 27.3 with bias ρ_v and variability ω_v^2 for pseudorater v. Therefore, the updated signal-detection model incorporating rater covariate effects is

$$p_{\xi a v} = P[X_{piv} = a \mid \xi_{pi} = \xi] \propto \exp\left\{-\frac{1}{2\omega_v^2}[a-(\xi+\rho_v)]^2\right\}. \tag{27.4}$$

The model structure can be specified in a design matrix Y with V rows (indicators for each pseudorater) and $Q + S$ columns (Q for raters and S for covariates). The Q columns contain binary indicators for each rater, and the S columns contain rater covariate values. For example, row $Y_v = (0, 1, 0, 0, 0, Z_{1v}, Z_{2v})$ corresponds to pseudorater v, which includes rater 2 with values for the two covariates, Z_{1v} and Z_{2v}. If there is no interest in controlling for individual rater effects, then, the first Q columns of the matrix may be omitted.

We may build a linear model for rating bias, depending on covariates Y, as

$$\rho_v = Y_v \boldsymbol{\eta}.$$

Here, the vector $\boldsymbol{\eta} = (\phi_1, \ldots, \phi_R, \eta_1, \ldots, \eta_S)^T$ gives the bias for the full rating process, where the first set of variables are the rater bias effects as defined below and the second set of variables are the covariate bias effects such that η_s represents the bias effect of covariate Z_s.

Similarly, we may build a linear model for rating variance (in the log scale), depending on covariates Y, as

$$\log \omega_v^2 = Y_v(\mathbf{log}\ \boldsymbol{\tau}^2).$$

where we define rater and covariate rater effects as $\mathbf{log}\ \boldsymbol{\tau}^2 = (\log \psi_1^2, \ldots, \log \psi_R^2, \log \tau_1^2, \ldots, \log \tau_S^2)^T$. Here, the first set of variables are individual rater variability parameters and the second set of variables are for the covariate variability such that τ_s represents the variability from covariate Z_s.

These considerations lead to the first version of the HRM with rater covariates discussed by Mariano and Junker (2007): their *fixed rating effects* model. Within this model, the augmented signal-detection model is given by

$$p_{\xi av} = P[X_{piv} = a \mid \xi_{pi} = \xi] \propto \exp\left\{-\frac{[a - (\xi + Y_v \eta)]^2}{2(\exp\{Y_v(\log \tau^2)\})}\right\}, \tag{27.5}$$

where the linear structures for rating bias and variability replace the original parameters ϕ_r and ψ_r that did not include covariates.

In the second version of the HRM with covariates, *random rating effects* model, the pseudorater bias ρ_v parameters in Equation 27.4 are treated as random draws from a normal distribution centered at a linear function of the covariates

$$\rho_v \sim N(Y_v \eta, \sigma_\rho^2) \tag{27.6}$$

and the pseudorater variability parameters ω_v^2 in Equation 27.4 are treated as random draws from a similarly centered log-normal distribution

$$\log \omega_v^2 \sim N(Y_v \log \tau^2, \sigma_\omega^2). \tag{27.7}$$

This results in a model very much like Equation 27.5, except that the fixed predictors $Y_v \eta$ and $Y_v(\log \tau^2)$ are replaced with random predictors $Y_v \eta + \varepsilon_{\rho,v}$ and $Y_v(\log \tau^2) + \varepsilon_{\omega^2,v}$, where $\varepsilon_{\rho,v}$ and $\varepsilon_{\omega^2,v}$ are suitable random normal draws for each pseudorater. In both the fixed and random rater effects models, care must be taken to ensure that the design matrix Y is full rank.

27.3 Estimation

Although the HRM has been estimated with maximum likelihood methods (Hombo and Donoghue, 2001), the literature has predominantly treated HRM as a Bayesian model (Patz et al., 2002), and it is treated as such here. In Section 27.5, we estimate an HRM using a Markov chain Monte Carlo (MCMC) algorithm, implemented in WinBUGS (Lunn et al., 2000). For an introductory discussion of MCMC, refer to Junker, Patz, and Vanhoudnos (Volume Two, Chapter 15); for more information on WinBUGS for IRT models, see Johnson (Volume Three, Chapter 21).

A basic HRM begins with the hierarchy in Equation 27.1, an IRT model such as the GPCM in Equation 27.2 for ideal ratings, and the signal-detection model in Equation 27.3 connecting observed ratings to ideal ratings. To use the Bayesian framework, we must also specify priors for seven sets of parameters: GPCM item parameters α_i, β_i, and γ_{ia}, for $i = 1, \ldots, I$ and $a = 0, \ldots, A-1$; rater parameters ϕ_r, ψ_r, where $r = 1, \ldots, R$; and examinee proficiency distribution parameters μ, σ^2. The prior distributions for the GPCM specified in Equation 27.2 should account for location indeterminacy by constraining either the latent proficiency mean μ, or the item difficulty parameters β_i. A similar scale indeterminacy problem can be addressed by constraining either the item discrimination parameters α_i or the latent proficiency variance σ^2. These constraints may be hard linear constraints, or soft constraints imposed through prior distributions.

Table 27.2 provides some noninformative parameters for the HRM parameters. In Table 27.2, the β_i are modeled as iid from the same Normal prior but constrained such that the negated sum of the first $I-1$ β's equals β_I, while the prior for μ is $N(0,10)$. Similarly, we give α's iid log-normal priors, except for the constraint that $\log \alpha_I = -\sum_{i=1}^{I-1} \log \alpha_i$. There is also indeterminacy for the γ_{ia} parameters; to solve this, the $A-1$ item step parameters γ_{ia} are assumed to be iid with $N(0,10)$ priors, except that the last γ_{ia} for each item is a linear function of the other such that $\gamma_{i(A-1)} = -\sum_{a=1}^{A-2} \gamma_{ia}$. Preliminary analyses for rater agreement may inform priors for the rater parameters; however, uninformative parameters such as those provided in Table 27.2 may be used when there is a lack of strong information. Similarly, we provide uninformative priors to reflect upon little prior knowledge about the examinee proficiency parameters.

Estimating the Bayesian HRM with MCMC is a straightforward extension of the MCMC approach to estimating a GPCM (Patz and Junker, 1999a,b); the extension must include the additional parameters for rater bias and variability and the ideal ratings (see Patz et al., 2002, for an in-depth discussion of MCMC for HRM). Additional modifications for rater covariates are discussed by Mariano and Junker (2007). The code for fitting the model in the next section is provided by Casabianca (2012).

The HRM has also been fitted with marginal maximum likelihood (MML, Hombo and Donoghue, 2001). DeCarlo et al. (2011) fitted the HRM–signal detection theory (SDT) model

TABLE 27.2

Some Uninformative Priors for the HRM Parameters as Presented in Patz et al. (2002)

Parameter	Uninformative Prior Distributions
α_i	i.i.d. from Log-normal (0,10)[a]
β_i	i.i.d. from $N(0,10)$[a]
γ_{ia}	i.i.d. from $N(0,10)$[a]
ϕ_r	$N(0,10)$
ψ_r	$\log(\psi_r) \sim N(0,10)$
μ	$N(0,10)$
σ^2	$1/\sigma^2 \sim Gamma\ (\alpha, \eta),\ \alpha = \eta = 1$

[a] Location indeterminacies for α_i, β_i, and γ_{ia} require sum-to-zero constraints to use these priors. See text for details.

using posterior modal estimation (PME) implemented with an expectation–maximization (EM) algorithm (Dempster et al., 1977; Wu, 1983).

27.4 Assessing Model Fit

Assessing model fit for the HRM is similar to assessing model fit for other complex hierarchically structured models. Overall indices of fit are usually based on the *model deviance*

$$D(\mathcal{M}) = -2 \times \log P_{\mathcal{M}}(\mathcal{X} \mid \hat{\Omega}),$$

where $P_{\mathcal{M}}(\mathcal{X} \mid \hat{\Omega})$ is the marginal likelihood (averaging over θ's) of the set of observed ratings $\mathcal{X}$ given estimates $\hat{\Omega}$ of all models parameters $\Omega = (\alpha_1, \alpha_2, \ldots, \beta_1, \beta_2, \ldots, \gamma_{11}, \gamma_{12}, \ldots, \ldots, \phi_1, \phi_2, \ldots, \psi_1, \psi_2, \ldots, \mu, \sigma^2)$, for the particular specification of interest $\mathcal{M}$ of the HRM.* If model $\mathcal{M}_1$ is nested within model $\mathcal{M}_2$, then, $D(\mathcal{M}_1) - D(\mathcal{M}_2)$ is the usual likelihood-ratio test statistic.

Non-nested models may be compared using the *Akaike information criterion* (AIC; Akaike, 1973)

$$AIC = D(\mathcal{M}) + 2k,$$

the *Bayesian information criterion* (BIC; Schwarz, 1978)

$$BIC = D(\mathcal{M}) + k \log(n),$$

where the model $\mathcal{M}$ has k locally free parameters and n observations, or some variation of AIC or BIC; see, for example, Burnham and Anderson (2004) or Claeskens and Hjort (2008). Patz et al. (2002) and Mariano and Junker (2007) provide examples of applying BIC to compare HRM fits.

A standard problem is on actually understanding how many free parameters or degrees of freedom are there in a complex hierarchical model. One approach is to use an estimate k_D derived from the curvature of the likelihood or posterior distribution near its mode. This leads to the *deviance information criterion* (DIC; Spiegelhalter et al., 2002)

$$DIC \approx D(\mathcal{M}) + 2k_D,$$

an estimate of which is provided in the output of WinBUGS and related software (see also Ando, 2012; Gelman and Hill, 2006, p. 525).

The BIC is itself an approximation to a more general model comparison tool, which is the Bayes factor (Kass and Raftery, 1995). Modern approaches to Bayes factor and marginal likelihood calculations based on MCMC and related methods (e.g., Chib, 1995; Chib and Jeliazkov, 2001; Gelman and Meng, 1998; Neal, 2001) are extensively surveyed and directly applied to the HRM, by Mariano (2002).

Another fruitful approach to model checking is the posterior predictive check (Gelman et al., 2003, Chapter 6), because it can be focused on particular model features of interest,

* Covariates are omitted for simplicity but could be included, to the right of the conditioning bar in $P_{\mathcal{M}}(\mathcal{X} \mid \hat{\Omega})$, without loss.

rather than providing only an omnibus test or comparison of fit. Posterior predictive checks are roughly analogous to model modification indices in other contexts. Posterior predictive checks tend to be somewhat conservative (see, e.g., the discussion of model checking in Gelman and Shalizi, 2013); so, when they suggest that a feature of the data is not well modeled, it likely really is not. Examples of posterior predictive checking in item response models related to the HRM include Sinharay et al. (2006), Levy et al. (2009), and Zhu and Stone (2011).

27.5 Example

To illustrate the HRM, we analyze data from a writing assessment conducted by CTB/McGraw-Hill (2012) that allow us to compare human and machine raters. We consider an extract of the data only for the first writing prompt. In this extract, 487 examinees responded to the prompt; their responses were then rated on five different writing features (such as mechanics, organization, etc.). Each feature was rated on a discrete 0–6 score scale (with 7 possible scores) and was rated by 4 raters (total number of ratings = 9740). Note that for any given examinee, the same 4 raters rated all 5 features. In total, there were 36 raters, including one machine that scored all features for all examinees. Hereafter, we will refer to these writing features as items.

We consider a single rater covariate, "gender." Our gender variable, however, includes three categories: male ($n = 18$), female ($n = 17$), and machine ($n = 1$). We are interested in learning about how a machine rater compares to males and females, and also how males compare to females, in terms of rating bias and variability.

The raters did not use the full scale (see Table 27.3, which gives the frequency distribution of ratings). Only one rater used the lowest score category, and similarly, only 23 raters (<1%) used the highest category. The most popular category for all items, a score of 3 (41%), was in the middle of the score scale. Mean overall ratings were not very different by the rater type: Males were the most lenient in terms of overall ratings (Mean = 3.87, SD = 0.997), then the machine (Mean = 3.83, SD = 0.751), and then females (Mean = 3.77, SD = 0.994).

To fit these data, we used an HRM (Equation 27.1) with random rater effects. The IRT model relating ideal ratings to the latent proficiency variable was the GPCM (Equation 27.2), and SDM used was as shown in Equation 27.4 with Equations 27.6 and 27.7 for bias

TABLE 27.3

Distribution of Ratings by Item

	Item					
Item Score	1	2	3	4	5	Percent
0	0	0	0	0	1	0.01
1	123	122	252	108	133	7.58
2	695	657	502	430	570	29.30
3	702	775	842	857	773	40.54
4	357	323	308	509	422	19.70
5	69	67	39	39	42	2.63
6	2	4	5	5	7	0.24

and variability. Therefore, in addition to the traditional HRM parameters, we are now interested in η and τ, which are the bias and variability parameters for the rater-type covariate. Since "rater type" did not change for a rater across ratings, the design matrix included $V = 36$ pseudoraters, which is the actual number of raters. Because we are fitting the random rater effects model, the design matrix Y for the fixed rater effects is 36 by 3—the 3 columns are rater-type indicators (male, female, and machine)—and there are 36 random intercepts, one for each rater.

We specified rater variability, ω_v^2, in the WinBUGS model differently than what is shown in Equation 27.4. The example's SDM included a transformed version of ω_v^2 such that $\zeta_v = \log(1/\omega_v^2)$. We estimated ζ_v for each rater and transformed the estimates back to the parameterization in Equation 27.4 using $\omega_v = \sqrt{1/\exp(\zeta_v)}$; this is what we report in Table 27.5 (later in the chapter). Additionally, because of the relationship between ω_v^2 and τ^2 in Equation 27.7, we similarly specified a transformed version of τ^2 such that $\kappa = \log(1/\tau^2)$. We estimated κ and then transformed it back with $\tau = \sqrt{1/\exp(\kappa)}$, reported in Table 27.6 (later in the chapter).

Prior distributions for the item and person parameters were specified as follows: $\beta_i \sim \text{Normal}(0,10)$, $\alpha_i \sim \text{Log Normal}(0,10)$, $\gamma_{ia} \sim \text{Normal}(0,10)$, $\theta \sim \text{Normal}(0,\zeta)$, and precision of θ, $\delta \sim \text{Gamma}(1,1)$. Rater and rater covariate parameters were specified as $\rho_v \sim \text{Normal}(\eta,\sigma_\rho^2)$, $\zeta_v \sim \text{Normal}(\kappa,\sigma_\zeta^2)$ where both η and κ have a Normal(0,10) prior, and both $1/\sigma_\zeta^2$ and $1/\sigma_\rho^2$ have a Gamma(1,1) prior. Note that the priors used for the rater and rater covariate variability parameters were appropriate for the way the WinBUGS model was specified as described above. Sum-to-zero constraints were placed on the α_i, β_i, and γ_{ia} parameters. Note that to reduce sparsity in the observed frequency distribution of ratings, we pooled rating categories 0–1 and recoded the pooled category as category 1, leading to an effect score scale running from 1 to 6.

This model was fitted to the CTB/McGraw-Hill (2012) data with an MCMC algorithm programmed with WinBUGS (Lunn et al., 2000) using the R (R Development Core Team, 2012) package RUBE (Seltman, 2010). Parameter estimates were based on 30,000 iterations from five chains, using a burn-in of 5000 iterations and thinning every fifth iteration. An R script file containing information on RUBE as well as the WinBUGS syntax to run this model is available from Casabianca (2012).

Posterior median and 95% credible interval estimates for the item parameters are given in Table 27.4. We note that all estimates of the discrimination parameters α_i are fairly similar to each other, except for $\hat{\alpha}_4$, indicating that Item 4 had a relatively high discrimination. Similarly, the β_i parameter estimates indicate fairly easy items, except for Item 3, which is somewhat more difficult than the others. The estimates of item category parameters γ_{ia} confirm the notion that the lowest category is not often used (all γ_{i1} estimates are far to the left of the other estimated γ_{ia}'s), and suggest some disordering of the other item category parameters, although most standard errors for $\hat{\gamma}$'s make it difficult to infer orderings with certainty.

Table 27.5 shows the posterior median estimates and 95% credible intervals for rater bias ρ_v and variability ω_v. Among the 36 raters, only 8 had bias estimates in absolute value greater than 0.50; 4 were severe (negative bias), and 4 were lenient (positive bias). Rater 20 had a very large negative bias compared to all other raters, $\hat{\rho}_{20} = -1.238$. The most lenient rater was Rater 11, whose bias was $\hat{\rho}_{11} = 0.81$. Aside from these exceptions, raters appeared to be in agreement with each other (i.e., $|\hat{\rho}_v| < 0.50$).

There appears to be no pattern in rater bias, in terms of gender (see also Table 27.6). Rater 36, the machine, however, had a positive bias ($\hat{\rho}_{36} = 0.069$). While this amount of bias is not practically significant, it was statistically significant (0.035, 0.102), likely due to

TABLE 27.4

Posterior Median and 95% Equal-Tailed Credible Interval (CI) Estimates of the Item Parameters

Parameter	Median	95% CI
α_1	0.900	(0.724, 1.143)
α_2	0.994	(0.798, 1.264)
α_3	0.888	(0.708, 1.130)
α_4	1.365	(1.063, 1.801)
α_5	0.909	(0.735, 1.149)
β_1	0.090	(−0.314, 0.549)
β_2	−0.232	(−0.661, 0.232)
β_3	1.135	(0.749, 1.588)
β_4	−0.973	(−1.416, −0.519)
β_5	−0.048	(−0.435, 0.386)
γ_{11}	−15.110	(−16.740, −13.610)
γ_{12}	5.721	(4.945, 6.593)
γ_{13}	4.189	(3.774, 4.642)
γ_{14}	3.252	(2.881, 3.694)
γ_{15}	1.933	(1.474, 2.450)
γ_{21}	−15.920	(−17.770, −14.240)
γ_{22}	6.077	(5.182, 7.108)
γ_{23}	4.451	(4.002, 4.932)
γ_{24}	3.325	(2.949, 3.771)
γ_{25}	2.043	(1.588, 2.543)
γ_{31}	−11.310	(−12.490, −10.210)
γ_{32}	2.897	(2.411, 3.429)
γ_{33}	3.801	(3.419, 4.215)
γ_{34}	3.653	(2.835, 7.661)
γ_{35}	0.947	(−3.029, 1.838)
γ_{41}	−15.140	(−17.100, −13.390)
γ_{42}	4.916	(3.969, 6.018)
γ_{43}	4.275	(3.829, 4.766)
γ_{44}	4.335	(3.537, 8.294)
γ_{45}	1.608	(−2.368, 2.510)
γ_{51}	−13.870	(−15.470, −12.450)
γ_{52}	4.508	(3.801, 5.346)
γ_{53}	4.091	(3.670, 4.552)
γ_{54}	4.438	(3.363, 8.394)
γ_{55}	0.826	(−3.165, 1.999)

the fact that the machine rated all items for all 487 examinees, driving down the standard error for estimating ρ_{36}. For comparison, the human raters rated 5 to 337 examinees, with raters 33, 34, and 35, each rating items for more than 300 examinees. As a whole, the rater bias estimates suggest little or no appreciable bias for the machine rater, relative to the humans.

In considering individual rater variability, recall that a small value of ω_v indicates a smaller chance of rater r rating in categories far from their most likely category, $\xi + \rho_v$

TABLE 27.5

Posterior Median and 95% Equal-Tailed Credible Interval (CI) Estimates of the Rater Bias ρ_v and Variability ω_v Effects for $v = 36$ (Pseudo) Raters

Rater Bias, ρ			Rater Variability, ω			
Parameter	**Median**	**95% CI**	**Parameter**	**Median**	**95% CI**	**Rater Type**
ρ_1	−0.020	(−0.361, 0.314)	ω_1	0.823	(0.635, 1.110)	M
ρ_2	0.150	(−0.047, 0.341)	ω_2	0.805	(0.688, 0.960)	M
ρ_3	−0.360	(−0.687, −0.058)	ω_3	0.729	(0.559, 0.993)	M
ρ_4	−0.345	(−0.462, −0.221)	ω_4	0.531	(0.436, 0.640)	M
ρ_5	0.171	(−0.101, 0.433)	ω_5	0.798	(0.616, 1.037)	F
ρ_6	−0.572	(−0.980, −0.219)	ω_6	0.942	(0.731, 1.253)	M
ρ_7	0.192	(0.009, 0.375)	ω_7	0.839	(0.724, 0.992)	M
ρ_8	0.078	(−0.152, 0.302)	ω_8	0.771	(0.638, 0.954)	M
ρ_9	−0.209	(−0.425, −0.011)	ω_9	0.670	(0.528, 0.854)	M
ρ_{10}	−0.203	(−0.356, −0.058)	ω_{10}	0.860	(0.755, 0.986)	F
ρ_{11}	0.810	(0.699, 0.922)	ω_{11}	0.723	(0.650, 0.811)	F
ρ_{12}	−0.220	(−0.492, 0.023)	ω_{12}	0.875	(0.706, 1.111)	F
ρ_{13}	−0.542	(−0.948, −0.210)	ω_{13}	0.932	(0.730, 1.235)	M
ρ_{14}	0.565	(0.384, 0.745)	ω_{14}	0.835	(0.714, 0.988)	F
ρ_{15}	−0.385	(−0.753, −0.059)	ω_{15}	0.951	(0.750, 1.263)	F
ρ_{16}	−0.222	(−0.301, −0.142)	ω_{16}	0.593	(0.539, 0.655)	F
ρ_{17}	0.361	(0.065, 0.650)	ω_{17}	0.818	(0.656, 1.057)	M
ρ_{18}	−0.652	(−0.884, −0.448)	ω_{18}	0.719	(0.588, 0.893)	F
ρ_{19}	0.271	(0.082, 0.466)	ω_{19}	0.586	(0.464, 0.750)	M
ρ_{20}	−1.238	(−1.435, −1.018)	ω_{20}	0.428	(0.313, 0.562)	F
ρ_{21}	−0.105	(−0.351, 0.113)	ω_{21}	0.978	(0.825, 1.183)	F
ρ_{22}	−0.074	(−0.294, 0.129)	ω_{22}	0.953	(0.819, 1.134)	M
ρ_{23}	−0.066	(−0.403, 0.229)	ω_{23}	1.006	(0.822, 1.275)	M
ρ_{24}	0.206	(−0.002, 0.408)	ω_{24}	0.752	(0.625, 0.928)	F
ρ_{25}	−0.200	(−0.496, 0.083)	ω_{25}	0.738	(0.574, 0.987)	M
ρ_{26}	−0.070	(−0.266, 0.114)	ω_{26}	0.688	(0.571, 0.851)	F
ρ_{27}	0.514	(0.352, 0.673)	ω_{27}	0.588	(0.475, 0.727)	F
ρ_{28}	−0.251	(−0.367, −0.140)	ω_{28}	0.819	(0.739, 0.914)	M
ρ_{29}	−0.124	(−0.430, 0.139)	ω_{29}	1.058	(0.875, 1.318)	F
ρ_{30}	0.382	(0.077, 0.675)	ω_{30}	0.862	(0.687, 1.114)	M
ρ_{31}	−0.122	(−0.507, 0.214)	ω_{31}	0.994	(0.789, 1.307)	M
ρ_{32}	0.551	(0.354, 0.742)	ω_{32}	0.749	(0.631, 0.906)	F
ρ_{33}	0.010	(−0.030, 0.049)	ω_{33}	0.489	(0.472, 0.508)	F
ρ_{34}	0.117	(0.082, 0.154)	ω_{34}	0.505	(0.487, 0.524)	F
ρ_{35}	0.027	(−0.011, 0.065)	ω_{35}	0.479	(0.462, 0.497)	M
ρ_{36}	0.069	(0.035, 0.102)	ω_{36}	0.516	(0.500, 0.533)	Machine

Note: M = males, F = females; Rater type is provided.

(ideal rating + bias); the probability approaches zero quickly as the chosen category *a* moves away from this most likely category (Patz et al., 2002). A large value of ω_v indicates a substantial probability of a rating in any categories near $\xi + \rho_v$.

Estimates of ω_v ranged from 0.428 (Rater 20) to 1.058 (Rater 29). Rater 20, Rater 35, and Rater 33 are the most reliable in their rating; they consistently score items in the category *a* nearest to $\xi + \rho_v$. Rater 31, Rater 23, and Rater 29 were the least reliable, suggesting that these raters were inconsistent in assigning ratings to similar quality work. None of the raters had very small variability estimates and so, the problem noted by Patz et al. (2002) that raters with a small rating variability would have a poorly estimated rater bias, did not occur in these data.

Rater 20 is particularly interesting because she had a large negative rating bias, but was among the most consistent raters in her scoring. An investigation of her raw ratings shows that she gave the same rating to all items for each of 11 examinees she rated. In other words, within a set of ratings for an examinee, she always used the same category. In addition to this high level of consistency, her ratings were always either one or two. This is a good example of how the HRM may be useful in identifying problematic raters; based on a qualitative review or check of these items, Rater 20's scores might be removed from contributing to estimates of proficiency.

Table 27.6 displays the posterior median and 95% credible interval estimates for the rater type covariate effects. The results show no difference in rating bias attributable to gender; both males and females had a tendency to rate close to the ideal ratings. The effect for the machine rater was −0.233, over 0.2 smaller than males and females, but not practically or statistically significant. Therefore, the machine effect yields ratings that are slightly severe compared to humans, but not enough to make a practical difference in ratings. The rater types were generally in agreement.

In terms of variability, the machine's ratings were significantly more consistent than males' ratings; there was a posterior median difference of 0.272 (0.066, 0.484). Male raters had the highest variability attributable to rater type and were the least consistent in their scoring. It may seem surprising that the machine exhibited any rater variability at all, since, given the same input, the machine should score the same way every time. If, however, the scoring algorithm fails to distinguish levels of the quality of student work that human raters do distinguish, then, the machine scoring will look inconsistent relative to the average human rating. Thus, the HRM can be helpful in identifying flaws in the scoring algorithm for machine rating, relative to human rating.

In summary, if we desired to compare human raters to the machine rater in this example, we might conclude that both humans and the machine are basically in agreement about ratings. Additionally, while there are differences in terms of consistency between males and the machine, it is the machine rater that is more reliable. This result provides evidence supporting the use of the machine for automated scoring.

TABLE 27.6

Posterior Median and 95% Equal-Tailed Credible Interval (CI) Estimates of the Rater-Type Covariate Effects

Effect		Bias			Variability	
		Median	95% CI		Median	95% CI
Male	η_1	−0.002	(−0.315, 0.309)	τ_1	0.871	(0.728, 1.046)
Female	η_2	−0.005	(−0.224, 0.210)	τ_2	0.733	(0.647, 0.834)
Machine	η_3	−0.233	(−0.629, 0.157)	τ_3	0.599	(0.476, 0.754)
Male–Female	η_1–η_2	0.003	(−0.367, 0.382)	τ_1–τ_2	0.139	(−0.036, 0.330)
Male–Machine	η_1–η_3	0.228	(−0.268, 0.735)	τ_1–τ_3	0.272	(0.066, 0.484)
Female–Machine	η_2–η_3	0.228	(−0.219, 0.670)	τ_2–τ_3	0.134	(−0.043, 0.293)

27.6 Discussion

The HRM addresses a fundamental flaw in information accumulation when IRT models (Linacre, 1989; Muraki, 1992) are used to model assessment items rated by multiple raters: these models incorrectly assert that perfect information can be obtained about the examinee's latent proficiency by increasing just the number of ratings on a fixed number of items, without also increasing the number of items (Mariano, 2002). It does so by building a structure similar to that of generalizability theory (GT) models (Brennan, 1992, 2001), but unlike GT, the HRM is designed for the nonlinear relationship between continuous latent proficiency variables and discrete observed ratings (Patz et al., 2002).

The IRT model for multiple raters (IRT-MMR) (Verhelst and Verstralen, 2001) was developed at about the same time as the HRM was being developed, and also combines IRT and GT ideas. Their model differs from the HRM in using a continuous latent variable for each examinee-by-item combination to represent the latent "quality of the response" of that examinee to that item, which is somewhat analogous to the HRM's ideal rating variables. Reasoning about rating student work (Baxter and Junker, 2001; DeCarlo et al., 2011, p. 335) often implicitly assumes discrete ideal ratings, but there is no reason on why one would have to do this, and in fact, arguments have been made that the number of ideal categories is unknown and perhaps different for different raters. A model with continuous latent "quality of response" variables, such as the IRT-MMR, may provide more flexibility in this direction.

The flaw in information accumulation, when IRT or facet models are applied to rated assessment data, arises because multiple ratings of the same items act like a dependent cluster of data. Both models of Patz et al. (2002) and Verhelst and Verstralen (2001) build hierarchical structures to model this cluster dependence, much as traditional GT does for continuous response data. Wilson and Hoskens (2001) developed an alternative IRT model for multiple ratings, in which the cluster dependence is modeled using "rater bundles," which are direct analogs of the "item bundles" or "testlets" of Wilson and Adams (1995), Bradlow et al. (1999), and Wainer et al. (2007). At approximately the same time, Bock et al. (2002) developed a GT-based correction for this cluster dependence, analogous to the "design effects" corrections that are often applied to clustered survey data. More recently, Briggs and Wilson (2007) proposed a direct grafting of GT structures onto an IRT framework.

There has also been some work in extending facet models to accommodate dependence in rated testing data. Wang and Wilson (2005) explicitly discuss local item dependence with many-faceted data, including situations of multiple raters. Their random-effects facets model treats rater bias as a random effect and incorporates a parameter for person and item interactions, thereby permitting intrarater and interrater variation in bias to be assessed. The generalized multilevel facets model (Wang and Liu, 2007) provides a framework for multilevel models with many facets and this framework was also extended to model longitudinal data (Hung and Wang, 2012). These extensions, however, account for the dependence between ratings from the same rater on multiple items, rather than between ratings of the same item by multiple raters. Muckle and Karabatsos (2009) explicitly show how to embed the facets model of Linacre (1989) in the hierarchical generalized linear model (HGLM) framework of Kamata (2001); see Mariano and Junker (2007, pp. 291–292) for a related proposal. Within this framework, many traditional GT coefficients have unambiguous definitions (though their interpretations may be different from that of traditional GT). Muckle and Karabatsos (2009, p. 205) point out that random effects analogous to the "quality of response" variables in the IRT-MMR could be incorporated into the HGLM framework. This would essentially reproduce the IRT-MMR, with a polytomous Rasch model for the discrete response data; see Mariano and Junker (2007, pp. 292–294) for some further details.

The HRM itself has been expanded in two significant ways. First, as outlined in Section 27.2.3, Mariano and Junker (2007) provided a full framework for incorporating covariates of the rating process into the HRM. As demonstrated by our earlier empirical example, this has been important for understanding how covariates affect rater bias and variability. Recent related work of Casabianca and Junker (2013) extends the HRM to longitudinally collected rating data. It is also obvious from the examples presented in Patz et al. (2002) and Mariano and Junker (2007) that the HRM allows for combining modeling of multiple ratings for some items with standard IRT modeling for other items. The less-noisy information from standard IRT models can be helpful for stabilizing the latent proficiency scale for rated responses.

The second expansion of the HRM is due to DeCarlo et al. (2011). They replace the original signal-detection model in Equation 27.3 with an alternate model from the SDT literature (DeCarlo, 2002, 2005). In the new model, each instance of rater r examining a response on item i in an ideal category k results in a normal latent "rater perception" variable whose location and scale is determined by r, i, and k. The rating category is then determined as a discretization of this latent variable with fixed category cutoffs. By varying the location and scale of the rater perception variable, effects such as rater bias and variability, and also other effects such as the tendency to favor central categories (DeCarlo, 2008; Wolfe and McVay, 2002), can be modeled.

In the past 30 years or so, a wide variety of statistical models have been proposed and used for the analysis of multiple ratings of rich examinee responses on assessments, and to account for various item and rater effects in fairly assigning scores. These range from the facets model (Linacre, 1989) to the HRM with covariates (Mariano and Junker, 2007) and HRM–SDT (DeCarlo et al., 2011). Earlier models did not accumulate information about multiple raters correctly from the data (Mariano, 2002). Although much can be learned even from a demonstrably wrong model, the HRM (Patz et al., 2002) and its relatives offer the possibility of correct inferences about raters, items, and examinees, accounting for all sources of dependence and uncertainty in the data.

Routine rating of complex student work may be reassigned from human raters to machines with little loss in reliability or bias, as a variety of studies (e.g., Attali and Burstein, 2005; Steedle and Elliot, 2012; J. Wang and Brown, 2012) show, and as is corroborated by our example in Section 27.5. However, a variety of other professional endeavors, such as teaching practice (Casabianca et al., 2013; Farrokhi et al., 2011; Junker et al., 2006), will continue to be evaluated by multiple human raters. Such data may be combined with simpler item-response-style data (e.g., a self-report or test of basic pedagogical or subject knowledge) to produce an estimate of professional quality. For these purposes, the HRM and its relatives will continue to be useful tools in the measurement toolkit.

References

Akaike, H. 1973. Information theory as an extension of the maximum likelihood principle. In B. N. Petrov and F. Csaki (Eds.), *Second International Symposium on Information Theory* (pp. 267–281). Budapest: Akademiai Kiado.

Ando, T. 2012. Predictive Bayesian model selection. *American Journal of Mathematical and Management Sciences, 31*, 13–38.

Attali, Y., and Burstein, J. 2005. *Automated Essay Scoring with e-Rater v. 2.0* (Technical Report No. RR-04-45). Princeton, NJ: Educational Testing Service.

Baxter, G. P., and Junker, B. W. 2001. Designing cognitive-developmental assessments: A case study in proportional reasoning. Paper presented at the *Annual Meeting of the National Council for Measurement in Education*, Seattle, Washington.

Bock, R., Brennan, R. L., and Muraki, E. 2002. The information in multiple ratings. *Applied Psychological Measurement, 26*(4), 364–375.

Bradlow, E. T., Wainer, H., and Wang, X. H. 1999. A Bayesian random effects model for testlets. *Psychometrika, 64*, 153–168.

Brennan, R. L. 1992. An NCME instructional module on generalizability theory. *Educational Measurement: Issues and Practice, 11*(4), 27–34.

Brennan, R. L. 2001. *Generalizability Theory*. New York: Springer-Verlag.

Briggs, D. C., and Wilson, M. 2007. Generalizability in item response modeling. *Journal of Educational Measurement, 44*(2), 131–155.

Burnham, K. P., and Anderson, D. R. 2004. Multimodel inference: Understanding AIC and BIC in model selection. *Sociological Methods and Research, 33*, 261–304.

Casabianca, J. M. 2012. *Software for the Chapter on HRM*. (Available on request from Jodi Casabianca <jcasabianca@austin.utexas.edu>.)

Casabianca, J. M., and Junker, B. 2013. Hierarchical rater models for longitudinal assessments. (*Annual Meeting of the National Council for Measurement in Education*, San Francisco, CA).

Casabianca, J. M., McCaffrey, D. F., Gitomer, D. H., Bell, C. A., Hamre, B. K., and Pianta, R. C. 2013. Effect of observation mode on measures of secondary mathematics teaching. *Educational and Psychological Measurement, 73*(5), 757–783.

Chib, S. 1995. Marginal likelihood from the Gibbs output. *Journal of the American Statistical Association, 90*, 1313–1321.

Chib, S., and Jeliazkov, I. 2001. Marginal likelihood from Metropolis–Hastings output. *Journal of the American Statistical Association, 96*(453), 270–281.

Claeskens, G., and Hjort, N. L. 2008. *Model Selection and Model Averaging*. New York: Cambridge University Press.

CTB/McGraw-Hill. 2012. *Multiple Rating Data from a Writing Assessment*. Monterey, CA: (Available on request from Jodi Casabianca <jcasabianca@austin.utexas.edu>.)

DeCarlo, L. T. 2002. A latent class extension of signal detection theory with applications. *Multivariate Behavioral Research, 37*, 423–451.

DeCarlo, L. T. 2005. A model of rater behavior in essay grading based on signal detection theory. *Journal of Educational Measurement, 42*, 53–76.

DeCarlo, L. T. 2008. *Studies of a Latent-Class Signal-Detection Model for Constructed Response Scoring* (Technical Report). Princeton, NJ (ETS Research Report No. RR-08-63).

DeCarlo, L. T., Kim, Y.-K., and Johnson, M. S. 2011. A hierarchical rater model for constructed responses with a signal detection rater model. *Journal of Educational Measurement Fall 2011, 48*(3), 333–356.

Dempster, A. P., Laird, N. M., and Rubin, D. B. 1977. Maximum likelihood from incomplete data via the EM algorithm. *Journal of the Royal Statistical Society Series B (Methodological), 39*(1), 1–38.

Donoghue, J. R., and Hombo, C. M. 2000. A comparison of different model assumptions about rater effects. Paper presented at the *Annual Meeting of the National Council on Measurement in Education*, New Orleans, LA.

Farrokhi, F., Esfandiari, R., and Vaez Dalili, M. 2011. Applying the many-facet Rasch model to detect centrality in self-assessment, peer-assessment and teacher assessment. *World Applied Sciences Journal (Innovation and Pedagogy for Lifelong Learning), 15*, 70–77.

Fischer, G. H. 1973. The linear logistic test model as an instrument in educational research. *Acta Psychologica, 37*, 359–374.

Gelman, A., Carlin, J. B., Stern, H. S., and Rubin, D. B. 2003. *Bayesian Data Analysis*, second edition. London: CRC Press.

Gelman, A., and Hill, J. 2006. *Data Analysis Using Regression and Multilevel/Hierarchical Models.* New York: Cambridge University Press.

Gelman, A., and Meng, X. L. 1998. Simulating normalizing constants: From importance sampling to bridge sampling to path sampling. *Statistical Science, 13,* 163–185.

Gelman, A., and Shalizi, C. R. 2013. Philosophy and the practice of Bayesian statistics. *British Journal of Mathematical and Statistical Psychology, 66,* 8–38.

Glas, C. A. W., and Pimentel, J. 2008. Modeling nonignorable missing data in speeded tests. *Educational and Psychological Measurement, 68,* 907–922.

Holman, R., and Glas, C. A. W. 2005. Modelling non-ignorable missing-data mechanisms with item response theory models. *The British Journal of Mathematical and Statistical Psychology, 58*(Pt 1), 1–17. Retrieved from http://www.ncbi.nlm.nih.gov/pubmed/15969835 doi: 10.1348/000711005X47168

Hombo, C., and Donoghue, J. R. 2001. Applying the hierarchical raters model to NAEP. Paper presented at the *Annual Meeting of the National Council on Measurement in Education,* Seattle, Washington.

Hung, L. F., and Wang, W. C. 2012. The generalized multilevel facets model for longitudinal data. *Journal of Educational and Behavioral Statistics, 37,* 231–255.

Junker, B. W., Weisberg, Y., Clare-Matsumura, L., Crosson, A., Wolk, M. K., Levison, A., and Resnick, L. B. 2006. *Overview of the Instructional Quality Assessment* (Technical Report). Los Angeles, CA. (CSE Report 671, http://www.cse.ucla.edu/products/reports.asp)

Kamata, A. 2001. Item analysis by the hierarchical generalized linear model. *Journal of Educational Measurement, 38,* 79–93.

Kass, R., and Raftery, A. 1995. Bayes factors. *Journal of the American Statistical Association, 90,* 773–795.

Levy, R., Mislevy, R. J., and Sinharay, S. 2009. Posterior predictive model checking for multidimensionality in item response theory. *Applied Psychological Measurement, 33*(7), 519–537.

Linacre, J. M. 1989. *Many-Faceted Rasch Measurement.* Chicago, IL: MESA Press.

Lunn, D. J., Thomas, A., Best, N., and Spiegelhalter, D. 2000. Winbugs—A Bayesian modelling framework: Concepts, structure, and extensibility. *Statistics and Computing, 10,* 325–337.

Mariano, L. T. 2002. *Information Accumulation, Model Selection and Rater Behavior in Constructed Response Student Assessments.* Unpublished doctoral dissertation, Department of Statistics, Carnegie Mellon University, Pittsburgh, PA.

Mariano, L. T., and Junker, B. W. 2007. Covariates of the rating process in hierarchical models for multiple ratings of test items. *Journal of Educational and Behavioral Statistics, 32,* 287–314.

Mislevy, R. J., and Wu, P. K. 1996. *Missing Responses and IRT Ability Estimation: Omits, Choice, Time Limits, and Adaptive Testing* (ETS Technical Report). Princeton, NJ: RR-96-30-ONR.

Muckle, T., and Karabatsos, G. 2009. Hierarchical generalized linear models for the analysis of judge ratings. *Journal of Educational Measurement, 46,* 198–219.

Muraki, E. 1992. A generalized partial credit model: Application of an EM algorithm. *Applied Psychological Measurement, 16,* 159–177.

Neal, R. M. 2001. Annealed importance sampling. *Statistics and Computing, 11,* 125–139.

Patz, R. J. 1996. *Markov Chain Monte Carlo Methods for Item Response Theory Models with Applications for the National Assessment of Educational Progress.* Unpublished doctoral dissertation, Department of Statistics, Carnegie Mellon University, Pittsburgh, PA.

Patz, R. J., and Junker, B. W. 1999a. Applications and extensions of MCMC in IRT: Multiple item types, missing data, and rated responses. *Journal of Educational and Behavioral Statistics, 24*(4), 342–366. Retrieved from http://jeb.sagepub.com/cgi/doi/10.3102/10769986024004342 doi: 10.3102/10769986024004342

Patz, R. J., and Junker, B. W. 1999b. A straightforward approach to Markov chain Monte Carlo methods for item response models. *Journal of Educational and Behavioral Statistics, 24*(2), 146–178. Retrieved from http://jeb.sagepub.com/cgi/doi/10.3102/10769986024002146 doi: 10.3102/10769986024002146

Patz, R. J., Junker, B. W., Johnson, M. S., and Mariano, L. T. 2002. The hierarchical rater model for rated test items and its application to large-scale educational assessment data. *Journal of Educational and Behavioral Statistics, 27,* 341–384.

R Development Core Team. 2012. *R: A Language and Environment for Statistical Computing* (Technical Report). Vienna, Austria: R Foundation for Statistical Computing. Retrieved from http://www.r-project.org/

Rubin, D. B., and Little, R. J. A. 2002. *Statistical Analysis with Missing Data.* Hoboken, NJ: John Wiley & Sons.

Schwarz, G. 1978. Estimating the dimension of a model. *Annals of Statistics, 6,* 461–464.

Seltman, H. 2010. *R Package RUBE (Really Useful Winbugs Enhancer).* http://www.stat.cmu.edu/~hseltman/rube/. (Version 0.2-13)

Sinharay, S., Johnson, M. S., and Stern, H. S. 2006. Posterior predictive assessment of item response theory models. *Applied Psychological Measurement, 30*(4), 298–321.

Spiegelhalter, D. J., Best, N. G., Carlin, B. P., and van der Linde, A. 2002. Bayesian measures of model complexity and fit (with discussion). *Journal of the Royal Statistical Society, Series B, 64,* 583–639.

Steedle, J., and Elliot, S. 2012. *The Efficacy of Automated Essay Scoring for Evaluating Student Responses to Complex Critical Thinking Performance Tasks* (Technical Report). New York, NY. (Collegiate Learning Assessment Whitepaper, obtained October 2012 from http://www.collegiatelearningassessment.org/research)

Verhelst, N. D., and Verstralen, H. H. F. M. 2001. An IRT model for multiple raters. In A. Boomsma, M. A. J. V. Duijn, and T. A. B. Snijders (Eds.), *Essays on Item Response Modeling* (pp. 89–108). New York: Springer-Verlag.

Wainer, H., Bradlow, E., and Wang, X. 2007. *Testlet Response Theory and Its Applications.* New York, NY: Cambridge University Press.

Wang, J., and Brown, M. S. 2012. Automated essay scoring versus human scoring: A comparative study. *Journal of Technology, Learning, and Assessment, 6*(2). (Retrieved October 2012 from http://www.jtla.org)

Wang, W., and Wilson, M. 2005. Exploring local item dependence using a facet random-effects facet model. *Applied Psychological Measurement, 29,* 296–318.

Wang, W. C., and Liu, C. Y. 2007. Formulation and application of the generalized multilevel facets model. *Educational and Psychological Measurement, 67,* 583–605.

Wilson, M., and Hoskens, M. 2001. The rater bundle model. *Journal of Educational and Behavioral Statistics, 26,* 283–306.

Wilson, M. R., and Adams, R. J. 1995. Rasch models for item bundles. *Psychometrika, 60,* 181–198.

Winke, P., Gass, S., and Myford, C. 2011. The relationship between raters' prior language study and the evaluation of foreign language speech samples. *ETS Research Report Series, 2011*(2), i–67.

Wolfe, E. W., and McVay, A. 2002. Application of latent trait models to identifying substantively interesting raters. *Educational Measurement: Issues and Practice, 31,* 31–37.

Wu, C. F. J. 1983. On the convergence properties of the EM algorithm. *Annals of Statistics, 11*(1), 95–103.

Yen, S. J., Ochieng, C., Michaels, H., and Friedman, G. 2005. The effect of year-to-year rater variation on IRT linking. Paper presented at the *Annual Meeting of the American Educational Research Association,* Montreal, Canada.

Zhu, X., and Stone, C. A. 2011. Assessing fit of unidimensional graded response models using Bayesian methods. *Journal of Educational Measurement, 48*(1), 81–97.

28

Randomized Response Models for Sensitive Measurements

Jean-Paul Fox

CONTENTS

28.1 Introduction 467
28.2 Presentation of the Models 468
28.2.1 Randomized IRT Models 469
28.2.2 Noncompliant Behavior 470
28.2.3 Structural Models for Sensitive Constructs 471
28.3 Parameter Estimation 471
28.4 Model Fit 473
28.5 Empirical Example 473
28.5.1 College Alcohol Problem Scale and Alcohol Expectancy Questionnaire 473
28.5.2 Data 474
28.5.3 Model Specification 474
28.5.4 Results 475
28.6 Discussion 477
Acknowledgments 478
Appendix 28A: CAPS-AEQ Questionnaire 478
References 479

28.1 Introduction

Research on behavior and attitudes typically relies on self-reports, especially when the infrequency of behavior and the research costs make it hardly impractical not to do so. However, many studies have shown that self-reports can be highly unreliable and actually serve as a fallible source of data.

Results from self-reports are often influenced by such factors as the question order, wording, or response format, even when they contain simple behavioral questions. In general, the psychology of asking questions has received considerable attention in the literature (e.g., Sudman et al., 1996; Tourangeau et al., 2000), and a growing body of research has provided sound methods of question development that do improve the quality of self-report data.

Finally, the quality of self-report data depends on the respondents' willingness to cooperate and give honest answers. Especially for sensitive topics, it is known that people tend to report in a socially desirable way; that is, in the direction of the researcher's expectations and/or what reflects positively on their behavior. Thus, the sensitivity of the questions easily leads to misreporting (i.e., under- or overreporting), even when anonymity and confidentiality of the responses is guaranteed. This observation is supported by considerable empirical evidence. For instance, survey respondents underreported socially undesirable behavior such as the use of illicit drugs (Anglin et al., 1993), the number of sex partners (Tourangeau and Smith, 1996), desires for adult entertainment (De Jong et al., 2010), welfare fraud (van der Heijden et al., 2000), and alcohol abuse and related problems (Fox and Wyrick, 2008).

Strategies have been developed to overcome respondents' tendencies to report inaccurately or even to refuse to provide any response at all. Typically, anonymity of the respondents is guaranteed and explicit assurances are given that each of their answers will remain completely confidential. Besides, questions are often phrased such that tendencies to provide socially desirable answers are diminished. Furthermore, respondents are motivated to provide accurate answers by stressing the importance of the research study.

Other ways of avoiding response tendencies to report inaccurately are based on innovative data collection methods that make it impossible to infer any identifying information from the response data. A general class of such methods for sensitive surveys is based on the randomized response technique (RRT) (Fox and Tracy, 1986), which involves the use of a randomizing device to mask individual responses. RRT has originated from Warner (1965), who developed a randomized response (RR) data collection procedure, where respondents are confronted with two mutually exclusive questions, for instance, "I belong to Group A," and "I do not belong to Group A." A choice is made between the two statements using a randomizing device (e.g., tossing of a die or use of a spinner). The randomization is performed by the respondent and the outcome is not revealed to the interviewer. The respondent then answers the question selected by the randomizing device. The interviewer only knows the response, not the question.

Because of this setup, the RR technique encourages greater cooperation from respondents and reduces socially desirable response behavior. The properties of the randomizing device are known, which still allows for population estimates of the sensitive behavior, for instance, proportions of the population engaging in a particular kind of behavior or, more generally, membership of Group A. Further analysis of the univariate RR data is limited to inferences at this aggregate data level.

Measurements of individual sensitive behaviors require support by multivariate randomized item-response data. The purpose of this chapter is to give an overview of item-response theory models modified such that they are suitable for the analysis of multivariate RR data. The general class of such models is referred to as randomized item-response theory (RIRT) models (Fox 2005; Fox and Wyrick 2008) or item randomized-response (IRR) models (Böckenholt and van der Heijden, 2007). Different RIRT models for binary, ordinal, and mixed responses are presented. Furthermore, it is shown how to extend these models to handle noncompliance, that is, when respondents do not follow the RR instructions (Clark and Desharnais, 1998) as well as to allow for measurement of multidimensional constructs. Our compensatory multidimensional modeling approach generalizes the noncompensatory model by Böckenholt and van der Heijden (2007), who considered multiple item bundles each measuring a specific construct given binary response data.

28.2 Presentation of the Models

In Warner's (1965) approach, a randomizing device (e.g., die, spinner) is required to make the choice between two logically opposite questions. The setup guarantees the confidentiality of each individual response, which cannot be related to either of the opposite questions.

Greenberg et al. (1969) proposed a more general unrelated question technique, where the outcome of the randomizing device controls the choice between a sensitive question and an irrelevant unrelated question.

Edgell et al. (1982) generalized the procedure by introducing an additional randomizing device to generate the answer on the unrelated question. The responses are then completely protected since it becomes impossible to infer whether they are answers to the sensitive question or forced answers generated by the randomizing device. Let the randomizing device select the sensitive question with probability ϕ_1 and a forced response with probability $1 - \phi_1$. The latter is supposed to be a success with probability ϕ_2. Let U_{pi} denote the RR of person $p = 1, \ldots, P$ to item $i, 1, \ldots, I$. Consider a success a positive response (score one) to a question and a failure a negative response (score zero). Then, the probability of a positive RR is represented by

$$P\{U_{pi} = 1; \phi_1, \phi_2\} = \phi_1 P\{\tilde{U}_{pi} = 1\} + (1 - \phi_1)\phi_2, \tag{28.1}$$

where $\tilde{U}_{pi}$ is the underlying response, which is referred to as the true response of person p to item i when directly and honestly answering the question.

For a polytomous RR, let $\phi_2(a)$ denote the probability of a forced response in category a for $a = 1, \ldots, A_i$ such that the number of response categories may vary over items. The probability of an RR of individual p in category a of item i is given by

$$P\{U_{pi} = a; \phi_1, \phi_2\} = \phi_1 P\{\tilde{U}_{pi} = a\} + (1 - \phi_1)\phi_2(a). \tag{28.2}$$

It follows that the forced RR model is a two-component mixture model, with the first component modeling the responses to the sensitive question and the second component modeling the forced responses. The mixture probabilities are controlled by the randomizing device. When $\phi_1 > 0.5$, the RR data contain sufficient information to make inferences from the responses.

28.2.1 Randomized IRT Models

In a multivariate setting, multiple items are used to measure an individual latent variable or construct (e.g., alcohol dependence; academic fraud) from multiple correlated randomized item responses. In this setting, the characteristics of the randomizing device are allowed to vary over items. For example, although they relate to the same sensitive latent variable, the sensitivity of items may vary. The variation in sensitivity can then be controlled by adjusting the randomizing device properties. But this option will not be further discussed here.

In randomized IRT modeling, the goal is to model the true item responses, $\tilde{U}$, which are latent because they are randomized before being observed. For dichotomous response data, the two-parameter (2PL) normal-ogive model defines the probability of a positive response given the sensitive latent construct θ_p and item discrimination and difficulty parameter a_i and b_i, respectively, which is given by

$$\pi_{pi} = P\{\tilde{U}_{pi} = 1; \theta_p, a_i, b_i\} = \Phi(a_i(\theta_p - b_i)), \tag{28.3}$$

where $\Phi(.)$ denotes the cumulative normal distribution function.

For polytomous responses, the probability of a response in category a of person p is supposed to be given by

$$\begin{aligned}\pi_{pi}(a) &= P\{\tilde{U}_{pi} = a; \theta_p, a_i, \mathbf{b}_i\} \\ &= \Phi(a_i(\theta_p - b_{i,(a-1)})) - \Phi(a_i(\theta_p - b_{i,a})),\end{aligned} \tag{28.4}$$

where vector b_i contains the threshold parameters of item i, which follow the order restriction: $b_i1 < \ldots < b_{iA}$ for response alternatives $a = 1, \ldots, A$ (for more on this type of graded response model, see Volume One, Chapter 6).

The last model can be extended to deal with questionnaires with items that measure multiple sensitive constructs. Let the multidimensional vector $\boldsymbol{\theta}_i$ of dimension D denote these constructs. Then, the probability of a true response in category a is

$$\begin{aligned}\pi_{pi}(a) &= P\{\tilde{U}_{pi} = a; \boldsymbol{\theta}_p, \boldsymbol{a}_i, \boldsymbol{b}_i\} \\ &= \Phi(\boldsymbol{a}_i^t(\boldsymbol{\theta}_p - b_{i,(a-1)})) - \Phi(\boldsymbol{a}_i^t(\boldsymbol{\theta}_p - b_{i,a})),\end{aligned} \tag{28.5}$$

where the vector of discriminations (factor loadings) of dimension D specifies the weights for each underlying dimension.

The models in Equations 28.3 through 28.5 can be embedded in an RR modeling framework. For example, for the two-parameter normal-ogive model for the true responses, the overall model becomes

$$P\{U_{pi} = 1; \theta_p, a_i, b_i\} = \phi_1 P\{\tilde{U}_{pi} = 1; \theta_p, a_i, b_i\} + (1 - \phi_1)\phi_2. \tag{28.6}$$

Thus, by combining the RRT with an IRT model, latent individual sensitive traits can be measured given observed randomized item responses. A major advantage of using an IRT model is its separation of item parameters and person parameters. Consequently, it can be used to interpret individual differences on the latent trait that is measured, allows for more complex test designs, and handles measurement error at the individual level.

28.2.2 Noncompliant Behavior

Despite the protection of privacy offered by randomized response techniques, some respondents may still show noncompliant behavior and consistently select the least self-incriminating response and completely ignore the RR instructions. Clark and Desharnais (1998) proposed a method to estimate the extent of noncompliance using two sampled groups each confronted with different RR designs. Böckenholt and van der Heijden (2007) and Cruyff et al. (2007) proposed the use of a two-component latent class model, where one group consists of respondents that follow the RR instructions and a second group of respondents that does not follow them.

RIRT modeling can be extended to account for noncompliance. To do so, let a binary latent class variable be $G_{pi} = 1$ when person p responds to item i in a noncompliant (self-protective) way and $G_{pi} = 0$ when p responds in a compliant way. Then, the randomized item-response model in Equation 28.6 is

$$P\{U_{pi} = 0\} = P\{G_{pi} = 0\} P\{U_{pi} = 0; \theta_p, a_i, b_i\} + P\{G_{pi} = 1\} I(U_{pi} = 0),$$

where $I(U_{pi} = 0)$ equals one when the answer to item i of respondent p is zero and equals zero otherwise. This mixture model consists of a randomized item-response model for the compliant class but a different model for the noncompliant class. Inferences are made from the responses by the compliant class, which requires information about the behavior of the respondents. That is, the assumption of an additional response model for G_{pi} is required (e.g., De Jong et al., 2010; Fox, 2010).

28.2.3 Structural Models for Sensitive Constructs

Respondents are usually independently sampled from a population, and a normal distribution is often used to describe the distribution of the latent variable. If so, the population model for the latent person variable is

$$\theta_p \sim N(\mu_\theta, \sigma_\theta^2).$$

For more complex sampling designs, respondents can be clustered, and the model for the population distribution needs to account for the dependencies between respondents in the same cluster. As described, among others, by Fox (2010) and Fox and Glas (2001; Volume One, Chapter 24), a multilevel population distribution for the latent person parameters needs to be defined. Let θ_{pj} denote the latent parameter of person p in group $j (j = 1, \ldots, J)$. The population distribution becomes

$$\theta_{pj} \sim N(\beta_j, \sigma_\theta^2),$$
$$\beta_j \sim N(\mu_\theta, \tau_{00}^2).$$

Or, for the multidimensional case,

$$\boldsymbol{\theta}_p \sim N(\mu_\theta, \Sigma_\theta),$$

where the covariance matrix of dimension D specifies the within-person correlations. This multidimensional model can also be extended to include a multilevel setting, but this case will not be discussed. Also, to explain variation between persons in latent sensitive measurements, explanatory variables at the level of persons and/or groups can also be included. Finally, variation in item parameters can also be modeled as described in De Boeck and Wilson (2004; Volume One, Chapter 33) and De Jong et al. (2010).

28.3 Parameter Estimation

A fully Bayesian estimation method with Markov chain Monte Carlo (MCMC) sampling from the posterior distribution of the parameters is presented. The method requires prior distributions for all model parameter. Noninformative inverse gamma priors are specified for the variance components. An inverse Wishart prior is specified for the covariance matrix. Normal and lognormal priors are specified for the difficulty and discrimination parameters, respectively. A uniform prior is specified for the threshold parameters while accounting for the order constraint.

Following the MCMC sampling procedure for IRR data in Fox (2005, 2010), Fox and Wyrick (2008), and De Jong et al. (2010), a fully Gibbs sampling procedure is developed which consists of a complex data augmentation scheme: (i) sampling of latent true responses, $\tilde{U}$; (ii) sampling latent continuous response data, $\boldsymbol{Z}$; and (iii) sampling latent class membership $\boldsymbol{G}$. The item-response model parameters and structural model parameters are sampled in a straightforward way given the continuous augmented data, as described by Fox (2010) and Johnson and Albert (2001).

Omitting conditioning on $G_{pi} = 0$ for notational convenience, the procedure is described for latent response data generated only for responses belonging to the compliant class. A probabilistic relationship needs to be defined between the observed RR data and the true response data. To do so, define $H_{pi} = 1$ when the randomizing device determines that person i answers item i truthfully and $H_{pi} = 0$ when a forced response is generated. It follows that the conditional distribution of a true response a given a RR a' is given by

$$P\{\tilde{U}_{pi} = a' \mid U_{pi} = a\} = \frac{P\{\tilde{U}_{pi} = a', U_{pi} = a\}}{P\{U_{pi} = a\}}$$

$$= \frac{\sum_{l \in \{0,1\}} P\{\tilde{U}_{pi} = a', U_{pi} = a \mid H_{pi} = l\} P\{H_{pi} = l\}}{\sum_{l \in \{0,1\}} P\{U_{pi} = a \mid H_{pi} = l\} P\{H_{pi} = l\}},$$

where $a, a' = \{0,1\}$ and $\{1,2, \ldots, A_i\}$ for binary and polytomous responses, respectively.

For binary responses, π_{pi} in Equation 28.3 defines the probability of a success. Subsequently, the latent responses are Bernoulli distributed,

$$\tilde{U}_{pi} \mid U_{pi} = 1, \pi_{pi} \sim B\left(\lambda = \frac{\pi_{pi}(p_1 + p_2(1-p_1))}{p_1\pi_{pi} + p_2(1-p_1)}\right),$$

$$\tilde{U}_{pi} \mid U_{pi} = 0, \pi_{pi} \sim B\left(\lambda = \frac{\pi_{pi}(1-p_1)(1-p_2)}{1-(p_1\pi_{pi} + p_2(1-p_1))}\right).$$

For polytomous response data, π_{pi} is defined in Equation 28.4 or 28.5, and $\tilde{U}_{pi}$ given $U_{pi} = a$ is multinomially distributed with cell probabilities

$$\Delta(a) = \frac{\pi_{pi}(a')p_1 I(a = a') + \pi_{pi}(a')(1-p_1)p_2(a)}{\pi_{pi}(a)p_1 + (1-p_1)p_2(a)}.$$

Following the data augmentation procedure of Johnson and Albert (2001) and Fox (2010), latent true response data are sampled given the augmented dichotomous or polytomous true response data.

The latent class memberships, G_{pi}, are generated from a Bernoulli distribution. Let $Y_{pi} = 0$ define the least self-incriminating response, then the success probability of the Bernoulli distribution can be expressed as

$$\frac{P\{G_{pi} = 1\} I(Y_{pi} = 0)}{P\{G_{pi} = 0\} P\{Y_{pi} = 0 \mid \theta_p, a_i, b_i\} + P\{G_{pi} = 1\} I(Y_{pi} = 0)},$$

where a Bernoulli prior is usually specified for the class membership variable G_{pi}.

Given the augmented data, class memberships, true responses, and latent true responses, all other model parameters can be sampled using a full Gibbs sampling algorithm. The full conditionals can be found in the MCMC literature for IRT (e.g., Junker et al., Volume Two, 2, Chapter 15).

28.4 Model Fit

A Bayesian residual analysis can be performed to evaluate the fit of the model. Residual analysis for binary and polytomous item-response models has been suggested by De Jong et al. (2010), Fox (2010), Geerlings et al. (2011), and Johnson and Albert (2001). Posterior distributions of the residuals can be used to evaluate their magnitude and make probability statements about them. Bayesian residuals are easily computed as by-products of the MCMC algorithm, and they can be summarized to provide information about specific model violations. For instance, sums of squared residuals can be used as a discrepancy measure for evaluating person or item fit. The extremeness of the observed discrepancy measure can be evaluated using replicated date generated under their posterior predictive distribution. Likewise, the assumption of local independence and unidimensionality can be checked using appropriate discrepancy measures. For an introduction to posterior predictive checks, see Sinharay (Volume Two, Chapter 19). Studies of different posterior predictive checks for Bayesian IRT models are reported in Glas and Meijer (2003), Levy et al. (2009), Sinharay et al. (2006), and Sinharay (2006).

28.5 Empirical Example

In a study of alcohol-related expectancies and problem drinking, responses to 13 items of the College Alcohol Problem Scale (CAPS; O'Hare, 1997) and four items of the Alcohol Expectancy Questionnaire (AEQ; Brown, Christiansen, and Goldman, 1987) were analyzed. The goal was to measure the sensitive constructs underlying both scales using multidimensional item-response theory. Furthermore, it was investigated whether the RRT improved the accuracy of the self-reports obtained by direct questions.

28.5.1 College Alcohol Problem Scale and Alcohol Expectancy Questionnaire

As an initial screening instrument, the College Alcohol Problem Scale (CAPS) instrument was developed to measure drinking problems among youth. Its items covered socio-emotional problems, such as hangovers, memory loss, nervousness, and depression, as well as community problems, such as driving under the influence, engaging in activities related to illegal drugs, and problems with the law. The questionnaire items are given in Appendix 28A. Self-reported information about negative consequences of drinking is likely to be biased due to socially desirable responding. Consequently, the survey was expected to lead to refusals to respond and responses given to conceal undesirable behavior. Therefore, an RRT was used to improve both the cooperation by the respondents and the accuracy of their self-reports.

The Alcohol Expectancy Questionnaire (AEQ) measures the degree of expectancies associated with drinking alcohol. Alcohol-related expectancies are known to influence alcohol use and behavior while drinking. The adult form of the AEQ consisted of 90 items and covers six dimensions. But in the study the focus was on alcohol-related sexual enhancement expectancies. The items covering sexual enhancement expectancies are given in Appendix 28A. The data were collected on a five-point ordinal scale, ranging from one (almost never) to five (almost always).

The CAPS data were reanalyzed by Fox and Wyrick (2008), who used a unidimensional randomized-item response model to measure general alcohol dependence. Although the model described the data well, CAPS was developed by O'Hare (1997) to measure different psychosocial dimensions of problem drinking among college students. Two of the dimensions, socio-emotional and community problems, were identified by analysis. Together, they explained more than 60% of the total variance of the responses. In the present study, a multidimensional modeling approach was carried out to investigate whether the CAPS data supported the measurement of multiple sensitive constructs given randomized responses. Besides, the multidimensional model was also used to jointly analyze the CAPS and AEQ data for the relationships between the multiple factors they measure. Finally, the effects of RRT on the measurement of its factors were jointly analyzed.

28.5.2 Data

A total of 793 students from four local colleges/universities, Elon University ($N = 495$), Guilford Technical Community College ($N = 66$), University of North Carolina ($N = 166$), and Wake Forest University ($N = 66$), participated in the survey study in 2002. Both the CAPS and AEQ items were administered to them and their age, gender, and ethnicity was recorded. It was logistically not possible to randomly assign students to the direct questioning (DQ) or the RR condition. However, it was possible to randomly assign classes of five to ten participants to one of the conditions.

A total of 351 students were assigned to the DQ condition. They served as the control group and were instructed to answer the questionnaire as they normally would. A total of 442 students in the RR condition received a spinner to assist them in completing the questionnaire. For each item, the spinner was used as a randomizing device, which determined whether to answer honestly or to give a forced response. According to a forced response design, the properties of the spinner were set such that an honest answer was requested with a probability of 0.60 and a forced response with a probability of 0.40. When a forced response was to be given, each of the five possible responses had a probability of 0.20.

28.5.3 Model Specification

The following multidimensional randomized-item response model was used to analyze the data,

$$\begin{aligned} P(Y_{pi} = a \mid \boldsymbol{\theta}_p, \boldsymbol{a}_i, \boldsymbol{b}_i) &= p_1 \pi_{pi} + (1 - p_1) p_2(a) \\ \pi_{pi} &= \Phi(\mathbf{a}_i^t(\boldsymbol{\theta}_p - b_{i,(a-1)})) - \Phi(\mathbf{a}_i^t(\boldsymbol{\theta}_p - b_{i,a})), \\ \boldsymbol{\theta}_p &\sim N(\boldsymbol{\mu}_{\theta,p}, \boldsymbol{\Sigma}_\theta) \\ \boldsymbol{\mu}_{\theta,p} &= \boldsymbol{\beta}_0 + \boldsymbol{\beta}_1 RR_p \end{aligned} \tag{28.7}$$

for $a = 1, \ldots, 5$ and $i = 1, \ldots, 17$. As just indicated, in the forced RR sampling design, $p_1 = 0.60$ and $p_2(a) = 0.20$, for $a = 1, \ldots, 5$, whereas for the direct-questioning conditioning $p_1 = 1$. The explanatory variable RR_p was equal to one when student p belonged to the RR group and equal to zero otherwise. The factor loadings, a, and item thresholds were assumed to be independent of the questioning technique.

Following Béguin and Glas (2001), the model was identified by fixing the mean score for each dimension, such that $\boldsymbol{\beta}_0 = \mathbf{0}$, while the variance components for each factor to set equal to one. To avoid the so-called rotational variance, one item was assigned uniquely to each of the Q dimensions.

The MCMC algorithm was used to estimate simultaneously all model parameters using 50,000 iterations, with a burn-in period of 10,000 iterations.

28.5.4 Results

In Table 28.1, the estimated factor loadings for a three-factor of the multidimensional RIRT model in Equation 28.7 are given. The factor loadings were standardized by dividing each of them by the average item loading. Furthermore, for each factor the sign of the loadings was set such that a higher latent score corresponded to a higher observed score. To avoid label switching, items 1, 5, and 14 were allowed to have one free nonzero loading, so that each of these items represented one factor.

Items 1–4, 6, 8, and 9 were positively associated with the first factor and had factor loadings higher than 0.60. This first factor represents drinking-related socio-emotional problems, including depression, anxiety, and troubles with family. These problems increased with alcohol consumption. Some of the items also loaded on the two other factors.

The second factor (community problems) covered items 5, 7, and 10–13, with loadings higher than 0.60, except for item 12. In the literature, item 12 has been associated with factor community problems, but in our analysis the item also related to the other factors, most strongly to the third. This second factor covers the acute physiological effects of drunkeness together with illegal and potentially dangerous activities (e.g., driving under the influence).

As expected, items 14–17 were associated with a third factor, which represented alcohol-related sexual enhancement expectancies. These expectancies increased with alcohol consumption but, given their negative loadings on the other two factors, slightly reduced the socio-emotional and community problems.

The multivariate latent factor model was extended with an explanatory variable denoted as RR, which indicated when a student was assigned to the RR (RR = 1) or the DQ condition (RR = 0). In addition, an indicator variable was included, which was set equal to one when the respondent was a female. Both explanatory variables were used for each factor. The RIRT model was further extended with a multivariate population model for all factors.

In Table 28.2, the parameter estimates of the three-factor and a two-factor model are given. For the latter, the loadings of items 1 and 14 were fixed to identify two factors, with one factor representing a composite measure of alcohol-related problems (i.e., socio-emotional and community problems) and the other alcohol-related sexual enhancement expectancies. A moderate positive correlation of 0.65 between the two factors was found.

The students in the RR condition scored significantly higher on both factors. For the RR group, the average latent scores were 0.20 and 0.22 on the composite problem and the alcohol-related expectancy factors, respectively, but both were equal to zero for the DQ group. The RR effect was slightly smaller than that of 0.23 reported by Fox and Wyrick (2008), who performed a unidimensional RIRT analysis using the CAPS items only. A comparable effect was found for the AEQ scale. Females and males showed comparable scores on both factors.

TABLE 28.1

CAPS-EAQ Scale: Weighted Factor Loadings for the Three-Component Analysis

	Three-Factor RIRT Model		
Subscale Items	**Factor 1**	**Factor 2**	**Factor 3**
Socio-Emotional Problems			
1. Feeling sad, blue, or depressed	1.00	0.00	0.00
2. Nervousness or irritability	1.00	0.01	−0.03
3. Hurt another person emotionally	0.96	0.27	0.10
4. Family problems related to drinking	0.82	0.56	0.14
6. Badly affected friendship	0.85	0.46	0.27
8. Other criticize your behavior	0.77	0.50	0.41
9. Nausea or vomiting	0.70	0.39	0.60
Community Problems			
5. Spent too much money on drugs	0.00	1.00	0.00
7. Hurt another person physically	0.48	0.84	0.26
10. Drove under the influence	0.43	0.74	0.53
11. Spent too much money	0.59	0.66	0.47
12. Feeling tired or hung over	0.57	0.41	0.71
13. Illegal activities	0.05	0.96	0.29
Sexual Enhancement			
14. I often feel sexier	0.00	0.00	1.00
15. I'm a better lover	−0.09	−0.12	0.99
16. I enjoy having sex more	−0.14	−0.06	0.99
17. I am more sexually responsive	−0.17	0.03	0.99

In the three-factor model, with the estimated loadings given in Table 28.1, the problems associated with drinking were represented by two factors (i.e., socio-emotional and community problems) and sexual enhancement expectancies by another factor. The RR effects were significantly different from zero for all three factors, while the effect on the factor representing community problems related to alcohol use was approximately 0.32. This was slightly higher than the effects of the other components, which were around 0.21. It seemed as if the students were less willing to admit to alcohol-related community problems and gave more socially desirable responses than for the other factors.

The male students scored significantly higher than the female students on the factor representing community problems related to alcohol use. That is, male students were more likely to experience alcohol-related community problems than females. This gender effect was not found for the other factors. The estimated effects indicated that the RR-group scored significantly higher in comparison to the DQ-group on each subscale. Although validation data are not available, the RR technique was expected to have led to an improved willingness of the students to answer truthfully, given their random assignment to the direct questioning and RR conditions.

Finally, the three factors yielded moderate positive correlations, as shown in Table 28.2. The factors community and socio-emotional problems correlated positively with sexual enhancement expectancies due to alcohol use. In-line with the alcohol expectancy theory, more positive expectancies of alcohol use lead to more positive drinking experiences, which in turn lead to more positive expectancies. Here, an increased expectancy of sexual

TABLE 28.2

CAPS-EAQ Scale: Parameter Estimates of Two- and Three-Component Randomized Item-Response Model

	Two Factor		Three Factor	
Parameter	**Mean**	**SD**	**Mean**	**SD**
Fixed Effects				
Socio-Emotional/Community				
γ_{11} (RR)	0.20	0.09	0.21	0.10
γ_{21} (Female)	0.01	0.06	0.05	0.07
Sexual Enhancement Expectancy				
γ_{12} (RR)	0.22	0.06	0.21	0.07
γ_{22} (Female)	0.03	0.04	0.06	0.05
Community				
γ_{13} (RR)			0.32	0.10
γ_{23} (Female)			−0.30	0.09
Variance Parameters				
$\Sigma_{\theta 11}$	0.96	0.05	0.98	0.05
$\Sigma_{\theta 12}$	0.65	0.07	0.55	0.06
$\Sigma_{\theta 13}$			0.38	0.08
$\Sigma_{\theta 22}$	0.98	0.05	1.06	0.05
$\Sigma_{\theta 23}$			0.42	0.08
$\Sigma_{\theta 33}$			0.99	0.07
Information Criteria				
-2log-likelihood	20,622		19,625	

enhancement stimulates alcohol use, which leads to more socio-emotional and community problems.

28.6 Discussion

Response bias is a serious threat to any research that uses self-report measures. Subjects are often not willing to cooperate or to provide honest answers to personal, sensitive questions. The general idea is that by offering confidentiality, respondents will become more willing to respond truthfully. Warner's (1965) RRT was developed to ensure such confidentiality.

Our multivariate extension of the technique still masks the responses to the items but enables us to estimate item characteristics and measure individual differences in sensitive behavior. The models can handle both dichotomous and polytomous responses to measure both unidimensional or multidimensional sensitive constructs. In the empirical example stated previously, a forced RR design was used to collect the data, but other options are available. Our RIRT models are easily adapted to a specific choice of response design.

To improve the cooperation of the respondents, both from an ethical and professional point of view, they should be informed about the levels of information that can and cannot be inferred from randomized item responses. The outcome of the randomization device is only known to the respondent, which protects them at the level of the individual items.

The RRT also has some disadvantages. The use of a randomization device makes the procedure more costly, and respondents have to trust the device. Respondents also have to understand the procedure to recognize and appreciate the anonymity they guarantee. Recently, Jann et al. (2012), Tan et al. (2009), and Coutts and Jann (2011) proposed nonrandomized response techniques to overcome the inadequacies of the RRT and tested their proposals empirically. The main idea of their so-called triangular and crosswise technique is to ask respondents a sensitive and a nonsensitive question and let them indicate whether the answers to the questions are the same (both 'yes' or both 'no') or different (one 'yes' and the other 'no'). Such a joint answer to both questions does not reveal the respondent's true status. The distribution of answers to the nonsensitive question has to be known and supports the measurement of the population prevalence on the sensitive question. These nonrandomized methods are designed to make inferences at an aggregate data level. Extensions are required to collect multivariate sensitive items responses that will support the measurement of sensitive constructs. In fact, more research is needed to explore the full potential of nonrandomized response techniques for the analysis of individual sensitive constructs.

Acknowledgments

The author thanks Cheryl Haworth Wyrick for providing the data from the study on alcohol use and abuse by college students.

Appendix 28A: CAPS-AEQ Questionnaire

CAPS: Socio-emotional and community problems

How often (almost always [5], often [4], sometimes [3], seldom [2], almost never [1]) have you had any of the following problems over the past years as a result of drinking too much alcohol?

1. Feeling sad, blue, or depressed
2. Nervousness or irritability
3. Hurt another person emotionally
4. Family problems related to your drinking
5. Spent too much money on drugs
6. Badly affected friendship or relationship
7. Hurt another person physically
8. Caused other to criticize your behavior
9. Nausea or vomiting

10. Drove under the influence
11. Spent too much money on alcohol
12. Feeling tired or hung over
13. Illegal activities associated with drug use

AEQ: Sexual enhancement

14. I often feel sexier after I've had a couple of drinks
15. I'm a better lover after a few drinks
16. I enjoy having sex more if I've had some alcohol
17. After a few drinks, I am more sexually responsive

References

Anglin, D., Hser, Y., and Chou, C. 1993. Reliability and validity of retrospective behavioral self-report by narcotics addicts. *Evaluation Review*, 17, 91–108.

Béguin, A. A., and Glas, C. A. W. 2001. MCMC estimation of multidimensional IRT models. *Psychometrika*, 66, 541–562.

Böckenholt, U., and van der Heijden, P. G. M. 2007. Item randomized–response models for measuring noncompliance: Risk–return perceptions, social influences, and self-protective responses. *Psychometrika*, 72, 245–62.

Brown, S. A., Christiansen, B. A., and Goldman, A. 1987. The alcohol expectancy questionnaire: An instrument for the assessment of adolescent and adult alcohol expectancies. *Journal of Studies on Alcohol*, 48, 483–491.

Clark, S. J., and Desharnais, R. A. 1998. Honest answers to embarrassing questions: Detecting cheating in the randomized response model. *Psychological Methods*, 3, 160–168.

Coutts, E., and Jann, B. 2011. Sensitive questions in online surveys: Experimental results for the randomized response technique (RRT) and the unmatched count technique (UCT). *Sociological Methods and Research*, 40, 169–193.

Cruyff, M. J. L. F., van den Hout, A., van der Heijden, P. G. M., and Böckenholt, U. 2007. Log-linear randomized-response models taking self-protective response behavior into account. *Sociological Methods and Research*, 36, 266–282.

De Boeck, P., and Wilson, M. 2004. *Explanatory Item Response Models: A Generalized Linear and Nonlinear Approach*. New York: Springer.

De Jong, M. G., Pieters, R., and Fox, J.-P. 2010. Reducing social desirability bias through item randomized response: An application to measure underreported desires. *Journal of Marketing Research*, 47, 14–27.

Edgell, S. E., Himmelfarb, S., and Duchan, K. L. 1982. Validity of forced responses in a randomized response model. *Sociological Methods and Research*, 11, 89–100.

Fox, J. A., and Tracy, P. E. 1986. *Randomized Response: A Method for Sensitive Surveys*. London: Sage.

Fox, J.-P. 2005. Randomized item response theory models. *Journal of Educational and Behavioral Statistics*, 30, 189–212.

Fox, J.-P. 2010. *Bayesian Item Response Modeling: Theory and Applications*. New York: Springer.

Fox, J.-P., and Glas, C. A. W. 2001. Bayesian estimation of a multilevel IRT model using Gibbs sampling. *Psychometrika*, 66, 269–286.

Fox, J.-P., and Wyrick, C. 2008. A mixed effects randomized item response model. *Journal of Educational and Behavioral Statistics*, 33, 389–415.

Geerlings, H., Glas, C. A. W., and van der Linden, W. J. 2011. Modeling rule-based item generation. *Psychometrika*, 76, 337–359.

Glas, C. A. W., and Meijer, R. R. 2003. A Bayesian approach to person fit analysis in item response theory models. *Applied Psychological Measurement*, 27, 217–233.

Greenberg, B. G., Abul-Ela, A.-L. A., Simmons, W. R., and Horwitz, D. G. 1969. The unrelated question randomized response model: Theoretical framework. *Journal of the American Statistical Association*, 64, 520–539.

Jann, B., Jerke, J., and Krumpal, I. 2012. Asking sensitive questions using the crosswise model: An experimental survey measuring plagiarism. *Public Opinion Quarterly*, 76, 1–18.

Johnson, V. E., and Albert, J. H. 2001. *Ordinal Data Modeling*. New York: Springer.

Levy, R., Mislevy, R. J., and Sinharay, S. 2009. Posterior predictive model checking for multidimensionality in item response theory. *Applied Psychological Measurement*, 33, 519–537.

O'Hare, T. 1997. Measuring problem drinking in first time offenders: Development and validation of the college alcohol problem scale (CAPS). *Journal of Substance Abuse Treatment*, 14, 383–387.

Sinharay, S. 2006. Bayesian item fit analysis for unidimensional item response theory models. *British Journal of Mathematical and Statistical Psychology*, 59, 429–449.

Sinharay, S., Johnson, M. S., and Stern, H. S. 2006. Posterior predictive assessment of item response theory models. *Applied Psychological Measurement*, 30, 298–321.

Sudman, S., Bradburn, N. M., and Schwarz, N. 1996. *Thinking about Answers: The Application of Cognitive Processes to Survey Methodology*. San Francisco, California: Jossey-Bass.

Tan, M. T., Tian, G.-L., and Tang, M.-L. 2009. Sample surveys with sensitive questions: A nonrandomized response approach. *The American Statistician*, 63, 9–16.

Tourangeau, R., Rips, L. J., and Rasinski, K. 2000. *The Psychology of Survey Response*. New York: Cambridge University Press.

Tourangeau, R., and Smith, T. W. 1996. Asking sensitive questions: The impact of data collection, question format, and question technique. *Public Opinion Quarterly*, 60, 275–304.

van der Heijden, P. G. M., van Gils, G., Bouts, J., and Hox, J. J. 2000. A comparison of randomized response, computer-assisted self-interview, and face-to-face direct questioning: Eliciting sensitive information in the context of welfare and unemployment benefit. *Sociological Methods and Research*, 28, 505–537.

Warner, S. L. 1965. Randomized response: A survey technique for eliminating evasive answer bias. *Journal of the American Statistical Association*, 60, 63–69.

29

Joint Hierarchical Modeling of Responses and Response Times

Wim J. van der Linden and Jean-Paul Fox

CONTENTS

29.1 Introduction 481
29.1.1 Levels of Modeling 484
29.2 Presentation of the Model 486
29.2.1 First-Level Models 487
29.2.2 Second-Level Models 487
29.2.3 Higher-Level Models 489
29.2.4 Identifiability 489
29.2.5 Alternative Plug-in Models 490
29.2.6 Dependency Structure of the Data 491
29.3 Parameter Estimation 491
29.4 Model Fit 493
29.4.1 Person Fit of the Response Model 494
29.4.2 Person Fit of RT Model 495
29.4.3 Item Fit 496
29.5 Empirical Example 496
29.6 Discussion 499
References 499

29.1 Introduction

In spite of its apparent simplicity, the event of a test taker responding to a test item is hard to disentangle, especially if the interest is both in the observed response and the time used to produce it. For one thing, the test taker's ability to solve an item is not the only factor that plays a role; the speed at which she/he operates should also be accounted for, which immediately raises the question of how the two are related. For some tests, there is more at stake for their test takers than for others, so motivation may have an impact on both as well. Similar questions arise with respect to the difficulty of the item and the amount of labor demanded by it. Besides, the nature of all these relationships is even more difficult to disentangle if the conditions change during testing; for example, due to increased fatigue toward the end of the test.

It is easy to confound or overlook the effects of some of these factors. In order to illustrate this point, Table 29.1 shows an extended version of the potentially confusing empirical example presented in an earlier chapter in this volume (Volume One, Chapter 16), which along with the first 10 response times (RTs) by an arbitrary pair of test takers on a 65-item

TABLE 29.1

Responses and RTs by Two Test Takers on the First 10 Items in a Cognitive Ability Test (Columns 1–5) Along with the Responses and RTs on Two Items by the First 10 Test Takers from the Same Data Set (Columns 6–10)

	Response		RT			Response		RT	
	Test Taker		Test Taker						
Item	$p=1$	$p=2$	$p=1$	$p=2$	Test Taker	Item $i=1$	Item $i=2$	Item $i=1$	Item $i=2$
1	0	0	22	26	1	0	1	10	14
2	0	0	19	38	2	1	1	27	56
3	1	0	40	101	3	0	1	22	40
4	1	1	43	57	4	0	0	26	101
5	1	1	27	37	5	1	1	29	42
6	0	0	21	27	6	0	1	18	8
7	1	1	45	116	7	1	0	18	37
8	0	1	23	44	8	0	1	12	36
9	1	1	14	10	9	1	0	20	51
10	1	1	47	117	10	0	1	21	22
R	0.89		0.20			0.54		0.19	
		0.21						−0.20	

cognitive ability test now also shows their responses (Columns 2–5). In addition, the table illustrates the RTs and responses on an arbitrary pair of items by the first 10 test takers in the dataset (Columns 7–10). Although the two test takers operated independently, their RTs appear to correlate $r = 0.89$. The same holds for their responses ($r = 0.20$). In fact, the data seem even more mysterious in that the responses by the first test taker appear to correlate with the RTs by the second ($r = 0.21$)! Similar patterns were observed for the pair of items in the dataset; both their response and RT vectors correlated positively ($r = 0.54$ and 0.19, respectively), while the responses on the first item also correlated with the RTs on the second, but this time negatively ($r = -0.20$). Why do these correlations not vary just randomly about zero? How could the responses by one test taker have been impacted by the RTs of the other? And why all of a sudden the negative correlation between the responses and RTs on the two items?

The only way to disentangle such apparently conflicting results is by careful modeling of the probability experiment of a test taker responding to an item. An important preliminary question, however, is if we should actually conceive of it as a single experiment or consider two distinct experiments—one for the response and the other for the RT, each driven by its own parameters. Exactly the same question bothered George Rasch when he introduced his two separate models for reading errors and reading speed (Rasch, 1960; Volume One, Chapter 15). Both of them had a parameter ξ_p for the test taker labeled as "ability" and an item parameter δ_i labeled as "difficulty." Rasch's question was whether these parameters were on the same reading scale or, in spite of their common names and notation, actually represented two distinct abilities and difficulties. His tentative solution was a combination of the two options. As for the item parameters, he thought it "reasonable that a text that gives rise to many mistakes—e.g., because it contains many unknown words or deals with a little known subject-matter—will also be rather slowly read." But for the ability parameters he did not expect it to be "a general rule that a slow reader also makes many mistakes" (Rasch, 1960, p. 42). For a further discussion of this issue, which touches on the very nature of the distinction between speed and power tests, see van der Linden (2009, pp. 250–251, 255–257; Volume Three, Chapter 12).

The position we take is the one of two independent experiments each with its own test-taker and item parameters, provided we can treat these parameters as fixed. The position is motivated by the long history of large-scale educational and psychological testing programs ignoring the RTs on their items, simply because the technical means necessary to record them had not yet arrived. The fit of the response models to the numerous collections of response data gathered during these days would have been impossible if the responses were actually driven by separate RT parameters as well. But, equally important, it should be noted that the independence is assumed to hold at the level of *fixed* parameters only, analogous to the assumption of local independence made throughout item response theory (IRT). Under this condition, specifying a relationship between any response and RT parameters would not make much sense. For instance, Rasch's statement that slow readers do not necessarily make many mistakes is without any observational meaning for readers each operating at constant speed and ability. The only thing our two models have to do at this level is adequately represent the probability distributions of the response and the RT for each combination of test taker and item. Imposing any relationship on their speed and ability parameters would unnecessarily constrain the models and therefore deteriorate their fit.

Continuing our second argument, in order to observe a possible relationship between ability and speed parameters, they thus have to vary over their range of possible values. They can do so in two entirely different ways: as parameters for the same test taker that change during the experiment or as parameters for different test takers that take different values. The former leads to the observation of a within-person relationship; the latter to a between-person relationship. Rasch's earlier quote on the speed of reading and the number of mistakes reminds us of the speed-accuracy tradeoff (SAT) established through extensive psychological research (Luce, 1986), which is an example of the former. It tells us that if someone decides to works faster she/he would do so at the expense of loss of accuracy (or in our current terminology: a lower effective ability). It is impossible to observe within-person relationships when test takers work at constant speed and ability. On the other hand, if the same quote is taken to refer to a between-person relationship, it would amount to an observation for a population of test takers reading at different speeds, with the slower readers tending to make fewer errors. The two different types of relationships do not imply each other at all, a point we will take up again when we introduce Figure 29.1.

The RT literature has suffered from a serious confounding of its levels of modeling—at least this is how we interpret its attempts to introduce relationships between response and RT parameters in single-level models, motivate these relationships by references to within-person or between-person phenomena, or more generally treat responses and RTs as dependent or with models that are cross-parameterized. The reason for this confounding may very well go back to be the fact that for most of educational and psychological testing, due to memory and/or learning effects, it is generally impossible to replicate the administration of an item to a test taker. Consequently, those who would want to study relationships between responses and RTs simply by "looking at the data" typically aggregate them across different items, test takers, and/or testing conditions. But such aggregations are just the operational equivalent of confounding the different levels of modeling for the underlying experiments that need to be considered.

The hierarchical framework of modeling responses and RTs reviewed in this chapter is an attempt to disentangle all relevant levels and model each of them appropriately. Although we make an obvious choice of models, the formal structure of the overall framework is the more important part of it. If necessary, each of our choices can be replaced

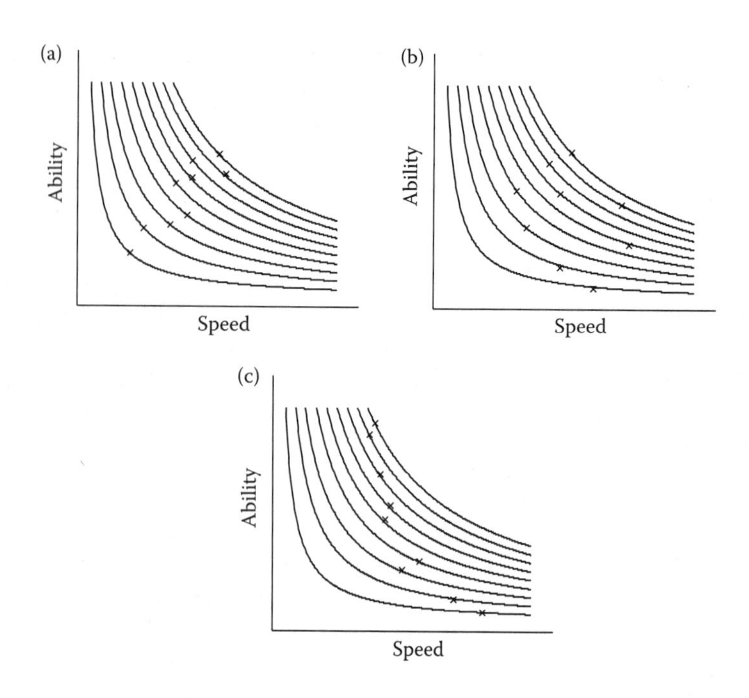

FIGURE 29.1
Three plots with SAT curves for the same test takers. Each test taker operates at a combination of speed and ability possible according to his personal tradeoff. The combination of choices leads to (a) positive, (b) zero, and (c) negative correlation between speed and ability among the test takers. (Reproduced with permission from van der Linden, W. J., *Journal of Educational Measurement*, 46, 247–272, 2009.)

by an alternative better suited to the specific application that has to be addressed. The Bayesian approach to parameter estimation and model evaluation presented below nicely complements the "plug-and-play approach" to the joint modeling of responses and RTs advocated in this chapter.

29.1.1 Levels of Modeling

The different levels of modeling in the framework are all empirical. A statistical level of prior distributions for the parameters at the highest level in the framework will be added later.

1. *Fixed test takers and fixed items.* At this level, both the person and item parameters have fixed values; the only randomness exist in the responses and the RTs. Typically, multiple test takers and items are modeled jointly in this way. But, as just noticed, it is unnecessary—and potentially even dangerous—to model any functional or statistical relationship between their response and RT parameters.
2. *Fixed items but changing test takers.* This within-person level applies when the conditions under which the test items are taken change or the test takers change their behavior for some more autonomous reason during the test. The assumption of fixed person parameters no longer holds and a relationship between speed and ability may now manifest itself. We expect these relations to have the shape of a "psychological law," that is, a constraint on the possible combinations of values for the speed and ability parameter established by psychological research. A prime

example is the SAT referred to earlier. Figure 29.1 illustrates the monotonically decreasing relationship between the speed and ability parameters implied by the constraint (otherwise the shape of the curves is arbitrary).

3. *Fixed items and a random sample from a population of fixed test takers.* The sampling creates a new level of randomness—that of a joint distribution for the speed and ability parameters in the two lower-level models. This case is typical of several large-scale educational assessments (Mazzeo, Volume Three, Chapter 14). The assumption of a population distribution of the ability parameters is also made in maximum marginal likelihood estimation of the item parameters in IRT (Glas, Volume Two, Chapter 11). It is hard to think of any psychological law that would define the distribution, however. As illustrated in Figure 29.1, possible laws at the within-person level certainly do not imply anything at the between-person level. The critical factor that moderates the two is the speed at which each individual test taker decides to operate within the range of possibilities left by his SAT—a decision likely to depend on such factors as their motivation, how they have interpreted the instructions, the time limit, strategic considerations, etc. At the current level of modeling, the responses and RTs for the given population of test takers may be further explored as a function of possible individual-level covariates of the person parameters (e.g., general intelligence, motivation, and physical fitness). Analogous to the idea of explanatory item response theory (Volume One, Chapter 33), these covariates can be incorporated as predictors in the modeling framework. Depending on their real-valued or categorical nature, the presence of the covariates introduces a regression or analysis-of-variance (ANOVA) structure in the population model.
4. *Fixed items and a stratified random sample from a population of fixed test takers with a hierarchical structure.* This case arises when the test takers are nested in groups and we observe a stratified random sample of them. If group membership does have an impact on the test takers' speed or abilities, the modeling framework should adopt an ANOVA structure with nested factors. But the hierarchical structure could also force us to add a higher level to the regression model above, namely when some of its parameters appear to vary as a function of group-level covariates.
5. *Similar levels for the items.* Each of the preceding alternative levels of modeling was specified for the test takers and their parameters only. The same classification makes sense for the items though. For instance, knowledge of the relationships between the items in a domain for a given subject area helps us to explain possible spurious correlations between responses and RTs due to their aggregation across items. Possible further dependencies between the response and RT parameters for the items could be explored by introducing item-level covariates for them (e.g., word counts, readability indices, and computational load). In the context of rule based-item generation, with families of items each generated by identical settings of the algorithm with some minor random variation added to it, it even makes sense to introduce a hierarchical structure for their parameters (Volume One, Chapter 26; Klein Entink et al., 2009b). Generally, the benefits of treating item parameters as random instead of fixed have been underestimated (De Boeck, 2008).

The framework reviewed in the next sections encompasses each of these possible levels both for the test takers and the items, with the exception of the within-person and within-item levels. Inclusion of the former would make sense for the analysis of experimental data with systematic manipulation of the test takers' speed and ability, but less so when the focus is on educational and psychological testing as in this chapter. For the latter, we refer to its treatment in the chapter by Glas, van der Linden, and Geerlings (Volume One, Chapter 26). Also, the extension of the framework to higher levels for the item domain is omitted. Our current focus will thus primarily be on the levels of fixed and randomly sampled test takers and items, with a possible individual-level regression structure to explain observed dependencies in their joint distributions. The option of test takers sampled from higher-level groups will be reviewed only briefly.

29.2 Presentation of the Model

At its lowest level, the framework has two distinct types of models for the responses and RTs. The next level consists of two models that specify the separate joint distributions of the person parameters for the population of test takers and item parameters for the domain of items, with possible individual-level explanatory variables for them. The possible extension of these distributions with a higher-level structure for the test-taker population and item domain will then be outlined. Our review of the framework draws heavily on earlier publications by Klein Entink et al. (2009a,b) and van der Linden (2007) (the reader should however be aware of the different parameterizations for the first-level models and their impact on the higher-level parameters in the first two of these publications).

It is instructive to follow the different types of separation and interaction between all parameters as we introduce the candidate models. At the lowest level, the response and RT parameters are not allowed to interact. But at the second level, the perspective is rotated by 90°. Now, the test-taker and item parameters are treated entirely separately; only the response and RT parameters are allowed to correlate. For a graphical illustration in the form of a causal diagram, see Figure 29.2. The same second-level pattern of separation and interaction continues when we add higher-level covariates of the parameters to the framework.

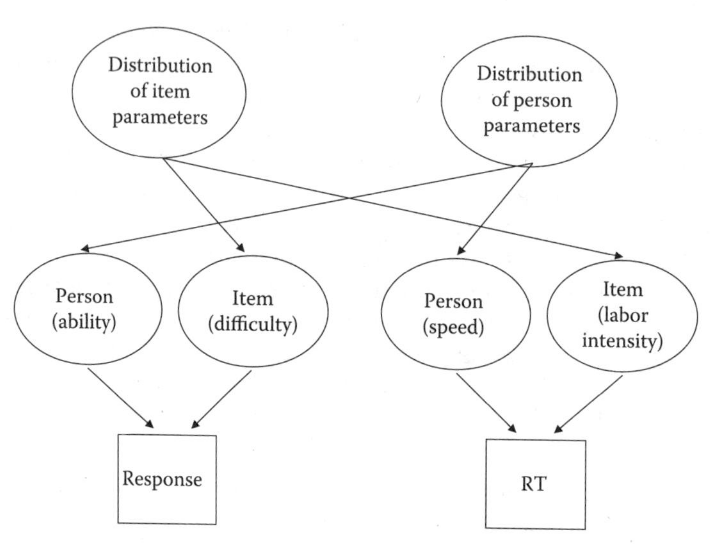

FIGURE 29.2
Hierarchical framework for joint modeling of responses and response times.

29.2.1 First-Level Models

Let U_{pi} and T_{pi} denote the random response and RT by test taker $p = 1, \ldots, P$ on item $i = 1, \ldots, I$.

As response model, the three-parameter logistic (3PL) model is adopted. That is, each response variable is assumed to be distributed as

$$U_{pi} \sim f(u_{pi};\theta_p, a_i, b_i, c_i), \tag{29.1}$$

where $f(\cdot)$ denotes a Bernoulli probability function with success parameter

$$p_p(\theta_j) \equiv c_i + (1 - c_i)\Psi(a_i(\theta_p - b_i)), \tag{29.2}$$

$\theta_p \in \mathbb{R}$ is the ability parameter for test taker p, $a_i \in \mathbb{R}^+$, $b_i \in \mathbb{R}$, and $c_i \in [0,1]$ are the discrimination, difficulty, and guessing parameters for item i, respectively, and $\Psi(\cdot)$ denotes the logistic distribution function. As the 3PL model is reviewed more completely in an earlier chapter (Volume One, Chapter 2), it does not need any further introduction here.

The response-time model specifies the distribution of T_{pi} as a lognormal with density function

$$T_{pi} \sim f(t_{pi};\tau_p, \alpha_i, \beta_i) = \frac{\alpha_i}{t_{pi}\sqrt{2\pi}} \exp\left\{-\frac{1}{2}[\alpha_i(\ln t_{pi} - (\beta_i - \tau_p))]^2\right\}, \tag{29.3}$$

where $\tau_p \in \mathbb{R}$ can be interpreted as a speed parameter for test taker p and $\beta_i \in \mathbb{R}$ and $\alpha_i \in \mathbb{R}^+$ as the time intensity and the discriminating power of item i, respectively. For a full treatment of the lognormal model, including its straightforward derivation from the definition of speed by a test taker on an item, see van der Linden (Volume One, Chapter 16).

Although the two discrimination parameters in Equations 29.2 and 29.3 have an analogous impact on the response and RT distribution (van der Linden, 2006, Figure 1; Volume One, Chapter 16) and the speed and time-intensity parameters do have similar antagonistic effects on the RTs as the ability and item difficulty parameters on the responses, the two models have important formal differences as well. For instance, the lognormal model in Equation 29.3 directly specifies a density function for the distribution of the RT. Its counterpart is the Bernoulli distribution in Equation 29.1, of which the 3PL model explains the success parameter. Also, Equation 29.2 has a guessing parameter that serves as a lower asymptote to this success parameter, whereas the RT distribution in Equation 29.3 is not constrained any further than by its natural lower bound at zero.

29.2.2 Second-Level Models

Let $\xi_p = (\theta_p, \tau_p)$ denote the vector with all parameters for test taker p in the response and RT model in Equations 29.2 and 29.3 and $\psi_i = (a_i, b_i, c_i, \alpha_i, \beta_i)$ the vector with all parameters for item i in the two models. The second-level models specify multivariate distributions of ξ_p and ψ_i for the population of test takers and the domain of items represented by the test, respectively.

An obvious choice of model for the distribution of ξ_p in a population $\mathcal{P}$ of test takers is a bivariate normal density function

$$f(\xi_p;\mu_{\mathcal{P}},\Sigma_{\mathcal{P}})=\frac{|\Sigma_{\mathcal{P}}^{-1}|^{1/2}}{2\pi}\exp\left[-\frac{1}{2}(\xi_p-\mu_{\mathcal{P}})^T\Sigma_{\mathcal{P}}^{-1}(\xi_j-\mu_{\mathcal{P}})\right] \tag{29.4}$$

with mean vector $\mu_{\mathcal{P}}=(\mu_\theta,\mu_\tau)$ and (2×2)-covariance matrix

$$\Sigma_{\mathcal{P}}=\begin{pmatrix}\sigma_\theta^2 & \sigma_{\theta\tau}\\ \sigma_{\theta\tau} & \sigma_\tau^2\end{pmatrix}. \tag{29.5}$$

Similarly, the item-domain model specifies the distribution of parameter vector ψ_i over a domain of items $\mathcal{I}$ as a multivariate normal density function

$$f(\psi_i;\mu_{\mathcal{I}},\Sigma_{\mathcal{I}})=\frac{|\Sigma_{\mathcal{I}}^{-1}|^{1/2}}{(2\pi)^{5/2}}\exp\left[-\frac{1}{2}(\psi_i-\mu_{\mathcal{I}})^T\Sigma_{\mathcal{I}}^{-1}(\psi_i-\mu_{\mathcal{I}})\right] \tag{29.6}$$

with mean vector $\mu_{\mathcal{I}}$ and (5×5)-covariance matrix $\Sigma_{\mathcal{I}}$ defined analogously to Equation 29.5.

The assumption of normality makes sense for an unrestricted population and item domain. However, the discrimination parameters in both first-level models and the guessing parameter in the response model have a restricted range. If the restriction appears to hamper the application, Equations 29.4 through 29.6 could be specified for transformed versions of the parameters. Obvious choices are log transformations for the discrimination parameters and logits for the guessing parameters; for the resulting lognormal and logit normal distributions, see Casabianca and Junker (Volume Two, Chapter 3).

Let $\mathbf{x}_p$ denote a $(Q+1)$-dimensional vector with the scores of test taker p on Q joint covariates of his ability and speed parameter and "1" as its first component, and assume that a linear regression model could be fitted to their relationships. It then holds that

$$\theta_p=\mathbf{x}_p'\gamma_\theta+\epsilon_{\theta_p} \tag{29.7}$$

and

$$\tau_p=\mathbf{x}_p'\gamma_\tau+\epsilon_{\tau_p}, \tag{29.8}$$

where γ_θ and γ_τ are the vectors with the regression weights and ϵ_{θ_p} and ϵ_{τ_p} are the residual terms in the regression of θ and τ on $\mathbf{x}$ for test taker p, respectively. Because of random sampling of the test takers, $\epsilon_{\theta_p}\sim N(0,\sigma_{\epsilon_\theta}^2)$ and $\epsilon_{\tau_p}\sim N(0,\sigma_{\epsilon_\tau}^2)$, where both are taken to be independent.

Likewise, it may be possible to predict the difficulty and time-intensity parameters of the items in the domain as

$$b_i=\mathbf{y}_i'\gamma_b+\epsilon_{b_i} \tag{29.9}$$

and

$$\beta_i=\mathbf{y}_i'\gamma_\beta+\epsilon_{\beta_i}, \tag{29.10}$$

where $\boldsymbol{\gamma}_b$ and $\boldsymbol{\gamma}_\beta$ are now the weights in the regression of b and β on common predictors $\mathbf{y}$ and $\epsilon_{b_i} \sim N(0,\sigma^2_{\epsilon_b})$ and $\epsilon_{\beta_i} \sim N(0,\sigma^2_{\epsilon_\beta})$ are their independent residual terms, respectively. Similar prediction of the other item parameters may be possible (although empirical research has not resulted in many serious predictors of them so far).

In the case of categorical covariates, dummy variables can be used to indicate their values. As the modification is straightforward, further details are omitted here. All covariates in Equations 29.7 through 29.10 were taken to jointly impact the test takers' ability and speed or the difficulties and time intensities of the item, albeit with possible different strengths of effects. The changes necessary to deal with separate covariates are obvious.

The regression structure explains the structure of the mean vectors and covariance matrices of $\mathcal{P}$ and $\mathcal{T}$. For instance, for $\mathcal{P}$ it now holds that

$$\mu_\theta = \mathcal{E}_P(\boldsymbol{\gamma}'_\theta \mathbf{x} + \epsilon_\theta) = \boldsymbol{\gamma}'_\theta \mathcal{E}_P(\mathbf{x}), \tag{29.11}$$

$$\mu_\tau = \mathcal{E}_P(\boldsymbol{\gamma}_\tau \mathbf{x} + \epsilon_\tau) = \boldsymbol{\gamma}'_\tau \mathcal{E}_P(\mathbf{x}) \tag{29.12}$$

and

$$\sigma_{\theta\tau} = \mathrm{Cov}(\boldsymbol{\gamma}'_\theta \mathbf{x} + \epsilon_\theta, \boldsymbol{\gamma}'_\beta \mathbf{x} + \epsilon_\beta) = \sum_u \sum_v \gamma_{\theta u} \gamma_{\tau_v} \mathrm{Cov}(x_u, x_v). \tag{29.13}$$

29.2.3 Higher-Level Models

For a population of test takers nested in groups, the introduction of (joint) group-level covariates may be meaningful. Let index $g = 1, \ldots, G$ denote the groups in a single-level structure of $\mathcal{P}$ and $\mathbf{y}_g$ the $(R + 1)$-dimensional vector of their scores on R of these covariates (again with a "1" in its first position). Assuming linear regression, we can treat the regression parameters in Equations 29.7 through 29.10 as

$$\gamma_{\theta g} = \mathbf{y}'_g \boldsymbol{\delta}_\theta + \epsilon_{\gamma_{\theta g}} \tag{29.14}$$

and

$$\gamma_{\tau g} = \mathbf{y}'_g \boldsymbol{\delta}_\tau + \epsilon_{\gamma_{\tau g}} \tag{29.15}$$

with all parameters and residual terms defined analogously to Equations 29.7 and 29.8. The option is not further explored here; it entirely parallels the case of multilevel response modeling addressed in Fox and Glas (Volume Two, Chapter 24), where further details can be found.

29.2.4 Identifiability

The response and RT models in Equations 29.2 and 29.3 are not yet identified. But, to obtain identification it suffices to set

$$\mu_\theta = 0; \quad \sigma^2_\theta = 1; \quad \mu_\tau = 0. \tag{29.16}$$

The first two restrictions are the ones generally used in item calibration studies for the 3PL model along with the assumption of a normal population distribution of θ. The third prevents the possible tradeoff between β_i and τ_j in Equation 29.3. Observe that $\mu_\tau = 0$ implies $\mu_\beta = \mathcal{E}(\ln(T))$, where the expectation is taken across all test takers and items in $\mathcal{P}$ and $\mathcal{T}$. Once the first-level models are identified, the same automatically holds for the proposed higher-level models.

It may come as a surprise that σ_τ^2 does not need to be fixed as well. But, as already indicated, although the proposed response and RT models may look to have completely analogous parameterizations at first sight, they are different: the 3PL model has latent parameter θ_p as its argument, whereas the argument of the RT model in Equation 29.3 is a manifest variable t_{pi}. Consequently, discrimination parameters α_i in the RT model are automatically fixed by the unit of time in which t_{ij} is measured (Volume One, Chapter 16).

29.2.5 Alternative Plug-in Models

The key idea in this chapter is a joint hierarchical modeling framework for the different types of interactions between the response and RT parameters at different levels. The individual models proposed here make sense because of earlier successful applications. But for the case of items with a polytomous response format, the 3PL model can easily be replaced by any of the popular polytomous models, for instance, the (generalized) partial credit (Volume One, Chapters 7 and 8) or graded response model (Volume One, Chapter 6). Options for models with still different response formats can be found in several of the other earlier chapters in this volume.

It is unclear how the response format of test items would impact their RT distributions. Generally, the amount of time spent on a test item seems to be more influenced by the nature of the problem formulated in it than how the test takers are assumed to submit their answers. Nevertheless, a more flexible alternative to the lognormal model is a normal Box–Cox type of model. However, a necessary condition for its compatibility with the current modeling framework is a common Box–Cox transformation of the RTs across all items; the more general case of item-specific transformations leads to item-specific scales for the RT parameters and therefore less plausible specification of the item-domain model as a multivariate normal (Klein Entink et al., 2009c). A rich family of models for RT distributions is the linear transformation model proposed by Wang et al. (2013). The family includes both parametric RT densities, including those based on the Box–Cox transformation, and the option of continuous nonparametric densities that are otherwise arbitrary.

Another example of a pair of response and RT models that fits the framework in this chapter quite naturally are Rasch's models for reading error and speed (van der Linden, 2009).

As already indicated, transformation of some of the first-level parameters may lead to improved fit of the second-level models or their possible regression specifications. This option may be a simple alternative to the choice of an entirely different family of models. Besides, another way to make the current choice of second-level models more robust is omitting any of the first-level parameters in whose behavior we are not interested at the higher levels of the framework. For instance, often the interest is primarily in the correlations between the test takers' abilities and speed and/or the difficulties and time intensities of the items. These parameters have obvious interpretations and knowledge of their dependencies would definitely help testing agencies to improve their programs (for instance, nonzero correlations between the difficulty and time intensity of the items tells them they would need to control their adaptive testing programs for possible differential test speededness). Statistically, leaving out any unnecessary first-level parameters at the higher levels reduces the number of parameters that have to be estimated and increases the likelihood of model fit, without introducing any bias in the estimates of the remaining parameters.

29.2.6 Dependency Structure of the Data

Table 29.1 illustrated the rather complex dependencies that seem to arise between the responses and RTs in an arbitrary dataset. However, responses and the RTs are always nested within test takers and items. Besides, for fixed test forms, the same set of items is administered to the same set of test takers. The result of both features is a multivariate dataset with a nested, cross-classified structure. To explain observed correlations between responses and RTs for such structures, different levels of modeling are required to correctly represent their underlying dependencies.

At the lowest level, both item and test-taker parameters need to be introduced. The item parameters are required to explain the dependencies between the responses and RTs across test takers. Likewise, test-taker parameters are necessary to account for the dependencies between the responses and RTs across items. However, given both sets of parameters, we assume the responses and RTs to be independently distributed—an assumption typically referred to as the local independence assumption in IRT. In the current context of multivariate data, the assumption applies in three different ways: between the responses of each test taker, between the RTs of each test taker, and between the responses and RTs for the same test takers (Glas and van der Linden, 2010; van der Linden and Glas, 2010).

At the next higher level, both the test taker and item parameters may show different degrees of dependency; hence, our choice of free parameters for their possible covariances. The covariances between the test-taker parameters explain the observed within-person dependencies between the responses and RT across the items. The same holds for the covariances between the item parameters and the observed within-item dependencies between the responses and RTs across the test takers.

When this cross-classified, nested structure of the data is ignored, it is easy to confuse causal and spurious correlations. For instance, the observed correlation of 0.21 between the responses and RTs of the two test takers in Table 29.1 should not be interpreted to have any causal meaning. It was spurious because the covariance between the difficulties and labor intensities of the items in the dataset was ignored. Similarly, the correlation of −0.20 between the responses and RTs on the two different items can be explained by a negative covariance between the test takers' speed and ability in the data set (just as in our empirical example below).

When the observed correlations are the result of a mixture of such higher-order effects, it becomes impossible to explain them intuitively. The statistical framework reviewed in this chapter is then required to shift our attention from the observed correlations to estimates of the more fundamental covariances between item and test-taker parameters.

29.3 Parameter Estimation

The proposed procedure is the Bayesian estimation of all parameters through Gibbs sampling of their joint posterior distribution. The procedure involves the division of all unknown parameters into blocks, with iterative sampling of the conditional posterior distributions of the parameters in each block given the preceding draws for the parameters in all other blocks (Fox, 2010, Chapter 4; Junker, Patz, and Vanhoudnos, Volume Two, Chapter 15). It is convenient to choose blocks that follow the individual models in the framework. This choice minimizes the number of changes necessary if we would replace one or more of them, and thus supports the idea of a "plug-and-play" approach proposed earlier in this chapter.

In addition to the number of levels and the specific models adopted at each of them, the nature of the Gibbs sampler also depends on possible additional choices one has made, such as the reparameterization of some of the models or use of data augmentation. Depending on these choices, the likelihoods associated with some of the blocks may combine with conjugate prior distributions for their parameters, with the benefit of efficient iterative updates of the conditional posterior distributions through simple adjustments of the values of their parameter. However, in the absence of conjugate prior distributions, an obvious alternative is the use of Metropolis–Hastings (MH) steps with a well-chosen proposal density (Junker, Patz, and Vanhoudnos, Volume Two, Chapter 15). Although tuning of its proposal density used to be bit of a waste of time, adaptive versions of MH sampling (Atchadé and Rosenthal, 2005; Rosenthal, 2007) have automated the job; for examples of applications based on Robbins–Monro stochastic approximation, see Cai (2010) and van der Linden and Ren (2015).

For the basic two-level framework with the standard parameterization in Equations 29.1 through 29.6, a convenient choice of Gibbs sampler is one with a mixture of MH steps and steps with direct sampling from conjugate posteriors in closed form. In each of these steps, the prior distribution for the first-level item and test-taker parameters is a conditional version of one of the joint second-level distributions in Equation 29.4 or 29.6. As these distributions are specified to be multivariate normal, each of their conditional distribution is also normal with a mean and variance that follow directly from their $\boldsymbol{\mu}$ and $\boldsymbol{\Sigma}$; see Casabianca and Junker (Volume Two, Chapter 3, Theorem 2.1) for their closed-form expressions. The prior distributions for the second-level parameters need to be specified separately, though.

Further, it is efficient to use the version of the lognormal RT model that postulates a normal distribution for the logtimes (Volume One, Chapter 16, Equation 8). Its choice implies a combination of a normal likelihood with a conjugate prior that yields a normal posterior for two of the three parameters in the RT model. Introductions to this normal–normal case are offered in nearly every Bayesian text (e.g., Gelman et al., 2013; Gill, 2015).

More specifically, the proposed Gibbs sampler has the following steps:

1. Item parameters (α_i, b_i, c_i) in the 3PL model are sampled using the MH algorithm with as prior distribution their second-level conditional distribution given the item parameters (α_i, β_i) in the lognormal model.
2. Item discrimination parameters α_i in the lognormal model are sampled using a similar MH step with as prior distribution their second-level conditional distribution given all other item parameters in the response and RT model.
3. Time-intensity parameters β_i have a normal likelihood for the observed logtimes in combination with a normal prior distribution given all other item parameters. They are sampled directly from normal posterior distributions with means and variances that have the typical Bayesian form of the combination of data and prior parameters for the normal–normal case. Because of the identifiability restriction $\mu_\tau = 0$ in Equation 29.16, the mean of the prior distribution of β_i can conveniently be set equal to $\mu_\beta = \overline{\ln t}$ (average logtime in the dataset).

4. Ability parameters θ_p are sampled using the MH algorithm with as prior distribution their second-level conditional distribution given τ_p. In fact, for this simple bivariate case, using the restrictions $\mu_\theta = 0$ and $\sigma_\theta = 1$ in Equation 29.16, the means and common variance of these prior distributions reduce all the way to $\mu_{\theta|\tau_p} = \sigma_{\theta\tau}\tau_p/\sigma_\tau^2$ and $\sigma^2_{\theta|\tau_j} = 1 - \sigma^2_{\theta\tau}/\sigma^2_\tau$.
5. Speed parameters τ_p have a normal likelihood and normal prior distributions given θ_p with means and a common variance that, because of the same restrictions, simplify to $\mu_{\tau|\theta_p} = \sigma_{\theta\tau}\theta_p/\sigma^2_\theta$ and $\sigma^2_{\tau|\theta_p} = \sigma^2_\tau - \sigma^2_{\theta\tau}$. Just as the time–intensity parameters, the speed parameters can thus be sampled from normal posterior distributions with means and variance that take the typical Bayesian form of the combination of data and prior parameters for the normal–normal case.
6. Choosing a member of the multivariate normal-inverse Wishart family as prior distribution, population and item-domain parameters $(\mu_\mathcal{P}, \Sigma_\mathcal{P})$ and $(\mu_\mathcal{I}, \Sigma_\mathcal{I})$ can be sampled from the same family with the typical Bayesian combination of data and prior parameters for multivariate normal data as parameters.

Alternatively, substituting the normal-ogive model with the slope-intercept parameterization for the 3PL model in Equation 29.2, the adoption of data augmentation allows for the replacement of the MH steps above by more efficient sampling from normal and beta posterior distributions. Details are provided in Fox (2010), Johnson and Albert (1999), Klein Entink et al. (2009a), and van der Linden (2007). However, the price to be paid for the reparameterization exists in the form of loss of the current interpretation of several of the parameters in the modeling framework. For instance, the intercept parameters cannot be interpreted to represent the difficulties of the items and, consequently, covariance matrix $\boldsymbol{\Sigma}_\mathcal{I}$ in Equation 29.6 no longer provides us with the correlation between item difficulties and time intensities.

If a regression structure is added to the population model, its former mean vector $\boldsymbol{\mu}_\mathcal{P} = (\mu_\theta, \mu_\tau)$ is further specified as in Equations 29.11 and 29.12. Likewise, the specification of $\boldsymbol{\Sigma}_\mathcal{P}$ then follows from Equation 29.13 and the residual variances are those in Equations 29.7 and 29.8. For this case, all regression parameters and residual variances can be sampled using the multivariate normal-inverse Wishart specification detailed in Fox (2010, Chapter 3) and Klein Entink et al. (2009a). The same references should be consulted for versions of the framework with a regression structure for the item-domain model or extensions to higher-level structures.

When the items have already been calibrated under the response and/or RT model of choice, an extremely efficient option is the use of MH steps for the item parameters with an independence sampler in the form of resampling of vectors of independent posterior draws saved for them during the calibration (for details, see Volume One, Chapters 3 and 16). The same option exists, for instance, when new item or person parameters have to be estimated but the higher-level structure can be assumed not to have changed.

29.4 Model Fit

In this section, new tools for evaluating the fit of the RT and response models are introduced following the procedure of Marianti et al. (2014). The basic idea is use of the log-likelihood statistic by Levine and Rubin (1979) in a Bayesian context to quantify the extremeness of responses and RTs under the model. In order to account for the uncertainty in each of the model parameters, a posterior probability of extremeness is computed by integrating the statistics across the prior distributions of all unknown model parameters. The approach corresponds to the prior predictive approach to hypothesis testing advocated by Box (1980). For a more complete review of possible fit analyses, see the chapters on the unidimensional logistic models (Volume One, Chapter 2) and the lognormal RT model (Volume One, Chapter 16).

29.4.1 Person Fit of the Response Model

Drasgow et al. (1985) proposed a standardized version of Levine–Rubin statistic. This standardized version has been shown to have statistical power to detect aberrant response patterns in educational testing (Karabatsos, 2003).

For two-parameter (2PL) response model, the original log-likelihood person-fit statistic is defined as

$$\begin{aligned} l_0(\mathbf{U}_p;\boldsymbol{\theta}_p,\mathbf{a},\mathbf{b}) &\equiv -\ln f(\mathbf{U}_p;\theta_p,\mathbf{a},\mathbf{b}) \\ &= -\sum_{i=1}^{I} u_{pi}\ln P(u_{pi}) + (1-u_{pi})\ln(1-P(u_{pi})), \end{aligned}$$

where $P(u_{pi}) \equiv \Pr\{U_{pi}=1 \mid \theta_p, a_i, b_i\}$. The statistic can be standardized using the following expressions for its mean and variance:

$$\mathcal{E}(l_0(\mathbf{U}_p;\theta_p,\mathbf{a},\mathbf{b})) = -\sum_{i=1}^{I} P(u_{pi})\ln P(u_{pi}) + (1-P(u_{pi}))\ln(1-P(u_{pi}))$$

$$Var(l_0(\mathbf{U}_p;\theta_p,\mathbf{a},\mathbf{b})) = \sum_{i=1}^{I} P(u_{pi})(1-P(u_{pi}))\ln\left(\frac{P(u_{pi})}{1-P(u_{pi})}\right)^2,$$

respectively. The result has an approximate standard normal distribution.

The person-fit test can be adapted for use under the 3PL model by introducing a dichotomous classification variable S_{pi} that classifies a correct response to be either a random guess with probability c_i ($S_{pi}=0$) or a response according to the two-parameter item response model ($S_{pi}=1$). Test statistic l_0 is then defined conditionally on $S_{pi}=1$ and evaluates the extremeness of non-guessed responses. The approach thus ignores the guessed responses, which are assumed to be random with the guessing parameters as probability of success.

A Bayesian version of the log-likelihood fit statistics is defined to adjust for the uncertainty in the model parameters. The result is a test that quantifies the extremeness of each response pattern as the probability of the statistic being greater than the threshold value C associated with a (frequentist) significance level of $\alpha = 0.05$. The (marginal) posterior probability of the statistic being greater than the threshold is

$$\begin{aligned} \Pr\{l_0(\mathbf{U}_p) > C\} &= \int \ldots \int \Pr\{l_0(\mathbf{U}_p;\boldsymbol{\theta}_p,\mathbf{a},\mathbf{b}) > C\} p(\boldsymbol{\theta}_p) p(\mathbf{a},\mathbf{b})\, d\boldsymbol{\theta}_p\, d\mathbf{a}\, d\mathbf{b} \\ &= \int \ldots \int \Phi(l_0(\mathbf{U}_p;\boldsymbol{\theta}_p,\mathbf{a},\mathbf{b}) > C) p(\boldsymbol{\theta}_p) p(\mathbf{a},\mathbf{b})\, d\boldsymbol{\theta}_p\, d\mathbf{a}\, d\mathbf{b} \\ &= p_l. \end{aligned}$$

Note that the likelihood statistic is integrated directly over the prior distributions of all parameters. The test can thus be interpreted as a prior predictive test as well.

Besides, it is also possible to compute the posterior probability of an aberrant response pattern under the model for the given significance level. Let F_p^u denote a random variable that takes the value one when an observed response pattern $\mathbf{U}_p = \mathbf{u}_p$ is marked as extreme and zero otherwise,

$$F_p^u = \begin{cases} 1, & \text{if } I(l_0(\mathbf{U}_p;\boldsymbol{\theta}_p,\mathbf{a},\mathbf{b}) > C), \\ 0, & \text{if } I(l_0(\mathbf{U}_p;\boldsymbol{\theta}_p,\mathbf{a},\mathbf{b}) \leq C), \end{cases}$$

where $I(.)$ is the indicator function. The posterior probability of $F_p^u = 1$ is computed by integrating over the model parameters,

$$\Pr\{F_p^u = 1\} = \int \ldots \int I(l_0(\mathbf{U}_p;\boldsymbol{\theta}_p,\mathbf{a},\mathbf{b}) > C) p(\boldsymbol{\theta}_p) p(\mathbf{a},\mathbf{b}) \, d\boldsymbol{\theta}_p \, d\mathbf{a} \, d\mathbf{b}.$$

A response pattern $\mathbf{u}_p$ can be identified as extreme when F_p^u equals one, for instance, with at least 0.95 posterior probability.

29.4.2 Person Fit of RT Model

Analogously, the log-likelihood of the RTs can be used to define a person-fit statistic to identify aberrant RT patterns. From the RT model in Equation 29.3, the log-likelihood statistic for the RT pattern of test taker p is

$$\ln f(\mathbf{T}_p;\tau_p,\alpha,\beta) = \sum_{i=1}^{I} \ln f(T_{pi};\tau_p,\alpha_i,\beta_i).$$

We use the sum of the squared standardized residuals in the exponent of the resulting expression as person-fit statistic for the RT pattern

$$l^t(\mathbf{T}_p;\tau_p,\alpha,\beta) = -\sum_{i=1}^{I} \alpha_i (\ln T_{pi} - (\beta_i - \tau_p))^2. \tag{29.17}$$

Given the person and item parameters, the person-fit statistic is chi-squared distributed with I degrees of freedom. A Bayesian significance test can be defined based on

$$\Pr\{l^t(\mathbf{T}_p;\tau_p,\alpha,\beta) > C\} = \Pr\{\chi_I^2 > C\}.$$

Another classification variable can be defined to quantify the posterior probability of an extreme RT pattern given a threshold value C. Let F_p^t denote the random variable which equals one when the RT pattern is flagged as extreme and zero otherwise; that is,

$$F_p^t = \begin{cases} 1, & \text{if } \Pr\{l^t(\mathbf{T}_p;\tau_p,\alpha,\beta)\} > C, \\ 0, & \text{if } \Pr\{l^t(\mathbf{T}_p;\tau_p,\alpha,\beta)\} \leq C. \end{cases}$$

Again, an observed RT pattern $\mathbf{t}_p$ is reported as extreme when the F_p^t equals one with a least 95 posterior probability.

Finally, classification variables F_p^u and F_p^t can be used jointly to flag test takers for their observed response and RT patterns. In order to do so, the joint posterior probability

$$\Pr(F_p^u = 1, F_p^t = 1 \mid \mathbf{t}_p, \mathbf{u}_p) = \int\int \Pr(F_p^u = 1, F_p^t = 1 \mid \xi_p, \boldsymbol{\psi}, \mathbf{t}_p, \mathbf{u}_p) f(\xi_p, \boldsymbol{\psi},) d\xi_p \, d\boldsymbol{\psi}$$

$$= \int \ldots \int \Pr(F_p^u = 1 \mid \theta_p, \mathbf{a}, \mathbf{b}, \mathbf{u}_p) \Pr(F_p^t = 1 \mid \tau_p, \boldsymbol{\alpha}, \boldsymbol{\beta}, \mathbf{t}_p)$$

$$\cdot f(\theta_p, \tau_p) f(\boldsymbol{\alpha}, \boldsymbol{\beta}, \mathbf{a}, \mathbf{b}) d\theta_p \, d\tau_p \, d\boldsymbol{\alpha} \, d\boldsymbol{\beta} \, d\mathbf{a} \, d\mathbf{b}$$

should be greater than 0.95, where $\boldsymbol{\psi} = (\psi_i)$. Observe how the local independence assumption between responses and RTs given speed and ability is used to represent the joint probability of aberrant patterns as a product of the probability of an aberrant response and of an aberrant RT pattern.

29.4.3 Item Fit

The Bayesian person-fit tests can easily be modified to evaluate the fit of responses and RTs to an item. The only necessary change is the definition of the log-likelihood statistic across the test takers for each item; everything else is similar. Just as the person-fit tests, the result is a more conservative test that is less likely to flag response and RT patterns on the items as aberrant due to its accounting for the uncertainty in the model parameters.

29.5 Empirical Example

Two versions of the joint model were estimated using a dataset consisting of the responses and RTs of $P = 454$ test takers on the $I = 60$ items of a certification exam. Both versions included the lognormal RT model in Equation 29.3. For a reason that will become clear below, one version included the two-parameter (2PL) response model, the other the three-parameter (3PL) model in Equation 29.2. The two response models were identified by restricting the population means of the ability and speed person parameters to be equal to zero and setting the product of all discrimination parameters equal to one. The parameters were estimated using the Gibbs sampler with the MH steps outlined above. As a software program, a modified version of the `cirt` package (Fox et al., 2007) was used to accommodate the parameterization in Equations 29.3 through 29.6. The total of number of iterations was 5000, with the first 1000 iterations used for burn in. Visual inspection of the chains showed good convergence.

Table 29.2 gives the mean and standard deviations of the expected a priori (EAP) estimates of the item parameters for the two versions of the joint model. The results appeared to be quite close. The mean time-intensity parameter of 3.87 corresponds to exp(3.87) = 47.94 s for test takers who operated at the average speed of $\tau = 0$. Basically, the only impact of the addition of the guessing parameters to the response model was a slight shift in the estimates of the difficulty parameters (from −0.61 to −0.47).

TABLE 29.2

Mean and SDs of the Estimated Item Parameters for the Two Versions of the Joint Model

	Version with 2PL Model		Version with 3PL Model	
Parameter	**Mean Estimate**	**SD**	**Mean Estimate**	**SD**
α_i	1.85	0.04	1.86	0.04
β_i	3.87	0.07	3.86	0.07
a_i	1.06	0.05	1.07	0.06
b_i	−0.61	0.09	−0.47	0.10
c_i	–	–	0.16	0.03

The correlation matrices between the speed and ability parameters for the two version of the model were equal to

$$\boldsymbol{\Sigma}_{\mathcal{P}} = \begin{pmatrix} 1.00 & -0.12 \\ -0.13 & 1.00 \end{pmatrix}$$

The two correlations (above and below the diagonal for the 3PL and 2PL model, respectively) were mildly negative differing only in their second decimal. Negative correlations between speed and ability are not uncommon; in fact, a similar correlation was already met in our explanation of the negative correlation between the responses and RTs for the two items in Table 29.1. A plausible explanation of them might be the better time-management skills of the more able test takers. Consequently, when they observe a less tight time limit, they might decide to slow down exploiting the opportunity to increase their scores, whereas less able test takers may be less inclined to do so (van der Linden, 2009).

The two correlation matrices for the item parameters are shown in Table 29.3. A comparison between them shows a minor impact of the presence of the guessing parameter on the estimates of the other model parameters only. It is especially instructive to note that the lack of impact by the guessing parameter on the correlation between the parameters in the RT model—a result that confirms our earlier observation of the possibility of a plug-and-play approach to the joint modeling of responses and RTs. Otherwise, the two patterns of correlation are typical of the results for other datasets observed by the authors. Generally, the correlation between the item difficulty and time-intensity parameters in these datasets

TABLE 29.3

Estimated Correlation Matrices for the Item Parameters for the Two Versions of the Joint Model

	Version with 2PL Model				Version with 3PL Model			
	α_i	β_i	a_i	b_i	α_i	β_i	a_i	b_i
α_i	1.00	0.20	−0.20	−0.09	1.00	0.20	−0.23	−0.07
β_i		1.00	0.12	0.61		1.00	0.28	0.61
a_i			1.00	0.02			1.00	0.15
b_i				1.00				1.00

tended to be positive and rather high while the correlations between all other item parameters were much smaller with a less predictable pattern. Intuitively, it does make sense for more difficult items to take more time. The usually relatively high correlation between these two item features is also the reason for the differential speededness observed in adaptive testing, where more able test takers tend to get more difficult items and then run the risk of experiencing considerably more time pressure (van der Linden, Volume Three, Chapter 12)

Figure 29.3 shows a plot of the estimated person-fit statistics against their Bayesian levels of significance. The upper plot represents the curves for the RT patterns, the lower plot for the response patterns under the 2PL model. The nominal significance level of $\alpha = 0.05$ led to 7.4% of the RT patterns flagged as aberrant for the original log-likelihood statistic, of which 7.0% were flagged with a posterior probability of at least 0.95. For the response patterns, the percentage flagged as extreme went down from 2.9% to 2.7% for the adopted posterior probability of at least 0.95. Finally, although the nominal significance level remained at 0.05, only 0.9% of test takers were flagged as aberrant for their joint RT and response pattern with the same posterior certainty.

We also checked the residual RTs for the test taker-item combinations. For the lognormal RT model, the residuals are normally distributed given the item and test-taker parameters. Following Fox (2010, p. 247), for a posterior probability of 0.95, 7.4% of the residuals were found to differ more than two standard deviations from their expected value of zero.

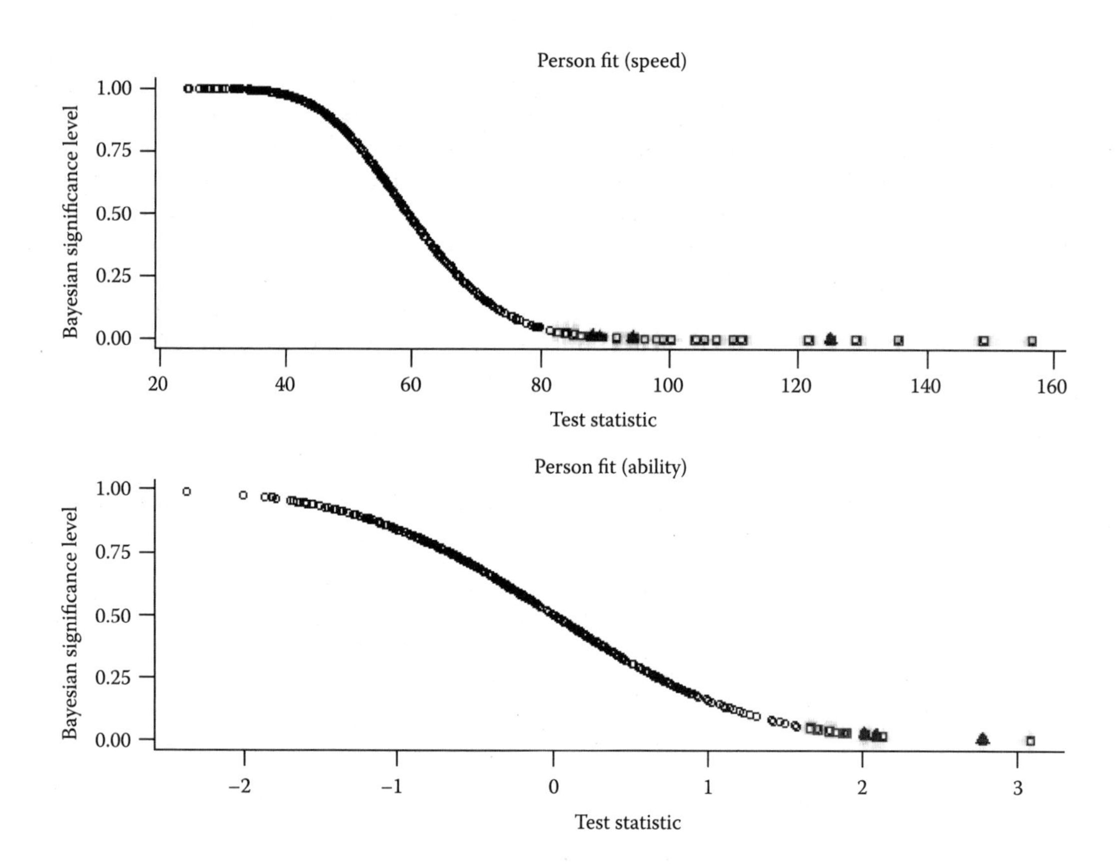

FIGURE 29.3
Plots of estimated person-fit statistics against their Bayesian significance level.

For the choice of the 3PL model, the percentage of RT patterns flagged as aberrant remained the same but the percentage of flagged response patterns decreased from 1.8% to 0.2% for the adopted minimum posterior probability of 0.95. Consequently, none of the test takers was flagged as extreme with respect to their joint RT and response pattern. The results suggest that one of main reasons of the previous flagging of model misfit for some of the test takers might have been the lack of accounting for their guessing by the 2PL model.

The item-fit statistics did not yield any significant results both for the responses and RTs. It was concluded that the 60 vectors of item responses and RTs observed in our dataset behaved according to the model.

29.6 Discussion

Our newly gained easy access to the RTs by test takers on items may force us to reconsider nearly every theoretical question and practical application addressed in the history of testing so far. Instead of pure speculation on the impact of the test taker's speed and the time features of the items on the observed behavior by the test takers, we are now able to base our conclusions on empirical data analyzed under statistical models checked for their fit.

As for practical applications, several of those related to the issue of test speededness and the selection of time limits are reviewed in van der Linden (Volume Three, Chapter 12). Other possible applications include support to the diagnosis of differential item functioning, improved analysis of item misfit, more refined assessment of the impact of test speededness on parameter linking and observed-score equating, improved standard setting procedures, and enhanced protection of item and test security.

In fact, the joint modeling of RTs and responses in this chapter offers the attractive statistical option to import the empirical collateral information collected in the RTs during testing to improve the power of all our current response-based inferences (and conversely). As an example, it allows for the use of the RTs collected during adaptive testing to improve the interim ability estimates—and hence item selection—in adaptive testing. The statistical foundation for doing so is a Bayesian approach with an empirical update of the initial prior distribution after each newly observed RT. The approach is novel in that, unlike a regular empirical Bayesian approach, it updates both the test taker's likelihood and the prior distribution (van der Linden, 2008). A similar new approach to item calibration is possible (van der Linden et al., 2010). For a discussion of the advantages of using RTs as collateral information in testing, see Ranger (2013).

References

Atchadé, Y. F., and Rosenthal, J. S. 2005. On adaptive Markov chain Monte Carlo algorithms. *Bernoulli, 20*, 815–828.

Box, G. E. P. 1980. Sampling and Bayesian inference in scientific modeling and robustness. *Journal of the Royal Statistical Society A, 143*, 383–430.

Cai, L. 2010. Metropolis–Hastings Robbins–Monro algorithm for confirmatory factor analysis. *Journal of Educational and Behavioral Statistics, 35*, 307–335.

De Boeck, P. 2008. Random item IRT models. *Psychometrika, 73,* 533–559.
Drasgow, F., Levine, M. V., and Williams, E. A. 1985. Appropriateness measurement with polytomous item response models and standardized indices. *British Journal of Mathematical and Statistical Psychology, 38,* 67–86.
Fox, J.-P. 2010. *Bayesian item Response Modeling: Theory and Applications.* New York: Springer.
Fox, J.-P., Klein Entink, R. H., and van der Linden, W. J. 2007. Modeling of responses and response times with the package CIRT. *Journal of Statistical Software, 20*(7), 1–14.
Gelman, A., Carlin, J. B., Stern, H. S., Dunson, D. B., Vehtari, A., and Rubin, D. B. 2013. *Bayesian Data Analysis* (3rd edn). Boca Raton, Florida: CRC Press.
Gill, J. 2015. *Bayesian Methods: A Social and Behavioral Sciences Approach* (3rd edn). Boca Raton, Florida: CRC Press.
Glas, C. A. W., and van der Linden, W. J. 2010. Marginal likelihood inference for a model for item responses and response times. *British Journal of Mathematical and Statistical Psychology, 63,* 603–626.
Johnson, V. E., and Albert, J. H. 1999. *Ordinal Data Modeling.* New York: Springer.
Karabatsos, G. 2003. Comparing the aberrant response detection performance of thirty-six person-fit statistics. *Applied Measurement in Education, 14,* 54–75.
Klein Entink, R. H., Fox, J.-P., and van der Linden, W. J. 2009a. A multivariate multilevel approach to simultaneous modeling of accuracy and speed on test items. *Psychometrika, 74,* 21–48.
Klein Entink, R. H., Kuhn, J.-T., Hornke, L. F., and Fox, J. P. 2009b. Evaluating cognitive theory: A joint modeling approach using responses and response times. *Psychological Methods, 14,* 54–75.
Klein Entink, R. H., van der Linden, W. J., and Fox, J.-P. 2009c. A Box–Cox normal model for response times. *British Journal of Mathematical and Statistical Psychology, 62,* 621–640.
Luce, R. D. 1986. *Response Times: Their Roles in Inferring Elementary Mental Organization.* Oxford, UK: Oxford University Press.
Marianti, S., Fox, J.-P., Avetisyan, M., Veldkamp, B. P., and Tijmstra, J. 2014. Testing for aberrant behavior in response time modeling. *Journal of Educational and Behavioral Statistics, 39,* 426–452.
Levine, M. V., and Rubin, D. B. 1979. Measuring the appropriateness of multiple-choice test scores. *Journal of Educational Statistics, 4,* 269–290.
Rasch, G. 1960. *Probabilistic Models for Some Intelligence and Attainment Tests.* Chicago: University of Chicago Press.
Ranger, J. 2013. A note on the hierarchical model of responses and response times on test of van der Linden 2007. *Psychometrika, 78,* 358–544.
Rosenthal, J. S. 2007. AMCMC: An R interface for adaptive MCMC. *Computational Statistics & Data Analysis, 51,* 5467–5470.
van der Linden, W. J. 2006. A lognormal model for response times on test items. *Journal of Educational and Behavioral Statistics, 31,* 181–204.
van der Linden, W. J. 2007. A hierarchical framework for modeling speed and accuracy on test items. *Psychometrika, 73,* 287–308.
van der Linden, W. J. 2008. Using response times for item selection in adaptive testing. *Journal of Educational and Behavioral Statistics, 33,* 5–20.
van der Linden, W. J. 2009. Conceptual issues in response-time modeling. *Journal of Educational Measurement, 46,* 247–272.
van der Linden, W. J., and Glas, C. A. W. 2010. Statistical tests of conditional independence between responses and/or response times on test items. *Psychometrika, 75,* 120–139.
van der Linden, W. J., Klein Entink, R. H., and Fox, J.-P. 2010. IRT parameter estimation with response times as collateral information. *Applied Psychological Measurement, 34,* 327–347.
van der Linden, W. J., and Ren, H. 2015. Optimal Bayesian adaptive design for test-item calibration. *Psychometrika, 80,* 263–288.
Wang, C., Chang, H.-H., and Douglas, J. A. 2013. The linear transformation model with frailties for the analysis of item response times. *British Journal of Mathematical and Statistical Psychology, 66,* 144–168.

Section VIII

Generalized Modeling Approaches

30

Generalized Linear Latent and Mixed Modeling

Sophia Rabe-Hesketh and Anders Skrondal

CONTENTS

30.1 Introduction 503
30.2 The GLLAMM Framework 505
30.2.1 Response Model 505
30.2.1.1 Links and Conditional Distributions 505
30.2.1.2 Linear Predictor 506
30.2.2 Structural Model 506
30.3 Response Processes 507
30.3.1 Dichotomous Responses 507
30.3.2 Ordinal Responses 508
30.3.3 Comparative Responses 508
30.3.4 Counts 510
30.3.5 Continuous Responses and Response Times 510
30.3.6 Some Other Response Processes 511
30.4 Person and Item Covariates 511
30.5 Multidimensional Models 512
30.5.1 Conventional MIRT 512
30.5.2 Random Coefficients of Item Characteristics 513
30.5.3 Growth IRT 513
30.6 Multilevel Models 514
30.6.1 Multilevel IRT 514
30.6.2 Multilevel SEM 516
30.7 Parameter Estimation 517
30.8 Model Fit 518
30.9 Example 519
30.9.1 PIRLS Data 519
30.9.2 Model 519
30.9.3 Results and Interpretation 520
30.10 Discussion 522
References 523

30.1 Introduction

It is evident from previous chapters in this handbook that conventional IRT models have been extended in many different ways, for instance, to accommodate different response processes, multidimensionality, and multilevel data. Similar developments have occurred in other areas, such as factor analysis, where multidimensional measurement models have a long tradition.

Unfortunately, it is often difficult to recognize the commonalities between different modeling traditions due to differences in terminology and notation—differences preserved and reinforced by academic compartmentalization. A striking example is the lack of cross-referencing between the literatures on IRT and categorical factor analysis, even after many of the connections were spelled out by Takane and de Leeuw (1987). A cost of this compartmentalization is that advances in one area are not automatically adopted in other areas.

A great advantage of unifying modeling frameworks is that they can help overcome such obstacles to progress. In the IRT setting, relevant frameworks include generalized linear models (GLMs), generalized linear IRT, structural equation models (SEMs), and generalized linear mixed models (GLMMs). The generalized linear latent and mixed model (GLLAMM) framework described in this chapter combines and extends all these models within a single framework.

GLMs (McCullagh and Nelder, 1989) remain a workhorse in applied statistics and include linear, logistic, Poisson, and other regression models as special cases. The benefit of such a coherent statistical framework is that theoretical results, estimation algorithms, diagnostics, and associated software can be developed for the entire model family instead of replicating this work for each of a multitude of special cases.

IRT models are typically specified via a response function, which can be thought of as the inverse link function of a GLM. Bartholomew (1987) made this connection explicit and Mellenbergh (1994) noted that most IRT models can be viewed as GLMs with latent explanatory variables. Mellenbergh's generalized linear IRT extended GLMs to include ordinal and nominal responses, in addition to continuous and binary responses and counts.

In the factor analysis tradition, a breakthrough was the development of SEMs with latent variables by Jöreskog (1973) and others. SEMs embed confirmatory factor models from psychometrics within structural equation models from econometrics, thus accommodating linear relationships among latent variables and between latent variables and observed covariates. Another important advance was due to Muthén (1984) and others who used a latent response formulation with thresholds to extend conventional SEMs to include probit measurement models (normal ogive models) for binary, ordinal, and censored responses. A limitation of this framework is that standard logistic IRT models for binary and ordinal items were not accommodated and that nominal responses and counts could not be modeled.

In statistics, GLMMs were developed for multilevel and longitudinal data (Breslow and Clayton, 1993). Recognizing that the Rasch model for binary responses and the parallel measurement model for continuous responses can be viewed as two-level GLMMs made it obvious how to formulate multilevel versions of these models for persons nested in clusters (Kamata, 2001; Maier, 2001; Raudenbush and Sampson, 1999). However, as Rijmen et al. (2003) pointed out, models with discrimination parameters do not fit within the GLMM framework.

The measurement models within the GLLAMM framework (Rabe-Hesketh et al., 2004; Skrondal and Rabe-Hesketh, 2004) include multilevel versions of Mellenbergh's generalized linear IRT. The framework also includes multilevel structural models that allow latent variables at one level to be regressed on latent and observed variables at the same and higher levels. Other formulations of multilevel structural models are limited to one level of clustering and to linear and probit measurement models (Asparouhov and Muthén, 2007; Fox and Glas, 2001, 2003).

30.2 The GLLAMM Framework

The GLLAMM framework consists of a response model, relating observed responses to latent and observed covariates, and a structural model, relating latent variables to each other and to observed covariates. The response model has three components: a linear predictor, a link, and a distribution.

We first define each model component in its general form. For brevity, we focus on continuous latent variables or latent traits in this chapter, although the framework also accommodates discrete latent variables (Skrondal and Rabe-Hesketh, 2004, 2007). For concreteness, we show how each of the components would be specified to obtain a simple IRT model, the two-parameter logistic (2PL) model (Birnbaum, 1968), possibly with a latent regression.

30.2.1 Response Model

30.2.1.1 Links and Conditional Distributions

The linear predictor ν (described in Section 30.2.1.2) is a function of a vector of latent variables $\boldsymbol{\theta}$ and vectors of covariates $\mathbf{x}$ and $\mathbf{z}$. We treat $\boldsymbol{\theta}$, $\mathbf{x}$, and $\mathbf{z}$ as random variables. The conditional expectation of the response U, given $\mathbf{x}$, $\mathbf{z}$, and $\boldsymbol{\theta}$, is "linked" to the linear predictor via a link function $g(\cdot)$,

$$g(\mathrm{E}[U|\,\mathbf{x},\mathbf{z},\boldsymbol{\theta}]) = \nu. \tag{30.1}$$

Here we have omitted subscripts for items and persons because general models will often have a multitude of other subscripts, for instance, for schools in multilevel IRT.

Link functions include the identity, log, logit, (scaled) probit, and complementary log-log link. For ordinal responses, cumulative logit, cumulative (scaled) probit, and cumulative complementary log-log links can be specified (see Section 30.3.2). For nominal and comparative responses, baseline-category logit and related links can be used (see Section 30.3.3). We refer to Albert (Volume Two, Chapter 1) for a general introduction to different link and response functions.

The conditional distribution of the response given the linear predictor is a member of the exponential family of distributions, including the binomial (or Bernoulli), normal, Poisson, and gamma distributions.

EXAMPLE 30.1: 2PL MODEL

The response U_{pi} for person p and item i is binary, the expectation is a probability, the link function is the logit link

$$\mathrm{logit}(\Pr\{U_{pi}=1\,|\,\theta_p\}) = \nu_{pi}$$

and the conditional response distribution is Bernoulli. See Section 30.2.1.2 on how the linear predictor $\nu_{pi} = a_i(\theta_p - b_i)$ is specified in the GLLAMM framework.

For distributions that have a dispersion parameter φ, such as the normal and gamma, a log-linear model can be specified for $\boldsymbol{\varphi}$

$$\ln\boldsymbol{\varphi} = \mathbf{v}'\boldsymbol{\alpha}, \tag{30.2}$$

where $\mathbf{v}$ is a vector of covariates and $\boldsymbol{\alpha}$ are regression coefficients.

Considerable flexibility in modeling response processes can be achieved using composite links (Thompson and Baker, 1981; Rabe-Hesketh and Skrondal, 2007). Defining several link functions $g_q(\cdot)$, $q = 1, \ldots, Q$, for each observation, the conditional expectation is written as a linear combination of inverse link functions with weights c_q,

$$E[U|\mathbf{x},\mathbf{z},\boldsymbol{\theta}] = \sum_{q=1}^{Q} c_q g_q^{-1}(\nu_q). \tag{30.3}$$

Section 30.3.6 provides examples demonstrating the power of composite link functions.

Different link functions and conditional distributions can furthermore be specified for different items and persons, for instance, for modeling items with mixed response formats and for modeling response times with censoring (see Section 30.3.5).

30.2.1.2 Linear Predictor

For multilevel settings where clusters are nested in superclusters, etc., the linear predictor contains latent variables varying at different levels $l = 2, \ldots, L$. To simplify notation, we will not use subscripts for the units of observation at the various levels. For a model with M_l latent variables at level l, the linear predictor then has the form

$$\nu = \mathbf{x}'\boldsymbol{\beta} + \sum_{l=2}^{L} \sum_{m=1}^{M_l} \theta_m^{(l)} \mathbf{z}_m^{(l)\prime} \boldsymbol{\lambda}_m^{(l)}. \tag{30.4}$$

The elements of $\mathbf{x}$ are covariates with "fixed" effects $\boldsymbol{\beta}$ and the mth latent variable $\theta_m^{(l)}$ at level l is multiplied by a linear combination $\mathbf{z}_m^{(l)\prime}\boldsymbol{\lambda}_m^{(l)}$ of covariates $\mathbf{z}_m^{(l)}$ with coefficients $\boldsymbol{\lambda}_m^{(l)}$.

EXAMPLE 30.2: 2PL MODEL

The model requires just $L = 2$ levels and $M_2 = 1$ latent variable at level 2. The covariate vectors are $\mathbf{x}_{pi} = \mathbf{z}_{pi} = \mathbf{s}_i$, where $\mathbf{s}_i$ is an I-dimensional selection vector with 1 in the ith position and 0 elsewhere. Omitting the superscript for levels (since there is only a latent variable at one level here), the linear predictor becomes

$$\nu_{pi} = \mathbf{s}_i'\boldsymbol{\beta} + \theta_p \mathbf{s}_i'\boldsymbol{\lambda} \tag{30.5}$$

$$= \beta_i + \theta_p\lambda_i \equiv a_i(\theta_p - b_i), \tag{30.6}$$

where $b_i = -\beta_i/\lambda_i$ and $a_i = \lambda_i$. While this item-intercept parameterization with intercepts β_i is more natural within the GLLAMM framework than the usual item-difficulty parameterization with difficulties b_i, the latter can be accomplished using the structural model described in the next section.

30.2.2 Structural Model

The vector of latent variables at level l is denoted $\theta^{(l)}$. The structural model for the vector of all latent variables $\boldsymbol{\theta} = (\boldsymbol{\theta}^{(2)\prime}, \ldots, \boldsymbol{\theta}^{(L)\prime})'$ has the form

$$\boldsymbol{\theta} = \mathbf{B}\boldsymbol{\theta} + \boldsymbol{\Gamma}\mathbf{w} + \boldsymbol{\zeta}, \tag{30.7}$$

where $\mathbf{B}$ is an $M \times M$ regression parameter matrix ($M = \sum_l M_l$), $\mathbf{w}$ is a vector of covariates, $\boldsymbol{\Gamma}$ is a regression parameter matrix and $\boldsymbol{\zeta}$ is a vector of disturbances.

EXAMPLE 30.3: 2PL MODEL WITH A LATENT REGRESSION

The unidimensional latent trait θ_p is regressed on covariates with regression coefficients in $\boldsymbol{\Gamma}$ (here a matrix with one row) with disturbance ζ_p,

$$\theta_p = \boldsymbol{\Gamma}\mathbf{w}_p + \zeta_p.$$

This model does not require a $\mathbf{B}$ matrix.

EXAMPLE 30.4: 2PL MODEL WITH ITEM-DIFFICULTY PARAMETERIZATION

We can think of a latent variable θ_{pi} representing ability minus difficulty, giving the following linear predictor and latent regression:

$$\begin{aligned} \nu_{pi} &= \theta_{pi}\mathbf{s}_i'\boldsymbol{\lambda}, \\ \theta_{pi} &= \boldsymbol{\Gamma}\mathbf{s}_i + \zeta_{p'}, \end{aligned} \tag{30.8}$$

where the elements of $\boldsymbol{\Gamma}$ are minus the item difficulties.

Latent variables cannot be regressed on latent variables varying at a lower level since such specifications do not make sense. Furthermore, the model is currently confined to recursive relations, not permitting feedback effects among the latent variables. The two restrictions together imply that the matrix $\mathbf{B}$ is strictly upper triangular if the elements of $\boldsymbol{\theta}^{(l)}$ are permuted appropriately, with elements of $\boldsymbol{\theta} = (\boldsymbol{\theta}^{(2)'}, \ldots, \boldsymbol{\theta}^{(L)'})'$ arranged in increasing order of l. Each element of $\boldsymbol{\zeta}$ varies at the same level as the corresponding element of $\boldsymbol{\theta}$ (In Equation 30.8 θ_{pi} is treated as a level-2 latent variable to obtain a person-level disturbance). Disturbances at a level l have a multivariate normal distribution with zero mean and covariance matrix Ψ_l and are independent of disturbances at other levels.

We now show how a wide range of IRT models can be specified within the GLLAMM framework. The focus is on exploring the scope of the framework, not in giving detailed descriptions or conceptual interpretations of the specific models. We start with measurement models for different types of responses, and then consider multidimensional measurement models, models with covariates, and finally multilevel models. Each section can be seen as a type of extension of traditional IRT models. Importantly, all these extensions can be combined to build a complex model.

30.3 Response Processes

30.3.1 Dichotomous Responses

As mentioned in Section 30.2.1.1, logistic IRT models for binary items are specified using a logit link and a Bernoulli distribution. Normal ogive models (e.g., Lawley, 1943; Lord, 1952) are obtained by using a probit link instead and Goldstein (1980) considered a complementary log-log link. Two-parameter versions of these models are accommodated in the GLLAMM framework but the three-parameter and four-parameter versions are not, unless the additional parameters are known (Rabe-Hesketh and Skrondal, 2007).

30.3.2 Ordinal Responses

In educational testing, ordinal responses arise when partial credit is awarded for partial success on items (Volume One, Chapter 7). In attitude measurement, Likert formats and other rating scales are often used.

Different kinds of logit links are typically used for ordinal responses. However, unlike the dichotomous case, the link function is not applied to the conditional expectation of the response as in Equation 30.1. For response alternatives $a = 1, \ldots, A$, a general ordinal logit model can be written as

$$\log\left[\frac{\Pr\{U \in \mathcal{A}_s | \theta_p\}}{\Pr\{U \in \mathcal{B}_s | \theta_p\}}\right] = \nu_p, \quad s = 1, \ldots, A-1, \tag{30.9}$$

where $\mathcal{A}_s$ and $\mathcal{B}_s$ are subsets of response alternatives.

The partial credit model (Volume One, Chapter 7) and rating scale model (Volume One, Chapter 5) use an adjacent-category logit link, where $\mathcal{A}_s = \{s + 1\}$ and $\mathcal{B}_s = \{s\}$, the sequential item response model (Tutz, 1990; Volume One, Chapter 9) uses a continuation-ratio logit link with $\mathcal{A}_s = \{s\}$ and $\mathcal{B}_s = \{s + 1, \ldots, A\}$, and the graded-response logit model (Volume One, Chapter 6) uses a cumulative logit link with $\mathcal{A}_s = \{s + 1, \ldots, A\}$ and $\mathcal{B}_s = \overline{\mathcal{A}}_s$.

The cumulative logit model can also be derived by assuming that the observed ordinal response results from applying thresholds to a latent continuous response U^*_{pi}, such that $U_{pi} = a$ if $\kappa_{a-1} \le U^*_{pi} < \kappa_a$ with $\kappa_0 = -\infty$ and $\kappa_A = \infty$. A linear model is then specified for the latent response, $U^*_{pi} = \nu_{pi} + \epsilon_{pi}$. If the error term ε_{pi} has a standard logistic distribution, the model for the observed response U_{pi} is a cumulative logit model; if the error has a standard normal distribution, we obtain a cumulative probit model; and if it has a Gumbel distribution, we get a complementary log-log model.

In the continuation-ratio logit model, the probability that $U_{pi} = a$ can also be expressed as the joint probability of a binary responses with an ordinary logit link, where the first $a - 1$ binary responses take the value 0 and the last response takes the value 1 (Cox, 1972). Läärä and Matthews (1985) moreover show that the continuation-ratio logit model is equivalent to a cumulative model with a complementary log-log link. However, other cumulative models require tailor-made link functions.

30.3.3 Comparative Responses

A problem with rating scales is that different persons may use different response styles, both in terms of location and dispersion (Brady, 1985). This problem can be addressed using comparative formats, sometimes referred to as forced choice items. For instance, persons could be asked which of two statements describes their attitude best, or which of several activities they enjoy the most, or to rank-order a set of political values by importance (Inglehart, 1971). Another way to handle different response styles is through anchoring vignettes and the model proposed by King et al. (2003) can be specified in the GLLAMM framework (Rabe-Hesketh and Skrondal, 2002).

In educational testing, archetypal nominal responses are the raw responses (not scored as correct or incorrect) to multiple-choice questions. These raw responses can be analyzed to exploit information revealed by the choice of "wrong" (or distractor) response. In the nominal response model (Bock, 1972; Volume One, Chapter 3), the response probabilities have the form

$$\Pr\{U_{pi} = a | \theta_p\} = \frac{\exp\left(\nu_{pi}^{[a]}\right)}{\sum_{s=1}^{A} \exp\left(\nu_{pi}^{[s]}\right)}, \tag{30.10}$$

where $\nu_{pi}^{[a]}$ is the linear predictor for response alternative a,

$$\nu_{pi}^{[a]} = c_{ia} + a_{ia}\theta_p, \quad \sum_{s=1}^{A} c_{ia} = \sum_{s=1}^{A} a_{ia} = 0$$

with item and alternative-specific parameters c_{ia} and a_{ia}.

The form of the probability in Equation 30.10 is the same as that of a conditional logit model. Specifically, it is the conditional joint probability for a set of binary responses $\{y_{pi}^{[a]}, a = 1, \ldots, A\}$, where the response $y_{pi}^{[a]}$ takes the value 1 if $U_{pi} = a$ and 0 otherwise, conditional on exactly one response being equal to 1 (or the sum of responses being equal to 1).

The GLLAMM framework allows the data to be expanded to A rows per item-person combination so that general linear predictors $\nu_{pi}^{[a]}$ can be specified as in Equation 30.4. Such an expanded form of the design matrix (but with less general linear predictors) is also used in Adams et al. (1997).

Using the conditional logit formulation for the nominal response model, the vector of linear predictors for two items ($i = 1, 2$) and three categories ($a = 1, 2, 3$), $\mathbf{v}_{p\{1,2\}} = \left(\nu_{p1}^{[1]}, \nu_{p1}^{[2]}, \nu_{p1}^{[3]}, \nu_{p2}^{[1]}, \nu_{p2}^{[2]}, \nu_{p2}^{[3]}\right)'$, is modeled as

$$\mathbf{v}_{p\{1,2\}} = \underbrace{\begin{bmatrix} 1 & 0 & 0 & 0 \\ 0 & 1 & 0 & 0 \\ -1 & -1 & 0 & 0 \\ 0 & 0 & 1 & 0 \\ 0 & 0 & 0 & 1 \\ 0 & 0 & -1 & -1 \end{bmatrix}}_{\mathbf{M}} \begin{bmatrix} c_{11} \\ c_{12} \\ c_{21} \\ c_{22} \end{bmatrix} + \theta_p \mathbf{M} \begin{bmatrix} a_{12} \\ a_{13} \\ a_{22} \\ a_{23} \end{bmatrix}.$$

For the generalized partial credit model (Muraki, 1992; Volume One, Chapter 8), the linear predictors have the forms

$$\nu_{pi}^{[a]} = \sum_{s=1}^{a} (a_i\theta_p - \delta_{is}) = -\sum_{s=1}^{a} \delta_{is} + \theta_p a \mathbf{s}_i'\boldsymbol{\lambda},$$

where δ_{is}/a_i are step difficulties with $\delta_{i1} = 0$, and $\sum_{s=1}^{1}(a_i\theta_p - \delta_{is}) = 0$. Using the conditional logit formulation (with $a_i \equiv \lambda_i$), the model for $\mathbf{v}_{p\{1,2\}}$ becomes

$$\mathbf{v}_{p\{1,2\}} = \begin{bmatrix} 0 & 0 & 0 & 0 \\ -1 & 0 & 0 & 0 \\ -1 & -1 & 0 & 0 \\ 0 & 0 & 0 & 0 \\ 0 & 0 & -1 & 0 \\ 0 & 0 & -1 & -1 \end{bmatrix} \begin{bmatrix} \delta_{12} \\ \delta_{13} \\ \delta_{22} \\ \delta_{23} \end{bmatrix} + \theta_p \begin{bmatrix} 1 & 0 \\ 2 & 0 \\ 3 & 0 \\ 0 & 1 \\ 0 & 2 \\ 0 & 3 \end{bmatrix} \begin{bmatrix} a_1 \\ a_2 \end{bmatrix}$$

(Zheng and Rabe-Hesketh, 2007). The conditional logit parameterization also makes it easy to model groups of items jointly and relax the local independence assumption by including interaction parameters among pairs of items (Hoskens and De Boeck, 1986).

A ranking of alternatives can be viewed as a first choice among all alternatives, followed by successive choices among remaining alternatives. Expanding the data for each of these successive choices (A rows, then $A-1$ rows, etc.) allows the baseline-category logit model to be applied in the usual way, giving the Luce–Plackett model for rankings (Luce, 1959; Plackett, 1975), sometimes called the "exploded logit model" (Chapman and Staelin, 1982). This model, as well as the nominal response model, can be derived by assuming that persons associate different (continuous) utilities with each alternative and choose the alternative with the highest utility (or rank-order alternatives according to their utilities), an idea dating back to Thurstone (1927, 1931). In the context of IRT, Bock (1972) referred to the utilities as "response tendencies." The utility formulation can also be used to model other kinds of comparative responses, such as pairwise comparisons (Böckenholt, 2001).

For more discussion about models for comparative responses within the GLLAMM framework, see Skrondal and Rabe-Hesketh (2003).

30.3.4 Counts

Rasch (1960) considered a Poisson model for the number of misreadings in a text where the rate parameter λ_{pi} for person p and test i is the product of a person (in) ability parameter and a test difficulty parameter. Other examples of counts in a test setting are the number of words read in a speed test (within a time limit) or the number of spelling errors.

Whereas Rasch treated ability as a fixed parameter, Owen (1969) proposed a Bayesian model in which ability had a gamma distribution, and Jansen and van Duijn (1992) considered such a model with covariates using a likelihood approach. The latter model, except with a log-normal distribution for ability, fits within the GLLAMM framework with

$$\log \mathrm{E}[U_{pi} | \theta_p] = \theta_p + \mathbf{s}_i'\boldsymbol{\beta} + \log t_i,$$

where t_i is the "exposure," such as the time limit in a speed test. It follows that the expectation of U_{pi} is multiplicative, $t_i \exp(\theta_p) \exp(\beta_i)$, where $\exp(\theta_p)$ is interpreted as the ability.

See Jansen (Volume One, Chapter 15) for more details on IRT models for counts.

30.3.5 Continuous Responses and Response Times

In conventional factor analysis and classical test theory, the responses are continuous, an identity link is specified, and a normal conditional distribution is sometimes assumed. The variance of the conditional distribution is typically allowed to be different for each item, which is implemented in the GLLAMM framework via a log-linear model for the dispersion parameter as shown in Equation 30.2.

The time to complete an item, often referred to as response time, contains important cognitive information and such information is readily available in computer-administered tests. A popular model for the response time is a log-normal model (van der Linden, 2006; Volume One, Chapter 16), which can be specified by log-transforming the response time and using an identity link and normal distribution.

When an item is attempted but not completed, the response time is right-censored in the sense that all that is known is that it exceeds the time t_{pi} spent on the task. Censoring can be accommodated in the GLLAMM framework using an identity link and normal distribution for observed (log-transformed) response times and a scaled probit link and Bernoulli distribution (for a binary variable taking the value 1) for censored response times. Specifically, the probability for the censored response becomes $\Phi((\nu_{pi} - \log t_{pi})/\sigma_i)$, where $-\log t_{pi}$ is an offset and σ_i is the standard deviation of the log-normal distribution for item i.

Some of the gamma models for response times discussed by Maris (1993) are GLMs and hence also GLLAMMs. Proportional hazard models for response times, including Cox regression (Cox, 1972), can also be specified in the GLLAMM framework using a log link, Poisson distribution, and appropriately expanded data (e.g., Clayton, 1988); see also Rabe-Hesketh and Skrondal (2012, Chapter 15).

Joint models for response times and item responses (van der Linden, 2007; Volume One, Chapter 29) can be specified as GLLAMMs using different links and conditional distributions depending on the response type. Other examples of modeling mixed response types are discussed in Skrondal and Rabe-Hesketh (2004, Chapter 14).

30.3.6 Some Other Response Processes

Randomized response designs can be used when it is feared that persons will not answer items truthfully, for example, when asked about illicit drug use. In the Warner design, a positive or negative wording (e.g., agreement with "I use heroin" or "I do not use heroin") is randomly chosen and revealed only to the respondent to preserve confidentiality.

If the (known) probability of the positive wording for item i is p_i, a randomized IRT model (Fox 2005; Volume One, Chapter 28) can be specified as

$$\Pr\{U_{pi} = 1 | \nu_{pi}\} = p_i \frac{\exp(\nu_{pi})}{1+\exp(\nu_{pi})} + (1-p_i)\left(1 - \frac{\exp(\nu_{pi})}{1+\exp(\nu_{pi})}\right).$$

Rabe-Hesketh and Skrondal (2007) point out that this expression corresponds to a composite link function as in Equation 30.3 where $Q = 2$, $g_1(\cdot)$ and $g_2(\cdot)$ are both logit links, $c_{1i} = p_i$, $c_{2i} = 1 - p_i$, and $\nu_{1pi} = -\nu_{2pi}$.

For attitude items, it may be unreasonable to assume that the response function is monotonic as in standard IRT. Instead, the response function may have a peak at an "ideal point" so that the probability of agreeing with the item decreases as the distance between the ideal point and latent trait increases. The generalized graded unfolding model proposed by Roberts and Laughlin (1996; Volume One, Chapter 22) can be specified as a GLLAMM with a composite link (Rabe-Hesketh and Skrondal, 2007).

30.4 Person and Item Covariates

In large-scale assessments and other surveys, a major goal is to compare latent traits between different groups of persons. A naive approach is to treat scores obtained from standard IRT models as the response variable in a linear model. However, it is well-known that such an approach can be problematic (Mislevy, 1991; Skrondal and Laake, 2001), and it is preferable to specify a latent regression for θ_p (using the structural model in the GLLAMM framework) and estimate all parameters simultaneously (e.g., Mislevy, 1987).

Such models have a long history in the SEM tradition where they are called multiple-indicator multiple-cause (MIMIC) models (e.g., Jöreskog and Goldberger, 1975). We demonstrated in Section 30.2.2 how the 2PL model with a latent regression can be specified in the GLLAMM framework.

In addition to person covariates, there may be item attributes, such as the number of steps required to solve an equation, which are often assembled in what is known as the Q matrix. In the linear logistic test model (LLTM) of Fischer (Fischer, 1973; Volume One, Chapter 13), the item difficulties are modeled as a linear function of item attributes, and in the 2PL-constrained model by Embretson (1999), both difficulty and discrimination parameters are modeled as linear functions of item attributes. In the GLLAMM framework, the 2PL-constrained model can be specified by replacing $\mathbf{s}_i$ in Equations 30.6 or 30.8 by a vector of item attributes $\mathbf{q}_i$ if the item-intercept parameterization is used.

Uniform differential item functioning (DIF) can be modeled by including a dummy variable for the focal group versus the reference group in the latent regression and interactions between the focal group dummy variable and dummy variables for the DIF items in the linear predictor, giving the models considered by Muthén and Lehman (1985) and Thissen and Steinberg (1988). Nonuniform DIF can be accommodated by including the same interactions in the covariate vector $\mathbf{z}$ that multiplies the latent variable. We refer to Gamerman, Gonçalves, and Soares (Volume Three, Chapter 4) for more information on DIF. See Rijmen et al. (2003) and De Boeck and Wilson (2004) for further discussion and examples of "explanatory" IRT models with covariates for items, persons, and for item-person combinations.

30.5 Multidimensional Models

30.5.1 Conventional MIRT

Often several latent traits $\theta_{1p}, \ldots, \theta_{M2p}$ are believed to affect item responses and multidimensional IRT (MIRT) models are therefore used. The most common MIRT models, sometimes called compensatory, are essentially confirmatory factor models for binary or ordinal items (e.g., Christofferson, 1975; McDonald, 1982; Reckase, 1985). See also Swaminathan and Rogers (Volume One, Chapter 11) and Reckase (Volume One, Chapter 12).

In the GLLAMM framework, such models can be specified by including $M_2 > 1$ latent variables in the linear predictor. Let $\mathbf{d}_m$ be a vector containing the indices i of the items that measure the mth trait or dimension. Also let $\mathbf{s}_i[\mathbf{d}_m]$ denote the corresponding elements of the selection vector $\mathbf{s}_i$. Then the linear predictor can be written as

$$\nu_{pi} = \mathbf{s}_i'\boldsymbol{\beta} + \sum_{m=1}^{M_2} \theta_{mp}\mathbf{s}_i[\mathbf{d}_m]'\boldsymbol{\lambda}_m, \quad \lambda_{1m} = 1.$$

For instance, for a between-item multidimensional test (Adams et al., 1997), often called an independent clusters model in factor analysis, each item measures only one dimension. If items 1 through 3 measure the first dimension and items 4 through 6 the second dimension, we would specify $\mathbf{d}_1 = (1,2,3)'$ and $\mathbf{d}_2 = (4,5,6)'$. For a within-item multidimensional test, $\mathbf{d}_1$ and $\mathbf{d}_2$ have some common elements corresponding to items that measure both dimensions.

30.5.2 Random Coefficients of Item Characteristics

Rijmen and De Boeck (2002) extend Fisher's (1973) LLTM model by including random coefficients of item attributes. Their random weights LLTM model can be specified in the GLLAMM framework using a linear predictor of the form

$$\nu_{pi} = \mathbf{q}_i'\boldsymbol{\beta} + \sum_{m=1}^{M_2} \theta_{mp} q_{mi},$$

where q_{mi} is the mth element of the vector $\mathbf{q}_i$ of item attributes, with $q_{1i} = 1$. Here θ_{1p} is the ability when all attributes are 0 (a random intercept) and θ_{mp} for $m > 1$ is the random coefficient of the mth attribute for person p.

If the attributes are dummy variables, the random part of the model is the same as a multidimensional (within-item) Rasch model. If the item attributes are dummy variables for items belonging to testlets (e.g., items sharing a common stimulus or stem), the model becomes the Rasch testlet model (Wang and Wilson, 2005). The 2PL version of this model is a bifactor model (Gibbons and Hedeker, 1992; Holzinger and Swineford, 1937; Rijmen, 2010).

Gibbons and Hedeker (1992) and Rijmen (2010) point out that estimation of these high-dimensional models can be greatly simplified using a sequence of two-dimensional integrations. In the GLLAMM framework, this is accomplished by defining the specific dimensions for testlets (or item clusters) j ($j = 1, \ldots, J$) to be the values of *one* latent variable $\theta_{pj}^{(2)}$ that varies between person-testlet combinations (level 2), instead of J latent variables that vary between persons (level 3). The linear predictor becomes

$$\nu_{pij} = \mathbf{s}_i'\boldsymbol{\beta} + \theta_{pj}^{(2)}\mathbf{s}_i'\boldsymbol{\lambda}^{(2)} + \theta_p^{(3)}\mathbf{s}_i'\boldsymbol{\lambda}^{(3)},$$

where $\theta_p^{(3)}$ is the general dimension that varies between persons.

30.5.3 Growth IRT

When the same persons are measured repeatedly over time, either using the same test, as is common in psychological assessment, or using different tests with linking items, as is common in educational testing, growth IRT models can be used to model growth in the latent trait simultaneously with the measurement models at the different time-points or occasions.

Andersen (1985) specified a 1PL model for each occasion (with time-constant item parameters) and allowed the latent traits at the different occasions to be freely correlated, given a between-item multidimensional IRT model. Embretson (1991) used a different parameterization where θ_{1pi} represents the latent trait at the first occasion and θ_{mpi}, $m > 1$, represent the changes in the latent trait from occasion 1 to occasion m.

In SEM, linear or polynomial growth models are typically used for latent variables measured by continuous items, sometimes called "curve-of-factors models" (McArdle, 1988). Pastor and Beretvas (2006) discuss a 1PL version of these models for binary responses. The linear predictor for the 2PL version can be written as

$$\nu_{tpi} = \mathbf{x}_{tpi}'\boldsymbol{\beta} + (\theta_{1p}^{(3)} + \theta_{2p}^{(3)} z_{tp} + \theta_{tp}^{(2)})\boldsymbol{\lambda}'\mathbf{s}_i,$$

where t indexes occasions, z_{tp} is the time associated with occasion t, $\theta_{1p}^{(3)}$ is a random intercept, $\theta_{2p}^{(3)}$ is a random slope, and $\theta_{tp}^{(2)}$ is a person and occasion-specific error term. Advantages of this multilevel formulation (compared with a multidimensional formulation where all latent variables are viewed as person-specific) include that it requires only three-dimensional integrations, regardless of the number of occasions, and that the timing and number of occasions can differ between persons.

30.6 Multilevel Models

Whenever units of observations fall into clusters or groups that are subdivided into subgroups that may be subdivided into finer subgroups, and so on, we have multilevel data where the levels (groupings) are labeled from 2 to L from the finest to the coarsest grouping.

Multilevel models include latent variables (or random effects) that vary at several different nested levels. An example considered earlier was the bifactor model where specific dimensions varied between person-testlet combinations (level 2) and the general dimension varied between persons (level 3). Another example was a growth IRT model where the error term varied between person-occasion combinations (level 2) and the random intercept and slope varied between persons (level 3). In this section, we focus on models that are multilevel because persons are nested in clusters such as schools.

In traditional multilevel regression models, there is one observed response variable, a set of observed covariates with fixed effects, and a set of covariates with random effects (corresponding to latent variables) varying at higher levels. GLLAMMs extends these models to allow several latent response variables (possibly varying at different levels) to be regressed on other latent or observed variables, making it possible to include intervening or mediating variables that are both responses and covariates.

For brevity, we focus on relatively simple models in this chapter. Section 30.6.1 discusses models where the response variable is latent, whereas Section 30.6.2 discusses models that include a latent covariate (the response variable may also be latent). The models are presented as path diagrams using the conventions introduced by Rabe-Hesketh et al. (2004), which we prefer to the within-between formulation that is often used for SEMs with two levels of nesting (see Rabe-Hesketh et al. [2012] for details of both approaches and a comparison).

30.6.1 Multilevel IRT

When persons are nested in clusters, for instance, students in schools, there may be between-cluster variability in the person-level latent traits. The simplest way of modeling such variability is using a variance-components IRT model (Rabe-Hesketh et al., 2004), depicted in Figure 30.1a. Here, the person-level latent trait $\theta_1^{(2)}$ (circle) is measured by three items (rectangles) and the short arrows pointing at the items represent residual variability (e.g., Bernoulli). All variables in this measurement model vary between persons and items (they have pk subscripts). The person-level latent variable $\theta_{1pk}^{(2)}$ is regressed on a cluster-level latent variable $\theta_{1k}^{(3)}$ (which does not vary between persons and is therefore inside the frame for clusters but not inside the frame for persons) with disturbance $\zeta_{1pk}^{(2)}$. In this model, $\theta_{1k}^{(3)}$ can be thought of as a cluster-level random intercept in a linear model for the person's latent trait. We do not generally enclose random intercepts (or slopes) in circles because

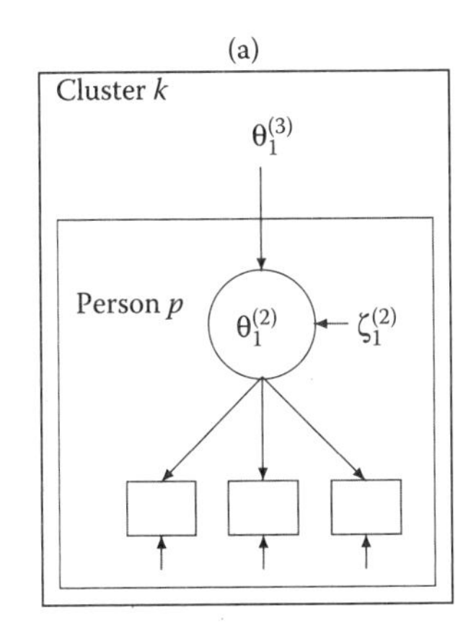

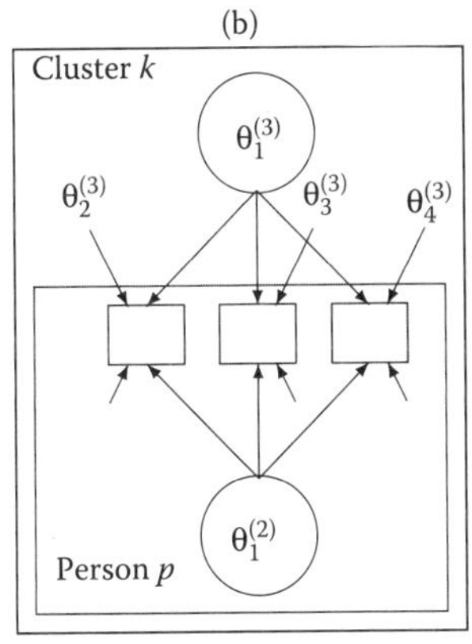

FIGURE 30.1
(a) A variance-components IRT model and (b) a general two-level IRT model. (Adapted from Rabe-Hesketh, S., Skrondal, A., and Pickles, A. 2004. *Psychometrika, 69,* 167–190.)

this helps to distinguish such random effects from latent traits measured by multiple items. The structural model is

$$\begin{bmatrix} \theta_{1pk}^{(2)} \\ \theta_{1k}^{(3)} \end{bmatrix} = \begin{bmatrix} 0 & 1 \\ 0 & 0 \end{bmatrix} \begin{bmatrix} \theta_{1pk}^{(2)} \\ \theta_{1k}^{(3)} \end{bmatrix} + \begin{bmatrix} \zeta_{1pk}^{(2)} \\ \zeta_{1k}^{(3)} \end{bmatrix}.$$

Fox and Glas (2001, Volume One, Chapter 24) called such a model a one-way random effects IRT ANOVA model.

The variance-components IRT model can easily be extended by regressing the person-level latent variable on observed covariates (e.g., Fox and Glas, 2001; Li et al., 2009). To allow DIF, models can include direct effects of covariates on items. Chaimongkol et al. (2006) and Stephenson (2011) specify a random slope for such a direct effect to allow DIF to vary randomly between schools.

A more complicated multilevel measurement model is given in Figure 30.1b. Here, a cluster-level latent variable $\theta_{1k}^{(3)}$ affects each item directly, with discrimination parameters $\lambda_i^{(3)}$ (discrimination parameters for the person-level latent variable are denoted $\lambda_i^{(2)}$). In addition, there are cluster-level error components or random intercepts $\theta_{i+1,k}^{(3)}$ for the items. Such cluster by item interactions can be viewed as random cluster-level DIF (De Jong and Steenkamp, 2007; Zheng, 2009).

A further generalization of the model in Figure 30.1 is achieved by specifying multidimensional measurement models at each level, possibly with different numbers of dimensions (Grilli and Rampichini, 2007; Longford and Muthén, 1992; Steele and Goldstein, 2006). In the GLLAMM framework, such a model can be written as

$$\nu_{pki} = \mathbf{s}_i'\boldsymbol{\beta} + \sum_{m=1}^{M_2} \theta_{mpk}^{(2)} \mathbf{s}_i [\mathbf{d}_m^{(2)}] \boldsymbol{\lambda}_m^{(2)} + \sum_{m=1}^{M_3} \theta_{mk}^{(3)} \mathbf{s}_i [\mathbf{d}_m^{(3)}] \boldsymbol{\lambda}_m^{(3)} + \sum_{m=1}^{I} \theta_{M_3+m,k}^{(3)} s_{im},$$

where $\mathbf{d}_m^{(l)}$ are the indices of the items measuring the mth dimension at level l ($l = 2, 3$), and the last term represents the cluster-level random intercepts.

30.6.2 Multilevel SEM

Figure 30.2, from Rabe-Hesketh et al. (2004), shows different multilevel SEMs with a latent covariate. Variables pertaining to the latent covariate have a C subscript, whereas those pertaining to the latent response variable have an R subscript.

In the model shown in Figure 30.2a, the latent covariate varies at the cluster level and is measured by cluster-level items. Such models were applied to PISA data by Rabe-Hesketh et al. (2007) and to PIRLS data by Rabe-Hesketh et al. (2012). The clusters were schools and the school-level latent trait was teacher excellence and school climate, respectively, with school-level item responses from the principals. The PIRLS application is briefly described in Section 30.9, where the parameter of main interest is the *cross-level effect* b_{12} of the school-level latent trait on student-level ability.

In the GLLAMM framework, person-level and cluster-level item responses are stacked into one vector and the corresponding linear predictors have the form

$$\nu_{pki} = \mathbf{x}'_{ijk}\boldsymbol{\beta} + \theta^{(2)}_{Rjk}\mathbf{z}'_{Rpki}\boldsymbol{\lambda}_R + \theta^{(3)}_{Cjk}\mathbf{z}'_{Cpki}\boldsymbol{\lambda}_C + \theta^{(3)}_{Rk} \times 0.$$

Here $\mathbf{z}_{Rpki}$ is a set of dummy variables for person-level items and $\mathbf{z}_{Cpki}$ is a set of dummy variables for cluster-level items. The structural model can be written as

$$\begin{bmatrix} \theta^{(2)}_{Rjk} \\ \theta^{(3)}_{Cjk} \\ \theta^{(3)}_{Rk} \end{bmatrix} = \begin{bmatrix} 0 & b_{12} & 1 \\ 0 & 0 & 0 \\ 0 & 0 & 0 \end{bmatrix} \begin{bmatrix} \theta^{(2)}_{Rjk} \\ \theta^{(3)}_{Cjk} \\ \theta^{(3)}_{Rk} \end{bmatrix} + \begin{bmatrix} \gamma_{11} & 0 \\ 0 & \gamma_{22} \\ 0 & 0 \end{bmatrix} \begin{bmatrix} w_{Rjk} \\ w_{Cjk} \end{bmatrix} + \begin{bmatrix} \zeta^{(2)}_{Rjk} \\ \zeta^{(3)}_{Cjk} \\ \zeta^{(3)}_{Rk} \end{bmatrix}.$$

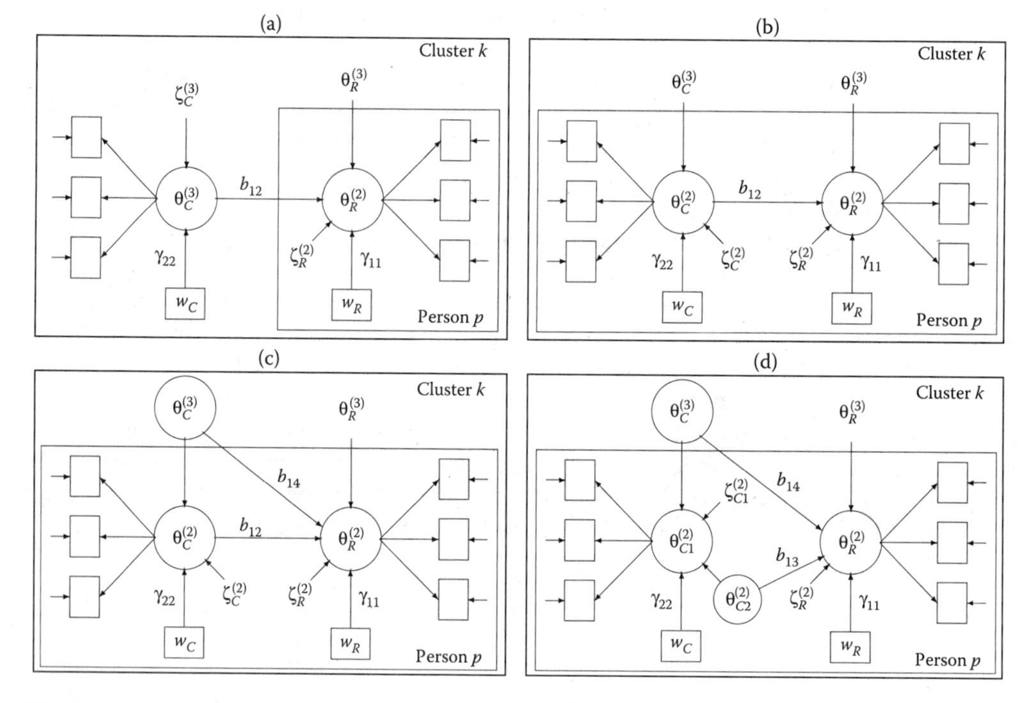

FIGURE 30.2
Path diagrams for structural equation models with latent covariates at different levels. (Adapted from Rabe-Hesketh, S., Skrondal, A., and Pickles, A. 2004. *Psychometrika*, 69, 167–190.)

In Figure 30.2b, the latent covariate and its items vary at the person level and the latent covariate has a cluster-level random intercept $\theta^{(3)}_{Ck}$. In Figure 30.2c, the latent covariate also varies between persons, but now the cluster-level random intercept is interpreted as a cluster-level latent trait that can affect the person-level outcome (via the path with regression coefficient b_{14}).

In multilevel models without latent variables, cluster-level latent traits are sometimes represented by aggregates (typically cluster-means) of person-level covariates. For example, a school's socioeconomic status (SES) is estimated by the mean of the students' SES (Raudenbush and Bryk, 2002). The contextual effect of SES is then defined as the effect of school SES after controlling for student SES. Marsh et al. (2009) pointed out that such "doubly manifest" models suffer from sampling error (due to a finite number of students per school) in addition to measurement error, whereas the "doubly latent contextual model" in Figure 30.2c addresses both sources of error. The idea of using doubly latent models for contexts or ecological settings was already proposed by Raudenbush and Sampson (1999), although their models treated factor loadings as known.

Another possibility is to consider the effect of deviations of persons latent traits from the cluster means of the latent traits. In Figure 30.2d, $\zeta^{(2)}_{C1pk}$ represents such a centered latent trait since it is the error term in the regression of the person's latent trait $\theta^{(2)}_{C1pk}$ on the cluster mean latent trait $\theta^{(3)}_{Ck}$. In the GLLAMM framework the disturbance $\zeta^{(2)}_{C1pk}$ cannot be a covariate and we therefore replace it by the latent variable $\theta^{(2)}_{C2pk}$ by setting the variance of $\zeta^{(2)}_{C1pk}$ to zero. The notion of such centered effects is common in sociological research, for instance, when studying relative deprivation.

30.7 Parameter Estimation

We express the likelihood in terms of the disturbances $\boldsymbol{\zeta}$ in the structural model Equation 30.7. When no structural model is specified, we simply have $\boldsymbol{\theta} = \boldsymbol{\zeta}$.

The likelihood is given by

$$L(\boldsymbol{\xi}) = \sum_k \ln f_k^{(L)}(\boldsymbol{\xi}),$$

where $f_k^{(L)}(\boldsymbol{\xi})$ is the likelihood contribution of a unit k at the highest level L and $\boldsymbol{\xi}$ is the vector of model parameters. Let $\boldsymbol{\zeta}_j^{(l+)} = (\boldsymbol{\zeta}_j^{(l)\prime}, \ldots, \boldsymbol{\zeta}_k^{(L)\prime})'$ be all the latent variables for a level-l unit j a levels l and above for $l \le L$. Exploiting conditional independence among level-($l-1$) units given the random effects $\boldsymbol{\zeta}_j^{(l+)}$ at levels l and above, the likelihood contribution of a given level-l unit j can be obtained recursively as

$$f_j^{(l)}(\boldsymbol{\xi}|\boldsymbol{\zeta}_k^{(l+1+)}) = \int g(\boldsymbol{\zeta}_j^{(l)}; \mathbf{0}, \boldsymbol{\Psi}^{(l)}) \prod_i f_i^{(l-1)}(\boldsymbol{\xi}|\boldsymbol{\zeta}_j^{(l+)}) d\boldsymbol{\zeta}_j^{(l)},$$

for $l = 1, \ldots, L-1$, where the product is over the units i at level $l-1$ that belong to level-l unit j, and unit j belongs to level $l+1$ unit k (only the lowest-level subscript is shown here). The integral is over the latent variables $\boldsymbol{\zeta}_j^{(l)}$ at level l with multivariate normal density $g(\boldsymbol{\zeta}_j^{(l)}; \mathbf{0}, \boldsymbol{\Psi}^{(l)})$, where $\boldsymbol{\Psi}^{(l)}$ is the covariance matrix at level l.

For example, in a multilevel item response model (for items i nested in persons j nested in schools k), when $l = 2$, $f_i^{(1)}(\boldsymbol{\xi}|\boldsymbol{\zeta}_j^{(2+)})$ is the conditional level-1 likelihood contribution $\Pr(U_i|\boldsymbol{\zeta}_j^{(2+)};\boldsymbol{\xi})$ for item i (and person j), the product is over the items, and the integral is over the person-level latent variables $\boldsymbol{\zeta}_j^{(2)}$.

For a unit k at the highest level L we obtain

$$f_k^{(L)}(\boldsymbol{\xi}) = \int g(\boldsymbol{\zeta}_k^{(L)};\mathbf{0},\boldsymbol{\Psi}^{(L)})\prod_j f_j^{(L-1)}(\boldsymbol{\xi}|\boldsymbol{\zeta}_k^{(L)})\mathrm{d}\boldsymbol{\zeta}_k^{(L)}.$$

In a multilevel item response model, where $L = 3$, $f_j^{(2)}(\boldsymbol{\xi}|\boldsymbol{\zeta}_k^{(3)})$ is the conditional likelihood contribution of person j in school k given the school-level latent variables $\boldsymbol{\zeta}_k^{(3)}$, the product is over all students within school k, and the integral is over the school-level latent variables.

The pseudo-likelihood for complex survey data is obtained by raising $f_i^{(l-1)}(\boldsymbol{\xi}|\boldsymbol{\zeta}_j^{(l+)})$ to the power $v_i^{(l-1)}$, where $v_i^{(l-1)}$ is the inverse probability that unit i at level $l-1$ was sampled given that the level-l unit it belongs to was sampled (Rabe-Hesketh and Skrondal, 2006).

Rabe-Hesketh et al. (2005) developed a novel adaptive quadrature method for evaluating likelihoods involving nested latent variables, giving a better approximation than ordinary Gauss–Hermite quadrature. The likelihood can be maximized using Newton–Raphson or an EM algorithm. Bayesian estimation via MCMC is also possible (e.g., Segawa et al., 2008). See Skrondal and Rabe-Hesketh (2004, Chapter 6) for an overview of estimation methods for latent variable models.

30.8 Model Fit

Here we make a few comments about assessment of model fit, but refer to Skrondal and Rabe-Hesketh (2004, Chapter 8) for a comprehensive treatment.

When a saturated model exists, for instance, if all observed variables are categorical, it is possible to test the global absolute fit of the model. However, due to sparseness, the test statistics can usually not be assumed to have a chi-square distribution under the null hypothesis. For this reason, tests of fit have been proposed based on lower-dimensional marginal tables (e.g., Glas, 1988, Volume Two, Chapter 11; Reiser, 1996).

Local absolute fit statistics such as "infit" and "outfit" abound in item response models (e.g., Hambleton and Wells, Volume Two, Chapter 20), but these are often somewhat ad-hoc, for instance, relying on a predicted value of ability to be plugged in. In a Bayesian setting, posterior predictive checking is more principled (e.g., Sinharay et al., 2006; Sinharay, Volume Two, Chapter 19) and provides test of local absolute fit that takes parameter uncertainty into account.

For comparing competing models that are nested, likelihood-ratio tests, Wald tests, or score tests can be used. However, care must be exercised when testing constraints for covariance matrices of latent variables where the null hypothesis is on the border of the parameter space, such as zero variances or perfect correlations (e.g., Self and Liang, 1987). To take into account the trade-off between model fit and complexity, and/or to compare non-nested models, information criteria are often employed, although their use for latent variable models is an ongoing area of research (e.g., Vaida and Blanchard, 2005).

30.9 Example

Here we consider an application to Progress in International Reading Literacy study (PIRLS) data from Zheng (2009) and Rabe-Hesketh et al. (2012). A similar model was applied to Program for International Student Assessment (PISA) data by Rabe-Hesketh et al. (2007).

30.9.1 PIRLS Data

PIRLS examines three aspects of reading literacy: purposes for reading, processes of comprehension, and reading behaviors and attitudes (Mullis et al., 2006). The purposes include reading for literacy experience and reading to acquire and use information. For each of these purposes, four types of comprehension processes are assessed, and we consider one of these: making straightforward inferences. In addition to the reading assessment, participating students and schools were asked to complete background questionnaires to provide information about reading behaviors and attitudes.

Here we analyze the American sample of the PIRLS 2006 assessment data, which comprises 5190 fourth grade students j from 183 U.S. schools k. Six items i about making inferences were chosen from a reading passage (Flowers); three multiple choice questions and three constructed response questions (where the student writes down the answer). All items were scored as 1: correct or 0: incorrect.

Four observed covariates from the student questionnaire are included in the analyses: student gender ([girl], 1: girl, 0: boy), language spoken at home ([English], 1: speak English at home, 0: no English at home), time spent reading ([reading], 1: some time spent reading every day, 0: otherwise), and number of books at home ([books], 1: more than 25 books, 0: otherwise).

At the school level, we focus on school climate measured by six survey questions in the school questionnaire, which is completed by school principals. The six questions concern teachers' job satisfaction, teachers' expectations for student achievement, parental support for student achievement, students' regard for school property, students' desire to do well in school, and students' regard for each other's welfare. The ordinal responses to the six questions originally had five categories that were collapsed into three categories due to sparseness.

The main research question motivating our analysis is whether there is a relationship between the school-level latent variable "school climate" and the student-level latent variable "reading ability."

30.9.2 Model

We number the six student assessment items as $i = 1, \ldots, 6$ and the six school-level items as $i = 7, \ldots, 12$. The student-level measurement model for reading item i for student p from school k is a two-parameter logistic IRT model,

$$\begin{aligned}\text{logit}(\Pr\{U_{pki} = 1 | \theta^{(2)}_{Rpk}\}) &= \beta_i + \lambda^{(2)}_i \theta^{(2)}_{Rpk} \\ &= \lambda^{(2)}_i [\theta^{(2)}_{Rpk} - (-\beta_i / \lambda^{(2)}_i)], \quad i = 1, \ldots, 6\end{aligned}$$

with discrimination parameter $\lambda_i^{(2)}$, difficulty parameter $-\beta_i/\lambda_i^{(2)}$ and student-level latent variable, reading ability $\theta_{Rpk}^{(2)}$. The first discrimination parameter $\lambda_1^{(2)}$ is set to 1 for identification (the residual ability variance is a free parameter, see below).

The measurement model for school climate is a logistic graded response model. In terms of a latent response U_{ki}^* for item i and school k, the model can be written as

$$U_{ki}^* = \lambda_i^{(3)}\theta_{Ck}^{(3)} + \epsilon_{ki}, \quad i = 7,\ldots,12,$$

where $\theta_{Ck}^{(3)}$ is the school-level latent variable, school climate, and $\lambda_i^{(3)}$ are discrimination parameters (with $\lambda_1^{(3)} = 1$ for identification). The error term ϵ_{ki} has a standard logistic distribution. The observed ordinal responses to the survey items are generated from the threshold model

$$U_{ki} = \begin{cases} 1 & \text{if} \quad -\infty \le U_{ki}^* < \alpha_1 + \tau_{i1} \\ 2 & \text{if} \quad \alpha_1 + \tau_{i1} \le U_{ki}^* < \alpha_2 + \tau_{i2} \quad \tau_{71} = \tau_{72} = 0 \\ 3 & \text{if} \quad \alpha_2 + \tau_{i2} \le U_{ki}^* < \infty. \end{cases}$$

In the threshold model α_s ($s = 1, 2$) represents the sth threshold for item 7 and τ_{is} ($i = 8, \ldots, 12$) the difference in the sth threshold between item i and item 7. Thus, $\alpha_s + \tau_{is}$ becomes the threshold parameter for the sth threshold of item i. The cumulative probabilities are given by

$$\begin{aligned} \text{logit}(\Pr\{U_{ki} > s | \theta_{Ck}^{(3)}\}) &= \lambda_i^{(3)}\theta_{Ck}^{(3)} - (\alpha_s + \tau_{is}) \\ &= \lambda_i^{(3)}[\theta_{Ck}^{(3)} - (\alpha_s + \tau_{is})/\lambda_i^{(3)}], \quad s = 1,2 \end{aligned}$$

so that $(\alpha_s + \tau_{is})/\lambda_i^{(3)}$ can be interpreted as the difficulty of rating item i above s.

The structural models are

$$\begin{aligned} \theta_{Rpk}^{(2)} &= b_{12}\theta_{Ck}^{(3)} + \gamma_1 w_{1pk} + \gamma_2 w_{2pk} + \gamma_3 w_{3pk} + \gamma_4 w_{4pk} + \theta_{Rk}^{(3)} + \zeta_{Rpk}^{(2)} \\ \theta_{Rk}^{(3)} &= \zeta_{Rk}^{(3)} \\ \theta_{Ck}^{(3)} &= \zeta_{Ck}^{(3)}, \end{aligned}$$

where b_{12} is the cross-level effect of school climate on student reading proficiency, w_{1pk} to w_{4pk} are the student-level covariates and γ_1 to γ_4 the associated regression coefficients. The disturbances are independently identically distributed as $\zeta_{Rpk}^{(2)} \sim N(0, \psi_R^{(2)})$, $\zeta_{Rk}^{(3)} \sim N(0, \psi_R^{(3)})$, and $\zeta_{Ck}^{(3)} \sim N(0, \psi_C^{(3)})$.

A path diagram for the model, showing only three items at the student and school levels and one student-level covariate, is given in Figure 30.2a, except that the covariate w_C should be omitted.

30.9.3 Results and Interpretation

Maximum-likelihood estimates for the model are given in Table 30.1. All estimates were obtained using gllamm (Rabe-Hesketh et al., 2004; Rabe-Hesketh and Skrondal, Volume Three, Chapter 29), which uses adaptive quadrature (Rabe-Hesketh et al., 2005) and runs in Stata.

TABLE 30.1

Maximum-Likelihood Estimates for Reading Achievement Study

Parameter		Estimate	(SE)	Difficulty
Reading Ability				
Measurement Model				$-\beta_i/\lambda_i^{(2)}$
β_1	(Item 1)	0.78	(0.31)	–0.78
β_2	(Item 2)	1.05	(0.52)	–0.60
β_3	(Item 3)	–0.36	(0.50)	0.21
β_4	(Item 4)	–0.18	(0.42)	0.13
β_5	(Item 5)	–0.06	(0.51)	0.04
β_6	(Item 6)	2.50	(0.66)	–1.13
$\lambda_1^{(2)}$	(Item 1)	1	–	
$\lambda_2^{(2)}$	(Item 2)	1.75	(0.39)	
$\lambda_3^{(2)}$	(Item 3)	1.71	(0.38)	
$\lambda_4^{(2)}$	(Item 4)	1.40	(0.31)	
$\lambda_5^{(2)}$	(Item 5)	1.75	(0.38)	
$\lambda_6^{(2)}$	(Item 6)	2.22	(0.56)	
Structural Model				
b_{12}	(School climate)	0.10	(0.04)	
γ_1	(Girl)	0.16	(0.08)	
γ_2	(Reading)	0.16	(0.06)	
γ_3	(English)	0.16	(0.09)	
γ_4	(Books)	0.38	(0.28)	
$\psi_R^{(3)}$	(Random intercept variance)	0.14		
$\psi_R^{(2)}$	(Residual variance)	0.76		
School Climate				
Measurement Model				$(\alpha_s + \tau_{is})/\lambda_i^{(3)}$
α_1	(Teachers' satisfaction)	–2.43	(0.29)	–2.43
$\tau_{8,1}$	(Teachers' expectations)	–0.64	(0.45)	–2.53
$\tau_{9,1}$	(Parental support)	2.39	(0.35)	–0.02
$\tau_{10,1}$	(Students' regard for property)	0.27	(0.43)	–1.11
$\tau_{11,1}$	(Students' ambition)	0.14	(0.53)	–0.83
$\tau_{12,1}$	(Students' regard for each other)	0.84	(0.44)	–0.65
α_2	(Teachers' satisfaction)	1.22	(0.22)	1.22
$\tau_{8,2}$	(Teachers' expectations)	–0.79	(0.27)	0.35
$\tau_{9,2}$	(Parental support)	1.51	(0.44)	1.44
$\tau_{10,2}$	(Students' regard for property)	1.26	(0.41)	1.27
$\tau_{11,2}$	(Students' ambition)	2.93	(0.75)	1.49
$\tau_{12,2}$	(Students' regard for each other)	2.96	(0.71)	1.71
$\lambda_7^{(3)}$	(Item 7)	1	–	
$\lambda_8^{(3)}$	(Item 8)	1.22	(0.27)	
$\lambda_9^{(3)}$	(Item 9)	1.90	(0.43)	
$\lambda_{10}^{(3)}$	(Item 10)	1.95	(0.44)	
$\lambda_{11}^{(3)}$	(Item 11)	2.78	(0.72)	
$\lambda_{12}^{(3)}$	(Item 12)	2.45	(0.60)	
Structural Model				
$\psi_C^{(3)}$	(School climate variance)	1.37		

Source: Adapted from Rabe-Hesketh, S., Skrondal, A., and Zheng, X. 2012. *Handbook of Structural Equation Modeling*. New York: Guilford.

Confirming our main hypothesis, the estimated effect of school climate on student reading ability is positive and significant at the 5% level ($p = 0.01$) after controlling for student-level covariates. Student gender, time spent reading, and number of books at home also have statistically significant effects on student reading performance. After controlling for the other variables, girls on an average perform better than boys; students who reported spending some time reading every day perform better than those who did not; and having more than 25 books at home is positively associated with reading achievement. Although students whose home language was English scored higher than students with a home language other than English, controlling for the other variables, the difference was not significant at the 5% level.

We can interpret the exponentiated regression coefficients as adjusted student-specific odds ratios for getting the first item right, as this item has a discrimination parameter of 1. The estimated odds ratios are 1.17 for girl, reading, and English, implying a 17% increase in the odds associated with these dummy variables, and 1.46 for books, implying a 46% increase in the odds of getting the first item right for those who read occasionally compared with those who do not, controlling for the other variables. For item 6 the corresponding odds ratios are 1.42 ($= \exp[2.22 \times 0.16]$) for girl, reading, and English and 2.32 ($= \exp[2.22 \times 0.38]$) for books. Regarding the effect of school climate, we consider a 0.69 increase in school climate, as this corresponds to doubling the odds that the principal rates question 7 (with discrimination parameter 1) above a given category (above 1 or above 2), since $\exp(0.69) = 2$. Such an increase in school climate is associated with a 7% ($= 100\%[\exp(0.69 \times 0.10) - 1]$) increase in the odds of getting the first item right and a 17% ($= 100\%[\exp(0.69 \times 0.10 \times 2.22) - 1]$) increase in the odds of getting the last item right, so the estimated effect size is rather small.

The school-level random intercept for reading proficiency has an estimated variance of 0.14 and the student-level residual variance is estimated as 0.76. Therefore, an estimated 16% of the residual variance in reading proficiency is between schools. The variance of school climate is estimated as 1.37.

30.10 Discussion

The GLLAMM framework unifies and extends a wide range of IRT models by combining a very general class of measurement models with multilevel structural models for the latent variables. We refer to Skrondal and Rabe-Hesketh (2004) and Skrondal and Rabe-Hesketh (2007) for extensive surveys of latent variable models and Skrondal and Rabe-Hesketh (2004) for a survey of estimation methods relevant for latent variable models.

A challenge with increasingly complex models is to keep them within a modeling framework that comes with a notation for writing the models down, corresponding software, syntax, and intuitive graphs. The LISREL model (e.g., Jöreskog, 1973) and software (e.g., Jöreskog and Sörbom, 1989) were great accomplishments in this regard. In a similar vein, the GLLAMM framework extends the range of IRT models while retaining a coherent modeling framework. This is important for efficient communication among researchers and between researchers and software, both in terms of correct model specification and sensible interpretation of results. Another advantage of a coherent statistical framework is that theoretical properties can be investigated and estimation algorithms developed and implemented for the entire model framework, instead of replicating this work for a multitude of special cases.

Statistical inference for all models in the GLLAMM framework can be performed using the associated gllamm program described in Chapter 29 of Volume Three of this Handbook.

References

Adams, R. J., Wilson, M., and Wang, W. 1997. The multidimensional random coefficients multinomial logit model. *Applied Psychological Measurement, 21,* 1–23.

Andersen, E. B. 1985. Estimating latent correlations between repeated testings. *Psychometrika, 50,* 3–16.

Andrich, D. 1978. A rating formulation for ordered response categories. *Psychometrika, 43,* 561–573.

Asparouhov, T., and Muthén, B. O. 2007. Computationally efficient estimation of multilevel high-dimensional latent variable models. In *Proceedings of the 2007 Joint Statistical Meetings* (pp. 2531–2535). Salt Lake City, Utah.

Bartholomew, D. J. 1987. *Latent Variable Models and Factor Analysis.* London: Griffin.

Birnbaum, A. 1968. Test scores, sufficient statistics, and the information structures of tests. In F. M. Lord and M. R. Novick (Eds.), *Statistical Theories of Mental Test Scores* (pp. 425–435). Reading: Addison-Wesley.

Bock, R. D. 1972. Estimating item parameters and latent ability when responses are scored in two or more nominal categories. *Psychometrika, 37,* 29–51.

Böckenholt, U. 2001. Hierarchical modeling of paired comparison data. *Psychological Methods, 6,* 49–66.

Brady, H. E. 1985. The perils of survey research: Inter-personally incomparable responses. *Political Methodology, 11,* 269–291.

Breslow, N. E., and Clayton, D. G. 1993. Approximate inference in generalized linear mixed models. *Journal of the American Statistical Association, 88,* 9–25.

Chaimongkol, S., Huffer, F. W., and Kamata, A. 2006. A Bayesian approach for fitting a random effect differential item functioning across group units. *Thailand Statistician, 4,* 27–41.

Chapman, R. G., and Staelin, R. 1982. Exploiting rank ordered choice set data within the stochastic utility model. *Journal of Marketing Research, 14,* 288–301.

Christofferson, A. 1975. Factor analysis of dichotomized variables. *Psychometrika, 40,* 5–32.

Clayton, D. G. 1988. The analysis of event history data: A review of progress and outstanding problems. *Statistics in Medicine, 7,* 819–841.

Cox, D. R. 1972. Regression models and life tables. *Journal of the Royal Statistical Society, Series B, 34,* 187–203.

De Boeck, P., and Wilson, M. (Eds.). 2004. *Explanatory Item Response Models: A Generalized Linear and Nonlinear Approach.* New York: Springer.

De Jong, M. G., and Steenkamp, J. 2007. Relaxing measurement invariance in cross-national consumer research using a hierarchical IRT model. *Journal of Consumer Research, 34,* 260–278.

Embretson, S. E. 1991. A multidimensional latent trait model for measuring learning and change. *Psychometrika, 56,* 495–515.

Embretson, S. E. 1999. Generating items during testing: Psychometric issues and models. *Psychometrika, 64,* 407–433.

Fischer, G. H. 1973. Linear logistic test model as an instrument in educational research. *Acta Psychologica, 37,* 359–374.

Fox, J. P. 2005. Randomized item response theory models. *Journal of Educational and Behavioral Statistics, 30,* 1–24.

Fox, J. P., and Glas, C. A. W. 2001. Bayesian estimation of a multilevel IRT model using Gibbs sampling. *Psychometrika, 66,* 271–288.

Fox, J. P., and Glas, C. A. W. 2003. Bayesian modeling of measurement error in predictor variables using item response theory. *Psychometrika, 68,* 169–191.

Gibbons, R. D., and Hedeker, D. 1992. Full-information item bi-factor analysis. *Psychometrika, 57,* 423436.

Glas, C. A. W. 1988. The derivation of some tests for the Rasch model from the multinomial distribution. *Psychometrika, 53,* 525–546.

Goldstein, H. 1980. Dimensionality, bias, independence and measurement scale problems in latent trait test score models. *British Journal of Mathematical and Statistical Psychology, 33,* 234–246.

Grilli, L., and Rampichini, C. 2007. Multilevel factor models for ordinal variables. *Structural equation Modeling, 14,* 1–25.

Holzinger, K. J., and Swineford, F. 1937. The bi-factor method. *Psychometrika, 2,* 41–54.

Hoskens, M., and De Boeck, P. 1986. A parametric model for local item dependence among test items. *Psychological Methods, 2,* 261–277.

Inglehart, R. 1971. The silent revolution in Europe: Intergenerational change in post-industrial societies. *The American Political Science Review, 65,* 991.

Jansen, M. G. H., and van Duijn, M. A. J. 1992. Extensions of Rasch's multiplicative Poisson model. *Psychometrika, 57,* 405–414.

Jöreskog, K. G. 1973. A general method for estimating a linear structural equation system. In A. S. Goldberger and O. D. Duncan (Eds.), *Structural Equation Models in the Social Sciences* (pp. 85–112). New York: Seminar.

Jöreskog, K. G., and Goldberger, A. S. 1975. Estimation of a model with multiple indicators and multiple causes of a single latent variable. *Journal of the American Statistical Association, 70,* 631–639.

Jöreskog, K. G., and Sörbom, D. 1989. *LISREL 7: A Guide to the Program and Applications.* Chicago, Illinois: SPSS Publications.

Kamata, A. 2001. Item analysis by hierarchical generalized linear model. *Journal of Educational Measurement, 38,* 79–93.

King, G., Murray, C. J. L., Salomon, J. A., and Tandon, A. 2003. Enhancing the validity of cross-cultural comparability of survey research. *American Political Science Review, 97,* 567–583.

Läärä, E., and Matthews, J. N. S. 1985. The equivalence of two models for ordinal data. *Biometrika, 72,* 206–207.

Lawley, D. N. 1943. On problems connected with item selection and test construction. *Proceedings of the Royal Society of Edinburgh, 61,* 273–287.

Li, D., Oranje, A., and Jiang, Y. 2009. On the estimation of hierarchical latent regression models for large-scale assessments. *Journal of Educational and Behavioral Statistics, 34,* 433–463.

Longford, N. T., and Muthén, B. O. 1992. Factor analysis for clustered observations. *Psychometrika, 57,* 581–597.

Lord, F. M. 1952. *A Theory of Test Scores.* Bowling Green, OH: Psychometric Monograph 7, Psychometric Society.

Luce, R. D. 1959. *Individual Choice Behavior.* New York: Wiley.

Maier, K. S. 2001. A Rasch hierarchical measurement model. *Journal of Educational and Behavioral Statistics, 26,* 307–330.

Maris, E. 1993. Additive and multiplicative models for gamma distributed random variables, and their application as psychometric models for response times. *Psychometrika, 58,* 445–469.

Marsh, H. W., Lüdtke, O., Robitzsch, A., Trautwein, U., Asparouhov, T., Muthén, B. et al. 2009. Doubly-latent models of school contextual effects: Integrating multilevel and structural equation approaches to control measurement and sampling error. *Multivariate Behavioral Research, 44,* 764–802.

Masters, G. N. 1982. A Rasch model for partial credit scoring. *Psychometrika, 47,* 149–174.

McArdle, J. J. 1988. Dynamic but structural equation modeling with repeated measures data. In J. R. Nesselroade and R. B. Cattell (Eds.), *Handbook of Multivariate Experimental Psychology* Volume II, (pp. 561–614). New York: Plenum.

McCullagh, P., and Nelder, J. A. 1989. *Generalized Linear Models* (2nd edition). London: Chapman & Hall.

McDonald, R. P. 1982. Linear versus nonlinear models in item response theory. *Applied Psychological Measurement, 6*, 379–396.

Mellenbergh, G. J. 1994. Generalized linear item response theory. *Psychological Bulletin, 115*, 300–307.

Mislevy, R. J. 1987. Exploiting auxiliary information about examinees in the estimation of item parameters. *Applied Psychological Measurement, 11*, 81–91.

Mislevy, R. J. 1991. Randomization-based inference about latent variables from complex samples. *Psychometrika, 56*, 177–196.

Mullis, I., Kennedy, A. M., Martin, M. O., and Sainsbury, M. 2006. *Pirls 2006 Assessment Framework and Specifications* (2nd edition). Chestnut Hill, MA: Boston College.

Muraki, E. 1992. A generalized partial credit model: Application of an EM algorithm. *Applied Psychological Measurement, 16*, 159–176.

Muthén, B. O. 1984. A general structural equation model with dichotomous, ordered categorical and continuous latent indicators. *Psychometrika, 49*, 115–132.

Muthén, B. O., and Lehman, J. 1985. A method for studying the homogeneity of test items with respect to other relevant variables. *Journal of Educational Statistics, 10*, 121–132.

Owen, R. J. 1969. *A Bayesian analysis of Rasch's multiplicative Poisson model for misreadings* (Tech. Rep. No. RB-69–64). Educational Testing Service, Princeton, New Jersey.

Pastor, D. A., and Beretvas, S. N. 2006. Longitudinal Rasch modeling in the context of psychotherapy outcomes assessment. *Applied Psychological Measurement, 30*, 100–120.

Plackett, R. L. 1975. The analysis of permutations. *Journal of the Royal Statistical Society, Series C, 24*, 193–202.

Rabe-Hesketh, S., and Skrondal, A. 2002. *Estimating chopit models in gllamm: Political efficacy example from king et al. (2002)* (Tech. Rep.). Institute of Psychiatry, King's College London. Available from http://www.gllamm.org/chopit.pdf

Rabe-Hesketh, S., and Skrondal, A. 2006. Multilevel modeling of complex survey data. *Journal of the Royal Statistical Society, Series A, 169*, 805–827.

Rabe-Hesketh, S., and Skrondal, A. 2007. Multilevel and latent variable modeling with composite links and exploded likelihoods. *Psychometrika, 72*, 123–140.

Rabe-Hesketh, S., and Skrondal, A. 2012. *Multilevel and Longitudinal Modeling Using Stata* (3rd edition), volume ii: *Categorical responses, counts, and survival.* College Station, Texas: Stata Press.

Rabe-Hesketh, S., Skrondal, A., and Pickles, A. 2004. Generalized multilevel structural equation modeling. *Psychometrika, 69*, 167–190.

Rabe-Hesketh, S., Skrondal, A., and Pickles, A. 2005. Maximum likelihood estimation of limited and discrete dependent variable models with nested random effects. *Journal of Econometrics, 128*, 301–323.

Rabe-Hesketh, S., Skrondal, A., and Zheng, X. 2007. Multilevel structural equation modeling. In S. Y. Lee (Ed.), *Handbook of Latent Variable and Related Models* (pp. 209–227). Amsterdam: Elsevier.

Rabe-Hesketh, S., Skrondal, A., and Zheng, X. 2012. Multilevel structural equation modeling. In R. H. Hoyle (Ed.), *Handbook of Structural Equation Modeling.* New York: Guilford.

Rasch, G. 1960. *Probabilistic Models for Some Intelligence and Attainment Tests.* Copenhagen: Danmarks Pædagogiske Institut.

Raudenbush, S. W., and Bryk, A. S. 2002. *Hierarchical Linear Models.* Thousand Oaks, California: Sage.

Raudenbush, S. W., and Sampson, R. 1999. Assessing direct and indirect effects in multilevel designs with latent variables. *Sociological Methods & Research, 28*, 123–153.

Reckase, M. D. 1985. The difficulty of test items that measure more than one ability. *Applied Psychological Measurement, 9*, 401–412.

Reiser, M. 1996. Analysis of residuals for the multinomial item response model. *Psychometrika, 61*, 509–528.

Rijmen, F. 2010. Formal relations and an empirical comparison between the bi-factor, the testlet, and a second-order multidimensional IRT model. *Journal of Educational Measurement, 47*, 361372.

Rijmen, F., and De Boeck, P. 2002. The random weights linear logistic test model. *Applied Psychological Measurement, 26*, 271–285.

Rijmen, F., Tuerlinckx, F., De Boeck, P., and Kuppens, P. 2003. A nonlinear mixed model framework for item response theory. *Psychological Methods, 8*, 185–205.

Roberts, J. S., and Laughlin, J. E. 1996. A unidimensional item response model for unfolding responses from a graded disagree-agree response scale. *Applied Psychological Measurement, 20*, 231–255.

Samejima, F. 1969. *Estimation of Latent Ability Using a Response Pattern of Graded Scores*. Bowling Green, OH: Psychometric Monograph 17, Psychometric Society.

Segawa, E., Emery, S., and Curry, S. J. 2008. Extended generalized linear latent and mixed models. *Journal of Educational and Behavioral Statistics, 33*, 464–484.

Self, S. G., and Liang, K.-Y. 1987. Asymptotic properties of maximum likelihood estimators and likelihood ratio tests under non-standard conditions. *Journal of the American Statistical Association, 82*, 605–610.

Sinharay, S., Johnson, M. S., and Stern, H. S. 2006. Posterior predictive assessment of item response theory models. *Applied Psychological Measurement, 30*, 298–321.

Skrondal, A., and Laake, P. 2001. Regression among factor scores. *Psychometrika, 66*, 563–576.

Skrondal, A., and Rabe-Hesketh, S. 2003. Multilevel logistic regression for polytomous data and rankings. *Psychometrika, 68*, 267–287.

Skrondal, A., and Rabe-Hesketh, S. 2004. *Generalized Latent Variable Modeling: Multilevel, Longitudinal, and Structural Equation Models*. Boca Raton, Florida: Chapman & Hall/CRC.

Skrondal, A., and Rabe-Hesketh, S. 2007. Latent variable modelling: A survey. *Scandinavian Journal of Statistics, 34*, 712–745.

Steele, F., and Goldstein, H. 2006. A multilevel factor model for mixed binary and ordinal indicators of women's status. *Sociological Methods & Research, 35*, 137–153.

Stephenson, S. Y. 2011. Extensions of Multilevel Factor Analysis to Detect School-Level Influences on Item Functioning. Ph.D. Thesis. University of California, Davis.

Takane, Y., and de Leeuw, J. 1987. On the relationship between item response theory and factor analysis of discretized variables. *Psychometrika, 52*, 393–408.

Thissen, D., and Steinberg, L. 1988. Data analysis using item response theory. *Psychological Bulletin, 104*, 385–395.

Thompson, R., and Baker, R. J. 1981. Composite link functions in generalized linear models. *Journal of the Royal Statistical Society, Series C, 30*, 125–131.

Thurstone, L. L. 1927. A law of comparative judgement. *Psychological Review, 34*, 278–286.

Thurstone, L. L. 1931. Multiple factor analysis. *Psychological Review, 38*, 406–427.

Tutz, G. 1990. Sequential item response models with an ordered response. *British Journal of Mathematical and Statistical Psychology, 43*, 39–55.

Vaida, F., and Blanchard, S. 2005. Conditional Akaike information for mixed effects models. *Biometrika, 92*, 351–370.

van der Linden, W. J. 2006. A lognormal model for response times on test items. *Journal of Educational and Behavioral Statistics, 31*, 181204.

van der Linden, W. J. 2007. A hierarchical framework for modeling speed and accuracy on test items. *Psychometrika, 72*, 287–308.

Wang, W.-C., and Wilson, M. 2005. The Rasch testlet model. *Applied Psychological Measurement, 29*, 126–149.

Zheng, X. 2009. Multilevel item response modeling: Applications to large-scale assessment of academic achievement. Unpublished doctoral dissertation, University of California, Berkeley, California.

Zheng, X., and Rabe-Hesketh, S. 2007. Estimating parameters of dichotomous and ordinal item response models with gllamm. *The Stata Journal, 7*, 313–333.

31

Multidimensional, Multilevel, and Multi-Timepoint Item Response Modeling

Bengt Muthén and Tihomir Asparouhov

CONTENTS

31.1 Introduction 527
31.2 Modeling 528
  31.2.1 Single-Level Modeling 528
  31.2.2 Two-Level Modeling 529
31.3 Estimation 530
  31.3.1 Weighted Least-Squares (WLSMV) 530
  31.3.2 Bayesian Estimation 531
31.4 Empirical Examples 531
  31.4.1 Item Bi-Factor Exploratory Factor Analysis 532
  31.4.2 Two-Level Item Bi-Factor Exploratory Factor Analysis 533
  31.4.3 Two-Level Item Bi-Factor Confirmatory Factor Analysis 533
  31.4.4 Two-Level Item Bi-Factor Confirmatory Factor Analysis with Random Factor Loadings 534
  31.4.5 Longitudinal Two-Level Item Bi-Factor Confirmatory Factor Analysis 535
31.5 Conclusions 537
References 538

31.1 Introduction

Item response modeling in the general latent variable framework of the Mplus program (Muthén and Muthén, 2012) offers many unique features including multidimensional analysis (Asparouhov and Muthén, 2012a); two-level, three-level, and cross-classified analysis (Asparouhov and Muthén, 2012b); mixture modeling (Muthén, 2008; Muthén and Asparouhov, 2009); and multilevel mixture modeling (Asparouhov and Muthén, 2008; Henry and Muthén, 2010). This chapter presents a subset of the Mplus item response modeling technique through the analysis of an example with three features common in behavioral science applications: multiple latent variable dimensions, multilevel data, and multiple timepoints. The dimensionality of a measurement instrument with categorical items is investigated using exploratory factor analysis with bi-factor rotation. Variation across students and classrooms is investigated using two-level exploratory and confirmatory bi-factor models. Change over grades is investigated using a longitudinal two-level model. The analyses are carried out using weighted least-squares, maximum-likelihood, and Bayesian analysis. The strengths of weighted least-squares and Bayesian estimation

TABLE 31.1

Response Percentages for Aggression Items of $n = 363$ Cohort 3 Males in the Fall of Grade 1

	Almost Never (Scored as 1)	Rarely (Scored as 2)	Sometimes (Scored as 3)	Often (Scored as 4)	Very Often (Scored as 5)	Almost Always (Scored as 6)
Stubborn	42.5	21.3	18.5	7.2	6.4	4.1
Breaks rules	37.6	16.0	22.7	7.5	8.3	8.0
Harms others and property	69.3	12.4	9.40	3.9	2.5	2.5
Breaks things	79.8	6.60	5.20	3.9	3.6	0.8
Yells at others	61.9	14.1	11.9	5.8	4.1	2.2
Takes others' property	72.9	9.70	10.8	2.5	2.2	1.9
Fights	60.5	13.8	13.5	5.5	3.0	3.6
Harms property	74.9	9.90	9.10	2.8	2.8	0.6
Lies	72.4	12.4	8.00	2.8	3.3	1.1
Talks back to adults	79.6	9.70	7.80	1.4	0.8	1.4
Teases classmates	55.0	14.4	17.7	7.2	4.4	1.4
Fights with classmates	67.4	12.4	10.2	5.0	3.3	1.7
Loses temper	61.6	15.5	13.8	4.7	3.0	1.4

as a complement to maximum-likelihood for this high-dimensional application are discussed. Mplus scripts for all analyses are available at www.statmodel.com.

As a motivating example, consider a teacher-rated measurement instrument capturing aggressive-disruptive behavior among a sample of U.S. students in Baltimore public schools (Ialongo et al., 1999). A total of 362 boys were observed in 27 classrooms in the fall of grade 1 and spring of grade 3. The instrument consisted of 13 items scored as 1 (almost never) through 6 (almost always). The items and the percentage in each response category are shown in Table 31.1 for fall grade 1. The item distribution is very skewed with a high percentage in the almost never category. The responses are modeled as ordered categorical. It is of interest to study the dimensionality of the instrument, to explore response variation due to both students and classrooms, and to study changes over the grades.

31.2 Modeling

31.2.1 Single-Level Modeling

Let U_{pi} be the response for person p on an ordered polytomous item i with categories $a = 1, 2, \ldots, A$, and express the item probabilities for this item as functions of D factors θ_{pd} $(d = 1, 2, \ldots, D)$ as follows,

$$P(U_{pi} = a \mid \theta_{p1}, \theta_{p2}, \ldots, \theta_{pD}) = F\left[\tau_{ia} - \sum_{d=1}^{D} \lambda_{id}\,\theta_{pd}\right] - F\left[\tau_{ia-1} - \sum_{d=1}^{D} \lambda_{id}\,\theta_{pd}\right], \qquad (31.1)$$

where F is either the logistic or standard normal distribution function, corresponding to logistic and probit regression. In statistics, this is referred to as a proportional odds model

(Agresti, 2002), whereas in psychometrics it is referred to as a graded response model. When $a = 0$, the threshold parameter $\tau_{i0} = -\infty$, resulting in $F = 0$, and when $a = A$, $\tau_{iA} = \infty$, resulting in $F = 1$. As usual, conditional independence is assumed among the items given the factors and the factors are assumed to have a multivariate normal distribution.

It is also useful to view the model as an equivalent latent response variable model, motivating the use of a threshold formulation. Consider I continuous latent response variables $U^*_{pi} (i = 1, 2, \ldots, I)$ for person p, following a linear factor model with D factors

$$U^*_{pi} = \nu_i + \sum_{d=1}^{D} \lambda_{id}\, \theta_{pd} + \epsilon_{pi} \tag{31.2}$$

with a threshold formulation such as for a three-category ordered categorical item,

$$U_{pi} = \begin{cases} 0 & \text{if } U^*_{pi} \le \tau_{i1} \\ 1 & \text{if } \tau_{i1} < U^*_{pi} \le \tau_{i2} \, . \\ 2 & \text{if } U^*_{pi} > \tau_{i2} \end{cases}$$

The intercept parameters ν are typically fixed at zero given that they cannot be separately identified from the thresholds τ. With normal factors θ and residuals ε the U^* variables have a multivariate normal distribution where the association among the items can be described via latent response variable (LRV) polychoric correlations (see, e.g., Muthén, 1978, 1984). This corresponds to a probit or normal ogive IRT model. With logistic density for the residuals, a logit IRT model is obtained.

31.2.2 Two-Level Modeling

The two-level model is conveniently expressed in terms of the continuous latent response variables U^*. For person p, item i, and cluster (classroom) j, consider the two-level, random measurement parameter IRT model (see, e.g., Fox, 2010; Volume One, Chapter 24)

$$U^*_{pij} = \nu_{ij} + \sum_{d=1}^{D} \lambda_{ijd}\, \theta_{pjd} + \epsilon_{pij}, \tag{31.3}$$

$$\nu_{ij} = \nu_i + \delta_{\nu ij}, \tag{31.4}$$

$$\lambda_{ijd} = \lambda_{id} + \delta_{\lambda ijd}, \tag{31.5}$$

$$\theta_{pjd} = \eta_{jd} + \zeta_{pjd}, \tag{31.6}$$

so that the factors have between- and within-cluster variation. Given that all loadings are free, one may set the within-cluster variance at one for identification purposes, $V(\zeta_{pjd}) = 1$. This model can be estimated in Mplus using Bayes as discussed in Asparouhov and Muthén (2012b). When the measurement parameters of ν (or τ) and λ are not varying across clusters but are fixed parameters, the variances of the δ residuals are zero.

It may be noted in the above IRT model that the same loading multiplies both within- and between-cluster parts of each factor. An alternative model has been put forward in the factor analysis tradition where this restriction is relaxed, acknowledging that some applications call for not only the loadings but also the number of factors to be different on the two levels (see, e.g., Cronbach, 1976; Harnqvist, 1978; Harnqvist et al., 1994). For the case of fixed loading parameters this model may be expressed as

$$U^*_{pij} = \nu_{ij} + \sum_{d=1}^{D_W} \lambda_{Wid}\, \theta_{Wpd} + \epsilon_{Wpij}, \tag{31.7}$$

$$\nu_{ij} = \nu_i + \sum_{d=1}^{D_B} \lambda_{Bid}\, \theta_{Bjd} + \epsilon_{Bij}, \tag{31.8}$$

which expresses the two-level factor model as a random intercepts (threshold) model. In this way, there is a separate factor analysis structure on each level. This type of modeling follows the tradition of two-level factor analysis for continuous responses; see, e.g., Goldstein and McDonald (1988), McDonald and Goldstein (1989), and Longford and Muthén (1992). For a discussion of two-level covariance structure modeling in Mplus, see Muthén (1994). For the case of $D_W = D_B$ and $\lambda_{Wid} = \lambda_{Bid}$, the model of Equations 31.7 and 31.8 is the same as the model of Equations 31.3 through 31.6 with zero variance for $\delta_{\lambda ijd}$, that is, with fixed as opposed to random loadings.

31.3 Estimation

Three estimators are considered in this chapter, maximum-likelihood, weighted least-squares, and Bayes. Maximum-likelihood (ML) estimation of this model is well-known (see, e.g., Baker and Kim, 2004) and is not described here. Weighted least-squares and Bayes are described only very briefly, pointing to available background references.

31.3.1 Weighted Least-Squares (WLSMV)

Let s define a vector of all the thresholds and LRV correlations estimated from the sample, let σ refer to the corresponding population correlations expressed in terms of the model parameters, and let W denote the large-sample covariance matrix for s. The weighted least-squares estimator minimizes the sums of squares of the differences between s and σ, where differences with larger variance are given less weight. This is accomplished by minimizing the following fitting function with respect to the model parameters (Muthén, 1978, 1984; Muthén and Satorra, 1995; Muthén et al., 1997)

$$F = (s - \sigma)'\, diag(W)^{-1}\, (s - \sigma). \tag{31.9}$$

Here, $diag(W)$ denotes the diagonal of the weight matrix, that is, using only the variances. The full weight matrix is, however, used for χ^2 testing of model fit and for standard error calculations (Asparouhov and Muthén, 2010a; Muthén et al., 1997). This estimator is referred to as WLSMV. Modification indices in line with those typically used for continuous items are available also with weighted least-squares (Asparouhov and Muthén, 2010a) and are useful for finding evidence of model misfit such as correlated residuals and noninvariance across groups.

Two-level analysis of categorical data can be carried out by the two-level weighted least-squares (WLSMV) estimator in Mplus developed by Asparouhov and Muthén (2007). This uses the model version of Equations 31.7 and 31.8. Two-level factor analysis may involve many latent variables and ML is therefore cumbersome due to many dimensions of numerical integration. Bayesian estimation is feasible, but two-level WLSMV is a simple and much faster procedure suitable for initial analysis. The computational demand is virtually independent of the number of latent variables because high-dimensional integration is replaced by multiple instances of one- and two-dimensional integration using a second-order information approach of WLSMV in line with the Muthen (1984) single-level WLSMV. This implies that residuals can be correlated and that model fit to the LRV structure can be obtained by chi-square testing.

31.3.2 Bayesian Estimation

As applied to the item response setting, the Bayes implementation in Mplus considers multivariate normal latent response variables as in Equation 31.2; see, for example, Johnson and Albert (1999) or Fox (2010). This relies on probit relations where the residual variances are fixed at one as in the maximum-likelihood parameterization using probit. In addition to parameters, latent variables, and missing data, the Bayes iterations consider the latent response variables as unknown to obtain a well-performing Markov Chain Monte Carlo approach. Posterior distributions for these latent response variables are obtained as a side product. Posterior predictive checking is available for the LRV structure (Asparouhov and Muthén, 2010b). The default for Bayesian estimation in Mplus is to use noninformative priors, but informative priors are easily specified. Like MLE, Bayesian estimation is a full-information approach and with noninformative priors Bayes gives asymptotically the same estimates as MLE. Bayesian estimation of latent variable models with categorical items as in IRT is, however, advantageous to MLE due to numerical integration required for the latter, which is slow or prohibitive with many dimensions of integration due to many latent variables. From a practical point of view, Bayesian analysis with noninformative priors can in such cases be seen as an approach to getting estimates close to those of MLE if ML estimates could have been computed. Bayesian estimation is particularly useful for two-level item response models due to many latent variable dimensions. For technical aspects, see Asparouhov and Muthén (2010b).

31.4 Empirical Examples

Returning to our motivating example, this section goes through a series of analyses to demonstrate the flexibility of the item response modeling in Mplus. For the software implementations of these analyses, see Muthén and Muthén (Volume One, Chapter 29).

31.4.1 Item Bi-Factor Exploratory Factor Analysis

Using ML, WLSMV, and Bayes estimation, Muthén et al. (2012) found that a three-factor exploratory factor analysis (EFA) model is suitable for the aggressive-disruptive items described in Section 31.1. The three factors correspond to verbally oriented, property-oriented, and person-oriented aggressive-disruptive behavior. An alternative EFA model with identical fit to the data is a bi-factor model that has both a general factor influencing all items and also specific factors, uncorrelated with the general factor, which influences sets of items. Drawing on Jennrich and Bentler (2011, 2012), it is possible to carry out a bi-factor EFA using an extension to categorical items implemented in Mplus. Assuming unidimensionality, a summed score of the 13 items of this measurement instrument has previously been used (Ialongo et al., 1999). The general factor of the bi-factor model is related to such an overall score, but it is of interest to also study additional specific dimensions.

The bi-factor EFA estimates in Table 31.2 pertain to the fall grade 1 data and are obtained using the WLSMV estimator with a bi-factor Geomin rotation allowing correlated specific factors. The table shows that all items load significantly on the general factor with approximately equal loadings. The first specific factor, labeled number 2, has significant and positive loadings for the items "stubborn" and "loses temper," which may be weakly indicative of a verbally-oriented aggressive-disruptive specific factor. There are, however, several large negative loadings, which make the interpretation less clear. The second specific factor, labeled number 3, has significant and positive loadings for the items "harms others," "fights," "teases classmates," and "fights with classmates," which may be indicative of a person-oriented aggressive-disruptive specific factor. The two specific factors have a small significant positive correlation. Modification indices suggest a few correlated residuals, but including them does not alter the original estimates in any important way.

TABLE 31.2

Bi-Factor EFA Solution Using WLSMV

	1	2	3
Bi-Geomin Rotated Loadings			
Stubborn	0.718[a]	**0.398**[a]	0.013
Breaks rules	0.796[a]	0.099	0.107
Harms others and property	0.827[a]	−0.197[a]	**0.198**[a]
Breaks things	0.890[a]	−0.330[a]	0.007
Yells at others	0.842[a]	0.180	−0.013
Takes others' property	0.848[a]	−0.242	−0.017
Fights	0.892[a]	−0.040	**0.367**[a]
Harms property	0.921[a]	−0.289	−0.020
Lies	0.906[a]	−0.049	−0.128[a]
Talks back to adults	0.870[a]	0.255	−0.116
Teases classmates	0.806[a]	0.008	**0.178**[a]
Fights with classmates	0.883[a]	0.060	**0.399**[a]
Loses temper	0.826[a]	**0.273**[a]	0.003
Bi-Geomin Factor Correlations			
1	1.000		
2	0.000	1.000	
3	0.000	0.115[a]	1.000

[a] Indicate significance at the 5% level.

31.4.2 Two-Level Item Bi-Factor Exploratory Factor Analysis

The data described in Section 31.1 are obtained from students in 27 classrooms. This multilevel structure was ignored in the previous analysis. It is, however, possible to carry out bi-factor EFA also for two-level categorical data using two-level weighted least-squares (WLSMV) estimation in Mplus developed by Asparouhov and Muthén (2007). This uses the model version of Equations 31.7 and 31.8.

Table 31.3 shows the two-level WLSMV solution for the student-level part of the model using a model with two specific factors on each level. The two-level pattern is much clearer than in the single-level analysis ignoring clustering shown in Table 31.2. The large negative loadings for the specific factors have largely disappeared. The first specific factor, labeled number 2, now has more significant positive loadings and thereby more clearly defines the factor as a verbally oriented aggressive-disruptive specific factor. The second specific factor, labeled number 3, is largely unchanged compared to the single-level analysis representing a person-oriented aggressive-disruptive specific factor.

The classroom-level loadings do not give an interpretable picture. Two of the three factors do not have any significant loadings so that it is not clear that the loadings are invariant across the two levels or that three factors are needed on the classroom level.

31.4.3 Two-Level Item Bi-Factor Confirmatory Factor Analysis

In this section, the analysis is changed from exploratory factor analysis to confirmatory factor analysis (CFA). Using the factor pattern shown in Table 31.3, a CFA bi-factor model is specified where the specific factors corresponding to verbal- and person-oriented

TABLE 31.3

Two-Level Analysis Using Bi-Factor EFA and the WLSMV Estimator

	1	2	3
Geomin Rotated Loadings			
Stubborn	0.699[a]	**0.360**[a]	−0.011
Breaks rules	0.829[a]	0.079	0.054
Harms others and property	0.876[a]	−0.053	−0.021
Breaks things	0.918[a]	−0.025	−0.211[a]
Yells at others	0.795[a]	**0.293**[a]	0.009
Takes others' property	0.875[a]	−0.134[a]	−0.015
Fights	0.927[a]	−0.043	**0.287**[a]
Harms property	0.944[a]	−0.003	−0.125[a]
Lies	0.894[a]	0.066	−0.070
Talks back to adults	0.837[a]	**0.349**[a]	−0.004
Teases classmates	0.814[a]	0.033	**0.187**[a]
Fights with classmates	0.919[a]	0.009	**0.314**[a]
Loses temper	0.780[a]	**0.390**[a]	0.009
Geomin Factor Correlations			
1	1.000		
2	0.000	1.000	
3	0.000	0.263[a]	1.000

[a] Student-level results.

TABLE 31.4

Two-Level Analysis Using Bi-Factor CFA and the Bayes Estimator

	G	Verbal	Person
Factor Loadings			
Stubborn	1.090[a]	0.633[a]	0.000
Breaks rules	1.519[a]	0.000	0.000
Harms others and property	1.839[a]	0.000	0.000
Breaks things	2.699[a]	0.000	0.000
Yells at others	1.543[a]	0.668[a]	0.000
Takes others' property	1.915[a]	0.000	0.000
Fights	3.525[a]	0.000	1.512[a]
Harms property	3.452[a]	0.000	0.000
Lies	2.166[a]	0.000	0.000
Talks back to adults	2.000[a]	0.884[a]	0.000
Teases classmates	1.511[a]	0.000	0.436[a]
Fights with classmates	4.534[a]	0.000	2.253[a]
Loses temper	1.689[a]	1.084[a]	0.000
Factor Variances			
Student-level variances	1.000	1.000	1.000
Classroom-level variances	0.322	0.375	0.547

[a] Significant at the 5% level.

aggressive-disruptive behavior are specified to be measured by only the factor loadings with asterisks.

Using two-level WLSMV estimation, a model is considered with the same loading pattern on the student and classroom levels, but not restricting the loadings to be equal across the two levels. Using this model, Wald testing of loading equality across levels is easily carried out in Mplus and it is found that equality cannot be rejected.

As an initial step in a series of further analyses, Bayesian estimation of the two-level bi-factor CFA model with loading invariance across levels is carried out and the estimates shown in Table 31.4 (the loadings are in a different metric than in Table 31.3 where the U^* variances are set to one). Because of the loading invariance across levels, it is possible to study the decomposition of factor variance for the two levels. The percentage due to the classroom variance may be seen as an indicator of heterogeneity of aggressive-disruptive classroom environment. It is seen that 24% ($100 \times 0.322/(1 + 0.322)$ of the general factor variance is due to variation across classrooms, with 27% for the verbal factor and 35% for the person factor.

31.4.4 Two-Level Item Bi-Factor Confirmatory Factor Analysis with Random Factor Loadings

This section changes the two-level model from the model version of Equations 31.7 and 31.8 to the model version of Equations 31.3 through 31.6. Because equality of factor loadings across the two levels is imposed in-line with the previous section, the key difference between the two model types is that the factor loadings are now allowed to vary across the clusters, in this case the classrooms. Technical aspects of this model are described in Asparouhov and Muthén (2012b) using several model variations. The specification makes an attempt to absorb as much of the factor loading variation as possible in factor variance differences across clusters.

The loadings are found to have substantial variation across the 27 classrooms. In other respects, however, the results are close to those of Table 31.4. The average loadings are similar for the general factor and the factor variances on the classroom level are similar: 0.348 for the general factor, 0.364 for the verbal factor, and 0.374 for the person factor. Given that the general conclusions are not altered, subsequent modeling holds loadings equal across clusters.

31.4.5 Longitudinal Two-Level Item Bi-Factor Confirmatory Factor Analysis

The 13 aggressive-disruptive behavior items were measured not only in grade 1 but also in the two subsequent grades. This section discusses longitudinal item response modeling with a focus on changes across time in factor means and variances. Joint analyses of grade 1 and grade 3 are carried out, while at the same time taking into account the classroom clustering. In this sense, three-level data are considered. The analyses presented here, however, will be carried out as two-level modeling because a wide format approach is taken for the longitudinal part of the model, formulating a model for the multivariate vector of 2×13 items.

As is typical in longitudinal studies, many students measured in grade 1 are missing in grade 3. In these data 28% of the students are missing. In such cases it is important to be able to draw on the missing data assumption of MAR (Little and Rubin, 2002), requiring the full-information estimation approaches of ML or Bayes. MCAR cannot be taken for granted, which is assumed by the WLSMV estimator due to using information from only pairs of variables. WLSMV is, however, without such a limitation when used together with a first step of multiple imputation of the missing data (for multiple imputation using Mplus, see Asparouhov and Muthén 2010c).

In the longitudinal setting, this application requires a special type of multilevel modeling of the classroom clustering. Classroom membership pertains to grade 1 classrooms, while the students are spread over many different classrooms by grade 3. An indication of the effect of grade 1 classroom clustering on grade 1 and grade 3 outcomes is obtained using intraclass correlations for the U^* variables behind the 2×13 items. Table 31.5 presents the intraclass correlations using the two-level WLSMV and Bayes estimators. This information is intractable to obtain with ML because of the requirement of an unrestricted model for the U^* variables on both the student and classroom level. Because a probit response function is used, these intraclass correlations are computed with a unit within-cluster variance as

$$icc = \frac{\sigma_B^2}{\left(1+\sigma_B^2\right)}, \tag{31.10}$$

where σ_B^2 is the between-level variance of the random intercept for the item.

Table 31.5 shows that the grade 1 items have sizeable intraclass correlations, but that by grade 3 the grade 1 classroomeffect has largely disappeared. Grade 3 classroom clustering is presumably still present because many students are in the same classroom, but the overall clustering effect in grade 3 is probably smaller due to fewer students of this cohort being in the same classroom in grade 3. Such a grade 3 clustering effect is ignored here.

TABLE 31.5

Intraclass Correlations for Grade 1 and Grade 3 Responses Estimated with WLSMV and Bayes

	Grade 1 WLSMV	Grade 1 Bayes	Grade 3 WLSMV	Grade 3 Bayes
Stubborn	0.110	0.099	0.000	0.080
Breaks rules	0.121	0.105	0.000	0.072
Harms others and property	0.208	0.138	0.000	0.075
Breaks things	0.380	0.222	0.015	0.104
Yells at others	0.215	0.142	0.000	0.070
Takes others' property	0.252	0.179	0.000	0.074
Fights	0.159	0.100	0.000	0.072
Harms property	0.314	0.202	0.001	0.083
Lies	0.211	0.172	0.000	0.070
Talks back to adults	0.143	0.122	0.000	0.068
Teases classmates	0.177	0.126	0.026	0.089
Fights with classmates	0.160	0.100	0.000	0.073
Loses temper	0.171	0.119	0.000	0.078

The longitudinal model to be used draws on the grade 1 model of Section 31.4.3 using a two-level item bi-factor confirmatory model with equal loadings across the two levels. The model is extended to include grade 3 responses as follows. For the student level the same bi-factor model is specified, holding loadings equal to those in grade 1. For the classroom level, the grade 3 items are influenced by the grade 1 general factor with no classroom-level factors added for grade 3. This is in-line with the small intraclass correlations for grade 3, where the expectation is that the classroom-level loadings for the grade 3 items on the general factor of grade 1 will be small. In this way, the longitudinal model has three student-level factors in grade 1, three student-level factors in grade 3, and three classroom-level factors in grade 1 for a total of nine factors. In addition, classroom-level residual variances for the items in grade 1 and grade 3 add further latent variable dimensions for a total of 35. The student-level factors are allowed to correlate across grades.

The longitudinal model is also extended to consider changes across time in the means of both the general and the specific factors. This is accomplished by also imposing measurement invariance for the item threshold across grade 1 and grade 3. Factor means are fixed at zero for grade 1 and estimated for grade 3.

For Bayes estimation, the fact that the model is high-dimensional does not present a problem. For ML estimation, however, this leads to intractable computations. Reducing the dimensions of the model to nine by fixing the classroom-level residual variances at zero, it is possible to carry out the ML computations using Monte Carlo integration (see Asparouhov and Muthén, 2012a) with a total of 100,000 integration points divided into 2174 points for the six student-level dimensions and 46 points for the three classroom-level dimensions. This computationally heavy ML analysis is, however, 10 times slower than the Bayes analysis of the full model.

The results of the Bayesian analysis are presented in Table 31.6. For simplicity, the items are dichotomized in this analysis. The items are dichotomized between the two most frequent item categories of almost never and rarely (see Table 31.1).

Table 31.6 shows that the student-level factor variances decrease from grade 1 to grade 3 so that student behavior becomes more homogeneous. The means of the general and verbal factors increase significantly from grade 1 to grade 3, while the increase for the person factor is not significant. In terms of grade 1 factor standard deviations, the increase

TABLE 31.6

Grade 1–Grade 3 Longitudinal Two-Level Analysis Using Bi-Factor CFA and the Bayes Estimator

	G	Verbal	Person
Factor Loadings			
Stubborn	1.270[a]	0.462[a]	0.000
Breaks rules	1.783[a]	0.000	0.000
Harms others and property	1.857[a]	0.000	0.000
Breaks things	1.838[a]	0.000	0.000
Yells at others	1.836[a]	0.327	0.000
Takes others' property	2.295[a]	0.000	0.000
Fights	3.106[a]	0.000	1.110[a]
Harms property	2.758[a]	0.000	0.000
Lies	2.815[a]	0.000	0.000
Talks back to adults	2.684[a]	1.571[a]	0.000
Teases classmates	1.649[a]	0.000	0.442[a]
Fights with classmates	3.397[a]	0.000	1.318[a]
Loses temper	1.708[a]	0.610[a]	0.000
Factor Variances			
Student-level Grade 1	1.000	1.000	1.000
Student-level Grade 3	0.895	0.427	0.620
Classroom-level Grade 1	0.203	0.318	0.418
Factor Means			
Grade 1	0.000	0.000	0.000
Grade 3	0.327[a]	0.646[a]	0.286
Factor Correlation			
G for Grade 1 with Grade 3	0.349[a]		

[a] Dichotomized items.

in factor mean from grade 1 to grade 3 is 0.298 for the general factor and 0.563 for the verbal factor, indicating that the increase in aggressive-disruptive behavior is mostly due to increased verbally oriented aggressive-disruptive behavior.

These analyses may form the basis for a multiple-indicator growth model across several grades where growth is considered for both the general and the specific factors. Previous growth analyses for these data have been carried out on the sum of the items assuming unidimensionality. Using growth mixture modeling, these analyses have uncovered different latent classes of trajectories for which an intervention has different effects (Muthén et al., 2002; Muthén and Asparouhov, 2009). Such a finite mixture generalization is also possible with the multi-dimensional, multi-level item response modeling considered here.

31.5 Conclusions

The analysis of the aggressive-disruptive behavior application exemplifies the flexibility of item response modeling in the Mplus framework. High-dimensional exploratory, confirmatory, multilevel, and longitudinal analyses are possible using a combination of weighted least-squares, maximum-likelihood, and Bayesian estimation.

Owing to lack of space, many more analysis possibilities relevant to this application are excluded from the discussion. Exploratory structural equation modeling (ESEM; Asparouhov and Muthén, 2009) can be used for multiple-group EFA of male and female students with varying degrees of invariance of measurement and structural parameters. ESEM can also be used in a longitudinal analysis to study measurement invariance across grades. Bayesian EFA and two-tier modeling can be carried out as discussed in Asparouhov and Muthén (2012a). Bayesian structural equation modeling (BSEM; Muthén and Asparouhov, 2012) can be used to allow cross-loadings in the confirmatory analysis, using informative zero-mean, small-variance priors for parameters that are not identified in maximum-likelihood analysis. Gender differences can be studied in two-level analysis that allows within-cluster groupings as discussed by Asparouhov and Muthén (2012c). Mplus applications of more general latent variable models with random subjects, random items, random contexts, and random parameters are also discussed by Asparouhov and Muthén (2012b).

References

Agresti, A. 2002. *Categorical Data Analysis.* Second edition. New York: John Wiley & Sons.

Asparouhov, T., and Muthén, B. 2007, July. Computationally efficient estimation of multilevel high-dimensional latent variable models. Paper presented at the Joint Statistical Meetings, Salt Lake City, Utah.

Asparouhov, T., and Muthén, B. 2008. Multilevel mixture models. In G. R. Hancock and K. M. Samuelsen (Eds.), *Advances in Latent Variable Mixture Models* (pp. 27–51). Charlotte, NC: Information Age Publishing, Inc.

Asparouhov, T., and Muthén, B. 2009. Exploratory structural equation modeling. *Structural Equation Modeling, 16*, 397–438.

Asparouhov, T., and Muthén, B. 2010a. Simple second order chi-square correction. Unpublished manuscript. Available at http://www.statmodel.com

Asparouhov, T., and Muthén, B. 2010b. Bayesian analysis using Mplus: Technical implementation. Unpublished manuscript. Available at http://www.statmodel.com

Asparouhov, T., and Muthén, B. 2010c. Multiple imputation with Mplus. Unpublished manuscript. Available at http://www.statmodel.com

Asparouhov, T., and Muthén, B. 2012a. Comparison of computational methods for high-dimensional item factor analysis. Manuscript submitted for publication. Available at http://www.statmodel.com

Asparouhov, T., and Muthén, B. 2012b. General random effect latent variable modeling: Random subjects, items, contexts, and parameters. Unpublished manuscript. Available at http://www.statmodel.com

Asparouhov, T., and Muthén, B. 2012c. Multiple group multilevel analysis. Mplus Web Notes: No. 16. November 15, 2012. Manuscript submitted for publication. Available at http://www.statmodel.com

Baker, F. B., and Kim, S. H. 2004. *Item Response Theory. Parameter Estimation Techniques.* New York: Marcel Dekker.

Cronbach, L. J. 1976. Research on classrooms and schools: Formulation of questions, design, and analysis. Unpublished manuscript. Available at http://www.statmodel.com

Fox, J. P. 2010. *Bayesian Item Response Modeling.* New York: Springer.

Goldstein, H., and McDonald, R. P. 1988. A general model for the analysis of multilevel data. *Psychometrika, 53,* 455–467.

Harnqvist, K. 1978. Primary mental abilities of collective and individual levels. *Journal of Educational Psychology, 70,* 706–716.

Harnqvist, K., Gustafsson, J. E., Muthén, B., and Nelson, G. 1994. Hierarchical models of ability at class and individual levels. *Intelligence, 18,* 165–187.

Henry, K., and Muthén, B. 2010. Multilevel latent class analysis: An application of adolescent smoking typologies with individual and contextual predictors. *Structural Equation Modeling, 17,* 193–215.

Ialongo, L. N., Werthamer, S., Kellam, S. K., Brown, C. H., Wang, S., and Lin, Y. 1999. Proximal impact of two first-grade preventive interventions on the early risk behaviors for later substance abuse, depression and antisocial behavior. *American Journal of Community Psychology, 27,* 599–641.

Jennrich, R. I., and Bentler, P. M. 2011. Exploratory bi-factor analysis. *Psychometrika, 76,* 537–549.

Jennrich, R. I., and Bentler, P. M. 2012. Exploratory bi-factor analysis: The oblique case. *Psychometrika, 77,* 442–454.

Johnson, V. E., and Albert, J. H. 1999. *Ordinal Data Modeling.* New York: Springer.

Little, R. J., and Rubin, D. B. 2002. *Statistical Analysis with Missing Data.* Second edition. New York: John Wiley and Sons.

Longford, N. T., and Muthén, B. 1992. Factor analysis for clustered observations. *Psychometrika, 57,* 581–597.

McDonald, R. P., and Goldstein, H. 1989. Balanced versus unbalanced designs for linear structural relations in two-level data. *British Journal of Mathematical and Statistical Psychology, 42,* 215–232.

Muthén, B. 1978. Contributions to factor analysis of dichotomous variables. *Psychometrika, 43,* 551–560.

Muthén, B. 1984. A general structural equation model with dichotomous, ordered categorical, and continuous latent variable indicators. *Psychometrika, 49,* 115–132.

Muthén, B. 1994. Multilevel covariance structure analysis. In J. Hox and I. Kreft (Eds.), *Multilevel Modeling, a Special Issue of Sociological Methods & Research, 22,* 376–398.

Muthén, B. 2008. Latent variable hybrids: Overview of old and new models. In G. R. Hancock and K. M. Samuelsen (Eds.), *Advances in Latent Variable Mixture Models* (pp. 1–24). Charlotte, NC: Information Age Publishing, Inc.

Muthén, B., and Asparouhov, T. 2009. Growth mixture modeling: Analysis with non-Gaussian random effects. In G. Fitzmaurice, M. Davidian, G. Verbeke, and G. Molenberghs (Eds.), *Longitudinal Data Analysis* (pp. 143–165). Boca Raton, FL: Chapman and Hall/CRC Press.

Muthén, B., and Asparouhov, T. 2012. Bayesian SEM: A more flexible representation of substantive theory. *Psychological Methods, 17,* 313–335.

Muthén, B., Brown, C. H., Masyn, K., Jo, B., Khoo, S. T., Yang, C. C., Wang, C. P., Kellam, S., Carlin, J., and Liao, J. 2002. General growth mixture modeling for randomized preventive interventions. *Biostatistics, 3,* 459–475.

Muthén, L. K., and Muthén, B. O. 1998–2012. *Mplus User's Guide.* Seventh Edition. Los Angeles, CA: Muthén and Muthén.

Muthén, B., Muthén, L., and Asparouhov, T. In press. *Regression Analysis, Factor Analysis, and Structural Equation Modeling Using Mplus.* Book in preparation.

Muthén, B., and Satorra, A. 1995. Technical aspects of Muthén's LISCOMP approach to estimation of latent variable relations with a comprehensive measurement model. *Psychometrika, 60,* 489–503.

Muthén, B., du Toit, S. H. C., and Spisic, D. 1997. Robust inference using weighted least squares and quadratic estimating equations in latent variable modeling with categorical and continuous outcomes. Unpublished manuscript. Available at http://www.statmodel.com

32

Mixed-Coefficients Multinomial Logit Models

Raymond J. Adams, Mark R. Wilson, and Margaret L. Wu

CONTENTS

32.1 Introduction 541
32.2 Presentation of the Models 542
32.2.1 Extended Mixed Coefficient Multinomial Logit Model 542
32.2.2 Simple Logistic Model 544
32.2.3 Models with Scoring Parameters 546
32.2.4 Population Model 547
32.2.5 Combined Model 548
32.2.6 Model Identification 548
32.3 Parameter Estimation 549
32.3.1 Maximum-Likelihood Estimation 549
32.3.2 Quadrature and Monte Carlo Approximations 550
32.3.3 Conditional Maximum-Likelihood Estimation 552
32.3.4 Latent Ability Estimation and Prediction 553
32.3.5 Estimating Functionals of the Population Distributions 553
32.4 Model Fit 555
32.4.1 Generalized Fit Test 555
32.4.2 Customized Fit Tests 556
32.4.3 Tests of Relative Fit 556
32.5 Empirical Example 557
32.5.1 Bundle Independence 557
32.5.2 Saturated Bundle Model 557
32.5.3 PCM Bundles 559
32.5.4 SLM Bundles 559
32.5.5 Customized Approaches 560
32.5.6 Empirical Comparisons 561
32.6 Discussion 562
References 563

32.1 Introduction

Over the past 30 years, a proliferation of item response models has emerged. In the Rasch logistic item response model family, notably, the simple logistic model (Rasch, 1980; Volume One, Chapter 3), the partial credit model (Volume One, Chapter 7), the rating scale model (Volume One, Chapter 5), the facets model (Linacre, 1989) and the linear logistic model (Fischer, 1973; Volume One, Chapter 13) have all played an important role in the analysis of item response data. Similarly, the two-parameter logistic model has been extended to polytomous items as the generalized partial credit model (Volume One, Chapter 8) and a variety of multidimensional models have been proposed as extensions for both the Rasch and two-parameter model families (Embretson, 1992; Volume One, Chapter 14; Reckase, 2009; Volume One, Chapter 12). Typically, the development of parameter estimation procedures is specific to each model, and the same some held for the development of dedicated software programs. Surveying the family of Rasch models, Adams and Wilson (1996) developed a unified approach to specifying the models and then consequentially estimating the parameters. There are at least two advantages of developing one single framework to encompass a family of models. First, the development of the estimation procedures and associated software for the implementation of the models can be streamlined within a single framework of models. That is, one only needs to develop one's set of estimation procedures and software program to carry out the estimation of the parameters in the models. Second, a generalized framework provides an opportunity for the development of new models that can fit within the framework. This allows for the flexible application of item response models to suit users' requirements.

This chapter describes a generalized framework for specifying a family of logistic item response models through the specification of design matrices. Their estimation procedures are also described. The idea of the use of design matrices is extended to the construction of a family of goodness-of-fit tests. Flexibility in the construction of fit tests allows the users to target specific hypotheses regarding the fit of the items to the model, such as the violation of local independence among subsets of items.

32.2 Presentation of the Models

32.2.1 Extended Mixed Coefficient Multinomial Logit Model

The extended mixed coefficients multinomial logit model is a categorical response model and, in most applications, the response patterns to a set of test items (the categorical outcomes) are modeled as the dependent variable. Under the model, the response patterns are predicted by logistic regression, where the independent variables are item difficulty characteristics and examinee abilities (both difficulty and ability are used as generic terms throughout this chapter).

The model is referred to as a mixed coefficients model because items are described by fixed sets of unknown parameters, $\boldsymbol{\xi}$ and $\boldsymbol{\tau}$, while the examinee outcome levels (the latent variable), $\boldsymbol{\theta}$, is a random effect. Later, the distributional assumptions for this random effect are discussed.

The model is specified as follows. Assume there are items indexed $i = 1, \ldots, I$ with each item admitting $A_i + 1$ response categories indexed $a = 0, 1, \ldots, A_i$. That is, a response to item i by an examinee can be allocated to one of $A_i + 1$ response categories. A vector valued random variable $\mathbf{U}_i = (U_{i1}, U_{i2}, \ldots, U_{iK_i})^T$, where for $a = 0, 1, \ldots, A_i$

$$U_{ia} = \begin{cases} 1 & \text{if response to item } i \text{ is in category } a \\ 0 & \text{otherwise} \end{cases}, \tag{32.1}$$

is used to indicate the $A_i + 1$ possible responses to item i.

A vector of zeroes denotes a response in category zero, making this category a reference category, which is necessary for model identification. Using this as the reference category is arbitrary, and does not affect the generality of the model.

Each $\mathbf{U}_i$ consists of a sequence of 0's and possibly one 1, indicating the examinee's response category for that item. For example, if the response category is 0 for an item with four categories (0,1,2,3), then $\mathbf{U}_i^T = (0,0,0)$. If the response category is 2, then $\mathbf{U}_i^T = (0,1,0)$.

The $\mathbf{U}_i$ can also be collected together into the single vector $\mathbf{U}^T = (\mathbf{U}_1^T, \mathbf{U}_2^T, \ldots, \mathbf{U}_I^T)$, called the response vector (or pattern). Particular instances of each of these random variables are indicated by their lower case equivalents; $\mathbf{u}$, $\mathbf{u}_j$, and u_{ia}.

Items are described through a vector $\boldsymbol{\xi}^T = (\xi_1, \xi_2, \ldots, \xi_Q)$ of Q parameters. Linear combinations of their components are used in the response probability model to describe the empirical characteristics of the response categories of each item. Design vectors $\mathbf{g}_{ik}$, ($i = 1, \ldots, I$; $a = 1, \ldots, A_i$), each of length Q, which can be collected to form a design matrix $\mathbf{G}^T = (\mathbf{g}_{11}, \mathbf{g}_{12}, \ldots, \mathbf{g}_{1A_1}, \mathbf{g}_{21}, \ldots, \mathbf{g}_{2A_2}, \ldots, \mathbf{g}_{IA_I})$, define these linear combinations.

The multidimensional form of the model assumes that a set of D latent abilities underlies the individuals' responses. These abilities define a D-dimensional latent space. The vector $\boldsymbol{\theta} = (\theta_1, \theta_2, \ldots, \theta_D)^T$ represents an individual's position in the D-dimensional latent space.

The model also contains a scoring function that allows the specification of the score or performance level assigned to each possible response category to each item. To do so, the notion of a response score b_{iad} is introduced, which gives the performance level of an observed response in category a to item i for dimension d. To allow flexibility in imposing constraints on b_{iad} they are in turn specified to be linear combinations of a set of s item scoring parameters $\boldsymbol{\tau}$, The scores for dimension d can be collected into a vector $\mathbf{b}_d = (b_{11d}, b_{12d}, \ldots, b_{1A_1d}, b_{21d}, \ldots, b_{2A_2d}, \ldots, b_{IA_Id})^T$ with $\mathbf{b}_d = \mathbf{C}_d \boldsymbol{\tau}$.

Under this approach the score for a response in the zero category is zero, but other responses may also be scored zero. This is an arbitrary identification constraint.

The regression of the response vector on the item and examinee parameters is

$$f(\mathbf{u};\boldsymbol{\xi},\boldsymbol{\tau} \mid \boldsymbol{\theta}) = \Psi(\boldsymbol{\theta},\boldsymbol{\xi},\boldsymbol{\tau}) \exp\left[\mathbf{u}^T\left(\sum_{d=1}^{D} \theta_d \mathbf{C}_d \boldsymbol{\tau} + \mathbf{G}\boldsymbol{\xi}\right)\right], \tag{32.2}$$

with

$$\Psi(\boldsymbol{\theta},\boldsymbol{\xi},\boldsymbol{\tau}) = \left\{\sum_{\mathbf{z}\in\Omega} \exp\left[\mathbf{z}^T\left(\sum_{d=1}^{D} \theta_d \mathbf{C}_d \boldsymbol{\tau} + \mathbf{G}\boldsymbol{\xi}\right)\right]\right\}^{-1}, \tag{32.3}$$

where Ω is the set of all possible response vectors.

In the case where there is no scoring, parameters τ would be a vector of 1's and if $\mathbf{B} = (\mathbf{b}_1, \mathbf{b}_2, \ldots, \mathbf{b}_D)$ then we can write

$$\mathbf{B}\boldsymbol{\theta} = \sum_{d=1}^{D} \theta_d \mathbf{C}_d \boldsymbol{\tau}, \tag{32.4}$$

so that Equations 32.2 and 32.3 become

$$f(\mathbf{u} \mid \boldsymbol{\xi}, \boldsymbol{\theta}) = \Psi(\boldsymbol{\theta},\boldsymbol{\xi})\exp[\mathbf{u}^T(\mathbf{B}\boldsymbol{\theta} + \mathbf{G}\boldsymbol{\xi})] \tag{32.5}$$

and

$$\Psi(\boldsymbol{\theta},\boldsymbol{\xi}) = \left\{\sum_{\mathbf{z}\in\Omega} \exp[\mathbf{z}^T(\mathbf{B}\boldsymbol{\theta} + \mathbf{G}\boldsymbol{\xi})]\right\}^{-1}. \tag{32.6}$$

32.2.2 Simple Logistic Model

With specific choices **A**, $\mathbf{C}_1$, $\boldsymbol{\xi}$, and $\boldsymbol{\tau}$, Equations 32.2 and 32.3 specialize to particular models. For example, a simple logistic model for three items is obtained using

$$\mathbf{A} = \begin{bmatrix} 1 & 0 & 0 \\ 0 & 1 & 0 \\ 0 & 0 & 1 \end{bmatrix}, \quad \mathbf{C}_1 = \begin{bmatrix} 1 \\ 1 \\ 1 \end{bmatrix}, \quad \boldsymbol{\xi} = \begin{bmatrix} \delta_1 \\ \delta_2 \\ \delta_3 \end{bmatrix}, \quad \boldsymbol{\tau} = [1]. \tag{32.7}$$

A partial credit item with three scoring categories 0, 1, and 2 is obtained using

$$\mathbf{G} = \begin{bmatrix} 1 & 0 & 0 & 0 \\ 1 & 1 & 0 & 0 \\ 0 & 0 & 1 & 0 \\ 0 & 0 & 1 & 1 \end{bmatrix}, \quad \mathbf{C}_1 = \begin{bmatrix} 1 \\ 2 \\ 1 \\ 2 \end{bmatrix}, \quad \boldsymbol{\xi} = \begin{bmatrix} \delta_{11} \\ \delta_{12} \\ \delta_{21} \\ \delta_{22} \end{bmatrix}, \quad \boldsymbol{\tau} = [1]. \tag{32.8}$$

To obtain a facets model (Linacre, 1989) where there are three raters, with each of them rating the same two dichotomous items six *generalized items* are defined as each of the possible combinations of a rater and an actual item. Generalized item one represents the response category given by rater one on item one. Generalized item two represents the response category given by rater one on item two, and so on. The following choices of **G**, $\mathbf{C}_1$, $\boldsymbol{\xi}$, and $\boldsymbol{\tau}$ will then give a facets model.

$$\mathbf{G} = \begin{bmatrix} 1 & 0 & 0 & 1 & 0 \\ 1 & 0 & 0 & 0 & 1 \\ 0 & 1 & 0 & 1 & 0 \\ 0 & 1 & 0 & 0 & 1 \\ 0 & 0 & 1 & 1 & 0 \\ 0 & 0 & 1 & 0 & 1 \end{bmatrix}, \quad \mathbf{C}_1 = \begin{bmatrix} 1 \\ 1 \\ 1 \\ 1 \\ 1 \\ 1 \end{bmatrix}, \quad \boldsymbol{\xi} = \begin{bmatrix} \rho_1 \\ \rho_2 \\ \rho_3 \\ \delta_1 \\ \delta_2 \end{bmatrix}, \quad \boldsymbol{\tau} = [1]. \tag{32.9}$$

The first row of **G** corresponds to category one of generalized item one (rater one item one); the second row corresponds to category one of generalized item two (rater one item two); the third row corresponds to category one of generalized item three (rater two item one); the fourth row corresponds to category one of generalized item four (rater two item two), and so on. The same row referencing applies to the matrix $\mathbf{C}_1$. The first three elements (ρ_1,ρ_2,ρ_3) of $\boldsymbol{\xi}$

are the severity parameters of raters one to three, respectively. The fourth and fifth element (δ_1, δ_2) of $\boldsymbol{\xi}$ are the difficulty parameters for the two dichotomous items. The single τ value is a common discrimination parameter that the facets model assumes for all item and raters.

Figure 32.1 shows two possible multidimensional models: a between-item multidimensional model and a within-item multidimensional model (Adams et al., 1997). In both cases, a hypothetical nine-item test is considered. In the between-item multidimensional case (left-hand side), each item is associated with a single dimension, but the collection of items covers three dimensions—three items are associated with each of the three latent dimensions. In the within-item case (right-hand side) some items are associated with more than one dimension. For example, item two is associated with both dimensions one and two.

If the items shown in Figure 32.1 are all dichotomous, then the matrices $\mathbf{G}$, $\mathbf{C}_1$, $\mathbf{C}_2$, $\mathbf{C}_3$, $\boldsymbol{\xi}$, and $\boldsymbol{\tau}$, as given in Equations 32.10 and 32.11, when substituted into Equation 32.2, will yield the between- and within-item multidimensional models shown in Figure 32.1, respectively.

Note that the only difference between Equations 32.10 and 32.11 are the $\mathbf{C}$ matrices. These scoring matrices are used to indicate the scores of the items on each of the three dimensions.

$$\mathbf{G}=\begin{bmatrix}1&0&0&0&0&0&0&0&0\\0&1&0&0&0&0&0&0&0\\0&0&1&0&0&0&0&0&0\\0&0&0&1&0&0&0&0&0\\0&0&0&0&1&0&0&0&0\\0&0&0&0&0&1&0&0&0\\0&0&0&0&0&0&1&0&0\\0&0&0&0&0&0&0&1&0\\0&0&0&0&0&0&0&0&1\end{bmatrix},\ \mathbf{C}_1=\begin{bmatrix}1\\1\\1\\0\\0\\0\\0\\0\\0\end{bmatrix},\ \mathbf{C}_2=\begin{bmatrix}0\\0\\0\\1\\1\\1\\0\\0\\0\end{bmatrix},\ \mathbf{C}_3=\begin{bmatrix}0\\0\\0\\0\\0\\0\\1\\1\\1\end{bmatrix},\ \boldsymbol{\xi}=\begin{bmatrix}\delta_1\\\delta_2\\\delta_3\\\delta_4\\\delta_5\\\delta_6\\\delta_7\\\delta_8\\\delta_9\end{bmatrix},\ \boldsymbol{\tau}=[1]. \quad (32.10)$$

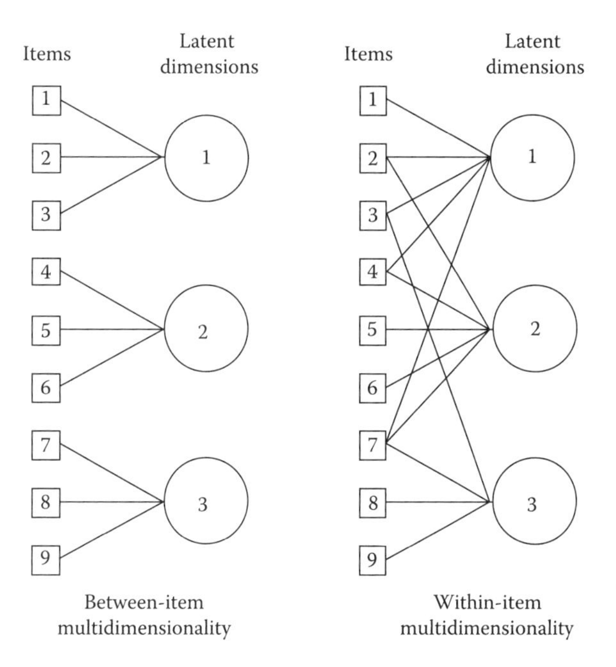

FIGURE 32.1
Between- and within-item multidimensionality.

$$\mathbf{G} = \begin{bmatrix} 1 & 0 & 0 & 0 & 0 & 0 & 0 & 0 & 0 \\ 0 & 1 & 0 & 0 & 0 & 0 & 0 & 0 & 0 \\ 0 & 0 & 1 & 0 & 0 & 0 & 0 & 0 & 0 \\ 0 & 0 & 0 & 1 & 0 & 0 & 0 & 0 & 0 \\ 0 & 0 & 0 & 0 & 1 & 0 & 0 & 0 & 0 \\ 0 & 0 & 0 & 0 & 0 & 1 & 0 & 0 & 0 \\ 0 & 0 & 0 & 0 & 0 & 0 & 1 & 0 & 0 \\ 0 & 0 & 0 & 0 & 0 & 0 & 0 & 1 & 0 \\ 0 & 0 & 0 & 0 & 0 & 0 & 0 & 0 & 1 \end{bmatrix}, \quad \mathbf{C}_1 = \begin{bmatrix} 1 \\ 1 \\ 1 \\ 1 \\ 0 \\ 0 \\ 1 \\ 0 \\ 0 \end{bmatrix}, \quad \mathbf{C}_2 = \begin{bmatrix} 0 \\ 1 \\ 0 \\ 1 \\ 1 \\ 1 \\ 1 \\ 0 \\ 0 \end{bmatrix}, \quad \mathbf{C}_3 = \begin{bmatrix} 0 \\ 0 \\ 1 \\ 0 \\ 0 \\ 0 \\ 1 \\ 1 \\ 1 \end{bmatrix}, \quad \boldsymbol{\xi} = \begin{bmatrix} \delta_1 \\ \delta_2 \\ \delta_3 \\ \delta_4 \\ \delta_5 \\ \delta_6 \\ \delta_7 \\ \delta_8 \\ \delta_9 \end{bmatrix}, \quad \boldsymbol{\tau} = [1]. \tag{32.11}$$

In the above examples, the items are dichotomous but the extension to polytomous items is relatively straightforward. In fact, in the polytomous case it is possible to allow different *steps* in an item to abilities from different dimensions. For example, the matrices in Equation 32.12 yield a two-dimensional partial credit model for two items in which the second step in each item draws upon a second ability dimension.

$$\mathbf{G} = \begin{bmatrix} 1 & 0 & 0 & 0 \\ 1 & 1 & 0 & 0 \\ 0 & 0 & 1 & 0 \\ 0 & 0 & 1 & 1 \end{bmatrix}, \quad \mathbf{C}_1 = \begin{bmatrix} 1 \\ 1 \\ 1 \\ 1 \end{bmatrix}, \quad \mathbf{C}_1 = \begin{bmatrix} 0 \\ 1 \\ 0 \\ 1 \end{bmatrix}, \quad \boldsymbol{\xi} = \begin{bmatrix} \delta_{11} \\ \delta_{12} \\ \delta_{21} \\ \delta_{22} \end{bmatrix}, \quad \boldsymbol{\tau} = [1]. \tag{32.12}$$

32.2.3 Models with Scoring Parameters

To extend Rasch's simple logistic model to the two-parameter logistic model (Volume One, Chapter 2), the set of matrices in Equation 32.7 is replaced with that in Equation 32.13. The changes are to $\mathbf{C}_1$ and τ. The elements of τ are the discrimination parameters for each of the items.

$$\mathbf{G} = \begin{bmatrix} 1 & 0 & 0 \\ 0 & 1 & 0 \\ 0 & 0 & 1 \end{bmatrix}, \quad \mathbf{C}_1 = \begin{bmatrix} 1 & 0 & 0 \\ 0 & 1 & 0 \\ 0 & 0 & 1 \end{bmatrix}, \quad \boldsymbol{\xi} = \begin{bmatrix} \delta_1 \\ \delta_2 \\ \delta_3 \end{bmatrix}, \quad \boldsymbol{\tau} = \begin{bmatrix} \tau_1 \\ \tau_2 \\ \tau_3 \end{bmatrix}. \tag{32.13}$$

To extend the partial credit model to the generalized partial credit model, the matrices in Equation 32.8 are replaced with those in

$$\mathbf{G} = \begin{bmatrix} 1 & 0 & 0 & 0 \\ 1 & 1 & 0 & 0 \\ 0 & 0 & 1 & 0 \\ 0 & 0 & 1 & 1 \end{bmatrix}, \quad \mathbf{C}_1 = \begin{bmatrix} 1 & 0 \\ 2 & 0 \\ 0 & 1 \\ 0 & 2 \end{bmatrix}, \quad \boldsymbol{\xi} = \begin{bmatrix} \delta_{11} \\ \delta_{12} \\ \delta_{21} \\ \delta_{22} \end{bmatrix}, \quad \boldsymbol{\tau} = \begin{bmatrix} \tau_1 \\ \tau_2 \end{bmatrix}. \tag{32.14}$$

It is also clear that this model is easily generalized to one in which the scoring parameters are not set at the item but at the category level. This would be achieved by replacing $\mathbf{C}_1$ with a 4×4 identity and making $\boldsymbol{\tau}$ a vector of length four. The facets models described above can similarly be formulated with scoring parameters, perhaps, allowing different raters or different items to have different scoring parameters.

32.2.4 Population Model

The item response model in Equation 32.2 can also be viewed as a conditional model, in the sense that it describes the process of generating item responses conditional on the realization of a random latent variable, $\boldsymbol{\theta}$. The complete version of this model requires the specification of a density, $f_\theta(\boldsymbol{\theta};\boldsymbol{\alpha})$ for the latent variable, $\boldsymbol{\theta}$. Let $\boldsymbol{\alpha}$ symbolize a set of parameters that characterize the distribution of $\boldsymbol{\theta}$. The most common practice, when specifying unidimensional marginal item response models, is to assume that examinees have been sampled from a normal population with mean μ and variance σ^2. That is,

$$f_{\boldsymbol{\theta}}(\boldsymbol{\theta};\boldsymbol{\alpha}) \equiv f_\theta(\theta;\mu,\sigma^2) = (2\pi\sigma^2)^{-1/2}\exp\left[-\frac{(\theta-\mu)^2}{2\sigma^2}\right], \tag{32.15}$$

or, equivalently,

$$\theta = \mu + E, \tag{32.16}$$

where $E \sim N(0,\sigma^2)$.

Adams et al. (1997) discuss a natural extension of Equation 32.15 with the replacement of the mean, μ, with the regression model, $\mathbf{Y}_p^T\boldsymbol{\beta}$ where $\mathbf{Y}_p$ is a vector of w fixed and known covariates for examinee p, and $\boldsymbol{\beta}$ is the corresponding vector of regression coefficients. For example, $\mathbf{Y}_p$ could be constituted of such examinee variables such as gender or socioeconomic status. Then, the model for examinee p becomes

$$\theta_p = \mathbf{Y}_p^T\boldsymbol{\beta} + E_p, \tag{32.17}$$

where E_p are assumed to be independently and identically normally distributed with mean zero and variance σ^2 across the examinees, so that Equation 32.15 can be generalized to

$$f_\theta(\theta_p \mid \mathbf{Y}_p,\boldsymbol{\beta},\sigma^2) = (2\pi\sigma^2)^{-1/2}\exp\left[-\frac{1}{2\sigma^2}\left(\theta_p - \mathbf{Y}_p^T\boldsymbol{\beta}\right)^T\left(\theta_p - \mathbf{Y}_p^T\boldsymbol{\beta}\right)\right], \tag{32.18}$$

that is, a normal distribution with mean $\mathbf{Y}_n^T\boldsymbol{\beta}$ and variance σ^2. The generalization needs to be taken one step further to apply it to the vector valued $\boldsymbol{\theta}$ (of length d) rather than the scalar valued θ. The extension results in the multivariate population model

$$f_{\boldsymbol{\theta}}(\boldsymbol{\theta}_p \mid \mathbf{W}_p,\boldsymbol{\gamma},\boldsymbol{\Sigma}) = (2\pi)^{-d/2}|\Sigma|^{-1/2}\exp\left[-\frac{1}{2}(\boldsymbol{\theta}_p - \boldsymbol{\gamma}\mathbf{W}_p)^T\boldsymbol{\Sigma}^{-1}(\boldsymbol{\theta}_p - \boldsymbol{\gamma}\mathbf{W}_p)\right], \tag{32.19}$$

where γ is a $d \times w$ matrix of regression coefficients, $\boldsymbol{\Sigma}$ is a $d \times d$ variance–covariance matrix and $\mathbf{W}_p$ is a $w \times 1$ vector of fixed covariates.

While, in most cases, the multivariate normal distribution in Equation 32.19 is assumed as the population distribution, other forms of the population distribution can also be considered. For example, Adams et al. (1997) considered a step distribution defined on a prespecified grid of ability values. They argued that this could be used as an approximation to an arbitrary continuous ability distribution. In this case $f_\theta(\boldsymbol{\theta};\boldsymbol{\alpha})$ is a discrete distribution defined on the grid. For the step distribution, a fixed vector of Q grid points

$(\Theta_{d1}, \Theta_{d2}, \ldots, \Theta_{dQ})$ is selected for each latent dimension, and the grid points of each of the $r = 1, \ldots, Q^D$ Cartesian coordinates are defined as

$$\boldsymbol{\Theta}_r = (\Theta_{1k_1}, \Theta_{2k_2}, \ldots, \Theta_{Dk_D}) \quad \text{with} \quad k_1 = 1, \ldots, Q; \quad k_2 = 1, \ldots, Q; \quad \cdots; \quad k_D = 1, \ldots, Q. \tag{32.20}$$

The parameters that characterize the distribution are $\boldsymbol{\Theta}_1, \boldsymbol{\Theta}_2, \ldots, \boldsymbol{\Theta}_{Q^D}$ and $w_1, w_2, \ldots, w_{Q^D}$ where $f_\Theta(\boldsymbol{\Theta}_r; \boldsymbol{\alpha}) = w_r$ is the density at point r. The grid points are specified a priori, and density values are estimated.

32.2.5 Combined Model

In the case of a normally distributed population, the conditional response model in Equation 32.2 and population model in Equation 32.19 are combined to obtain the following marginal model:

$$f_{\mathrm{u}}(\mathbf{u}; \boldsymbol{\gamma}, \boldsymbol{\Sigma}, \boldsymbol{\xi}, \boldsymbol{\tau}) = \int_{\theta} f_{\mathrm{u}}(\mathbf{u}; \boldsymbol{\xi}, \boldsymbol{\tau} \mid \boldsymbol{\theta})\, f_\theta(\boldsymbol{\theta}; \boldsymbol{\gamma}, \boldsymbol{\Sigma})\, d\boldsymbol{\theta}. \tag{32.21}$$

It is important to recognize that, under this model, the locations of the examinees on the latent ability variables are not estimated. The parameters of the model are $\boldsymbol{\gamma}$, $\boldsymbol{\Sigma}$, $\boldsymbol{\xi}$, and $\boldsymbol{\tau}$, where $\boldsymbol{\gamma}$, $\boldsymbol{\Sigma}$ are the population parameters and $\boldsymbol{\xi}$ and $\boldsymbol{\tau}$ are the item parameters.

Likewise, for the discrete case,

$$\begin{aligned} f_{\mathrm{u}}(\mathbf{u}; \boldsymbol{\xi}, \boldsymbol{\tau}, \boldsymbol{\alpha}) &= \sum_{r=1}^{Q} f_{\mathrm{u}}(\mathbf{u}; \boldsymbol{\xi}, \boldsymbol{\tau} \mid \boldsymbol{\Theta}_r)\, f_\Theta(\boldsymbol{\Theta}_r; \boldsymbol{\alpha}) \\ &= \sum_{r=1}^{Q} f_{\mathrm{u}}(\mathbf{u}; \boldsymbol{\xi}, \boldsymbol{\tau} \mid \boldsymbol{\Theta}_r)\, w_r. \end{aligned} \tag{32.22}$$

32.2.6 Model Identification

When there are no scoring parameters, the identification of Equation 32.21 requires certain constraints be placed on the design matrices **G** and **B** (in fact, the design matrices used in our earlier examples did not yield identified models). Volodin and Adams (1995) showed that the following are conditions for the identification of Equation 32.21 with D latent dimensions and the length of parameter vector $\boldsymbol{\xi}$ equal to Q:

Proposition 32.1

If $A_i + 1$ is the number of response categories for item i, and $A = \sum_{i \in \mathbf{I}} A_i$, then the model in Equation 32.6, if applied to the set of items **I**, can only be identified if $Q + D \le A$.

Proposition 32.2

The model in Equation 32.6 can only be identified if $rank(\mathbf{G}) = Q$, $rank(\mathbf{B}) = D$, and $rank([\mathbf{G}\ \mathbf{B}]) = Q + D$.

Proposition 32.3

If $A_i + 1$ is the number of response categories for item i, and $A = \sum_{i \in \mathbf{I}} A_i$, then model (32.6), if applied to the set of items **I**, can be identified if and only if $rank([\mathbf{G}\ \mathbf{B}]) = Q + D \leq A$.

32.3 Parameter Estimation

In Section 32.3.1, a maximum-likelihood approach to estimating the parameters is sketched (Adams et al., 1997), and when there are no scoring parameters the possibility of a conditional maximum-likelihood (Andersen, 1970) approach is discussed.

32.3.1 Maximum-Likelihood Estimation

The maximum-likelihood approach to estimating the parameters in Equation 32.21 proceeds as follows. Let $\mathbf{u}_p$ be the response pattern of examinee p and assume independent observations are made for $p = 1, \ldots, P$ examinees. It follows that the likelihood for the P sampled examinees is

$$\Lambda = \prod_{p=1}^{P} f_{\mathbf{u}}(\mathbf{u}_p; \boldsymbol{\gamma}, \boldsymbol{\Sigma}, \boldsymbol{\xi}, \boldsymbol{\tau}). \tag{32.23}$$

Differentiating with respect to each of the parameters and defining the marginal density of θ_n as

$$h_{\theta}(\boldsymbol{\theta}_p; \mathbf{W}_p, \boldsymbol{\gamma}, \boldsymbol{\Sigma}, \boldsymbol{\xi}, \boldsymbol{\tau} \mid \mathbf{u}_p) = \frac{f_{\mathbf{x}}(\mathbf{u}_p; \boldsymbol{\xi}, \boldsymbol{\tau} \mid \boldsymbol{\theta}_p)\, f_{\theta}(\boldsymbol{\theta}_p; \mathbf{W}_p, \boldsymbol{\gamma}, \boldsymbol{\Sigma})}{f_{\mathbf{u}}(\mathbf{u}_p; \mathbf{W}_p, \boldsymbol{\gamma}, \boldsymbol{\Sigma}, \boldsymbol{\xi}, \boldsymbol{\tau})}, \tag{32.24}$$

the following system of likelihood equations is derived:

$$\mathbf{G}^T \sum_{p=1}^{P} \left[\mathbf{u}_p - \int_{\boldsymbol{\theta}_p} E(\mathbf{z}, \boldsymbol{\theta}_p, \boldsymbol{\xi}, \boldsymbol{\tau})\, h_{\theta}(\boldsymbol{\theta}_p; \mathbf{W}_p, \boldsymbol{\gamma}, \boldsymbol{\Sigma}, \boldsymbol{\xi}, \boldsymbol{\tau} \mid \mathbf{u}_p)\, d\boldsymbol{\theta}_p \right] = \mathbf{0}, \tag{32.25}$$

$$\sum_{p=1}^{P} \int_{\boldsymbol{\theta}_p} \left[\sum_{d=1}^{D} \boldsymbol{\theta}_{pd} \mathbf{C}_d^T \mathbf{u}_p - E\left(\sum_{d=1}^{D} \boldsymbol{\theta}_{pd} \mathbf{C}_d^T \mathbf{z}, \boldsymbol{\theta}_p, \boldsymbol{\xi}, \boldsymbol{\tau} \right) \right] h_{\theta}(\boldsymbol{\theta}_p; \mathbf{W}_p, \boldsymbol{\gamma}, \boldsymbol{\Sigma}, \boldsymbol{\xi}, \boldsymbol{\tau} \mid \mathbf{u}_p)\, d\boldsymbol{\theta}_p = \mathbf{0}, \tag{32.26}$$

$$\hat{\gamma} = \left(\sum_{p=1}^{P} \bar{\boldsymbol{\theta}}_p \mathbf{W}_p^T \right) \left(\sum_{p=1}^{P} \mathbf{W}_p \mathbf{W}_p^T \right)^{-1} \tag{32.27}$$

and

$$\hat{\boldsymbol{\Sigma}} = \frac{1}{P}\sum_{p=1}^{P}\int_{\boldsymbol{\theta}_p}(\boldsymbol{\theta}_p-\boldsymbol{\gamma}\mathbf{W}_p)(\boldsymbol{\theta}_p-\boldsymbol{\gamma}\mathbf{W}_p)^T h_{\boldsymbol{\theta}}(\boldsymbol{\theta}_p;\mathbf{W}_p,\boldsymbol{\gamma},\boldsymbol{\Sigma},\boldsymbol{\xi},\boldsymbol{\tau}\,|\,\mathbf{u}_p)\,d\boldsymbol{\theta}_p, \tag{32.28}$$

where

$$\mathrm{E}(y,\boldsymbol{\theta}_p,\boldsymbol{\xi},\boldsymbol{\tau}) = \Psi(\boldsymbol{\theta},\boldsymbol{\xi},\boldsymbol{\tau})\sum_{\mathbf{z}\in\Omega} y\,\exp\left[\mathbf{z}^T\left(\sum_{d=1}^{D}\theta_d\mathbf{C}_d\boldsymbol{\tau} + \mathbf{G}\boldsymbol{\xi}\right)\right] \tag{32.29}$$

and

$$\bar{\boldsymbol{\theta}}_p = \int_{\boldsymbol{\theta}_p}\boldsymbol{\theta}_p h_{\boldsymbol{\theta}}(\boldsymbol{\theta}_p;\mathbf{W}_p,\boldsymbol{\gamma},\boldsymbol{\Sigma},\boldsymbol{\xi},\boldsymbol{\tau}\,|\,\mathbf{u}_p)\,d\boldsymbol{\theta}_p \tag{32.30}$$

(Adams et al., 1997). The system of equations defined by Equations 32.25 through 32.28 can be solved using an EM algorithm (Dempster et al., 1977) following the approach of Bock and Aitkin (1981).

32.3.2 Quadrature and Monte Carlo Approximations

The integrals in Equations 32.25 and 32.28 can be approximated numerically using quadrature or Monte Carlo methods. Either case proceeds by defining $\boldsymbol{\Theta}_q$, $q = 1, \ldots, Q$, a set of Q D-dimensional vectors (referred to as grid points) and, for each point, a weight $W_q(\gamma,\Sigma)$. The vector of response probability in Equation 32.21 can then be approximated using

$$f_{\mathbf{u}}(\mathbf{u};\xi,\alpha) \approx \sum_{p=1}^{Q} f_{\mathbf{u}}(\mathbf{u}_p;\boldsymbol{\xi},\boldsymbol{\tau}\,|\,\boldsymbol{\Theta}_p)W_p(\boldsymbol{\gamma},\boldsymbol{\Sigma}) \tag{32.31}$$

and the marginal density in Equation 32.24 is approximated as

$$h_{\boldsymbol{\Theta}}(\boldsymbol{\Theta}_q;\mathbf{W}_p,\boldsymbol{\xi},\boldsymbol{\tau},\boldsymbol{\gamma},\boldsymbol{\Sigma}\,|\,\mathbf{u}_p) = \frac{f_{\mathbf{u}}(\mathbf{u}_p;\boldsymbol{\xi},\boldsymbol{\tau}\,|\,\boldsymbol{\Theta}_q)\,W_q(\boldsymbol{\gamma},\boldsymbol{\Sigma})}{\sum_{p=1}^{Q} f_{\mathbf{u}}(\mathbf{u}_p;\boldsymbol{\xi},\boldsymbol{\tau}\,|\,\boldsymbol{\Theta}_p)W_p(\boldsymbol{\gamma},\boldsymbol{\Sigma})} \tag{32.32}$$

for $q = 1, \ldots, Q$.

The EM algorithm then proceeds as follows:

Step 1. Prepare a set of grid points and weights depending upon $\gamma^{(t)}$ and $\Sigma^{(t)}$, which are the estimates of $\boldsymbol{\gamma}$ and $\boldsymbol{\Sigma}$ at iteration t.

Step 2. Calculate the discrete approximation of the marginal density of $\boldsymbol{\theta}_n$, given $\mathbf{x}_n$ at iteration t, using

$$h_{\boldsymbol{\Theta}}(\boldsymbol{\Theta}_q;\mathbf{W}_p,\boldsymbol{\xi}^{(t)},\boldsymbol{\tau}^{(t)},\boldsymbol{\gamma}^{(t)},\boldsymbol{\Sigma}^{(t)}|\,\mathbf{u}_p) = \frac{f_{\mathbf{u}}(\mathbf{u}_p;\boldsymbol{\xi}^{(t)},\boldsymbol{\tau}^{(t)}|\,\boldsymbol{\Theta}_q)\,W_q(\boldsymbol{\gamma}^{(t)},\boldsymbol{\Sigma}^{(t)})}{\sum_{p=1}^{Q} f_{\mathbf{u}}(\mathbf{u}_p;\boldsymbol{\xi}^{(t)},\boldsymbol{\tau}^{(t)}|\,\boldsymbol{\Theta}_p)W_p(\boldsymbol{\gamma}^{(t)},\boldsymbol{\Sigma}^{(t)})}, \tag{32.33}$$

where $\boldsymbol{\xi}^{(t)}$, $\boldsymbol{\tau}^{(t)}$, $\boldsymbol{\gamma}^{(t)}$, and $\boldsymbol{\Sigma}^{(t)}$ are estimates of $\boldsymbol{\xi}$, $\boldsymbol{\gamma}$, and $\boldsymbol{\Sigma}$ at iteration t.

Step 3. Compute

$$\bar{\boldsymbol{\Theta}}_p^{(t)} = \sum_{r=1}^{Q} \boldsymbol{\Theta}_r \, h_{\boldsymbol{\Theta}}\left(\boldsymbol{\Theta}_r; \mathbf{W}_p, \boldsymbol{\xi}^{(t)}, \boldsymbol{\gamma}^{(t)}, \boldsymbol{\Sigma}^{(t)} \mid \mathbf{u}_p\right) \tag{32.34}$$

and

$$\mathbf{B}^{(t)} = \left[\sum_{p=1}^{P} \sum_{d=1}^{D} \bar{\Theta}_p^{(t)} \mathbf{C}_d\right]^T. \tag{32.35}$$

Step 4. Use the Newton–Raphson method to solve the following for ξ to produce estimates of $\hat{\xi}^{(t+1)}$

$$\mathbf{G}^T \sum_{p=1}^{P} \left[\mathbf{u}_p - \sum_{r=1}^{Q} E(\mathbf{z}, \boldsymbol{\Theta}_r, \boldsymbol{\xi}, \boldsymbol{\tau}^{(t)}) h_{\boldsymbol{\Theta}}(\boldsymbol{\Theta}_r; \mathbf{W}_p, \boldsymbol{\xi}^{(t)}, \boldsymbol{\tau}^{(t)}, \boldsymbol{\gamma}^{(t)}, \boldsymbol{\Sigma}^{(t)} \mid \mathbf{u}_p)\right] = \mathbf{0}. \tag{32.36}$$

Step 5. Use the Newton–Raphson method to solve the following for $\boldsymbol{\tau}$ to produce estimates of $\hat{\boldsymbol{\tau}}^{(t+1)}$

$$\mathbf{B}^{(t)} \sum_{p=1}^{P} \left[\mathbf{u}_p - \sum_{r=1}^{Q} E(\mathbf{z}, \boldsymbol{\Theta}_r, \boldsymbol{\xi}^{(t+1)}, \boldsymbol{\tau}) h_{\boldsymbol{\Theta}}(\boldsymbol{\Theta}_r; \mathbf{W}_p, \boldsymbol{\xi}^{(t)}, \boldsymbol{\tau}^{(t)}, \boldsymbol{\gamma}^{(t)}, \boldsymbol{\Sigma}^{(t)} \mid \mathbf{u}_p)\right] = \mathbf{0}. \tag{32.37}$$

Step 6. Estimate $\boldsymbol{\gamma}^{(t+1)}$ and $\boldsymbol{\Sigma}^{(t+1)}$ using

$$\hat{\boldsymbol{\gamma}}^{(t+1)} = \left(\sum_{p=1}^{P} \bar{\boldsymbol{\Theta}}_p^{(t)} \mathbf{W}_p^T\right)\left(\sum_{n=1}^{N} \mathbf{W}_p \mathbf{W}_p^T\right)^{-1} \tag{32.38}$$

and

$$\hat{\boldsymbol{\Sigma}}^{(t+1)} = \frac{1}{P} \sum_{p=1}^{P} \sum_{r=1}^{Q} \left(\boldsymbol{\Theta}_r - \boldsymbol{\gamma}^{(t+1)} \mathbf{W}_p\right)\left(\boldsymbol{\Theta}_r - \boldsymbol{\gamma}^{(t+1)} \mathbf{W}_p\right)^T h_{\boldsymbol{\Theta}}\left(\boldsymbol{\Theta}_r; \mathbf{Y}_p, \boldsymbol{\xi}^{(t)}, \boldsymbol{\tau}^{(t)}, \boldsymbol{\gamma}^{(t)}, \boldsymbol{\Sigma}^{(t)} \mid \mathbf{u}_p\right). \tag{32.39}$$

Step 7. Return to step 1.

The difference between the quadrature and Monte Carlo methods lies in the way the grid points and weights are prepared. For the quadrature case, begin by choosing a fixed set of Q points, $\left(\Theta_{d1}, \Theta_{d2}, \ldots, \Theta_{dQ}\right)$, for each latent dimension d and then define the set of Q^D grid points indexed as $r = 1, \ldots, Q^D$ with Cartesian coordinates

$$\boldsymbol{\Theta}_r = \left(\Theta_{1j_1}, \Theta_{2j_2}, \ldots, \Theta_{Dj_D}\right) \text{with } j_1 = 1, \ldots, Q;\ j_2 = 1, \ldots, Q; \ldots;\ j_D = 1, \ldots, Q.$$

The weights are then chosen to approximate the continuous multivariate latent population density in Equation 32.19 as

$$W_r = K(2\boldsymbol{\pi})^{-d/2} |\boldsymbol{\Sigma}|^{-1/2} \exp\left[-\frac{1}{2}(\boldsymbol{\Theta}_r - \gamma\mathbf{W}_p)^T \boldsymbol{\Sigma}^{-1}(\boldsymbol{\Theta}_r - \boldsymbol{\gamma}\mathbf{W}_p)\right], \tag{32.40}$$

where K is a scaling factor to ensure that the sum of the weights is equal to one.

In the Monte Carlo case, the grid points are drawn from a multivariate normal distribution with mean $\boldsymbol{\gamma}\mathbf{W}_n$ and variance Σ, which leads to weights $1/Q$ for all grid points.

For further information on the quadrature approach, see Adams et al. (1997), whereas Volodin and Adams (1995) should be consulted for details of the Monte Carlo method.

If a discrete distribution is used, there are no W variables or parameters $\boldsymbol{\gamma}$ and $\boldsymbol{\Sigma}$, and step 6 is replaced with

$$\hat{w}_q^{(t+1)} = \sum_{p=1}^{P} \boldsymbol{\Theta}_q h_{\boldsymbol{\Theta}}(\boldsymbol{\Theta}_q; \boldsymbol{\xi}^{(t)} | \mathbf{u}_p). \tag{32.41}$$

32.3.3 Conditional Maximum-Likelihood Estimation

In the case where there no scoring parameters conditional maximum-likelihood (CML) estimation is possible, the first step in the derivation of the CML estimators is to compute the probability of a response pattern conditional on that pattern yielding a specific score. More formally, let $\mathbf{X}$ be a vector-valued random variable that is the vector of scores of a response pattern, then a realization of this variable is $\mathbf{x} = \mathbf{u}^T\mathbf{B}$, where $\mathbf{u}^T$ and $\mathbf{B}$ are as used in Equation 32.5, and the probability of a response pattern conditional on $\mathbf{X}$ taking the value $\mathbf{x}$ is

$$\begin{aligned}
f(\mathbf{u}; \boldsymbol{\xi}, \boldsymbol{\gamma}, \boldsymbol{\Sigma} \mid \mathrm{X} = \mathbf{x}) &= \frac{f(\mathbf{u}; \boldsymbol{\xi}, \boldsymbol{\gamma}, \boldsymbol{\Sigma}, \mathrm{X} = \mathbf{x})}{\sum_{\mathbf{z} \in \Omega_x} f(\mathbf{z}; \boldsymbol{\xi}, \boldsymbol{\gamma}, \boldsymbol{\Sigma}, \mathrm{X} = \mathbf{x})} \\
&= \frac{\int f_{\mathrm{x}}(\mathbf{u}; \boldsymbol{\xi}, \mathrm{X} = \mathbf{x} \mid \boldsymbol{\theta}) f_{\boldsymbol{\theta}}(\boldsymbol{\theta}; \boldsymbol{\gamma}, \boldsymbol{\Sigma}) d\boldsymbol{\theta}}{\sum_{\mathbf{z} \in \Omega_x} \int f_{\mathrm{x}}(\mathbf{z}; \boldsymbol{\xi}, \mathrm{X} = \mathbf{x} \mid \boldsymbol{\theta}) f_{\boldsymbol{\theta}}(\boldsymbol{\theta}; \boldsymbol{\gamma}, \boldsymbol{\Sigma}) d\boldsymbol{\theta}} \\
&= \frac{\int \boldsymbol{\Psi}(\boldsymbol{\theta}, \boldsymbol{\xi}) \exp(\mathbf{x}\boldsymbol{\theta} + \mathbf{u}^T \mathrm{G}\boldsymbol{\xi}) f_{\boldsymbol{\theta}}(\boldsymbol{\theta}; \boldsymbol{\gamma}, \boldsymbol{\Sigma}) d\boldsymbol{\theta}}{\sum_{\mathbf{z} \in \Omega_x} \int \boldsymbol{\Psi}(\boldsymbol{\theta}, \boldsymbol{\xi}) \exp(\mathbf{x}\boldsymbol{\theta} + \mathbf{z}^T \mathrm{G}\boldsymbol{\xi}) f_{\boldsymbol{\theta}}(\boldsymbol{\theta}; \boldsymbol{\gamma}, \boldsymbol{\Sigma}) d\boldsymbol{\theta}} \\
&= \frac{\exp(\mathbf{u}^T \mathbf{G}\boldsymbol{\xi}) \int \boldsymbol{\Psi}(\boldsymbol{\theta}, \boldsymbol{\xi}) \exp(\mathbf{x}\boldsymbol{\theta}) f_{\boldsymbol{\theta}}(\boldsymbol{\theta}; \boldsymbol{\gamma}, \boldsymbol{\Sigma}) d\boldsymbol{\theta}}{\sum_{\mathbf{z} \in \Omega_x} \exp(\mathbf{z}^T \mathbf{G}\boldsymbol{\xi}) \int \Psi(\boldsymbol{\theta}, \boldsymbol{\xi}) \exp(\mathbf{x}\boldsymbol{\theta}) f_{\boldsymbol{\theta}}(\boldsymbol{\theta}; \boldsymbol{\gamma}, \boldsymbol{\Sigma}) d\boldsymbol{\theta}} \\
&= \frac{\exp(\mathbf{u}^T \mathbf{G}\boldsymbol{\xi})}{\sum_{\mathbf{z} \in \Omega_x} \exp(\mathbf{z}^T \mathbf{G}\boldsymbol{\xi})},
\end{aligned} \tag{32.42}$$

where Ω_x is the set of all possible response patterns when the vector of scores is $\mathbf{x}$.

The Equation 32.42 shows that the probability of a response pattern conditional on **X** taking the value **x** neither depends on the ability $\boldsymbol{\theta}$, nor its distribution. The consequential advantage of the CML approach is that it provides the same estimates for the item parameters regardless of the choice of the population distribution. As such, the CML item parameter estimator is not influenced by any assumption about the population distribution. The disadvantage is that the population parameters are not estimated. If, as is often the case, the population parameters are of interest, they must be estimated in a second step.

The second step involves solving the system of equations in (32.38) and (32.39) while assuming the item parameters are known. Apart from underestimating the uncertainty in the population parameter estimates, the consequences of using the CML item parameter estimates, in this second step, as if they were true values, are not clear. In contrast, the maximum-likelihood approach provides direct estimates of both item parameters and population parameters. However, it suffers from the risk that if the population distributional assumption is incorrect, the item parameters may be biased.

32.3.4 Latent Ability Estimation and Prediction

The marginal response model in Equation 32.21 does not include parameters for the latent abilities $\boldsymbol{\theta}_n$; and hence, the estimation algorithm does not result in any estimates of them. Although this may not be of concern when modeling is undertaken for the purposes of estimating population parameters $\boldsymbol{\gamma}$ and $\boldsymbol{\Sigma}$, it does cause inconveniences when there is an interest in estimates of the latent abilities of examinees.

There are a number of standard approaches that can be applied to provide estimates, or perhaps more accurately predictions, of the latent values. Perhaps the most common approach is to go Bayesian, using the population distributions as a common prior distribution of θ_n for all examinees, and calculate the expectation of its posterior distribution (Bock and Aitkin, 1981). This expected a posteriori estimate (EAP) of θ_n is

$$\boldsymbol{\theta}_p^{EAP} = \sum_{r=1}^{Q} \boldsymbol{\Theta}_r \, h_{\boldsymbol{\Theta}}(\boldsymbol{\Theta}_r; W_p, \hat{\boldsymbol{\xi}}, \hat{\boldsymbol{\tau}}, \hat{\boldsymbol{\gamma}}, \hat{\boldsymbol{\Sigma}} \mid \mathbf{u}_p) \quad (32.43)$$

whereas the variance of the posterior distribution of θ_n can be estimated using

$$\mathrm{var}\left(\boldsymbol{\theta}_p^{EAP}\right) = \sum_{r=1}^{Q} \left(\boldsymbol{\Theta}_r - \boldsymbol{\theta}_p^{EAP}\right)\left(\boldsymbol{\Theta}_r - \boldsymbol{\theta}_p^{EAP}\right)^T h_{\boldsymbol{\Theta}}\left(\boldsymbol{\Theta}_r; W_p, \hat{\boldsymbol{\xi}}, \hat{\boldsymbol{\tau}}, \hat{\boldsymbol{\gamma}}, \hat{\boldsymbol{\Sigma}} \mid \mathbf{u}_p\right). \quad (32.44)$$

An alternative to the EAP is the maximum a-posteriori (MAP) (Bock and Aitkin, 1981), which requires finding the modes rather than the expectations of the posterior distributions.

A maximum-likelihood approach to the estimation of the ability estimates can also be used, for example, the weighted likelihood approach of Warm (1985, 1989).

32.3.5 Estimating Functionals of the Population Distributions

The model presented in Equation 32.21 provides estimates of the $\boldsymbol{\gamma}$ and $\boldsymbol{\Sigma}$ parameters of the population, but of course there are many other characteristics of the population that may be of interest. It may be tempting to estimate these functionals from point estimates of the $\boldsymbol{\theta}_p$ parameters. It is well-known; however, that the use of point estimates, such as the

EAP, MLE, and WLE, leads to biased estimates due to negligence of the estimation error in them.

As an alternative to using point estimates, the statistical literature recommends a Bayesian approach based on imputed values for the θ_n parameters, that is, multiple random draws from their posterior distributions. In the educational assessment literature, the random draws have become widely known as plausible values (Mislevy, 1991; Mislevy et al., 1992). Observe that, upon assumption of the population distribution as the common prior distribution for the $\boldsymbol{\theta}$ parameters of all examinees, Equation 32.24 becomes the posterior distribution of $\boldsymbol{\theta}_p$.

The following describes an inverse transformation method for drawing plausible values from the posterior distribution $\boldsymbol{\theta}_n$. Recall from Equation 32.24 the marginal posterior is

$$h_{\boldsymbol{\theta}}(\boldsymbol{\theta}_p;\mathbf{W}_p,\boldsymbol{\xi},\boldsymbol{\gamma},\boldsymbol{\Sigma}\mid\mathbf{u}_p)=\frac{f_{\mathbf{u}}(\mathbf{u}_p;\boldsymbol{\xi}\mid\boldsymbol{\theta}_p)f_{\boldsymbol{\theta}}(\boldsymbol{\theta}_p;\mathbf{W}_p,\boldsymbol{\gamma},\boldsymbol{\Sigma})}{\int_{\theta}f_{\mathbf{u}}(\mathbf{u}_p;\boldsymbol{\xi}\mid\boldsymbol{\theta})f_{\boldsymbol{\theta}}(\boldsymbol{\theta},\boldsymbol{\gamma},\boldsymbol{\Sigma})d\boldsymbol{\theta}}. \tag{32.45}$$

Unlike previously described methods for drawing plausible values (Beaton, 1987; Mislevy et al., 1992), there is no need whatsoever to approximate Equation 32.2 by a normal distribution. Rather, we draw directly using the following steps: first draw M vector-valued random deviates, $\{\boldsymbol{\varphi}_{mp}\}_{m=1}^{M}$, from the estimated multivariate normal distribution, $f_{\boldsymbol{\theta}}(\boldsymbol{\theta}_p;\mathbf{W}_p,\boldsymbol{\gamma},\boldsymbol{\Sigma})$ for each p. These vectors are used to compute an approximation to the integral in the denominator of Equation 32.45, using the Monte Carlo integration

$$\int_{\theta}f_{\mathbf{u}}(\mathbf{u};\boldsymbol{\xi}\mid\boldsymbol{\theta})f_{\boldsymbol{\theta}}(\boldsymbol{\theta},\boldsymbol{\gamma},\boldsymbol{\Sigma})d\boldsymbol{\theta}\approx\frac{1}{M}\sum_{m=1}^{M}f_{\mathbf{u}}(\mathbf{u};\boldsymbol{\xi}\mid\boldsymbol{\varphi}_{mp})\equiv\Im. \tag{32.46}$$

At the same time, the values

$$s_{mp}=f_{\mathbf{u}}(\mathbf{u}_p;\boldsymbol{\xi}\mid\boldsymbol{\varphi}_{mp})f_{\boldsymbol{\theta}}(\boldsymbol{\varphi}_{mp};\mathbf{W}_p,\boldsymbol{\gamma},\boldsymbol{\Sigma}) \tag{32.47}$$

are calculated, and the set of pairs $(\boldsymbol{\varphi}_{mp},s_{mp}/\Im)_{m=1}^{M}$ is obtained. This set of pairs can be used as an approximation of the posterior density (32.45); and the probability that φ_{jp} could be drawn from this density is given by

$$q_{pj}=\frac{s_{jp}}{\sum_{m=1}^{M}s_{mp}}. \tag{32.48}$$

At this point, L random deviates, $\{\eta_i\}_{i=1}^{L}$, are independently generated from the standard uniform distribution on the interval from 0 to 1; and for each deviate, that is, for each $i=1,\ldots,L$, the vector, $\boldsymbol{\varphi}_{pi_0}$, that satisfies the condition

$$\sum_{s=1}^{i_0-1}q_{sp}<\eta_i\le\sum_{s=1}^{i_0}q_{sp} \tag{32.49}$$

is selected as a plausible vector.

32.4 Model Fit

32.4.1 Generalized Fit Test

A convenient way to assess the fit of the model is to follow the residual-based approach of Wright and Stone (1979) and Wright and Masters (1982). Wu (1997) extended this approach for application with the marginal model used here, and more recently Adams and Wu (2004) generalized it further to tests the range of specific hypotheses.

If $\mathbf{G}_r$ is used to indicate the r-th column of design matrix $\mathbf{G}$, the Wu fit statistic is based upon the standardized residuals

$$z_{pr}(\theta_p) = \left(\mathbf{G}_r^T \mathbf{u}_p - E_{pr}\right) \Big/ \sqrt{V_{pr}}, \tag{32.50}$$

where $\mathbf{G}_r^T\mathbf{u}_p$ is the contribution of examinee p to the sufficient statistic for parameter r, and E_{pr} and V_{pr} are, respectively, the conditional expectation and variance of $\mathbf{G}_r^T\mathbf{u}_p$.

To construct an unweighted fit statistic, the square of this residual is averaged over the cases and then integrated over the marginal ability distributions to obtain

$$Fit_{out,r} = \int_{\theta_1}\int_{\theta_2}\cdots\int_{\theta_P}\left[\frac{1}{P}\sum_{p=1}^{P} \hat{z}_{pr}^2(\theta_p)\right]\prod_{p=1}^{P} h_\theta\left(\theta_p;\mathbf{W}_p,\,\hat{\boldsymbol{\xi}},\,\hat{\boldsymbol{\tau}},\,\hat{\boldsymbol{\gamma}},\,\hat{\boldsymbol{\Sigma}}\,|\,\mathbf{u}_p\right)d\theta_P\, d\theta_{P-1}\ldots d\theta_1. \tag{32.51}$$

Alternatively, a weighted average of the squared residuals is used as follows:

$$Fit_{in,r} = \int_{\theta_1}\int_{\theta_2}\cdots\int_{\theta_P}\left[\frac{\sum_{p=1}^{P}\hat{z}_{pr}^2(\theta_p)V_{pr}(\theta_p)}{\sum_{p=1}^{P}V_{pr}(\theta_p)}\right]\prod_{p=1}^{P} h_\theta(\theta_p;\mathbf{W}_p,\hat{\boldsymbol{\xi}},\hat{\boldsymbol{\tau}},\hat{\boldsymbol{\gamma}},\hat{\boldsymbol{\Sigma}}\,|\,\mathbf{u}_p)d\theta_P\, d\theta_{P-1}\ldots d\theta_1. \tag{32.52}$$

It is convenient to use the Monte Carlo method to approximate the integrals in Equations 32.51 and 32.52. Wu (1997) has shown that the statistics produced by Equations 32.51 and 32.52 have approximated scaled chi-squared distributions. These statistics are transformed to approximate normal deviates using the Wilson–Hilferty transformations

$$t_{out,r} = \left(Fit_{out,r}^{1/3} - 1 + 2/(9\rho P)\right) \Big/ (2/(9\rho P))^{1/2} \tag{32.53}$$

and

$$t_{in,r} = \left[Fit_{in,r}^{1/3} - 1\right] \times \frac{3}{\sqrt{Var(Fit_{in,r})}} + \frac{\sqrt{Var(Fit_{in,r})}}{3}, \tag{32.54}$$

where ρ is the number of draws used in the Monte Carlo approximation of the integrals in Equations 32.51, 32.52 and

$$Var(Fit_{in,r}) = \left(1 \Big/ \sum_p V_{pr}\right)^2 \left(\sum_p \left(E\left(\left(\mathbf{G}_r^T\mathbf{u}_p - E_{pr}\right)^4\right) - V_{pr}^2\right)\right). \tag{32.55}$$

The derivation and justification for these transformations is given in Wu (1997).

32.4.2 Customized Fit Tests

The fit testing approach described above works at the parameter level; that is, it provides a fit statistic for each of the estimated item parameters. A more general approach was introduced by Adams and Wu (2004), who suggested replacing matrix **A** used in Equation 32.50 by an alternative matrix **F** of the same length, which they called a fit matrix. The use of **F** relaxes the restriction of linear combinations of the item responses being limited to $\mathbf{G}_r^T\mathbf{u}_p$, where $\mathbf{G}_r$ is the design vector for the parameter ξ_r. One can compute the expectation and variance of $\mathbf{F}^T\mathbf{u}_p$, and construct a fit statistic in exactly the same way as for $\mathbf{G}_r^T\mathbf{u}_p$.

The following is an example of a user-defined fit test of local independence for a simple dichotomous Rasch model. Consider a test consisting of 10 dichotomous items; the design matrix, **G**, for the simple logistic model for such a test would be a 10 by 10 identity matrix.

Using the notation defined earlier, the first column of **G** is $\mathbf{G}_1^T = (1\ 0\ 0\ 0\ 0\ 0\ 0\ 0\ 0\ 0)$. The product $\mathbf{G}_1^T\mathbf{u}_p$ gives the item response of examinee n on item one. This is the contribution of examinee p to the sufficient statistic for the first item parameter. Similarly, $\mathbf{A}_2^T = (0\ 1\ 0\ 0\ 0\ 0\ 0\ 0\ 0\ 0)$, and $\mathbf{A}_2^T\mathbf{u}_p$ is the contribution of examinee p to the sufficient statistic for the second item parameter, and so on.

For a user-defined fit test, consider the matrix

$$\mathbf{F} = \begin{bmatrix} 1 & 0 & 1 \\ 0 & 1 & 1 \\ 0 & 0 & 1 \\ 0 & 0 & 1 \\ 0 & 0 & 1 \\ 1 & 0 & 0 \\ 0 & 1 & 0 \\ 0 & 0 & 0 \\ 0 & 0 & 0 \\ 0 & 0 & 0 \end{bmatrix}. \tag{32.56}$$

A fit test based upon the first column of this matrix tests whether items one and six are both answered correctly as often as would be expected under the model. Similarly, the second column provides a test of whether item two and seven are both answered correctly as often as would be expected under the model. As such, these are tests of the local independence between these pairs of items. The third column compares the score on the first five items with its expectation; that is, whether the subtest consisting of the first five items fits with the rest of the items as predicted by the model.

32.4.3 Tests of Relative Fit

Where models are hierarchically ordered, they may be compared in terms of statistical significance using the likelihood-ratio statistic

$$G^2 = -2\sum f_{xt} \ln\left(\frac{m_{xt}}{f_{xt}}\right), \tag{32.57}$$

where f_{xt} is the observed frequency of examinees with response pattern x and score t, m_{xt} is the modeled frequency, and the summation is taken overall nonzero cells. When models are not hierarchically ordered, we could use Akaike's information criterion (AIC) (Akaike, 1973; Cohen and Cho, Volume Two, Chapter 18),

$$AIC = G^2 + 2p + T, \tag{32.58}$$

where p is the number of estimated parameters, G^2 is as above, and the value T is an arbitrary constant. For example, one might choose T such that the AIC for the saturated model is 0.00.

32.5 Empirical Example

This section discusses the application of a series of particular cases of the EMCML model to situations where test items are grouped into subsets and the common attributes of items within these subsets brings into question the usual assumption of conditional independence. Grouped subsets of items are called *item bundles* (Rosenbaum, 1988) or testlets (Wainer and Kiely, 1987). The use of the models is illustrated using item bundles constructed in the framework of the SOLO taxonomy of Biggs and Collis (1982).

32.5.1 Bundle Independence

Following Rosenbaum (1988), define bundle independence as

$$\Pr(u;\delta \mid \theta) = \prod_{c=1}^{C} \Pr(u_c;\xi \mid \theta), \tag{32.59}$$

where the probability of a bundle response u_c is conditional on a vector of parameters ξ associated with bundle c, and (possibly) with other bundles.

We begin by considering a bundle c that consists of I_c dichotomous items. Such a bundle will admit 2^{I_c} bundle response patterns. If an ordering for these patterns is determined, then they can be referenced using the integers $k = 0, \ldots, K_c$, where $K_c = 2^{I_c} - 1$. In our ordering $k = 0$ always corresponds to the pattern $(0, \ldots, 0)^T$, which will sometimes be referred to as the reference pattern.

Linear combinations of parameters in the vector ξ are used in modeling the bundle response patterns. These linear combinations are defined by the design vectors $\mathbf{g}_{ik}$, as above. These design vectors can be gathered into bundle design matrices $\mathbf{D}_c$, which, in turn, can be gathered into the test design matrix $\mathbf{G}$ in (32.2).

32.5.2 Saturated Bundle Model

We are now in a position to consider some specific models for item bundles. One model that will be a possibility in any bundle context is the model that includes a different parameter for each bundle response vector for each item (except for the reference response vector). This is called the saturated bundle model.

Consider, for example, a set of seven item bundles, with three dichotomous items in each bundle, where the items within each bundle are linked by common stimulus material. In the case of a strict bundle independence model, that is, a model without parametric dependence between bundles, the maximum number of parameters that can be estimated for each bundle is $2^3 - 1 = 7$; that is, one for each possible response pattern to a three-item bundle composed of dichotomous items, except for the one reference vector, (0, 0, 0). The bundle response vectors are then 3-vectors, where the elements are zero (for an incorrect response to an item) and 1 (for a correct response to an item). Table 32.1 shows all such vectors (in the second column) for a generic bundle c. The order shown in Table 32.1 is used throughout this paper. The third column gives the score of each category, $r(\mathbf{u}_c)$. The remaining columns of the table shows the components of the bundle design vectors that contain any nonzero elements. The test design matrix **G**, then takes the following form:

$$\mathbf{G} = \begin{bmatrix} \mathbf{D}_1 & \mathbf{0} & \mathbf{0} & \mathbf{0} & \mathbf{0} & \mathbf{0} & \mathbf{0} \\ \mathbf{0} & \mathbf{D}_2 & \mathbf{0} & \mathbf{0} & \mathbf{0} & \mathbf{0} & \mathbf{0} \\ \mathbf{0} & \mathbf{0} & \mathbf{D}_3 & \mathbf{0} & \mathbf{0} & \mathbf{0} & \mathbf{0} \\ \mathbf{0} & \mathbf{0} & \mathbf{0} & \mathbf{D}_4 & \mathbf{0} & \mathbf{0} & \mathbf{0} \\ \mathbf{0} & \mathbf{0} & \mathbf{0} & \mathbf{0} & \mathbf{D}_5 & \mathbf{0} & \mathbf{0} \\ \mathbf{0} & \mathbf{0} & \mathbf{0} & \mathbf{0} & \mathbf{0} & \mathbf{D}_6 & \mathbf{0} \\ \mathbf{0} & \mathbf{0} & \mathbf{0} & \mathbf{0} & \mathbf{0} & \mathbf{0} & \mathbf{D}_7 \end{bmatrix}, \qquad (32.60)$$

where **0** is an 8×7 matrix of zeroes. In the convention adopted here, we have chosen to fix the location of the scale by setting the mean of the examinee parameters to zero. This is a particularly straightforward model, including just one bundle parameter for each possible bundle response vector. It is a saturated model and as such provides a *best fit* in the strict bundle independence class, against which to compare other models with fewer parameters. For the saturated strict bundle model, the response probability model in Equation 32.2 takes on a particularly simple form if we index the bundle response vectors as in Table 32.1. Then,

$$\Pr(k, \xi \mid \theta) = \frac{\exp(r(k)\theta + \delta_{ck})}{\sum_{j=1}^{r_c} \exp(r(j)\theta + \delta_{cj})}. \qquad (32.61)$$

TABLE 32.1

Scores and Design Vectors for a Bundle of Three Dichotomous Items for the Saturated Bundle Model

Category Number (k)	$\mathbf{U}_c$	$r(\mathbf{U}_c)$	$\mathbf{D}_c$						
			p = 1	p = 2	p = 3	p = 4	p = 5	p = 6	p = 7
0	(0,0,0)	0	0	0	0	0	0	0	0
1	(1,0,0)	1	1	0	0	0	0	0	0
2	(0,1,0)	1	0	1	0	0	0	0	0
3	(0,0,1)	1	0	0	1	0	0	0	0
4	(1,1,0)	2	0	0	0	1	0	0	0
5	(1,0,1)	2	0	0	0	0	1	0	0
6	(0,1,1)	2	0	0	0	0	0	1	0
7	(1,1,1)	3	0	0	0	0	0	0	1

where k is the index, and $r(k)$ is the score (given in Table 32.1) of response k to bundle c and δ_{ck} is the parameter associated with response k in cluster c. We note, in passing, that this is a special case of the ordered partition model (Wilson, 1992; Wilson and Adams, 1993).

32.5.3 PCM Bundles

One approach to the issue of item bundles has been used classically: utilize the sum score from the items in each bundle as though the bundle was simply a partial credit item. We call this the *PCM Bundle* approach. This approach was used implicitly in the analyses of the original developers of the items used in our example (Romberg et al., 1982a,b). This approach is based on the assumption that what is most important about the bundle is its total score, ignoring the specific items that make up the score. Among the Rasch family of models, this approach is epitomized by Masters' partial credit model (PCM; Volume One, Chapter 7), which can be written using Equations 32.10 through 32.12 previously shown. However, there is an alternative way to express the parameters of the PCM using a bundle model. In Table 32.1 substitute

$$\mathbf{D}_c = \begin{bmatrix} 0 & 0 & 0 \\ 1 & 0 & 0 \\ 1 & 0 & 0 \\ 1 & 0 & 0 \\ 1 & 1 & 0 \\ 1 & 1 & 0 \\ 1 & 1 & 0 \\ 1 & 1 & 1 \end{bmatrix}. \tag{32.62}$$

Note that this expression constitutes the PCM using Andersen's parameterization (Andersen, 1973, 1983)—see Wilson and Adams (1993) for a detailed explanation of this point. Although the two parameterizations are different, they are observationally equivalent, and hence, the model fit statistics are the same.

32.5.4 SLM Bundles

One criticism of the PCM bundles approach is that it ignores the relative difficulties of the items within each bundle. The second approach to bundle situations has been described (Wilson and Adams, 1993). It preserves the role of the item difficulty parameters, and also observes the bundle dependencies. Consider the test design matrix **G** built as in Equation 32.62 from the following bundle design matrices:

$$\mathbf{D}_c = \begin{bmatrix} 0 & 0 & 0 \\ 1 & 0 & 0 \\ 0 & 1 & 0 \\ 0 & 0 & 1 \\ 1 & 1 & 0 \\ 1 & 0 & 1 \\ 0 & 1 & 1 \\ 1 & 1 & 1 \end{bmatrix}. \tag{32.63}$$

This model, which we call *SLM Bundles*, has an interesting relationship with the SLM. For a response vector $\mathbf{u}_c$ we can write

$$\Pr(u_c, \boldsymbol{\xi}_c \mid \theta) = \prod_{j=1}^{3} \frac{\exp(u_{cj}\theta + \xi_{cj})}{1 + \exp(\theta + \xi_{cj})}, \tag{32.64}$$

which indicates that the probability of a bundle response vector is modeled in exactly the same way as it is for the SLM.

32.5.5 Customized Approaches

The major point we wish to make is that a *standard* approach, such as in the PCM bundles approach should not be seen as limiting one's thinking about and estimation of models appropriate to bundle situations. Using a generalized model allows or even encourages sensitivity to the real-world context and the construction of models built specifically for particular applications. We survey some possible custom models that arise in our particular example. Many more models could as easily be constructed; the ones we display are intended to be expository rather than exhaustive.

The score-based PCM bundles model as in Equation 32.62 represents each bundle using three parameters. The saturated model uses seven in this case. It is quite straightforward to construct models that are positioned between these two as for their numbers of parameters. Suppose, for example, that we had substantive reasons to believe that the items in the bundle formed an approximation to a Guttman scale. That is, in the example stated previously, let us assume that we had reason to believe that the Guttman response patterns (1, 0, 0) and (1, 1, 0) (assuming the items go from easiest to the most difficult) predominated within their respective score groups. We do not need to model this for (0, 0, 0) and (1, 1, 1), because they are the only members of their score groups. Instead, we might consider augmenting the bundle design matrix as follows:

$$\mathbf{D}_c = \begin{bmatrix} 0 & 0 & 0 & 0 \\ 1 & 0 & 0 & 1 \\ 1 & 0 & 0 & 0 \\ 1 & 0 & 0 & 0 \\ 1 & 1 & 0 & 1 \\ 1 & 1 & 0 & 0 \\ 1 & 1 & 0 & 0 \\ 1 & 1 & 1 & 0 \end{bmatrix}. \tag{32.65}$$

This formulation could be interpreted as modeling an adjustment to the probabilities for the Guttman response patterns, especially, if the estimated extra parameter is negative (and significantly different from zero), which indicates that the Guttman response patterns do indeed have a higher probability than the others. This model assumes that the Guttman adjustment (in the metric of the parameters) is the same for score category 1 and score category 2. This may be too restrictive; perhaps the Guttman nature of the scale is more pronounced within one score category compared to the other. In that case we could consider the model represented by the following bundle design matrix:

$$\mathbf{D}_c = \begin{bmatrix} 0 & 0 & 0 & 0 & 0 \\ 1 & 0 & 0 & 1 & 0 \\ 1 & 0 & 0 & 0 & 0 \\ 1 & 0 & 0 & 0 & 0 \\ 1 & 1 & 0 & 0 & 1 \\ 1 & 1 & 0 & 0 & 0 \\ 1 & 1 & 0 & 0 & 0 \\ 1 & 1 & 1 & 0 & 0 \end{bmatrix}. \tag{32.66}$$

This model expresses the possibility that each of the two Guttman responses (1, 0, 0) and (1, 1, 0) requires a different adjustment. There are many other models between matrix (32.66) and the saturated model. Many have no clear interpretation. Some, such as those above, will be found interpretable in certain circumstances, and may prove useful in particular applications.

32.5.6 Empirical Comparisons

In this section, we consider the relative fit of several of the models described above to the particular data gathered by Romberg and his colleagues. Note that in the following discussion we report results for more models than one would be likely to consider in a real-world data analysis. The purpose of this example is only to illustrate the method and so we have tried several models. Table 32.2 gives the equation number for each model, $2 \times$ log-likelihood, the number of bundle parameters, and the AIC.

Consider first the models based on the score-based PCM bundles approach. A comparison between the models in Equation 32.62 and yielded Equation 32.65, $X^2 = 2032.72$ ($df = 7$), between the two levels of Guttman adjustment, $X^2 = 330.63$ ($df = 7$); and between the model given in Equation 32.66 and the saturated model (32.62), $X^2 = 118.34$ ($df = 14$). All the comparisons are significant at the 0.01 level. Clearly, for this example the model given in Equation 32.62 is not a model of choice. Its performance is improved dramatically by the uniform Guttman adjustment but can still be improved by applying the adjustment differentially in the 1 and 2 score groups. Equally, the saturated model also added statistically significant information to the model given in Equation 32.66.

TABLE 32.2

Scores and Design Vectors for a Bundle of Three Dichotomous Items for the Saturated Bundle Model

Model Equation Number	Bundle Parameters	$-2 \times$ Loglikelihood	AIC
(32.62)	20	6828.08	2425.69
(32.65)	27	4795.36	408.97
(32.66)	34	4464.73	92.34
(32.63)	21	4559.42	159.03
(32.60)	48	4346.39	0.00

Comparing the score-based approach to the item-based SLM bundles approach, the major difference between the two sets of results is that the item-based model in Equation 32.63 is doing a great deal better than the score-based model in Equation 32.62. These two models are not hierarchically ordered, so we cannot compare them using a likelihood-ratio test, but we can use the AIC. Choosing an arbitrary constant so that the AIC for the saturated model is 0.00, the AICs for the models are shown in the last column of Table 32.2. Note the difference between the AIC for the models in Equations 32.62 and 32.63. It reflects the impact of the item difficulty ordering within the bundles brought about by the SOLO item generation procedure. The order of fit remains the same when we make the uniform and independent Guttman adjustments. It takes two extra parameters per item (the two Guttman adjustments), as in the model in Equation 32.66, to make both the score-based and the item-based model fit.

The class of models introduced in this paper was designed to express some possible ways in which conditional dependence can be modeled by subsets of items within a test, which has common attributes not shared with any of the other items. While models such as the SLM bundles model can be seen as having generic importance, we should expect them to need adaptation to differing circumstances. The required adaptation can range from something as straightforward as having different numbers of items per bundle, to more complicated a priori structures, such as the Guttman-like possibilities described above. The EMCML model allows one to specify complications such as these within a common framework, although the specification of the design matrix can become quite tedious in cases with large numbers of bundles and with a nonuniform pattern of bundle parameterization. There are a number of important possibilities that have not been considered above. One is that the items in the bundle might themselves be polytomous. This situation is mathematically no more complex than the examples discussed here, but it does introduce some interpretational complexities. A second, and potentially more important possibility, is that a unidimensional model is insufficient to model the complexity introduced by item bundles. For example, one might associate the overall bundle scores with one dimension, and differential item probabilities with a second. This could also be modeled in the EMCML framework.

32.6 Discussion

This paper demonstrated the flexible power of the use of design matrices to specify a family of item response models. Not only can standard models such as the partial credit, rating scale, and facets models be included under one single framework of models, but many other models can also be specified through user-defined design matrices.

The estimation procedures described in this paper allow for a joint (or one-step) calibration of both item parameters and population parameters, as opposed to a two-step process where individual examinee abilities are first estimated and then aggregated to form population parameter estimates. The advantages of a joint calibration of parameters include more accurate standard errors for the estimates of the population parameters and less bias of some population parameters.

Similarly, user-defined fit design matrices allows for more focused test of goodness-of-fit of the data to the model. In many cases, such focused fit tests are statistically more powerful in detecting misfit in the data.

However, the theoretical elegance of the use of design matrices can be overshadowed by the tediousness of the construction of these matrices in practice. A software package, ConQuest (Adams et al., 2012; Volume Three, Chapter 27), has been developed where users can specify various item response models through a command language. The design matrices are then automatically built by ConQuest, but it also allows users to import a design matrix should the need arise. Thus, the advantages of a unified framework of item response models can be easily exploited in practice for the analysis of a vast range of different data sets.

References

Adams, R. J., and Wilson, M. R. 1996. A random coefficients multinomial logit: A generalized approach to fitting Rasch models. In G. Engelhard and M. Wilson (Eds.), *Objective Measurement III: Theory into Practice* (pp. 143–166). Norwood, New Jersey: Ablex.

Adams, R. J., Wilson, M. R., and Wang, W. C. 1997. The multidimensional random coefficients multinomial logit model. *Applied Psychological Measurement, 21*, 1–23.

Adams, R. J., Wilson, M. R., and Wu, M. L. 1997. Multilevel item response modelling: An approach to errors in variables regression. *Journal of Educational and Behavioral Statistics, 22*, 47–76.

Adams, R. J., and Wu, M. L. 2004, June. The construction and implementation of user-defined fit tests for use with marginal maximum likelihood estimation and generalised item response models. Paper presented at the *International Meeting of the Psychometric Society*, Pacific Grove, California, USA.

Adams, R. J., Wu, M. L., and Wilson, M. R. 2012. ACER ConQuest Version 3: Generalised Item Response Modelling Software [Computer program]. Camberwell: Australian Council for Educational Research.

Andersen, E. B. 1970. Asymptotic properties of conditional maximum likelihood estimators. *Journal of the Royal Statistical Society, Series B, 32*, 283–301.

Andersen, E. B. 1973. Conditional inference for multiple choice questionnaires. *British Journal of Mathematical and Statistical Psychology, 26*, 31–44.

Andersen, E. B. 1983. A general latent structure model for contingency table data. In H. Walner and S. Messick (Eds.), *Principals of Modern Psychological Measurement* (pp. 117–138). Hillsdale, New Jersey: Lawrence Erlbaum.

Andrich, D. 1978. A rating formulation for ordered response categories. *Psychometrika, 43*, 561–573.

Beaton, A. E. 1987. Implementing the new design: The NAEP 1983-84 Technical Report. (Report No.15-TR-20). Princeton, New Jersey: Educational Testing Service.

Biggs, J. B., and Collis, K. F. 1982. *Evaluating the Quality of Learning: The SOLO Taxonomy*. New York: Academic Press.

Bock, R. D., and Aitkin, M. 1981. Marginal maximum likelihood estimation of item parameters: An application of the EM algorithm. *Psychometrika, 46*, 443–459.

Dempster, A. P., Laird, N. M., and Rubin, D. B. 1977. Maximum likelihood estimation with incomplete data via the EM algorithm. *Journal of the Royal Statistical Society, Series B, 39*, 1–38.

Fischer, G. H. 1973. The linear logistic model as an instrument in educational research. *Acta Psychologica, 37*, 359–374.

Linacre, J. M. 1989. *Many-Faceted Rasch Measurement*. Chicago: MESA Press.

Masters, G. N. 1982. A Rasch model for partial credit scoring. *Psychometrika, 47*, 149–174.

Mislevy, R. J. 1991. Randomization-based inference about latent variables from complex samples. *Psychometrika, 56*, 177–96.

Mislevy, R. J., Beaton, A. E., Kaplan, B., and Sheehan, K. M. 1992. Estimating population characteristics from sparse matrix samples of item responses. *Journal of Educational Measurement, 29*, 133–61.

Muraki, E. 1992. A generalized partial credit model: Application of an EM algorithm. *Applied Psychological Measurement, 16,* 159–176.

Rasch, G. 1980. *Probabilistic Models for Some Intelligence and Attainment Tests.* Chicago: University of Chicago Press (original work published 1960).

Reckase, M. 2009. *Multidimensional Item Response Theory.* New York: Springer.

Romberg, T. A., Collis, K. F., Donovan, B. F., Buchanan, A. E., and Romberg, M. N. 1982a. The development of mathematical problem solving superitems (Report of NIE/EC Item Development Project). Madison, Wisconsin: Wisconsin Center for Education Research.

Romberg, T. A., Jurdak, M. E., Collis, K. F., and Buchanan, A. E. 1982b. Construct validity of a set of mathematical superitems (Report of NIE/ECS Item Development Project). Madison, Wisconsin: Wisconsin Center for Education Research.

Rosenbaum, P. R. 1988. Item bundles. *Psychometrika, 53,* 349–359.

Volodin, N., and Adams, R. J. 1995. Identifying and estimating a D-dimensional Rasch model. Unpublished manuscript, Australian Council for Educational Research, Camberwell, Victoria, Australia.

Wainer, H., and Kiely, G. L. 1987. Item clusters and computerized adaptive testing: A case for testlets. *Journal of Educational Measurement, 24,* 185–201.

Warm, T. A. 1985. Weighted maximum likelihood estimation of ability in item response theory with tests of finite length (Technical Report CGI-TR-85-08). Oklahoma City: U.S. Coast Guard Institute.

Warm, T. A. 1989. Weighted likelihood estimation of ability in item response theory. *Psychometrika, 54,* 427–450.

Wilson, M. 1992. The ordered partition model: An extension of the partial credit model. *Applied Psychological Measurement,* 16, 309–325.

Wilson, M., and Adams, R. J. 1993. Marginal maximum likelihood estimation for the ordered partition model. *Journal of Educational Statistics, 18,* 69–90.

Wright, B. D., and Masters, G. N. 1982. *Rating Scale Analysis.* Chicago: MESA Press.

Wright, B. D., and Stone, M. H. 1979. *Best Test Design.* Chicago: MESA Press.

Wu, M. L. 1997. The development and application of a fit test for use with marginal maximum likelihood estimation and generalised item response models. Unpublished master's thesis, University of Melbourne.

33

Explanatory Response Models

Paul De Boeck and Mark R. Wilson

CONTENTS

33.1 Introduction 565
33.1.1 Measurement and Explanation 566
33.1.2 Explanatory and Descriptive Measurement 567
33.1.3 Review of Literature 567
33.2 Presentation of the Model 568
33.2.1 Long Data Form 568
33.2.2 Initial Model Formulation 568
33.3 General Model 569
33.4 Types of Covariates 569
33.5 Parameter Estimation 570
33.6 Model Fit 570
33.7 Empirical Example 571
33.7.1 Data 571
33.7.2 Dendrification of the Responses 572
33.7.3 Model Description 572
33.8 Parameter Estimates 575
33.8.1 Model Fit 576
33.9 Discussion 577
References 578

33.1 Introduction

Regression models have a familiar suitability as explanatory models: a criterion or dependent variable is explained in terms of predictors or independent variables. In the explanatory item response modeling (EIRM) approach, item responses are seen as dependent variables to be explained by independent variables, such as properties of persons, items, and pairs of persons and items. Examples of person properties are gender, age, socioeconomic status, or perhaps treatment status, or location on another test variable. Examples of item properties depend on the domain of testing. For example, in a spatial ability test, the angle of the required rotation is such a property, while in a mathematical test, the number and the different kinds of numerical operations are interesting explanatory properties. Because of its explanatory nature, the EIRM approach has the intrinsic qualities to function as a vehicle for research based on substantive theories. It can help to meet a basic concern expressed by Sijtsma (2012) "Ask what psychometrics can do for psychology."

(a)

$$\begin{array}{c} i=123 \\ \begin{array}{l} p=1 \\ p=2 \end{array} \begin{bmatrix} 100 \\ 101 \end{bmatrix} \end{array}$$

(b)

$$\begin{array}{ll} p=1 & i=1 \\ p=1 & i=2 \\ p=1 & i=3 \\ p=2 & i=1 \\ p=2 & i=2 \\ p=2 & i=3 \end{array} \begin{bmatrix} 1 \\ 0 \\ 0 \\ 1 \\ 0 \\ 1 \end{bmatrix}$$

(c) $k=0123$

$$\begin{bmatrix} 1010 \\ 1021 \\ 1030 \\ 1111 \\ 1121 \\ 1130 \end{bmatrix}$$

(d) $k=0123$

$$\begin{bmatrix} 1100 \\ 1010 \\ 1001 \\ 1100 \\ 1010 \\ 1001 \end{bmatrix}$$

FIGURE 33.1
Data format with wide and long form, as well as covariates. (a) Example $\boldsymbol{U}$ wide form, (b) example $\boldsymbol{U}$ long form, (c) example $\boldsymbol{Z}$ covariates, (d) $\boldsymbol{Z}$ for Rasch model.

However, familiar as the regression formulation is, one cannot simply apply the linear regression model to item responses for two reasons. First, the data to be explained have a person-by-item array format (see Figure 33.1a), not the one of a single dependent variable. Second, the data are categorical so that a linear regression is not appropriate. As explained in Section 33.1.1, generalized linear mixed modeling (GLMM) (McCulloch and Searle, 2001; Volume One, Chapter 30) can be used to solve both problems. First, the data are reorganized into a long form (Figure 33.1b), so that they can serve as a dependent variable. Second, a linear combination of the independent variables is connected to the categorical dependent variable through a link function and a stochastic component.

The focus of this chapter is on explanatory models of the GLMM type. Models outside the GLMM framework, such as mixture models and models with products of parameters, are not discussed. A well-known item response model with a product of parameters is the two-parameter model. A well-known item response model that can be interpreted as a mixture model is the three-parameter model. (It can be seen as a mixture model for pairs of persons and items. Also, the discrimination and the guessing parameters can be explained through covariates, see Section 33.7; but this is not a topic of this chapter.) The description and discussion of EIRM in De Boeck and Wilson (2004) relies on the GLMM as a starting point, while it also deals with bilinear and mixture extensions. A broader approach, which does include nonlinear mixed models, is described by Rijmen et al. (2003).

33.1.1 Measurement and Explanation

Psychometrics is commonly seen as the science of measuring latent variables, not directly observable variables, also referred to as "constructs." Their latent character requires the availability of observable responses provided by items. Hence, the primary purpose of items is as an instrument to measure the latent variable. Also, measurement itself is instrumental, and its purposes are diverse: testing theories, monitoring, evaluation, and decision-making.

In an *explanatory approach*, the data and not the latent variables are the focus of interest. The observed responses are the explanandum—something to be explained. In this view, latent variables are a by-product of data explanation, needed in case the regression weights vary across persons. From an explanatory viewpoint, a latent variable is a (randomly) varying regression weight, for example, a random intercept; or, possibly more interesting, random regression weights for the covariates used to explain the responses. Thus, in this sense, the explanatory power of the model does not come from the latent variable, as has been criticized by, among others, Borsboom et al. (2003), but from the person and item properties that serve as independent variables in the model.

Measurement and explanation are not mutually exclusive. They can be combined and be mutually reinforcing in the form of *explanatory measurement* (De Boeck and Wilson, 2004). Explanatory measurement contributes to the internal validity of a test (Embretson, 1983), it makes one understand better the meaning of test scores.

33.1.2 Explanatory and Descriptive Measurement

Measurement is *descriptive* if numbers ("measures") are assigned to the units of observation, such as persons and items, without linking the numbers to known properties of the units. Measurement is *explanatory* if the numbers are linked indeed to known properties of the units (possibly with a residual term). Crossing the two dimensions of descriptive versus explanatory and items versus persons lead to four different types of measurement (Wilson and De Boeck, 2004): doubly descriptive (no properties as covariates), person explanatory (with person properties), item explanatory (with item properties), and doubly explanatory (with person and item properties) measurement.

33.1.3 Review of Literature

EIRM dates back to the early history of item response theory. The linear logistic test model (LLTM) (Scheiblechner, 1972; Fischer, 1973, 1983; Volume One, Chapter 13) is based on explanatory item covariates, while the latent regression Rasch model (Adams et al., 1997; Zwinderman, 1991) is based on explanatory person covariates. Early applications of EIRM addressed cognitive tests (Whitely, 1976, 1977). Also, more recently various types of intelligence and achievement test response have been approached with EIRM: mathematical problem solving (Daniel and Embretson, 2010), language and reading comprehension (Gorin and Embretson, 2006; Wilson and Moore, 2011), spatial ability (Embretson and Gorin, 2001; Ivie and Embretson, 2010), cognitive diagnosis (Wilson, 2008), learning progressions (Wilson, 2009), longitudinal data (Wilson et al., 2012), and teacher judgments (Draney and Wilson, 2008). To understand rather specific task processes, EIRM approaches have been used for numerical division problems (Hickendorff et al., 2009, 2010), phoneme segmentation (Bouwmeester et al., 2011), and word decoding (Gilbert et al., 2011). Other recent applications of EIRM involve automated item generation (Embretson, 1999; Daniel and Embretson, 2010; Geerlings et al., 2011; Volume One, Chapter 26), modeling response times (Jansen, 1997; Volume One, Chapter 15) as well as response times in combination with response accuracy (Klein-Entink et al., 2009).

A special type of explanatory model is the model with an internal restriction on item difficulty (MIRID) (Butter et al., 1998; Bechger et al., 2002). In an MIRID model, the item difficulties of a set of items are explained in terms of the difficulties of the item subtasks. This type of model has recently been generalized in different ways (Hung, 2011; Wang and Jin, 2010). Cognitive diagnostic models (Rupp et al., 2010) are also explanatory, because they use (binary) item-related covariates to explain the responses. Since MIRID involves multiplication of parameters (item difficulties having weights) and cognitive diagnostic models are mixture models, they will not be further discussed in this chapter.

33.2 Presentation of the Model

33.2.1 Long Data Form

To see an EIRM as the extension of a regression model, the standard person-by-item data array must be reformatted as a data vector $\boldsymbol{U}$ with $P \times I$ entries, where P is the number of persons and I the number of items. The first two parts of Figure 33.1 illustrate the reformatting for a small binary data array for two persons by three items (wide form) into a vector of length six with the same data (long form).

This reformatting makes it possible to see the item responses as dependent variables and the covariates as their predictors. The covariates in Figure 33.1 are further explained in Section 33.2.2.

33.2.2 Initial Model Formulation

In the GLMM, the expected values of the data points, $E(U_{pi})$ with $p = 1, \ldots, P$ for the persons and $i = 1, \ldots, I$ for the items, are transformed by a link function, which lead to $\eta_{pi} = f_{link}[E(U_{pi})]$, where $E(U_{pi})$ is the probability of observing a correct response. In line with the current practice, we assume the logit link, but the probit link is a good alternative. The transformed value, η_{pi}, is modeled as a linear function of covariates Z_k, with $k = 1, \ldots, K$. All pairs (p,i) have a value on the covariates, Z_{pik}. Figure 33.1c shows a $(P \times I) \times K$ matrix $\mathbf{Z}$ of covariates. Its first column defines an intercept, the second represents a covariate depending on the persons, the third is a covariate depending on the items, and the fourth is a person-by-item covariate that depends both on the persons and items. The parameters of the model are the effects (regression weights) of the covariates. These effects can be fixed or random. It is commonly assumed that the random effects have a (multivariate) normal distribution, but other assumptions are possible, both of a parametric kind (Azevedo et al., 2011) and nonparametric kind (Bock and Aitkin, 1981; Jara et al., 2011). Jara et al. (2011) provide an example of an EIRM application based on a Dirichlet process.

When there are only random person but no random item effects, the linear function is

$$\eta_{pi} = \sum_{k=0}^{K} \beta_k Z_{pik} + \sum_{k=0}^{K} \theta_{pk} Z_{pik} \tag{33.1}$$

with β_k as fixed effects, and θ_{pk} as random effects, and $\theta_p \sim MVN(\mathbf{0}, \Sigma_\theta)$.

Note that matrix $\mathbf{Z}$ represents both covariates with fixed effects and random effects. The model in Equation 33.1 is a Rasch model if (a) Z_k, $k = 1, \ldots, K$ are item indicators (Figure 33.1d) and $b_i = \sum_{k=1}^{k} \beta_k Z_{ik}$ is thus the item parameter and (b) the only person effect stems from Z_{pi0} with random person intercept θ_{p0}, commonly referred to as ability in a unidimensional cognitive test.

Equation 33.1 does not have an error term whereas typical regression models do. But an error term can be introduced by adopting an unobserved variable V_{pi}, such that $V_{pi} = \eta_{pi} + \varepsilon_{pi}$, with ε_{pi} a term with a standard logistic distribution (or the standard normal for the probit link) and where $V_{pi} \geq 0 \rightarrow U_{pi} = 1$ and $V_{pi} < 0 \rightarrow U_{pi} = 0$. This is a formulation equivalent to the model for $P\{U_{pi} = 1\}$. Constraining the error variance is typical of item response models, while in factor models the variance of V_{pi} per item i is usually constrained to one.

33.3 General Model

To formulate a more general model, the notion of a *stratum* must be introduced. A stratum is a dimension of a data array. For a person-by-item array, the two strata are persons and items. We use an ordered set of indices to represent both strata and their combinations. For the simple case of data in a person-by-item array, this leads to a vector of three indices, $(p,i,[p,i])$, which we will denote by a meta-index ν ($\nu = 1, \ldots, N$) used as a superscript for random effects. For example, ability in a Rasch model is $\theta^{(1)}_{p0}$. Also as a general notation, ω will be used to collect all relevant strata (e.g., as in $\omega = [p,i]$). Hence, our covariate notation becomes $Z_{\omega k}$. Suppose the data array has three different strata: persons, items, and measurement occasions (indexed by o [$o = 1, \ldots, O$]). Then, $\omega = (p,i,o)$, and the set of indices for the random effects is $\{p,i,o,(p,i),(p,o),(i,o),(p,i,o)\}$, which would yield superscripts $\nu = 1, \ldots, 7$.

The general model can now be written as

$$\eta_\omega = \sum_{k=0}^{K} \beta_k Z_{\omega k} + \sum_{k=0}^{K} \sum_{\nu=1}^{N} \theta^{(\nu)}_{\nu^* k} Z_{\omega k} \tag{33.2}$$

where ν^* is replaced with the corresponding index, for example, with i if $\nu = 2$, and with pi if $\nu = 4$. The two parts in the right-hand side correspond to the fixed effects and the random effects, respectively. Again, the $Z_{\omega k}$ cover all covariates needed for either the fixed effects or the random effects. For example, $\theta^{(2)}_{ik} = 0$, for all values of i and k, in a model with fixed item effects only. For each value of ν, it holds for the vector of random effects that $\boldsymbol{\theta}^{(\nu)}_{\nu^*} \sim MVN(\mathbf{0}, \Sigma_{\boldsymbol{\theta}(\nu)})$, with, for example, $\boldsymbol{\theta}^{(1)}_{p} \sim MVN(\mathbf{0}, \Sigma_{\boldsymbol{\theta}(1)})$ and $\boldsymbol{\theta}^{(2)}_{i} \sim MVN(\mathbf{0}, \Sigma_{\boldsymbol{\theta}(2)})$.

33.4 Types of Covariates

Common covariates depend on persons or items. Covariates dependent on person-item pairs are less common, although useful, for example, to model local dependence (Meulders and Xie, 2004) or the effect of strategies (Hickendorff et al., 2009). In general, covariates can be dependent on any stratum or their combinations. Other examples than items and persons are covariates depending on measurement occasions created by situations or time (Cattell, 1966). The latter include the cases of linear and quadratic time used in growth-curve item response modeling (McArdle, 2009). Examples of situation-related covariates are private versus public and imposed versus freely chosen character in the study of behavior.

Strata can be crossed or nested. Persons and items are usually crossed, while nested strata in the form of persons within groups lead to *multilevel* designs. Also, covariates may be useful at different levels, such as in the multilevel regression model, for example, with demographic characteristics of individual students and the goal orientation of their teachers as covariates.

We use two different types of strata covariates: indicators and properties. Use of the indicators as covariates leads to descriptive measurement, whereas the properties are required for explanatory measurement. Binary group indicators such as gender indicators can be useful to model *heteroscedasticity*, for instance, in the form of differences in variance between men and women.

For the response categories, *category covariates* can be used, either in the form of category indicators or properties. Examples of category properties are the extremity of a category or one or more score functions (Revuelta, 2008; Thissen and Steinberg, 1986). As soon as more than one category covariate is involved, like in the partial-credit model (Volume One, Chapter 7) and the graded-response model (Volume One, Chapter 6), the models become multivariate (Fahrmeir and Tutz, 2001), except when they are of the sequential type (Volume One, Chapter 9). Our empirical example in Section 33.7 illustrates the sequential type.

33.5 Parameter Estimation

Parameter estimation for models with crossed random effects is reviewed here. As an example consider the case of crossed random person (i.e., the latent trait) and item effects. Random item effects are useful as residuals in an item explanatory model (De Boeck, 2008; Janssen et al., 2004). But crossed effects create a serious computational burden because of their high-dimensional integrals. Three approaches can be roughly followed (Bellio and Varin, 2005): (a) simulation-based algorithms such as MCMC, (b) maximization based on approximate methods such as Laplace, and (c) likelihood-based maximization divided into components, with possible alternation between them. Examples of the last approach are the pairwise likelihood method by Bellio and Varin (2005) based on pairs of observations with at least one common random term and the posterior imputation estimation method of Cho and Rabe-Hesketh (2011), which alternates between the items and persons.

The most commonly used methods, however, are Bayesian methods (Fox, 2010) and the Laplace approximation, as implemented in the lme4 package in R (Bates et al., 2012). The latter easily handles large numbers of random effects and is computationally fast, but has the disadvantage of a downward bias of its variance estimators (Joe, 2008), although the bias is still limited for models with no more than 10 units in a stratum (Hofman and De Boeck, 2011). A comparison of these three methods for crossed random effects (i.e., Bayesian, Laplace, division) can be found in Cho et al. (2012); see also Rijmen, Jeon, and Rabe-Hesketh (Volume Two, Chapter 14) where a variational estimation approach is described.

33.6 Model Fit

An excellent general review of goodness-of-fit (GOF) methods is given by Maydeu-Olivares and García-Forero (2010): indices versus statistics, absolute versus relative, and overall versus piecewise. Because item response models commonly suffer from small expected frequencies, limited information methods based on marginal probabilities are helpful, or one can rely on resampling (bootstrap) methods. Relative GOF indices (AIC, BIC; Cohen and Cho, Volume Two, Chapter 18) and likelihood-ratio (LR) tests (Glas, Volume Two, Chapter 17) are perhaps the most popular GOF approaches. LR tests are assumed to give very good results also for models estimated with the Laplace method (Baayen et al., 2008), the method used in our example below.

A special GOF issue concerns testing of the random effects of covariates, since the null hypothesis of a zero variance is located at the boundary of the parameter space. Baayen et al. (2008) recommend the LR test nevertheless, but as a conservative test. Verbeke and Molenberghs (2003) discuss the issue and propose three methods, one of which is a mixture-χ^2 LR test for the comparison of J and J + 1-correlated random effects, to be tested with a p-value obtained from a mixture of χ^2, $0.5\chi_J^2 + 0.5\chi_{J+1}^2$ (Stram and Lee, 1994, 1995). The standard LR test would be a χ_{J+1}^2 (one additional variance plus J covariances). For the test of $J = 0$ versus $J = 1$, the p-value should be divided by two.

From an explanatory point of view, the well-known R^2 is an interesting statistic because it is a measure of the explanatory power of the model rather than a GOF index. The use of an incremental R^2 statistic has been proposed for a linear mixed model with fixed effects, as an indication of effect size (Edwards et al., 2008). Edwards (2012) extends the R^2 for the GLMM, and thus also for binary data: $R^2 = 1 - e^{-LR\chi^2/n_{obs}}$, where $LR\chi^2$ is the likelihood-ratio χ^2 and n_{obs} is the total number of observations. For common null hypothesis testing of fixed effects, an LR test, score test, or Wald test can be used (Glas, Volume Two, Chapter 17).

33.7 Empirical Example

33.7.1 Data

Our empirical example is based on the verbal aggression data set from De Boeck and Wilson (2004). The data were from 316 respondents, 243 females, and 73 males, presented with a verbal aggression inventory with 24 items. The inventory was constructed with a substantive interest in mind concerning the source of verbal aggression (type of situation), the kind of verbally aggressive behavior, and its possible inhibition. This is why each item consists of (a) one of four frustrating situations, two of which are other-to-blame and two of which are self-to-blame, followed by (b) one of three verbally aggressive behaviors (cursing, scolding, shouting), phrased in (b) one of two behavioral modes (wanting or doing). A sample item is "A bus fails to stop for me. I would want to curse" (other-to-blame, cursing, wanting). The doing version is "I would actually curse." The items had three response categories: "no," "perhaps," and "yes." The long form of the data set was taken from the lme4 package (Bates et al., 2012).

Earlier, two simple explanatory models for the dichotomized data (e.g., "yes" and "perhaps" versus "no") were used. First, the item difficulties were constrained to be a linear function of the three kinds of item covariates (type of situation, behavior, and behavior mode) in an LLTM (Wilson and De Boeck, 2004), so that Equation 33.2 is reduced to

$$\eta_{pi} = \sum_{k=0}^{K} \beta_k Z_{pik} + \theta_{p0}^{(1)} Z_{pi0} \tag{33.3}$$

where $\theta_{p0}^{(1)} Z_{pi0} = \theta_p$ is the latent trait. Second, a random item residual was added to the LLTM for the case the explanation was not yet perfect. This option of an LLTM with error was applied by Janssen et al. (2004); see also Janssen (Volume One, Chapter 13). Formally, the model is

$$\eta_{pi} = \sum_{k=0}^{k} \beta_k Z_{pik} + \theta_{p0}^{(1)} Z_{pi0} + \theta_{i0}^{(2)} Z_{pi0} \tag{33.4}$$

where $\theta_{i0}^{(2)} Z_{pi0} = \varepsilon_i$ is the residual item difficulty.

33.7.2 Dendrification of the Responses

Because, there were three ordered response categories, a choice had to be made regarding the logits to be considered: adjacent logits (partial credit), cumulative logits (graded response), or continuation logits (sequential); for these choices, see Masters (Volume One, Chapter 7), Samejima (Volume One, Chapter 6), and Tutz (Volume One, Chapter 9). Although all three could have been used, we worked with continuation logits in order to stay within the univariate GLMM family; the other two approaches require a multivariate GLMM (Fahrmeir and Tutz, 2001). Following the continuation logit approach, each response is decomposed into two sub-responses, based on the tree in Figure 33.2 (De Boeck and Partchev, 2012).

A substantive interpretation of the nodes in Figure 33.2 can be based on the semiotic square concept in Greimas (1983), which relates four kinds of statements: assertion, nonassertion, nonnegation, and negation. The top node defines a contrast between *non-negation* ("perhaps" and "yes," and $U_{pi1} = 1$) and *negation* ("no" and $U_{pi1} = 0$), and will be called the *non-negation node*. The bottom node defines a contrast between *assertion* ("yes," $U_{pi2} = 1$) and *non-assertion* ("perhaps" and $U_{pi2} = 0$) given a nonnegation (given the response is not a "no" and $U_{pi1} = 1$). The bottom node will be called the *assertion node*. We expect an assertion not only to be stronger than a nonnegation but also qualitatively different. Of course, different assumptions could lead to a different dendrification as well as model formulation.

The tree shown in Figure 33.2 corresponds to a continuation ratio model. On the basis of the tree, the dendrification of the responses leads to two sub-responses per original response: (0, NA) for "no," (1, 0) for "perhaps," and (1, 1) for "yes," with NA for missing (there cannot be node 2 observations regarding a "no"). This dendrification is illustrated by the response vector in Figure 33.3 for the responses "no," "perhaps," and "yes," respectively.

33.7.3 Model Description

The three strata in the model are $\omega = (p,i,m)$, with m as an index for the nodes and thus for the two subitems. The covariates of the model are:

(0) 1-constant to define the intercept

(1) Gender (women versus men), person covariate with effect coding

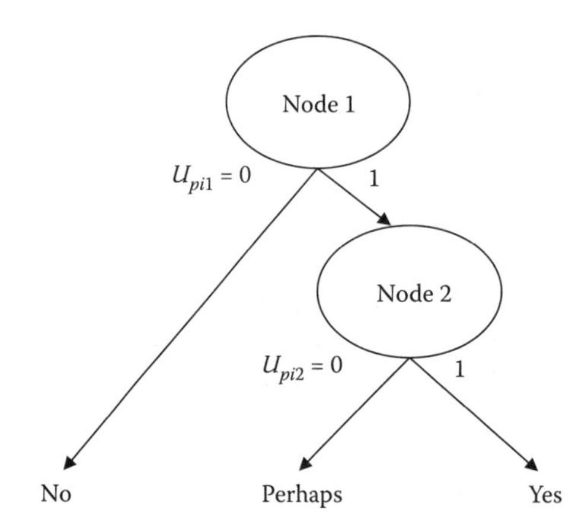

FIGURE 33.2
Dendrification of item responses into subitem responses.

(2) Behavior mode (do versus want), item covariate with effect coding

(3) Type of situation (other-to-blame versus self-to-blame), item covariate with effect coding

(4)–(5) Behavior (scolding versus cursing, shouting versus cursing and scolding), item covariates with Helmert coding

(6) Node (node 1 versus node 2), category covariate, effect coding

Figure 33.3 gives sample values of these covariates. Also, the inclusion of all pairwise products of covariates 2 to 6 (but not the product of covariates 4 and 5) allowed us, for example, to investigate gender and node differences in the effects of the item covariates. In total, there were 21 covariates, each with a fixed effect; see Equation 33.5 and Table 33.1.

The random effects of the model were defined as follows. Since we did not expect the fixed covariate effects to explain all variations, random person and item intercepts were included in the model: $\theta_{p0}^{(1)}$ and $\theta_{i0}^{(2)}$. Because of our interest in possible qualitative differences between the nodes (nonnegation and an assertion given a nonnegation), random person, and random item effects were added for node (covariate 6, effect coded): $\theta_{p6}^{(1)}$ and $\theta_{i6}^{(2)}$. Finally, in order to investigate individual differences in the inhibition of verbal aggression, the random person effect of behavior mode, $\theta_{p2}^{(1)}$, was part of the model (covariate 2, effect coded).

The full logistic model considered was

$$
\begin{aligned}
\eta_{\omega} = & \sum_{k=0}^{20} \beta_k Z_{\omega k} \\
& + \theta_{p0}^{(1)} Z_{\omega 0} + \theta_{p2}^{(1)} Z_{\omega 2} + \theta_{p6}^{(1)} Z_{\omega 6} \\
& + \theta_{i0}^{(2)} Z_{\omega 0} + \theta_{i6}^{(2)} Z_{\omega 2},
\end{aligned} \tag{33.5}
$$

	Subresponse	Covariates 0	1	2	3	4	5	6
person 1 item 1 node 1	0	1	−1	1	−1	1	−1	1
person 1 item 1 node 2	NA	1	−1	1	−1	1	−1	−1
person 1 item 2 node 1	1	1	−1	1	1	−1	−1	1
person 1 item 2 node 2	0	1	−1	1	1	−1	−1	−1
person 1 item 3 node 1	1	1	−1	−1	1	0	2	1
person 1 item 3 node 2	1	1	−1	−1	1	0	2	−1

FIGURE 33.3
First six lines of the dendrified long form data with covariates. "Subresponse" refers to the node (subitem) responses, with NA for missing. The original responses are "no," "perhaps," and "yes," respectively. The column numbers are 0 for the column with 1s, 1 for gender (woman = 1, man = −1), 2 for behavior mode (do = 1, want = −1), 3 for type of situation (other-to-blame = 1, self-to-blame = −1); 4 and 5 for behavior (curse = −1, −1; scold = 1, −1; shout = 0, 2); 6 for node (node 1 = 1, node 2 = −1).

TABLE 33.1

Fixed Effect Estimates

	Effect	S.E.	z
0. Intercept	–0.591	0.107	–5.511***
1. Gender			
Women versus men	–0.163	0.089	–1.837**
2. Behavior mode			
Do versus want	–0.317	0.072	–4.404***
3. Situation			
Other versus self	0.573	0.066	8.711***
4. Behavior			
Scold versus curse	–0.414	0.079	–5.255***
5. Behavior			
Shout versus other	–0.491	0.048	–10.161***
6. Node			
Node 1 versus 2	0.502	0.062	8.109***
Gender*mode			
1*2	–0.189	0.040	–4.864***
Gender*situation			
1*3	0.040	0.028	1.424
Gender*behav			
1*4	0.005	0.032	0.144
Gender*behav			
1*5	0.108	0.022	4.896***
Gender*node			
1*6	–0.001	0.061	–0.020
Mode*situation			
2*3	0.018	0.064	0.282
Mode*behav			
2*4	–0.022	0.077	–0.292
Mode*behav			
2*5	–0.106	0.046	–2.317**
Mode*node			
2*6	0.044	0.027	1.621
Situation*behav			
3*4	0.081	0.077	1.051
Situation*behav			
3*5	0.037	0.046	0.804
Situation*node			
3*6	–0.046	0.026	–1.729*
Behav*node			
4*6	–0.142	0.029	–4.882***
Behav*node			
5*6	–0.112	0.021	–5.341***

* $p \leq 0.10$, ** $p \leq 0.05$, *** $p \leq 0.001$.

where the first line is for the fixed effects, β_k, the second for the random person effects, $(\theta^{(1)}_{p0},\theta^{(1)}_{p2},\theta^{(1)}_{p6}) \sim MVN(0,\Sigma_{\theta(1)})$, and the third for the random item effects, $(\theta^{(2)}_{p0},\theta^{(2)}_{p6}) \sim MVN(0,\Sigma_{\theta(2)})$.

The model allows both for multidimensionality of individual differences ($\theta^{(1)}_{p6}$) and item differences ($\theta^{(2)}_{i6}$) in the use of ordered response categories. Although with different approaches, such individual difference were studied earlier by Johnson (2003, 2007) and Wang and Wu (2011) using a divide-by-total model.

33.8 Parameter Estimates

The model was estimated using Laplace approximation (`glmer` function of the R package `lme4`; Bates et al., 2012). A preliminary check on the dimensionality of the data structure was made through a comparison within two model sets. The first set contained models in which one, two, or three random person effects (of covariates 0; 0 and 2; 0, 2 and 6) were combined with one or two random item effects (of covariates 0, 0 and 6), each time with fixed effects of gender and node, and their interaction (six models in total). In the second set of models, the two random item effects were replaced by fixed effects, either of item indicators (item specific difficulties) or of subitem indicators (subitem specific difficulties) (again six models in total).

Using AIC and BIC as criteria, the best fitting models of both sets have three random person effects (0, 2, and 6), with two random item effects (0 and 6) in the first set, and fixed subitem effects (and thus item-by-node interactions) in the second set. These preliminary results indicated that the nodes differed in a qualitative way, and also that wanting and doing were two different dimensions. We therefore continued with the model in Equation 33.5, labeled the *full-covariate 3RP-2RI model* because of its three random person effects (3RP) and two random item effects (2RI).

The estimated structure of the *random effects* is shown in the left side of Table 33.2. For a comparison, the right side of Table 33.2 shows the estimated structure for the case of no item covariates , the *no-item-covariate 3RP-P2I* model. The results confirmed the multidimensionality of the random person effects, although inhibition (do versus want) and differential use of the nodes (node 1 versus node 2) had a variance clearly smaller than the

TABLE 33.2

Structure of the Random Effects

	With Item Covariate Effects				Without Item Covariate Effects			
Persons	**Var**		**Cor**		**Var**		**Cor**	
Intercept (0)	1.538	–			1.537	–		
Behavior mode (2)	0.187	0.429	–		0.206	0.411	–	
Node (6)	0.596	–0.034	–0.298	–	0.588	–0.021	–0.290	–
Items								
Intercept (0)	0.081	–			1.060	–		
Node (6)	0.000*	1.000	–		0.054	0.577	–	

* The estimate is 0.000098, so that the correlation of 1.000 should not be interpreted.

intercept variance (12% and 39%, respectively, compared to the intercept variance). Also, it was clear that nodes 1 and 2 could be represented by one random item residual term, given that item covariates were used to explain the difference between the two nodes (left side of Table 33.2).

Table 33.1 shows the estimates of the fixed single and pairwise covariate effects in the full-covariate 3RP–2RI model. Let us first look at the single *person covariate* (gender) and its interaction with the item and node covariates. Women appeared to be slightly less verbally aggressive on average (effect 1), they showed a larger discrepancy between doing and wanting and thus more inhibition (effect 1*2), and they shouted relatively more frequently than they cursed and scolded compared to men (effect 1*5).

All *item covariate effects* appeared to have a significant main effect. The subjects seemed to refrain from their verbally aggressive wants (effect 2), they were more aggressive when someone else had caused the frustrating situation (effect 3), they scolded less than they cursed (effect 4), and they shouted less than they cursed and scolded (effect 5). Only one interaction effect was significant, indicating that the discrepancy between doing and wanting was larger for shouting than for cursing and scolding (effect 2*5). In other words, cursing and scolding were less inhibited than shouting.

Finally, there was a main effect of *node* (effect 6) as well as some interaction effects of node and item covariates (effects 3*6, 4*6, 5*6). The main effect refers to nonnegation being more popular than assertion. The difference was smaller for other-to-blame situations than self-to-blame situations (effect 3*6), smaller for scolding than for cursing (effect 4*6), and smaller for shouting than for cursing and scolding (effect 5*6). Using a divide-by-total model and adjacent logits for the same data set, Johnson (2007) also found that the three response categories required two dimensions, primarily because of the type of situation.

33.8.1 Model Fit

The relative goodness of fit was tested separately for the random-effect and fixed-effect parts of the model. The need for *fixed covariate effects* was tested comparing the full-covariate 3RP–2RI model with a low-covariate 3RP–2RI model with neither person nor item covariates but only two fixed effects: a fixed intercept (effect 0 in Table 33.1) and a fixed node effect (effect 6 in Table 33.1). The AIC and BIC values were 12,265 and 12,485 for the full-covariate model, and 12,365 and 12,446 for the low-covariate model. The LR test yielded a $\chi^2_{19} = 137.61$, $p < 0.001$, so that the low-covariate model was rejected. The incremental R^2 for the gain of introducing fixed covariate effects on top of random item effects was 0.018, while the random effects already take care of much of the variation in the responses. When the random item effects were omitted, the incremental R^2 was 0.190—a substantial effect size given the fact that much of the variation in the responses was person variation.

To test the need for *random effects*, the full-covariate 3RP–2RI model was compared with the 3RP–FSI model, which replaces the two random item effects (2RI) with fixed subitem effect (FSI), so that the item and node covariates also had to be omitted. The AIC and BIC values of the 3RP–FSI model were 12,270 and 12,672. These values were higher than those for the full-covariate 3RP–2RI model, so that we continued with the latter. Apparently, it was better to use explanatory item covariates plus a random residual than just fixed subitem effects (equivalent with node and item effects and their interaction).

Further, it was tested whether each of the five random effects (residuals) in the full-covariate 3RP–2RI model were needed. We tested the fit of the following sequence of full-covariate models using the mixture χ^2 approach: 1RP–1RI (two random intercepts),

2RP–1RI (plus random person effect of behavior mode), 3RP–1RI (plus random person effect of node), and 3RP–2RI (plus random item effect of node). The respective mixture χ^2-test results were LR $\chi^2 = 400.22$, $p < 0.001$, LR$\chi^2 = 93.89$, $p < 0.001$, and LR $\chi^2 = 0.05$, $p > 0.50$. Apparently, the use of more random person effects yielded better fit than the use of fewer, but the model with one random item residual was not rejected against the one with two residuals. This last result was not surprising given the small variances and the perfect correlation in Table 33.2. However, the 3RP–0RI model was rejected against the 3RP–1RI (LR mixture $\chi^2 = 81.28$, $p < 0.001$), so we decided to keep one random item effect.

33.9 Discussion

From an explanatory point of view, the results from our empirical example inform us about factors that determine verbal aggression tendencies. For example, women showed more inhibition than men; the least common behavior, shouting, was a relatively more popular among women; and, interestingly, this was also the type of behavior with the largest discrepancy between wanting and doing. From a measurement point of view, we were informed that the inventory was multidimensional, and that the ordered-response scale did not seem to be ordinal as for the traits intended to be measured (because of the two node dimensions). Using the appropriate covariates with fixed and random effects made it possible to capture the intended latent traits (verbal aggression and inhibition) and to better understand what was actually measured.

Our explanatory model was primarily an item explanatory model, which was a deliberate choice since the item side is a somewhat neglected side. Generally, the dimensionality of psychometric models and the existence of latent variables depend on individual differences between persons. But while the meaning of latent variables depends on the items, the item parameters are mostly left unexplained, so that the full potential for an interpretation of the latent variables is not realized. Of course, person explanatory models are important for an explanation of the individual differences, as well as their distribution. When the distribution deviates from normality, one possible explanation is lack of appropriate person covariates. In fact, the use of person covariates is the manifest equivalent of an approach where a mixture of normals is used to approximate a nonnormal distribution.

In addition to the use of persons and items as data strata, other interesting models can be defined: multilevel explanatory item response models assume person or item groups and covariates for these groups. Growth-curve explanatory item response models assume the existence of measurement occasions. Obvious covariates for measurement occasion are time and functions of time, and an appropriate selection of item and person covariates can help to explain differential growth.

Our empirical example did not illustrate many other possible uses of EIRM. One option is the use of pairwise covariates from two strata, such as person-by-item covariates, which seem to be quite useful to model, and thus also explain, differential item functioning or local item dependencies (De Boeck et al., 2011; Meulders and Xie, 2004). Another option is the explanation of item discriminations (Daniel and Embretson, 2010; Embretson, 1999; Klein Entink et al., 2009). Also, the treatment in this chapter was limited to item-response models of the (univariate) GLMM type and neither includes the two-parameter model, nor its multidimensional extensions.

We consider it as ideal to understand what is measured as much as possible. Explanation is a major goal of research. Measurement is an important tool to that end, in addition to being of high practical importance, for instance, to monitor progress, for evaluation, and for decision-making. The combination of both goals in explanatory measurement creates opportunities for mutual benefits.

References

Adams, R. J., Wilson, M., and Wang, W. 1997. The multidimensional random coefficients multinomial logit model. *Applied Psychological Measurement, 21*, 1–23.

Azevedo, C. L., Bolfarine, H., Andrade, D. F. 2011. Bayesian inference for a skew-normal IRT model under the centered parameterization. *Computational Statistics and Data Analysis, 55*, 353–365.

Baayen, R. H., Davidson, D. J., and Bates, D. M. 2008 Mixed-effects modeling with crossed random effects for subjects and items. *Journal of Memory and Language, 59*, 390–412.

Bates, D., Maechler, M., and Bolker, B. 2012. lme4: Linear mixed-effects models using S4 classes. R package, URL http://CRAN.R-project.org/package=lme4.

Bechger, T. M., Verstralen, H. H. F. M., and Verhelst, N. 2002. Equivalent linear logistic test models. *Psychometrika, 67*, 123–136.

Bellio, R., and Varin, C. 2005. A pairwise likelihood approach to generalized linear models with crossed random effects. *Statistical Modeling, 5*, 217–227.

Bock, R. D., and Aitkin, M. 1981. Marginal maximum likelihood estimation of item parameters: Application of an EM algorithm. *Psychometrika, 46*, 443–459.

Borsboom, D., Mellenbergh, G. J., and van Heerden, J.2003. The theoretical status of latent variables. *Psychological Review, 110*, 203–219.

Bouwmeester, S., van Rijen, E. H. M., and Sijtsma, K. 2011. Understanding phoneme segmentation performance by analyzing abilities and word properties. *European Journal of Psychological Assessment, 27*, 95–102.

Butter, R., De Boeck, P., and Verhelst, N. 1998. An item response model with internal restrictions on item difficulty. *Psychometrika, 63*, 1–17.

Cattell, R. B. 1966. The data box: Its ordering of total resources in terms of possible relational systems. In R. B. Cattell (Ed.), *Handbook of Multivariate Experimental Psychology* (pp. 67–128). Chicago: Rand McNally.

Cho, S.-J., Partchev, I., and De Boeck, P. 2012. Parameter estimation of multiple item response profiles model. *British Journal of Mathematical and Statistical Psychology, 65*, 438–466.

Cho, S.-J., and Rabe-Hesketh, S. 2011. Alternating imputation posterior estimation of models with crossed random effects. *Computational Statistics and Data Analysis, 55*, 12–25.

Daniel, R. C., and Embretson, S. E. 2010. Designing cognitive complexity in mathematical problem-solving items. *Applied Psychological Measurement, 34*, 348–364.

De Boeck, P. 2008. Random item IRT models. *Psychometrika, 73*, 533–559.

De Boeck, P., Bakker, M., Zwitser, R., Nivard, M., Hofman, A., Tuerlinckx, F., and Partchev, I. 2011. The estimation of item response models with the lmer function from the lme4 package in R. *Journal of Statistical Software, 39(12)*, 1–28.

De Boeck, P., and Partchev, I. 2012. IRTrees: Tree-based item response models of the GLMM family. *Journal of Statistical Software, 48*, Code snippet 1.

De Boeck, P., and Wilson, M. (Eds.). 2004. *Explanatory Item Response Models*. New York: Springer.

Draney, K., and Wilson, M. 2008. A LLTM approach to the examination of teachers' ratings of classroom assessments. *Psychology Science Quarterly, 50*, 417–432.

Edwards, L. J. 2012. A brief note on a R^2 measure for fixed effects in the generalized linear mixed model (arXiv:1011.0471v1). Retrieved April 20, 2012, from http://arxiv.org/abs/1011.0471.

Edwards, L. J., Muller, K. E., Wolfinger, R. D., Qaqish, B. F., and Schabenberger, O. 2008. A R^2 statistic for fixed effects in the linear mixed model. *Statistics in Medicine, 27,* 6137–6157.
Embretson, S. E. 1983. Construct validity: Construct representation versus nomothetic span. *Psychological Bulletin, 93,* 179–197.
Embretson, S. E. 1999. Generating items during testing: Psychometric issues and models. *Psychometrika,* **64**, 407–433.
Embretson, S. E., and Gorin, J. S. 2001. Improving construct validity with cognitive psychology principles. *Journal of Educational Measurement, 38,* 343–368.
Fahrmeir, L., and Tutz, G. 2001. *Multivariate Statistical Modeling Based On Generalized Linear Models* (2nd ed.). New York: Springer.
Fischer, G. H. 1973. The linear logistic test model as an instrument in educational research. *Acta Psychologica, 3,* 359–374.
Fischer, G. H. 1983. Logistic latent trait models with linear constraints. *Psychometrika, 48,* 3–26.
Fox, J.-P. 2010. *Bayesian Item Response Modeling: Theory and Applications.* New York: Springer.
Geerlings, H., Glas, C. A. W., and van der Linden, W. J. 2011. Modeling rule-based item generation. *Psychometrika, 76,* 337–359.
Gilbert, J. K., Compton, D. L., and Kearns, D. M. 2011. Word and person effects on decoding accuracy: A new look at an old question. *Journal of Educational Psychology, 103,* 489–507.
Gorin, J. S., and Embretson, S. E. 2006. Item difficulty modeling of paragraph comprehension items. *Applied Psychological Measurement, 30,* 394–411.
Greimas, A. J. 1983. *Structural Semantics: An Attempt at a Method.* Lincoln: University of Nebrasks Press.
Hickendorff, M., Heiser, W. J., van Putten, C. M., and Verhelst, N. D. 2009. Solution strategies and achievement in Dutch complex arithmetic: Latent variable modeling of change. *Psychometrika, 74,* 331–350.
Hickendorff, M., van Putten, C. M., Verhelst, N., and Heiser, W. J. 2010. Individual differences in strategy use on division problems: Mental versus written computation. *Journal of Educational Psychology, 102,* 438–452.
Hofman, A., and De Boeck, P. 2011. Distributional assumptions and sample size using crossed random effect models for binary data: A recovery study based on the lmer function. Unpublished paper.
Hung, L.-F. 2011. Formulating an application of the hierarchical generalized random-situation random-weight MIRID. *Multivariate Behavioral Research, 46,* 643–668.
Ivie, J. L., and Embretson, S. 2010. Cognitive process modeling of spatial ability: The assembling objects task. *Intelligence, 38,* 324–335.
Jansen, M. G. H. 1997. Rasch's model for reading speed with manifest explanatory variables. *Psychometrika, 62,* 393–409.
Janssen, R., Schepers, J., and Peres, D. 2004. Models with item and item group predictors. In P. De Boeck and M. Wilson (Eds.), *Explanatory Item Response Models.* (pp. 189–212). New York: Springer.
Jara, A., Hanson, T., Quintana, F. A., Müller, P., and Posner, G. L. 2011. DPackage: Bayesian semi- and nonparametric modeling in R. *Journal of Statistical Sofware, 40(5),* 1–30.
Joe, H. 2008. Accuracy of Laplace approximation for discrete response mixed models. *Journal of Computational Statistics and Data Analysis, 52,* 5066–5074.
Johnson, T. R. 2003. On the use of heterogeneous thresholds ordinal regression models to account for individual differences in response style. *Psychometrika, 68,* 563–583.
Johnson, T. R. 2007. Discrete choice models for ordinal response variables: A generalization of the stereotype model. *Psychometrika, 72,* 489–504.
Klein Entink, R. H., Kuhn, J.-T., Hornke, L. F., and Fox, J.-P. 2009. Evaluating cognitive theory: A joint modeling approach using responses and response times. *Psychological Methods, 14,* 54–75.
Maydeu-Olivares, A., and García-Forero, C. 2010. Goodness of fit testing. In P. Peterson, E. Baker, and B. McGaw (Eds.), *International Encyclopedia of Education* (3rd ed.) (pp. 190–196). Oxford: Elsevier.
McArdle, J. J. 2009. Latent variable modeling of differences and change with longitudinal data. *Annual Review of Psychology, 60,* 577–605.

McCulloch, C. E., and Searle, S. R. 2001. *Generalized, Linear, and Mixed Models*. New York: Wiley.

Meulders, M., and Xie, Y. 2004. Person-by-item predictors. In P. De Boeck, and M. Wilson (Eds.), *Explanatory Item Response Models* (pp. 213–240). New York: Springer.

Revuelta, J. 2008. The generalized logit-linear item response model for binary-designed items. *Psychometrika, 73*, 385–405.

Rijmen, F., Tuerlinckx, F., De Boeck, P., and Kuppens, P. 2003. A nonlinear mixed model framework for item response theory. *Psychological Methods, 8*, 185–205.

Rupp, A. A., Templin, J., and Henson, R. A. 2010. *Diagnostic Measurement*. New York: Guilford Press.

Scheiblechner, H. 1972. Das Lernen und Lösen komplexer Denkaufgabben [Learning and solving complex thinking tasks]. *Zeitschrift für Experimentelle und Angewandte Psychologie, 19*, 476–506.

Sijtsma, K. 2012. Future of psychometrics: Ask what psychometrics can do for psychology. *Psychometrika, 77*, 4–20.

Stram, D. O., and Lee, J. W. 1994. Variance components testing in the longitudinal mixed-effects model. *Biometrics, 50*, 1171–1177.

Stram, D. O., and Lee, J. W. 1995. Correction to: Variance components testing in the longitudinal mixed-effects model. *Biometrics, 51*, 1196.

Thissen, D.,and Steinberg, L. 1986. A taxonomy of item response models. *Psychometrika, 51*, 567–577.

Tutz, G. 1990. Sequential item response models with ordered response. *British Journal of Mathematical and Statistical Psychology, 43*, 39–55.

Verbeke, G., and Molenberghs, G. 2003. The use of score tests for inference on variance components. *Biometrics, 59*, 254–262.

Wang, W.-C., and Jin, K.-Y. 2010. Multilevel, two-level and random-weights generalizations of a model with internal restrictions on item difficulty. *Applied Psychological Measurement, 34*, 46–65.

Wang, W.-C., and Wu, S.-L. 2011. The random-effect generalized rating scale model. *Journal of Educational Measurement, 48*, 441–456.

Whitely, S. E. 1976. Solving verbal analogies: Some cognitive components of intelligence test items. *Journal of Educational Psychology, 68*, 234–242.

Whitely, S. E. 1977. Information processing on intelligence test items: Some response components. *Applied Psychological Measurement, 1*, 465–476.

Wilson, M. 2008. Cognitive diagnosis using item response models. *Zeitschrift für Psychologie, 216*, 74–88.

Wilson, M. 2009. Measuring progressions: Assessment structures underlying a learning progression. *Journal for Research in Science Teaching, 46*, 716–730.

Wilson, M., and De Boeck, P. 2004. Descriptive and explanatory item response models. In P. De Boeck, and M. Wilson (Eds.), *Explanatory Item Response Models* (pp. 43–74). New York: Springer.

Wilson, M., and Moore, S. 2011. Building out a measurement model to incorporate complexities of testing in the language domain. *Language Testing, 28*, 441–462.

Wilson, M., Zheng, X., and McGuire, L. 2012. Formulating latent growth using an explanatory item response model approach. *Journal of Applied Measurement, 13*, 1–22.

Zwinderman, A. H. 1991. A generalized Rasch model for manifest predictors. *Psychometrika, 56*, 589–600.

Index

A

Acceleration model, 101, 102, 104, 106; *see also* Graded response model (GRM)
ACL, *see* Adjective Check List (ACL)
Adjacent categories logits, 143
Adjacent category logit link, 508
Adjective Check List (ACL), 159
AEQ, *see* Alcohol Expectancy Questionnaire (AEQ)
Aggregation bias, 407; *see also* Multilevel response models
AIC, *see* Akaike information criterion (AIC)
Akaike information criterion (AIC), 57
Alcohol Expectancy Questionnaire (AEQ), 473
  CAPS-AEQ questionnaire, 478–479
Analysis-of-variance (ANOVA), 485
ANOVA, *see* Analysis-of-variance (ANOVA)
Arc-length metric, 342–343
Ashworth spasticity scale, 87; *see also* Rasch rating-scale model (RSM)
Attitude measurement, *see* Item difficulty; latitude of acceptance
Automated item generation, 211, 214, 446, 567
Automated item-selection procedure, 314, 315, 319
Automated machine-scoring, 449, 460
Average speed of moving object, 265

B

Bartlett's least squares, 157
Bayesian estimation, 217, 471, 491, 518, 527, 531, 534, 538
Bayesian information criterion (BIC), 57
Bayesian model fit methodology, 26; *see also* Unidimensional logistic response models
Bayesian nonparametric modeling approach (BNP modeling approach), 324; *see also* Bayesian nonparametric response models
Bayesian nonparametric response models, 323
  basic, 326
  conditional likelihood, 330
  covariates, 323
  discrete-time Harris ergodic Markov chain, 330
  discussion, 333–334
  empirical example, 332–333
  extensions, 329
  mixture IRT and BNPs, 325
  model fit, 331
  model presentation, 326–330
  parameter estimation, 330–331
Bayesian structural equation modeling (BSEM), 538; *see also* Mplus item response modeling technique
Bernoulli distribution, 14, 160, 268, 472
BIC, *see* Bayesian information criterion (BIC)
Bifactor dimension reduction, 425; *see also* Two-tier item factor analysis modeling
Bifactor model, 183, 184, 421, 422, 423, 433; *see also* Two-tier item factor analysis modeling
Binet, Alfred, 1–3; *see also* Item response theory (IRT)
  parameter separation application, 8
Binomial distribution, *see* Bernoulli distribution
BNP-IRT model, *see* Bayesian nonparametric response models
BNP modeling approach, *see* Bayesian nonparametric modeling approach (BNP modeling approach)
Bootstrap test, 400, 570
Boundary separation, 285, 287, 289, 298
BSEM, *see* Bayesian structural equation modeling (BSEM)
Bundle independence, 557; *see also* Mixed-coefficients multinomial logit models

C

CAPS-AEQ questionnaire, 478–479; *see also* Randomized response models
CAPS, *see* College Alcohol Problem Scale (CAPS)
Category
  order, 66, 529
  threshold, 128, 371, 372, 375, 378, 379, 386
Category boundary discrimination (CBD), 68
Category characteristic curves (CCCs), 90
CBD, *see* Category boundary discrimination (CBD)
CCCs, *see* Category characteristic curves (CCCs)

cdf, *see* Cumulative distribution function (cdf)
Censoring, 506, 510
CFA, *see* Confirmatory factor analysis (CFA)
Cheating detection, 284
Chebychev–Hermite polynomials; *see also* Hermite–Chebychev polynomials
Classical test theory (CTT), 1
CML, *see* Conditional maximum-likelihood (CML)
Coefficient omega, 159
Cognitive
  complexity model, 225, 226, 228, 236, 238, 240
  component, 195, 226, 236
  diagnosis, 7, 221, 236, 567
College Alcohol Problem Scale (CAPS), 473
  CAPS-AEQ questionnaire, 478–479
Compensatory MIRT model, 191
Computer adaptive testing (CAT), 110
Conditional association, 310, 311, 313
Conditional logit model, 509
Conditional maximum-likelihood (CML), 17, 552; *see also* Partial credit model (PCM)
  procedure, 115–116
Confirmatory factor analysis (CFA), 533; *see also* Mplus item response modeling technique
Confirmatory item factor analysis model, 421; *see also* Two-tier item factor analysis modeling
Congeneric model, 161; *see also* Models for continuous responses
Consecutive steps, 139
Consistency, 91
Contextual effect, 517
Continuation-ratio logit link, 508
Covariates, 323
  item, 511
  rater, 452–453
  types, 569–570
Cross level effect, 408, 516, 520, 574
CTT, *see* Classical test theory (CTT)
Cumulative distribution function (cdf), 324
Cumulative probability models, 144; *see also* Sequential mechanism
Curve-of-factors model, 513
Customized fit test, 556
Cut score, 307, 319

D

D-Diffusion Mode, 287
DDP, *see* Dependent Dirichlet process (DDP)
Decision making, 110, 285, 566, 578
Dependent Dirichlet process (DDP), 326
Design matrix, 439, 452
  item-, 213, 214, 215, 220, 221
Deterministic-input noisy-AND (DINA), 236
Deviance information criterion (DIC), 414, 415
D-factor model, 155; *see also* Models for continuous responses
DFF, *see* Differential facet functioning (DFF)
DIC, *see* Deviance information criterion (DIC)
Differential facet functioning (DFF), 216; *see also* Linear logistic models
Differential item functioning (DIF), 161, 216, 512
Diffusion-based response-time models, 283, 284–287
  D-diffusion model, 287–289
  discussion, 297–298
  distance-difficulty hypothesis, 288
  empirical example, 292–297
  ICCs, 289, 290
  model fit, 292
  parameter estimation, 291
  parameters, 285
  Q-diffusion model, 289
  relations to other models and ideas, 289–291
  for two-choice decisions, 285
Diffusion model, 284–287, *see* Diffusion-based response-time models
DIF, *see* Differential item functioning (DIF)
DINA, *see* Deterministic-input noisy-AND (DINA)
Direct questioning (DQ), 474
Dirichlet process (DP), 325
Distance-difficulty hypothesis, 288
Double monotonicity model, 305, 306, 307; *see also* Mokken models
Doubly latent contextual model, 517
DP, *see* Dirichlet process (DP)
DQ, *see* Direct questioning (DQ)
Drift rate, 285, 287, 289, 290, 292, 294, 297, 298

E

EAP, *see* Expected a posteriori (EAP)
Ecological fallacy, *see* Aggregation bias
Educational research data, 407
EFA, *see* Exploratory factor analysis (EFA)
EIRM, *see* Explanatory item response modeling (EIRM)
Elementary symmetric function, 17, 83
EM algorithm, *see* Expectation–maximization algorithm (EM algorithm)

EPPCs, *see* Extended posterior predictive checks (EPPCs)
EPPP, *see* Extended posterior predictive *p*-value (EPPP)
ESEM, *see* Exploratory structural equation modeling (ESEM)
EVC, *see* Expected value curve (EVC)
Expectation–maximization algorithm (EM algorithm), 18, 103
Expected a posteriori (EAP), 84
Expected value curve (EVC), 89
Explanatory item response modeling (EIRM), 565; *see also* Explanatory response models
Explanatory item response theory, 486
Explanatory response models, 565
  covariate types, 569–570
  data format with wide and long form, 566
  discussion, 577–578
  empirical example, 571–575
  explanatory and descriptive measurement, 567
  general model, 569
  initial model formulation, 568
  latent variable, 566
  literature review, 567
  long data form, 568
  measurement and explanation, 566–567
  model fit, 570–571, 576–577
  model presentation, 568
  parameter estimation, 570, 575
  psychometrics, 566
Exploded logit model, 510
Exploratory factor analysis (EFA), 532; *see also* Mplus item response modeling technique
Exploratory structural equation modeling (ESEM), 538; *see also* Mplus item response modeling technique
Exponential family distribution, 505
Extended competency-demand hypothesis, 217
Extended mixed coefficient multinomial logit model, 542–544; *see also* Mixed-coefficients multinomial logit models
Extended posterior predictive checks (EPPCs), 441
Extended posterior predictive *p*-value (EPPP), 441
Extra-Poisson variation, *see* Overdispersion variation
Extreme responding, 52, 56, 70

F

Facets model, 212, 449, 461, 541, 544, 545, 546, 562
Factor loading, 155, 175, 179, 475, 534, 537
Fit matrix, 556; *see also* Mixed-coefficients multinomial logit models
Fixed effect, 14, 17, 18, 20, 215, 253, 477
Fixed rating effects model, 453; *see also* Hierarchical rater models (HRM)
Fourier basis, 55

G

Gamma distribution, 505
General component latent trait model (GLTM), 226; *see also* Multicomponent models
Generalizability theory (GT), 461
Generalized graded unfolding model (GGUM), 369
  discussion, 386
  EAP estimate, 380
  empirical example, 382–385
  GGUM2004 program, 377, 381
  item characteristic curve, 374, 375
  latent response, 370
  model fit, 380–382
  model presentation, 370–375
  observed response, 370
  parameter estimation, 375–380
  parameter-specific derivative, 386–388
  probability curves, 371, 373
  probability function, 372
  thresholds, 372
Generalized latent variable (GLV), 290
Generalized least squares (GLS), 178
Generalized linear latent and mixed model framework (GLLAMM framework), 504; *see also* Generalized linear latent and mixed modeling
Generalized linear latent and mixed modeling, 503
  comparative responses, 508–510
  continuous responses and response times, 510–511
  conventional MIRT, 512
  counts, 510
  dichotomous responses, 507
  discussion, 522–523
  example, 519–522
  GLLAMM framework, 505
  growth IRT, 513–514

Generalized linear latent and mixed modeling (*Continued*)
- linear predictor, 506, 513
- links and conditional distributions, 505–506
- model fit, 518
- multidimensional models, 512
- multilevel IRT, 514–515
- multilevel models, 514
- multilevel SEM, 516–517
- ordinal responses, 508
- parameter estimation, 517–518
- path diagrams for structural equation models, 516
- person and item covariates, 511
- random coefficients of item characteristics, 513
- response model, 505
- response processes, 507, 511
- structural model, 506–507, 515, 516
- two-level IRT model, 515
- variance-components IRT model, 515

Generalized linear latent model, 147, 328, 461, 504
Generalized linear latent modeling (GLLAM), 26
Generalized linear mixed models (GLMMs), 216, 504; *see also* Generalized linear latent and mixed modeling
Generalized linear models (GLMs), 504
Generalized partial credit model (GPCM), 55, 101, 127; *see also* General-purpose multidimensional nominal model
- discussion, 136
- empirical example, 134–136
- E-step, 131–132
- expected value of scoring function, 130
- goodness of fit, 133–134
- ICRFs, 128
- model presentation, 127
- M-step, 132–133
- parameter estimation, 130–131
- polytomous item response models, 129

Generalized partial credit model, 147, 408, 504, 566
General-purpose multidimensional nominal model, 54; *see also* Nominal categories models
- Fourier version for linear effects and smoothing, 55
- identity-based **T** matrix for equality constraints, 55
- nominal MIRT model, 54
- nominal model and IRT models, 56

GGUM, *see* Generalized graded unfolding model (GGUM)
Gibbs sampler, 492–493; *see also* Joint hierarchical modeling
GLLAMM framework, *see* Generalized linear latent and mixed model framework (GLLAMM framework)
GLLAM, *see* Generalized linear latent modeling (GLLAM)
GLMs, *see* Generalized linear models (GLMs)
GLS, *see* Generalized least squares (GLS)
GLTM, *see* General component latent trait model (GLTM)
GLV, *see* Generalized latent variable (GLV)
GOF methods, *see* Goodness-of-fit methods (GOF methods)
Goodness-of-fit methods (GOF methods), 570; *see also* Explanatory response models
Graded response model (GRM), 77, 95, 114; *see also* Partial credit model (PCM)
- acceleration model, 101
- asymptotic basic function, 98
- Bock's nominal response model, 100
- category response function, 96
- cumulative score category response function, 97
- discussion, 106
- general, 96–97
- goodness of fit, 104–105
- heterogeneous case, 100–102, 105–106
- homogeneous case, 98–100
- LPEF models, 105–106
- model presentation, 95–96
- multilog, 103
- parameter estimation, 102–104
- score category response function, 99
- step acceleration parameter, 101

Growth model, 513
GT, *see* Generalizability theory (GT)
Gumbel distribution, 508
Guttman vector, 80, 86

H

HADS, *see* Hospital Anxiety and Depression Scale (HADS)
Harmonic analysis, 178
HCM, *see* Hyperbolic cosine model (HCM)
Hermite-Chebychev polynomials, 171, 173, 174, 176; *see also* Normal-Ogive multidimensional models
Hermite polynomials, *see* Hermite–Chebychev polynomials

HGLM, *see* Hierarchical generalized linear model (HGLM)
Hidden structure, 393
Hierarchical generalized linear model (HGLM), 461
Hierarchical rater models (HRM), 449
  augmented signal-detection model, 453
  discussion, 461–462
  estimation, 453–455
  example, 456–460
  fixed rating effects model, 453
  ideal rating, 451
  IRT model for item responses, 451
  model fit assessment, 455–456
  random rating effects model, 453
  rater accuracy, 451–452
  rater covariates, 452–453
  rating probability table for signal-detection process, 451
  signal-detection model, 452
  spread parameter, 452
  uninformative priors for HRM parameters, 454
Highest posterior density (HPD), 416
Hospital Anxiety and Depression Scale (HADS), 315; *see also* Mokken models
HPD, *see* Highest posterior density (HPD)
HRM, *see* Hierarchical rater models (HRM)
Hyperbolic cosine model (HCM), 353
  constraint on parameter, 360–361
  deterministic unfolding and cumulative and response patterns, 354
  discussion, 365
  example, 362–364
  goodness of fit, 361–362
  inconsistency of parameter estimates, 360
  initial estimates, 360
  latitude of acceptance, 357–358
  model presentation, 353–357
  parameter estimation, 358–359
  parameterization, 355
  person's ideal point, 354
  response functions, 355
  solution algorithm, 359

I

ICCs, *see* Item characteristic curves (ICCs)
ICRF, *see* Item-category response function (ICRF)
Ideal rating, 451, 461
Identity link, 510
ID, *see* Interthreshold distance (ID)
IEA, *see* International Association for the Evaluation of Educational Achievement (IEA)
Incomplete block design, 38
Intelligence, 2
Intelligence quotient (IQ), 2
International Association for the Evaluation of Educational Achievement (IEA), 110
International Personality Item Pool (IPIP), 400
Interthreshold distance (ID), 375
Invariant item ordering, 307
IPIP, *see* International Personality Item Pool (IPIP)
IQ, *see* Intelligence quotient (IQ)
IRF, *see* Item response function (IRF)
IRR, *see* Item randomized-response (IRR)
IRT-MMR, *see* Item response theory model for multiple raters (IRT-MMR)
IRT, *see* Item response theory (IRT)
Item; *see also* Item-family models; Two-tier item factor analysis modeling
  bundle, 461, 557, 558, 559, 562
  covariates, 511
  curve, 167
  difficulty, 177
  difficulty model, 236
  factor analysis, 421
  families, 437
Item-category response function (ICRF), 128
Item characteristic curves (ICCs), 167, 289
Item-family models, 437
  discussion, 446
  empirical example, 442–446
  family information, 442
  general model formulation, 438–439
  model fit, 440
  model presentation, 438
  optimal test design, 441–442
  parameter estimation, 440
  posterior predictive *p*-value, 441
  restrictions on model, 439
Item randomized-response (IRR), 468
Item response function (IRF), 337
Item response theory (IRT), 1
  Binet, Alfred, 1–3
  contributions to, 6–7
  diagrams underlying IRT models, 8
  Lord, Frederic, 5
  models, 504
  multilevel modeling perspective on, 408–409
  parameter separation principle, 8
  Rasch, George, 5–6
  Thurstone, Louis, 3–5
  unifying features of, 8–9

Item response theory model for multiple raters (IRT-MMR), 461; *see also* Hierarchical rater models (HRM)

J

JML, *see* Joint maximum likelihood (JML)
Joint hierarchical modeling, 481
  alternative plug-in models, 490–491
  dependency structure of data, 491
  discussion, 499
  empirical example, 496–499
  first-level models, 487
  Gibbs sampler, 492–493
  hierarchical framework, 486
  higher-level models, 489
  identifiability, 489–490
  item fit, 496
  levels of modeling, 484–486
  model fit, 493
  model presentation, 486
  parameter estimation, 491–493
  person fit of response model, 494–495
  person fit of RT model, 495–496
  person-fit test, 494
  responses and RTs by two test takers, 482
  response-time model, 487
  SAT curves, 484
  second-level models, 487–489
Joint maximum likelihood (JML), 17, 36
Joint model, 291, 486, 496, 497, 499

L

Lagrange multiplier (LM), 26
  tests, 275
Laplace approximation, 147, 570, 575
Latent class analysis (LCA), 399, 400
Latent invariant item ordering, 312–313
Latent profile model, 154, 155, 156, 161
Latent response variable (LRV), 529
Latent structure, 80, 394, 395, 399
Latent variable, 2, 566; *see also* Models for continuous responses
  models, 153
Latitude of acceptance, 357–358; *see also* Hyperbolic cosine model (HCM)
LCDM, *see* Loglinear cognitive diagnosis model (LCDM)
LID, *see* Local dependencies (LID)
Likelihood-ratio (LR), 570
Likert scale, 75
Linear logistic models, 211, 213; *see also* Linear logistic test model (LLTM)
  discussion, 221
  empirical example, 217–221
  extended competency-demand hypothesis, 217
  item characteristics estimation, 214–215
  item variation incorporation, 214–216
  including local dependencies between items, 215
  logit expression for focal group, 216
  MIRID, 214–215
  model fit, 217
  model presentation, 213
  parameter estimation, 216
  RWLLTM, 215–216
Linear logistic model with relaxed assumptions (LLRA), 213
Linear logistic test model (LLTM), 211, 512, 567; *see also* Linear logistic models
  applications of, 212
  cognitive operation, 212
  2PL-constrained model, 213
  random-effects version of, 214
Linear predictor, 505, 506, 509, 512, 516
LLCA, *see* Located latent classes (LLCA)
LLRA, *see* Linear logistic model with relaxed assumptions (LLRA)
LLTM, *see* Linear logistic test model (LLTM)
LM, *see* Lagrange multiplier (LM)
Local dependencies (LID), 215
Local independence, 14, 154, 168, 265, 305
Located latent classes (LLCA), 36
Logistic mixture-distribution response models, 393
  discussion, 401–402
  EM algorithm, 397
  empirical example, 400
  estimating mixture IRT models, 398–399
  hidden structure, 393
  likelihood function, 396
  local independence assumption, 394
  MCMC estimation of mixture IRT models, 397
  model fit, 399–400
  model presentation, 394–396
  parameter estimation, 396–399
  Saltus model, 401
Logistic model, 20, 22, 25, 26, 28, 29, 133, 160, 204
Logistic multidimensional models, 189; *see also* MIRT models; Multidimensional item response theory (MIRT)

assumption, 191
discussion, 207–208
empirical example, 205–207
model fit, 204–205
model presentation, 190
parameter estimation, 203–204
Logistic positive exponent family model (LPEF model), 105–106; *see also* Graded response model (GRM)
Logit link, 411, 505, 507, 508, 568
Loglinear cognitive diagnosis model (LCDM), 236
Lognormal distribution, 247, 257, 267, 275, 291
Lognormal response-time model, 261; *see also* Response times (RTs)
adjusted sampler, 274
assumptions, 264–266
discrimination parameter, 268
discussion, 278–279
empirical example, 276–278, 279, 280
formal model, 266–267
model fit, 274–276
model presentation, 264
moments of RT distributions, 270–271
moving object speed, 265
parameter estimation, 272–274
parameter identifiability and linking, 268–270
parameter interpretation, 267–268
probability integral transform theorem, 275
relationships with other models, 271
Roskam's model, 263
RT patterns, 262
speed, 266
Lord, Frederic, 5; *see also* Item response theory (IRT)
two-parameter model, 6
LPEF model, *see* Logistic positive exponent family model (LPEF model)
LR, *see* Likelihood-ratio (LR)
LRV, *see* Latent response variable (LRV)

M

MAEs, *see* Mean absolute errors (MAEs)
MAP, *see* Maximum a-posteriori (MAP)
Marginal distribution, 395
Marginal maximum *a posteriori* (MMAP), 375
Markov chain Monte Carlo methods (MCMC methods), 25; *see also* Unidimensional logistic response models
Martin-Löf test, 42
MASI, *see* Multiple-attempt single-item (MASI)
Maximum a-posteriori (MAP), 553
Maximum likelihood (ML), 56
Maximum likelihood estimation (MLE), 44
Maximum marginal likelihood (MML), 18, 36, 116–117; *see also* Partial credit model (PCM)
MCAR, *see* Missing completely at random (MCAR)
MCCIs, *see* Monte Carlo confidence intervals (MCCIs)
MCMC methods, *see* Markov chain Monte Carlo methods (MCMC methods)
MDISC, *see* Multidimensional discrimination (MDISC)
Mean absolute errors (MAEs), 444
Measurement precision, 407
Mehler identity, 174
Memorial symptom assessment scale—short form (MSAS-SF), 66
Mental functions, 2
Metropolis–Hastings (MH), 492
MH, *see* Metropolis–Hastings (MH)
MIMIC, *see* Multiple indicator multiple-cause (MIMIC)
MIRT models, *see* Multidimensional item response theory models (MIRT models)
Missing completely at random (MCAR), 450
Mixed-coefficients multinomial logit models, 541
between-and within-item multidimensionality, 545
bundle independence, 557
combined model, 548
conditional maximum-likelihood estimation, 552–553
customized approaches, 560–561
customized fit tests, 556
discussion, 562–563
empirical comparisons, 561–562
empirical example, 557
estimating functionals of population distributions, 553–554
extended mixed coefficient multinomial logit model, 542–544
generalized fit test, 555
item bundles, 557
latent ability estimation and prediction, 553
maximum-likelihood estimation, 549–550
model fit, 555
model identification, 548–549

Mixed-coefficients multinomial logit models (*Continued*)
- model presentation, 542
- models with scoring parameters, 546
- parameter estimation, 549
- PCM bundles, 559
- plausible values, 554
- population model, 547–548
- quadrature and Monte Carlo approximations, 550–552
- saturated bundle model, 557–559
- simple logistic model, 544–546
- SLM bundles, 559–560
- tests of relative fit, 556–557

Mixture IRT models, 399; *see also* Logistic mixture-distribution response models

MLE, *see* Maximum likelihood estimation (MLE)

ML, *see* Maximum likelihood (ML)

MLTM-D, *see* Multicomponent latent trait model for diagnosis (MLTM-D)

MLTM, *see* Multicomponent latent trait model (MLTM)

MMAP, *see* Marginal maximum *a posteriori* (MMAP)

MML, *see* Maximum marginal likelihood (MML)

Model
- Luce-Plackett, 510
- Luce's choice, 51
- MIMIC, 512
- mixed coefficients multinomial logit, 541
- mixture, 470–471
- mixture generalized partial credit, 396
- Mokken, 303
- monotone homogeneity, 305
- multidimensional mixture, 395
- multilevel, 407, 408
- nominal response, 101
- normal-ogive, 167, 184–185
- one-parameter logistic, 13, 42, 215, 235
- partial credit, 109, 559
- polytomous, 129
- polytomous mixture, 427
- proportional hazard, 511
- Q-diffusion, 289
- random weights linear logistic test model, 215–216
- ranking, 510
- Rasch, 5–6, 31, 306, 323
- rating scale, 56, 75, 129
- response-time, 487
- RWLLTM, 215
- Saltus, 401
- sequential model, 139
- sequential sampling, 283
- signal detection, 451
- Spearman's one-factor, 155
- step, 140
- structural equation, 504
- testlet, 422
- unfolding, 369
- unidimensional, 13, 189

Models for continuous responses, 153
- congeneric model, 161
- *D*-factor model, 155
- discussion, 160–162
- empirical example, 159–160
- item parameters, 157
- latent variable models, 153–154
- model fit, 159
- model presentation, 154–157
- observed test-score precision, 158–159
- parameter estimation, 157
- person parameters, 157
- precision of estimates, 157–158
- response scale, 153
- Spearman's one-factor model, 155

Model with an internal restriction on item difficulty (MIRID), 214–215, 567; *see also* Explanatory response models; Linear logistic models

Mokken models, 303
- conditional association and local independence, 310–311
- discussion, 319
- double-monotonicity model, 306, 307
- empirical example, 315–319
- invariant item ordering, 307
- item selection, 313–314
- manifest and latent invariant item ordering, 312–313
- manifest monotonicity and latent monotonicity, 311–312
- measurement of attributes, 304
- model fit, 310
- model presentation, 304–308
- monotone homogeneity model, 305
- MSP, 311
- nonparametric IRT models, 305
- parameter estimation, 308–309
- person fit and reliability, 314–315
- polytomous items, 307
- Rasch model, 306
- reliability of total score, 314
- scalability coefficient, 309

Mokken Scale analysis for Polytomous items (MSP), 311; *see also* Mokken models
Monotone homogeneity model, 305; *see also* Mokken models
Monotonicity
latent response function, 311–312
manifest response function, 312–313
Monte Carlo confidence intervals (MCCIs), 331
Monte Carlo integration, 442
Mosier, 4
Mplus item response modeling technique, 527, 537–538
Bayesian estimation, 531
Bi-factor EFA solution using WLSMV, 532
empirical examples, 531
estimation, 530
item bi-factor confirmatory factor analysis, 535–536
item bi-factor exploratory factor analysis, 532, 533
modeling, 528
response percentages for aggression items, 528
single-level modeling, 528
two-level modeling, 529–530
WLSMV, 530–531
MSAS-SF, *see* Memorial symptom assessment scale—short form (MSAS-SF)
MSP, *see* Mokken Scale analysis for Polytomous items (MSP)
Multicomponent latent trait model (MLTM), 226; *see also* Multicomponent models
Multicomponent latent trait model for diagnosis (MLTM-D), 226, 228; *see also* Multicomponent models
full component model, 235
null model, 235
potential for, 236
restricted component model, 235
Multicomponent models, 225
advantages of, 240
applications of, 226
cognitive components in mathematical achievement items, 236–239
discussion, 239–240
empirical examples, 234
fit index, 233
goodness of fit, 233
history of models, 226–228
model fit, 232–234
model presentation, 228
parameter estimation, 231–232
regression of total item response probability, 229
standards-based mathematical skills, 234–236
for subtask data, 228–230
tasks for mathematical word problem, 227
types of, 228
variables in revised cognitive model, 237
for varying components, 230–231
Multidimensional discrimination (MDISC), 197
index, 177
Multidimensional item response theory (MIRT), 54, 189–190; *see also* Logistic multidimensional models
estimation procedures for, 203
Multidimensional item response theory models (MIRT models), 173, 189–190, 225, 226; *see also* Logistic multidimensional models; Multicomponent models
compensatory, 191
directional derivative, 199
forms of, 194–195
information functions for item, 199
item vector, 198, 202
logistic form of, 192
multidimensional discrimination, 197
partially compensatory model, 194
pseudo-guessing parameter, 193
reference composite, 201, 202
test characteristic surface, 201
test information functions, 201
test item and, 196
vector plot of items in three dimensions, 198
Multilevel modeling, 408; *see also* Multilevel response models
Multilevel response models, 407
Bayesian multilevel IRT modeling, 409
DIC, 414
discussion, 417
empirical example, 415–417
GLMM presentation, 411
level-2 model, 410
mixture IRT model, 412
model fit, 414
model presentation, 410
multilevel IRT model, 410–411
multilevel IRT with random item parameters, 412–413
multilevel modeling perspective on IRT, 408–409
multilevel Rasch model reformulation, 411
multiple-group IRT model, 411–412
parameter estimation, 413
two-parameter IRT model, 410

Multilog, 103; *see also* Graded response model (GRM)
Multiple-attempt single-item (MASI), 245
Multiple indicator multiple-cause (MIMIC), 512
Multiple population sampling, 393

N

NAEP, *see* National Assessment of Educational Progress (NAEP)
National Assessment of Educational Progress (NAEP), 134
Newton method, 40
Newton-Raphson method, 551
NLMM, *see* Nonlinear mixed models (NLMM)
NOHARM program, 178, 203, 205; *see also* Normal-Ogive multidimensional models
Nominal categories models, 51
  checking behavior of item responses, 68–70
  current research, 70
  determination of response category order, 66–68
  discussion, 65
  empirical examples, 58
  general-purpose multidimensional nominal model, 54–56
  model fit, 57–58
  model presentation, 52–53
  multidimensional example, 62–65
  original nominal categories model, 53–54
  parameter estimation, 56–57
  response pattern testlets, 65–66
  unidimensional example, 58–62
  uses of nominal model, 65
Nominal MIRT model, 54; *see also* General-purpose multidimensional nominal model
Noncompliance, 468, 470
Nondecision time, 285
Nonlinear factor analysis approach, 168
Nonlinear mixed models (NLMM), 217
Nonnested clustering, 217
Nonparametric IRT models, 305; *see also* Mokken models
Normal distribution, 27, 36, 276, 291, 312, 319
Normal-Ogive multidimensional models, 167, 184–185
  asymptotic distribution of estimators, 178–181
  empirical example, 183–184
  goodness of fit, 182–183
  Hermite–Chebychev polynomials, 171
  item characteristic curve, 167
  item curve, 167
  item difficulty, 177
  Mehler identity, 174
  model presentation, 170–177
  multidimensional item response models, 173
  NOHARM program, 178
  nonlinear factor analysis approach, 168
  parameter estimation, 177–178
  strong principle of local independence, 168
  theorem, 179–181
  wide-sense linear model, 169

O

OECD, *see* Organisation for Economic Co-operation and Development (OECD)
Offset, 411, 511
One-parameter logistic model (OPLM), 13, 42, 215, 235
1PL model, *see* One-parameter logistic model (OPLM)
Operating density characteristic, 99
Optimal test design, 441–442
Organisation for Economic Co-operation and Development (OECD), 415
Original nominal categories model, 53–54; *see also* Nominal categories models
Overdispersion variation, 247

P

Paired comparisons, 35, 51
Pairwise comparison, 115, 510
Parameter; *see also* Item response theory (IRT); Unidimensional logistic response models
  acceleration, 101
  discrimination, 268
  dispersion, 110
  estimation methods, 25–26
  identifiability, 268–270
  invariance, 393
  linking, 268–270
  random item, 157, 412–413
  separability, 8
  separation principle, 8
Partial credit model (PCM), 109, 559; *see also* Mixed-coefficients multinomial logit models
  applications, 110

bundle approach, 559
calibrating markers, 120–123
CML estimation, 115–116
to define regions of measurement continuum using, 112
discussion, 123
empirical example, 119
goodness of fit, 118
marginal maximum-likelihood estimation, 116–117
measuring essay quality, 120
model presentation, 111–112
parameter estimation, 115
parameter interpretation, 113–114
person ability parameter, 113
relations to other models, 114–115
response functions, 112–113
unweighted mean-square, 118–119
weighted mean-square, 119
Partial credit parameterization, 84; *see also* Rasch rating-scale model (RSM)
Path diagram, 422, 516, 520
PCM, *see* Partial credit model (PCM)
pdf, *see* Probability density function (pdf)
Performance
index, 341
manifold, 344, 348
Personality assessment, 288
Person fit
coefficient, 314
index, 42
Person-fit test, 494; *see also* Joint hierarchical modeling
Person's ideal point, 354; *see also* Hyperbolic cosine model (HCM)
PGSI, *see* Problem gambling severity index (PGSI)
PIRLS, *see* Progress in International Reading Literacy study (PIRLS)
PISA, *see* Programme for International Student Assessment (PISA)
Pitman–Koopman–Darmois theorem, 6; *see also* Item response theory (IRT)
Plausible values, 554
PME, *see* Posterior modal estimation (PME)
pmf, *see* Probability mass function (pmf)
PM, *see* Posterior mean (PM)
Poisson; *see also* Rasch model for speed; Rasch Poisson counts model (RPCM)
and gamma models, 245
distribution, 246, 250, 211
process models, 245
Posterior mean (PM), 219
Posterior modal estimation (PME), 455
Posterior predictive checks (PPC), 219
Posterior predictive *p*-value, 441
Posterior standard deviation (PSD), 219
PPC, *see* Posterior predictive checks (PPC)
Preference measurement, 354
Probability density function (pdf), 324
Probability integral transform theorem, 275
Probability mass function (pmf), 324
Probit link, 507, 511, 568
Problem gambling severity index (PGSI), 68
Processing function, 96, 101
Proficiency levels, 221
Programme for International Student Assessment (PISA), 110, 415, 519
Progress in International Reading Literacy study (PIRLS), 519
Propensity distribution, 154
PSD, *see* Posterior standard deviation (PSD)
Psychometrics, 566
models for item response times, 246

Q

QM, *see* Quotient model (QM)
QQ-plot, 296
Quasi Newton method, 40, 343, 398
Quotient model (QM), 290; *see also* Diffusion-based response-time models

R

Random effect, 17, 18, 19, 20, 146, 257, 461, 514
Randomized item-response theory (RIRT), 468
Randomized response (RR), 468
Randomized response models, 467
CAPS-AEQ questionnaire, 478–479
disadvantages, 478
discussion, 477–478
empirical example, 473–477
forced RR model, 469
mixture model, 470–471
model fit, 473
model presentation, 468–469
noncompliant behavior, 470–471
parameter estimation, 471–473
randomized IRT models, 469–470
randomizing device, 468
response bias, 477
self-reports, 467
structural models for sensitive constructs, 471
Randomized response technique (RRT), 468

Random rating effects model, 453; *see also* Hierarchical rater models (HRM)
Random weights linear logistic test model (RWLLTM), 215–216; *see also* Linear logistic models
RAN, *see* Rapid automated naming (RAN)
Rapid automated naming (RAN), 255
Rasch, George, 5–6
Rasch model, 5–6, 31, 306, 323; *see also* Item response theory (IRT); Mokken models
  discussion, 44
  empirical example, 42–44
  model fit, 41–42
  model presentation, 32–35
  parameter estimation, 35–40
  parameterizations of, 23
  parameter separability, 35
Rasch model for speed, 249; *see also* Rasch Poisson counts model (RPCM)
  discussion, 257
  examples, 253–257
  factorization drawback, 250
  Gellman–Rubin statistics Brus test, 255
  LM-statistic, 252–253
  marginal likelihood function, 251
  model derivation, 249
  model extensions, 251
  model fit, 252–253
  parameter estimation, 250–251
Rasch Poisson counts model (RPCM), 246; *see also* Rasch model for speed
  discussion, 257
  expectation and variance, 247
  joint maximum likelihood estimation, 247
  maximum marginal likelihood estimation, 248–249
  model derivation, 246
  overdispersion or extra-Poisson variation, 247
  parameter estimation, 247
  probability of specific observed count, 246
Rasch rating-scale model (RSM), 75
  class of Rasch models, 76–77
  coefficients and scoring functions for, 77
  contiguous intervals for ordered categories, 76
  discussion, 92–93
  empirical example, 87–90
  empirical ordering of categories, 78–82
  Guttman Subspace, 80
  Guttman vector, 80, 86
  identical threshold parameters, 81
  interpretation of response structure and threshold disorder, 91
  interpreting threshold distances, 91–92
  item fit, 86
  item parameters, 82–84
  model fit, 85
  model presentation, 76
  noncollapsed adjacent categories, 77–78
  ordered category assessments in rating-scale format, 76
  parameter estimation, 82
  partial credit parameterization, 84
  person fit, 86–87
  person parameters, 85
  resolved-independent design, 79
  standard design, 78
  threshold order, 90–91
Rater
  covariates, 452–453
  human, 456–460
  machine, 456–460
Reading speed, 246; *see also* Rasch model for speed
Regression
  Cox, 511
  latent, 505, 507
  models, 565; *see also* Explanatory response models
Reliability, 314–315
Residual
  dependence, 430, 433
  time, 285
Response
  bias, 477
  bounded-continuous, 171
  caution, 285, 287
  comparative, 508–510
  dichotomous, 507
  direct, 362, 364
  forced-choice, 508
  graded, 77, 95
  ordinal, 508
  polytomous, 129
  randomized, 467
  style, 401, 462
  subjective, 215
  time, 510–511
Response data modeling, 337, 349–350
  arc-length metric, 342–343
  functional test analysis algorithm, 346–348
  introductory psychology data, 346–347
  metrics, 343
  modeling item response manifolds, 338–341

performance-indexing systems, 341
performance manifold estimation, 344
plotting performance manifolds, 348
projection into 3D principal-component subspace, 348
space curve, 337
sum score improvisation by weighting, 344–346
sum scores and weighted counterparts, 345
Thurstone tradition, 341–342
TSβ distribution with density, 343
uniform measure or rank, 342
Response function (RF), 354; *see also* Hyperbolic cosine model (HCM)
category, 95, 96
cumulative score category, 97
logistic, 13, 169
multivariate logistic, 52, 53
nonparametric, 323
single-peaked, 354
Response scale, 153; *see also* Models for continuous responses
Response times (RTs), 261, 481; *see also* Lognormal response-time model
density, 264
modeling, 263–264
patterns, 262
RF, *see* Response function (RF)
RIRT, *see* Randomized item-response theory (RIRT)
RMSEA, *see* Root mean squared error of approximation (RMSEA)
Root mean squared error of approximation (RMSEA), 57–58
RPCM, *see* Rasch Poisson counts model (RPCM)
RR, *see* Randomized response (RR)
RRT, *see* Randomized response technique (RRT)
RSM, *see* Rasch rating-scale model (RSM)
RTs, *see* Response times (RTs)
Rule-based item generation, 7, 439
RWLLTM, *see* Random weights linear logistic test model (RWLLTM)

S

SATs, *see* Speed–accuracy trade-offs (SATs)
Saturated bundle model, 557–559; *see also* Mixed-coefficients multinomial logit models
Scalability coefficient, 309
Scale
ordinal, 306, 474
rating, 56, 75, 129
Thurstone, 4
School's socioeconomic status (SES), 517
Score
sum, 344–346
total, 314
weighted sum, 345
Scoring
function, 77, 130
parameters, 546
rule, 2, 284
SDM, *see* Signal-detection model (SDM)
Self-reports, 467
SEMs, *see* Structural equation models (SEMs)
Sensitive constructs, 471
Sequential mechanism, 140; *see also* Sequential models
category probabilities, 142
cumulative probability models, 144
modeling of steps, 141–143
partial credit model, 143–144
sequential model and graded response model, 144
Sequential models, 139; *see also* Sequential mechanism
discussion, 149
empirical example, 148–149
joint maximum likelihood, 145–146
log-likelihood ratio statistic, 147
marginal maximum likelihood, 146–147
model fit, 147–148
model presentation, 140
parameter estimation, 144
SES, *see* School's socioeconomic status (SES)
Short portable mental status questionnaire (SPMSQ), 65
Signal-detection model (SDM), 451; *see also* Hierarchical rater models (HRM)
Simple logistic model (SLM), 77, 544–546; *see also* Mixed-coefficients multinomial logit models
bundles, 559–560
Simplex structure, 341, 380
SLM, *see* Simple logistic model (SLM)
Software
GGUM, 370
GLLAM, 26
IRTPRO, 56, 59, 204
Latent Gold, 26, 56
LEM, 398
Mdltm, 43
NOHARM, 176, 178, 205
SAS PROC MIXED, 56
WINMIRA, 45

Spearman's one-factor model, 155; *see also* Models for continuous responses
Specific objectivity, 35
Speed–accuracy trade-offs (SATs), 262, 483
Speededness, 265, 498
Speed of moving object, average, 265
SPMSQ, *see* Short portable mental status questionnaire (SPMSQ)
Statistic
  limited goodness-of-fit, 148, 233, 240
  localized t-statistics, 381
  M_2, 57, 295
Statistical test
  graphical model, 41
  likelihood-ratio, 57, 59
  Martin-Löf test, 42
  multilog, 103
  Person fit test, 494
Step model, 140; *see also* Sequential models
Stochastic process, 264, 286
Structural equation models (SEMs), 504
Subtasks, 215, 226
Sufficiency
  factorization theorem of sufficiency, 37
  sufficient statistic, 20

T

TCCs, *see* Threshold characteristic curves (TCCs)
Test administration effects, 212
TESTFACT, 203
Testlet response theory model, 422; *see also* Two-tier item factor analysis modeling
Tetrachoric series, 167, 174
Thissen's model, 246
Three-parameter logistic (3PL), 6, 13, 341, 487, 496
Three-parameter logistic model (3PL model), 6, 13; *see also* Item response theory (IRT)
  equations, 14
3RP, *see* Three random person effects (3RP)
3PL model, *see* Three-parameter logistic model (3PL model)
Three random person effects (3RP), 575; *see also* Explanatory response models
Threshold characteristic curves (TCCs), 90
Thurstone, Louis, 3–5; *see also* Item response theory (IRT)
  scale for pacifism–militarism, 4
  scaling of Binet test items, 3
Tilted scaled beta (TSβ), 338; *see also* Response data modeling
Time pressure, 287
TSβ, *see* Tilted scaled beta (TSβ)
Two-level model, 529; *see also* Mplus item response modeling technique
Two-parameter logistic model (2PL model), 13
Two-parameter model, 6; *see also* Item response theory (IRT)
2RI, *see* Two random item effects (2RI)
2PL-constrained model, 213; *see also* Linear logistic test model (LLTM)
2PL model, *see* Two-parameter logistic model (2PL model)
Two random item effects (2RI), 575; *see also* Explanatory response models
Two-tier item factor analysis modeling, 421
  conditional probability of response pattern, 427
  discussion, 433
  empirical example, 430–433
  marginal probability of response pattern, 428
  model fit, 429–430
  model presentation, 422–426
  parameter estimation, 426–428
  polytomous MIRT models, 427
  probability mass function, 427
  Q-point quadrature rules, 428
  two-tier model applications, 423–426
Two-tier item factor model, 422; *see also* Two-tier item factor analysis modeling
Typological variable, 393

U

UIRT, *see* Unidimensional item response theory (UIRT)
Unfolding models, 369; *see also* Generalized graded unfolding model (GGUM)
Unidimensional item response theory (UIRT), 189; *see also* Multidimensional item response theory (MIRT)
Unidimensional logistic response models, 13
  different parameterizations of Rasch model, 23
  discussion, 28–29
  empirical example, 26–28
  fixed-effects models, 14–17
  item map for test of body height, 28
  model fit, 26
  model identifiability, 19–21
  model interpretation, 21–25
  parameter estimation, 25–26

parameter linking, 21
presentation of models, 14
random-effects models, 17–19
reformulation of fixed-effects three-parameter normal ogive model, 17
response functions for set of 40 arithmetic items, 15
3PL model equations, 14

V

Validity
construct, 221, 225, 239
content, 221, 236
Variance components, 408, 515

W

Weighted Least-Squares (WLSMV), 530; *see also* Mplus item response modeling technique
Weighted likelihood estimation (WLE), 43
Wide-sense linear model, 169; *see also* Normal-Ogive multidimensional models
WLE, *see* Weighted likelihood estimation (WLE)
WLSMV, *see* Weighted Least-Squares (WLSMV)

◎ 编辑手记

著名数学家贾吉特·辛格(Jagjit Singh)曾指出:

> 正是由于数学方法的万能性和广泛性,使得它能够处理种类众多的问题,如空间的和运动的,机会的和概率的,统计学的和社会科学的,艺术的和文学的,逻辑学的和哲学的,音乐的和建筑的,战争的和政治的,食物的和医药的,伦琴射线的和晶体的,遗传的和继承的,人类思维的和电脑的.

本书是一部大型的英文版的应用统计学著作,是社会与行为科学中的统计学系列中的一本,中文书名或可译为《项目反应理论手册.第一卷,模型》.

本书借鉴了该领域国际知名专家的工作,并且介绍了所有主要的项目反应模型.本书为《项目反应理论手册》三卷书中的第一卷,涵盖了近20年内在项目反应理论中许多模型的发展,描述了不同反应模式或反应过程的模型,对更深入的参数化的需要是由于反应数据的多级或层次结构,以及其他拓展内容和见解.

在第一卷中,所有章节都有一个共同的格式,每一章都聚焦于一组模型或者建模方法.每章的介绍部分都包含一些模型的历史,以及它的重要的目的.之后的部分更为形式地展示了模型,对其参数进行了估计,展示了如何评估其与经验数据的拟合程度,通过一个经验性的例子解释模型的应用,并且讨论了其更加深远的应用和待解决的研究问题.

本书的特点为:

(1)包含了国际知名专家的贡献,包括在他们各自领域中的原创的和成功的工作历史;

(2)在所有章节中提供了扩展性的交叉参考和通用符号表示;

(3)覆盖了项目反应理论的主要模型,包括二分法和多节反应模型、反应时间、多维能力以及分层反应;

(4)强调了将所有模型视为统计严格方法的重要性.

本书的作者为维姆·J.范·德·林登(Wim J. van der Linden).他是太平洋度量公司的一位著名的科学家和研究创新的指导者,同时也是特文特大学测量和数据分析专业的退休教授.他的主要研究方向包括测试理论、计算机化自适性测试、优化测试集合、参数环绕、测试等值、反应时间模型以及决策理论与其在教育决策问题中的应用.

本书的版权编辑佟雨繁女士为了使读者能快速了解本书的大致内容,特翻译了本书的目录,如下:

统计工具目录

应用目录

前言

贡献者

1 简介
(Wim J. van der Linden)

第1部分 二分法模型

2 一维逻辑斯谛反应模型
(Wim J. van der Linden)

3 Rasch 模型
(Wim J. van der Linden)

第2部分 名义模型与顺序模型

4 名义分类模型
(David Thissen 和 Li Cai)

5 Rasch 评分-尺标模型
(David Andrich)

6 分次反应模型

(Fumiko Samejima)

7 部分信用模型

(Geoff N. Masters)

8 一般部分信用模型

(Eiji Muraki and Mari Muraki)

9 有序反应的顺序模型

(Gerhard Tutz)

10 连续反应模型

(Gideon J. Mellenbergh)

第 3 部分 多维与多分量模型

11 正规分布累积曲线多维模型

(Hariharan Swaminathan 和 H. Jane Rogers)

12 逻辑斯谛多维模型

(Mark D. Reckase)

13 线性逻辑斯谛模型

(Rianne Janssen)

14 多分量模型

(Susan E. Embretson)

第 4 部分 反应时间模型

15 读取速度与错误的泊松与 γ 模型

(Margo G. H. Jansen)

16 对数正态反应时间模型

(Wim J. van der Linden)

17 扩散有基反应时间模型

(Francis Tuerlinckx, Dylan Molenaar 和 Han L. J. van der Maas)

第 5 部分 非参数模型

18 Mokken 模型

(Klaas Sijtsma 和 Ioo W. Molenaar)

19 贝叶斯非参数反应模型

(George Karabatsos)

20 模型化反应数据的泛函方法

(James O. Ramsay)

第 6 部分 非单调项目模型

21 开折反应的双曲余弦模型

(David Andrich)

22 一般分级开折模型

(James S. Roberts)

第 7 部分 分层反应模型

23 逻辑斯谛混合分布反应模型
(Matthias von Davier 和 Jürgen Rost)

24 具有协变量和多重群的多级反应模型
(Jean-Paul Fox 和 Cees A. W. Glas)

25 双层项目因子分析建模
(Li Cai)

26 项目族模型
(Cees A. W. Glas, Wim J. van der Linden 和 Hanneke Geerlings)

27 分层评分模型
(Jodi M. Casabianca, Brian W. Junker 和 Richard J. Patz)

28 灵敏测量的随机反应模型
(Jean-Paul Fox)

29 反应和反应时间的联合分层建模
(Wim J. van der Linden 和 Jean-Paul Fox)

第 8 部分 一般建模方法

30 一般线性潜在和混合建模
(Sophia Rabe-Hesketh 和 Anders Skrondal)

31 多维,多级,多时间点项目反应模型
(Bengt Muthén 和 Tihomir Asparouhov)

32 混合系数多项对数单位模型
(Raymond J. Adams, Mark R. Wilson 和 Margaret L. Wu)

33 解释反应模型
(Paul De Boeck 和 Mark R. Wilson)

索引

统计工具目录

第 1 部分 基本工具

1 对数单位,概率单位与其他反应函数
(James H. Albert)

2 离散分布
(Jodi M. Casabianca 和 Brian W. Junker)

3 多元正态分布
(Jodi M. Casabianca 和 Brian W. Junker)

4 与项目反应理论相关的指数族分布

(Shelby J. Haberman)

5 观察得分分布的对数线性模型

(Tim Moses)

6 非全同随机变量和的分布

(Wim J. van der Linden)

7 信息论与其在测试上的应用

(Hua-Hua Chang, Chun Wang 和 Zhiliang Ying)

第 2 部分 建模问题

8 项目反应理论的识别

(Ernesto San Martín)

9 具有干扰和附带参数的模型

(Shelby J. Haberman)

10 项目反应理论中的缺失反应

(Robert J. Mislevy)

第 3 部分 参数估计

11 极大似然估计

(Cees A. W. Glas)

12 期望最大算法与推广

(Murray Aitkin)

13 贝叶斯估计

(Matthew S. Johnson 和 Sandip Sinharay)

14 变分近似法

(Frank Rijmen, Minjeong Jeon 和 Sophia Rabe-Hesketh)

15 项目反应理论的马尔可夫链蒙特·卡洛方法

(Brian W. Junker, Richard J. Patz 和 Nathan M. VanHoudnos)

16 统计优化设计理论

(Heinz Holling 和 Rainer Schwabe)

第 4 部分 模型拟合与比较

17 频率模型拟合检验

(Cees A. W. Glas)

18 信息标准

(Allan S. Cohen 和 Sun-Joo Cho)

19 贝叶斯模型拟合与模型比较

(Sandip Sinharay)

20 残差分析的模型拟合

(Craig S. Wells 和 Ronald K. Hambleton)

索引

应用目录

第 1 部分　项目校准与分析

1　项目–校准设计
(Martijn P. F. Berger)
2　参数环绕
(Wim J. van der Linden 和 Michelle D. Barrett)
3　维度分析
(Robert D. Gibbons 和 Li Cai)
4　差异项目功能化
(Dani Gamerman, Flávio B. Gonçalves 和 Tufi M. Soares)
5　校准技术–增强项目
(Richard M. Luecht)

第 2 部分　人员拟合与评分

6　人员拟合
(Cees A. W. Glas 和 Naveed Khalid)
7　得分报告与解释
(Ronald K. Hambleton 和 April L. Zenisky)
8　项目反应理论观察得分等化
(Wim J. van der Linden)

第 3 部分　测试设计

9　最优测试设计
(Wim J. van der Linden)
10　自适性测试
(Wim J. van der Linden)
11　标准制定
(Daniel Lewis 和 Jennifer Lord-Bessen)
12　测试速度与时间极限
(Wim J. van der Linden)
13　项目与测试安全性
(Wim J. van der Linden)

第 4 部分　应用领域

14　大规模集体得分评估

(John Mazzeo)

15 心理测验

(Paul De Boeck)

16 认知诊断评估

(Chun Wang 和 Hua-Hua Chang)

17 健康测量

(Richard C. Gershon, Ron D. Hays 和 Michael Kallen)

18 市场研究

(Martijn G. de Jong 和 Ulf Böckenholt)

19 用 Rasch 模型测量变化

(Gerhard H. Fischer)

第 5 部分 计算机程序

20 R 语言中的项目反应理论包

(Thomas Rusch, Patrick Mair 和 Reinhold Hatzinger)

21 项目反应理论模型的运用吉布斯抽样的贝叶斯推论(BUGS)

(Matthew S. Johnson)

22 BILOG-MG

(Michele F. Zimowski)

23 PARSCALE

(Eiji Muraki)

24 IRTPRO

(Li Cai)

25 Xcalibre 4

(Nathan A. Thompson 和 Jieun Lee)

26 EQSIRT

(Peter M. Bentler, Eric Wu 和 Patrick Mair)

27 ACER ConQuest

(Raymond J. Adam, Margaret L. Wu 和 Mark R. Wilson).

28 Mplus

(Bengt Muthén 和 Linda Muthén)

29 GLLAMM

(Sophia Rabe-Hesketh 和 Anders Skrondal)

30 Latent GOLD

(Jeroen K. Vermunt)

31 WinGen

(Kyung (Chris) T. Han)

32 Firestar

(Seung W. Choi)
33 jMetrik
(J. Patrick Meyer)
索引

正如本书作者在前言中所指出的:

项目反应理论(IRT)起源于路易斯·瑟斯顿(Louis Thurstone)在20世纪20年代的开创性工作,少数作者(如Lawley和Richardson)在20世纪40年代的工作,以及20世纪50年代至20世纪60年代由Alan Birnbaum,Frederic Lord和George Rasch完成的更为关键的工作.本书所展现的最大突破在于解决了一个从经典测试理论继承而来的基本缺点——我们的测量对象和用来测量它的测试项目之间的系统混淆.

试验管理是观测性的研究,在其中被测试者会收到一系列项目,然后我们观察他们的反应.反应是一种项目性质和被测试者能力结合的效应.与在其他观测性研究中一样,将这种效应仅仅归结为其中一个潜在的因果因素是方法论上的错误.但是看起来我们是被迫这么做的.如果一个项目是现场测验的,那么人们感兴趣的问题仅在于他们的性质,而对于在研究中使用的大部分任意选取的被测试者的能力的任何混淆都将会使我们对他们的推断产生偏差.同样地,如果测试者收到测试,我们所感兴趣的仅仅是他们的能力,而不想让他们的得分因为项目的附属性质而产生偏差.经典测试理论确实会产生这样的偏差,比如它将项目的p值视为他们的难度参数,但是这些值同样依赖于现场测试的被测试者样本的能力.尽管术语不同,但这样的问题同样会在项目识别参数和测试可靠性定义中存在.另一方面,经典测试理论中数值修正得分通常是一种同等于反映测试难度和被测试者能力的得分.实际上,传统的指标例如项目或者被测试者的参数和得分仅仅是系统上地隐藏了这种混淆.

项目反应理论解决了这个问题,它将每个反应视为不同的可以被模型化为将项目和被测试者参数分开的概率实验的结果.因此它的项目参数允许我们在估计能力时修正项目效应.同样地,能力参数的出现允许在我们估计项目参数时修正其效应.对于这种范式转变的最好的介绍之一是Rasch(1960,第一章),这本书对于任何对这个问题感兴趣的人都是一本必读的阅读材料.本章给出了行为科学和社会科学研究在传统的更广泛的背景下的一个新范式,其对模糊定义的被测试者"群体"的主题有持续的兴趣,除了一些随机噪声外,将其视为可交换的,并使用统计技术,如相关系数、方差分析和假设检验,假设从中"随机抽样".

项目反应理论自从原始概念化以来,它的发展是迅速的.当Ron Hambleton和我编辑更早的一本《项目反应理论手册》(Wim J. van der Linden和Hambleton, 1997)时,我们有印象这本书的28章总结了极多的我们对这个话题的讨论. 但是20年之后,这三卷章节数几乎相同的书,每一卷看起来都是很有必要的. 并且我对所有对项目反应理论的巨大课题做出贡献,但是其贡献没有被收录在这本新的手册中的研究者和实践者感到抱歉. 不仅有对二分法反应的原始模型及不同反应模式和反应过程的数值模型的补充,并且现在清楚的是,例如,测试的反应时间的模型需要同样类型的对项目和被测试者效应负责的参数. 另一个重要的发展是对于由于反应数据的多层和多级结构而产生的对深度参数化的需求的认识. 这个发展引导了解释,如解释性协变量,影响项目或能力参数的群结构,反应过程的混合,反应和反应时间之间的高级关系或项目域的特殊结构. 例如,由于使用基于规则的项目生成的结构的可能性. 同时,如何将项目反应理论嵌入到一般潜在变量模型的更为广阔的发展中变得更加清晰. 并且为了了解和推广这些新的内容的结果,我们现在更热衷于在将模型的参数看作固定的或随机的中做出选择. 本套书的第一卷覆盖了这些发展的大部分. 每一章基本上都回顾了一个模型. 但是,所有章节都有一个共同的介绍部分,包括模型的历史以及其相关的目的,后续的部分更为形式化地对这个模型进行了介绍,将估计视为它的参数,展示如何计算其与经验数据的拟合,并且通过经验案例解释模型的使用. 最后一部分将会讨论更加深远的应用和有待解决的研究问题.

和任何其他类型的统计学模型一样,项目反应理论非常依赖于使用统计工具处理其模型和应用. 但是与项目反应理论系统有关的介绍和重点的回顾很难在统计学文献中找到. 第二卷则是填补了这个空白. 第二卷的相关章节包括了在项目反应理论中经常使用的概率分布,具有目的参数和干扰参数模型的问题,信息准则的使用,处理缺失数据的方法,模型识别问题,以及一些参数估计和模型拟合与比较的问题. 尤其在最后两个领域中,最近的发展是十分迅速的. 例如,当前一卷《项目反应理论手册》出版时,贝叶斯方法已经取得了一些进展,但是还没有普及. 但是得益于马尔可夫链蒙特·卡洛方法在计算上的成功,这些方法现在已经成为标准方法,尤其是那些在第一卷后半部分中的复杂模型.

第三卷的章节复习了一些项目反应理论在日常测试的应用. 虽然在项目校准和分析领域,人员匹配和得分领域及测验设计领域的每一个选择的问题在许多测试理论文献中有足够的资源,但本书的章节还是专门强调了项目反应理论给这些问题带来的贡献. 这一卷的章节回顾了项目反应理论是如何在大规模教育评估、心理学测试、认知诊断、健康测试、市场调研或者更一般的变化测量领域展现优势的. 同时也汇集了大量对可用于运行第一卷和第三卷中任何应用和模型的计算机软件程序的介绍.

我希望这本《项目反应理论手册》可以作为在项目反应理论领域的研究者和实践者的日常信息来源,同时也可以作为新人的教材. 为了更好地服务于这些目的,所有的章节都是独立的. 但是它们共同的核心符号以及广泛的交叉引用,使得某一章的读者可以在借鉴和查阅其他章节以获得背景信息时不会有很多的困扰.

著名数学家M.加德纳(M. Gardner)曾指出:

> 数学不仅在摧毁着物理科学紧锁的大门,而且正在侵入并摇撼着生物科学、心理学和社会科学.会有这样的一天,经济的争执能够用数学以一种没有争吵的方式来解决.现在想象这一天的到来不再是荒谬的了.

当然这是数学家天真的预测.事实上,许多时候人都是情绪化的、非理性的.单凭数学和科学是不够的,也是不行的!

刘培杰
2021年10月10日
于哈工大

刘培杰数学工作室
已出版(即将出版)图书目录——原版影印

书　　名	出版时间	定　价	编号
数学物理大百科全书.第1卷(英文)	2016－01	418.00	508
数学物理大百科全书.第2卷(英文)	2016－01	408.00	509
数学物理大百科全书.第3卷(英文)	2016－01	396.00	510
数学物理大百科全书.第4卷(英文)	2016－01	408.00	511
数学物理大百科全书.第5卷(英文)	2016－01	368.00	512
zeta函数,q-zeta函数,相伴级数与积分(英文)	2015－08	88.00	513
微分形式:理论与练习(英文)	2015－08	58.00	514
离散与微分包含的逼近和优化(英文)	2015－08	58.00	515
艾伦·图灵:他的工作与影响(英文)	2016－01	98.00	560
测度理论概率导论,第2版(英文)	2016－01	88.00	561
带有潜在故障恢复系统的半马尔柯夫模型控制(英文)	2016－01	98.00	562
数学分析原理(英文)	2016－01	88.00	563
随机偏微分方程的有效动力学(英文)	2016－01	88.00	564
图的谱半径(英文)	2016－01	58.00	565
量子机器学习中数据挖掘的量子计算方法(英文)	2016－01	98.00	566
量子物理的非常规方法(英文)	2016－01	118.00	567
运输过程的统一非局部理论:广义波尔兹曼物理动力学,第2版(英文)	2016－01	198.00	568
量子力学与经典力学之间的联系在原子、分子及电动力学系统建模中的应用(英文)	2016－01	58.00	569
算术域(英文)	2018－01	158.00	821
高等数学竞赛:1962—1991年的米洛克斯·史怀哲竞赛(英文)	2018－01	128.00	822
用数学奥林匹克精神解决数论问题(英文)	2018－01	108.00	823
代数几何(德文)	2018－04	68.00	824
丢番图逼近论(英文)	2018－01	78.00	825
代数几何学基础教程(英文)	2018－01	98.00	826
解析数论入门课程(英文)	2018－01	78.00	827
数论中的丢番图问题(英文)	2018－01	78.00	829
数论(梦幻之旅):第五届中日数论研讨会演讲集(英文)	2018－01	68.00	830
数论新应用(英文)	2018－01	68.00	831
数论(英文)	2018－01	78.00	832

刘培杰数学工作室
已出版(即将出版)图书目录——原版影印

书　　名	出版时间	定　价	编号
湍流十讲(英文)	2018－04	108.00	886
无穷维李代数:第3版(英文)	2018－04	98.00	887
等值、不变量和对称性(英文)	2018－04	78.00	888
解析数论(英文)	2018－09	78.00	889
《数学原理》的演化:伯特兰·罗素撰写第二版时的手稿与笔记(英文)	2018－04	108.00	890
哈密尔顿数学论文集(第4卷):几何学、分析学、天文学、概率和有限差分等(英文)	2019－05	108.00	891
偏微分方程全局吸引子的特性(英文)	2018－09	108.00	979
整函数与下调和函数(英文)	2018－09	118.00	980
幂等分析(英文)	2018－09	118.00	981
李群,离散子群与不变量理论(英文)	2018－09	108.00	982
动力系统与统计力学(英文)	2018－09	118.00	983
表示论与动力系统(英文)	2018－09	118.00	984
分析学练习.第1部分(英文)	2021－01	88.00	1247
分析学练习.第2部分,非线性分析(英文)	2021－01	88.00	1248
初级统计学:循序渐进的方法:第10版(英文)	2019－05	68.00	1067
工程师与科学家微分方程用书:第4版(英文)	2019－07	58.00	1068
大学代数与三角学(英文)	2019－06	78.00	1069
培养数学能力的途径(英文)	2019－07	38.00	1070
工程师与科学家统计学:第4版(英文)	2019－06	58.00	1071
贸易与经济中的应用统计学:第6版(英文)	2019－06	58.00	1072
傅立叶级数和边值问题:第8版(英文)	2019－05	48.00	1073
通往天文学的途径:第5版(英文)	2019－05	58.00	1074
拉马努金笔记.第1卷(英文)	2019－06	165.00	1078
拉马努金笔记.第2卷(英文)	2019－06	165.00	1079
拉马努金笔记.第3卷(英文)	2019－06	165.00	1080
拉马努金笔记.第4卷(英文)	2019－06	165.00	1081
拉马努金笔记.第5卷(英文)	2019－06	165.00	1082
拉马努金遗失笔记.第1卷(英文)	2019－06	109.00	1083
拉马努金遗失笔记.第2卷(英文)	2019－06	109.00	1084
拉马努金遗失笔记.第3卷(英文)	2019－06	109.00	1085
拉马努金遗失笔记.第4卷(英文)	2019－06	109.00	1086
数论:1976年纽约洛克菲勒大学数论会议记录(英文)	2020－06	68.00	1145
数论:卡本代尔1979:1979年在南伊利诺伊卡本代尔大学举行的数论会议记录(英文)	2020－06	78.00	1146
数论:诺德韦克豪特1983:1983年在诺德韦克豪特举行的Journees Arithmetiques数论大会会议记录(英文)	2020－06	68.00	1147
数论:1985－1988年在纽约城市大学研究生院和大学中心举办的研讨会(英文)	2020－06	68.00	1148

刘培杰数学工作室
已出版(即将出版)图书目录——原版影印

书名	出版时间	定价	编号
数论:1987年在乌尔姆举行的Journees Arithmetiques数论大会会议记录(英文)	2020-06	68.00	1149
数论:马德拉斯1987:1987年在马德拉斯安娜大学举行的国际拉马努金百年纪念大会会议记录(英文)	2020-06	68.00	1150
解析数论:1988年在东京举行的日法研讨会会议记录(英文)	2020-06	68.00	1151
解析数论:2002年在意大利切特拉罗举行的C.I.M.E.暑期班演讲集(英文)	2020-06	68.00	1152
量子世界中的蝴蝶:最迷人的量子分形故事(英文)	2020-06	118.00	1157
走进量子力学(英文)	2020-06	118.00	1158
计算物理学概论(英文)	2020-06	48.00	1159
物质,空间和时间的理论:量子理论(英文)	2020-10	48.00	1160
物质,空间和时间的理论:经典理论(英文)	2020-10	48.00	1161
量子场理论:解释世界的神秘背景(英文)	2020-07	38.00	1162
计算物理学概论(英文)	2020-06	48.00	1163
行星状星云(英文)	2020-10	38.00	1164
基本宇宙学:从亚里士多德的宇宙到大爆炸(英文)	2020-08	58.00	1165
数学磁流体力学(英文)	2020-07	58.00	1166
计算科学:第1卷,计算的科学(日文)	2020-07	88.00	1167
计算科学:第2卷,计算与宇宙(日文)	2020-07	88.00	1168
计算科学:第3卷,计算与物质(日文)	2020-07	88.00	1169
计算科学:第4卷,计算与生命(日文)	2020-07	88.00	1170
计算科学:第5卷,计算与地球环境(日文)	2020-07	88.00	1171
计算科学:第6卷,计算与社会(日文)	2020-07	88.00	1172
计算科学.别卷,超级计算机(日文)	2020-07	88.00	1173
代数与数论:综合方法(英文)	2020-10	78.00	1185
复分析:现代函数理论第一课(英文)	2020-07	58.00	1186
斐波那契数列和卡特兰数:导论(英文)	2020-10	68.00	1187
组合推理:计数艺术介绍(英文)	2020-07	88.00	1188
二次互反律的傅里叶分析证明(英文)	2020-07	48.00	1189
旋瓦兹分布的希尔伯特变换与应用(英文)	2020-07	58.00	1190
泛函分析:巴拿赫空间理论入门(英文)	2020-07	48.00	1191
卡塔兰数入门(英文)	2019-05	68.00	1060
测度与积分(英文)	2019-04	68.00	1059
组合学手册.第一卷(英文)	2020-06	128.00	1153
*一代数、局部紧群和巴拿赫*一代数丛的表示.第一卷,群和代数的基本表示理论(英文)	2020-05	148.00	1154
电磁理论(英文)	2020-08	48.00	1193
连续介质力学中的非线性问题(英文)	2020-09	78.00	1195
多变量数学入门(英文)	2021-05	68.00	1317
偏微分方程入门(英文)	2021-05	88.00	1318
若尔当典范性:理论与实践(英文)	2021-07	68.00	1366
伽罗瓦理论.第4版(英文)	2021-08	88.00	1408

刘培杰数学工作室
已出版(即将出版)图书目录——原版影印

书　　名	出版时间	定　价	编号
典型群,错排与素数(英文)	2020—11	58.00	1204
李代数的表示:通过 gln 进行介绍(英文)	2020—10	38.00	1205
实分析演讲集(英文)	2020—10	38.00	1206
现代分析及其应用的课程(英文)	2020—10	58.00	1207
运动中的抛射物数学(英文)	2020—10	38.00	1208
2—纽结与它们的群(英文)	2020—10	38.00	1209
概率,策略和选择:博弈与选举中的数学(英文)	2020—11	58.00	1210
分析学引论(英文)	2020—11	58.00	1211
量子群:通往流代数的路径(英文)	2020—11	38.00	1212
集合论入门(英文)	2020—10	48.00	1213
酉反射群(英文)	2020—11	58.00	1214
探索数学:吸引人的证明方式(英文)	2020—11	58.00	1215
微分拓扑短期课程(英文)	2020—10	48.00	1216
抽象凸分析(英文)	2020—11	68.00	1222
费马大定理笔记(英文)	2021—03	48.00	1223
高斯与雅可比和(英文)	2021—03	78.00	1224
π 与算术几何平均:关于解析数论和计算复杂性的研究(英文)	2021—01	58.00	1225
复分析入门(英文)	2021—03	48.00	1226
爱德华·卢卡斯与素性测定(英文)	2021—03	78.00	1227
通往凸分析及其应用的简单路径(英文)	2021—01	68.00	1229
微分几何的各个方面.第一卷(英文)	2021—01	58.00	1230
微分几何的各个方面.第二卷(英文)	2020—12	58.00	1231
微分几何的各个方面.第三卷(英文)	2020—12	58.00	1232
沃克流形几何学(英文)	2020—11	58.00	1233
仿射和韦尔几何应用(英文)	2020—12	58.00	1234
双曲几何学的旋转向量空间方法(英文)	2021—02	58.00	1235
积分:分析学的关键(英文)	2020—12	48.00	1236
为有天分的新生准备的分析学基础教材(英文)	2020—11	48.00	1237
数学不等式.第一卷.对称多项式不等式(英文)	2021—03	108.00	1273
数学不等式.第二卷.对称有理不等式与对称无理不等式(英文)	2021—03	108.00	1274
数学不等式.第三卷.循环不等式与非循环不等式(英文)	2021—03	108.00	1275
数学不等式.第四卷.Jensen 不等式的扩展与加细(英文)	2021—03	108.00	1276
数学不等式.第五卷.创建不等式与解不等式的其他方法(英文)	2021—04	108.00	1277

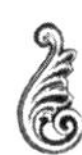

刘培杰数学工作室
已出版(即将出版)图书目录——原版影印

书　　名	出版时间	定　价	编号
冯·诺依曼代数中的谱位移函数:半有限冯·诺依曼代数中的谱位移函数与谱流(英文)	2021-06	98.00	1308
链接结构:关于嵌入完全图的直线中链接单形的组合结构(英文)	2021-05	58.00	1309
代数几何方法. 第1卷(英文)	2021-06	68.00	1310
代数几何方法. 第2卷(英文)	2021-06	68.00	1311
代数几何方法. 第3卷(英文)	2021-06	58.00	1312

书　　名	出版时间	定　价	编号
代数、生物信息和机器人技术的算法问题. 第四卷,独立恒等式系统(俄文)	2020-08	118.00	1199
代数、生物信息和机器人技术的算法问题. 第五卷,相对覆盖性和独立可拆分恒等式系统(俄文)	2020-08	118.00	1200
代数、生物信息和机器人技术的算法问题. 第六卷,恒等式和准恒等式的相等 问题、可推导性和可实现性(俄文)	2020-08	128.00	1201
分数阶微积分的应用:非局部动态过程,分数阶导热系数(俄文)	2021-01	68.00	1241
泛函分析问题与练习:第2版(俄文)	2021-01	98.00	1242
集合论、数学逻辑和算法论问题:第5版(俄文)	2021-01	98.00	1243
微分几何和拓扑短期课程(俄文)	2021-01	98.00	1244
素数规律(俄文)	2021-01	88.00	1245
无穷边值问题解的递减:无界域中的拟线性椭圆和抛物方程(俄文)	2021-01	48.00	1246
微分几何讲义(俄文)	2020-12	98.00	1253
二次型和矩阵(俄文)	2021-01	98.00	1255
积分和级数. 第2卷,特殊函数(俄文)	2021-01	168.00	1258
积分和级数. 第3卷,特殊函数补充:第2版(俄文)	2021-01	178.00	1264
几何图上的微分方程(俄文)	2021-01	138.00	1259
数论教程:第2版(俄文)	2021-01	98.00	1260
非阿基米德分析及其应用(俄文)	2021-03	98.00	1261
古典群和量子群的压缩(俄文)	2021-03	98.00	1263
数学分析习题集. 第3卷,多元函数:第3版(俄文)	2021-03	98.00	1266
数学习题:乌拉尔国立大学数学力学系大学生奥林匹克(俄文)	2021-03	98.00	1267
柯西定理和微分方程的特解(俄文)	2021-03	98.00	1268
组合极值问题及其应用:第3版(俄文)	2021-03	98.00	1269
数学词典(俄文)	2021-01	98.00	1271
确定性混沌分析模型(俄文)	2021-06	168.00	1307
精选初等数学习题和定理. 立体几何. 第3版(俄文)	2021-03	68.00	1316
微分几何习题:第3版(俄文)	2021-05	98.00	1336
精选初等数学习题和定理. 平面几何. 第4版(俄文)	2021-05	68.00	1335

刘培杰数学工作室
已出版(即将出版)图书目录——原版影印

书　　名	出版时间	定　价	编号
狭义相对论与广义相对论:时空与引力导论(英文)	2021—07	88.00	1319
束流物理学和粒子加速器的实践介绍:第2版(英文)	2021—07	88.00	1320
凝聚态物理中的拓扑和微分几何简介(英文)	2021—05	88.00	1321
混沌映射:动力学、分形学和快速涨落(英文)	2021—05	128.00	1322
广义相对论:黑洞、引力波和宇宙学介绍(英文)	2021—06	68.00	1323
现代分析电磁均质化(英文)	2021—06	68.00	1324
为科学家提供的基本流体动力学(英文)	2021—06	88.00	1325
视觉天文学:理解夜空的指南(英文)	2021—06	68.00	1326
物理学中的计算方法(英文)	2021—06	68.00	1327
单星的结构与演化:导论(英文)	2021—06	108.00	1328
超越居里:1903年至1963年物理界四位女性及其著名发现(英文)	2021—06	68.00	1329
范德瓦尔斯流体热力学的进展(英文)	2021—06	68.00	1330
先进的托卡马克稳定性理论(英文)	2021—06	88.00	1331
经典场论导论:基本相互作用的过程(英文)	2021—07	88.00	1332
光致电离量子动力学方法原理(英文)	2021—07	108.00	1333
经典域论和应力:能量张量(英文)	2021—05	88.00	1334
非线性太赫兹光谱的概念与应用(英文)	2021—06	68.00	1337
电磁学中的无穷空间并矢格林函数(英文)	2021—06	88.00	1338
物理科学基础数学.第1卷,齐次边值问题、傅里叶方法和特殊函数(英文)	2021—07	108.00	1339
离散量子力学(英文)	2021—07	68.00	1340
核磁共振的物理学和数学(英文)	2021—07	108.00	1341
分子水平的静电学(英文)	2021—08	68.00	1342
非线性波:理论、计算机模拟、实验(英文)	2021—06	108.00	1343
石墨烯光学:经典问题的电解解决方案(英文)	2021—06	68.00	1344
超材料多元宇宙(英文)	2021—07	68.00	1345
银河系外的天体物理学(英文)	2021—07	68.00	1346
原子物理学(英文)	2021—07	68.00	1347
将光打结:将拓扑学应用于光学(英文)	2021—07	68.00	1348
电磁学:问题与解法(英文)	2021—07	88.00	1364
海浪的原理:介绍量子力学的技巧与应用(英文)	2021—07	108.00	1365
多孔介质中的流体:输运与相变(英文)	2021—07	68.00	1372
洛伦兹群的物理学(英文)	2021—08	68.00	1373
物理导论的数学方法和解决方法手册(英文)	2021—08	68.00	1374
非线性波数学物理学入门(英文)	2021—08	88.00	1376
波:基本原理和动力学(英文)	2021—07	68.00	1377
光电子量子计量学.第1卷,基础(英文)	2021—07	88.00	1383
光电子量子计量学.第2卷,应用与进展(英文)	2021—07	68.00	1384
复杂流的格子玻尔兹曼建模的工程应用(英文)	2021—08	68.00	1393
电偶极矩挑战(英文)	2021—08	108.00	1394
电动力学:问题与解法(英文)	2021—09	68.00	1395
自由电子激光的经典理论(英文)	2021—08	68.00	1397

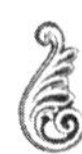

刘培杰数学工作室
已出版(即将出版)图书目录——原版影印

书　　名	出版时间	定　价	编号
曼哈顿计划——核武器物理学简介(英文)	2021—09	68.00	1401
粒子物理学(英文)	2021—09	68.00	1402
引力场中的量子信息(英文)	2021—09	128.00	1403
器件物理学的基本经典力学(英文)	2021—09	68.00	1404
等离子体物理及其空间应用导论. 第1卷,基本原理和初步过程(英文)	2021—09	68.00	1405
拓扑与超弦理论焦点问题(英文)	2021—07	58.00	1349
应用数学:理论、方法与实践(英文)	2021—07	78.00	1350
非线性特征值问题:牛顿型方法与非线性瑞利函数(英文)	2021—07	58.00	1351
广义膨胀和齐性:利用齐性构造齐次系统的李雅普诺夫函数和控制律(英文)	2021—06	48.00	1352
解析数论焦点问题(英文)	2021—07	58.00	1353
随机微分方程:动态系统方法(英文)	2021—07	58.00	1354
经典力学与微分几何(英文)	2021—07	58.00	1355
负定相交形式流形上的瞬子模空间几何(英文)	2021—07	68.00	1356
广义卡塔兰轨道分析:广义卡塔兰轨道计算数字的方法(英文)	2021—07	48.00	1367
洛伦兹方法的变分:二维与三维洛伦兹方法(英文)	2021—08	38.00	1378
几何、分析和数论精编(英文)	2021—08	68.00	1380
从一个新角度看数论:通过遗传方法引入现实的概念(英文)	2021—07	58.00	1387
动力系统:短期课程(英文)	2021—08	68.00	1382
几何路径:理论与实践(英文)	2021—08	48.00	1385
论天体力学中某些问题的不可积性(英文)	2021—07	88.00	1396
广义斐波那契数列及其性质(英文)	2021—08	38.00	1386
对称函数和麦克唐纳多项式:余代数结构与 Kawanaka 恒等式(英文)	2021—09	38.00	1400
杰弗里·英格拉姆·泰勒科学论文集:第1卷. 固体力学(英文)	2021—05	78.00	1360
杰弗里·英格拉姆·泰勒科学论文集:第2卷. 气象学、海洋学和湍流(英文)	2021—05	68.00	1361
杰弗里·英格拉姆·泰勒科学论文集:第3卷. 空气动力学以及落弹数和爆炸的力学(英文)	2021—05	68.00	1362
杰弗里·英格拉姆·泰勒科学论文集:第4卷. 有关流体力学(英文)	2021—05	58.00	1363

刘培杰数学工作室
已出版(即将出版)图书目录——原版影印

书　　名	出版时间	定　价	编号
非局域泛函演化方程:积分与分数阶(英文)	2021－08	48.00	1390
理论工作者的高等微分几何:纤维丛、射流流形和拉格朗日理论(英文)	2021－08	68.00	1391
半线性退化椭圆微分方程:局部定理与整体定理(英文)	2021－07	48.00	1392
非交换几何、规范理论和重整化:一般简介与非交换量子场论的重整化(英文)	2021－09	78.00	1406
数论论文集:拉普拉斯变换和带有数论系数的幂级数(俄文)	2021－09	48.00	1407
挠理论专题:相对极大值,单射与扩充模(英文)	2021－09	88.00	1410
强正则图与欧几里得若尔当代数:非通常关系中的启示(英文)	2021－10	48.00	1411
拉格朗日几何和哈密顿几何:力学的应用(英文)	2021－10	48.00	1412
时滞微分方程与差分方程的振动理论:二阶与三阶(英文)	2021－10	98.00	1417
卷积结构与几何函数理论:用以研究特定几何函数理论方向的分数阶微积分算子与卷积结构(英文)	2021－10	48.00	1418
经典数学物理的历史发展(英文)	2021－10	78.00	1419
扩展线性丢番图问题(英文)	2021－10	38.00	1420
一类混沌动力系统的分歧分析与控制:分歧分析与控制(英文)	即将出版		1421
伽利略空间和伪伽利略空间中一些特殊曲线的几何性质(英文)	即将出版		1422
数论与密码学导论:第二版(英文)	即将出版		1423
一阶偏微分方程:哈密尔顿—雅可比理论(英文)	2021－11	48.00	1424
各向异性黎曼多面体的反问题:分段光滑的各向异性黎曼多面体反边界谱问题:唯一性(英文)	2021－11	38.00	1425
项目反应理论手册.第一卷,模型(英文)	2021－11	138.00	1431
项目反应理论手册.第二卷,统计工具(英文)	2021－11	118.00	1432
项目反应理论手册.第三卷,应用(英文)	2021－11	138.00	1433
二次无理数:经典数论入门(英文)	即将出版		1434
数,形与对称性:数论,几何和群论导论(英文)	即将出版		1435
有限域手册(英文)	2021－11	178.00	1436
计算数论(英文)	2021－11	148.00	1437
拟群与其表示简介(英文)	2021－11	88.00	1438

联系地址:哈尔滨市南岗区复华四道街10号　哈尔滨工业大学出版社刘培杰数学工作室
网　　址:http://lpj.hit.edu.cn/
邮　　编:150006
联系电话:0451－86281378　　13904613167
E-mail:lpj1378@163.com